World Regional Geography Book Series

Series Editor
E. F. J. De Mulder
DANS, NARCIS, Utrecht, Noord-Holland, The Netherlands

What does Finland mean to a Finn, Sichuan to a Sichuanian, and California to a Californian? How are physical and human geographical factors reflected in their present-day inhabitants? And how are these factors interrelated? How does history, culture, socio-economy, language and demography impact and characterize and identify an average person in such regions today? How does that determine her or his well-being, behaviour, ambitions and perspectives for the future? These are the type of questions that are central to The World Regional Geography Book Series, where physically and socially coherent regions are being characterized by their roots and future perspectives described through a wide variety of scientific disciplines. The Book Series presents a dynamic overall and in-depth picture of specific regions and their people. In times of globalization renewed interest emerges for the region as an entity, its people, its landscapes and their roots. Books in this Series will also provide insight in how people from different regions in the world will anticipate on and adapt to global challenges as climate change and to supra-regional mitigation measures. This, in turn, will contribute to the ambitions of the International Year of Global Understanding to link the local with the global, to be proclaimed by the United Nations as a UN-Year for 2016, as initiated by the International Geographical Union. Submissions to the Book Series are also invited on the theme 'The Geography of…', with a relevant subtitle of the authors/editors choice. Proposals for the series will be considered by the Series Editor and International Editorial Board. An author/editor questionnaire and instructions for authors can be obtained from the Publisher.

This book series is published in cooperation with the International Geographical Union (IGU). The IGU is an international, non-governmental, professional organization devoted to the development of the discipline of Geography. The purposes of the IGU are primarily to promote Geography through initiating and coordinating geographical research and teaching in all countries of the world.

More information about this series at http://www.springer.com/series/13179

Igor Jelen • Angelija Bučienė
Francesco Chiavon • Tommaso Silvestri
Katie Louise Forrest

The Geography of Central Asia

Human Adaptations, Natural Processes and Post-Soviet Transition

Springer

Igor Jelen
Department of Political and Social Sciences
University of Trieste
Trieste, Italy

Angelija Bučienė
Study Centre of Social Geography and Regional Studies
Klaipėda University
Klaipėda, Lithuania

Francesco Chiavon
Freelance, independent analyst
Udine, Italy

Tommaso Silvestri
Department of Political and Social Sciences
University of Trieste
Trieste, Italy

Katie Louise Forrest
Liceo linguistico
Istituto Gaspare Bertoni
Udine, Italy

ISSN 2363-9083 ISSN 2363-9091 (electronic)
World Regional Geography Book Series
ISBN 978-3-030-61265-8 ISBN 978-3-030-61266-5 (eBook)
https://doi.org/10.1007/978-3-030-61266-5

Picture Credit: Anton Balazh – Fotolia.com

This Springer imprint is published by the registered company Springer Nature Switzerland AG
The registered company address is: Gewerbestrasse 11, 6330 Cham, Switzerland

Preface

It is not easy neither obvious to get updated information about Central Asia, especially for those expressing environment- and social-sensitive character. The presentation of arguments, which in such narrative have evidently a synthetic character, is more difficult and subject to the risk of intolerable simplification or even of becoming praecox obsolete. Geographers often find themselves in such risky situations, undertaking such task, and they have to be sufficiently rapid in travelling and observing, reading and interpreting as well as writing, and even more rapid in controlling results and findings, possibly presenting them in a readable manner. Evidently, for all these reasons, the authors are the only ones responsible for errors, mistakes and any kind of imperfections that exist in this text. As a possible excuse, beyond any further criteria, we have to say we tried to write in a way that could be understandable for students and for those, in general, approaching such kind of studies: after 25 years of experiencing, observing, talking and listening to Central Asian protagonists, we tried to write what we deemed could be useful for interpreting such rapidly changing environments and societies, in order to consider the significances of such changes at all the scales, in any form.

Trieste, Italy — Igor Jelen
Klaipėda, Lithuania — Angelija Bučienė
Udine, Italy — Francesco Chiavon
Trieste, Italy — Tommaso Silvestri
Udine, Italy — Katie Louise Forrest

July 2018

Acknowledgement

I would like to express cordial thanks to my family members – husband Dr. Saulius Bučas, sons Prof. Dr. Martynas Bučas and Algirdas Bučas for moral support and assistance in selection, preparing and improving quality of some figures, which were necessary for this book. Great thanks to the initiator of this book Prof. Dr. Igor Jelen for inviting me as a co-author. Angelija Bučienė.

Thanks to my wife and my daughter: as promised, after this book, I will spend more time at home. And deepest thanks to countless Central Asian friends, testimonials, just ordinary peoples I met here and there: they are the anonymous co-authors of this book, who showed exquisite hospitality under any circumstance – possibly the most important resource of these marvellous countries. Igor Jelen

Contents

About the Authors

Igor Jelen University of Trieste, Piazzale Europa 1, 34100 Trieste, Italy. Email: igor.jelen@dispes.units.it.

Research interests: globalization-induced transformation on local realities, political-economical and cultural postmodern geographies.

Igor Jelen wrote the Chaps. 1, 8, 9, 11, 12, 14 and 15.

Angelija Bučienė Klaipėda University, Study Centre of Social Geography and Regional Sciences, Faculty of Social and Humanitarian Sciences, address: S.Nėries 5 Klaipėda, LT-92227, Lithuania. Email: angelija.buciene@ku.lt.

Research interests: changes in rural and agrarian landscape, impact of different cropping systems on agroecosystems and their components, and diffuse pollution problems in watersheds.

Angelija Bučienė was the main contributor to Chaps. 2, 3, 4, 5, 6 and 15.

Tommaso Silvestri was the main contributor to Chap. 7, and prepared the tables for Chaps. 8, 9, 10, 11 and 12.

Francesco Chiavon was the main contributor to Chap. 13.

Katie Forrest was the main contributor to Chap. 10, and supervised the editing for Chaps. 1, 8, 9, 11, 12, 13, 14 and 15.

Abbreviations

AO	Avtonomnaja Oblast'
AR	Avtonomnaja Respublika
CA	Central Asia
CBD	central business district
CIS	Commonwealth of Independent States
EBRD	European Bank for Reconstruction and Development
EU	European Union
FDI	foreign direct investments
FI	foreign investments
FSU	Former Soviet Union
GCS	global civil society
GNI	gross national income
HC	hydrocarbon
HSHC	high speed high capacity (train)
ID	irrigation and drainage
IDB	Islamic Development Bank
IO	international organization
MNC	multinational companies
NATO	North Atlantic Treaty Organization
NIC	new independent countries
NIS	new independent states
s.c.	so-called
m. osl.	metres over sea level
RD	research and development
SCO	Shanghai Cooperation Agreement
SR	Silk Road
SSR	Sovetskaja Socialističeskaja Respublika
SU	Soviet Union
USA	United States of America
WWI	First World War
WWII	Second World War

List of Figures

List of Tables

1 Premise: A Land of Extremes

Abstract

The human and ecologic dimensions of Central Asia (CA) are rapidly changing; this fact is due, to some extent, to the stress affecting the physical environment, characterized by endorheic conditions and continentalism, and by other elements, deriving from political-cultural changes and transformations; above all, the position of this region, considering the technological, geo-political and geo-economic situation, makes these countries appear as central or peripheral compared to the wider Eurasiatic spaces. In fact, the significance of this region has been changing continuously. In its history it has often just been a remote and neglected land. In other cases, it has been the target of expansionism, a space and cache of resources to be exploited. Seldom has it been regarded as the cradle of original cultures, which would exert their influence on the wider world.

Keywords

Central Asia ecology · Geopolitics · Human adaptations · Environmental changes · Main geographical characteristics

1.1 Several Narrative Axes

The most difficult task is that of having a conceptual vision of what is currently going on, namely of what is effectively in front of our eyes: this is because it is difficult—or impossible—to have an idea of the extent to which processes are materially and culturally developing.

This is a conclusion, rather than a starting point, yet it is easy to verify. As far as the last few decades are concerned, the geography, both human and physical, of Central Asia (CA) appears so shifting that a book—intended as a narrative and with an organic character, fixed on paper—might appear as an inadequate instrument. It would seem to be necessary to write and re-write its chapters continuously, then, after having finished one, start on the updating of the others. It is evidently an archaic way of describing such complicated topics and discussions.

In fact, the CA region, as it appears today, is articulated in five spaces made up of the five post-Soviet republics (Fig. 1.1, the "stans": Kazakhstan, Kyrgyzstan, Uzbekistan, Turkmenistan and Tajikistan), resulting from the combination of innumerable elements that are continuously overlapping, sometimes confused beyond the mere concept of borderlines, in various scenarios.

Therefore, it is necessary to assume somewhat reductive criteria, in order to describe what otherwise might appear just as an uninterrupted flow of snapshots and anecdotes. This is not an arbitrary choice, evidencing some particular elements to the detriment of others. It is simply expedient so as to give some order to the narrative that—whatever the aim of the writing is—should result from an ordinate sequence of descriptions.

This is in principle the well-meant intention; in proceeding with the work, selection among such criteria proved to be too limitative, simply too difficult; therefore, we just decided to carry on, without binding the narrative to a scheme, rather, continuing with a "pluralistic" method, leaving a certain dose of improvisation to the "pen."

Indeed, the resulting summary should evidence—intended as general narrative axes—some priorities, among which: the evolution from the natural to the human, from the ancient to the recent, from the supposedly "original" to the current (and artificial) reality, from the original community equilibrium to confusion, from the social-simple dimension to the economic complex one, or to the hyper-specialized actuality, and occasionally some others too.

It is an attempt to escape naïve assumptions, among which those that describe the break-up of a supposed primordial ecological equilibrium provoked by modernism,

I. Jelen et al., *The Geography of Central Asia*, World Regional Geography Book Series,
https://doi.org/10.1007/978-3-030-61266-5_1

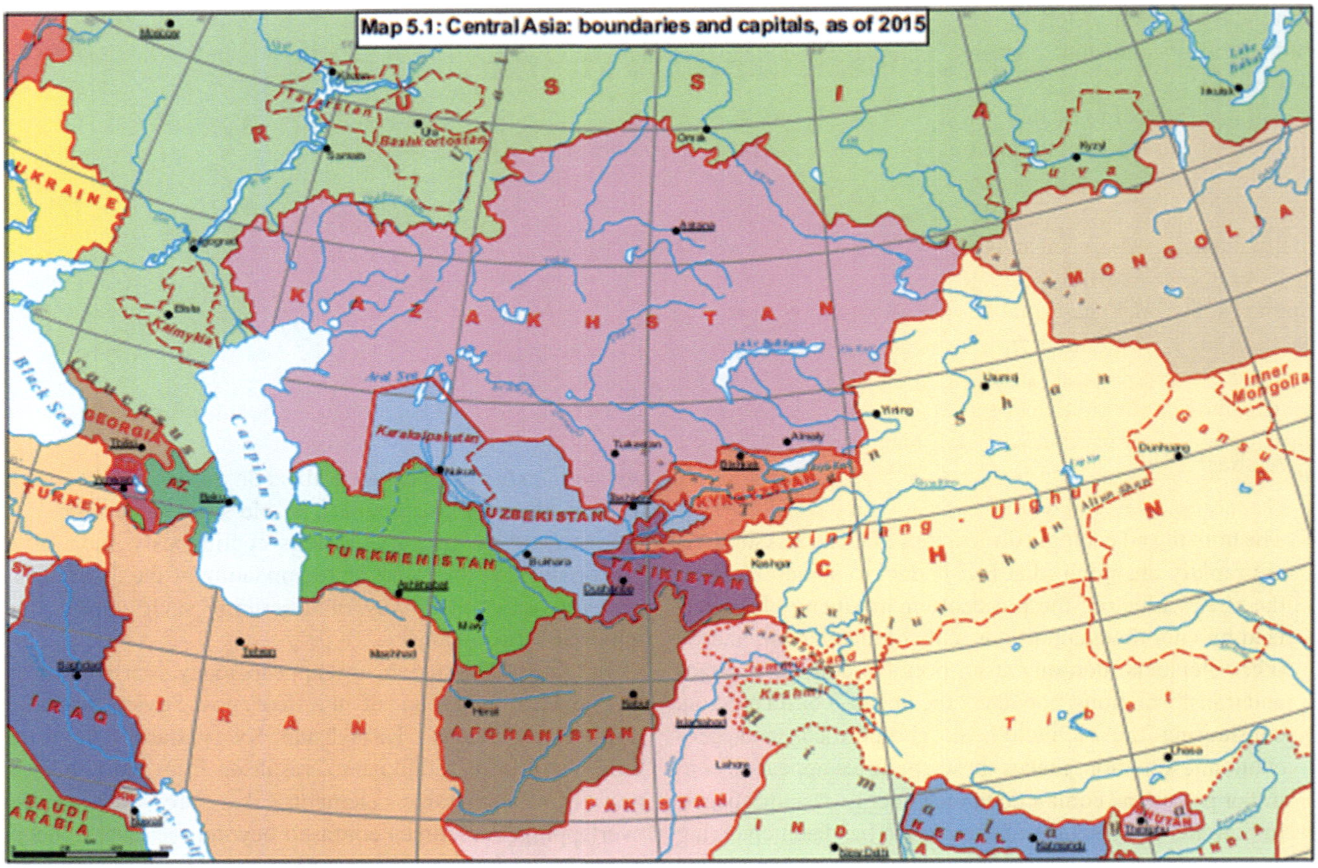

Fig. 1.1 Central Asia: boundaries and capitals. (Source: Courtesy, Barisitz Stephan (2017:280))

and the consequent efforts to restore an original situation (into a current post-modern era, once a centennial cycle is exhausted).

On the contrary, the intention is to highlight the interlacement among natural and cultural evolutions—and regressions—with corresponding transformations that are usually confusing and that mutually feed-back into each other, on different scales. In order to describe these evolutions (that we do not assume as deterministically oriented), we chose a limited set of spots, namely of critical passages, trying to describe (rather than understand) situations and to evidence contradictions.[1]

1.2 Intentions and Goals

The starting point of this work is the description of reality (the basic job of the geographer); on this basis and consequently developing the observations, it will be possible to derive further definitions and explanations, eventually explicative statements, while applying a cautionary principle, as usual in such situations.

All these are to be combined with further elements, sometimes apparently extraneous to the evolution game affecting that reality, namely independent variables. It is the case of the sequence of unforeseeable inventions, of the spread of specific technologies, of external interference, or of cultural creations, as well as of new ideologies or innovative solidarity patterns, aggregating individuals and groups. It could be the case, typically for today, of information and communications technology (ICT), and of mass- and social-media, to some extent substituting traditional ways of evidencing identity (belonging to a group) and to organize communication on any scale.

All these would change the territorial "circumscriptions"—and the corresponding symbols and references—in which the populations try continuously to identify their selves, and that are to be considered finally as the typical "ideals" of a certain epoch. It is the case of village communities and of nomadic warrior tribes, of city- or nation-states, of colonial empires or of de-territorialized globalized aggregations: all these have to be intended as elements of a dis-

[1]Valikhanov Č.Č., 1985; Valikhanov capt and Veniukof M., 1865; Scharr K., Steinicke E., (hg.), 2012; Gumilev L.N., 1972; Barisitz Stephan, 2018; Thorez Julien, 2018; Edgar Adrienne Lynn, 2004; Rasanayagam Johan and Beyer Judith, 2014; Khazanov A.M., 1994.

crete classification of territories, prefiguring discontinuities and possibly tension, and of human organizational processes.

All this in order to reconstruct the nexus that periodically forms between humans, organized in more or less numerous groups, and between humans and natural elements (and resources) and environments, which occasionally—not really in principle—reach an equilibrium.

Therefore, it is surely limitative to present the five post-Soviet "stans" like individual bodies (namely "organisms"), each on their own, either from a natural or from a human and socio-technological point of view. But it is also necessary to consider that the life of each of these countries occasionally melts and confuses with the others. There are adaptations and compromises that assume different forms, transcending border and political categories that the elite try, from to time to time, to superimpose on the spontaneous development of the societies. It is the case of both natural and human processes, that occasionally overcome some threshold; it is the case of both cultural and technological elements, that exert an increasing but differentiated impact on local territories and populations; it is the case of internal revolutionary innovations or of exogenous influences, that may affect the new CA independent countries today, as has often happened in the past from the south, to the north, the west or the east.

It is a critical point; cultural changes have different impacts in different moments and in different places. It is the case of some fundamental divisions in the CA history, that derive both from natural and cultural causes, which are sometimes difficult to distinguish from each other.

Among such events is it possible to solicit the original urbanization and the formation of trade routes that cross the CA space from ancient times; the early wave of Islamization and Arabization in the seventh century; the Medieval Mongol and Turkish invasions, resulting in major waves of destructions, but as well as resulting in the construction of major empires, that, starting from the areas being dealt with, would conquer still wider areas. So too for the conquests that happened in modernity subsequently, in times of colonial and Soviet rule in the nineteenth and twentieth centuries; it is possible to say the same about the current era of Globalization, that forces local societies to open and integrate to some extent in a wider world.

Such processes exert different impacts, sometimes producing adaptations, sometimes destructive waves, inducing consequent cultural shocks; sometimes they have generated new civilizations from scratch, sometimes delaying and pulverizing their effects in the long term and onto wider surfaces (as happened in the environment of remote steppes, deserts and high mountains).

Some elements occasionally did converge, or occasionally diverge, among the different populations, being perceived in different and also opposite ways; sometimes they would suddenly lose their innovative potential, being assimilated into the cultural body, or, on the contrary, would be violently rejected.

Sometimes a certain phenomenon, generated from outside and arriving in this peripheral space, changed its intrinsic original significance. This happened after the Bolshevik Revolution and it happened with the post-Soviet independence, that was initially perceived as a risk, rather than an opportunity, both by the elite and the general population.

Evidently, in order to understand the power of certain transformations, it is necessary to collocate such processes on the "longue durée" scale, but this would not really be in the possibility of our "instant" book.

1.3 Early Forms of Settlements

Starting from the origins, the first obstacle, and also the first resource, for human survival is that of wild nature in its different forms, that manifests itself as something impossible to dialogue with, but also as something necessary to use in order to satisfy functional needs.

In such circumstances, such needs prefigured different tendencies, sometimes inducing the rarefied populations to aggregate (in hordes, tribes, oasis towns, city-states, periodical bazaars or caravan itineraries); sometimes inducing them to fight each other in order to get the resources they needed for survival; in other cases, just inducing the populations to migrate, trying to escape, sometime for thousands of miles, then assuming a kind of permanent mobile "genre de vie" (becoming "nomads").

The primordial conditions of these places shape opposite outcomes, oscillating periodically between different extremes, or on the contrary inducing new barriers and fractures, consolidating old centres or new peripheries, namely remote areas that are impossible to settle and to control.

Only on a few epochal occasions did the chance appear to elaborate a common cultural element, unifying the whole space, that of the oasis and of the steppe, as may have happened in occasion of some of the already mentioned moments—Islamization, Altaic invasions and modernity revolutions as occurred in both colonial and Soviet times, and possibly with current globalization too. Such processes did sometimes coincide with major exogenous waves that exerted a persistent impact, with changes being imposed rather than assimilated and interpreted by the different local populations. In this context, any classification method would appear insufficient and incoherent.

Nevertheless, we need some descriptive key in order to try to understand some long-term lines of changes, that a single segment of research may not evidence, with the fundamental aim of recognizing the basic element of a certain evolution.

1.4 Undifferentiated Space, Transit Territory or Continental Pivot?

The first consideration, as evidenced by literature—which is mainly elaborated by sedentary cultures, namely biased by a "territorialized" point of view—assumes the territory in all its dimensions: structure, resources, extension, shape. Among these, it assumes the position the area occupies with regard to all the other positions, and then the perception of this collocation in the wider context: depending on the points of view, CA appears as a "means" or as an "end" a remote periphery, a transit corridor or—rarely, indeed, in history—as a centre of power (Figs. 1.4, 1.6 and 1.7.)

This is due to wider scale functioning, to the presence of traffic roads crossing the area or to the presence of powerful neighbours; more in general, it depends on the technological paradigm, namely to the availability of instruments for travelling long distances, for controlling wide surfaces and remote areas, and for using the corresponding resources—which are not usually easy to access.

The "world of the steppe" is a space where, in pre-modern history, it is impossible to build up towns and infrastructures, where it is apparently impossible to accumulate cultures in any sense—economic wealth, political organization, material codes like written languages, techniques and knowledge, arts or production units—in which it is in principle impossible to extend any sense of statehood. Namely, it is impossible to start any political appropriation (or institutionalization) process.

In general, this region is occupied by large unproductive areas, characterized by the alternance of sandy regions and steppes, deserts and—in the eastern and southern regions—mountain ridges, that represent areas in which any kind of organized life results as being extraordinarily difficult.[2]

Such configuration means the absence of relevant geographical forms (namely of natural circumscription, in Carneiro's sense): it does not present significant discontinuities where a definite group would be able to develop security or further functions that may be considered the basis for starting a civilization cycle, in which primordial agrarian or commercial organizations would then have the possibility to consolidate themselves.

From a certain point of view, it appears as an undifferentiated space, or as a mere transit area, describing a certain continental corridor; from another, it is just a huge extension of uncontrollable territories; from a further perspective, and just because of its remoteness, it may be seen as the "pivot" of world politics, namely the uncontrollable "heartland" in the continental landmass (as defined by the founder of geopolitics, Sir Halford Mackinder).

1.5 The Local Point of View

Such opposite interpretations do not impact the destiny of native populations very much, as they are usually outside of their capability of perception; it depends, then, on a subjective point of view, on circumstances and in particular on their strength and on the cultural attitude of their organizations (village networks, city-states or confederations of tribes); and it has obviously depended on the intentions of their neighbours, sometimes organizing themselves in contiguous areas, powerful states with pretensions of expanding their dominions.

Then, it depends on further elements, and above all on the importance that local resources assume in different epochs, eventually becoming the target for aggression; this is the case of natural and human resources, of agro-pastoral soil, water, raw materials, or currently, hydrocarbons (HC); and it is the case of free and unknown territories in which the nomadic populations were used to escaping from invasion or from despotic power.

The most evident aspect of CA environment is the unbalanced distribution of such survival resources; this fact determines different consequences, above all the capability of settling such territories permanently. Even today, the main criterion in order to classify these places lies in the possibility of organizing a static or dynamic "genre de vie," namely between nomadic and sedentary ways of organizing life.

A question about adaptation possibilities was already observed by many scholars in ancient times: Strabo produced a systemic description of CA,[3] while Ibn Khaldun[4] defined nomadic and sedentary civilization cycles; much earlier, Herodotus[5] originally described the Skyft tribes moving in the Eurasian steppe, stimulating the curiosity and the fear of sedentary populations.

Such questions have interested the mainstream of geographical studies over time, until the moment in which they were theorized by some of the founders of geo-political

[2]Ramazanova N., Berdenov Zh., Ramazanov S., Kazangapova N., Romanova S., Toksanbaeva S., Wendt Jan, 2019.

[3]Biraschi A.M., 2000.

[4]Ibn Khaldun, 1980.

[5]Erodoto, 1988.

thought, inspiring and conditioning the centuries of modernity; they indeed often degenerated into ideologies, often just positivist interpretations, but useful for justifying some pretensions (already in the colonialist epoch).

Among them, the one already cited is of Mackinder, who considered the inner side of the continent in a visionary perspective, as the impregnable fortress, around which entire world politics would pivot (since "whoever rules the world island may rule the world").

It is the case of Lev Gumilev,[6] who theorized the nomadic migration waves as deterministically dependent on environment cycles; and it is the case, earlier, but in the same context, of Ferdinand Von Richthofen,[7] who idealized the transcontinental movement, producing a kind of mythical but fortunate definition of the "silk road" (SR, Seidenstrasse).

This road, in particular, had connected the extreme sides of the Eurasian continent since ancient times, crossing inhospitable environments, plateaus and arid deserts, "disseminating," along the same road, commercial activities, bazaars and cities, that over time would become fabulously rich: people for millennia ran these routes, without having a definite perception of their dimension.

Such reasoning finally inspired a theory of the "oriental despotism," following the same western tradition, tracing to Illuminism, and even to the same Herodotus the original observations. It is the case of Karl Wittfogel, who tried to elaborate a materialistic way to understand the process of civilization formation, even if his assumptions are often just interpreted as a geo-determinist way for producing stereotypes about the "orient" and the other "outside" worlds (eventually set against the "occident").

His theory derives from direct observations of territories, of resources and of the organization method the population were used to practicing. In such circumstances of structural scarcity, it seems that power had the chance to use vital resources in a manipulatory way (water and agricultural soil in the past, HC in times of modernity), finally accumulating an infinite control of its own population: a situation that would represent the exact opposite of a democratic idea of power—responsible and reversible—traceable to the ancient Greeks, and having spread since those times over the occidental hemisphere and subsequently worldwide.

But these are just few of the schemes, ideas and theories, that today have to be applied in order to interpret evolution, that only give the example of a comprehension method, that assume modernist geographical discoveries (the "new worlds" prospected by colonial expansionism) as a sequence of opposite realities; these indeed would represent what "we are not," namely the "other" in opposition of the "self" (namely the western "ego"), applying a schema that would also produce a sequence of dramatic misunderstandings.

1.6 Classification in Geo-Political Units

All these periods are characterized by specific ideologies—or political technologies—that the corresponding elites would apply in order to make some organized units emerge from these huge extensions.

These vary greatly in the different epochs; in current times they usually coincide with a sequence of nation-states, delimited by linear borders. Relying on this classification, the strict geographical definition of the area comprehends the cited five post-Soviet republics that share a set of common cultural and geo-environmental characteristics and much of the recent and ancient past of their history.

From a more general point of view, CA results in a wider geographical definition, comprehending the surrounding regions defined with territorial labels such as Afghanistan, Mongolia, Xinxiang, as well as Central-Southern Siberia, Tibet and further contiguous areas, assuming a concentric interpretation.

Such definitions result from variables that are assumed on the basis of wider geographical characteristics—such as natural barriers and continentalism, the presence of evident discontinuities such as rivers and inland waters, desert depressions and mountain ridges, namely landscape homogeneous units—as well as on the basis of human elements, like the presence of signs of material civilization (towns, infrastructures, fortifications) and—possibly the most popular definition—the development of long-range roads, which have connected the west and the east, the north and the south of Eurasian landmass since ancient times.

It is the case of the most ancient and glorious of them, the SR, and also of further caravan routes, each one characterized by the specific product that was traded and transported on it, over continental distances, requiring incredible unearthly fatigue and difficulty. It is the case of incense, gold, spices, fur, tea and of further products characterized by similar specificities—being light, relatively precious and rare, easy to conserve and to transport. They define by themselves certain routes, characterized by junctions, passes and connections that would be the reference for the further formation of clusters of outposts, bazaars and towns.

Such transcontinental roads were segmented into innumerable itineraries, side and lateral variants; sometimes they were connected by pre-existent landscape signposts (signalled by a holy rock, a sacred tree, a traditional cemetery, a

[6]Gumilev L.N., 1972; Gvozdetsky N.A., 1974.

[7]A concept and a definition invented by Richthofen in the 1870s describing a set of traffic itineraries, that the merchants crossed possibly each one running for small segments, but connecting long distances for millennia; so similarly for many other variants in all the directions, overlying and combining with it (Barisitz Stephan, 2017:10).

"kurgan"), useful for orienting the travellers; sometimes they were connected by settlements called Chinese Towns, "kitay gorod,"[8] by Tatar and Russian merchants, that had been using these roads since early modern times: towns traditionally located at the distance defined by the duration of a day-light trip—the obvious measurement unit assumed by caravans—that had always meant for merchants, pilgrims, or travellers, the expectation of finding shelter at the end of the day in unknown countries.

In fact, in such borderless spaces, an oasis would appear on the horizon—to the eyes of the tireless travellers—as an explosion of humanity and nature; then a place of plenty for food, life, freshness and shade (or comfortable heat in winter), voices and colours, and above all water.

Sometimes, such oases would become important centres over time, namely the hub for powerful civilizations, developing irrigated agriculture, manufactures and bazaars; they would represent, in the imagination of the nomads, the exact opposite of the landscapes they were used to wandering, then wide extensions of deserts, steppes and frozen plateaus (a kind of representation of the paradise, exactly what they were looking for).

1.7 World History Primordial Engine

With time these spaces would design a kind of continental corridor stretching across the steppe longitudinally, between the Taiga forest (in the north) and the southern barrier-belt represented by CA mountains and deserts, connecting on an east–west axis the extreme parts of Eurasia (Figs. 1.2, 1.3 and 1.4).[9]

In this space powerful "nomadic empires" would arise periodically, appearing as aggressive hordes invading and destroying, often opposed to oasis towns, defining the terms of a confrontation that would inspire several geographical-political theories.

The most popular of them is the one assuming the east–west migration as one of the most powerful "geo-political engine" of all the times—at least until modernity. In fact, this space was crossed by apparently never-ending waves of populations that were pushing westwards, and occasionally in other directions: hordes and tribes appear periodically in this space, indeed following, not always intentionally, a path delineated since ancient times, and prefigured by a set of geographical signposts.

Such populations would melt, collaborate or sometime fight each other; they mainly arrived from the eastern side of the CA steppe, from Kazakh and the Mongolian prairies, where they originally formed.

In general, this movement originates from the typical continental disequilibria, namely by climatic and environmental conditions that, after periods of abundance, and of growth in any sense, can suddenly change (because of a drought, a sequence of bad seasons, epidemics or innumerable further causes), forcing the hordes to move in search of new pastures (Fig. 1.5).

Then an alternation of good and bad periods: good pasturages and a mild climate would signify in the long term, after just a few generations, a demographic boom for cattle, horses and consequently for human groups. Such situations, such continental "traps," would provoke a crisis, forcing the tribes to start a migration, occupying other territories and struggling for resources indistinctly with anyone they would eventually encounter on their path.

These changes had an evident impact on the destiny of the sedentary civilizations that were flourishing at the edge of this continental arid and semi-arid area, as a kind of necklace of bazaars, and kingdoms. The attitude of the nomads towards these settlements was actually different: sometimes they would dialogue with them, exchanging products and obtaining supplies in food or water, or whatever they needed, but in times of necessity, they would attack them mercilessly, starting a game of conquests, settling or destroying cities and infrastructures, or any kind of sedentary civilization.

This is the case of ancient oasis cultures—known by the classic names of Sogdiana, Bactria and Transoxiana, from Oxus, the old Greek name for Amu Darya—at the edge of the steppes, but also of further spaces, such as the Chinese and Persian empires, the Indian and Hellenistic kingdoms, the Roman and eastern Mediterranean civilizations.

In many cases, indeed, after having destroyed such civilizations, the nomads demonstrated that they were much more inclined to integration into those same institutions than expected. However, they never changed—and never could change—their basic territorial behaviour. Usually, instead of consolidating the conquest, after the incursions, in order to avoid the retaliations, they returned to the steppe, to the desert or the mountains; in this way they configured a kind of a

[8]Namely, outposts on the long way to China; oral source Maryashev A.N., Kazak Academy; Mariyashev A.N., 1994.

[9]Yilmaz Harun, 2013:49; the but reason is evident, just with a glance to the map at a continental scale, since the other directions were blocked by natural obstacles: the North and the East ones by climatic and natural barriers (the Taiga, big rivers such as Amur and Aigun, frozen soil winters as in Eastern Siberia); the South direction was blocked by deserts (Gobi, Taklimakan) and later by Chinese Big Wall, that since the second century B.C. started to pose a significant barrier to nomad movements – that nevertheless were capable of periodically overcoming such wall, conquering and destroying the Chinese states. The better (and often only) chance was to take the westwards direction, invading and riding (literally) along the steppe corridor, in the (relatively) narrow continental strip between mountains, forest Taiga (in the north), "internal seas" and deserts, "opening the door" fighting and attacking occasionally the oppositions they were meeting on the road.

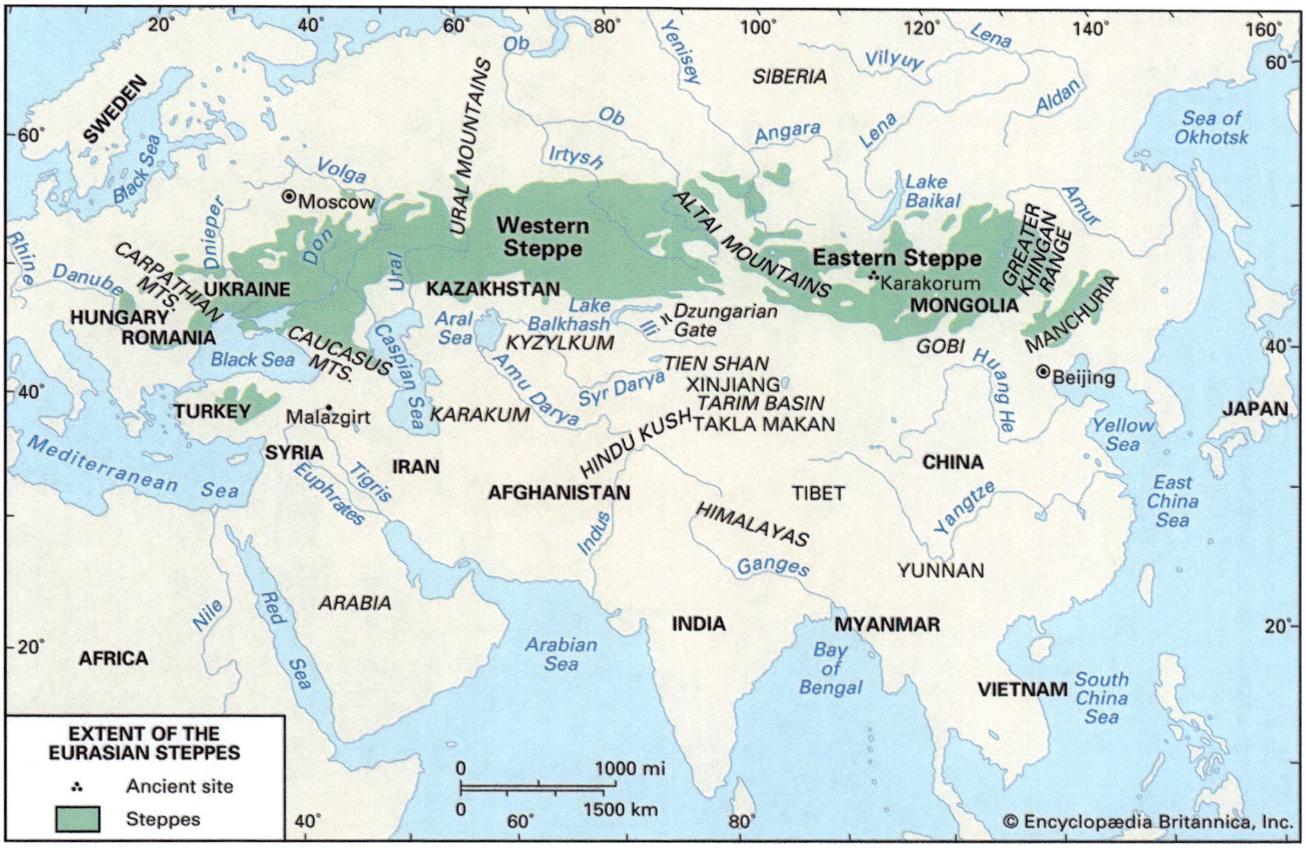

Fig. 1.2 Representation of the steppe corridor, connecting the east and the west, along what developed into the main millennial flows of Altaic populations westwards. (Source: Encyclopædia Britannica, Inc, https://www.britannica.com/place/the-Steppe/images-videos#/media/1/565551/3658, from McNeill, William H.. "The Steppe". Encyclopedia Britannica, https://www.britannica.com/place/the-Steppe. Accessed 4 July 2018)

pendulum movement between oasis city-states and the arid outback.

From a wider point of view, such migrations configure a prevalently westwards movement: the same geographical constraints forced the nomadic populations ahead into a sequence of pushes, in a cascade manner towards the west.

In this movement, and because of such invasions, the same role of the whole CA region changed, reverting occasionally from a peripheral position (dispersedly inhabited and an unknown area, impossible to settle) to the seat of impressive horsemen armies capable of conquering almost the whole world—as it was known in those times. Its destiny was that of becoming alternatively the centre of huge empires, ruled by valorous horsemen, and a remote land of desperate shepherds, respectively the subject or the object of the Big History played out on a wider scale in other places.

1.8 The Affirmation of Modern Territoriality

This lasted until the "last invasion"[10] of the Dzungarian Mongols in the eighteenth century, that found the Kazakh nomads in a critical moment: in fact they had already began, exactly in that moment, a sedentarization process; furthermore they were already under the pressure of the western modern states that had already begun their continental expansion, limiting their available space in the steppe.

The invasion may already be considered as something anachronistic, since at this time the destiny of nomadic organizations, versus the modern states, had already been defined: the figure of the destructive Dzungar hordes (or Oirate, or "tout court" Mongolian, depending on the definition) remained in the local images as the symbol of the barbarian enemy, the last one standing before modernization.

Such elaborations were ironically made by the same already sedentarizing Turkish tribes, just in the moment that they were abandoning nomadic living: they would record

[10]Bussagli M., 1970; Barisitz Stephan, 2017.

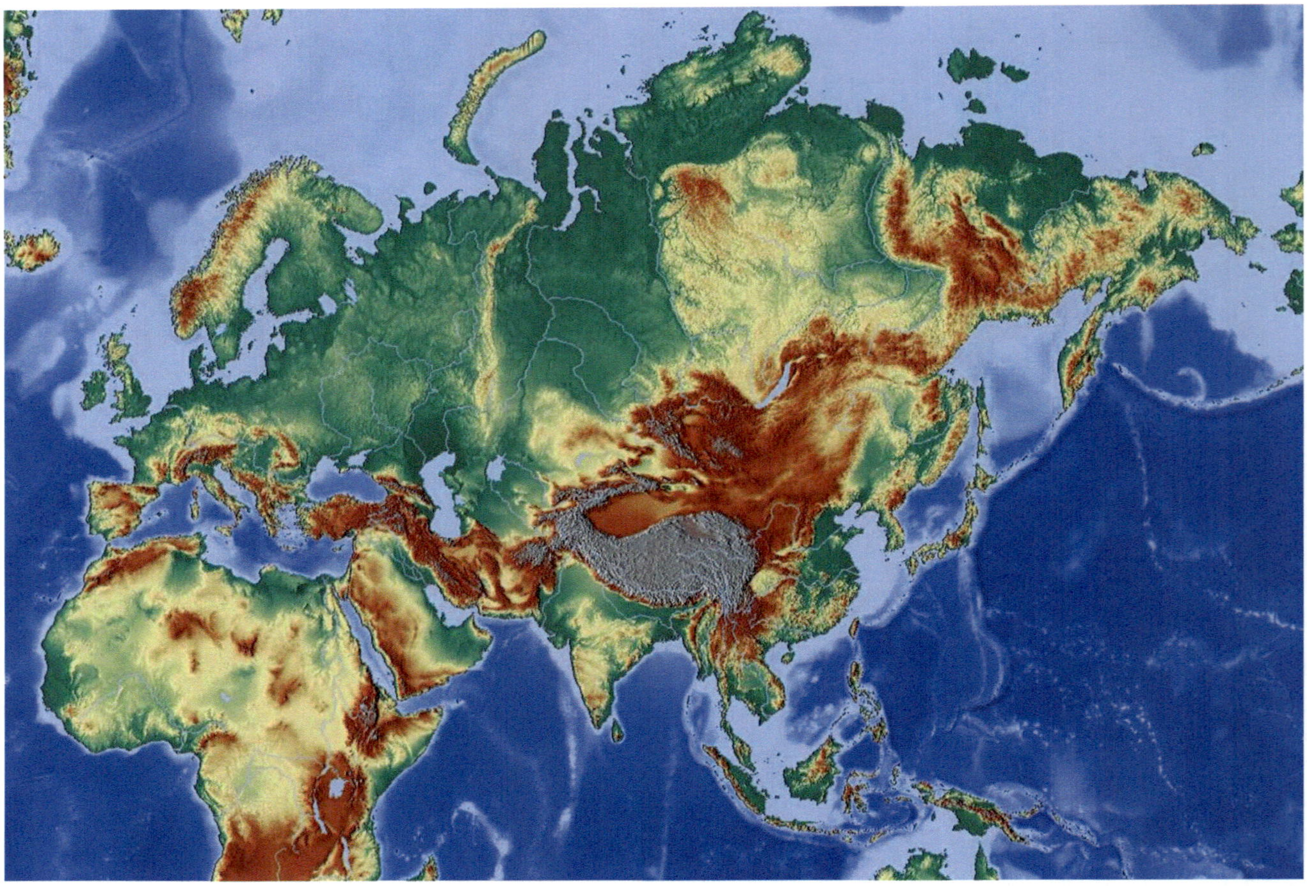

Fig. 1.3 The central collocation of Central Asia in the world physical map: similarly, as in Fig. 1.2, it is possible to observe the steppe corridor in the combination of plateaus, flatlands and deserts. (Source: Wikipedia commons https://commons.wikimedia.org/wiki/File:Eurasia_location_map_-_Physical.jpg, accessed 4.7.2018)

Fig. 1.4 United Nations geographical subregions. (Source: https://commons.wikimedia.org/wiki/File:United_Nations_geographical_subregions.png, accessed 4.7.2018)

Fig. 1.5 Kazakhstan, Dzungarian Alatau, 1997, shepherds summer camp. (Photo: Igor Jelen)

these "last wars" in their tales, then soon after they were codified in a written manner. It is the evidence of the fulfilment of a cycle, then the transformation of the traditional "genre de vie" into nothing else than a narrative made up of legends, tales, epics and oral literature (as usual for nomadic cultures).

It is the case of the extraordinary Kyrgyz epos Manas and of the Kazakh "batyr" mythologies that describe and transmit such memories, namely values and events of the legendary past, which is generally the case of CA classic narratives (what children usually study in school today).[11]

Such final invasions, with the consequent sedentarization, signified a definitive end to what was once a people of fierce horsemen, escaping the new coming barbarian hordes and looking to settle down in a definite area (possibly a kind of state limited by a perimeter of recognized and protected borders).

1.9 Stereotypes, Namely Proud Nomads, Flourishing Oases and Caravans on the Silk Road

The destiny of CA seems to be shaped by its same geography. The core of the Eurasiatic continent evidences such characteristics that make it almost impossible to imagine any kind of stability, or that can be managed in the frame of some political continuity: the undifferentiated spaces designed by deserts, plateaus and steppes are simply too far, and too difficult to organize for any structured civilization.

In this space, any semblance of stable culture seems to represent an extraordinary exception. It is the case of the bazaar cities that became the target of the same nomads: such appearance of richness is accentuated by the same characteristics of the area, with dramatic discontinuities predisposing the arising of cultural differences; simply such disequilibria may feed imaginations of the different populations, that would incite more and more traffic, in search of social relations, economic goods or for political predominance.

The geography of CA results basically from a condition of opposites, that are continuously mirroring each other: fierce warriors opposed to the richness of Samarqand and Bukhara, a struggle for survival as opposed to lazy life in artistic decorated palaces and mosques; and then, the free life of the steppe warriors as opposed to feudal institutions, slavery and exploitation—a further set of stereotypes that over time would become the main knowledge devices concerning this area and the "orient" in general.

Such confrontations characterized the CA for millennia: vanishing populations, riding camels and horses, appearing here or there in the wide spaces, seemingly poor and deprived of any richness, yet carrying with them sophisticated and "transportable" culture, namely precious jewellery, carpets and textile manufactures, and an extraordinary, even if invisible, patrimony of oral narrative.

Such figures—of nomadic riders, appearing and disappearing periodically like a natural phenomenon—will be reworked in different manners by the sedentary cultures

[11]Yilmaz Harun, 2013:53.

that had the chance to codify such images in a written code (a chance that the nomadic warriors did not have), finally representing the perfect opposite of their "own" point of view (in fact the perfect representation of the "other").

So, the image with which the nomads entered into world culture (and consciousness) was already originally something biased, namely the result of the elaboration of partisan writers representing their sedentary antagonists; they were generally missionaries, ambassadors or merchants (later explorers and journalists), that had the ability to reproduce such figures, eventually presenting them as infidels, cruel robbers or desperate migrants, used to facing the wild nature with their own naked hands (not aware of the richness their country represented, as the illuminist Von Humboldt stated in the early nineteenth century).[12]

And so, furthermore, images represented the fierce outlaws, the warrior and the hunter, or eventually the poetic shepherd "counting the stars in the endless steppe night sky" (as the poet Leopardi represented definitively in Occidental literature).[13] Then images of the "wild" that would have been at the time of early modernity "tout court" assimilated to the "primitive," as they would represent something residual of an original, full of energy but rule-less, life: just the opposite of modern (and western) supposed rationality, when such cultural models started to become prevalent.

Such representations soon became subject to manipulation, finally being reworked into cultural stereotypes.

It is the case of typical "resignation" to despotism, the fanatical believer, definitively incapable of integrating into modern culture.

It is the case of the images of the tirelessly roaming nomad, capable of survival in an almost instinctive and shamanic relation within their territory; or in its opposite images of the vicious emirs, feudally ruling and enslaving a wide population of serfs, made to do the heavy work on the land, blackmailing them with the menace of cutting off the little water available for irrigation and for any use—actually the perfect despotism opposed to the perfect nomadism, both indeed susceptible of being interpreted as symbols of immaturity.

Such images would soon become the object of a stigma that it would prove very difficult to emancipate from. As for the native culture in any part of the world, such scenarios represent the contraposition (that would become one of the most crucial for the entire colonial scenario) between structured civilizations and the never-changing natives, subject to a life that would "repeat always the same,"[14] as a kind of underdevelopment trap (therefore destined to be destroyed by the more "civilized" systems).

Finally, such considerations inspired not just literature and poets, but ideologists and politicians as well as entire populations—who would later eventually migrate and invade those same countries. They were possibly predisposed to some interested interpretation, namely to an ideology of their superiority inducing a new cycle of conquests and expropriations of domestic resources: an ideology that would legitimize colonization, and the imposition of a modernization cycle.

1.10 CA Civilizations in World Perception Until the Current Days

All these figures were soon classified into some corresponding branch of "orientalism,"[15] enlisting a sequence of myths and practices, of picturesque landscapes and human habits.

So, in a similar way, for both natural conditions and human constructions, and for the huge but rapidly disappearing nomadic empires—that spread out into the "nothingness" of the steppes—originally "opposed" to oases and cities, that would sometimes become paradoxically the cradle of new civilization cycles at the end of a destruction campaign.

This is the case of universally celebrated cultures, enriched by the presence of protagonists such as Ibn Sina (Latinized in Avicenna), Ulugh Beg and further famous scientists and writers (Fig. 7.7).

For the steppe nomadic tribes these tales, representing the rudimentary shepherd's life, also may have opposed the sophisticated atmosphere of the Persian courts, their magnificent literature, architecture, sense of decoration and aesthetics.

It is particularly the case of the "Thousand Nights" and of the tales of Scheherazade, of the Samarqand literature heritage, which could be traced back to the original traditionally orally transmitted tales. These consisted in a set of variants adapted over the centuries: "[t]he golden age of Persian (Tajik) literature is also closely associated with these cities and many Persian-speaking poets, astronomers and Islamic

---

12 Von Humboldt A., 1975.

13 In his nineteenth-century poem "Canto notturno di un pastore errante per l'Asia", Giacomo Leopardi cites the travel of baron Meyendorff, namely his writing "Voyage d'Orenbourg à Boukhara fait en 1820", inspired by such style of life, always attracting the interest of writers of the sedentary societies, representing the phenomenon of people without a home, continuously tirelessly wandering; Leopardi G. (1828).

14 Braudel Fernand, 1979.

15 Said E.W., 1991, but also Barthold V.V., 1947; Barthold W., 1913, 1945, 1968.

theologians lived and worked here, giving Samarqand and Bukhara a halo of glory."[16]

Such images also have corresponding meanings in social and in political patterns, based on the same rhetoric of the "extremes": the free tribes are represented as "warrior democracies,"[17] from the beginning opposed to the "civilized" organizations, where power is exhibited in some city, or in some palace or ritual. Such representations signify the opposition between material power—the rule of force—and institutional power, namely its symbols. But sometimes such contraposition finds a compromise, eventually becoming the site of some wider authority, starting the accumulation of power in some wider territory or even an empire.

So too, and in a similar way, for further representations of life, for societal organizations, religious beliefs, values and habits; so are the prevalent figures representing power, the apparently senseless liturgies, the fanatic faith of the believer, practicing rituals beyond any limit, until sacrifice and martyrdom, or until the master is transfigured into an omnipotent being (as a sacred divinity).

The perception of Asia, and of CA culture, is based on such mythical descriptions—in fact representing just a part of the reality—of absolute power and of absolute vacuum. In such images, from the time Herodotus theorized the confrontation between Orient and Occident, the former became the place of excess in any field, in culture, art, religion, aesthetics, behaviours and values, moulding all aspects of social and individual life. And this especially counted for politics, with the image of power being sacralised (the definition of the "oriental despotism"): a representation that is still current, having been adapted to recent interpretations.

In fact, such elements have been reproduced in times of modernity, in different circumstances but sometimes in a paradoxical manner (considering that the modernity has been defined as the epoch of the triumph of functionality and of a rationalistic interpretation of the human role in reality).

However, as would soon become evident, such changes result in further confusion between the dimensions of the "traditional" and of the "modern," in the context of colonial and of Soviet oppression (and eventually of the current globalization), of new technologies presented as "magic," of the imposition of new powers, prefiguring a confusion between real and metaphysical interpretations.

[16]Dagiev Dagikhudo, 2014:110–111.

[17]Or "military democracy". Esenova Saulesh, 1998; Khazanov A. M., 1994; Glavnaja Redakcija Kirgizskoj Enciklopedii, 1994, 1995; Zhirmunski V.M., 1995; Gyozalov F., 1995.

It is the case (among the others, that this book will try to evidence) of spreading the idea of the President (or of the Unique Party) as that of a sacred figure, and of the modernist technique as the miraculous demonstration of power, finally legitimizing the "revolution" in the relationship between human dimension and the balance in relations with nature.

1.11 Evolutions Coming to Anything?

Today, after a cycle of tragedies, with the supposed upstaging of modernity, and with the collapse of the modernist "empires," local society has had the chance to become aware of the damage, sometimes irreversible, provoked by such changes. They were indeed so severe that they brought the western and indeed the whole world civilization, at a certain point of their history, close to self-destruction.

Many signs indicate a change in the attitude of elites and populations, although the practices do not always truly reflect this change, remaining often at the level of formal declaration, of manifested ideologies and propaganda—concerning democratization, exhibition of environmentally friendly policies and peaceful international relations.

The change in attitude seems to have started a new cycle of adaptations, of mutual recognition, of attempts to restore what in nature and in culture has been destroyed by an indiscriminate idea of progress: peoples and institutions are aware that modernity provoked a sequence of shocks, culminating in the experience of totalitarian power that probably changed the relationship between power and societies in a definitive manner, disseminating a persistent fear in the face of authority.

Today is just the beginning of a new chapter; the forces of geopolitics, the drift of technological innovations, or the transformations induced by a global culture, which is self-imposing and which penetrates all societies all over the world, seem to exert an influence on local evolution, independently from intentions.

At the same time, the post-Soviet populations—finally gaining independence, taking advantage of a set of cultural and material improvements—seem to have a chance. This means the possibility of starting a new cycle, recovering what they had lost in the previous centuries; finally, they have all the resources they need, not just natural and energetic resources but also an easily convertible cultural richness necessary to face this passage (Figs. 1.6 and 1.7).

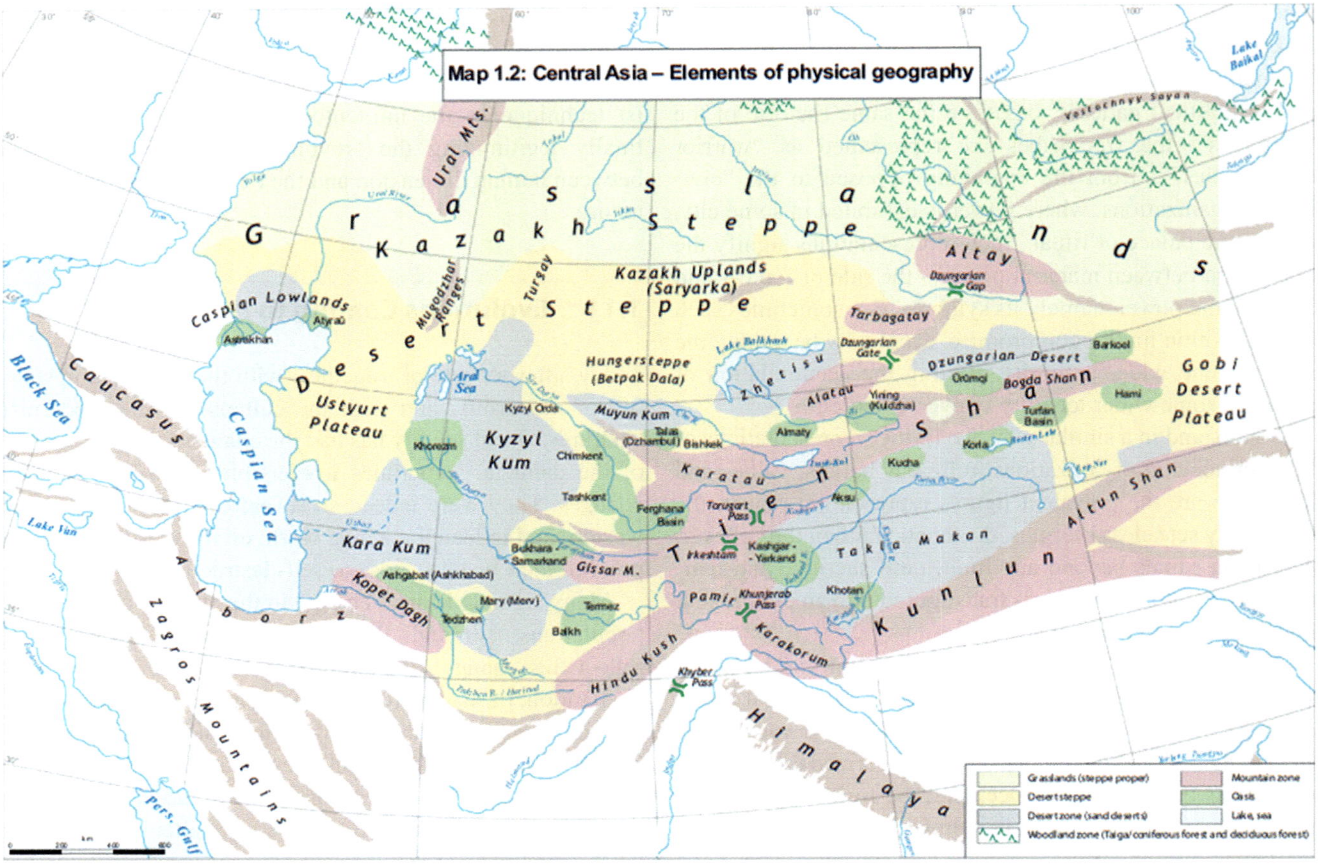

Fig. 1.6 Elements of physical geography in Central Asia. (Source: Courtesy, Barisitz Stephan (2017:5))

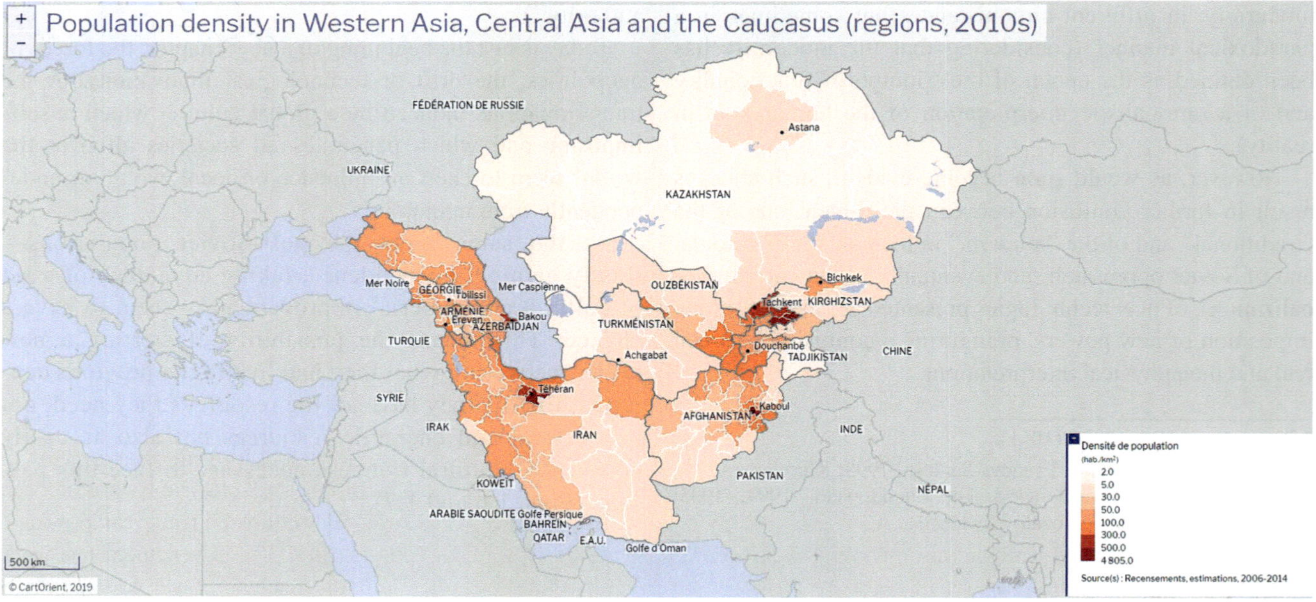

Fig. 1.7 Central Asia and contiguous countries' population density. (Source: Courtesy, Thorez Julien, www.cartorient.cnrs.fr)

References

Barisitz Stephan (2017) Central Asia and the Silk Road, economic rise and decline over several millennia. Springer, Cham

Barisitz Stephan (2018) Myth and actuality of the Silk Road: geo-economic opportunity vs. and geo-political convenience, ASIAC annual conference 2018, 5–7 December 2018, University of Trieste – Gorizia campus (manuscript)

Barthold VV (1947) La Découverte de l'Asie: Historie de l'orientalisme en Europe et en Russie. Payot, Paris

Barthold W (1913) Die geographische und historische Erforschung des Orients mit besonderer Berücksichtigung der russischen Arbeiten. Wigand, Lipsia

Barthold W (1945) Histoire des Turcs d'Asie centrale. Adrien-Maisonneuve, Paris

Barthold W (1968) Turkestan down to the Mongol invasion. E.J.W. Gibb Memorial, Luzac, London

Biraschi AM (2000) Introduzione a Strabone, Geografia. BUR, Milano

Braudel Fernand (1979) Civilisation matérielle, économie et capitalisme, XVe-XVIIIe siècle 1. Les Structures du quotidien – 2. Les Jeux de l'échange – 3. Le Temps du monde, Paris, Armand Colin

Bussagli M (1970) Asia Centrale e mondo dei nomadi. In: Storia universale dei popoli e delle civiltà. Asia Centrale e Giappone. UTET, Torino

Dagiev Dagikhudo (2014) Regime transition in Central Asia. Routledge, New York

Edgar Adrienne Lynn (2004) Tribal nation: the making of Soviet Turkmenistan. Princeton University Press, Princeton

Erodoto (1988) Storie. Mondadori, Milan

Esenova Saulesh (1998) 'Tribalism' and identity in contemporary circumstances: the case of Kazakhstan. Cent Asia Surv 17(3):443–462

Glavnaja Redakcija Kirgizskoj Enciklopedii (1994) Kyrgyzstan

Glavnaja Redakcija Kirgizskoj Enciklopedii (1995) Encyclopaedical phenomenon of epos "Manas", article collection, Biškek

Gumilev LN (1972) Gli Unni. Einaudi, Torino

Gvozdetsky NA (1974) Soviet geographical explorations and discoveries. Progress Publishers, Moscow

Gyozalov F (1995) The problem ideological schemes in the "Manas" epos. Kyrgyz Republic National Academy of Sciences 1995, Bishkek, pp 67–68

Ibn Khaldun (1980) The Maqaddimah (trans: Rosenthal F), 2nd edn. Princeton University Press, Princeton

Khazanov AM (1994) Nomads and the outside world, 2nd edn. The University of Wisconsin Press, or. ed. in Russian 1983

Leopardi G (1828) Canto notturno di un pastore errante per l'Asia, v. anche lo Zibaldone, 4400 in data 3 ottobre 1828

Mariyashev AN (1994) Petroglyphs of South Kazakhstan and Semirechye. A.H. Margulan Archeology Institute, Pilgrim Firm, Zaman Company, Almaty

Ramazanova N, Berdenov Z, Ramazanov S, Kazangapova N, Romanova S, Toksanbaeva S, Jan W (2019) Landscape-geochemical analysis of steppe zone basin Zhaiyrk. News Natl Acad Sci Rep Kaz 4(436):33–41. https://doi.org/10.32014/2019.2518-170X.95

Rasanayagam Johan, Beyer Judith (2014) Ethnographies of the State in Central Asia Performing Politics. Indiana University Press, Bloomington. ISBN: 978-0-253-01141-1

Said EW (1991) Orientalismo. Bollati Boringhieri, Torino

Scharr K, Steinicke E (hg.) (2012) Vom euphorischen Aufbruch in die Realitaet des Alltages, 1898–2000 – zwei Jahreszenten Transformationsforschuing, University of Innsbruck Press

Thorez Julien (2018) Central Asia: the state of the research, Asian annual conference 2018, 5–7 December 2018, University of Trieste – Gorizia campus (manuscript)

Valikhanov ČČ (1985) Sobranje sočinij v pjatu tomach. Glavnaja Redakcija Kazachskoj Sovetskoj enciklopedii, Alma-Ata

Valikhanov capt, Veniukof M., et alii (1865) The Russians in Central Asia (trans: John, Robert Mitchell, Edward Stanford). London

Von Humboldt A (1975) a cura di Milanesi M. – Visconti Viansson A., La geografia, i viaggi. F. Angeli, Milano

Yilmaz Harun (2013) The Soviet construction of Kazakh batyrs. In: Sevket A, Richard C (eds) Social and cultural change in Central Asia. The soviet legacy. Routledge, London and New York, pp 45–62

Zhirmunski VM (1995) Introduction in learning of epos "Manas". In: Glavnaja Redakcija Kirgizskoj Enciklopedii, pp 330–374

The Geographical Setting and Physical Environment: A Literature Review

2

Abstract

The geographical setting of Central Asia (CA), as its *place* in the space of Northern Hemisphere, can be described by the inter-relations of the main components and phenomena of physical environment, its geology, relief/geomorphology, climatology, hydrology and biology, not only recent but also of the past. The total area of the Central Asia region with five former Soviet republics (Kazakhstan, Uzbekistan, Tajikistan, Kyrgyzstan and Turkmenistan) occupies 4,003,451 km^2. It extends between 45 and 90° E and 36 and 55° N; the prevailing surface elevation ranges between 0 and 500 m above sea level (asl), with the lowest point at Turpan Depression, –154 m, and the highest Ismoil Somoni Peak at Pamirs, 7495 m. The region is mainly endorheic with small drainage area in the northern part, belonging to Kazakhstan, towards the Arctic Ocean.

The continental core of Asia was assembled during the late Palaeozoic (about 500 million years ago) by numerous collisions between small continental blocks and by accretionary growth along convergent margins. The recent tectonic activity is associated with the zones of faulting.

The Caspian Sea and the Aral Sea are the remnants of the Tethys Sea, which extended between the fragments of Europe and Asia. The Aral Sea depression was formed towards the end of the Neogene Period. Both the Aral and Caspian seas fluctuated in area and volume many times through their history.

The geological structures of the region are rich in different deposits and resources. It accounts for about 4% of global energy deposits: the oil reserves in Central Asia and along the Caspian Sea coast are comparable to that of Qatar.

Tajikistan, which possesses 53% of Central Asia hydropower resources, is one of the world leaders in hydropower potential. The renewable energy sources (RES) have a large potential in the Central Asia region; however for the moment-being their usage is low, except for hydropower stations in Tajikistan and Kyrgyzstan.

All the countries in the region are rich in different minerals: metallic ores, non-ferrous, rare earth metals. Kazakhstan had the second largest uranium reserves in the world in 2016 (after Canada).

According to Köppen, there are 7 climatic zones in Central Asia region, and the general trend during the last 30 years showed an increase in air temperature and aridity. Among 21 agroclimatic zones in the CA, 2 of them are dominated in the area: (1) semi-arid with cold winter and warm summer and (2) arid with cold winter and warm summer.

Keywords

Geographical setting · Geologic history · Natural deposits and minerals · Climate features

2.1 The Geographical Setting

Even starting with human geography, we have to focus on the physical background of this large region of Central Asia (hereinafter CA), extending from the Caspian Sea in the west to the border of western China in the east and from the Siberian Russia in the north to the mountains and deserts in the south bordering with Iran, Afghanistan and China. It is extremely large and rich in objects of physical geography, including high passes and mountains (mostly in Tajikistan and Kyrgyzstan), vast deserts as Kyzyl Kum and Kara Kum, extending in Turkmenistan and Uzbekistan, and especially treeless, grassy steppes of Kazakhstan (Fig. 2.1).

I. Jelen et al., *The Geography of Central Asia*, World Regional Geography Book Series,
https://doi.org/10.1007/978-3-030-61266-5_2

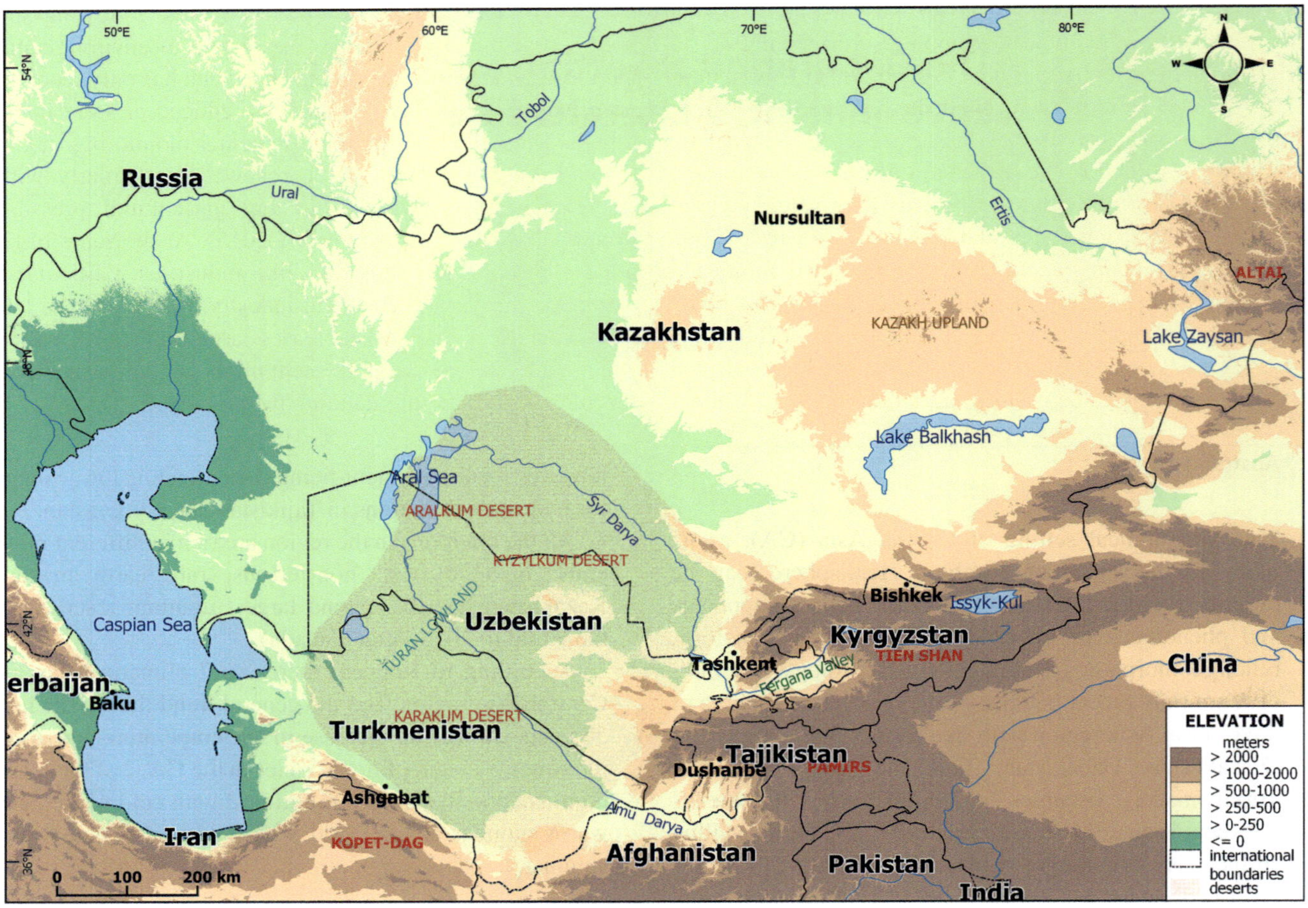

Fig. 2.1 The physical map of CA region with the main surface elevation groups. (Drawn by Martynas Bučas, Klaipėda University)

Two main rivers of CA region—the Syr Darya (older name Jaxartes) and the Amu Darya (older name Oxus)—after rise in the Tian Shan and Pamir mountain ranges wind north-westward through the different landscapes flowing into the Aral Sea. Their watersheds make the largest area in total Aral Sea basin and provide most of the region's water resources. The northern Kazakhstan is drained by rivers flowing north through Russia into the Arctic Ocean. It is necessary to mark that the western part of Kazakhstan geographically belongs to Europe (about 10% of country's area; Bruk et al. 1999; Elr. "Central Asia"… 2017).

The high mountain systems extend in the east, such as Altai Mountains, or in the south of CA, like the Tian Shan ranges. The highest peaks in the Tian Shan are Victory Peak (Kyrgyz, Jengish Chokusu; Russian, Pik Pobedy), which reaches 7439 m, and Khan Tengri Peak (Kyrgyz, Kan-Too Chokusu), with 6995 m. In the eastern Tian Shan, the deepest Turpan depression (–154 m below sea level) is found indicating large variation in the territory's altitudes (Bruk et al. 1999).

Another high mountain system of CA is the Pamir Mountains (Fig. 2.2).

From the Pamirs, also called **Pamir**, several south-central Asian mountain ranges radiate, including the Hindu Kush, the Karakoram Range, the Kunlun Mountains and the Tian Shan. Most of the Pamirs extend within Tajikistan, where the peaks higher than 7000 m are found (Allan and Zakharova 2017). The highest is Ismoil Somoni Peak (known from 1932 to 1962 as Stalin Peak and from 1962 to 1998 as Communism Peak), 7495 m. Few other high peaks are: Ibn Sina Peak (still unofficially known as Lenin Peak), 7134 m, and Korzhenevskaja Peak, 7105 m (Gvozdetskij 1970).

The steppe areas of CA belong to the Eurasian Steppe geographical zone.

Major surface bodies of water include the Aral Sea and Lake Balkhash, both of which are the part of the huge west-central Asian endorheic (closed, not having drainage) basin that also includes the Caspian Sea. The Aral Sea and the Lake Balkhash have shrunk significantly in recent decades due to decrease in runoff from rivers that feed them for irrigation and other purposes. Water is an extremely valuable

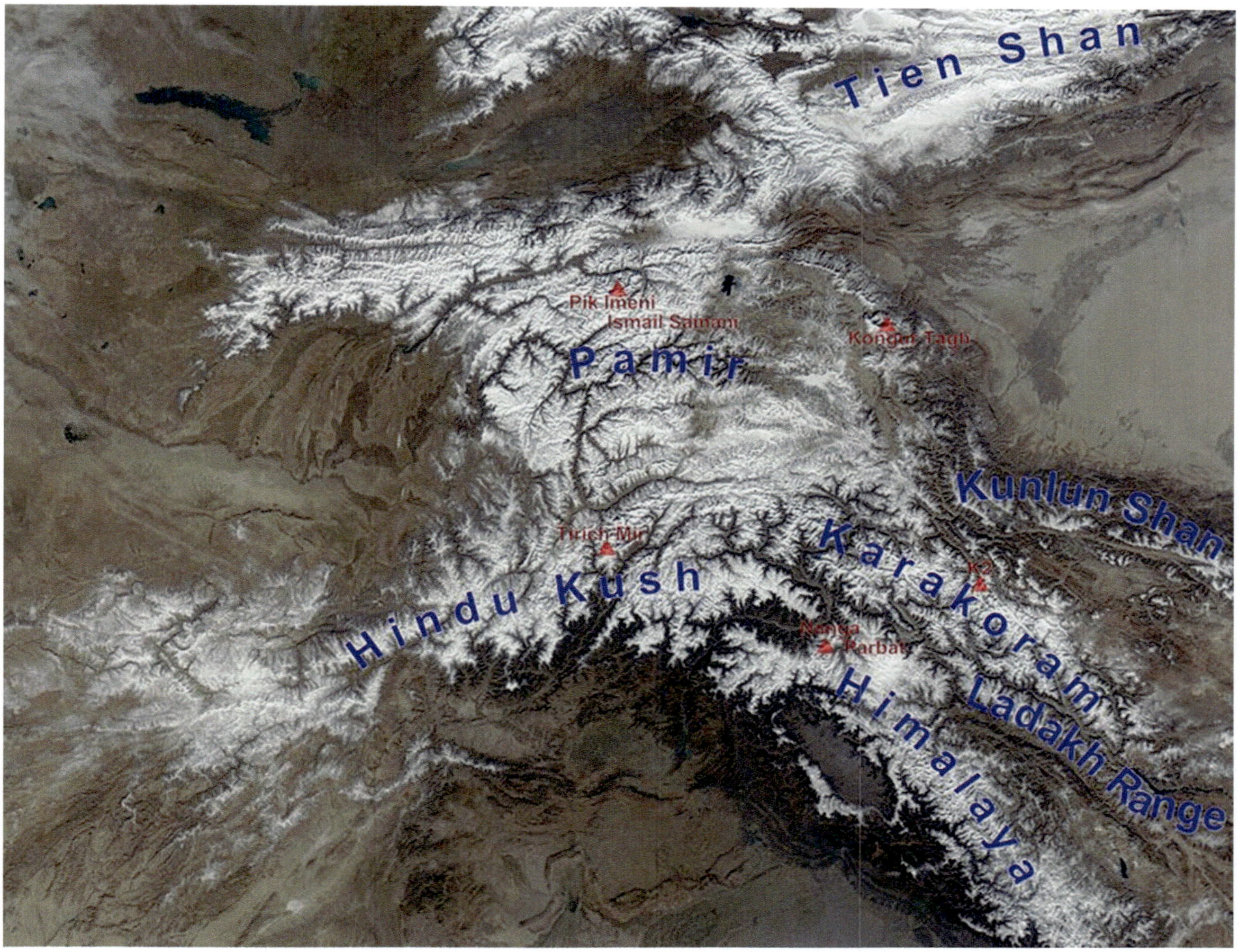

Fig. 2.2 Satellite image of CA region mountains (27 November 2012). (Source: Derived from Hindu Kush satellite image.jpg:, Public Domain, https://commons.wikimedia.org/w/index.php?curid=22886435)

resource in arid CA and can lead to rather significant international disputes (Central Asia 1998–2014).

2.2 Geological Conditions: Main Events and Structures During the Geological History of the Region

After the collapse of the Soviet Union, new and modern research data on geology and physical geography of CA region have become available for the researchers in different parts of the world. In the twenty-first century, more scientific papers, prepared by the younger generation of researchers in CA countries as well as in Russia and in collaboration with foreign partners, have been published in English. Thus, the region becomes more opened for the world's academic society nowadays.

The recent geological, geochemical, geophysical and Earth space related research, carried out by scientists from different countries, has revealed more precise picture of the formation history of Eurasia continent. According to Heubeck (2001), a core of Asia was assembled during the late Palaeozoic (about 500 million years ago) by numerous collisions between small continental blocks and by accretionary growth along convergent margins (Fig. 2.3).

But the early formation of geological structures of recent CA region was closely connected with global tectonic processes of Precambrian timeline. It started with the breakup of Rodinia around 970–850 Ma and the following plate tectonic events, resulting in formation of new oceans and orogenic belts as well as the accretion of island arcs in the recent territory of CA region (Dobretsov and Buslov 2007; Dobretsov et al. 2003; Buslov et al. 2004; Rogers and Santosh 2004; Yarmolyuk et al. 2013). According to Windley et al. (2007, p. 44), "it is reasonable to use 'modern' accretionary models

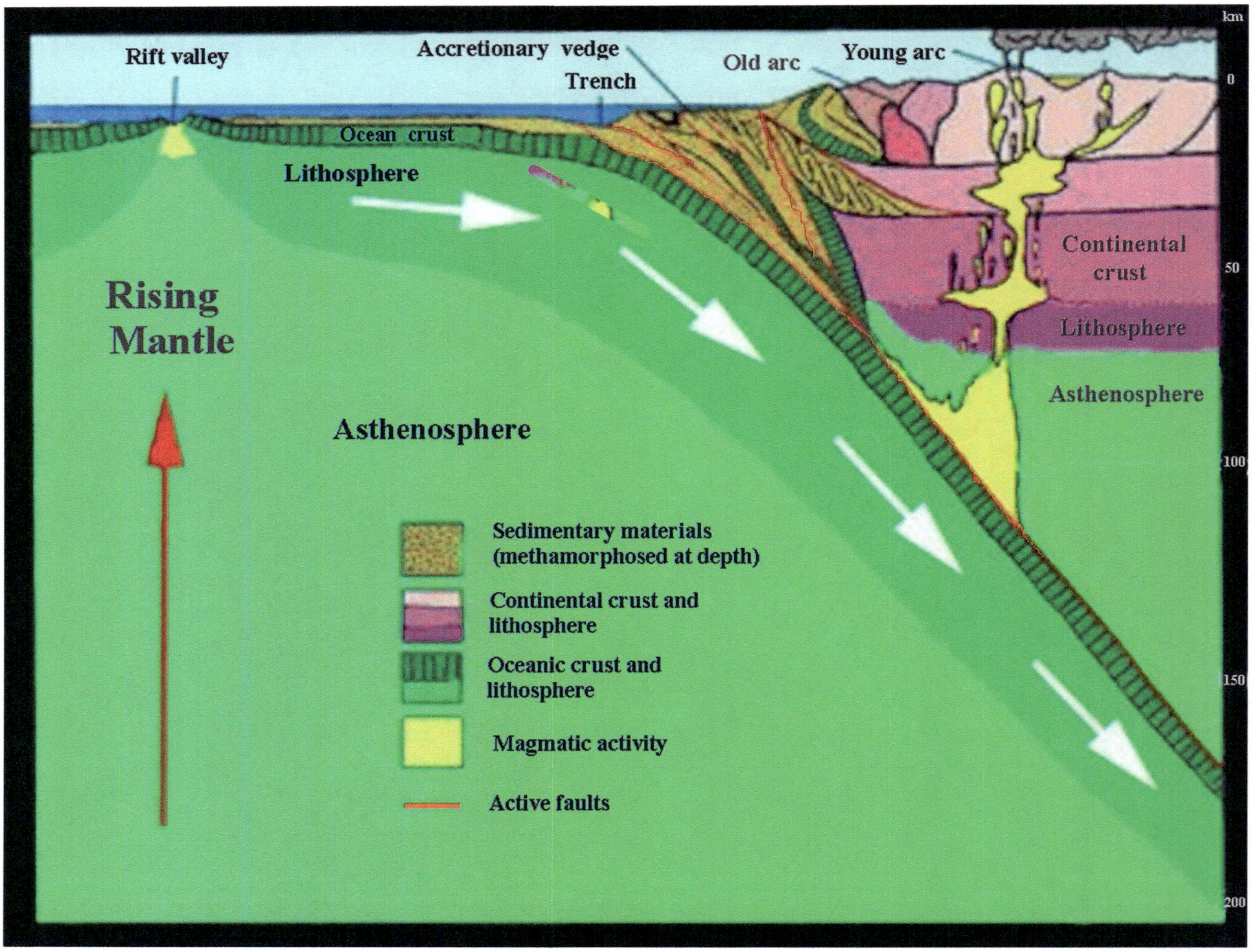

Fig. 2.3 Illustration of accretionary process with old and young arc convergence. (Source: USGS_Visual_Glossary_Accretionary_wedge.gif from Wikimedia Commons)

to explain the tectonic evolution of the Central Asian Orogenic Belt. The question arises: which model is the most viable? Data point towards an archipelago-type (Indonesian) model."

Other authors, Safonova (2009) and Safonova and Santosh (2014), in their publications have focused on different cases of accretion of these geological structures. The tectonic history of the CA mountains is given in Zonenshain et al. (1990). Two main contrasting hypotheses for the evolution of the CA orogenic belt (hereinafter CAOB) have been worked out. The supporters of the first hypothesis (Şengör et al. 1993, Şengör and Natal'in 1996) recognize the CAOB as an originally single and long chain of the Palaeozoic intra-oceanic island arc systems. Another, supported by Zonenshain et al. (1990), Mossakovsky et al. (1993), Dobretsov et al. (1995), Kröner et al. (2014), Windley et al. (2007) and Safonova (2009), suggests that the CAOB comprises a collage of microcontinents and oceanic arcs that collided with one another and eventually accreted to the Siberian, Tarim and North China cratons. The recent research on CAOB has shown that these processes were more complicated: some structures like the Kazakhstan arc chain were characterized by multiple subduction, whereas others like the northern fringe of the Tarim Craton remained as a passive margin (Xiao and Santosh 2014).

Being the world's largest continent, Eurasia is the youngest continent with diverse geological composition and particularly orogenic belts. The coexistence of Pacific-type (P-type) and collision-type (C-type) orogenic belts in the CAOB was revealed, where the P-type is dominant (Safonova and Maruyama 2014).

The geologic history of CAOB structures is evident in the minerology of their rocks. For example, the Tien Shan ranges are composed in the main of crystalline and sedimentary rocks of the Palaeozoic Era (about 540–250 million years ago). During that period, at least two foldings took place. The basins between the mountains are filled with younger sediments that were formed by the erosive action of the rivers. Granitic rocks are found mostly in the north and east of the Tien Shan, and sedimentary metamorphosed rock with some volcanic and intrusive patterns are found in the southern-western part of the ranges. Loose substrates of weathered and eroded rocks have accumulated into the valleys such as the Fergana Valley making almost 8 km thick layers. Large salty deposits from the evaporated lakes are found here too (Bruk et al. 1999; Bossu et al. 1996).

The history of formation of the Caspian Sea began from the Triassic following the interaction of the Eurasia, India and Arabia plates, and numerous microplates and closure of Tethys Sea. The South and North Caspian underwent reorganization during the Oligocene–Neogene (Sagers 1994). Since the Pliocene, its area and volume has been changing many times (Fig. 2.4).

Svitoch (1991) subdivided the Caspian Sea level cycles in five classes. From the time of sea formation as a closed basin in the Pliocene, at least six major transgressional phases have been distinguished. According to Rychagov (1977, 1993), up to five transgressional phases have been described and ^{14}C dated around 8000 BP, 7000 BP, 6000 BP, 3000 BP and 200 BP. Seemingly the reasons of these phases might be complex, affected by both climatic and geological factors. The last can be considered as a consequence of plate tectonic subduction, as epeirogenic processes in the lithosphere crust (Alden 2020). These processes continue nowadays overlapping with sedimentary deposits and dynamics in the southern part of Sea basin as it is described in Firoozfar (2014). The active sedimentation is taking place also in other parts of the Caspian Sea. According to Lilienberg (1985, 1993), levelling record of the last century shows that the Volga delta area is subject to uplift up to 3 mm/year. During the twentieth century, sharp changes in the sea water level have been taking place too: it dropped by 3 m between 1929 and 1977 and then largely recovered with a 2.6 m rise from 1978 to 1995. The causes of such frequent fluctuations are not yet entirely clear. Some studies and their authors distinguish the climate change and anthropogenic factors as the main, while others tend to explain the Sea level cycles by the intensity of sedimentary dynamics and geological factors. The recent tectonic dynamism of the CA region from north-west to southeast is connected with orientation of the main faults, being formed during Palaeozoic timeline. The registered earthquake data approve the importance of tectonic factors in the formation of both terrestrial and underwater geological structures of CA region (Thomas et al. 1999).

Fig. 2.4 Maximum and minimum levels of the Caspian Sea. (Source: https://commons.wikimedia.org/wiki/File:Caspian_sea_history.jpg#filelinks NASA)

The Aral Sea is also the remnant basin of the large Tethys Sea. The plate drift, uplift of continental blocks and huge sedimentation caused the Tethys to disappear. The sediments found in the basin of the Aral Sea are of Jurassic and Cretaceous periods mostly. According to Breckle and Geldyeva (2012, p. 14), they "consist of limestone, clay marls, sandstones and rather often evaporites, such as salt (NaCl), gypsum and even potassium salts, indicating various strong phases of desiccation, regressions and transgressions." These authors also share the opinion that combination of different factors, like tectonic-geological (including sedimentary dynamics) and climatic, accompanied by anthropogenic activities have caused big changes in the hydrological cycles of the Caspian and the Aral seas during the last several thousand years (Breckle and Geldyeva 2012).

The recent plate movements around Eurasia are dominated by northward directions, which according to the opinion of Safonova and Maruyama (2014) will result in Africa–Eurasia collision in the near future. Thus, Eurasia continent is further going to be accreted and likely to become in a 200–250 Ma future the supercontinent "Amasia" or "Afeurasia".

2.3 Main Deposits and Minerals in CA Region; Oil, Gas, Coal—Main Energy Resources

CA countries dispose the rich deposits of the most known metals, minerals and other resources (Table 2.1).

CA accounts for about 4% of global energy deposits. The oil reserves in CA and along the Caspian Sea coast amount

Table 2.1 Main deposits, minerals and other resources found in CA region

CA countries	Resources
Kazakhstan	Coal, oil, gas, iron, chromium, gold, copper, zinc, lead, manganese, uranium, diamond
Kyrgyzstan	Deposits of metals including gold, antimony and rare earth metals; locally exploitable coal, oil and gas
Uzbekistan	Oil, gas, coal, gold, uranium, lead, zinc, copper, wolfram, molybdenum
Turkmenistan	Oil, gas, sulphur, salt
Tajikistan	Lignite, some oil, aluminium, uranium, rare earth metals

Source: The CIA World Factbook (2017)

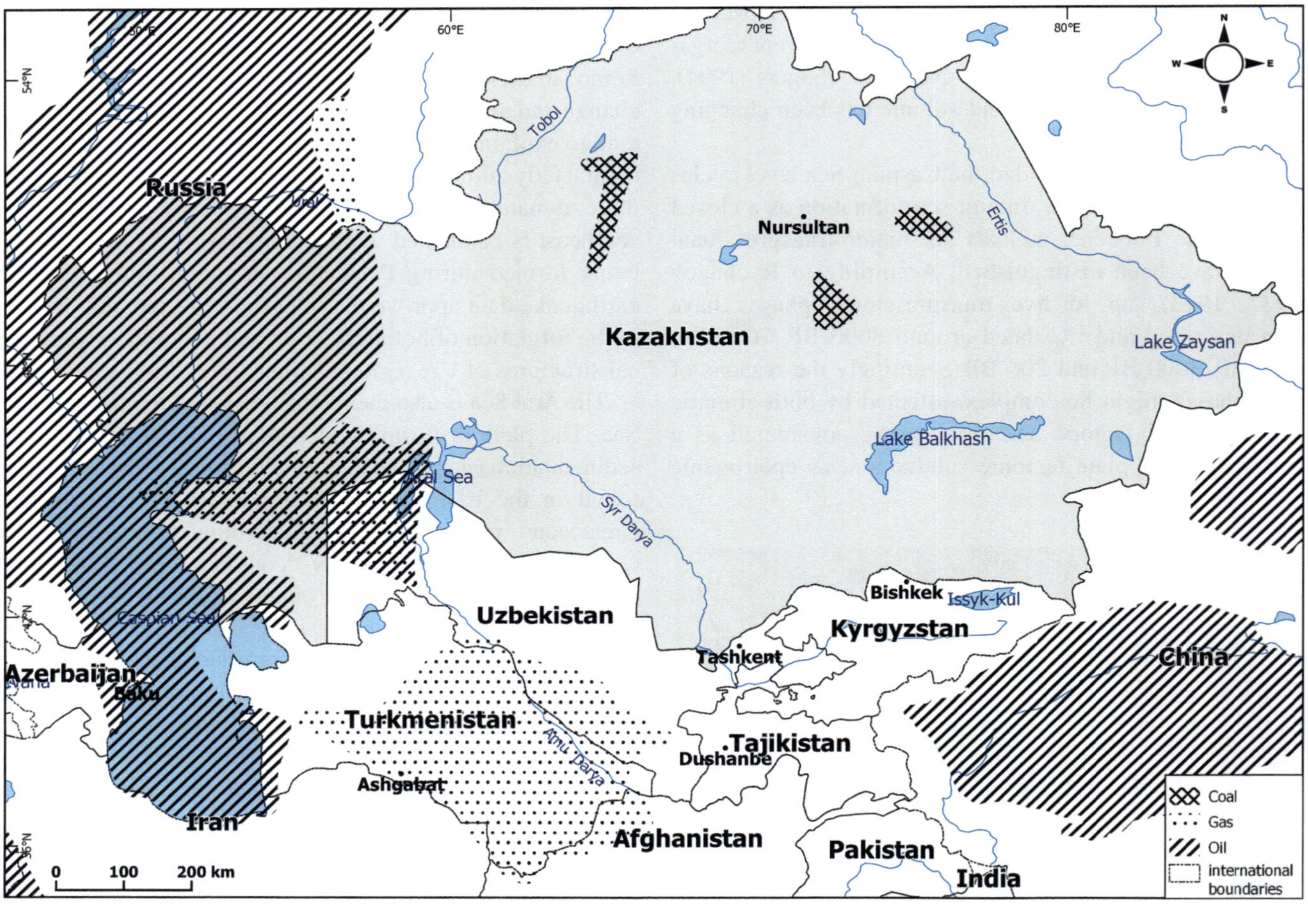

Fig. 2.5 Distribution of oil and gas fields and coal deposits in CA region. (Made by Martynas Bučas (Klaipėda University) using (Sources: Li (2011); Central Asia Atlas of Natural Resources (2010))

to 17–33 bbl/d, which are comparable to that of Qatar (Thomas et al. 1999). The location and connection of oil fields to territorial tectonic structures is outlined by Samsonow (1994), Persits et al. (1997) and Wandrey and Law (1999). Figure 2.5 shows the distribution of the main oil, gas and coal deposits in the CA region.

Major fields of oil and gas containing rocks are associated with the North Afghan block, Amu Darya Basin and Afghan-Tajik Basin provinces. Almost all these structures are related to the Neogene wrench faulting and are similar to the traps along the San Andreas fault (Harding 1976). Oil and gas traps of CA region are found mainly in the Upper Jurassic carbonates and Lower Cretaceous sandstones or in mudstones within carbonaceous shales and coal, and in Palaeogene (Eocene) marine or possibly lacustrine shale (Kingston and Clarke 1995; Klett et al. 2006). According to

Table 2.2 Energy production by CA country in 2000–2006 and 2014–2016

Commodity	Kazakhstan	Kyrgyzstan	Tajikistan	Turkmenistan	Uzbekistan
2000–2006					
Oil (ths.b/d)*	1370[f]	1.4[d]	2.5[d]	187[f]	114[f]
Gas (bcm)	11.6[g]	0.016[g]	0.05[g]	63.1[g]	4.852[b]
Coal (mtoe)	43.2[de]	0.508[a]	0.0127[a]	0	2.096[a]
Electricity (billion kWh)	66.8[c]	11.7[b]	15.1[b]	11.4[g]	47.7[b]
2014–2016					
Oil (ths.b/d)	1672[i]	1.0[g]	1.8[i]	261[i]	55.0[i]
Gas (bcm)	21.4[ei]	30.0[eh]	20.0[eh]	83.7[ei]	55.7[eh]
Coal (mtoe)	106.5[h]	1.6[g]	0.6[g]	0	4.0[h]
Electricity (billion kWh)	100.4[eh]	12.8[eh]	17.0[eh]	21.2[eh]	59.0[ei]

Note: *Includes crude oil, shale oil, oil sands and natural gas liquids (NGLs – the liquid content of natural gas where this is recovered separately) and excludes liquid fuels from other sources such as biomass and derivatives of coal and natural gas; ths.b/d, thousands of barrels per day; bcm, billion cubic metres; mtoe, million tonnes of oil equivalent; kWh, kilowatt-hours; [a]2000; [b]2002; [c]2003; [d]2004; [e]estimated; [f]2006; [g]2014; [h]2015; [i]2016.

Sources: The CIA World Factbook (2005, 2018); British Petroleum (2004); https://www.worldenergy.org/data/resources/region/south-central-asia/coal/; https://www.worldenergy.org/wp-content/uploads/2017/03/WEResources_Coal_2016.pdf

the U.S. Geological Survey, there are large volumes of undiscovered technically recoverable, conventional oil and gas resources, particularly in the Amu Darya Basin Province and the Afghan–Tajik Basin Province (Sagers 1994). The future potential of oil and gas accumulations is related with structures beneath the Upper Jurassic evaporite section and slope and basin gravity-flow deposits (Assessment … 2011).

Changes in energy production by CA country from 2000–2006 to 2014–2016 are shown in Table 2.2.

Kazakhstan was the leading producer of oil, and Turkmenistan was the leader in gas production at the beginning of the twenty-first century in the CA region. In recent years, the extraction of oil in Kazakhstan and gas production in Turkmenistan showed the tendency for increase by 22 and 40% consequently as compared with the situation during 2000–2006. In 2018, the oil exploration still increased by 4.7% in Kazakhstan as compared with 2017, which was achieved mostly due to the stabile increase in exploration in the Kashagan, Tengiz and Karachaganak oil fields. But it is expected that there would be a decrease in oil exploration in Kazakhstan in 2019 due to the general reconstruction activities in the mentioned fields and emptying of reserves in Kyzylorda and Aktyubinsk areas. Exploration of gas in Kazakhstan in 2018 made 55.5 billion m^3, which was 4% more than in 2017. The plan for 2019 is to keep on the same 55.5 billion m^3 (Godovoy otciot 2018).

The production of coal doubled in Kazakhstan and in Uzbekistan in 2015 as compared with 2000–2004. The production of electricity in 2015–2016 mostly increased in Turkmenistan, followed by Kazakhstan and Uzbekistan, by 86.0, 50.3 and 23.7%, respectively. The later estimates of exploration and production of oil and gas show almost unchanged situation from the beginning of the twenty-first century, however, with small changes: Uzbekistan was in the second position after Turkmenistan according to the produced gas (Li 2011).

In Tajikistan, the coal deposits are one of the main resources, while natural gas and oil are less found (Sinor et al. 1999). Uzbekistan is rich in natural gas, oil and coal resources, with major oil reserves in the Fergana Valley (Allworth 1999). Turkmenistan is rich in natural gas and oil resources in the western part, both in terrestrial areas and underwater along the Caspian Sea (Zhmuida et al. 1998).

Thus, CA is predominantly the oil, gas and coal producing region, and the potential for the development particularly of gas is even higher than of oil since it is more evenly spread throughout the region. But natural gas from here is generally high in sulphur and must be treated before being transported through pipelines.

In total the estimates of undiscovered technically recoverable, conventional oil and gas resources only for the Amu Darya Basin Province are about 962 million barrels (MMB) of crude oil, 52,025 billion cubic feet (BCF) of natural gas and 582 MMB of natural gas liquids. The estimated mean volumes of these resources for the Afghan–Tajik Basin Province are even higher (Assessment … 2011). Thus, the CA region will continue exploration of these fossil-fuel based resources for their own needs and on the world-scale in near future.

In the twenty-first century, with global climate warming, the ideas leading towards more sustainable usage of different resources were initiated and more research fulfilled on the measurement and estimation of greenhouse gas emissions and lower-carbon economy. It was estimated that worldwide in 2004 about 7.5 Pg C was emitted by fossil-fuel burning, and this was almost four times higher than the rate of CO_2 emission from land use change. With 107 Tg C yr^{-1}, CA countries contributed only about 1.4%

of CO_2 that was set free worldwide by fossil-fuel burning (Sommer and De Pauw 2011). Kazakhstan was the largest and Uzbekistan the second largest emission producer. With the development of economy, CA countries will need to generate more energy in future. However, uneven geographic distribution of energy resources makes some of them more dependent on the imported energy sources or electricity.

2.4 Recent Energy Consumption in CA Countries

According to electricity consumption by sectors, the industry is one of the largest users in all CA countries, except Kyrgyzstan, where the residential and agriculture sectors are using most (Electricity in Central… 2007). The industry, agricultural and residential sectors use the most of electricity in Uzbekistan, which can be explained by the largest number of population and as most important sectors of economics. In addition to this, transport can be mentioned as one of the most electricity consuming sectors in Kazakhstan, since the country covers the largest area in the CA region. Coal in Kazakhstan is still the mostly used resource for electricity production (more than 75% of electro-heating stations of the country are based on coal burning). It is planned to be reduced in the country's energy balance by less polluted energy sources. Recently some changes in final energy consumption in Kazakhstan have appeared: it has grown fast in industry and particularly in transport and service sectors from 2012 to 2015 (Nacionalny doklad... 2016).

Oil and natural gas comprise 97% of Uzbekistan's energy balance, and this country has increased its electricity production and consumption (Electricity in Central… 2007; IEA 2009; https://energypedia.info/wiki/Uzbekistan_Energy_Situation). In contrast to neighbouring countries, Kyrgyzstan has much less of non-renewable energy resources. Instead it possesses huge water resources and uses the hydropower mostly (90% of total country's energy). It was the first country in the Commonwealth of Independent States (CIS) to develop an independent regulatory agency for economic regulation of the energy sector. Approximately 95% of the population is connected to the grid; however losses in the distribution system range from 40 to 50% and reliability is poor. About 30% of the electricity distribution systems need to be replaced. The residential and agricultural sectors with forestry are the largest electricity consumers in Kyrgyzstan (https://energypedia.info/wiki/Kyrgyzstan_Energy_Situation).

In Tajikistan, after the end of the Soviet Union, electricity subsidies were terminated and that revealed the problematic situation with usage of energy resources in this country. Tajikistan has enormous hydropower potential as it possesses 4% of the world's hydropower resources and 53% of CA resources; hydropower provides 94% of electricity, but at the same time this potential is used only by 5%. The country faces an energy deficit of 3.0–3.5 GWh, resulting in regular blackouts from October to April. The industry (like aluminium), agriculture with forestry and residential sectors are the largest consumers of electricity in Tajikistan (https://energypedia.info/wiki/Tajikistan_Energy_Situation).

Turkmenistan's energy market is mainly controlled by the State. The electricity is produced in thermal stations and the most of the population receives natural gas and electricity for free. Those who do pay, enjoy the world's lowest energy prices, but this policy remains to be inefficient and not sustainable. The electricity consumption by sector is the following: agriculture and forestry 31.8%, industry 36%, residential 21% and transport 2.6% (Turkmenistan Energy Situation https://energypedia.info/wiki/Turkmenistan_Energy_Situation#Energy_Situation

2.4.1 Renewable Energy Resources

The energy sector in the economically developed countries is becoming diverse with rapid increase in share of renewable resources. Some of the CA region countries have made substantial commitments towards the development of non-traditional or renewable energy (RE) too. The Kyun government agency in Kyrgyzstan was established and devoted entirely to examining alternative energy supplies, notably solar, wind, geothermal and coalbed methane gas. The solar energy has a potential in the Kazakhstan's southern and northern tiers, which receive abundant sunshine, while the potential wind and biogas resources have been identified in the west of the country. Both the southern and north-eastern regions of Kazakhstan have the potential for the geothermal energy (Dorian 2006). Kazakhstan also has experience in hydropower use. Another suitable place for wind turbines with a capacity of 700 kW was found on the border of Kyrgyzstan and Kazakhstan. It is planned to increase the share of RES (Renewable Energy Sources) in Kazakhstan's energy balance to 5% by 2024 (UNDP 2014). In Tashkent, with international support, the Solar Energy Institute was founded. The solar energy is very promising for small remote outback settlements. Turkmenistan has more than 300 sunny days a year. Universities of the country started training Bachelors and Masters with the major in RES. It is planned to create in Turkmenistan a scientific centre "Renewable Energy Systems." In Tajikistan, the use of local, autonomous solar collectors and small- and micro-hydropower plants seems very actual and promising for the country's needs. Also within the framework of the project "Advancement of Renewable Energy Plants in Mountain Areas" (in cooperation with Mountain Societies Development Support Programme [MSDSP] of Aga Khan Foundation, with funding from the EU) 30 solar photovoltaic power

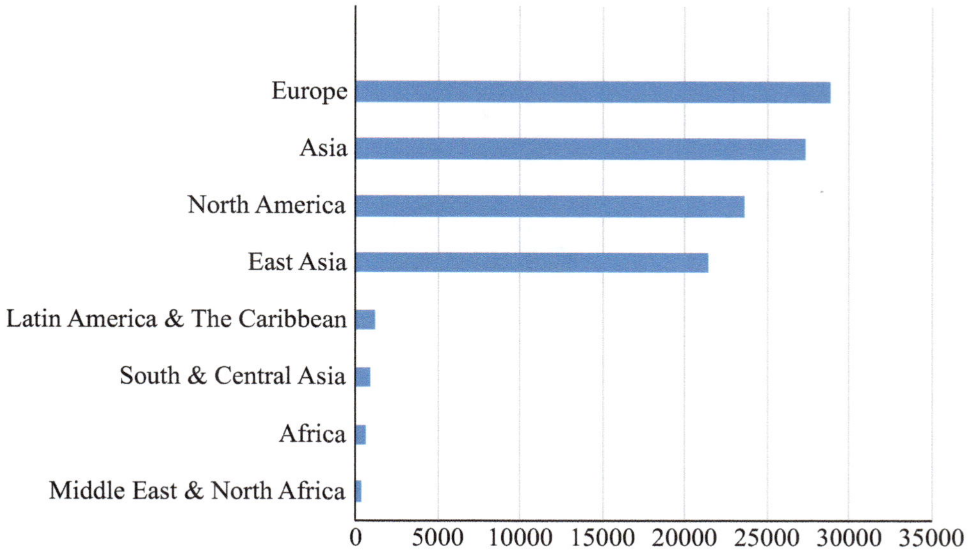

Fig. 2.6 Geothermal installed capacity in MW by region. (Source: World Energy Issues Monitor (2017))

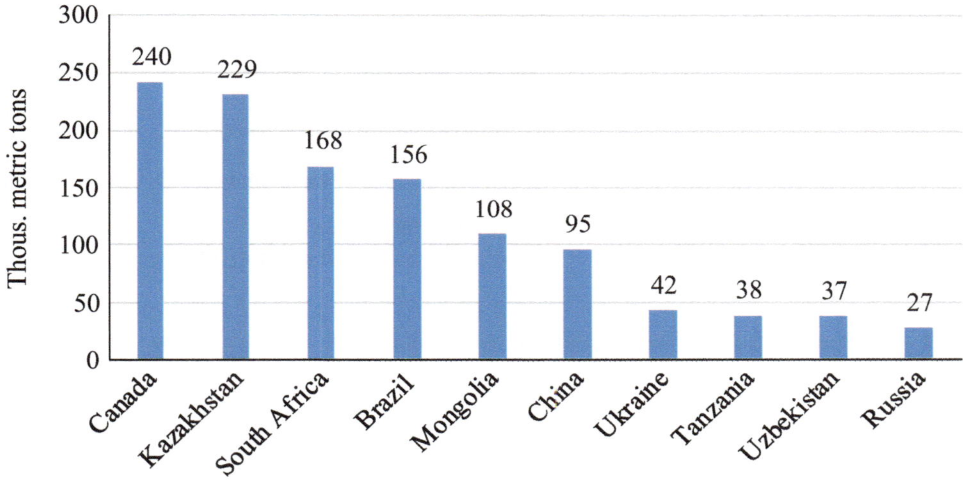

Fig. 2.7 World's leading countries based on uranium reserves in 2016 (in 1000 metric tonnes). (Source: Countries with the largest uranium… (2018))

plants, seven wind power plants and four plants for biogas production from biomass have been installed (Winner 2013).

As for the geothermal energy, it is mostly used as geothermal water for household needs, health restoration, medical treatment and the like. However, CA countries have a large potential for usage of geothermal energy sources more in near future (Fig. 2.6).

The following can be mentioned as barriers to renewable energy (RE) deployment in Central Asia (Nabiyeva 2015):

- High fossil-fuel subsidies and low electricity prices make RE uncompetitive
- Limited or no access to affordable bank loans
- High initial investment costs and risk of return of investment
- Limited number of local RE technology providers and specialists
- Lack of feasibility studies on RE perspectives
- Little public discussion of RE benefits

2.5 Minerals

Tajikistan possesses rich mineral deposits, among them iron, lead, zinc, antimony etc. (Table 2.1).

Uzbekistan's resources include metallic and polymetallic ores. Turkmenistan in the western coast along the Caspian Sea is rich with deposits of mirabilite, iodine, bromine, sulphur, potassium and salt (Zhmuida et al. 1998). Kyrgyzstan had become a source for non-ferrous metals, notably of antimony and mercury ores, and gold (Sinor and Allworth 1999). The most important minerals found in Kazakhstan are copper, lead, zinc, iron ore, silver, gold etc. The country distinguishes by the rich resources of uranium. In 2006, KazAtomProm, Kazakhstan's state-owned nuclear holding company, produced 8% of the world's mined uranium. In 2016, Kazakhstan was the second largest country in the world in terms of uranium reserves (Fig. 2.7).

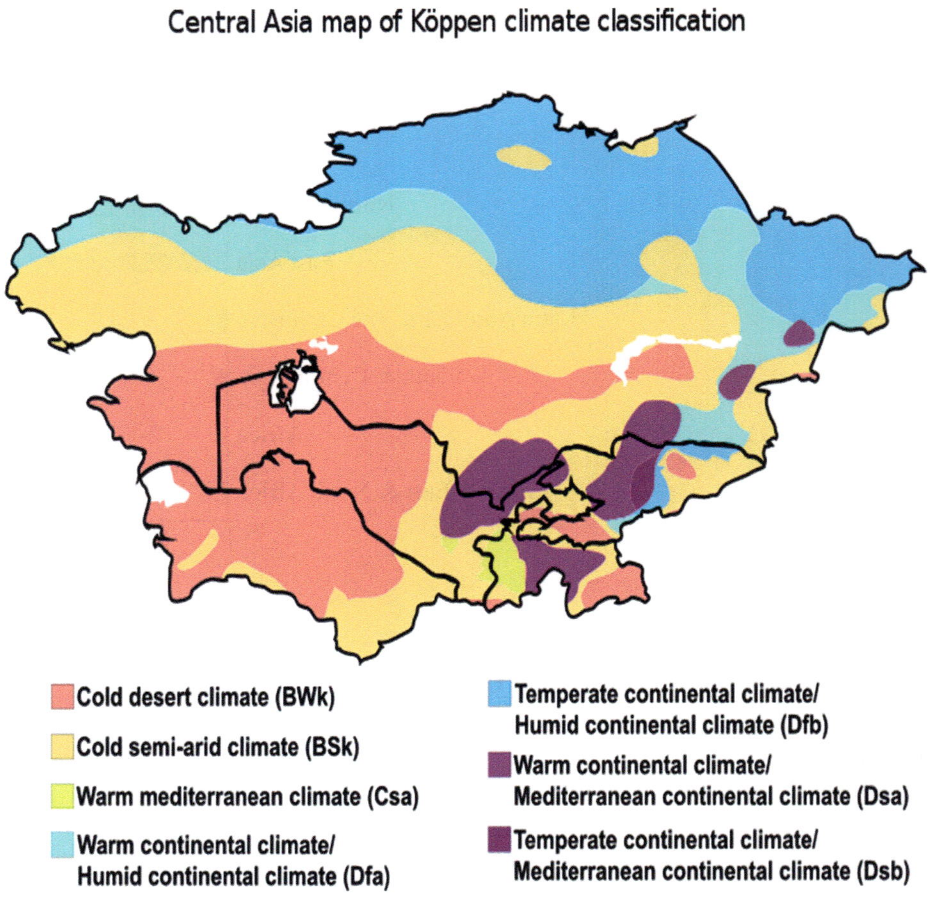

Fig. 2.8 The climatic zones in CA region according to Köppen. (Source: Derived from World Köppen Classification.svg., CC BY-SA 4.0, https://commons.wikimedia.org/wiki/File:Central_Asia_map_of_K%C3%B6ppen_climate_classification.svg)

Uzbekistan, with 2% of world reserves, was second in the CA region and ninth among world leaders in 2016.

2.6 Climate and Climate Zones in CA Region and Countries

According to Köppen (Grieser et al. 2006), there are seven climatic zones in the CA region (Fig. 2.8). Two of them are dominating in the area: (1) cold desert climate (Bwk) and (2) cold semi-arid climate (Bsk). The third largest climatic zone with temperate continental climate (Dfb) is spread in the north and the northeast. The climate of the mountainous region of CA is diverse and variable, influenced by the complex topography and position of ranges in accordance to the main air masses' circulation paths and seasons. The maximum precipitation rate in western Tien Shan occurs in winter season mostly, while in the rest of Tien Shan and eastern Pamir the highest in precipitation rate are the spring and summer months (Siegfried et al. 2012).

The changes in seasonal and monthly air temperatures show some differences between CA countries, but the general trend of changes in air temperature in CA countries showed a significant regional increase of 0.36–0.42°C per decade in the last 33 years (Hu et al. 2014).

The largest increase in air temperature is characteristic to the spring season. This phenomenon is mostly evident in the middle of the CA region, which can be considered as a warming centre. However, seasonal situations with air temperature rise in separate CA countries can be different. In Kazakhstan, for example, positive trend of average monthly air temperatures was observed almost everywhere for the same time-period (from 1941 to 2012). But the temperature increase here was higher in cold season (November–March) than in warm (April–October; Petrova and Ilyakova 2013).

The data on distribution of mean annual precipitation in the CA region have shown that the highest precipitation is distributed mostly in the eastern part of the region, which is known as the region of high mountain systems. The minimal rates of precipitation are observed in the central, southern and western parts of the CA region, and round the Aral Sea.

Turkmenistan distinguishes by the lowest mean annual precipitation rate. In larger countries of CA, like Kazakhstan, the changes in precipitation are more diverse. In some regions, precipitation increased slightly, whereas in other regions it decreased for the period 1941–2012. The annual potential evapotranspiration in the lowlands usually exceeds the annual precipitation. Due to this, constant irrigation is required for vegetation period of cultivated agricultural crops (Hamidov et al. 2016). Thus, the obtained data illustrate the

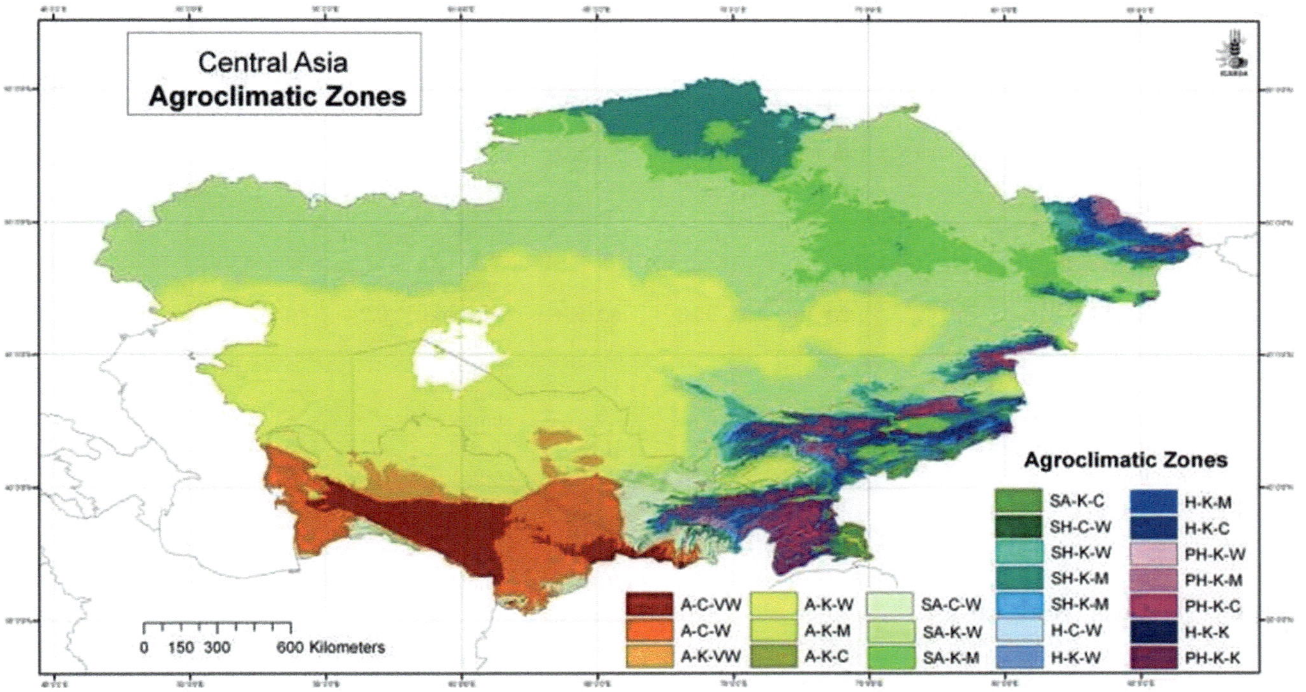

Fig. 2.9 Agroclimatic zones of CA region. (Source: De Pauw (2008))

strong continentality of the region's annual climatic conditions with wide amplitudes between the minimum and maximum values of air temperature, precipitation and evapotranspiration, whose indices change while going from the north to south and according to the territory's altitude.

The climate makes direct impact on agricultural systems in CA: spatial distribution of 21 zone reflects the diversity of agroclimatic conditions (Fig. 2.9; Table 2.3).

Two widest agroclimatic zones, occupying more than 68% of whole CA area, are as follows: (1) semi-arid, cold winter, warm summer, and (2) arid, cold winter, warm summer. Both distinguish by the high aridity index.

Besides physical factors, the anthropogenic factor makes the impact on the environment and through different activities on the climate too.

2.7 Climate by Countries

2.7.1 Kazakhstan

Being the largest country in CA region, Kazakhstan distinguishes by large fluctuations of seasonal air temperature and distribution of precipitation, but the general feature of country's climate is its strong continentality. Average January temperatures in northern and central regions range from −19° to −16°C; in the south, temperatures are milder, ranging from −5° to −1.4°C. Average July temperatures range from 20°C in the northern to 29°C in the southern regions (Allworth and Sinor 1998). Precipitation rate varies depending on the location, relief/altitude and season. It reaches 200–300 mm in the northern and central regions and increases to 400–1600 mm a year at the foothills and mountain regions in the south. Southwest Kazakhstan with semi-deserts and deserts and Balkhash region in the southeast receive 100–200 mm (in some years <100 mm) per year. In the north and in the centre, the maximum precipitation falls during the summer months, while it occurs in the early spring in the south. It is one of the most windy countries in the CA region (State of environment... 2000).

In Altai Mountains the regional climate is severely continental, but in western part precipitation rate is high: they vary between 500 and 2000 mm per annum depending on the elevation. The summers are hot, but short. Average January temperature ranges between –14°C in the foothills and –32°C in the sheltered hollows of the east, with minimum −60°C recorded in the Chu steppes (Altai-Mountains 2007).

2.7.2 Kyrgyzstan

Kyrgyzstan's climate is strongly continental too, with even higher impact of the surface elevation than in Kazakhstan. Average January temperatures in the valleys vary between –1.5 and –8°C, whereas in the mountains

Table 2.3 Description of agroclimatic zones of CA: Details to the map's (Fig. 2.9) legend

Code of zone	Description	Aridity index	Temperature range of the coldest month (°C)	Temperature range of the warmest month (°C)	% in total
SA-K-W	Semi-arid, cold winter, warm summer	0.2–0.5	≤0	20–30	37.9
A-K-W	Arid, cold winter, warm summer	0.03–0.2	≤0	20–30	30.8
SA-K-M	Semi-arid, cold winter	0.2–0.5	≤0	10–20	6.6
SH-K-M	Sub-humid, cold winter	0.5–0.75	≤0	10–20	5.9
A-C-W	Arid, cool winter, warm summer	0.03–0.2	0–10	20–30	4.9
A-C-VW	Arid, cool winter, very warm summer	0.03–0.2	0–10	>30	2.9
PH-K-C	Per-humid, cold winter, cool summer	>1	≤0	0–10	2.0
H-K-M	Humid, cold winter, mild summer	0.75–1	≤0	10–20	1.6
SA-C-W	Semi-arid, cool winter, warm summer	0.2–0.5	010	20–30	1.5
SH-K-W	Sub-humid, cold winter, warm summer	0.5–0.75	≤0	20–30	1.4
A-K-VW	Arid, cold winter, very warm summer	0.03–0.2	≤0	>30	1.2
PH-K-M	Per-humid, cold winter, mild summer	>1	≤0	10–20	1.2

(continued)

Table 2.3 (continued)

Code of zone	Description	Aridity index	Temperature range of the coldest month (°C)	Temperature range of the warmest month (°C)	% in total
SH-K-C	Sub-humid, cold winter-cool summer	0.5–0.75	≤0	0–10	0.5
SA-K-C	Semi-arid, cold winter, cool summer	0.2–0.5	≤0	0–10	0.5
H-K-C	Humid, cold winter, cool summer	0.75–1	≤0	0b–b10	0.5
H-K-W	Humid, cold winter, warm summer	0.75–1	≤0	20–30	0.2
SH-C-W	Sub-humid, cold winter, warm summer	0.5–0.75	0–10	20–30	0.1
A-K-M	Arid, cold winter, cool summer	0.03–0.2	≤0	10–20	0.1
PH-K-K	Per-humid, cold winter, cold summer	>1	≤0	≤0	0.1
PH-K-W	Per-humid, cold winter, warm summer	>1	≤0	20–30	0.0
A-K-C	Arid, cold winter, cold summer	0.03–0.2	≤0	0–10	0.0

Source: De Pauw (2008)

between −8 and −28°C. As to the minimum temperature, even −50°C was registered at the Ak-Say valley of the inner Tien Shan (Bruk et al. 1999). Average July temperatures in the valleys reach 15–28°C and in the high mountains are by 5 degrees lower. Yearly precipitation varies from 180 mm in the eastern to 1000 mm on the northern-western slopes of the Tien Shan (Sinor and Allworth

1999). The main precipitation in the mountains is in the form of snow. The most of precipitation falls in autumn, winter and spring, and summer is usually dry (State of environment... 2000).

2.7.3 Tajikistan

The climate of Tajikistan is sharply continental with little precipitation, dry air and little cloud cover, but changes with altitude drastically (State of environment... 2000). Average January temperatures are mild, ranging between –1 and 2°C in the warm-temperate valley areas with hot and dry summers (the mean temperature in July is 27–30°C). But in the highlands, climatic conditions are different: in the Pamirs in January it reaches −20°C on the average, and in July fluctuates around 0°C. The extremal temperatures here in winter months can drop even to –46°C (Sinor et al. 1999). The annual precipitation rate ranges within large scale too: the smallest (till 70 mm a year) is recorded in the eastern Pamirs, the highest reaches 1600 mm on the southern slopes of Gissar ridge, whereas in the valleys and plains it ranges between 100 and 200 mm. The maximum precipitation falls in March–April, except in eastern Pamir, where it occurs in July–August (State of environment... 2000).

2.7.4 Uzbekistan

The territory of Uzbekistan is the third largest by area in the CA region; thus, its climate is diverse with features of temperate zone continental and subtropical climate. The average January temperatures range from 2–3°C in the south-eastern region to –10°C in the north-western. The winter lasts from 1.5–2 months in the extreme south to 5 months in the far north (State of environment... 2000). The average July temperature reaches 32°C, but the extremum can surpass 40°C (Allworth 1999). Much sunshine and high aridity is typical for Uzbekistan. Annual precipitation ranges from <100 mm in the plains, increasing till 150–200 mm on foothills with maximum 1000 mm in the mountain areas. More than 70% of precipitation comes in winter and spring, and the snow cover is not permanent on the plains (State of environment... 2000).

2.7.5 Turkmenistan

Turkmenistan's territory is the second largest in the CA region; thus, its climate is varying from strongly continental of temperate zone to the subtropical, but the main feature is high aridity. The average January temperatures range between 4 and –5°C, while the average July temperatures reach 28–34°C. The absolute maximum high temperature of 50°C was registered in the southeast Karakum in the shade. Contrastly, in winter the temperature on the border with Afghanistan drops to −33°C (Zhmuida et al. 1998). Precipitation varies between <100 mm a year in the desert regions and 300–400 mm in the mountains. The strong winds of northern direction are prevailing in most regions of the country.

2.8 Topics for Further Reading and Discussions on the Content of this Chapter

- The newest discoveries in research of the paleo-history of CA region
- Practical analysis of aridity in CA countries and their inner regions using meteorological data and climatic maps

Recommended References

De Pauw E (2008) Climatic and soil datasets for the ICARDA wheat genetic resource collections of the Eurasia region: explanatory notes. Technical Note, GIS Unit, ICARDA, Syria

Kröner A (2015) The Central Asian Orogenic Belt – present knowledge and comparison with the SW Pacific. In: Kröner A (ed) The Central Asian Orogenic Belt. Geology, evolution, tectonic, and models, Borntraeger Science Publishers, Stuttgart:1–6

Safonova IY, Santosh M (2014) Accretionary complexes in the Asia Pacific region: tracing archives of ocean plates stratigraphy and tracking mantle plums. Gondw Res 2 (5): 126–158

Siegfried T, Bernauer T, Guiennet R, Sellars S, Robertson AW, Mankin J, Bauer-Gottwein P, Yakovlev A (2012) Will climate change exacerbate water stress in Central Asia? Clim Change 112: 881–899

Xiao WJ, Santosh M (2014) The western Central Asian Orogenic Belt: a window to accretionary orogenesis and continental growth: Gondw Res, 18: 1429–1444

References

Alden A (2020) Epeirogeny: understanding vertical continental drift. ThoughtCo, Feb. 11, 2020, thoughtco.com/what-is-epeirogeny-1440831, https://www.thoughtco.com/what-is-epeirogeny-1440831. Accessed 29 Feb 2020

Allan NJR, Zakharova TK with contribution of Gaur A (2017) Pamirs. Mountain regions. Asia. Encyclopædia Britannica. https://www.britannica.com/place/Pamirs. Accessed 10 June 2017

Allworth E (1999) Uzbekistan. In: Encyclopaedia Britannica online Databases. https://www.britannica.com/place/Uzbekistan. Accessed 6 June 2017

Allworth E, Sinor D (1998) Kazakhstan. In: Encyclopaedia Britannica. https://www.britannica.com/place/Kazakhstan. Accessed 6 June 2017

Altai Mountains (2007) The Editors of Encyclopaedia Britannica. https://www.britannica.com/place/Altai-Mountains. Accessed 15 July 2017

Assessment of Undiscovered Oil and Gas Resources of the Amu Darya Basin and Afghan–Tajik Basin Provinces, Afghanistan, Iran, Tajikistan, Turkmenistan, and Uzbekistan (2011) World Petroleum Resources Project, USGS science for a changing world, https://pubs.usgs.gov/fs/2011/3154/report/FS11-3154.pdf. Accessed 11 July 2017

Bossu R, Grasso JR, Plotnikova LM, Nurtaev B, Fréchet J, Moisy M (1996) Complexity of intracontinental seismic faulting: the Gazli, Uzbekistan, sequence. Bull Seismol Soc Am 86:959–971

Breckle SW, Geldyeva GV (2012) Dynamics of the Aral Sea in geological and historical times (chapter 2). In: Aralkum-a man-made desert: the Desiccated floor of the Aral Sea (Central Asia). Springer, Berlin/Heidelberg, pp 13–37

British Petroleum, BP Statistical Review of World Energy (2004) London US Department of Energy, Energy Information Administration, May 2002, "Central Asia". Country Analysis Brief, Washington, DC

Bruk SI, Rantsman YY, Allan NJR (1999) Tien Shan. In: Encyclopaedia Britannica. https://www.britannica.com/place/Tien-Shan. Accessed 6 June 2017

Buslov MM, Watanabe T, Fujiwara Y, Iwata K, Smirnova LV, Safonova IY, Semakov NN, Kiryanova AP (2004) Late Paleozoic faults of the Altai region, Central Asia: tectonic pattern and model of formation. J Asian Earth Sci 23:655–671, Central Asia i. Geographical Survey. Encyclopædia Iranica (2012) 2:159–161, available online at http://www.iranicaonline.org/articles/central-asia-i. Accessed 10 June 2017

Central Asia written by the Editors of Encyclopædia Britannica (1998–2014) Encyclopædia Britannica, Inc. https://www.britannica.com/place/Central-Asia. Accessed 6 July 2017

Central Asia Atlas of Natural Resources (2010) Central Asian Countries Initiative for Land Management, Asian Development Bank, Manila, Philippines. https://think-asia.org/handle/11540/155. Accessed 10 June 2020

CIA The World Factbook (2005). https://www.cia.gov/news-information/press-releases-statements/press-release-archive-2005/pr04282005.html. Accessed 15 July 2017

CIA The World Factbook (2017). https://www.cia.gov/library/publications/the-world-factbook/geos/uz.html. Accessed 15 July 2017

CIA The World Factbook (2018). https://www.cia.gov/library/publications/the-world-factbook/geos/uz.html. Accessed 15 March 2018

Countries with the largest uranium reserves worldwide 2016. Statista (2018). https://www.statista.com/statistics/264781/countries-with-the-largest-uranium-reserves/. Accessed 18 Feb 2018

De Pauw E (2008) Climatic and soil datasets for the ICARDA wheat genetic resource collections of the Eurasia region: explanatory notes. Technical Note, GIS Unit, ICARDA, Syria

Dobretsov NL, Buslov MM (2007) Late Cambrian-Ordovician tectonics and geodynamics of Central Asia. Russ Geol Geophys 48:71–82

Dobretsov NL, Berzin NA, Buslov MM (1995) Opening and tectonic evolution of the Paleo- Asian ocean. Intern Geol Rev 37:335–360

Dobretsov NL, Buslov MM, Vernikovsky VA (2003) Neoproterozoic to Early Ordovician evolution of the Paleo-Asian Ocean: implications to the break-up of Rodinia. Gondw Res 6:143–159

Dorian JP (2006) Central Asia: A major emerging energy player in the 21st century. Energy Policy 34:544–555

Electricity in Central Asia market and investment opportunity report (2007) World Energy Council. https://www.worldenergy.org/wpcontent/uploads/2012/10/PUB_Asia_Regional_Report_Electricty_Market_And_Investment_Opportunity_2007_WEC.pdf. Accessed 30 Jan 2018

Firoozfar A (2014) Beach profiles and sediments, a case of Caspian Sea. Int J Mar Sci 4(43):1–9. http://ijms.biopublisher.ca, https://doi.org/10.5376/ijms.2014.04.0043. Accessed 29 Feb 2020

Godovoy otciot 2018 (2018) AO „KazMunayGaz" (Annual raport 2018, AC „KazMunayGaz") Accessed 01 Mar 2020

Grieser J, Gommes R, Cofield S, Bernardi M (2006) New gridded maps of Koeppen's climate classification. FAO of the UN, Viale delle Terme di Caracalla, 00100 Rome, Italy http://www.fao.org/nr/climpag/globgrids/kc_classification_en.asp. Accessed 05 Dec 2017

Gvozdetskij N (1970) Fizicheskaya geografiya SSSR. Aziatskaya chast' (Physical geography of SSSR, part:Asia), Moskva

Hamidov A, Helming K, Balla D (2016) Impact of agricultural land use in Central Asia: a review. Agron Sustain Dev 36(23):6. https://doi.org/10.1007/s13593-015-0337-7. Accessed 30 Apr 2018

Harding TP (1976) Tectonic significance and hydrocarbon trapping consequences of sequential folding synchronous with San Andreas faulting. San Joaquin Valley, California. Am Assoc Pet Geolo Bull 60:356–378

Heubeck C (2001) Assembly of central Asia during the middle and late Paleozoic. In: Hendrix MS, Davis GA (eds) Paleozoic and Mesozoic tectonic evolution of Central and Eastern Asia, Mem Geol Soc Am 194:1–22

Hu Z, Zhang C, Hu Q (2014) Temperature Changes in Central Asia from 1979 to 2011 based on multiple datasets. J Clim 27(3):1143–1167. https://doi.org/10.1175/JCLI-D-13-00064.1

IEA (International Energy Agency). World Energy Outlook (2009). http://www.worldenergyoutlook.org/media/weowebsite/2009/WEO2009.pdf. Accessed 18 Dec 2017

Kingston J, Clarke W (1995) Petroleum Geology and Resources of Afghanistan. Int Geol Rev 37(2):111–127

Klett TR, Ulmishek GF, Wandrey CJ, Agena WF (2006) U.S. Geological survey assessment of undiscovered technically recoverable conventional petroleum resources of Northern Afghanistan, Afghanistan Ministry of Mines and Industry Joint oil and gas resource assessment team, United States Geological Survey, Open-File Report. http://pubs.usgs.gov/of/2006/1253/. Accessed 1 Sept 2017

Kröner A, Kovach V, Belousova E, Hegner E, Armstrong R, Dolgopolova A, Seltmann R, Alexeiev DV, Hoffmann JE, Wong J, Sun M, Cai K, Wang T, Tong Y, Wilde SA, Degtyarev KE, Rytsk E (2014) Reassessment of continental growth during the accretionary history of the Central Asian Orogenic Belt. Gondw Res 25:103–125

Kyrgyzstan Energy Situation. Energypedia Info. https://energypedia.info/wiki/Kyrgyzstan_Energy_Situation#Energy_Situation. Accessed 30 Jan 2018

Li G (2011) World atlas of oil and gas basins. Wiley. https://books.google.lt/books?id=GFXkR7i7t8YC&dq=map+of+oil+and+gas+in+Central+Asia+countries&lr=&hl=lt&source=gbs_navlinks_s. Accessed 11 July 2017

Lilienberg DA (1985) Recent geodynamics of the Alpine oro- genie belt of Southern Europe. Geomorphology 1985(4):16–29. (in Russian)

Lilienberg DA (1993) Current trends in the endodynamics of the Caspian Sea and changes in its water level. Dokl Ross Akad Nauk 33 1(6):745–750. (in Russian)

Mossakovsky AA, Ruzhentsev SV, Samygin SG, Kheraskova TN (1993) The Central Asian foldbelt: geodynamics evolution and the history of formation. Geotektonika 6:3–33. (in Russian)

Nabiyeva K (2015) First steps towards energy transition in Central Asia: opportunities for German and EU engagement, 15-06 Istanbul. http://www.iea.org/media/workshops/2015/platformistanbul2015/First_Steps_towards_Energy_Transition_in_Central_Asia__Opportunities_for_German_and_EU_Engagement.pdf. Accessed 13 July 2017

Nacionalny doklad o sostoyanii okruzhayuscey sredy i ob ispolzovanyi prirodnyh resursov za 2015 god. Energetika (2016) (The National report on state of environment and use of natural resources for year 2015. Energetics). http://ecogosfond.kz/wp-content/uploads/2018/03/140320171.pdf.1. Accessed 30 Jan 2018

Persits FM, Ulmishek GF, Steinshouer DW (1997) Maps showing geology, oil and gas fields and geologic provinces of the former Soviet Union (CD-ROM): U.S. Geological Survey, Open File Report 97-470

Petrova EE, Ilyakova RM (eds) Annual bulletin of climate change monitoring in Kazakhstan: 2012 (2013) Ministry of Environmental Protection of the Republic of Kazakhstan, Republican State Enterprise "Kazhydromet", Astana, https://kazhydromet.kz/upload/pdf/climat/en_1490785559.pdf. Accessed 29 Apr 2018

Rogers JJW, Santosh M (2004) Continents and supercontinents. Oxford Press, New York

Rychagov GI (1977) The Pleistocene history of the Caspian Sea. Unpublished D.Sc. thesis, Moscow State University, 252 pp. Autoreferate 62 p (in Russian)

Rychagov GI (1993) The sea-level regime of the Caspian Sea during the last 10,000 years. Bull. Moscow State University Ser. 5, 1993 (2):3819 (in Russian)

Safonova IY (2009) Intraplate magmatism and oceanic plate stratigraphy of the Paleo-Asian and Paleo-Pacific Oceans from 600 to 140 Ma. Ore Geol Rev 35:137–154

Safonova IY, Maruyama S (2014) Asia: a frontier for a future supercontinent Amasia. Int Geol Rev 56(9):1051–1071

Safonova IY, Santosh M (2014) Accretionary complexes in the Asia Pacific region: tracing archives of ocean plates stratigraphy and tracking mantle plums. Gondw Res 2(5):126–158

Sagers MJ (1994) Oil and Gas production in the Amu Dar'ya Basin of Western Uzbekistan and Eastern Turkmenistan. Int Geol Rev 60:416–434

Samsonow JW (1994) Afghanistan. In: Kulke H (ed) Regional Petroleum Geology of the World: Part 1. Gebruder Borntraeger, Brlin. Beitrage der Regionale Geologie der Erde, vol. 21: 595–606

Şengör AMC, Natal'in BA (1996) Palaeotectonics of Asia: fragments of a synthesis. In: Yin A, Harrison M (eds) Tectonic Evolution of Asia. Cambridge University Press, Cambridge, pp 486–640

Şengör AMC, Natal'in BA, Burtman VS (1993) Evolution of the Altaid tectonic collage and Paleozoic crustal growth in Eurasia 364:299–307

Siegfried T, Bernauer T, Guiennet R, Sellars S, Robertson AW, Mankin J, Bauer-Gottwein P, Yakovlev A (2012) Will climate change exacerbate water stress in Central Asia? Clim Change 112:881–899

Sinor D, Allworth E (1999) Kyrgyzstan. In.: Encyclopaedia Britannica online Database. https://www.britannica.com/place/Kyrgyzstan. Accessed 6 June 2017

Sinor D, Allworth E, Imshenetsky AI (1999) Tajikistan In.: Encyclopaedia Britannica online, Database. https://www.britannica.com/place/Tajikistan. Accessed 6 June 2017

Sommer R, De Pauw E (2011) Organic carbon in soils of Central Asia–status quo and potentials for sequestration. Plant Soil 338:273–288. https://www.researchgate.net/publication/225342880_Organic_carbon_in_soils_of_Central_Asia-status_quo_and_potentials_for_sequestration/figures. Accessed 08 Aug 2017

State of Environment of the Aral Sea Basin (2000) Regional report of the Central Asian States. http://enrin.grida.no/htmls/aralsoe/aralsea/index.htm. Accessed 26 Aug 2017

Svitoch AA (1991) Fluctuations of the Caspian sea-level in the Pleistocene (Classification and systematic description). In: The Caspian Sea. Paleogeography and Geomorphology of the Caspian Area in the Pleistocene. Nauka, Moscow, pp 3–100. (in Russian)

Tajikistan Energy Situation. Energypedia Info, https://energypedia.info/wiki/Tajikistan_Energy_Situation#Energy_Situation. Accessed 30 Jan 2018

Thomas J-C, Grasso J-R, Bossu R, Martinod J, Nurtaev B (1999) Recent deformation in the Turan and South Kazakh Platforms, Western Central Asia, and its relation to Arabia-Asia and India-Asia collisions. Tectonics 18(2):201–214

Turkmenistan Energy Situation. Energypedia Info. https://energypedia.info/wiki/Turkmenistan_Energy_Situation#Energy_Situation. Accessed 30 Jan 2018

UNDP renewable energy snapshots (2014). http://www.eurasia.undp.org/content/rbec/en/home/library/environment_energy/renewable-energy-snapshots.html. Accessed 1 Sept 2017

Wandrey CJ, Law BE (1999) Map showing geology, oil and gas fields, and geologic provinces of South Asia: U.S. Geological Survey Open File Report 97-470C, version 2, 1 CD-ROM

Windley BF, Alexeiev D, Xiao W, Kröner A, Badarch G (2007) Tectonic models for accretion of the Central Asian Orogenic Belt. 164 1 Journal of the Geological Society, London: 31–47

Winner V (2013) Central Asia countries increase potential in renewable energy use. EAN News portal of Eurasia countries. http://eurasianews.info/en/news/central-asia-countries-increase-potential-in-renewable-energy-use.html. Accessed 27 June 2017

World Energy Issues Monitor (2017) Exposing the new energy realities (2017) World Energy Council. https://www.worldenergy.org/assets/downloads/1.-World-Energy-Issues-Monitor-2017-Full-Report.pdf. Accessed 6 June 2018

Xiao WJ, Santosh M (2014) The western Central Asian Orogenic Belt: a window to accretionary orogenesis and continental growth. Gondw Res 18:1429–1444

Yarmolyuk VV, Kuz'min MI, Vorontsov AA (2013) West Pacific-type convergent boundaries and their role in the formation of the Central Asian fold belt. Russ Geol Geophys 54(12):1427–1441. Publishing House of Siberian Branch of the Russian Academy of Sciences, Novosibirsk, Russia

Zhmuida, VB, Allworth E, Sinor D (1998) Turkmenistan. In: Encyclopaedia Britannica online Database. https://www.britannica.com/place/Turkmenistan. Accessed 6 June 2017

Zonenshain LP, Kuzmin MI, Natapov LM (1990) Geology of the USSR: a Plate Tectonic Synthesis, Geodyn. Ser., 21, 242 p, AGU, Washington, DC

3 Geobiology, Bioresources and Biodiversity

Abstract

The main ecosystems and biomes of CA are of global significance, and include deserts, steppes, montane temperate forests, montane coniferous forests, alpine and sub-alpine meadows, river deltas and riparian forests, flooded Tugay type of forests and freshwater wetlands. The region is rich in biodiversity since it extends in the core of Eurasia continent with natural crossroads of species migrating in different directions depending on natural obstacles and corridors. Particularly rich in biodiversity are the mountains of CA. As to the forested landscapes, the region during 1990–2015 experienced the smallest gain/reduction in the forested land area; however, forested area monitoring is conducted with some disturbances. The illegal cutting of trees and fuelwood harvesting as well as conversion of forest land to agricultural farming all led towards decrease in the forested areas of the CA region. On the other hand, the anthropogenic factor and its different activities are becoming more and more important in CA countries: long-term land use under agriculture, irrigation or intensive grazing supported the increase in soil salinization, contamination and the drastic changes towards the worsening of biodiversity of natural ecosystems there. These negative tendencies are evident in nearly half of all natural pastures in whole arid zone. The attempts to protect endangered species and entire ecosystems have been evident during the last years in the CA region. Tajikistan is distinguished by the highest percentage of total protected areas. The problems of desertification and degradation in relation to climate warming and due to the increased anthropogenic impact on biomes and ecosystems can lead to a large number of ecological problems and disasters in CA countries. The Aral Sea area and its surroundings are one of these hotspots.

Keywords

Biomes · Ecoregions · Ecosystems · Forests · Anthropogenic impact · Desertification

3.1 From Biomes, Ecoregions and Ecosystems to Biodiversity, Plant Resources and Nature Protected Areas

The recent ecological situation with vegetation complexes, biomes and ecosystems, in the different parts of the CA region, is closely related to the whole history of the landscape changes, driven by many physical, biological and anthropogenic factors.

According to the definition of biome given in world biomes site (http//www.blueplanetbiomes.org/world_biomes.htm), "a biome is a large geographical area of distinctive plant and animal groups, which are adapted to that particular environment. Each biome consists of many ecosystems whose communities have adapted to the small differences in climate and the environment inside the biome". The CA region is distinguished by vast temperate grasslands/steppe, semi-desert and desert with xeric shrubland, montane grasslands with shrubland and few temperate conifer forest biomes. These biomes are characterized by large diversity of different ecosystems; however, some of them became degraded under the pressure of anthropogenic factor meanwhile. For example, CA desert pastures have historically and traditionally been used for developing cattle, sheep, camel and herd horse breeding and will probably be maintained in the near future. The well-being of these areas, as well as people who live in desert areas of CA countries, largely depends on the condition and productivity of natural fodder lands.

I. Jelen et al., *The Geography of Central Asia*, World Regional Geography Book Series,
https://doi.org/10.1007/978-3-030-61266-5_3

Among the physical factors, the climatic conditions, particularly those related to the hydrothermic regime of environment, are the most important in this region. Long dry periods, sharp seasonal contrasts in temperature and precipitation are responsible for the distribution of representatives of subtropical flora and fauna in the southern part of CA, including among them the relicts from warm Paleogene and Neogene.

Two main geobotanical provinces of Turan and Kazakhstan are to be found in the territory of CA, the border line between them stretches in the ranges of Karatau, Talas Alatau, Fergan, Kyzyl Kum desert (approximately along the line from Kyzyl-Orda to Kara Bogaz Gol). The province of Turan is a part of larger Iran–Arabic geobotanical range; thus, its subtropical vegetation is close by forms and species to the Middle East countries of Euroasia continent. The Kazakhstan province has much in common with flora of the temperate zone of CA. The differences between provinces flora can be observed also in the mountain areas; however, the altitude makes these differences less (Biomes and Regions 2002). The limit factor for vegetation growth in high altitudes is air temperature. The northern part of Kazakhstan is characteristic for the wide steppe zone with different grasses, which is gradually shifted by semi-desert zone with typical vegetation, while going from north to south. At 47–48° latitude, the desert flora zone is prevailing, and it can be divided to northern and southern (subtropical) desert (Babaev 1996).

According to the list given in World Heritage (2005), CA terrestrial biomes and ecosystems presented in Table 3.1 are recognized as being of global importance.

Sandy deserts are spread within three countries of the CA region (Kazakhstan, Turkmenistan and Uzbekistan) with largest area in Turkmenistan (Rachkovskaya 1995). The climate of these deserts has features of a transition from temperate to subtropical one. The main tree representatives are the white saxaul (*Haloxylon persicum*) and the black saxaul (*H. aphyllum*). They grow together with many endemic species of shrubs like *Calligonum leucocladum, C. eriopodum and C. setosum*. For example, the vegetation in the sand desert of Karakum is dominated by such shrub species as saxaul, cherkez and kandym, with an herbaceous cover of sand sedge (*Carex physodes*) and ephemerous plants (Rustamov 1994). The fauna of sandy deserts consists of representatives of insects and reptiles including rare and endangered species.

The steppes biome with related ecosystems is spread in the west-northern and northern part of the CA region, mainly in Kazakhstan, and transformed for the agricultural needs mostly by cultivation of winter wheat and other crops. The World Wildlife Fund for Nature (WWF) divides the CA steppe zone into seven ecoregions with distinct communities of fauna and flora: (1) Alai-Western Tien Shan; (2) the Kazakh upland; (3) the Altai steppe and semi-desert; (4) the Emin Valley steppe in Kazakhstan; (5) the Kazakh steppe; (6) the Kazakh forest steppe; and (7) the Tien Shan foothill arid steppe (Larger than tigers 2019). Natural representatives of CA steppes are different species of perennial grasses like *Stipa zalesskii, Festuca rupicola, Artemisia marschalliana*, etc (The Kazakh Steppe 2011).

The Alpine and Sub-alpine Meadows with different species of perennial grasses like *Fescue*, *Poa relaxa*, *Puccinella subspicata*, *N.cocanica Nepeta podostachys* stretch at 3100–3400 m asl in the highest mountain chains of the CA region (https://www.worldwildlife.org/ecoregions/pa0808).

Freshwater wetlands are very important from the point of increased biodiversity and habitats for endangered species. The 64 Ramsar sites are recognized in CA countries, and the total area of wetlands extends over approximately 7.7 million ha (Larger Than Tigers 2019).

Table 3.1 Main biomes and ecosystems of CA region of global importance

Main biomes and/or ecosystems	Location, country/ies	Representatives of flora and fauna
Sandy Deserts	Karakum and Kyzylkum in Kazakhstan, Turkmenistan, Uzbekistan	*Haloxylon* shrubs/saxaul; herpetofauna
Steppes	In Kazakhstan mainly	Daffodils and tulips; saiga antelope (*Saiga tatarica*) and Asian wild ass (*Equus hemionus)*
Montane Temperate Forests	In all five countries	Wild forebears of apple-trees, pears, apricots, almonds, pistachios and walnuts, red tulips (*Tulipa micheliana*); Siberian ibex, Bukhara urial, etc.
Montane Coniferous Forests	In Kazakhstan mainly	Juniper and spruce forests
Alpine and Sub-alpine Meadows	In all five countries	Pikas, marmots, wild goat, wild sheep and threatened snow leopard
River Deltas and Riparian Forests	Along the Amu Darya, Syr Darya and in the Volga Delta in Kazakhstan, Tajikistan, Turkmenistan, Uzbekistan	The "Tugay" riparian forests of poplars, *Elaeagnus* and *Tamarix;* Bukhara deer, sturgeon
Freshwater wetlands	Lakes and marshes in Kazakhstan, Kyrgyzstan, Turkmenistan, Uzbekistan	Migrating waterfowl species, endangered species like Dalmatian pelican and white-headed duck

Source: Adopted from World Heritage (2005)

The main forest biomes and ecosystems of the CA region will be presented in the following sub-chapter "Distribution of forest resources globally and in CA countries".

3.1.1 Biodiversity

The CA region is located on a biological crossroads, with species representative of three of the world's biogeographic realms: the Palearctic, Indo-Malayan and Afrotropic (Larger Than Tigers 2019). Various topographic and climatic factors also have resulted in the region's highly diverse fauna and flora. According to the Global 200 ecoregions organized by biomes and biogeographic realms (Olson and Dinerstein 2002), the estimated conservation and biodiversity status of those biomes stretching in the CA region can be assessed as vulnerable or critical/endangered. Conservation status assessments of the Global 200 ecoregions were based on landscape or aquascape-level criteria, such as total habitat loss, the degree of fragmentation, water quality and estimates of future threat (Shahgedanova 2002; Aladin et al. 2008).

The number of threatened species in separate country and in the total CA region is shown in Table 3.2. The recent data were taken from the IUCN Red List of threatened species, made in June 2016 (http://www.iucnredlist.org/).

Among all terrestrial Global 200 ecoregions (142 in total), 75 ecoregions (53%) are considered critical or endangered. Of these at least two are located in the CA region (the Central Asian Deserts and Aral Sea basin as the intrazonial ecoregion). Other three terrestrial ecoregions are considered as of vulnerable status. The richer biodiversity and endemism are typical for montane ecoregions and ecosystems due to the isolation and higher mosaic of landscape and habitat patterns.

According to Krever et al. (1998) in the CA region, "the angiosperm flora comprises some 7,000 species; there are more than 900 species of vertebrates, including 172 mammals, 540 birds, 106 species of reptiles, 14 species of amphibians and approximately 150 species of fish; and more than 20,000 invertebrate species have been recorded". The endemism in the CA region is high, particularly among higher plants (World Heritage 2005). But the threat of increase in the species extinction rate nowadays is recognized also very high (Chivian and Bernstein 2008; Lazkov and Sennikov 2014; Sennikov et al. 2016; Tojibaev et al. 2017). In order to improve knowledge on the situation with bio-resources and ecosystem services, more research on biodiversity, habitats quality and associated anthropogenic efforts needed to be coordinated and the results exchanged. This is important not only on the national but also on the international level since the ecosystems and bio-resources have no borders (Gambette et al. 2014). According to Zakirov (1989), the vegetation of CA is subdivided into 9 types and 20 subtypes. The desert flora in the CA region is also diverse depending on the substrate: psammophytes grow on the sands, while hydrophytes—mostly on stony substrates—and halophytes—on saline soils, solonchaks. The mixed *Artemisia pauciflora* and *Salsola* species are spread on the clay and loess-clay soils of northern part of the desert zone, and the ephemeral flora in the south. The microrelief of desert also makes the impact on the biodiversity of plants growing there: with water accumulation in the small hollows of the ground, the vegetation cover is more complex than in vast flat areas. Another factor enhancing biodiversity in the desert is higher level of ground water and/or nearness of river. The formation of tugaj flora (flooded wood ecosystems) near the banks of Amu Darya in Turkmenistan can be one of such examples. In the tugaj ecosystems of Amu Darya Nature reserve, 97 species of plants are registered (Sokolova 2014).

The anthropogenic factor and its different activities are becoming more and more important and influence biodiversity in the CA region: long-term land use under agriculture, irrigation or intensive grazing supported the increase in soil salinization, contamination and drastic changes towards the worsening of biodiversity of natural ecosystems there. These negative tendencies are evident in nearly half of all natural pastures in whole arid zone. Thus, the environmental restoration of degraded pasture and other ecosystems has to be done applying developed biogeocenotic methods based on the use of dominant species of fodder plants with violent and patient features and their phytocenotically balanced combinations, which would ensure the formation of a biological frame of nodal, zonally typical self-renewing pasture ecosystems

Table 3.2 Total number of threatened species (marine and terrestrial), per CA country and total region

Country	Critically endangered	Endangered	Vulnerable	Total
Kazakhstan	17	19	35	71
Kyrgyzstan	7	13	24	44
Tajikistan	8	10	25	43
Turkmenistan	10	9	3	52
Uzbekistan	11	16	26	53
Total	53	67	113	333

Table made using Annex 2 from the source: Larger than tigers (2019, p. 558)

within a short period (2–3 years) (Shamsutdinov and Shamsutdinov 2012).

Being the largest country in CA by area, Kazakhstan is distinguished by number of species of biota. Only paleorelict species are about 2000, more than 400 species of plants, which are endemic or relict and 300 of vertebrates might extinct soon. One can be mentioned, *Malus sieversii* (Lebed.) M. Roem as the world-known wild progenitor specie of cultivated apple-tree (Dzhangaliev et al. 2003). These plants grow in the forests of Western Tien Shan and other mountains and are unique world-known genetic plant resources.

In general, the mountains of CA and Kazakhstan are one of the four centres of biological diversity within the territory of the former Soviet Union (Agakhanjanz and Breckle 1995).

The largest number of plant species is typical for the Tajikistan and Kyrgyzstan mountain regions. However, in these countries, problems of preservation of biodiversity also occur, for example one of endemic species *Otostegia nikitinae*, found in Kyrgyzstan, is threatened (Prioritety sohranenya 2014). The Kopet Dag Mountains laying in southern Turkmenistan—northern Iran—have a rich and diverse flora; of which 332 species are endemic. Unfortunately a large number of these rare species (more than 25%) have been lost due to habitat decline (https://www.worldwildlife.org/ecoregions/pa1008). The critical situation with biodiversity and degradation of ecosystems is documented in 66% of territory of Kazakhstan, particularly in the zone of desertification and steppes, where the land is ploughed or overgrazed (Nacionalny doklad 2016).

3.2 Nature Protected Areas by Country

According to the International Union for Conservation of Nature (UNEP-WCMC and IUCN 2016), a protected area is "a clearly defined geographical space, recognized, dedicated and managed, through legal or other effective means, to achieve the long-term conservation of nature with associated ecosystem services and cultural values". They can be totally or partly protected, depending on their status and form, which vary from natural monuments, wildlife sanctuaries, protected landscapes to nature reserves and national parks. There are currently 286 protected areas in CA (see Table 3.3). The overwhelming majority of them have been established in mountain ecosystems.

Table 3.3 Nature protected areas in CA countries in km^2 and in % of total country area

Country	km^2	%
Tajikistan	3105	22.0
Kazakhstan	240,188	8.8
Kyrgyzstan	147.6	7.4
Uzbekistan	20,520	5.0
Turkmenistan	19,750	4.0

Sources: Nacionalny doklad (2016); Underutilised species (2006); https://www.worldwildlife.org/ecoregions/pa0808; Prioritety sohranenya (2014)

In most CA countries, except Tajikistan, nature protected territories occupy less than 8% of their total area. The data in this table was collected by using different national or international sources, provided with data from different years (2005 for Uzbekistan and Turkmenistan, 2012 for Tajikistan, 2015 for Kazakhstan and 2017 for Kyrgyzstan).

In Tajikistan the total area of nature protected territories covers 3,104,839 ha or 22% of the territory of the republic. The total area of the state natural reserves makes up 173,200 ha or 1.2% of the territory of the republic. The area state natural refuges—313,260 ha or 2.2% of the territory of the republic, Natural parks—6805 ha and the territory of National parks covers 18% of the territory of the republic (Tajikistan—terrestrial protected areas 2015). Recently, the National Park of Tajikistan in Pamir Mountains, which is one of the highest in the world, was announced as the first natural World Heritage site. The park stretches in the area with more than 2.5 million ha in the east of the country, "at the centre of the so-called "Pamir Knot", a meeting point of the highest mountain ranges on the Eurasian continent. It consists of high plateaux in the east and, to the west, rugged peaks, some of them over 7000 meters high, and features extreme seasonal variations of temperature". This site is famous not only for the longest valley type of glacier and the impressive number of 1085 inventoried mountain glaciers, but also as a shelter for the rare species of flora and fauna. Anthropogenic impact here is rather small, what is also desirable for such type of parks. In addition, due to the frequent earthquakes, the park territory is like a natural lab for those who study tectonics.

The protected areas of Kyrgyzstan cover 1,476,121.6 ha and account for 7.4% of the country's total area (2017), among them: 10 state nature reserves with 509,952.7 ha. There are 13 nature national parks occupying 724,670.2 ha, 1 biosphere area and 66 other protected areas with a total area of 241,499 ha (Prioritety sohranenya 2014).

The main categories of specially nature protected areas in Uzbekistan are two national parks, nine state reserves, one national centre on breeding of rare species of animals and three nature monuments. Among them, the largest in area is the national parks (Underutilised species 2006).

Currently, there are several specially protected areas (nature reserves) in the transboundary ecoregion of Gissaro-Alai mountains: Kugitang in Turkmenistan; Zaamin and Gissar in Uzbekistan; and Ramit in Tajikistan (Central Asia: Kyrgyzstan, Tajikistan and Uzbekistan https://www.worldwildlife.org/ecoregions/pa0808).

There are 9 nature reserves and 13 sanctuaries in Turkmenistan with a total area of 19,750 km^2 or more than 4% of Turkmenistan's territory (Turkmenistan—terrestrial protected areas 2015).

The area under the nature conservation reached 240,188 km^2 or 8.8% of total area of Kazakhstan in 2015 (Table 3.4).

The reserved zones and sanctuaries occupy the largest area in protected nature territory of Kazakhstan. In terms of protected territory area, nature conservation territories in four of five CA countries are not as high as in other European or American countries; only Tajikistan is a long step ahead even worldwide.

The distribution of vegetation belts, zones and patches in different landscapes depends on many factors, but two of the most important are climate and soil cover conditions. Climate scenarios on the global and regional scale have received much attention from the researchers of many countries, including those from the CA region. However, landscape-scale vegetation response across time and space has received far less emphasis. The studies focused on the vegetation dynamics and driving variables in three regional CA landscapes (desert, steppe and mountainous) were conducted using remote sensing methods in 1981–2008 (Kariyeva and van Leeuwen 2011). This research has revealed some tendencies in the spatial and year-to-year variation in vegetation dynamics; according to this parameter, the steppe regional landscapes were mostly affected by temperature. Also these landscapes showed relatively low temporal variability of vegetation dynamics, and that suggested higher resistance to change. Vegetation dynamics in the mountainous regional landscape were correlated primarily to climate variables, especially precipitation observed along elevation gradients. Vegetation dynamics in the desert regional landscape were affected by the combination of temperature regimes, winter precipitation patterns and soil carbon content. Thus, this research has revealed that dynamics of vegetation in a given ecosystem are influenced not only by climate but also by topography and soil properties of the system. If we look at the case of Kazakhstan, the tendencies of Normalized Difference Vegetation Index (NDVI) variation during 1982–2015 were such: the highest average NDVI was obtained in the cropland area (0.59) followed by grassland (0.45) and shrubland (0.26) (Luo Liang et al. 2017).

Table 3.4 Distribution of nature protected areas in Kazakhstan by status in 2015

Nature protected areas	km^2	%
Nature reserves	16,114	6.7
Reservations	23,041	9.6
National parks	25,242	10.5
Botanical gardens	4.24	0.002
Regional parks	1891	0.78
Reserved zones	113,124	47.1
Sanctuaries	60,763	25.3
Nature monuments	4.03	0.002
Total	240,188	100.0

Source: Nacionalny doklad (2016)

3.3 Distribution of Forest Resources Globally and in CA Countries

In 1990, the world had 4128 million ha of forest; by 2015, this area had decreased to 3999 million ha. Global forest land area declined from 31.6% in 1990 to 30.6% in 2015. The calculations made by the Food and Agriculture Organization of the United Nations (FAO) showed that the forest area changes differ in accordance with the world climatic-vegetation zones and time periods (Global forest resources 2015). In tropical zone, the largest reduction in forested area occurred during 1990–2000; and in temperate zone the largest decrease in forested area was observed during 2000–2010. In boreal and subtropical zones the changes in forest area were the least. The CA region during 1990–2015 experienced very small changes (gain/reduction) in the forested land area, and in 2015 reached about 6% of total CA region on the average (Table 3.5).

Table 3.5 shows that Turkmenistan has become the leader in the region on the basis of the share of forested and other wooded land area with 8.7% of total country's area. Concerning conservation policy, the largest forest areas designated for that purpose are to be found in Tajikistan (26.1% of forested country's area). As per the total forest harvest in 2015, Kazakhstan remained the leader in the CA region with 371,000 m^3. Comparatively low production of timber biomass in this region is limited by climatological/meteorological conditions. But the forested area monitoring in the CA countries is conducted with some disturbances. Some researchers (Yin et al. 2017) and organizations, like the Food and Agriculture Organization of the United Nations (FAO), state that forest statistics provided by these CA countries are mostly outdated or missing (State of the World Forests 2007).

Notwithstanding their relatively small coverage area, forests in CA play a key role in the regional hydrology by helping maintain a steady river discharge from the high mountains to the irrigated lowlands (Akhmadov 2008). The forests particularly in the mountainous territories serve as the shelters for endangered and endemic flora and fauna species and as carbon sinks. Calculated carbon stock in living forest biomass of all forests in the CA region and separate countries is presented in Fig. 3.1.

Table 3.5 Distribution of forest and other wooded land in 1000 ha and % of country's total area, share of forest area conserved for biodiversity in % and estimated total harvest in 1000 m³ in the CA countries in 2015

	Forest and other wooded land		Share of forest area conserved for biodiversity	Estimated total harvest
Country	1000 ha	%	%	1000 m³
Kazakhstan	12,904	4.7	20.0	371
Kyrgyzstan	1663	8.3	6.2	18
Tajikistan	563	4.0	26.1	9
Turkmenistan	4264	8.7	2.4	10
Uzbekistan	3369	7.5	12.0	36

Adopted using source: Forests and forest sector overviews (State of Forests 2019)

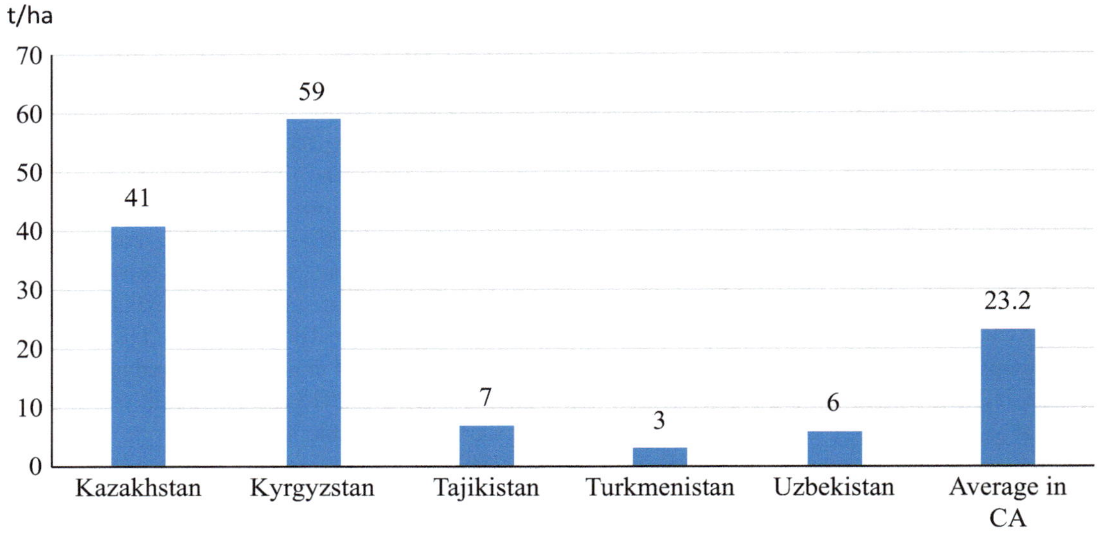

Fig. 3.1 Calculated carbon stock in tonnes ha^{-1} in living forest biomass in CA countries in 2010. (Source: Forests and climate change (2010))

3.3.1 Kazakhstan

According to the Forest Code of the Republic of Kazakhstan, forests are defined as "a natural complex formed on a certain territory based on the aggregate of trees and shrubs vegetation and other wildlife components interacting with environment and having environmental, economic and social importance" (State of the World Forests 2007). In 2015, forests accounted for almost 5% of Kazakhstan's territory (Fig. 3.2).

The main concentrations of forests in Kazakhstan are in the fertile forest-steppe zone extending from Russia along the northern border, at the foothills and slopes of the Altai, Alatau and Tien Shan mountains along the eastern and south-eastern borders, and along the Syr Darya and other main rivers in the southern deserts. About 300,000 people in the country are directly dependent on the sector, while an estimated 2.5 million rely on its resources for their living (State of the World Forests 2007). Kazakhstan as a country in its forest resources management strategy recognizes the multi-use functions of its forests, and this encourages for the future more forested land to be designated for conservational and recreational targets. Particularly, afforestation is vital for the dry Aral Sea bed and surroundings and Lake Balkhash basin, where the saxaul and other shrub type forests were growing widely, but with desertification increase and overgrazing, they need to be restored and maintained in a more sustainable way (Babaev and Kharin 1999; Kushlin et al. 2004) (Fig. 3.3).

In general, Kazakh forests are very diverse; they have 68 species of trees and 266 species of bushes, and provide habitats for 835 species of vertebrates. The coniferous forests account for 15.5%, hardwoods for 61.8% and bushes for 22.7% of forest areas (Kushlin et al. 2004). In addition, the riparian forests along major rivers and in treeless western Kazakhstan are very important agents of water and local climate regulation in the southern floodplains (Tugay forest) (Global Forest 2014a).

There are many reasons for forest degradation in Kazakhstan: the clear cutting, illegal logging, forest fires, overgrazing. In the forest steppe zones, intensive agricultural activities are competing with the forest management for the land use: one alarming example along the Irtysh River in the north-eastern Kazakhstan confirms this (Sehring 2012).

Fig. 3.2 Forests on the slopes of south-eastern Kazakhstan near Almaty and Ile-Alatau National park (about 1500 m asl). (Made by A. Bučienė in October 2017)

Fig. 3.3 Saxaul and tamarisk shrubs with the herd of camels in the area of Barsakelmes reservation (Kazakhstan) near Small Aral. (Made by Saulius Bučas in October 2017)

3.3.2 Kyrgyzstan

Kyrgyzstan is comparatively rich in forest resources in the CA region, with more than 8% of forested and other wooded area. Kyrgyz forests, which are all state owned, are also highly diverse. They can be grouped in four main types: (1) spruce forests; (2) walnut-fruit forests; (3) juniper forests; and (4) riverside forests. Apart from natural forests of the above types, there are plantations (Carter et al. 2001; Fisher et al. 2004; Integrated Assessment 2010; Global Forest 2014b). The first group forests are found in the west, in the centre and in the highest mountain ranges, between 1700 and 3000 m asl. The main trees are *Picea schrenkiana* Fisch. et May, also small areas of stands with the endemic Semenov fir (*Abies semenovii* B. Fedtsch) can be found in the very west of the country. The second group of forests are found in the northern and north-eastern slopes of the Fergana valley. They are represented by the ecosystems with walnut (*Juglans regia* L.), apple (*Malus* spp.), hawthorn (*Crataegus* spp.), plumb (*Prunus* spp.), rose species (*Rosa* spp.) almond (*Prunus amygdalus* Stokes) and pistachio (*Pistacia vera* L.). According to Carter et al. (2003), this group of forests of Kyrgyzstan are considered to be the biggest remaining areas of this particular forest type worldwide and therefore to be of global significance for biodiversity conservation. The third group of forests grow under arid conditions or in very high altitudes up to 3500 m asl and represented by Juniper (*Juniperus* spp.). Riverside forests can be found in all parts of the country along streams and rivers with typical representatives like willow (*Salix*), poplar (*Populus*), birch (*Betula*) and tamarix (*Tamarix*), sometimes also with sea buckthorn (*Hippophae rhamnoides* L.) (Carter et al. 2003).

The idea of promotion of collaborative forest management (CFM) in accordance with the programme, run through Swiss collaboration, has been supporting the forestry sector in Kyrgyzstan since 1995. However according to Ulybina (2015), the capacity of local communities should be supported by cross-sectoral national or even cross-border efforts and joint actions with foreign experts in order to adopt the most suitable for locals concepts.

3.3.3 Tajikistan

The total area of lands of state forest fund makes about 4% of the territory of the republic. Various tree-bushy breeds are significant varieties of vegetative communities: wide deciduous and small-leaved forests, unique juniper groves and Tugay, nut and pistachio forests, and also deserted sparse growth of trees from saxauls, kandym, cherkez and other sandy breeds (Global Forest Resources 2010a). The best message about the forest sector of Tajikistan is its highest percentage under the conservational status for biodiversity purpose (26.1% of all forest area). All Tajikistan forests are mountainous, and divided into five types: (1) broadleaved mesophilous forests, (2) hard-leaved xerophilous light forests (shibliak), (3) small-leaved microthermous mountain forests, (4) juniperus forests and (5) Tugay forests. Despite state efforts to maintain conservative status of the country's forests, the depletion of forest resources is taking place mostly due to the population poverty. The grazing of livestock in the forested areas, illegal cutting of timber, expending of agricultural land use and other routine use of forest resources lead to their depletion (Akhmadov 2008). Among these factors, the overgrazing must be considered an essential factor in forest degradation and destruction in Tajikistan. Thus, more sustainable pasture management and/or improvement of pastures is another solution for the stopping of forest degradation by livestock (Kirchhoff and Fabian 2010).

3.3.4 Turkmenistan

According to the last year's official statistics data, the forest land area in Turkmenistan is one of the highest—8.7%. The dominating species on the areas classified as forest are Saxauls (*Haloxylon spp.*) with very low growing stock harvest (Global Forest Resources Assessment 2010b). But still the forests distributed on the plains and highlands of the country are diverse in tree-species, habitats/communities, etc.

Kopet Dag woodlands found from the lowland plane 300 m up to 2500 m above the sea level can be characterized by several types: (1) xeric shrublike woodlands, (2) juniper woodlands and (3) mesophytic riparian forest. The first group is known as *shiblyak*, distinctive by the Turkmen maple (*Acer tucomanicum*), hawthorns (*Crataegus spp.*) or by the Christ's thorn (*Paliurus spinachristi*). Kopet Dag forests are known for the high endemism reaching 18% in flowering plants. They serve also as the center of origin and genetic diversity for wild relatives of cultivated plants such as grapes, pomegranates, figs, almonds, walnuts, wheat, barley and many others (https://www.worldwildlife.org/ecoregions/pa1008). At the same time, these woodlands are vulnerable to be disturbed under the increasing of logging and overgrazing because overgrazing in the upper parts of the mountain lands leads to serious destruction of the grasslands and thus to mudflows, disturbing woodlands on the lower slopes. Dry summers of the last years have dried out vegetation (Pereladova et al. 1998). Thus, for Turkmenistan, the afforestation and sustainable management of forested areas continues to be a great challenge.

Urban areas in this country are also designed to be afforested or greened. For example about 24,000 ha representing 30 million seedlings have been planted in total with modern techniques using drip irrigation close to Ashkhabad (Global Forest Resources Assessment 2010b).

3.3.5 Uzbekistan

About 7.5% of Uzbekistan's territory is covered by forests and shrubland, but the majority of forests are found in the desert zone. Besides desert, forest is growing in the valleys of rivers (Tugay-type) and in mountains (Botman 2008). The current forests are very different from the original forests in terms of composition of forest species, density and area distribution primarily due to changing of climatic conditions towards aridity (Global Forest Resources 2014a, b). The state recognized the importance of multi-functionality of forest ecosystems with focus on the conservation of biological diversity and combating against desertification (Botman 2008). Still the main problems of forestry development in Uzbekistan are as follows: (1) lack of qualified specialists working in the forestry enterprises; (2) no National Forest Program for country; (3) no common system of inventory of forests between all stakeholders; (4) out-of-date methods of forest management and forest resources monitoring; (5) weak relations between the forestry sector and manufacturers using wood products and (6) uncontrolled overgrazing negatively influencing natural afforestation. Among the measures to start solving these problems, it is important to develop the National Forestry Program and participate in the international projects-initiated afforestation (Vildanova 2006).

Forest ecosystems produce more different products than timber or firewoods. In Table 3.6, main non-wood products and services supplied by forests in CA countries are presented.

If grazing in the forest land is recognized as one of the forest services for Kyrgyzstan and Uzbekistan rural areas, other non-wood products and services like hunting, nuts, forest fruits, berries, medicinal herbs, seedlings and honey are agreed to be important for all CA countries. Yet novel products and services, like handicrafts, aromatic plants, mushrooms and hay, are important in a separate country.

3.4 Biomes, Vegetation and Their Impact on the Climate System

Different biomes, ecoregions and ecosystems are not only passive observers of climate changes; they are rather active participants of processes of energy and matter cycles on the earth. The researchers Thomas and Roundtree (1992) were first to notice that land cover change in the boreal forest has a significant influence on the climate system mostly through changes to the surface radiative energy budget. Changes in the surface albedo due to snow versus vegetation cover and the variability of cloud cover combine to strongly influence the radiative balance at the surface. They simulated different scenarios boreal forest ecosystems. Bonan et al. (1992, 1995) elaborated further these ideas working with coupled atmosphere-biosphere model, CCM3-IBIS and analysing the climatic processes. It was confirmed that removal of the boreal forest vegetation has a large effect on the surface radiative balance because of the strong snow/ice/albedo feedback. Later research (Snyder et al. 2004a) has revealed that boreal forests are the strongest agents influencing on the surface albedo as compared with temperate forests or grassland/steppe biomes (Table 3.7).

Table 3.6 Main non-wood products and services supplied by forests in CA countries

Country	Main non-wood products and services
Kazakhstan	Hunting, forest fruits, berries, nuts, mushrooms
Kyrgyzstan	Grazing (meat and dairy), nuts, berries, honey, medicinal herbs
Tajikistan	Hunting, forest fruits, berries, nuts, medicinal plants, honey, seedlings
Turkmenistan	Nuts (pistachio, walnut), berries, fruits, medical herbs, hay, seedlings
Uzbekistan	Honey, grazing, hunting, nuts (pistachio, walnut), medicinal and aromatic plants, fruits, handicrafts

Adopted from the source: Forests and forest sector overviews in (State of Forests 2019)

Vegetation cover impacts were analysed also through air temperature and evapotranspiration effects within different biomes. Boreal forests have shown weak effects on evapotranspiration as compared with temperate forest and steppe, which showed moderate effects. The removal of temperate forest vegetation leads to a warming and drying during the

Table 3.7 Summary of each biome's influence on the climate due to albedo and evapotranspiration effects. Ranking is from "strong" to "moderate" and "weak" and is qualitatively based on the results from the individual biomes

Biome	Effects of changing albedo	Effects of changing evapotranspiration (ET)	Description
Boreal forest	Strong	Weak	Large surface albedo inscrease causes a large decrease in net radiation and the surface cools. Reduction in ET limits near-surface moisture and precipitation only during summer growing season. Surface cooling leads to a reduction in planetary boundary layer height and a large increase in cloud cover.
Temperate forest	Moderate	Moderate	Increase in albedo decreases net radiation during all seasons. Winter and spring increase causes a cooling while summer and fall increase causes a warming when combined with a moderate reduction in ET. Decrease in ET reduces near-surface moisture and precipitation (mostly in summer)
Grassland and steppe	Moderate	Moderate	Moderate surface albedo increase causes a reduction in net radiation. In winter and spring this cools the surface while in summer it warms the surface when combined with a moderate reduction in ET.

Source: Adopted from Snyder et al. (2004a)

summer and fall months due to a reduction in latent cooling with removal of the normally transpiring vegetation and a cooling in the winter and spring as the surface albedo of snow is higher than the vegetation (Snyder et al. 2004a). In Uzbekistan, Botman (2008) identified that under future climate change scenarios the mountain coniferous forest ecosystems mostly represented by juniper (*Juniperus spp.*) will be the most affected. The new research has shown that from a dendroclimatic perspective, some species have higher potential for the extremal situations than others. For example, *J. turkistanica* at its distribution limit at the treeline is a suitable proxy of summer temperature (Seim et al. 2015).

The situation in grassland/steppe biome is forecasted differentially for central US and Asia biomes: the reduction in evapotranspiration will lead to climate warming in North America continent, while increase in albedo will cause the climate cooling in CA grassland/steppe biomes (Snyder et al. 2004b).

Still there are many unanswered questions regarding the role of different biomes in influencing the climate on global and regional scale; thus the research on that theme has to be continued with more variables included.

3.5 Problems of Protection Steppe/ Grassland Biome

All biomes are important reservoirs of carbon. Forests are the largest terrestrial carbon stock, but they are under the risk of deforestation, degradation and the long-time impact on climate change. Grasslands are potentially the largest carbon sink, but loss and degradation are currently releasing large amounts of carbon (Yildiray 2010). Steppe/grassland biomes are found in about 20 countries, mainly in mid-latitudes where seasonal climates and soils favour the dominance of perennial grasses covering about 8% of the earth's surface. Biodiversity in these habitats includes a number of large grazing mammals and associated predators in addition to burrowing mammals, numerous bird species and insects. However, only 3.4% of the world's temperate grasslands are protected. Different human practices, particularly agriculture, have led and are continuing to lead to habitat loss, declining biodiversity, desertification and fragmentation. The recent climate change brings the increasing concern too. In CA countries, particularly in Kazakhstan, there are five steppe ecological zones: forest steppe, meadow steppe, dry steppe, desertified steppe and steppe-semidesert, which cover about 59% of the country's territory. Habitat fragmentation and loss have had a major impact on the grassland and steppe biome (Sklyarenko 2010; Ibragimov 2016). Habitat types within temperate steppe biome in Kazakhstan allowed a rich diversity of flora and fauna to exist. Notable mammals in this ecoregion are herds of saiga, wild boar, lynx and badger, and the region's avifauna is particularly rich: about 300 different species are found, of which 30 are considered rare. During seasonal migrations as many as 3.5 million geese visit the area (Rachkovskaja, Pereladova Central Asia: Kazakhstan https://www.worldwildlife.org/ecoregions/pa0811). Thus this biome requires urgent measures to be taken for the protection and more sustainable management. Scientists and practitioners have identified that traditional grazing regimes and management practices are critical to maintaining vegetation, soils and composition of natural communities. This can be illustrated by the inter-linkages: decline of Saiga in Kazakhstan = decline of grasslands = decline of grassland-dependent species. In general steppes are the home for many migratory species; thus, this feature also has to be respected. The sustainable steppe management requires strong community relations and landscape approach (Jamison 2010).

3.6 Climate Warming and Land Cover Changes in Different Biomes in CA Region

The great attention has to be given to the tendencies of global climate change. Recent climate warming impacts are leading to longer periods with stable hot weather in summer, which increases the probability of years with bad grazing conditions (Sklyarenko 2010). Using the remote sensing methods, one can see the evident changes or shifts of different land use/land cover patterns in biomes of the CA region from 1982 to 1999 (Fig. 3.4).

From 1982 to 1999 four land cover change trends were documented in the CA region: (1) areas with changes towards intensification of agriculture; (2) areas with trend to naturalization of vegetation cover; (3) areas with retrenchment of agriculture; and (4) areas with retrenchment of natural vegetation. The main factors of these trends are mixed: the climate change and anthropogenic impact. As the calculations show (Sommer and De Pauw 2011), the conversion of native land into agricultural land and the degradation of rangelands reduced the soil organic carbon by 4.1% of the total stocks on the average, while the most of the carbon losses resulted from the past conversion of rangelands into rainfed or irrigated agricultural land in the north of Kazakhstan.

But let us look more in detail for natural and anthropogenic factors, influencing the desertification process in some localities like the Aral Sea surroundings.

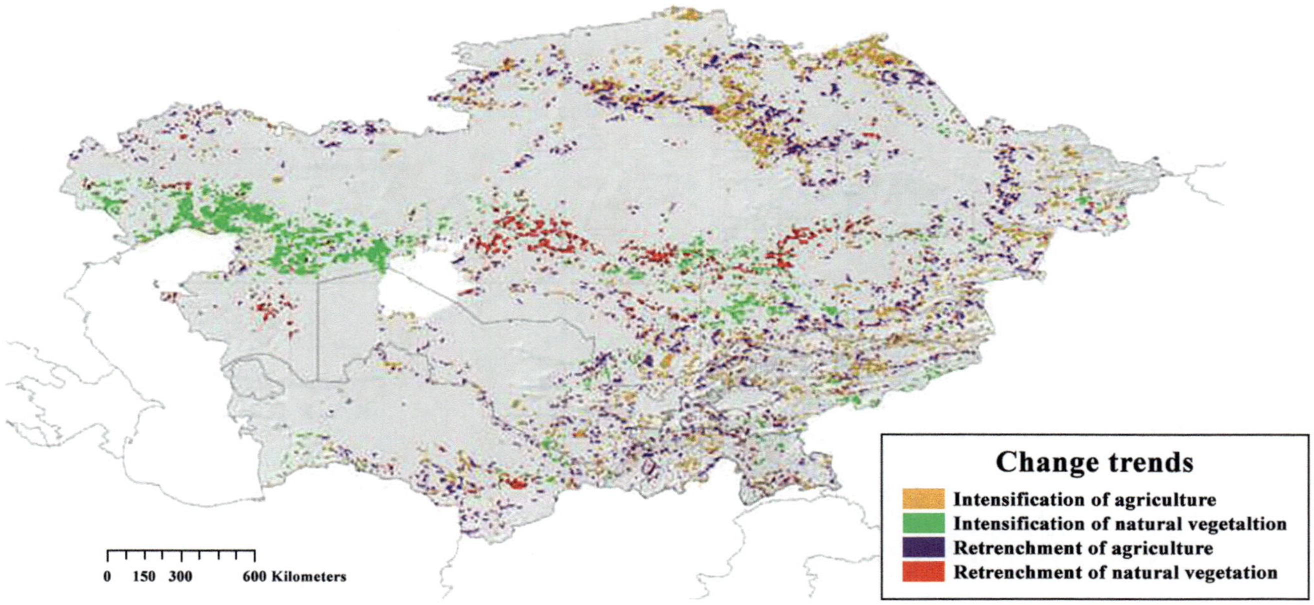

Fig. 3.4 Changes in different land use cover in CA from 1982 to 1999. (Source: Sommer and De Pauw (2011))

3.7 Recent Desertification and Its Consequences for the Biomes and Ecosystems

3.7.1 Case Study: Situation in Aral Sea Basin

Among the first-order drivers of land degradation and desertification the climatic characteristics, such as aridity, continentality, wind action, etc., which cause such phenomena as drought, hot winds, deflation, and atmospheric transportation of sand, salts and dust to the long distances have to be ranked. Among the drivers of second importance different human or anthropogenic activities have to be mentioned. The interaction of those natural and anthropogenic forces can be evidently studied in nowadays region surrounding the Aral Sea. The total basin of the Aral Sea includes the area of two southern oblasts (region) of Kazakhstan (Kyzyl-Orda and Southern Kazakhstan), three oblasts of the Kyrgyz Republic, the whole territory of Tajikistan and Uzbekistan and four velayats (regions) of Turkmenistan (without Balkhan), in total 155 million ha (Dukhovny et al. 2005) (Fig. 3.5).

The agricultural land in the Aral Sea basin is distributed as shown in Fig. 3.6.

The largest area of agricultural land in the basin is used as a pasture or rangeland (about 83%). Though the percentage of ploughed land is not high (17.0%), the irrigated land made 78.7% of that; thus, this number makes to be concerned on the rational water use for the cultivated crops.

The climate was very contrasting during different historical periods in the Aral Sea basin; thus, it might have impacted the area of the sea too. The analyses of sediment cores taken from the NW large Aral Sea area have revealed shifts in vegetation development from subdesertic to steppe vegetation in the Aral Sea basin during the late Holocene. Researchers have reconstructed climate in CA during the past 2000 years and came to the conclusion that there were at least three cold and arid periods. Conversely, the most suitable climate conditions occurred between ca. AD 400 and 900, and AD 1150 and 1450, when steppe vegetation was enriched by plants requiring moister conditions (~ 250–500 mm/yr) and some trees developed. These results are fairly consistent with other late Holocene records from the eastern Mediterranean region and the Middle East, showing that regional rainfall in CA is predominantly controlled by the eastern Mediterranean cyclonic system when the North Atlantic Oscillation (NAO) is in a negative phase (Sorrel et al. 2007; Breckle and Geldyeva 2013; Geldyeva et al. 2013). Thus seemingly the Aral Sea might have had a number of cycles of recession and transgression in Holocene.

The irrigated agriculture in the CA region was a target for the Soviet government to decide in the 1960s to divert the Syr Darya and Amu Darya rivers and later their tributaries so

Fig. 3.5 The Aral Sea basin area. (Source: Micklin (2007); Landsat satellite imagery from USGS/NASA; Digital Elevation Model from USGS EROS; visualisation by UNEP/GRID-Sioux Falls https://na.unep.net/geas/articleimages/Jan-14-figure-1.png)

Fig. 3.6 Distribution of agricultural land cover in the Aral Sea basin. (Sources: The Aral Sea (2008a, b))

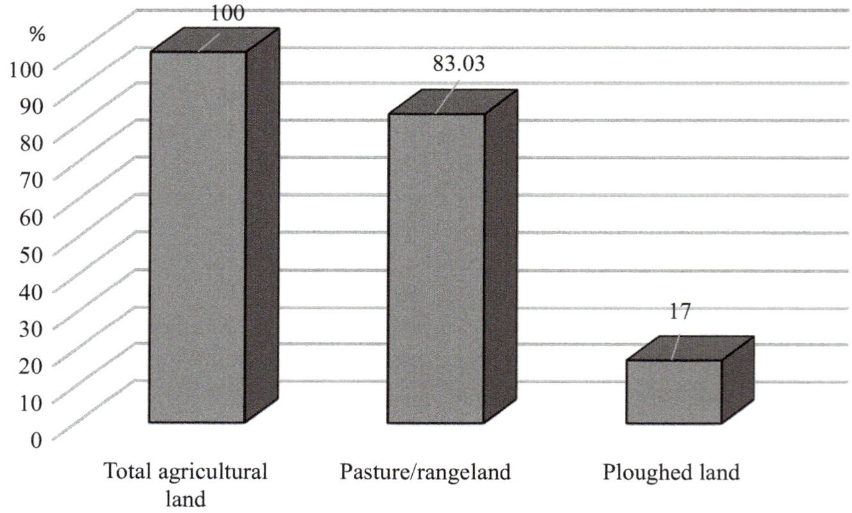

that they could irrigate the desert region surrounding the Sea in order to favour agriculture, especially for cotton. Most of surface water was regulated by canals system at the expense of the Aral Sea supply, but another part was being soaked up by the desert or wasted (Micklin 2016). As a result, since 1960s the water level in the Aral Sea started drastically decreasing. It has dropped approximately 23 meters since the onset of its primary sources of water being diverted (Zavialov

2005). In normal conditions the water budget in the basin is determined by several components: inflow from the river, evaporation, precipitation rates and groundwater inflow. Thus the Aral Sea gets approximately one-fifth of its water supply through rainfall, while the rest is delivered to it by the Amu Darya and Syr Darya rivers. But the intensification of irrigation has disturbed these processes. Net evaporation defined as the difference between evaporation and precipitation at the surface started to decrease, while the river inflow has been drastically decreasing since 1960 in the basin. Yet the groundwater inflow has remained approximately the same, the overall situation resulted in a net deficit of water to the sea. Evaporation causes the water level to decrease by the same amount that flows into the Sea, making it sustainable as long as inflow is equal to evaporation on average. Therefore, the diversion of rivers is at the origin of the imbalance that caused the sea to slowly desiccate over the last four decades. More precipitation in the mountains and some reduction in irrigation withdrawals increased river discharge during the 1990s and reduced the water balance deficit, slowing the sea's recession (Micklin 2007, 2016; The Aral Sea Crisis 2008a, c).

Up until the third quarter of the 20th century the Aral Sea was the world's fourth largest saline lake, and contained 10 grams of salt per litre. Nowadays under both climate warming and still continued anthropogenic pressure expressed mostly via irrigated agriculture, the desertification and secondary soil salinization is taking place in the Aral Sea basin. Since 1956, researchers at Kazakhstan's U.U. Uspanov Research Institute of Soil Science and Agrochemistry have conducted monitoring studies which have determined that the drying up and desertification of hydromorphic soils at the current delta of the Syr Darya river are accompanied by increased salinization processes, a sharp decrease in non-saline soils and increases in soil salinity of various degrees, including the formation of saline marsh soils (Saparov 2014). As a result salt is intensively accumulated on the surface, and the solonchak is to be formed and spread in previous agricultural (mostly irrigated or grazed) landscape (Fig. 3.7a, b).

The desiccation of the Aral Sea greatly increased the number of dust and salt storms in the area. Satellite images have revealed that there are up to ten major storms occurring annually in the region, most of them occurring between April and July (Glantz 2007; The Aral Sea crisis 2008a). According to Saparov (2014) at the Kazakhstan part of the Aral Sea, three large sources emitting sand-salt aerosols into the atmosphere have formed and their influence can be observed at a distance of 200–250 km and more.

The salinization of the Aral Sea water also took place due to the decrease in precipitation rate and river inflows, and that led to the disappearance of freshwater fish and other invertebrates. Anthropogenic regression and salinization of the Aral Sea also took place. The Aral Sea was divided into a northern Small Aral and a southern Large Aral at the end of the 1980s, with the two having different hydrological regimes (Plotnikov et al. 2016; Aladin et al. 1995). Later starting from September 2009, the Aral Sea was separated into four waterbodies. Up to now discussions among scientists from around the world and the politicians of CA countries are going on whether it is possible to restore the Sea or not. At least the partial restorations seem to be feasible. CA countries encountered difficulties in supporting their scientific institutions after the collapse of the Soviet Union and during the period of profound transformations. As scientific knowledge and technical know-how are the basis for all future development, UNESCO saw its first task as supporting the scientists to enable them to continue their research related to the Aral Sea. Thus international projects such as "Ecological research and monitoring of the Aral Sea Deltas" began in 1994 with extra-budgetary funds provided by the German Federal Ministry of Research and Technology. Since 1998, UNESCO has intensified its efforts to address water-related problems, including social aspects of water use in the Aral Sea basin, and established a Scientific Advisory Board on Aral Sea Basin Problems (SABAS), whose members are representatives of the scientific community of the region (Rasmussen 2014). UNESCO also launched its "Aral Sea Initiative" outlining future activities in the area focusing on the Aral Sea Deltas environment (Water Related Vision 2000).

3.7.2 Attempts to Stop Decrease of Aral Sea Area and Water Salinity

Among those measures, at first the Kokaral Dam should be mentioned (Fig. 3.8a, b).

The dam was constructed in 1992, but needed to be improved and finally it was finished in 2004. The level of water of the Small Aral Sea increased with gradual decline in salinity. Today the Small Aral has again become brackish (Plotnikov et al. 2016).

According to the satellite images from 1960-08-22 and 2017-08-22, the Aral Sea area and volume has been changing dramatically, even after some measures to stop decrease of Small (Northern) Aral were undertaken (Fig. 3.9a, b, Table 3.8).

The upper image was made using an Argon reconnaissance satellite on 22 August 1964, before the drop in water levels altered the shoreline and devastated surrounding communities. The name of the Aral Sea means "Sea of Islands"; thus they were many in this sea earlier. One of these islands, Vozrozhdeniye, translated as "renaissance" or "rebirth" island, was the home of biological weapons lab in 1966 (A U.S.-Uzbek team decontaminated the former island in 2002). Today this island is a part of a peninsula reconnected to the mainland in 2001. The lower image was captured by the

a

b

Fig. 3.7 Sevier desertification and salenization is taking place in the south of Aral town near the Aralkum settlement (**a**): the sand and salt storms frequency has increased here few times from the end of the 20th century, and in the bottoms of previous sea lagoons and streams, the solonchak (**b**) is forming. Photos made by S. Bučas in October 2017

Moderate Resolution Imaging Spectroradiometer (MODIS) on NASA's Terrasatellite, (https://earthobservatory.nasa.gov/IOTD/view.php?id=90857&eocn=related_to&eoci=related_image).

The two largest rivers, Syr Darya and Amu Darya, act as biological refugia for endemic species of the Aral Sea (Micklin 2010). The environmental monitoring including hydroecological data measuring station network is a necessary tool for further ecological stabilization of the region. The landscape of the delta of Syr Darya is shown in Fig. 3.10.

It was estimated the needed reduction in irrigation areas and intensity in order to stop decrease of water level in the recent Aral Sea fragments. For example, to cover the deficit of water runoff would require by 42% reducing the irrigated area of 8.2 million ha as it was in 2010 (Micklin 2014). The land areas surrounding of Aral Sea, which are now free of water, compose the so-called new deserts where ecosystems need to be developed. This can be the only way to stop dust and salt dissemination, and wind erosion of the soils. Development of a layer of salt-resistant vegetation and further succession of plant communities can be the only way to create a healthy environment for the local people (https://www.worldwildlife.org/ecoregions/pa1310).

Fig. 3.8 The Kokaral dam (**a**); at the Berg's Strait bellow the Kokaral dam (**b**) October 2017. (Made by S. Bučas)

3.7.3 Impact of Recent Desertification in the Aral Sea Basin on Human Social and Physical Health

With a population greater than 32 million occupying 1.5 million km of land in the Aral Sea basin, profound societal changes are to be expected as desertification increases (Smith 1994). It has to be recognized that drop in economics will lead to the social instability. Many people in the Aral basin were employed either directly or indirectly by Sea- or River-dependent jobs. As more agricultural land is taken out of production, the poor socioeconomic conditions will only increase. Further, the quality of life has been deteriorating, due to a variety of severe health problems caused by worsening of environment: salty storms, occurrence of pesticides from agricultural fields, poor quality of drinking water leading to the increase in infectious diseases. Infant mortality and morbidity rates across the entire region are increasing (Micklin 1993). Restoring the Aral Sea is of minor importance, compared to the radical improvements needed in public health and socioeconomic situation (Allen 1997).

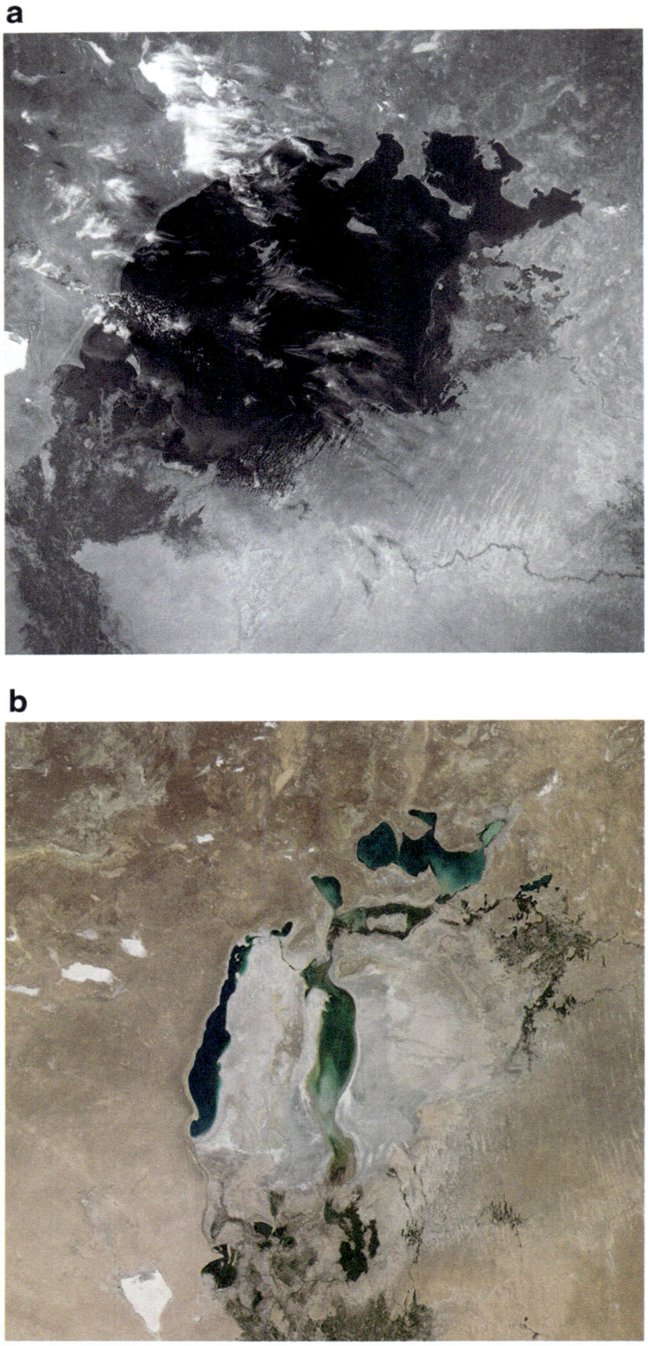

Fig. 3.9 Two satellite images of the Aral Sea (a and b) remoted one from another by 53 years. (Source: https://earthobservatory.nasa.gov/IOTD/view.php?id=77193)

Sakiev and Batyrbekova (2015) have noted that the most vulnerable groups of society in accordance to the increased disease risks are children, youngsters and women.

The newest research has revealed some tendencies in the structure of oncological diseases: the dominated haematological malignancies, including organ tumours—tumours of the musculoskeletal system and the skin, digestive system, brain and central nervous system are taking place. In areas of the Aral Sea region (disaster, crisis and pre-crisis condition) primary oncological morbidity of child population was higher than the control area, but the degree of excess was not entirely consistent with environmental stress zones. General morbidity in the disaster area was somewhat lower than in the crisis area, which may be associated with a disease process undulating dynamic. Gender peculiarities of malignant tumours in the investigated area have not been revealed. The dynamics of the primary oncological morbidity in children over a 10-year period have been unstable; incidence peaks were observed in the zones (Rybalkina et al. 2017).

Table 3.8 The main hydrological data on the Aral Sea from 1960 to 2011

Year	Sea level m (asl)	Sea area (1000 km^2)	Sea volume (km^3)	Salinity (g/l)
1960	53.4	67.5	1,089	10
1971	51.1	60.2	1,007	10
1976	48.3	55.7	763	14
1989	39.6	39.7	364	30
2009 Large Sea	26.5	4.9	56	>100-200
2009 Small Sea	42.0	3.6	27	10-14
2009 Whole	30.8	8.5	83	10-200
2011 Large Sea	27.7	6.7	57	50->100
2011 Small Sea	42.0	3.6	27	8
2011 Whole	31.5	10.3	84	8->100

Source: Adopted from Micklin (2014)

Analysis of demographic variables in the studied areas of the Aral Sea region showed an increase in the population, is characterized by high birth rates and a decrease in total mortality, which provides high levels of natural increase. Despite positive shifts in the demographic situation, demographic burden on the working-age population of persons older than working age increased in all regions studied. For Priaralie currently mortality has been caused mainly by exogenous factors. Analysis of the structure of the causes among the population of all the areas studying mortality showed that the dominant causes of the enlarged group were cardiovascular diseases, tumours, injuries and poisoning, accidents and respiratory diseases.

The average life expectancy on the coastal zone is relatively low compared to European rates and other regions of Kazakhstan. During the period under the review there was a positive dynamics of growth in life expectancy; there is a significant difference in life expectancy between men and women (Salimbayeva et al. 2017).

Fig. 3.10 Landscape of the delta of Syr Darya at Small (North) Aral Sea, made by S. Bučas in October 2017

3.8 Topics for Further Reading and Discussions on the Content of this Chapter

- Recent problems of maintaining biodiversity in desert biome of CA
- Unique nature protected monuments and complexes in CA countries: state overview

Recommended References

Aladin N, Micklin P, Plotnikov I (2008) Biodiversity of the Aral Sea and its importance to the possible ways of rehabilitating and conserving its remnant water bodies. In: Qi J, Evered KT (eds) Environmental problems of Central Asia and their economic, social and security impacts, NATO Science for Piece and Security Series C: Environmental Security. Springer, Dordrecht

Biodiversity in Central Asia. A visual synthesis (2018). https://zoinet.org/wp-content/uploads/2018/02/Biodiversity-CA-EN.pdf

Geldyeva GV, Breckle SW, Gel'dyev BV (2013) Geography and geomorphological and lithological characteristics of the Aralkum (Chapter 3)/Aralkum-a man-made desert: the desiccated floor of the Aral Sea (Central Asia). Springer, Berlin/Heidelberg, pp 37–50

Mount Etna and the Mountains of Pamir inscribed on World Heritage List alongside El Pinacate and Gran Desierto de Altar (2013) UNESCO. Culture. World Heritage Centre. News and Events. News. http://whc.unesco.org/en/news/1042. Accessed 26 June 2018

Sennikov AN, Tojibaev KSh, Khassanov FO, Beshko NYu (2016) The flora of Uzbekistan project. Phytotaxa 282:107–118

World Heritage Thematic Study for Central Asia – August 2005. http://cmsdata.iucn.org/downloads/central_asia.pdf. Accessed 3 Apr 2018

References

Agakhanjanz O, Breckle SW (1995) Chapter 5: Origin and evolution of the mountain flora in Middle Asia and neighbouring mountain regions. In: Chapin FS, Körner C (eds) Arctic and Alpine Biodiversity, Ecological Studies, vol 113. Springer, Berlin/Heidelberg

Akhmadov K (2008) Forest and forest products. Country profile: Tajikistan. Geneva Timber and Forest Discussion Paper 46, New York. United Nations FAO, Geneva

Aladin NV, Plotnikov IS, Potts WTW (1995) The Aral Sea desiccation and possible ways of rehabilitating and conserving its northern part. Environmetrics 6:17–29. https://doi.org/10.1002/env.3170060104

Aladin N, Micklin P, Plotnikov I (2008) Biodiversity of the Aral Sea and its importance to the possible ways of rehabilitating and conserving its remnant water bodies. In: Qi J, Evered KT (eds) Environmental problems of Central Asia and their economic, social and security impacts, NATO Science for Peace and Security Series C: Environmental Security. Springer, Dordrecht

Allen TH (1997) The Aral Sea crisis: desiccation and perspectives on recovery, 10 p. http://employees.oneonta.edu/allenth/Class-Readings-Password/aral%20sea%20crisis.pdf. Accessed 13 Dec 2017

Babaev AG (1996) Problemy osvojenya aridnyh zemel (Problems of Arid Land Development). Moscow State University Press, Moscow, 282 p

Babaev AG, Kharin NG (1999) The monitoring and forecast of desertification processes. Desert problems and desertification in Central Asia. Springer, pp 59–75

Biodiversity in Central Asia. A visual synthesis (2018). https://zoinet.org/wp-content/uploads/2018/02/Biodiversity-CA-EN.pdf

Biomes and Regions of Northern Eurasia (2002) Russian nature. In: Shahgedanova M (ed) The physical geography of northern Eurasia. Oxford University Press. http://www.rusnature.info/reg/. Accessed 13 July 2017

Bonan GB, Pollard D, Thompson SL (1992) Effects of boreal forest vegetation on global climate. Nature 359:716–718

Bonan GB, Chapin FS III, Thompson SL (1995) Boreal forest and tundra ecosystems as components of the climate system. Clim Change 29:145–167
Botman E (2008) Vulnerability of juniper formations to climate change on the territory of Uzbekistan. In: SANIGMI (ed) Climate change consequences in Uzbekistan, adaptation issues. Information on fulfillment of commitments to the UNFCCC by Uzbekistan, 7th edn, Tashkent, pp 68–75
Breckle SW, Geldyeva GV (2013) Dynamics of the Aral Sea in geological and historical times (chapter 2)/Aralkum-a Man-Made Desert: the desiccated floor of the Aral Sea (Central Asia). Springer, Berlin/Heidelberg, pp 13–37
Carter J, Haldimann E, Kamytov M (2001) From top down Soviet planning to local forestry decision making: coping with change in Kyrgyzstan. Forests, Trees and People Newsletter No. 44
Carter J, Steenhof B, Haldimann E and Akenshaev N (2003) Collaborative forest management in Kyrgyzstan: moving from top-down to bottom-up decision-making, iied International Institute for Environment and Development Natural Resources Group and Sustainable Agriculture and Rural Livelihoods Programme, SIDA, Gatekeeper series No 108 (3)
Central Asia: Kyrgyzstan, Tajikistan, and Uzbekistan. Temperate grasslands, savannas and shrublands. World Wild Life. https://www.worldwildlife.org/ecoregions/pa0808. Accessed 1 Feb 2018
Central Asia: Southern Turkmenistan and Northern Iran. World Wild Life. https://www.worldwildlife.org/ecoregions/pa1008. Accessed 1 Nov 2017
Chivian E, Bernstein A (2008) Sustaining life: How human health depends on biodiversity. Center for Health and the Global Environment. Oxford University Press, New York
Dukhovny VA, Umarov PD, Yakubov KE, Tuchin AI (2005) Drainage in the Aral Sea Basin – Towards a strategy for sustainable irrigated agriculture. In: International conference report on Towards a Strategy for Sustainable Irrigated Agriculture with Feasible, Investment in Drainage, Aral Sea Basin, Central Asia, 10–13 March 2004, Tashkent, Uzbekistan, pp 157–160
Dzhangaliev AD, Salova TN, Turekhanova PM (2003) The wild fruit and nut plants of Kazakhstan. In: Janick J (ed) Horticultural reviews, Wild apple and fruit trees in Central Asia, vol 29. Wiley, pp 311–312
Fisher RJ, Schmidt K, Steenhof B, Akenshaev N (2004) Poverty and forestry. A Case Study of Kyrgyzstan with Reference to Other Countries in West and Central Asia, FAO LSP Working Paper 13. http://www.fao.org/docrep/007/j2603e/j2603e06.htm. Accessed 1 Nov 2017
Forests and climate change in Eastern Europe and Central Asia (2010) Working paper 8 (Dr. Csaba Mátyás, ed), FAO, Rome, 188 p. http://www.fao.org/docrep/014/k9589e/k9589e.pdf. Accessed 4 Apr 2018
Gambette P, Eggermont H, Le Roux X (2014) Temporal and geographical trends in the type of biodiversity research funded on a competitive basis in European countries. BiodivERsA report, 33 p. https://www.researchgate.net/publication/272179557_Temporal_and_geographical_trends_in_the_type_of_biodiversity_research_funded_on_a_competitive_basis_in_European_countries. Accessed 12 July 2017
Geldyeva GV, Breckle SW, Gel'dyev BV (2013) Geography and Geomorphological and Lithological Characteristics of the Aralkum (chapter 3)/Aralkum-a Man-Made Desert: the desiccated floor of the Aral Sea (Central Asia). Springer, Berlin/Heidelberg, pp 37–50
Glantz MH (2007) Aral Sea basin: a sea dies, a sea also rises. Ambio 36(4):323. Stockholm
Global forest resources assessment. Country report. Tajikistan (2010a) FAO, Rome. http://www.fao.org/docrep/013/al640E/al640e.pdf. Accessed 31 Oct 2017
Global forest resources assessment 2010. Country report. Turkmenistan (2010b) FAO, Rome. http://www.fao.org/docrep/013/al650E/al650E.pdf. Accessed 31 Oct 2017
Global forest resources assessment 2015. Country report. Kazakhstan (2014a) FAO, Rome. http://www.fao.org/3/a-az250e.pdf. Accessed 31 Oct 2017
Global forest resources assessment 2015. Country report. Uzbekistan (2014b) FAO, Rome. http://www.fao.org/3/a-az370e.pdf. Accessed 31 Oct 2017
Global forest resources assessment (2015) How are the world's forests changing? (2016) Food and agriculture organization of the United Nations, Rome, 2nd edn. http://www.fao.org/3/a-i4793e.pdf. Accessed 31 Oct 2017
Ibragimov F (2016) Evaluation report on Land Degradation in Kazakhstan. https://www.linkedin.com/pulse/evaluation-report-land-degradation-kazakhstan-firuz-ibragimov. Accessed 7 Jan 2018
Integrated Assessment of Natural Resources of Kyrgyzstan 2008–2010 (2010) FAO, Bishkek. http://www.fao.org/forestry/30655-067a616376e5bf5ebac056446ec010d1f.pdf. Accessed 31 Oct 2017
IUCN Red List of threatened species. http://www.iucnredlist.org/. Accessed 14 Mar 2020
Jamison E (2010) Working group results: grasslands and steppes. In: Stolton S, Dudley N (eds) Managing for climate change-developing strategies for protected area managers. Results of Seminar, organised by BfN and the United Nations Development Programme at the International Academy for Nature Conservation on the island of Vilm, Germany, August, 2010, pp 49–52
Kariyeva J, van Leeuwen WJD (2011) Environmental Drivers of NDVI-Based Vegetation Phenology in Central Asia. Remote Sens 3(2):203–246
Kirchhoff JF, Fabian A (2010) Forestry sector analysis of the Republic of Tajikistan, Eschborn, Germany: GTZ [Deutsche Gesellschaft für Technische Zusammenarbeit]
Krever V, Pereladova O, Williams M, Jungius H (eds) (1998) Biodiversity conservation in Central Asia. An analysis of biodiversity and current threats and initial investment Portfolio. World Wide Fund for Nature, Gland, Switzerland, 112 p
Kushlin A, Van Veen TS, Sutton W (2004) Kazakhstan – Forest sector in transition: the resource, the users and sustainable use. World Bank, Washington, DC. http://documents.worldbank.org/curated/en/251201468041112065/Kazakhstan-Forest-sector-in-transition-the-resource-the-users-and-sustainable-use. Accessed 13 Mar 2020
Larger than tigers. Inputs for a strategic approach to biodiversity conservation in Asia – Regional reports (2019). European Commission Directorate-General for Communication Publications 1049 Brussels BELGIUM. https://www.academia.edu/37956274/Larger_than_Tigers. Accessed 11 Mar 2020
Lazkov G, Sennikov A (2014) New records in vascular plants alien to Kyrgyzstan. Biodivers Data J 2:e1018. Published 2014 Jan 21. https://doi.org/10.3897/BDJ.2.e1018
Lazkov GA, Sultanova BA (2011) Checklist of vascular plants of Kyrgyzstan. Norrlinia 24:1–166. https://www.ncbi.nlm.nih.gov/pmc/articles/PMC4030214/#B419352. Accessed 12 Mar 2020
Liang L, Wenpeng D, Huimin Y, Lin Z, Yuet D (2017) Spatio-temporal Patterns of Vegetation Change in Kazakhstan from 1982 to 2015. J Resour Ecol 8(4):378–384. https://doi.org/10.5814/j.issn.1674-764x.2017.04.009. www.jorae.cn
Micklin PP (1993) The shrinking Aral Sea. Geotimes:14–18
Micklin P (2007) The Aral sea disaster. Annu Rev Earth Planet Sci 35:47–72
Micklin P (2010) The past, present, and future Aral Sea. Lakes Reserv Res Manag 15:193–213. https://doi.org/10.1111/j.1440-1770.2010.00437.x
Micklin P (2014) Chapter 15: Efforts to revive the Aral Sea. In: Micklin P et al (eds) The Aral Sea, Springer Earth System Sciences, pp 361–380
Micklin P (2016) The future Aral Sea: hope and despair. Environ Earth Sci 75:844. https://doi.org/10.1007/s12665-016-5614-5

Mount Etna and the Mountains of Pamir inscribed on World Heritage List alongside El Pinacate and Gran Desierto de Altar (2013) UNESCO. Culture. World Heritage Centre. News and Events. News. http://whc.unesco.org/en/news/1042. Accessed 26 June 2018

Nacionalny doklad o sostoyanii okruzhayuscey sredy i ob ispolzovanyi prirodnyh resursov za 2015 (2016) (National Report on the state of environment and use of natural resources for 2015) Bioraznoobrazye. 1.3. http://doklad.ecogosfond.kz/biodiversity. Accessed 30 Jan 2018

Olson DM, Dinerstein E (2002) The global 200: priority ecoregions for global conservation. Ann Mo Bot Gard 89:199–224. https://www.researchgate.net/publication/233729606. Accessed 12 Mar 2020

People, forests and trees in West and Central Asia (2007) Outlook for 2020. FAO, Rome. http://www.fao.org/docrep/009/a0981e/a0981e00.htm. Accessed 1 Nov 2017

Pereladova O, Krever V, Williams M (1998) Biodiversity conservation in Central Asia – analysis of modern situation and project portfolio. WWF, Moscow

Plotnikov IS, Ermakhanov ZK, Aladin NV, Micklin P (2016) Modern state of the Small (Northern) Aral Sea fauna. Lakes Reserv Res Manage 21:315–328. https://doi.org/10.1111/lre.12149

Prioritety sohranenya biologicheskogo raznoobrazya KR na period do 2024 goda (2014) (The priorities of protection of biodiversity of Republic of Kyrgyzstan for the period till 2024) Postanovlenye pravitelstva Kyrgizskoy Respubliki. http://ecology.gov.kg/page/view/id/25#. Accessed 15 Feb 2018

Rachkovskaya EI (1995) Vegetation of Kazakhstan and Middle Asia (Desert region). Vegetation Map of Kasakhstan and Middle Asia. Scale 1:2 500 000. Komarov Botanic Institute, Russian Academy of Sciences, Saint Petersburg

Rachkovskaja K, Pereladova O. Central Asia: Kazakhstan, World Wild Life (PA0811). https://www.worldwildlife.org/ecoregions/pa0811. Accessed 1 Feb 2018

Rasmussen C (2014) NASA data find some hope for water in Aral Sea Basin, NASA Earth Science News Team. https://www.nasa.gov/jpl/news/aral-sea-20140214/. Accessed 30 Jan 2018

Rustamov IG (1994) Vegetation of the deserts of Turkmenistan. In: Fet V, Atamuradov KI (eds) Biogeography and ecology of Turkmenistan, Monographiae Biologicae, vol 72. Springer, Dordrecht

Rybalkina DH, Zhanbasinova NM, Salimbayeva BM (2017) Oncological morbidity of children in the region of Aral sea. Bulletin of the Karaganda University. Biol Med Geogr Ser 1(85):75–79

Sakiev KZ, Batyrbekova LS (2015) Vlijanie faktorov okruzhajuscey sredy na sostojanie hepatobiliarnoy sistemy naselenya, prozhivajuscego v ekologicheski neblagoprijatnych regionah (Influence of environmental factors on the hepatobiliary system condition of people living in the ecologically unfavorable regions), Karagandinskiy gosudarstvenniy medicinski universitet, Karaganda

Salimbayeva BM, Sraubaev EN, Sergaliev TS, Uresaev AO, Baltaeva ZE, Ash M (2017) Medical and demographic situation in the Aral sea region. In: YeK K (ed) Bulletin of the Karaganda University, Biology. Medicine. Geography Series, 1(85), pp 80–86

Saparov A (2014) Soil resources of the Republic of Kazakhstan: current status, problems and solutions. In: Mueller L, Saparov A, Lischeid G (eds) Novel measurement and assessment tools for monitoring and management of land and water resources in agricultural landscapes of Central Asia, vol XXIII. Springer, pp 61–74. http://www.springer.com/978-3-319-01016-8. Accessed 6 June 2017

Sehring J (2012) Forests in the context of climate change in Kazakhstan. Zentrum für internationale Entwicklungs- und Umweltforschung der Justus-Liebig-Universität Gießen, No. 60

Seim A, Tulyaganov T, Omurova G, Nikolyai L, Botman E, Linderholm HW (2015) Dendroclimatological potential of three juniper species from the Turkestan range, northwestern Pamir-Alay Mountains, Uzbekistan. Trees, Springer, Berlin/Heidelberg. https://doi.org/10.1007/s00468-015-1316-y

Sennikov AN, Tojibaev KS, Khassanov FO, Beshko NY (2016) The flora of Uzbekistan project. Phytotaxa 282:107–118

Shahgedanova M (ed) (2002) The physical geography of Northern Eurasia, Oxford University Press. http://www.rusnature.info/reg/16_7.htm. Accessed 15 July 2017

Shamsutdinov ZS, Shamsutdinov NZ (2012) Biogeocenotic principles and methods of environmental restoration of desert pasture ecosystems in Central Asia. Arid Ecosyst 2(3):139–149

Sklyarenko S (2010) Introduction: grasslands and steppes. In: Stolton S, Dudley N (eds) Managing for climate change-developing strategies for protected area managers. Results of Seminar, organised by BfN and the United Nations Development Programme at the International Academy for Nature Conservation on the island of Vilm, Germany, August, 2010, pp 46–48

Smith D (1994) Change and variability in climate and ecosystem decline in Aral Sea basin deltas. Post-Sov Geogr 35:142–165

Snyder PK, Delire C, Foley JA (2004a) Evaluating the influence of different vegetation biomes on the global climate. Clim Dyn 23:279–302

Snyder PK, Foley JA, Hitchman MH, Delire C (2004b) Analyzing the effects of complete tropical forest removal on the regional climate using a detailed three-dimensional energy budget: an application to Africa. J Geophys Res 109:D21102. https://doi.org/10.1029/2003JD004462

Sokolova N (2014) Unikalnye tugai Amudarjy – sokrovische mirovogo znachenya (Unic Tugay forests of Amu Darya – the treasure of world importance) Priroda i ekologia. http://infoabad.com/priroda-i-yekologija-turkmenistana/unikalnye-tugai-amudari- sokrovische-mirovogo-znachenija.html. Accessed 8 Aug 2017

Sommer R, De Pauw E (2011) Organic carbon in soils of Central Asia–status quo and potentials for sequestration. Plant Soil 338:273–288. https://www.researchgate.net/publication/225342880_Organic_carbon_in_soils_of_Central_Asia-status_quo_and_potentials_for_sequestration/figures. Accessed 08 Aug 2017

Sorrel P, Popescu S, Klotz S, Suc J, Oberhänsli H (2007) Climate variability in the Aral Sea basin (Central Asia) during the late Holocene based on vegetation changes. Q Res 67(3):353–370

State of forests of the Caucasus and Central Asia (2019) Geneva timber and forest study Paper. Overview of forests and sustainable forest management in the Caucasus and Central Asia region, UNECE Food and Agriculture Organization of the United Nations, New York and Geneva. https://digitallibrary.un.org/record/3814608. Accessed 13 Mar 2020

State of the world forests 2007 (2007) FAO (Food and Agriculture Organization of the United Nations). Rome. http://www.fao.org/3/a--a0773e.pdf. Accessed 8 July 2017

Tajikistan – terrestrial protected areas (% of total land area) (2015) United Nations Environmental Program and the World Conservation Monitoring Centre, as compiled by the World Resources Institute, based on data from national authorities, national legislation and international agreements. https://www.indexmundi.com/facts/tajikistan/indicator/ER.LND.PTLD.ZS. Accessed 8 July 2017

The Aral Sea Crisis. Environment impacts (2008a). http://www.columbia.edu/~tmt2120/environmental%20impacts.htm. Accessed 26 Aug 2017

The Aral Sea Crisis. Impacts to life in the region (2008b). http://www.columbia.edu/~tmt2120/impacts%20to%20life%20in%20the%20region.htm. Accessed 26 Aug 2017

The Aral Sea Crisis. Introduction (2008c). http://www.columbia.edu/~tmt2120/introduction.htm. Accessed 26 Aug 2017

The Aral Sea, Once the Fourth Largest Lake in the World is Gone (2014). http://environmentalfuture.org/the-aral-sea-once-the-fourth-largest-lake-in-the-world-is-gone/. Accessed 26 Aug 2017

The Kazakh Steppe (2011) Encyclopaedia britannica blogs. Posted by Kara Rogers. http://blogs.britannica.com/2011/09/kazakh-steppe/. Accessed 11 Mar 2020

Thomas G, Roundtree PR (1992) The boreal forests and climate. Q J R Meteorol Soc 118:469–497

Tojibaev KS, Beshko NY, Batashov AR, Karimov FI, Lee D-H, Turginov OT, Usmonov MX, Kodirov UH, Tajeddinova D (2017) Ten new records of vascular plants for the flora of Uzbekistan (Asteraceae). Korean J Plant Taxon 47(3):171–179. https://doi.org/10.11110/kjpt.2017.47.3.171

Turkmenistan – terrestrial protected areas (2015) United Nations Environmental Program and the World Conservation Monitoring Centre, as compiled by the World Resources Institute, based on data from national authorities, national legislation and international agreements. https://www.indexmundi.com/facts/turkmenistan/terrestrial-protected-areas. Accessed 8 July 2017

Ulybina O (2015) Participatory forest management: the experience of foreign-funded programmes in the Kyrgyz Republic. Environ Policy Govern 25(1):70–82. https://doi.org/10.1002/eet.1648

Underutilised species, policies and strategies (2006) Analysis of existing national policies and legislation that enable or inhibit the wider use of underutilized plant species for food and agriculture in Uzbekistan. Report presented by Uzbek Scientific and Production Centre of Agriculture, Ministry of Agriculture and Water resources of the Republik of Uzbekistan, Tashkent, 17 p

UNEP-WCMC and IUCN (2016) Protected planet report 2016. UNEP-WCMC and IUCN, Cambridge/Gland. https://portals.iucn.org/library/sites/library/files/documents/2016-051.pdf. Accessed 12 Mar 2018

Vildanova G (2006) Forest and forest products, country profile Uzbekistan, Geneva timber and forest discussion paper 45. FAO United Nations, New York/Geneva. http://www.unece.org/fileadmin/DAM/timber/docs/dp/dp-45.pdf. Accessed 19 Dec 2017

Water related vision for the Aral sea basin for the year 2025 (2000) UNESCO, France, 237 p. http://unesdoc.unesco.org/images/0012/001262/126259mo.pdf. Accessed 27 Aug 2017

World biomes. http://www.blueplanetbiomes.org/world_biomes.htm. Accessed 13 July 2017

World Heritage Thematic Study for Central Asia – August 2005. http://cmsdata.iucn.org/downloads/central_asia.pdf. Accessed 3 Apr 2018

Yildiray L (2010) Overview: biome management. In: Stolton S, Dudley N (eds) Managing for climate change-developing strategies for protected area managers. Results of Seminar, organised by BfN and the United Nations Development Programme at the International Academy for Nature Conservation on the island of Vilm, Germany, August, 2010

Yin H, Khamzina A, Pflugmacher D, Martius C (2017) Forest cover mapping in post-Soviet Central Asia using multi-resolution remote sensing imagery. Sci Rep 7:1375. https://www.nature.com/articles/s41598-017-01582-x#Tab1. Accessed 3 May 2018

Zakirov KZ (1989) Classification of Central Asian vegetation. Fan, Tashkent

Zavialov P (2005) Physical oceanography of the dying Aral Sea. Praxis Publishing Ltd, Chichester

Ecological Base and Environmental Constraints

4

Abstract

Soils of CA are diverse: Xerosols and Yermosols, followed by Kastanozems, Solonetz and Lithosols with rock outcrops, occupy the largest area in CA, the smaller patterns are covered by Gleysols, Histosols, Gypsisols, Solonchaks etc. Soil cover in the forest-steppe biome distinguishes by Chernozem type of soils, and the steppe biome by Kastanozem type of soil: these two are the most fertile soil types in CA region, divided in several subtypes. Summarizing the climate change impact on agriculture in CA, it is necessary to state that it has positive and negative consequences. Among the negatives, the expected increase in aridity across entire region, but mostly in the western parts of Turkmenistan, Uzbekistan and Kazakhstan. This situation will be possible to control under the controlled, coordinated and rational water resources management only. As the positive outcome of climate change impact, a modest increase in winter period precipitation rate is expected across the region, especially in the eastern part of Kazakhstan and in Kyrgyzstan and Tajikistan, which is benevolent for growing of winter wheat. The soils of irrigated lands experience strong impact of anthropogenic factors which modify their morphology, pedogenesis and properties leading towards secondary salinization. Other types of human economic activities, like overgrazing, industry, mining and military use, also strengthen the desertification and degradation processes of CA region landscape via increased erosion and soil cover contamination.

Keywords

Soil resources · Agricultural soil properties · Irrigation · Secondary salinization

4.1 Soil Resources in CA Region; Climate Change Challenges for Agriculture and Soil; Irrigation and Secondary Salinization

4.1.1 Soil Resources Overview

Genetic typology of soils depends on the classification criteria, thus traditionally in CA region two main soil classifications are used: national, based on Russian soil classification system and FAO-UNESCO classification (FAO-UNESCO 1974) (Fig. 4.1).

Soils of CA are diverse and follow the main vegetation and climatic zones. The soil formation processes are directly linked with the dominating geomorphological, hydrological and climatological variables, changes in altitudes and bioclimatic conditions. The largest areas of the region are covered with Xerosols and Yermosols, followed by Kastanozems, Solonetz and Lithosols. The map shows that sand dunes prevail in the Kyzylkum and Karakum deserts of the south, and the Lithosols extend into the eastern and south-eastern part of CA region.

The air and soil temperatures and soil moisture regime direct development of soil types or pedons within different biomes and ecoregions. Soil cover in the forest-steppe biome is formed under the year round equilibrium conditions among precipitation, evaporation and biomass productivity. In such a condition, humus accumulation in the soil is considerable, and dark-coloured soils Chernozems are formed (Pocvy Rossii 2001–2016a). These are the most fertile of all Eurasia soils, also they are common in North American prairies. In the CA, they are distributed mainly in Northern Kazakhstan up to 52° N latitude. When followed from north to south (from forest-steppe to dry southern steppe), few subtypes of these soils are found: 1. Leached and Podzolized Chernozems; 2. Usual Chernozems (Thick and Typical); 3. Southern Chernozems (Table 4.1).

I. Jelen et al., *The Geography of Central Asia*, World Regional Geography Book Series,
https://doi.org/10.1007/978-3-030-61266-5_4

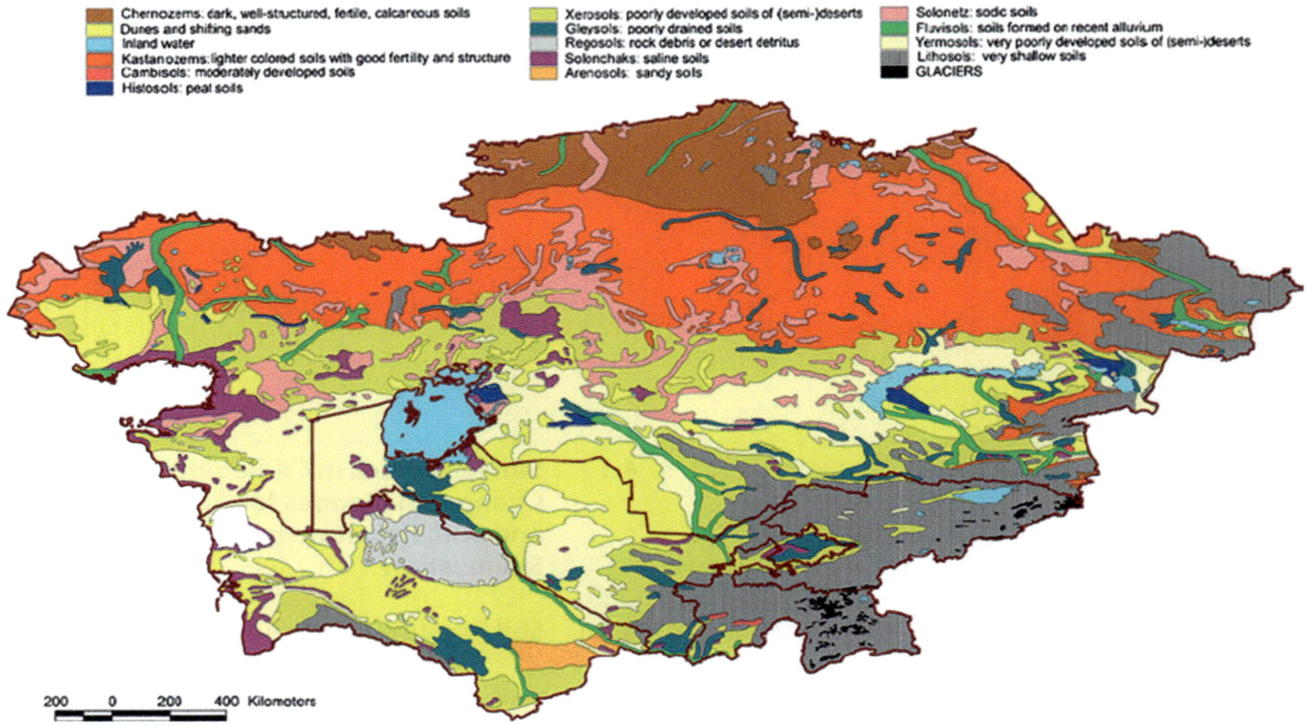

Fig. 4.1 Soil map (major units) of Central Asia (FAO-UNESCO 1974). (Source: Sommer and De Pauw (2011))

These sub-types of Chernozems differ mostly by the thickness of humus-containing layer: it varies from 40–50 cm in the Southern Chernozems to 85–120 cm in the Thick Chernozems.

Chestnut soils or Kastanozems are found south of the Chernozems between 52° and 48° N latitude within the temperate grassland zone, and those soil types distinguish by the highest natural productivity. They are subdivided into Dark Kastanozems of the moderately dry steppe (10.5%), Typical Kastanozems of the dry steppe (9.6%) and Light Kastanozems soils of the semi-desert (14.2%). Kastanozems are used mostly for grazing, pastures, hey making and for such arable crops like wheat, maize, *Panicum* species and sunflower (Shahgedanova 2002).

In Kazakhstan, there are 25.7 million ha of Chernozems, 90.4 million ha of Kastanozems, 119.2 million ha of brown and grey-brown soils (Calcisols), and 37 million ha of Mountain soils (Faizov et al. 2006).

The main threats to soils in this zone are water and wind erosion and carbon mineralization. Sandy soils with sparse vegetation under the hot and arid climate are most prone to wind erosion: this can be observed in Turkmenistan where about 80% of the soils are sandy (Gupta et al. 2009). Concerning the organic matter/humus content, the Chernozems are the richest, and the Kastanozems are the second-richest soil type in CA region. That is distinguished from brown topsoil of 20–40 cm thick in which the organic matter content ranges between 2 and 6%. The topsoil of Kastanozems has brown granular or fine blocky structure which is favourable for plants. The deeper layers are lighter in colour and are characterized by the secondary accumulation of calcite. Kastanozems are found in relatively dry climatic zones. They are rather rich in main nutrients, Ca is also available and mostly are used for irrigated farming and grazing (World reference base… 2015). But the frequent droughts occurring in the region hinder their potential use in agriculture (Encyclopaedia of Soil Science 2008). Another serious problem on Kastanozems is that overgrazing on light-textured soils often produces deflation and destroys the topsoil (Wang et al. 2010).

South of 48° N latitude, Brown and Brownish grey tracts of Semi-desert and Desert soils are prevalent; they alternate with large tracts of Sandy Desert soils and Takyr-like soils. Brown soils of the northern desert sub-zone (21.6%) and Grey-brown soils confined to the middle and southern desert sub-zones (22%) dominate in this area. The saline soil is also found there, and agriculture is possible only with the use of irrigation, which gives rise to specific cultivated sub-type of Sierozems used for the cultivation of cotton. These soils are characteristic by low humus and high carbonate, organic phosphorus, potassium as well as comparatively high content of lightly hydrolysed nitrogen compounds (World reference base… 2015). In the mountains of the Western and Northern Tien Shan the Grey-brown soils give way to

Table 4.1 The main sub-types and features of Chernozems and Kastanazems of CA region

Soil type	Soil sub-type	Humus content in the topsoil in %	Average thickness of humus horizon in cm	Natural zone and sub-zones
Chernozems	Podzolized	5–8	50–80/120	Forest steppe zone with northern forest-steppe sub-zone
	Leached	6–10	50–80	Forest steppe zone with typical forest-steppe sub-zone
	Thick	8–12/15	85–120	Forest steppe zone with southern forest-steppe sub-zone
	Typical	6–10	65–80	Steppe zone with northern steppe sub-zone
	Southern	4–6	40–50	Steppe zone with typical steppe sub-zone
Kastanozems (Chestnut soil)	Dark	3.5–5.0	30–50	Steppe zone with dry steppe sub-zone
	Typical	2.5–4.0	20–25	Semi-desert zone with northern semi-desert sub-zone
	Light	1.5–2.5	15–20	Semi-desert zone with typical semi-desert sub-zone
	Brown semi-desert	1.0–2.5	10–15	Semi-desert zone with southern semi-desert sub-zone

The table was adopted from the Pocvy Rossii 2001–2016a, b based on two sources:
Physical Geography of Northern Eurasia (2002). Edited by Maria Shahgedanova, Oxford University Press and The Natural History of the USSR by Algirdas Knystautas (1987). Swallow Editions Limited; figures after/show the upper values
http://www.ecosystema.ru/08nature/soil/094t.htm

Sierozems and Light Chestnut soils of the plain and foothill sub-zone. Few sub-types of Brown, Kastanozems and Chernozems can be found in the mountains of Western and Northern Tien Shan, Southern Tarbagatai and Western Altai. In the Kazakhstan territory, mountain soils make up 12.6% (Pachikin et al. 2014).

The Xerosols according to FAO classification are the soils having a weak ochric top A horizon (after 1998 they were renamed as Durisols) and the Yermosols with very weak A horizon, both are developing on an aridic moisture regime, lacking permafrost within 200 cm of the surface (World reference base… 2015).

Gypsisols are characterized by a significant secondary accumulation of calcium sulphate. They are distinguished by the accumulation of gypsum crystal aggregates in to the cemented impermeable layer, as the petrogypsic horizon. These soils extend in the driest part of the arid climatic zone in Kazakhstan, Turkmenistan and Uzbekistan. They are used mostly for extensive grazing because under the irrigation, the dissolution of gypsum results in the irregular subsidence of the land surface, caving in canal walls, and in the corrosion of concrete structures (World reference base… 2015).

Solonetz are the soils having a natric B horizon, and the lithosols are the soils which are limited in depth by continuous coherent and hard rock within 10 cm of the surface (Legend of the Soil map… 1974). Salt bearing rocks represent the major initial source of salts in the deserts of CA. Differences in the distribution of saline grey brown soils are mainly related to the geological history of particular territories. Another source of salts in the automorphic soils of CA is related to eolian processes. Salts of the eolian genesis are particularly wide spread on coastal plains of the Aral Sea region (Pankova and Konyushkova 2013). Soil salinization is a major process of land degradation that decreases soil fertility and is a significant component of desertification processes in the world's drylands (Middleton and Thomas 1997). The largest percentage of saline and salinized soils of the total irrigated areas is concentrated in the Uzbekistan (about 60%) and in the central part of Fergana valley even 80% (Glazovsky 1995).

The different processes within soils of CA region also are distributed according to the climatic zones, biomes and the absolute height of the terrestrial localities. The temperate forest zone in CA region stretches only in the north-eastern part and in the mountain areas. Characteristic of soil formation in that forest zone is the leaching process, expressed by vertical transport of soil solutions from the upper to the deeper layers, thus forming the infertile light grey soil layer called podzols (Russian: "under ash"). Beneath the podzols, the rusty brown soil layer with iron and other metal oxides is accumulated. The podzolization is characteristic also for the mountain forest-steppe soils (Asia. Soils… https://www.britannica.com/place/Asia/Soils).

Climate change can lead to the changes in species composition and the rate of litter fall. But these processes are also related to the feedbacks from biomes and ecoregions, thus inter-relations between climatic, soil and vegetation variables are the most important for further studies (White 1987; Tóth et al. 2007; Lenton and Huntingford 2003; Otto et al. 2005; Otto Simpson 2006; Thevenot et al. 2010; Grunwald 2009). It was determined that the SOM (soil organic matter) decomposition is more temperature-sensitive at lower than at higher temperature (Kirschbaum 1995, 2000, 2006). High soil temperatures and moist conditions accelerate soil respiration and thus increase CO_2 emissions (Brito et al. 2005). Another research has shown that the SOM decomposition rate is also influenced by physicochemical protection mechanisms such as occlusion within aggregates and by association with mineral surfaces (Conrad et al. 2011; Freedman 2014). Thus, knowledge about the temperature sensitivity of diverse SOM and SOC (soil organic carbon) fractions, and their change in the soil under climate change, remains important direction for future research (Hakkenberg et al. 2008; Status of world soil resources… 2015).

4.2 Climate Change and Challenges for Agriculture

Grazing land and pastures occupy the largest areas of agricultural land in CA countries. It is recognized, that grassland soils typically result in the accrual of soil C "owing to lower soil disturbance and reduced C removal in harvested products" (Smith et al. 2008, p. 792). But the overgrazing of pastures can also lead to the losses of SOC. Thus, it is important for CA countries to learn more and adopt such agricultural and land use practices that could lead to reduced GHG (greenhouse gas) emissions taking place in their local agroecosystems. There are different scenarios for the impact of climate change on farming for the shorter or longer time period. Let us look into the future climate scenarios and agriculture in three CA region cases.

4.2.1 Case Study: Turkmenistan

The comparison of meteorological parameters, like mean air temperature, from the second part of twentieth century up to the beginning of twenty-first century, has indicated the slight warming tendency (from 0.2°–0.3° C to 0.6 °C, particularly in the northern part of the country. The analysis of repetitions and continuity of the periods with air temperature of 40 °C and higher according to the data obtained in three Meteorological Stations (Gyzylarvat, Tahtabazar and Ashgabad) has confirmed the increase of such cases since 1983. From 1960 to 1990 the amount of annual precipitation in the country slightly increased in winter season, but decreased in warm seasons. This climate change influences pasture's vulnerability, reduces soil moisture and consequently leads to the decrease of pasture's productivity. Few scenarios predict the increase of average annual air temperature from 4.2 ° to 6.1 °C, and the amount of precipitation with zero changes or a decrease from 15 to 56% (Annaklycheva 2002).

4.2.2 Case Study: Uzbekistan

For Uzbekistan, meteorological records since 1950s confirm the rising air temperature trend, thus the climate warming. Monitoring has shown that in Khorezm 10-year average of summer temperatures were by 0.2° and 0.5° C higher than the long-term average. The highest shift was recorded in the vicinity of the Karakum desert (Conrad et al. 2011). The days with maximum temperatures (higher than 40°C) increased twofold in Prearalie, while in other parts of the country this increase was much less. On the contrary, the number of days with low temperatures decreased in the whole country. Some of the researchers are optimistic for the near future (till 2040): they conclude that farmers in Uzbekistan will benefit from climate change due to more favourable weather conditions for crop growth. However, revenues are expected to decline in the late future (2070–2100) due to increasing temperatures and increasing risk of water deficit (Bobojonov and Aw-Hassan 2014). In general, the growing rate of salinization is considered as one of the main causes for land degradation in Uzbekistan, which is increasing under the influence of both climatic and anthropogenic factors (World Bank 2018). Among them, the irrational water use, physical aging of irrigation and drainage systems, ineffective methods of irrigation, absence of crop rotation, and as a consequence the low humus content in soil have to be mentioned.

4.2.3 Case Study: Kazakhstan

The Kazakhstan is one of the largest wheat producers in the world. The moderate climate warming is benevolent for that crop, but too hot temperatures during flowering (flower sterility) will become a problem in the long-term future in some, mostly southern, areas and in the spring wheat areas of Northern Kazakhstan. But in general, this crop tolerates higher temperatures. Though the irrigation and water requirements do not increase under climate change, the management of this resource has to be improved. The irrigation is closely connected to drainage management too, thus it has to be paid more attention due to the risk of increased N (nitrogen) loss out of the soil by leaching in response to higher and more intensive precipitation in some periods (Sommer et al. 2013).

Summarizing on climate change in CA (Lioubimtseva and Henebry 2009), it is projected to become warmer during the coming decades with increased aridity, especially in the western parts of Turkmenistan, Uzbekistan and Kazakhstan. The annual precipitation rate will not change much, but seasonal differences are forecasted which will lead to their shortage during the warm seasons (mostly summer and fall). Thus, if the additional measures (transboundary agreements on the surface water resources use and management) are not taken by all five CA countries, it can make the situation with irrigated agriculture more risky, both environmentally and economically, than it is today.

4.3 Anthropogenic Impact of Agriculture on Soil Hydrology: The Secondary Salinization and Land Degradation

Soil is the most active and integral axis of agroecosystem directly and indirectly integrated within growing or ever grown plant cover. Thus, changes in vegetation can also change the water and salt balances in soil, the vadose zone, and shallow groundwater. In the less continental deserts of Uzbekistan and Turkmenistan, no regular soil freezing is observed, thus the leaching of soluble salts from the upper soil horizons into the groundwater takes place in the spring. This results in the increased groundwater salinity and in the depletion of salts from the surface soil horizons (Pankova and Konyushkova 2013). But in general, the worsened situation with maintenance of irrigation-drainage systems (I&D) has been continued from the end of Soviet times up to now. The water usage in agriculture in CA countries is very high: it comprised from 81.8% in Kazakhstan to 97.5% in Turkmenistan of total water use in each CA country in 2001 (FAO 2002, 2011). Nevertheless, in 2010 the average use of water resources in agriculture of CA countries was even higher than in 2001. Still the irrigation systems are spread widely within the main agricultural land in CA countries. Having in mind the poor functioning of I&D and increased demand of water in many western and south-western agricultural regions of CA, ecological situation of soils in these regions has worsened too resulting in rising water tables and secondary salinization. Yields have dropped, and also the farmers' incomes (Bucknall et al. 2003).

The groundwater resources are strongly hydraulically connected to the surface runoff due to the prevailing of coarse sediments. The research has shown that ground shallow water can be salinized by both the natural processes and secondary salinization with increased groundwater level caused by irrigation (Ostrovsky 2007). The groundwater table and its composition depend on the location of irrigated areas too: in general water quality upstream is better than downstream due to the less water mineralization and the higher surface runoff (FLERMONECA Regional… 2015).

Groundwater resources were not widely used for irrigated agriculture in Central Asian Republics during the soviet period due to sufficient amount of surface water, reliable water supply, and irrigation infrastructure delivered to the farmers. Thus, the groundwater resources were used primarily for livestock sector and very site-specific purposes, for example, drinking water supply in both urban and rural areas.

During the Soviet period groundwater resources were not widely used for irrigated agriculture in CA countries. However, they might be an alternative source for different sectors of agriculture especially in the events of frequent droughts and in the irrigated fields, where the surface water is strongly contaminated (Rakhmatullaev et al. 2010). According to the CAWater-Info (2018), approximately 30% of regional groundwater is stored within a trans-boundary aquifer in the Aral Sea Basin, and the figure is much higher in the mountainous regions (e.g. Tian Shan Mountains) (Yu et al. 2019). Still there is urgent need to coordinate and agree on the specific regulations of the groundwater resources usage in CA region since they are there for common welfare.

Thus, salt-induced land and degradation of water resources in CA is the consequence of both naturally occurring phenomena (primary salinity) and anthropogenic activities causing secondary salinity. The contribution from anthropogenic activities is greater than primary salinity (Qadir et al. 2009; Arabov 2014).

Summarizing different findings obtained from the arid areas of CA, the main anthropogenic factors affecting land resources, soil and water quality can be distinguished as it is illustrated in the Fig. 4.2.

Research and practices have shown that many other activities than land conversion, for example frequent ploughing, intensive fertilization, disturbing peatland, installing drainage, monocultures instead of crop rotations, can be mentioned as leading towards the SOM reduction (Kibblewhite et al. 2005; Qadir et al. 2009; Status of the world soil resources… 2015).

Bellow three cases with irrigated land use and changes in soil properties are presented and discussed.

4.3.1 Case Study: Uzbekistan

If the measures to improve the irrigation management will not be undertaken, with climate warming it will continue to increase in the irrigated lands soil salinity and desertification. Approximately 40% of all irrigated land is strongly salinizated in the lands belonging to Aral Sea basin (Lubin

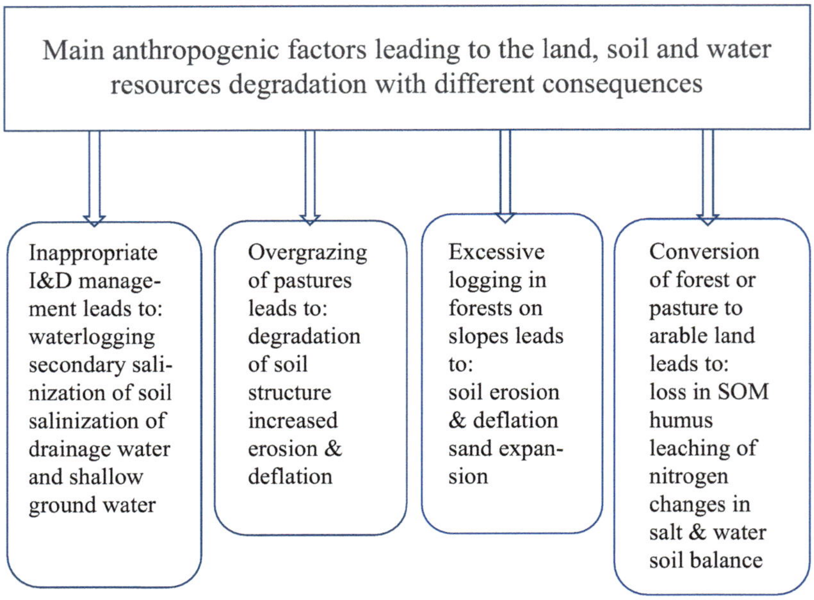

Fig. 4.2 Effect of anthropogenic factors on land resources and soil degradation in the arid agricultural and wooded areas of CA region. (Prepared by author after compilation of different sources: Almaganbetov and Grigoruk 2008; Arabov 2014; Babaev and Kharin 1999; Faizov et al. 2006; Kibblewhite et al. 2005; Qadir et al. 2009; Saparov 2014; Pankova and Konyushkova 2013; Status of the world soil resources… 2015)

1989). Secondary salinization, due to excess irrigation, is rendering much agricultural land unusable, and that might increase in the future and all that together will impact negatively on the development of agricultural production (Precoda 1991).

The results of the research from three irrigational sites, Khiva, Khonka and Yangibazar in Khorezm region, suggest that the elevated groundwater levels forced soil salinization by annually adding 3.5–14 t ha^{-1} of salts depending on the position and salinity of the groundwater table (Ibrakhimov et al. 2007).

About 80% of Uzbekistan territory are the flat plains, and the agricultural lands constitute about 46.1% of the country's total area. The distribution of soils in Uzbekistan reflects a complex system of pedo-geographical regularities, and due to the climate warming and not rational usage of agricultural measures in I&D systems management, the degraded soil area might even increase (Arabov 2010).

Among measures to stop or reduce soil salinity on arable irrigated lands, few conservation agricultural measures in the Khorezm region were tested. The reduced tillage, establishing of permanent raised sowing beds, enriching of land with plant residues under the cotton-wheat-maize rotation were studied and have shown the positive results (Devkota et al. 2015).

Pastures in the desert zone of Uzbekistan mostly are used without a management plan, thus the animals are grazing all year round on the same pastures leading to degradation. In order to improve the situation, the proposed technology was developed and introduced within the framework of the UNDP/GEF and Government of Uzbekistan project "Achieving ecosystem stability on degraded lands in Karakalpakstan and Kyzylkum desert" in 2007–2009. Based on traditional methods and approaches to pasture use, the assessment of pastures included the type, forage value and pasture capacity, followed by the calculation of the forage balance, pasture productivity, number of animals, water wells functioning. Thus, the pastures were divided into three rotation areas of 1308 ha each, to be grazed sequentially (Pasture rotation… 2018). The purpose of this technology is to prevent further pasture degradation, loss of biodiversity and to create prerequisites for regeneration of vegetation and pasture improvement.

4.3.2 Case Study: Kazakhstan

The anthropogenic factors strongly modify the morphology and genesis of the soils. Among these factors, the pasturing, agricultural farming and industrial & mining activities are the main types of country's economic development (Ibragimov 2016).

Irrigated soils are mainly transformed into Anthrosols. For Kazakhstan, with existing drylands in sandy and saline soils and their rapid extension, the most important environmental and economic problem is to find measures and stop processes of land degradation and desertification, which is evident in more than 60% of the total country's area (Almaganbetov and Grigoruk 2008). The mentioned anthropogenic factors act almost in all natural landscapes, but particularly in the vicinity of Aral Sea. In the dried-up bottoms of the sea and the Syr Darya River, the increase in saline marsh soils and solonchaks is taking place (Saparov 2014). About 10–15% of irrigated lands fell out of agricultural use as a result of soil salinization and swamping in the Syr Darya river basin (Qadir et al. 2009).

Though there was high drop in the cropland and livestock number after the end of Soviet Era, some places, particularly those located near large rural settlements, experienced an increase in anthropogenic impacts by intensification of agricultural use and grazing (Propastin and Kappas 2008). Land degradation and desertification processes are severe near the Caspian Sea and in the Lake Balkhash region too (Almaganbetov and Grigoruk 2008).

In the pastures with high livestock number and overgrazing, soil fertility decreased in two to three folds, mostly due to the dehumidification, and crop harvesting declined. According to the data from the Agency of the Republic of Kazakhstan on land management, salinized and solonetz lands occupy 94.9 million ha, which is equal to 42.1%. Soil erosion covers 30.5 million ha or 13.0%. Soil salinization of dried lakes and other reservoirs is another specific feature of Kazakhstan (Qadir et al. 2009).

The irrigated water, particularly in long term used for rice growing, enhances the migration and accumulation of heavy metals within soil profile. Research experiments carried in Shely site, have shown, that Pb, Cd and Ni are the mostly migrated metals in the topsoil of irrigated area (Otarov 2014).

The data obtained show that in the studied soils the observed concentrations of heavy metals were not stable, but rather mobile within soil horizons. Clarke value in this soil was excessively high for Cd and for Pb. Zn and Cu have Clarke values close to the land crust values, but their accumulation coefficient in plough horizon is one of the highest, particularly of Cu. Based on above results, it can be concluded, that the irrigation of rice-marsh soils by means of permanent flooding has a significant influence on the distribution of mobile forms of heavy metals studied.

Among the solutions concerning the re-usage of the abandoned salt-affected lands, the cultivation of multipurpose tree and shrub plantations for renewable energy purposes has to be considered (Khamzina et al. 2008).

As the research showed (Liang et al. 2017) during 1982–2015, mostly degraded areas in Kazakhstan accounted for about 24.0% of the total land area, and is the result of various interacting factors. The authors suggested more research efforts to be given to development of soil and ecosystem restoration strategies.

4.4 Case Study: Turkmenistan

Deserts occupy about 79% of the Turkmenistan area. According to the research conducted by the Desert Institute of Turkmenistan, 66.5% of the country's territory is affected by the processes of desertification with varied extent of severity (Babaev and Kharin 1999). These processes mainly are induced by anthropogenic influence.

Under conditions of continuous massive irrigation in Turkmenistan, considerable attention has to be given to collectors and other drainage facilities intended for the removal of excess water from the soil. The research has revealed that soil in Turkmenistan was waterlogged in 14% and salinized in 73% of total irrigated land (Stanchin and Lerman 2008; Kostianoy et al. 2011).

The research on the degradation of agricultural land and soil has revealed some patterns of degraded land and distinguished the factors responsible for this. Estimated results were based on the usage of space photos, statistical materials, thematic maps and stationary observations. The research revealed main causes of desertification: besides climate, those were overgrazing, cutting down of desert vegetation, inappropriate agricultural methods and uncontrolled driving over the desert (Annaklycheva 2002).

4.5 Other Sources of Land Degradation

Not only agricultural activities, but other economic activities like industry and mining, municipal and domestic waste management, transport emissions, and unreclaimed mining contribute to land degradation as well. There is great difference between the effect of natural and anthropogenic factors. If the natural processes of land degradation and desertification develop slowly, human activities accelerate them, and the impacts of anthropogenic degradation appear in a short period of time.

The land degradation and desertification can be caused also by the contamination with oil and oil products, heavy metals and radionuclides (Almaganbetov and Grigoruk 2008) as well as other pollutants from industrial, transport, military and other social activities which may pollute air, soil, surface and groundwater and damage the living resources (Ibragimov 2016).

4.6 Air and Ground Pollution by Anthropogenic Factors

Air and soil pollution hazard has sharply increased since the development of Industrial Revolution with intensive mining industry, but as the global/regional ecological problem was recognized in the twentieth century only and continues to be so in the beginning of twenty-first century. The main pollutants spreading through the mobile and stationary industrial, transport and other human living environmental sources are sulphur, nitrogen, dust etc. (Kļaviņš et al. 2014). The CA region is not exception for the appearance of these hazards too, particularly in the main cities and industrial regions.

Within Kazakhstan, there are some territories under stronger air pollution depending mostly on the industry and transport sources. What is the level of pollution of atmosphere by stationary sources in Kazakhstan can be seen from the Table 4.2.

From 2012 to 2015, there is a tendency to decrease the emissions from this source of air pollution. However, two regions, Pavlodarsk and Karaganda, the largest industrial regions of Kazakhstan, are distinguished by the highest total emission from stationary sources. The main pollutants of this kind of source were SO_2 and CO_2. The monitoring data of 2015 has shown the highest atmospheric pollution by all sources in towns such as Stepnogorsk, Ust-Kamenogorsk, Almaty, Ridder, Zheskazgan, Shymkent, Temirtau and Karaganda. In total, in 2015 there were 248 of cases of high atmosphere pollution and 47 extreme cases with accidents and dust storms (Nacionalny doklad… 2016).

Stationary industrial sources are the major contributors to emissions of such air pollutants as sulphur dioxide (SO_2) and hydrocarbons in Uzbekistan (Fig. 4.3).

The main source of other pollutants is vehicular transport. There are ten cities in Uzbekistan where stationary industrial sources account for over 50% of total emissions, including almost 70% of particulates and 55% of gaseous pollutants. The highest concentrations of chemical substances emitted from chemical and metal industries remain in the cities of Almalyk, Chirchik, Samarkand and Fergana (Nacionalny doklad 2013). The large industrial enterprises of Uzbekistan are mostly distributed in narrow mountain valleys, thus this situation resulted in air interchange of emission products between cities and settlements and the higher polluted background (Uzbekistan Human… 1995).

Table 4.2 Emissions from the stationary sources of pollution in thousand tonnes in Kazakhstan administrative regions and largest cities, 2012–2015

Regions, cities	2012	2013	2014	2015
Kazakhstan republik	2384.3	2282.7	2256.7	2179.2
Akmolinsk region	105.7	83.8	84.6	85.6
Aktyubinsk region	123.9	125.4	121.8	134.2
Almaty region	64.2	68.4	51.6	55.0
Atyrau region	133.1	138.4	109.1	110.6
Western Kazakhstan region	62.1	60.4	44.7	42.4
Zhambyl region	40.7	33.6	38.2	41.9
Karaganda region	641.4	572.6	603.6	569.3
Kostanay region	100.6	115.4	103.8	91.6
Kyzylorda region	31.1	31.3	30.8	30.1
Mangystau region	64.2	77.5	88.3	72.5
Southern Kazakhstan region	48.6	56.3	59.9	68.9
Pavlodarsk region	676.0	650.4	610.2	552.9
Northern Kazakhstan region	75.7	71.4	71.9	74.8
Eastern Kazakhstan region	140.0	124.9	129.6	127.1
Astana city	64.9	60.5	65.1	56.3
Almaty city	12.1	12.4	43.5	39.1

Source: Nacionalny doklad… (2016) http://doklad.ecogosfond.kz/atmosfernyi-vozdux

The basic sources of industrial air pollution in Turkmenistan are oil refining, oil and gas production, chemical, machinery building industries, industry of construction materials and energy generation. Distribution of them is uneven: most of the oil production is found in the west, chemical industry is spread in the west, central and eastern parts and gas production – in different parts of Karakum desert. Still the main air pollutant in the cities of Turkmenistan is the transport (Nacionalny doklad… 2006).

Main stationary sources of air pollution in Tajikistan are mining, metallurgy, chemical industries, heat power plants and agriculture. Large stationary sources of atmospheric air pollution are Tajik aluminium plant, Isfara metallurgy plant, Dushanbe cement plant, Gold company, Anzob etc. (Atmospheric air: Pressure… 2000).

As the data of two last decades show (Air pollution trends… 2016), some heavy metal emissions, like cadmium and mercury, were slightly reduced in the EECCA countries (Eastern Europe, Caucasus and Central Asia).

4.7 Contamination with Radioactive Materials

Many of the industries, particularly mining, chemical including the extraction of oil and gas, mill and processing of rare-earth and uranium, leave lots of solid wastes enriched with natural radioactive elements like uranium-238, radium-226, radium-228, thorium-230 (Managing industrial solid wastes https://www.princeton.edu/~ota/disk1/1992/9225/922506.PDF).

The mining is one of the most developed industrial activities in Kazakhstan and other CA countries, thus the radioactive wastes of natural origin are common in their environment. Old and unmanaged tailings from uranium mining in Kazakhstan, Kyrgyzstan, Uzbekistan and Tajikistan make threats to the environment and society. The Koshkar-Ata uranium tailing in Kazakhstan, near the Aktau city, contains more than 100 million tonnes of radioactive wastes (Uranium and Nuclear Power… 2018; Koshkar Ata Tailings Pond http://ecocitizens.kz/en/hotspots/koshkar-ata-tailings-pond). In other regions of Kazakhstan, the Stepnogorsk and Ust-Kamenogorsk tailing deposits contain about 50 million tonnes of such wastes (Hays 2016).

During the Soviet times the Mailuu-Suu site in Western Kyrgyzstan was also a significant uranium mining area. Though the mining ceased nowadays, the site with large uranium tailings remained as well as a high risk of cancer for local people (Uranium in Kyrgyzstan 2018).

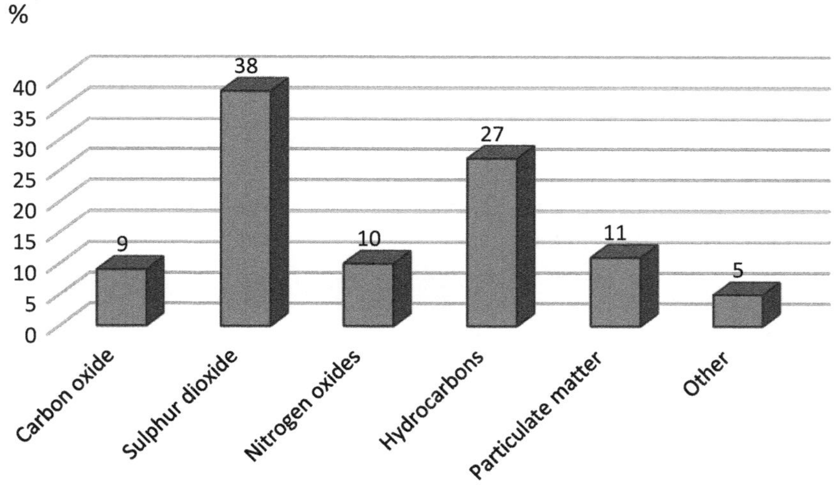

Fig. 4.3 Emissions of pollutants from stationary pollution sources in Uzbekistan. (According to source: Kļaviņš et al. (2014))

In Tajikistan, the uranium was mined and milled mostly during the Soviet times. Several mines were explored in Taboshar (40 kms north of Khujand), Adrasman, Mailisui, Uigur and Tyuya-Muyun, and the mining continued till 1992. Since then 32 million cubic metres of tailings (33.7 Mt) and 2.7 million cubic metres (1.8 Mt) of below-ore-grade material was accumulated in Tajikistan, which was not safely disposed (Uranium in Tajikistan 2017). The European Bank for Reconstruction and Development (EBRD) in 2015 set up a fund to deal with radioactive contaminated material resulting from Soviet-era uranium mining and processing in the Kyrgyz Republic, Tajikistan and Uzbekistan, thus the international financing has pushed the mentioned states to start the rehabilitation of numerous uranium tailing sites in the densely populated Fergana valley and other urbanized areas. In Uzbekistan, today most uranium is mined in the middle of the country, at Navoi, but all significant uranium resources are located in the Central Kyzylkum area (Uranium in Uzbekistan 2020).

Kazakhstan, Turkmenistan, Uzbekistan and Tajikistan also have the radioactive tailings and wastes related to the oil-gas industry. In addition, the Kazakhstan was for a long time involved in the nuclear weapon testing: from 1947 to 1990, when the country was a part of the Soviet Union, hundreds of nuclear tests were conducted at the 19,000 km^2 Semipalatinsk test site. Five of the surface tests were not successful and resulted in the dispersion of plutonium into the environment. In 1991, the testing was terminated. A joint US–Russian project with Kazakh assistance over 1996–2012 removed a significant quantity of high-enriched uranium and plutonium from the Semipalatinsk site, and encased more material in concrete. But the land of testing site still in some areas is heavily contaminated with radioactive materials and not recommended to be used for agriculture and as residential area, but only for commercial use.

Kazakhstan had also one functioning nuclear reactor in Aktau for water desalinization and electricity production, which in 1999 was closed (Uranium and nuclear power… 2018). For the near future (till 2030), government has foreseen to build smaller units of nuclear power for country's energy needs. It was announced in 2015 that a reactor, likely a Russian one, would be built at Kurchatov, and a second one would be at Balkhash if energy demand justified it. For the last reactor, the debates were arranged with foreign companies from USA, Japan and South Korea. The Ulken (near Balkhash Lake) and Kurchatov sites are preferred to be chosen for this purpose. There are plans in Uzbekistan to build the first new nuclear power station for the rapidly increasing country's energy demand. For this, in September 2018 an inter-governmental agreement was signed for construction by Rosatom (Russia Federation) of two VVER-1200 reactors to be commissioned about 2028. The region for this station after long debates was proposed near Lake Tuzkan, in the Farishsky district, and the number of reactors increased from two to four (Uranium in Uzbekistan 2020).

To sum up, CA countries have well-developed environmental policies and regulations. However, following the break-up of the Soviet Union, the system of environmental monitoring and conservation collapsed and has only recently been partially restored, soil surveys reduced or were interrupted. Thus, the soil and environmental monitoring in CA region countries needed to be renewed, particularly with changing situation of climate and agricultural pressure (Status of world soil resources… 2015). At the same time, other anthropogenic activities, like industries, particularly extraction, heavy and chemical industry, transport have left from Soviet-Era and still increases large tailings of wastes which under the poor management might become dangerous for living environment including the humans.

4.8 Topics for Further Reading and Discussions on the Content of this Chapter

- Situation analysis with air pollution in the largest CA cities.
- Ecological situation in the Caspian Sea and on-shore in relation to oil exploration.

Recommended References

FLERMONECA Regional Environmental Centre for Central Asia, Environmental Agency of Austria, Zoi Environment Network (2015) The state of the environment in Central Asia: illustrations of selected environmental themes and indicators. http://www.zoinet.org/web/sites/default/files/publications/SOE-regional-eng.pdf. In: Nugumanova L, Frey M, Yemelina N, Yugay S (2017) Environmental problems and policies in Kazakhstan: air pollution, waste and water, IOS Working Papers, No. 366, Leibniz-Institut für Ost- und Südosteuropaforschung (IOS), Regensburg. Accessed 22 Feb 2020

Kļaviņš M, Azizov A, Zaļoksnis J (2014) Environment, pollution, development: the case of Uzbekistan. UL Press, Riga. https://uzwater.ktu.lt/index.php/download-files/compendia-and-textbooks/module-a-environmental-science/129-environment-pollution-development/file. Accessed 22 Mar 2020

Koshkar Ata Tailings Pond. http://ecocitizens.kz/en/hotspots/koshkar-ata-tailings-pond

Managing industrial solid wastes from manufacturing, mining, oil and gas production, and utility coal combustion. Chapter 4: Oil and gas wastes. https://www.princeton.edu/~ota/disk1/1992/9225/922506.PDF. Accessed 3 Feb 2018

Nacionalny doklad o sostoyanii okruzhayuscey sredy i ob ispolzovanyi prirodnyh resursov za 2015 god. Atmosfernyi vozdux (2016) (National Report on the state of environment and use of natural resources for year 2015). http://doklad.ecogosfond.kz/atmosfernyi-vozdux. Accessed 1 Feb 2018

References

Air pollution trends in the EMEP region between 1990 and 2012 (2016) Joint report EMEP co-operative programme for monitoring and evaluation of the long-range transmission of air pollutants in Europe. Norwegian Institute for Air Research, NILU, pp 92–93

Almaganbetov N, Grigoruk V (2008) Degradation of soil in Kazakhstan: problems and challenges. In: Simeonov L, Sargsyan V (eds) Environmental security-soil chemical pollution, risk assessment, remediation and security, NATO Science for Peace and Security Series C. Springer, Dordrecht, pp 309–320

Annaklycheva J (2002) Combating desertification in Turkmenistan on the grass roots level example of the Central Karakum desert, Inaugural-Dissertation zur Erlangung des Doktorgrades der Mathematisch-Naturwissenschaftlichen Fakultät der Universität zu Köln, Aschgabad, 155 p

Arabov SA (ed) (2010) Atlas of soil resources of the Republic of Uzbekistan. Tashkent, Goskomzemgeodezkadastr of the Republic of Uzbekistan, State Scientific Production Enterprise Kartografiya

Arabov SA (ed) (2014) National report on the state of land resources of the Republic of Uzbekistan. Tashkent, Goskomzemgeodezkadastr of the Republic of Uzbekistan, State Scientific Production Enterprise Kartografiya

Asia – soils. Encyclopaedia Britannica. https://www.britannica.com/place/Asia/Soils. Accessed 15 Oct 2017

Atmospheric air: Pressure (2000) Tajikistan state of the environment report. http://enrin.grida.no/htmls/tadjik/soe2/eng/htm/air/press.htm. Accessed 11 July 2017

Babaev AG, Kharin NG (1999) The monitoring and forecast of desertification processes. In: Desert problems and desertification in Central Asia. Springer, pp 59–75

Bobojonov I, Aw-Hassan A (2014) Impacts of climate change on farm income security in Central Asia: an integrated modeling approach. Agric Ecosyst Environ 188:245–255

Brito LF, La Scala N Jr, Merques J Jr, Pereira GT (2005) Temporal variability of CO2 soil emissions and its relation with soil temperature in different positions of the landscape passing to areas farmed with sugarcane. In: Simpósio sobre Plantio direto e Meio ambiente; Seqüestro de carbono e qualidade da agua, pp 210–212. Anais. Foz do Iguaçu, 18–20 de Maio 2005

Bucknall J, Klytchnikova I, Lampietti J, Lundell M, Scatasta M, Thurman M (2003) Irrigation in Central Asia. Social economic and environmental considerations. The World Bank

CAWater-Info (2018) Water resources of the Aral Sea Basin. Groundwater: reserves and uses. www.cawater-info.net/aral/groundwater_e.htm

Conrad C, Schorcht G, Tischbein B, Davletov S, Sultonov M, Lamers JPA (2011) Agro-meteorological trends of recent climate development in Khorezm and implications for crop production. In: Martius C, Rudenko I, Lamers JPA, Vlek PLG (eds) Cotton, water, salts and soums: economic and ecological restructuring in Khorezm, Uzbekistan. Springer, pp 25–36

Devkota M, Martius C, Gupta RK, Devkota KP, McDonald AJ, Lamers JPA (2015) Managing soil salinity with permanent bed planting in irrigated production systems in Central Asia. Agric Ecosyst Environ 202:90–97

Encyclopaedia of Soil Science (2008) Chesworth Ward (ed). http://www.springer.com/gp/book/9781402039942?wt_mc=ThirdParty.SpringerLink.3.EPR653.About_eBook#otherversion=9781402039959. Accessed 24 Feb 2018

Faizov KS, Kenenbaev SB, Mamutov ZU, Esimbekov MB (2006) Geografia i ekologia pocv Kazakhstana (Soil geography and ecology of Kazakhstan), Almaty, 348 p

FAO-UNESCO Soil map of the world (1974) UNESCO, Paris

FAO (2002) Water resources, development and management service. AQUASTAT Information System on Water in Agriculture: Review of Water Resource Statistics by Country. http://www.fao.org/waicent/faoinfo/agricult/agl/aglw/aquastat/water_res/index.htm. Accessed 1 Nov 2017

FAO (2011) The state of the world's land and water resources for food and agriculture (SOLAW) – managing systems at risk. Food and Agriculture Organization of the United Nations, Rome and Earthscan, London. http://www.fao.org/docrep/017/i1688e/i1688e.pdf. Accessed 17 Jan 2018

FLERMONECA Regional Environmental Centre for Central Asia, Environmental Agency of Austria, Zoi Environment Network (2015) The state of the environment in Central Asia: illustrations of selected environmental themes and indicators. http://www.zoinet.org/web/sites/default/files/publications/SOE-regional-eng.pdf. In: Nugumanova L, Frey M, Yemelina N, Yugay S (2017) Environmental problems and policies in Kazakhstan: air pollution, waste and water, IOS Working Papers, No. 366, Leibniz-Institut für Ost- und Südosteuropaforschung (IOS), Regensburg. Accessed 22 Feb 2020

Freedman B (2014) Soil organic matter dynamics, climate change effects. Handb Glob Environ Pollut 1:317–323

Glazovsky NF (1995) The Aral Sea basin: environmental changes human driving forces societal and recognition of the Aral Sea problem, possible solutions and rescue scenarios, the Aral Sea basin trajectory, saving the Aral Sea. In: Kasperson JX, Kasperson E, Turner BL (eds) Regions at risk: comparison of threatened environments. The U.N. University, Tokyo/New York/Paris, pp 92–139

Grunwald S (2009) Multi-criteria characterization of recent digital soil mapping and modeling approaches. Geoderma 152:195–207. http://citeseerx.ist.psu.edu/viewdoc/download?doi=10.1.1.459.3655&rep=rep1&type=pdf. Accessed 24 Feb 2018

Gupta R, Kienzler K Martius C, Mirzabaev A, Oweis T, De Pauw E, Qadir M, Shideed K, Sommer R, Thomas R, Sayre K, Carli C, Saparov A, Bekenov M, Sanginov S, Nepesov M, Ikramov R (2009) Research prospectus: a vision for sustainable land management research in Central Asia. ICARDA Central Asia and Caucasus Program. Sustainable Agriculture in Central Asia and the Caucasus Series No.1. CGIAR-PFU, Tashkent, Uzbekistan. 84 p. http://geoagro.icarda.org/downloads/publications/geo/Sustainable_Agriculture_1.pdf. Accessed 5 May 2018

Hakkenberg R, Churkina G, Rodeghiero M, Börner A, Steinhof A, Cescatti A (2008) Temperature sensitivity of the turnover times of soil organic matter in forests. Ecol Appl 18:119–131

Hays J (2016) Environmental issues in Kazakhstan. http://factsanddetails.com/central-asia/Kazakhstan/sub8_4f/entry-4681.html. Accessed 7 Jan 2018

Ibragimov F (2016) Evaluation report on Land Degradation in Kazakhstan. https://www.linkedin.com/pulse/evaluation-report-land-degradation-kazakhstan-firuz-ibragimov. Accessed 7 Jan 2018

Ibrakhimov M, Khamzina A, Forkutsa I, Paluasheva G, Lamers JPA, Tischbein B, Vlek PLG, Martius C (2007) Groundwater table and salinity: spatial and temporal distribution and influence on soil salinization in Khorezm region (Uzbekistan, Aral Sea Basin). Irrig Drain Syst 21:219–236

Khamzina A, Lamers JPA, Vlek PLG (2008) Tree establishment under deficit irrigation on degraded agricultural land in the lower Amu Darya River region, Aral Sea Basin. For Ecol Manag 255:168–178

Kibblewhite M, Jones RJA, Baritz R, Huber S, Arrouays D, Michéli E, Dufour MJD (2005) Environmental assessment of soil for monitoring. European Commission Desertification meeting. Brussels, 12–13 Oct 2005

Kirschbaum MUF (1995) The temperature dependence of soil organic matter decomposition, and the effect of global warming on soil organic storage. Soil Biol Biochem 27:753–760

Kirschbaum MUF (2000) Will changes in soil organic carbon act as a positive or negative feedback on global warming ? Biogeochemistry 48:21–51

Kirschbaum MUF (2006) The temperature dependence of organic-matter decomposition – still a topic of debate. Soil Biol Biochem 38:2510–2518

Kļaviņš M, Azizov A, Zaļoksnis J (2014) Environment, pollution, development: the case of Uzbekistan. Riga, UL Press. https://uzwater.ktu.lt/index.php/download-files/compendia-and-textbooks/module-a-environmental-science/129-environment-pollution-development/file. Accessed 22 Mar 2020

Koshkar Ata Tailings Pond, Environment of Mangystau Province, Kazakhstan. http://ecocitizens.kz/en/hotspots/koshkar-ata-tailings-pond. Accessed 13 Oct 2017

Kostianoy AG, Lebedev SA, Solovyov DM (2011) Satellite monitoring of water resources in Turkmenistan. Int Water Technol J IWTJ I(1):4–15

Legend of the Soil Map of the World (1974) UNESCO. In: Paris Middleton NJ, Thomas DSG (eds) World Atlas of desertification. Edward Arnold, London. 1997

Lenton TM, Huntingford C (2003) Global terrestrial carbon storage and uncertainties in its temperature sensitivity examined with a simple model. Glob Chang Biol 9(10):1333–1352

Liang L, Wenpeng D, Huimin Y, Lin Z, Yuet D (2017) Spatio-temporal patterns of vegetation change in Kazakhstan from 1982 to 2015. J Resour Ecol 8(4):378–384. https://doi.org/10.5814/j.issn.1674-764x.2017.04.009. www.jorae.cn

Lioubimtseva E, Henebry GM (2009) Climate and environmental change in arid Central Asia: impacts, vulnerability, and adaptations. J Arid Environ 73:963–977

Lubin N (1989) Uzbekistan: the challenges ahead. Middle East J 43:619–634

Managing industrial solid wastes from manufacturing, mining, oil and gas production, and utility coal combustion. Chapter 4: Oil and gas wastes. https://www.princeton.edu/~ota/disk1/1992/9225/922506.PDF. Accessed 3 Feb 2018

Middleton NJ, Thomas DSG (1997) World atlas of desertification. Edward Arnold, London

Nacionalny doklad o sostoyanii okruzhayuscey sredy i ob ispolzovanyi prirodnyh resursov za 2015 god. Atmosfernyi vozdux (2016) (National Report on the state of environment and use of natural resources for year 2015). http://doklad.ecogosfond.kz/atmosfernyi-vozdux. Accessed 1 Feb 2018

Nacionalny doklad o sostoyanii okruzhajuscey sredy i ispolzovanyi prirodnyh resursov v respublike Uzbekistan (2008–2011 g.g.) (2013) (National Report on the state of environment and use of natural resources in the Republic of Uzbekistan for years 2008–2011) Tashkent, Izdatelstvo "CHINOR ENK". http://docplayer.ru/58059280-Gosudarstvennyy-komitet-respubliki-uzbekistan-po-ohrane-prirody-nacionalnyy-doklad.html. Accessed 7 May 2018

Nacionalny doklad. Vypolnenye konvencii OON po borbe s opustynivanyem v Turkmenistane (2006) (National Report. Execution of Convention of UN on the measures against desertification in Turkmenistan), Ashgabat. www.unccd-prais.com. Accessed 7 May 2018

Ostrovsky VN (2007) Comparative analysis of groundwater formation in arid and super-arid deserts (with examples from Central Asia and Northeastern Arabian Peninsula). Hydrogeol J 15:759–771

Otarov A (2014) Concentration of heavy metals in irrigated soils in Southern Kazakhstan. In: Mueller L, Saparov A, Lischeid G (eds) Novel measurement and assessment tools for monitoring and management of land and water resources in agricultural landscapes of Central Asia, Environmental Science and Engineering. Springer, pp 641–652

Otto A, Simpson MJ (2006) Evaluation of CuO oxidation parameters for determining the source and stage of lignin degradation in soil. Biogeochemistry 80:121–142

Otto A, Shunthirasingham C, Simpson M (2005) A comparison of plant and microbial biomarkers in grassland soils from the Prairie Ecozone of Canada. Org Geochem 36:425–448

Pachikin K, Erokhina O, Funakawa S (2014) Soils of Kazakhstan, their distribution and mapping. In: Mueller L, Saparov A, Lischeid G (eds) Novel measurement and assessment tools for monitoring and management of land and water resources in agricultural landscapes of Central Asia, Environmental Science and Engineering. Springer, pp 519–534

Pankova EI, Konyushkova MV (2013) Climate and soil salinity in the deserts of Central Asia. Eurasian Soil Sci 46(7):721–727

Pasture rotation in the desert areas of Uzbekistan (2018) (CACILM) – Central Asian Countries' initiative for land management. https://qt.wocat.net/qt_summary.php?qt_id=624. Accessed 07 Jan 2018

Pocvy Rossii i Sovetskogo Soyuza (2001–2016a) cast' III. Sistematicheskye opisanya pocv. Pocvy centralnoy lesostepnoy i stepnoy oblasti. Tip Chernozemnyh pocv (Soils of Russia and Soviet Union, part III. Systemic description of soils. Soils of central forest-steppe and steppe zones. Chernozem type of soils) In: Field

Ecology Center “Ecosystem”™ Alexander Bogolyubov. http://www.ecosystema.ru/08nature/soil/094t.htm. Accessed 20 Aug 2017
Pocvy Rossii i Sovetskogo Soyuza (2001–2016b) cast' III. Sistematicheskye opisanya pocv
Pocvy centralnoy lesostepnoy i stepnoy oblasti. Tip Kashtanovyh pocv (Soils of Russia and Soviet Union, part III. Systemic description of soils. Soils of central forest-steppe and steppe zones. Kashtanozem type of soils) In: Field Ecology Center “Ecosystem”™, Alexander Bogolyubov. http://www.ecosystema.ru/08nature/soil/101t.htm. Accessed 20 Aug 2017
Precoda N (1991) Requiem for the Aral Sea. Ambio 20(3–4):109–114
Propastin P, Kappas M (2008) Spatio-temporal drifts in AVHRR/NDVI-precipitation relationships and their linkage to land use change in Central Kazakhstan. EARSeL eProc 7(1):30–45
Qadir M, Noble AD, Qureshi AS, Gupta RK, Yuldashev T, Karimov A (2009) Salt-induced land and water degradation in the Aral Sea basin: a challenge to sustainable agriculture in Central Asia. Nat Res Forum 33:134–149
Rakhmatullaev S, Huneau F, Kazbekov J, Le Coustumer P, Jumanov J, El Oifi B, Motelica-Heino M, Hrkal Z (2010) Groundwater resources use and management in the Amu Darya River Basin (Central Asia). Environ Earth Sci 59(6):1183–1193
Saparov A (2014) Soil resources of the Republic of Kazakhstan: current status, problems and solutions. In: Mueller L, Saparov A, Lischeid G (eds) Novel measurement and assessment tools for monitoring and management of land and water resources in agricultural landscapes of Central Asia, vol XXIII. Springer, pp 61–74. http://www.springer.com/978-3-319-01016-8. Accessed 6 June 2017
Shahgedanova M (ed) (2002) The physical geography of northern Eurasia. Oxford University Press. http://www.rusnature.info/reg/16_7.htm. Accessed 15 July 2017
Smith P, Martino D, Cai Z, Gwary D, Janzen HH, Kumar P, McCarl B, Ogle S, O'Mara F, Rice C, Scholes RJ, Sirotenko O, Howden M, McAllister T, Pan G, Romanenkov V, Schneider U, Towprayoon S, Wattenbach M, Smith JU (2008) Greenhouse gas mitigation in agriculture. Philos Trans R Soc B 363:789–813
Sommer R, De Pauw E (2011) Organic carbon in soils of Central Asia–status quo and potentials for sequestration. Plant Soil 338:273–288. https://www.researchgate.net/publication/225342880_Organic_carbon_in_soils_of_Central_Asia-status_quo_and_potentials_for_sequestration/figures. Accessed 08 Aug 2017
Sommer R, Glazirina M, Yuldasheva T, Otarov A, Ibraeva M, Martynova L, Bekenov M, Kholov B, Ibragimov N, Kobilov R, Karaev S, Sultonov M, Khasanova F, Esanbekov M, Mavlyanov D, Isaev S, Abdurahimov S, Ikramov R, Shezdyukova L, De Pauw E (2013) Impact of climate change on wheat productivity in Central Asia. Agric Ecosyst Environ 178:78–99
Stanchin I, Lerman Z (2008) Water in Turkmenistan. In: Spoor M, Arsel M (eds) The last drop: water, security, and sustainable development in Central Eurasia. Routledge, London. State Department of Natural Protected Areas. https://web.archive.org/web/20120331085424/http://www.tajikpark.tj/en/index.php?option=com_content&task=view&id=1&Itemid=2. Accessed 1 Feb 2018
Status of the world soil resources (2015) FAO Main report, prepared by Intergovernmental Technical Panel on Soils (ITPS). http://www.fao.org/3/a-i5199e.pdf. Accessed 1 July 2017
Thevenot M, Dignac M-F, Rumpel C (2010) Fate of lignins in soils: a review. Soil Biol Biochem 42:1200–1211
Tóth JA, Lajtha K, Kotroczó Z, Krakomperger Z, Caldwell B, Bowden R, Papp M (2007) The effect of climate change on soil organic matter decomposition. Acta Silv Lingaria Hung 3:1–11
Uranium and Nuclear Power in Kazakhstan (2018) World Nuclear Association. http://www.world-nuclear.org/information-library/country-profiles/countries-g-n/kazakhstan.aspx. Accessed 20 Feb 2018
Uranium in Kyrgyzstan (2018). https://www.world-nuclear.org/information-library/country-profiles/countries-g-n/kyrgyzstan.aspx. Accessed 24 Mar 2020
Uranium in Tajikistan (2017). https://www.world-nuclear.org/information-library/country-profiles/countries-t-z/tajikistan.aspx. Accessed 24 Mar 2020
Uranium in Uzbekistan (2020). https://www.world-nuclear.org/information-library/country-profiles/countries-t-z/uzbekistan.aspx. Accessed 24 Mar 2020
Uzbekistan Human development report (1995) Tashkent, publ. house UNDP, JV ILTECH-DIMAGE (Uzbekistan/Turkey)
Wang L, Wang D, He Z, Liu G, Hodgkinson KC (2010) Mechanisms linking plant species richness to foraging of a large herbivore. J Appl Ecol 4:868–875
White R (1987) Introduction to the principles and practice of soil science. Blackwell Scientific Publication
World Bank (2018) Uzbekistan – sustainable agriculture and climate change Mitigation Project (English). World Bank Group, Washington, DC. http://documents.worldbank.org/curated/en/631671539872712334/Uzbekistan-Sustainable-Agriculture-and-Climate-Change-Mitigation-Project. Accessed 18 Mar 2020
World reference base for soil resources 2014 (2015) FAO World Soil Resources Report 106, Rome. http://www.fao.org/3/i3794en/I3794en.pdf. Accessed 26 June 2018
Yu Y, Pi Y, Yu X, Ta Z, Sun L, Disse M, Zeng F, Li Y, Chen X, Yu R (2019) Climate change, water resources and sustainable development in the arid and semi-arid lands of Central Asia in the past 30 years. J Arid Land 11(1):1–14. https://doi.org/10.1007/s40333-018-0073-3. Accessed 22 Feb 2020

Modernization and Corresponding Ecological/Human Ruptures

5

Abstract

Agriculture remains one of the most important economic sectors in the CA region, creating more than 20% of GDP and providing jobs for about 40% of the working age population. Animal husbandry is dominant among the agricultural sectors in the region. However, agricultural land use as well as maintenance of different land, soil, water, crop and animal resources have experienced many changes from Soviet Union times. Agricultural land use decreased slightly from 1992 to 2015 on the average, but the changes were different in separate CA countries. The largest cropland is characteristic to Kazakhstan, followed by Uzbekistan and Turkmenistan, while the largest areas of irrigated cropland are found in Uzbekistan. There were essential changes in the farming structure in CA countries. Nowadays enterprises, peasant farms and household plots are the main subjects of agricultural farming in CA region. The cultivation methods make the impact on the soil properties and not always positive: the traditional intensive agriculture decreases the soil organic matter (SOM) as a result of accelerated microbial decomposition activated by any mechanical treatment of soil. The most traditional in CA countries is the conventional cropping system under the irrigation, but the infrastructure of irrigation systems has become more degraded nowadays and inefficient, causing the secondary salinization and drop in crop yields. The modern agriculture confirms the necessity to select crops and cropping systems to fit the local environment and climate in order to overcome risks of soil compaction, erosion, salinization, degradation and water deficit. One of such measures applicable in CA is the promotion of conservation agriculture, which allows to rehabilitate soil fertility and organic matter content. Recovering and further sustainable development of inland fishery in CA region is perspective too.

Keywords

Agriculture · Land use · Cropping and farming systems · Sustainable development · Animal husbandry · Fishery

5.1 Modern Agriculture, Use of Land Resources, Soil Properties, Farming and Cropping Systems, Agroecosystems, Conservation and Organic Agriculture, Animal Husbandry, Development of Fishery

5.1.1 Agricultural Sectors and Cultivated Land

Land resources and topsoil layer are the most important components of biosphere and mankind's welfare depends on the skill how to make the best use of them. The World's cultivated and irrigated land area has increased significantly during the last 50 years resulting in two and half to three times higher production of major agricultural crops (FAO 2011). According to Food and Agricultural Organization (FAO) data at present 10.4% of the total land area or 2.95% of the total terrestrial globe surface are cultivated. The importance of agriculture varies among the countries. The five leading countries according to the percentage area under agricultural use in European and CA countries are Kazakhstan, Moldova, United Kingdom, Ukraine and Turkmenistan, and the five least agricultural countries are Greenland, Norway, Finland, Sweden and the Russian Federation (Status of world soil… 2015). But this is true for total agricultural land area which includes vast rangelands in semi-deserts too. The agricultural land use in CA countries has experienced many changes from the Soviet Union times up to recent years (Fig. 5.1).

The agricultural area in the CA countries decreased slightly from 1992 to 2015 on the average, but the changes

I. Jelen et al., *The Geography of Central Asia*, World Regional Geography Book Series,
https://doi.org/10.1007/978-3-030-61266-5_5

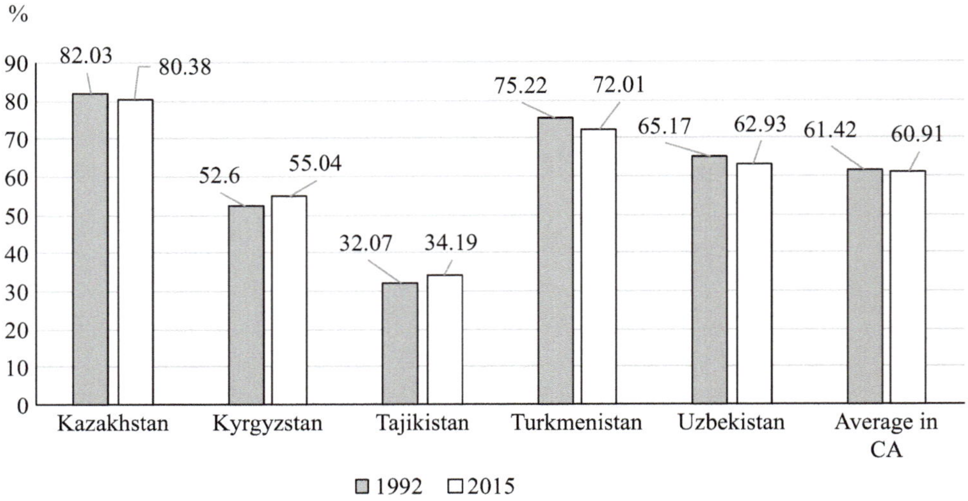

Fig. 5.1 Changes in agricultural land area in percentage of all land in CA countries from 1992 to 2015. (Sources: FAO (2011; FAOSTAT Land use…; CIA The World Factbook 2017))

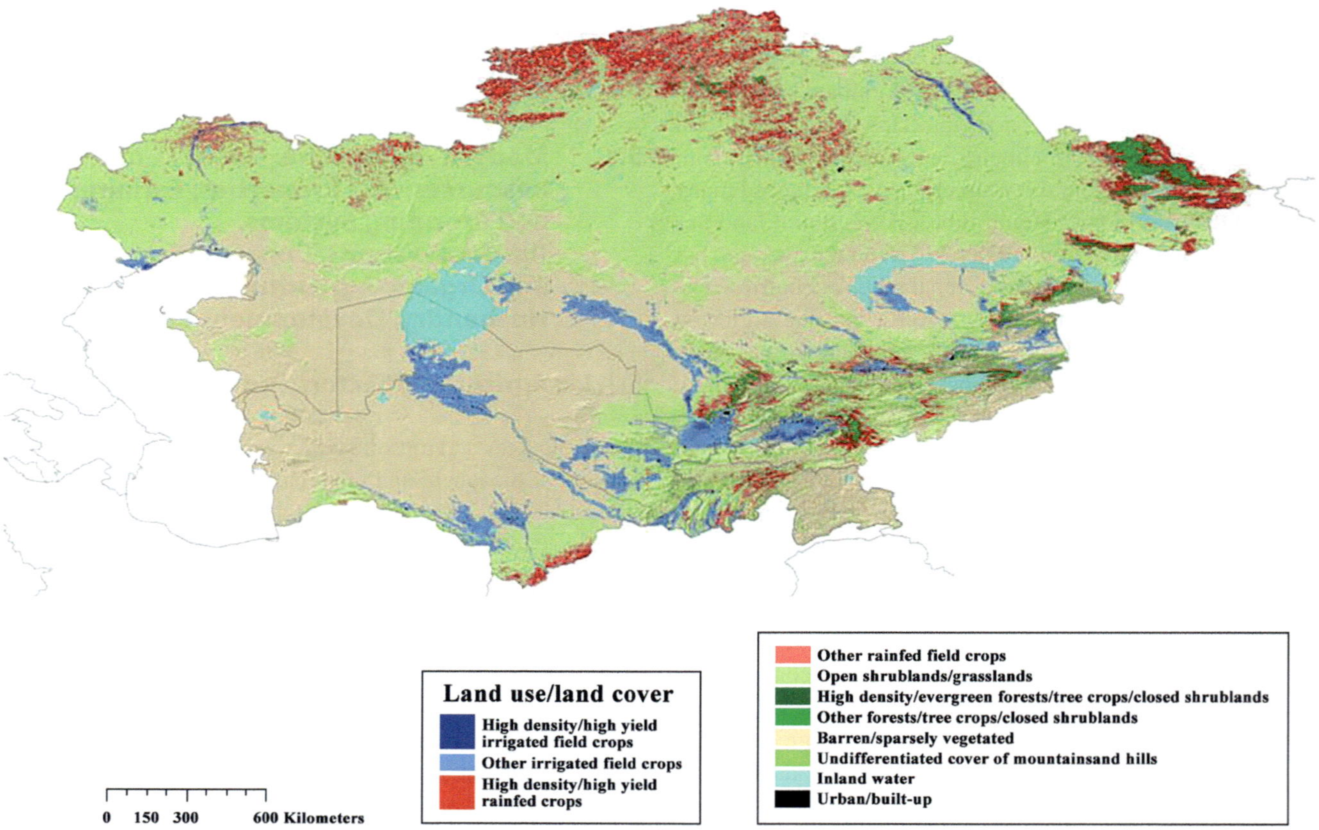

Fig. 5.2 The spatial distribution of land cover/land use of CA in 1997–1999 (Celis et al. 2007). (Source: Sommer and De Pauw (2011))

were different in separate countries: varied from −3.21% in Turkmenistan to +2.44% in Kyrgyzstan.

The spatial distribution of different land cover and use in CA region in 1997–1999 is demonstrated in Fig. 5.2.

Though CA region extends in the area of 400 million ha, only 20% of that is suitable for farming (Bobojonov and Aw-Hassan 2014). Figure 5.2 shows that irrigated agricultural land occupied rather large areas along the main Syr Darya and Amu Darya streams and their tributaries, channels, in the valleys and plain landscape of central, eastern and south-eastern parts of CA region. However, the largest areas are covered by the open shrublands/grassland and the bare/sparsely vegetated arid areas, mostly used as a rangeland for grazing.

Agriculture is one of the most important economic sectors in the CA region, resulting in more than 20% of GDP on the average and creating jobs for about 40% of total labour force (Qushimov et al. 2007). Of the branches, the animal husbandry is the dominant among the agricultural sectors in the region, and it was the dominant also during the Soviet governance. Cattle, sheep and poultry are the main cultivated animal species, and among them famous local breeds like race horses of Turkmenistan, the Karakul sheep and the Akhal-Teke horse have to be mentioned. Some regions, like Fergana Valley in Uzbekistan and Tajikistan and many smaller farms in Turkmenistan, also cultivate mulberry trees and breed silkworms which was the traditional branch for development of silk weaving (Meller 2013).

According to Celis et al. (2007), the dominant cultivated land type of CA region is irrigated field crops and rainfed crops. The irrigation started with 991,000 ha in the mid-1920s, mostly in the Syr Darya River Basin, had increased more than five times by the 1960s (Qadir et al. 2009). Though the irrigation and drainage benefited from massive investment during the Soviet era, but water was not well managed. From the independence, more problems have occurred for farmers, thus in many cases the infrastructure of irrigation systems became even more degraded. Recently with 9.61 million ha under irrigated agriculture, CA is one of the largest region in the world (The state of food... 2008).

Data in Tables 5.1 and 5.2 show the distribution of cultivated cropland, permanent pastures and meadows and irrigated land area in the CA region in 1999 and 2017.

According to the available statistics, the cultivated cropland and permanent meadows and pastures area decreased in CA region in 2017 as compared with situation in 1999. But that tendency is characteristic to Kazakhstan and Uzbekistan, two largest countries in the region. In Kyrgyzstan, there were no changes in land use of both types, and in Tajikistan there registered an increase in the area of permanent meadows and pastures, while in Turkmenistan a slight increase in cropping land and decrease in permanent meadows and pastures area was observed.

Though the cropland decreased from 1999 to 2017 in CA region, the area of irrigated land slightly increased by area and by percentage of cropland. As to countries, the increase of irrigated land area was registered in Tajikistan and Turkmenistan, while in Kyrgyzstan and Uzbekistan it remained the same. The decrease of irrigated land area in ha was registered only in Kazakhstan, but the country has ambitious plans to increase it in near future (Gosudarstvennaya programma... 2017; Satubaldina 2019). Still the largest areas of irrigated cropland are located in Uzbekistan. The most problematic is the high losses of water due to the inefficient I&D systems in all CA countries (The United Nations Economic Commission... 2015). The ineffective irrigation practices, particularly during droughts, can lead to acute water scarcity (Mogilevskii et al. 2017; Asian Development Bank (ADB) 2013; CIAT; World Bank 2018). Currently, irrigated farming is the main source of water and soil pollution by salinization (Novel Measurement and Assessment... 2014). Over-irrigation and lack of proper crop rotations on the irrigated land also lead to the higher losses in nutrients and crop yields (Reddy et al. 2013; Devkota et al. 2013; Xenarios et al. 2019; Irrigatcya: Kyrgyzstan... 2017)

Thus CA constitutes a particularly vulnerable region because of the physical geography (dominated by temperate deserts and semi-deserts) and due to its relative specialization and cultivation of monocultures before 1991, also the drastic social-economic changes after 1991, which have postponed the modernization and more effective usage of water I&D systems. The recent climate changes even increased region's vulnerability from the ecological point of view (Lioubimtseva and Henebry 2009). The land reforms, which immediately took place after 1991 when CA countries gained independence, on the one hand have slowed the modernization of agricultural sectors, but from the longer perspective have made the essential changes concerning land and other resources and property issues. The fast increase in the peasant farm number and area as well as decrease of the role of large agricultural enterprises inherited from the Soviet planning economics were observed in CA, however, with country specifics. For example, there are three main farming forms adopted and spread in Tajikistan and Uzbekistan nowadays: enterprises, peasant farms and household plots. The role of enterprises decreased, and the one of peasant farms increased from the state's independence period in 1991 up to 2006–2007 in both countries, but there are differences in management with other than arable land (rangeland): more peasant farms with rangeland is typical for Tajikistan, than for Uzbekistan where the enterprises are still more popular. According to the agricultural production output, such a tendency was observed in Uzbekistan and Tajikistan: total agricultural production and crop production increased evidently in the peasant farms and

Table 5.1 The area of cultivated cropland and permanent meadows and pastures in million ha in CA countries and region in 1999 and 2017

	1999		2017	
	Cultivated cropland	Permanent meadows and pastures	Cultivated cropland	Permanent meadows and pastures
Kazakhstan	30.1	18.3	29.5	18.8
Kyrgyzstan	1.4	9.2	1.4	9.2
Tajikistan	0.9	3.6	0.9	3.9
Turkmenistan	1.9	33.6	2.0	31.8
Uzbekistan	4.8	22.5	4.4	21.1
Total in CA	39.1	87.2	38.2	84.8

Source: FAOSTAT Land use http://www.fao.org/faostat/en/#data/RL/visualize

Table 5.2 The area of irrigated land in million ha and in pecentage of cropland in CA countries in 1999 and 2017

	1999		2017	
	million ha	% of cropland	million ha	% of cropland
Kazakhstan	2.3	7.1	2.1	7.1
Kyrgyzstan	1.0	69.7	1.0	73.3
Tajikistan	0.7	79.0	0.9	89.6
Turkmenistan	1.8	95.2	2.0	100.0
Uzbekistan	4.2	86.9	4.2	95.1
Total in CA	10.0	25.6	10.2	26.7

Sources: FAOSTAT Land use http://www.fao.org/faostat/en/#data/RL/visualize; FAO AQUASTAT… (2018)

household plots and decreased in the large enterprises from 1997 to 2007. At the same time, the livestock production remained as it was concentrated and increased mostly in the household plots (Lerman and Sedik 2009a, b). With all the reforms, the agricultural production as well as land productivity started to grow from 1997 in Uzbekistan and from 1999 in Tajikistan. It was observed, that peasant or individual farms were not only cosmetically reorganized (Lerman 2007).

In Kyrgyzstan in 2008, there was the sharp decrease in number of large agricultural enterprises and increase in the individual farms. Thus, the total number of farms increased, and livestock today is concentrated almost exclusively in household plots and peasant farms. The shift of productive resources—land and livestock—from enterprises to the individual sector has resulted in a significant increase in the share of individual farms in agricultural production (Lerman and Sedik 2009a).

As to the crops, there was an increase in production of cereals, legumes, vegetables, melons and potatoes, but decrease in cotton, barley and pastures between 1952 and 2002 in CA countries. The area under the cereal crops increased by 38– 47% except in Kazakhstan (Qushimov et al. 2007). The wheat became a strategic crop among other cereals, it was cultivated even on marginal salt-affected soils under large scale irrigation. The highest shift to wheat was observed in Turkmenistan and Uzbekistan (Fig. 5.3).

Crop productivity mostly has increased in the household plots from 1999 to 2007. The average enterprise in Kyrgyzstan in 2002 had 220 ha of arable land as compared with average peasant farm with 3.80 ha and a household plot of 0.10 ha. The leasing of land is popular mostly among the individual farms. Concerning the crops grown in Kyrgyzstan, there was the drop in forage crop cultivation area in order to expand cereals, cotton and other crops (Lerman and Sedik 2009a, b).

In Kazakhstan, the agrarian reform also has gone through many ups and downs from 1991 and later, privatization has been uneven (Spoor 2007; Kazakhstan: country fact sheet… 2017). Wheat is the main agricultural crop cultivated in Kazakhstan. More than 70% of the wheat grain comes from the fields of the large enterprises. Second popular crop, the potato, is contrarily produced mostly in the individual farms and household plots of the country (Kazakhstan Agricultural Overview 2010). The grape production is still problematic in the CA countries (Spoor 2006, 2007).

According to Lerman and Stanchin (2005, p. 4), "Turkmenistan has liberalized much of its agricultural production and food trade, but the main strategic commodities—cotton and wheat (as well as the much less important rice)—remain subject to state orders". Privatization and restructuring of farmland in this country are limited, and the agricultural producers are controlled by state administration much. Only the livestock sector operates on a more private basis (Turkmenistan: agricultural… 2012).

Other problems, related to agricultural land use in CA countries are the abundant cultivated areas, which have lost their previous importance due to the desertification and degradation of soil. They were investigated from 2003 to 2016 by using Moderate Resolution Imaging Spectroradiometer (MODIS) method in the Aral Sea Basin (ASB) area. It was determined that cropland abandonment occurred mostly in the downstream regions of the Amu Darya and the Syr Darya rivers and was associated with degraded land and areas prone to water stress (Löw et al. 2018). But this research has shown also some abandoned cropland in areas with good farming conditions, thus it is necessary to look for other causes leading to the abandoned cropland still.

5.2 Soils in CA Region Under the Agricultural Land Use, Agroecosystems and Farming Intensity

Agriculture is cultivated almost in all biomes of CA; however, the plant husbandry, both rainfed and irrigated, is spread in the areas with mostly fertile soils supplied with water. Such are the Chernozems in the northern part and Kashtanozems in other parts of the region. The Chernozems and Dark Kastanozems (Chestnut soils) have been mostly cultivated, in some places up to 90–95% of the total land area (Status of world soils… 2015).

Degradation of soil organic matter (SOM) occurs as a result of accelerated microbial decomposition activated by any mechanical treatment of soil and serves as the main determinant and indicator of soil biological activity, fertility and C-sequestration. Thus, understanding factors controlling carbon retention in the ecosystems is particularly important maintaining environmental quality and ecosystem sustain-

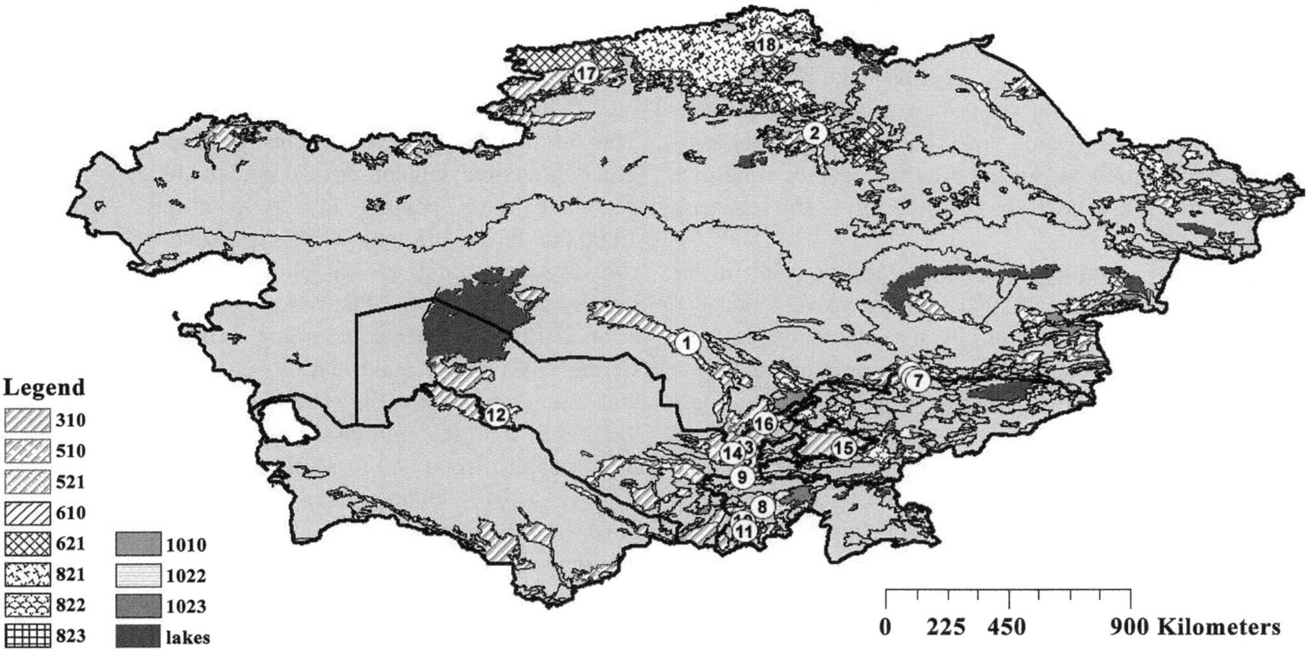

Brief description of agro-ecological zones of CA according to De Pauw (2010) suitable for cultivation of wheat

AEZ	Description	Countries (in order of abundance)
310	Irrigated wheat in arid climate with cold winter and hot summer	Uzbekistan, Kazakhstan, Tajikistan
510, 521	Irrigated or rainfed wheat in a semi-arid climate with cold winter and mostly warm summer	Kazakhstan, Uzbekistan, Tajikistan, Kyrgyzstan
610, 621	Irrigated or rainfed wheat in mostly semi-arid climate with cold winter and mild summer	Kazakhstan, Kyrgyzstan
821, 822, 823	Rainfed wheat in sub-humid climate with cold winter and mild summer	Kazakhstan, Kyrgyzstan, Uzbekistan
1010, 1022, 1023	Irrigated or rainfed wheat in a humid climate with cold winter and mild summer	Kazakhstan, Tajikistan, Kyrgyzstan, Uzbekistan

Fig. 5.3 Agroecological zones suitable for cultivation of wheat in Central Asia. (Sources: (Sommer et al. 2013; De Pauw 2010))

ability. Soil aggregation, water and nutrient retention characteristics, as well as resistance against erosion by wind and water, all these increase with increasing content of organic carbon (Mamilov 2010).

Some soil properties and fractions, like mobile fractions of organic matter, available phosphorus (P) and potassium (K), microbial and enzyme activity, react quickly to changes in land use and management, whereas passive pools of SOM, notably their organic nitrogen (N org) and humus contents, are more stable (Bučienė et al. 2003). Thus, SOC (soil organic carbon) can be selected as an indicator of soil fertility and ecological soil status for long term.

The compilation of different research data has shown that ten most important soil types in CA region contain different loads of SOC in the upper 30 cm layer. The SOC still is the highest in Chernozem-type of soils, followed by the Chestnut or Kastanozem soils and Lithosol. However, the data on the content of SOC in the last type is from the source of 1968, and it needs to be tested again (Sommer and de Pauw 2011).

Findings of many researchers show (Bučienė 2012) that agroecosystems arise on the interface of different farming/cropping systems and natural ecosystems with their resources. The main agroecosystems in the CA region are the fields of arable land crops both irrigated and rainfed, permanent crops (fruit and berry plantations), different types of rangelands (from cultivated grassland and rainfed pastures to the steppe and mountain meadows and the arid shrub-land). The farming systems can be of different input level: from low to high or so-called intensive farming systems (Gomiero et al. 2006; Wolf and Allen 1995). In CA countries, the most traditional is the conventional farming/cropping system under the irrigation. Since the Soviet period till now, there have been large changes with cropping and crop rotation systems in CA countries. For example, the typical crop rotation in Uzbekistan was three years of alfalfa followed by six years or more of cotton, but after independence in 1991 the share of alfalfa and especially cotton decreased in favour of winter wheat (Kienzler 2010; FAO Food and Agriculture…;

Khalikov 2010). In southern Kazakhstan, the official recommendations (Japan International… 2010; Khalikov and Tillaev 2006; Umirsakov et al. 2011; Kienzler 2010; Conrad et al. 2016) may vary following the soil quality. Here after two years of rice cultivation, fields are temporarily set aside from rice cultivation and alfalfa or other legumes are grown for up to three consecutive years, which allows soil regeneration (Löw et al. 2018).

The intensive farming systems not always result in the highest output. This is evident from the intensively managed crops in the irrigated fields of CA: here the increased (secondary) salinization is following by a corresponding drop in crop yields. Such a situation forced the farmers to apply ever-greater quantities of water in order to flush the salt out of the soil, making water application even more wasteful, and also increased waterlogging (Bucknall et al. 2003). It was estimated, that the loss in cotton yields is 20–30% on slightly salinized land, 40–60% on moderately salinized land and up to 80% and beyond on severely salinized land (Royal Haskoning… 2001). The soil salinity via regulation of osmotic pressure inhibits the ability of plants to absorb water, also some ion toxicity can occur. The plants differ according to their adaptation to different levels of salinity in soil–water environment. Among the crops grown in CA, the most salt-tolerant to higher salinity are barley and sugar beets; moderately tolerant crops are alfalfa, rice, cotton, wheat, corn, potatoes, carrots, onion, cucumbers, pomegranates, figs, melons and grapes. The least salt-tolerant crops are stone fruits, almonds, peas and beans (World Bank 2001). In specific conditions, like in Priaralye, at Kyzylorda province, the experiments testing adaptation of different vegetables sown under the conditions of high soil salinity showed promising results. The soils in this region are rich in potassium, contain average to below average amounts of phosphorus, and low levels of nitrogen. The percentage of humus varies from 0.5 to 4%, seldom reaching 6%. The arable land, which was previously used for rice growing, was chosen for this experiment with vegetables tolerant to salts. The results showed that the most salt-stable seeds were from the nightshade and mustard families. The performance of carrots was also impressive (Tyan et al. 2005).

Another end-product of SOM decomposition is nitrate (NO_3^-), which is very essential in plant nutrition. However, a number of environmental problems occur when nitrification rate in soil exceeds demands of crops. Contamination of drainage and groundwaters with nitrates is one of them. It happens even in extremely arid regions. The other process is microbial denitrification, and as result accelerated emissions of nitrous oxide (N_2O)—an important green-house gas, which is 150 times more efficient than CO_2 retaining heat, and having ten times longer residence in atmosphere. The contribution of modern agriculture to global N_2O emissions is about 90%. Achievements of modern agriculture confirm the necessity to select crops and cropping systems to fit the local environment and climate in order to overcome risks of soil compaction, salinization and water deficit (Mamilov 2010).

So called conservation agriculture, according to FAO (2010, 2013), is increasing in many countries of the world. This farming direction which relies on reduced or zero tillage, dense cover of plant residues and viable crop rotation is becoming more popular in CA countries. There was 1,600,000 ha of agricultural land in Kazakhstan, where the conservation agriculture was adopted in 2011 though the country started this way of farming only since 2000 (Friedrich et al. 2012; FAO 2016). According to the FAO Investment Centre mission to Kazakhstan, the adoption of zero tillage and conservation agriculture had raised domestic wheat production by almost 2 million tonnes, and in parallel brought more qualified food and resulted in higher sequestration of CO_2 (FAO 2012). Taking into account that CA region distinguishes by the large variety of agroclimatic zones, soil types, cropping specialization, scarcity of water resources, population density etc., the conservation agriculture cannot be expected to be the only solution, but rather one of the solutions (Kienzler et al. 2012). Among them, the Organic Centre of Kazakhstan can be mentioned (Organic Centre… www.organiccenter.kz). It was founded in 2008 as a result of cooperation between the institutions of Kazakhstan and the Netherlands. Recently in Kazakhstan the total area controlled by this Organic Centre was 133,562 ha and seven companies were certified (Klimov 2010). In Tajikistan, the organic agriculture is not so developed in comparison with other CA countries. But recently few projects together with one institute from Netherlands on bio-cotton growing and processing in Sougd oblast started (Sultanova 2010).

The first cooperative of organic farmers "BioFarmer" was founded in Kyrgyzstan in 2007 with the aim to produce organic cotton fibre, organic wheat and pulses. Besides, production of medicinal and aromatic Plants (calendula and chamomile) and dried apricots has started at pilot level. For the domestic market, the cooperative produces canned vegetables and fruits and sells production to the cities (Borkoshev 2010).

In Uzbekistan, the organic production is concentrated in the Qashqadaryo province. The low input soil and crop management combined with diversified crop rotation allow to get qualitative crop outcomes, but for modernization of technologies higher state support is necessary (Nurbekov et al. 2018).

Novel methods used for crop production and soil management allow faster and better adoptation in site-specific practices. For example, the method of agroecological zoning using GIS procedures is proposed to CA countries to define areas that can be considered homogeneous in their biophysical characteristics at a relatively fine resolution (1–5 km^2) and can thus serve as a first basis for area-specific agricultural (research) planning, particularly under the conditions of climate change (De Pauw 2010). The research confirmed that the overall impact of climate change on wheat productivity in CA is positive except for very arid southern territories of

Kazakhstan and some northern areas for spring wheat growing. Irrigation water requirements do not increase under climate change if technologies are proper (Sommer et al. 2012). However, looking from the current situation. it is necessary to concentrate more on the improvement of I&D systems and their proper management.

5.3 Animal Husbandry

After the collapse of the Soviet Union, the animal husbandry, like the other branches of agricultural sector in CA region, has experienced the large drop in the productivity and incomes on the one hand, and on the other hand with restructuring land property, the shift between the main producers has been taking place. Still from 1991 up to now the development of this industry is characterized by unbalanced development. The problem is that disproportion between livestock and feed resources is growing rapidly, the area under the fodder crops on the cultivated land is decreasing. The fodder capacity of summer and winter pastures decreases as well. Pasture land is degraded, and in many places this becomes irreversible. This requires the transfer of livestock to animal housing systems to give pastures the opportunity for natural restoration of green cover (Umarov 2019). The data in Fig. 5.4 show as to what was the situation in 2018 with the main livestock groups and numbers of live animals in CA countries.

About 70% of Kazakhstan's total land area is permanent pastureland, thus it was traditional to raise sheep, goats, horses, cattle, pigs and camels. Today the largest in number are the sheep and cattle herds. The new country's Livestock Programme intends to support smaller scale farms of 100–200 head cattle or 600 head sheep, expanding the number of farms from today's 20 thousand on 58 million ha to 100 thousand on 100 million ha in 2027 (Vanderberg 2018).

The main livestock products in Kyrgyzstan include meat (50% beef, 29% sheep meat, 10% horse meat, 11% other meat), cow's milk, wool and eggs, but the productivity of branch is low, it satisfies mostly owner needs. Livestock numbers fell in the first half of the 1990s, and then recovered to a different extent: the number of cattle and horses are now much higher than in 1990, while sheep, goats and poultry are still well below their 1990 level (Mogilevskii et al. 2017). There is disproportion between animal number and grazing areas, thus the over-grazing and the continued pressure on pastures, particularly those at lower altitudes is taking place. Also there is noticed a drop in productivity of the grasslands as compared with Soviet times, some parts of the country even experienced a 50% drop (Fitzherbert 2006; Van Berkum 2015).

The largest area of Tajikistan's agricultural land is suitable for pastoralism; however, the situation with animal husbandry is not benevolent. After a sharp drop in the number of animals after 1991, the animal husbandry sector, which rears mostly sheep, cows and goats, began to grow again after 2000. Nowadays a major problem is a sufficient supply of winter fodder and the large number of small-scale producers with low incomes and restricted in credits they need (EBRD 2012; Swanson et al. 2011).

The raising of sheep in Turkmenistan is the third highest among the CA countries following Kazakhstan and Uzbekistan. In contrast to the majority of former Soviet republics, Turkmenistan experienced a growth in its animal husbandry sector as early as the 1990s, due in part to the growth in milk yields per cow (Lerman et al. 2012). But the productivity is low, the milk and meat are processed and consumed for the farmers' own requirements. The majority of pastures and grass steppes of the country are located in the northern part.

The vast majority of Uzbekistan is made up of desert pastures, suitable only for sheep, however, raising of cattle is becoming important branch of animal husbandry, with growing in number from year to year (USDA FAS 2011).

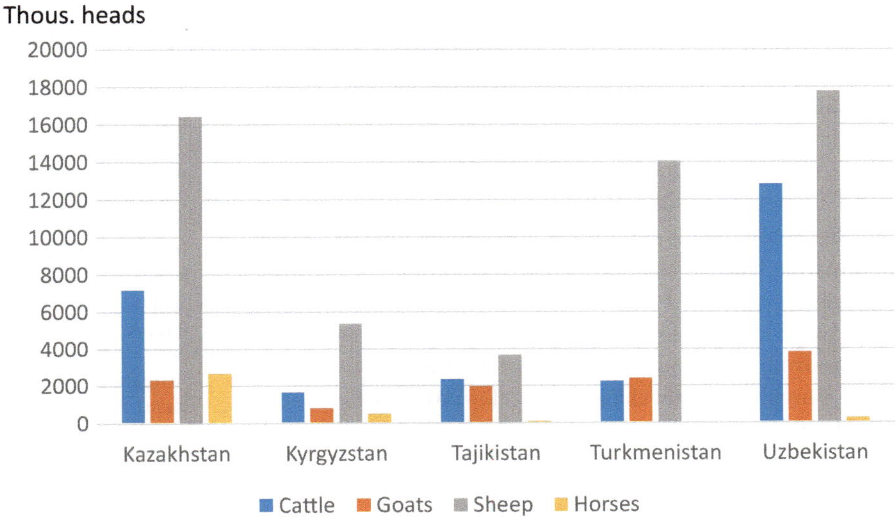

Fig. 5.4 Number of live animals (in thousand. heads) of the main livestock groups in CA countries in 2018. (Source: FAOSTAT Live animals http://www.fao.org/faostat/en/#data/QA/visualize)

In order to make more sustainable grazing system in CA countries, local institutional innovation ensuring the participatory range management of pastoral communities has to be strengthened (Mirzabaev et al. 2016).

5.4 Situation with Fishery

Fishery in this arid CA region was more intensive during the Soviet Union period than today, but attention was not paid about the fish habitats ecology, environmental pollution etc. Thus both the uncontrolled fishery and other anthropogenic loads, like irrigated agriculture and hydro-power systems, together with climate warming and drying, resulted in the shrinking and salinization of the largest lakes and drastic drop in the aboriginal fish resources (FAO 2010–2014; 2007; FAO The State… 2016; Timirkhanov et al. 2010; Ermakhanov et al. 2012; Mendikulova 2008; Savvaitova and Petr 1995; Petr and Mitrofanov 1998). Seemingly fishery was not the priority sector in CA countries after they have gained independence in 1991. By 2012, however, few countries like Uzbekistan and Kazakhstan, had made a progress. The Uzbekistan, for example, is focusing on the aquaculture development (Graham 2016; FAO 2010–2014, 2011), and the Kazakhstan has done significant attempts for fish recovery in the Small Aral. After building of Kokaral dam which has stabilized the water volume in the northern part of the sea and reduced the excessive water salinity, fresh water fish resources started to recover (Plotnikov et al. 2016). Apart of commercial fishing, there is growth in recreational fishing.

The Caspian Sea is world–known sea of six species of sturgeon, but the illegal catches have led to their dramatic reduction (Pourkazemi 2006; Ermolin and Svolkinas 2018). The Volga–Caspian region is of greatest significance for the sturgeon fishery with respect to numbers migrating to spawning grounds and the size of commercial catch (75–78% of the total sturgeon harvest in the basin) (Ruban and Khodorevskaya 2011). Recently by the initiative of Russian Federation, it was decided to develop a joint programme for the Caspian states to restore sturgeon stocks through the coordinated efforts in scientific research and fishery regulations (Caspian countries extend… 2019). The Kazakhstan and Turkmenistan will join this programme too

FAO recommends for CA countries to include the inland and marine fishery resources management in the state priority tasks (FAO 2011). This sector can recover only with the state assistance for legal and environmental regulation and modernization of aquaculture including such valuable fish like sturgeon (Timirkhanov et al. 2010; Ruban and Khodorevskaya 2011)

Topics for Further Reading and Discussions on the Content of This Chapter

- Sustainable agriculture in CA countries: from theory to practical solutions.
- World countries mostly advanced in irrigation technics and technologies.

Recommended References

Conrad C, Lamers JPA, Ibragimov N, Löw F, Martius C (2016) Analysing irrigated crop rotation patterns in arid Uzbekistan by the means of remote sensing: a case study on post-Soviet agricultural land use. J Arid Environ 124:150–159

Conway GR (1987) The properties of agro-ecosystems. Agric Syst 24:95–117

De Pauw E (2010) Agro-ecological zoning of the CWANA region. In: El-Beltagy A, Saxena MC (eds) Sustainable development in drylands. Meeting the Challenge of Global Climate Change. Proceedings, pp 335–348

Devkota M, Martius C, Lamers JPA, Sayre KD, Devkota KP, Gupta RK, Egamberdiev O, Vlek PLG (2013) Combining permanent beds and residue retention with nitrogen fertilization improves crop yields and water productivity in irrigated arid lands under cotton, wheat and maize. Field Crops Res 149:105–114

FAO (2010) Conservation agriculture spreads further a field. http://www.fao.org/agriculture/crops/agp-in-action/en/. Accessed 21 Jan 2018

FAO (2012) Conservation agriculture in Central Asia: status, policy, institutional support, and strategic framework for its promotion, FAO Sub-Regional Office for Central Asia (FAO-SEC). http://www.fao.org/docrep/017/aq278e/aq278e.pdf. Accessed 26 Jan 2018

FAO (2013) Conservation agriculture in Central Asia. Status, policy and institutional support and strategic framework for its promotion, FAO Sub-Regional Office for Central Asia (FAO-SEC) Ankara. http://www.fao.org/docrep/019/i3275e/i3275e.pdf. Accessed 22 Jan 2018

Friedrich T, Derpsch R, Kassam A (2012) Overview of the global spread of conservation agriculture. Field actions science reports. Reconciling poverty eradication and protection of the environment. Special issue, 6. http://factsreports.revues.org/1941. Accessed 25 Feb 2018

Giampietro M (1994) Using hierarchy theory to explore the concept of sustainable development. Futures 26(6):616–625

Giampietro M (1997) Socioeconomic pressure, demographic pressure, environmental loading and technological changes in agriculture. Agric Ecosyst Environ 65:201–229

Gomiero T, Giampietro M, Mayumi K (2006) Facing complexity on agro-ecosystems: a new approach to farming system analysis. Int J Agric Resour Gover Ecol 5(2–3):116–144

Hart RD (1984) The effect of inter level hierarchical system communication on agricultural system input-output relationships. International Association for Ecology Series Study. In: Options Mediterraneennes Ciheam IAMZ-84-1

Ikerd JE (1993) The need for a system approach to sustainable agriculture. Agric Ecosyst Environ 46:147–160

Irrigatcya: Kyrgyzstan vyhodit na novyj vitok razvitya (2017) [Irrigation: Kyrgyzstan appears in the new round of the development] (2017). http://kabar.kg/news/irrigatciia-pravitel-stvo-kr-vykhodit-na-novyi- vitok-razvitiia/. Accessed 29 Mar 2020

Kienzler KM, Lamers JPA, McDonald A, Mirzabaev A, Ibragimov N, Egamberdiev O, Ruzibaev E, Akramkhanov A (2012) Conservation agriculture in Central Asia – what do we know and where do we go from here ? Field Crops Res 132:95–105

Klimov J (2010) Development of organic agriculture in Kazakhstan. Information on the Organic-Center. Conference proceedings, Astana, Kazakhstan, pp 31–33. www.conference.organiccenter.kz. Accessed 27 Jan 2018

Novel Measurement and Assessment Tools for Monitoring and Management of Land and Water Resources in Agricultural Landscapes of Central Asia (2014). Lothar M, Abdulla S, Lischeid G (eds). Springer. https://books.google.lt/booksid=0T4nAQAAQBAJ&printsec=frontcover&dq=Novel+Measurement+and+Assessment+Tools+for+Monitoring+and+Management+of+Land+and+Water+Resources+in+Agricultural+Landscapes+of+Central+Asia/. Accessed 27 Mar 2020

Nurbekov A, Aksoy U, Muminjanov H, Shukurov A (2018) Organic agriculture in Uzbekistan: status, practices and prospects. FAO, Tashkent. http://www.fao.org/3/i8398en/I8398EN.pdf. Accessed 31 Mar 2020

Reddy JM, Jumaboev K, Matyakubov B, Eshmuratov D (2013) Evaluation of furrow irrigation practices in Fergana valley of Uzbekistan. Agric Water Manag 117:133–144

References

Asian Development Bank (ADB) (2013) Kyrgyz Republic: developing water resources sector strategies in Central and West Asia Financed by the Multi-Donor Trust Fund under the Water Financing Partnership Facility and ADB TASF-other sources. The consultants' report. ADB, Bishkek. https://www.adb.org/

Bobojonov I, Aw-Hassan A (2014) Impacts of climate change on farm income security in Central Asia: an integrated modeling approach. Agric Ecosyst Environ 188:245–255

Borkoshev J (2010) "Organic farming enriches your life" – example of the "BioFarmer" Cooperative in Kyrgyzstan. In: Conference proceedings, Astana, Kazakhstan, p 67. www.conference.organiccenter.kz. Accessed 27 Jan 2018

Bučienė A (2012) On the sustainability of different farming systems: conventional, organic and integrated systems. In: Jakobsson C (ed) Sustainable agriculture. Baltic University Programms, Uppsala University, pp 17–26

Bučienė A, Šlepetienė A, Šimanskaitė D, Svirskienė A, Butkutė B (2003) Changes in soil properties under high and low input cropping systems in Lithuania. Soil Use Manag 19(4):291–297

Bucknall J, Klytchnikova I, Lampietti J, Lundell M, Scatasta M, Thurman M (2003) Irrigation in Central Asia. Social economic and environmental considerations. The World Bank

Caspian countries extend sturgeon moratorium (2019) World fishing and aquaculture. https://www.worldfishing.net/news101/industry-news/caspian-countries-extend-sturgeon-moratorium. Accessed 30 Apr 2020

Celis D, De Pauw E, Geerken R (2007) Assessment of land cover and land use in central and West Asia and North Africa. Part 1. Land cover/land use – base year 1993. International Center for Agricultural Research in Dry Areas (ICARDA), Aleppo, Syria, 54 p

CIA The World Factbook (2017). https://www.cia.gov/library/publications/the-world-factbook/geos/uz.html. Accessed 15 July 2017

CIAT; World Bank (2018) Climate-Resilient Agriculture in the Kyrgyz Republic. CSA Country Profiles for Asią Series. International Center for Tropical Agriculture (CIAT). World Bank, Washington, DC, 28 p. www.worldbank.org https://www.csa_Profile_The_Kyrgyz_Republic.pdf. Accessed 28 Mar 2020

Conrad C, Lamers JPA, Ibragimov N, Löw F, Martius C (2016) Analysing irrigated crop rotation patterns in arid Uzbekistan by the means of remote sensing: a case study on post-Soviet agricultural land use. J Arid Environ 124:150–159

De Pauw E (2010) Agro-ecological zoning of the CWANA region. In: El-Beltagy A, Saxena MC (eds) Sustainable development in drylands. Meeting the Challenge of Global Climate Change. Proceedings, pp 335–348

Devkota M, Martius C, Lamers JPA, Sayre KD, Devkota KP, Gupta RK, Egamberdiev O, Vlek PLG (2013) Combining permanent beds and residue retention with nitrogen fertilization improves crop yields and water productivity in irrigated arid lands under cotton, wheat and maize. Field Crops Res 149:105–114

EBRD (2012) Strategy for Tajikistan. EBRD, London

Ermakhanov ZK, Plotnikov IS, Aladin NV, Micklin P (2012) Changes in the Aral Sea ichthyofauna and fishery during the period of ecological crisis. Lakes Reserv Res Manag 17:3–9

Ermolin I, Svolkinas L (2018) Assessment of the sturgeon catches and seal bycatches in an IUU fishery in the Caspian Sea. Mar Policy 87:284–290

FAO AQUASTAT for Kyrgyzstan between 2018–2013 FAO (2018) Rome. http://www.fao.org/nr/water/aquastat. Accessed 26 Jan 2020

FAO (2007) Fishery country profile: Kyrgyz Republic. FAO, FID/CP/KGZ, Rome

FAO (2010) Conservation agriculture spreads further a field. http://www.fao.org/agriculture/crops/agp-in-action/en/. Accessed 21 Jan 2018

FAO (2011) The State of the World's Land and Water Resources for Food and Agriculture (SOLAW) – managing systems at risk. Food and Agriculture Organization of the United Nations/Earthscan, Rome/London. http://www.fao.org/docrep/017/i1688e/i1688e.pdf. Accessed 17 Jan 2018

FAO (2012) Conservation agriculture in Central Asia: status, policy, institutional support, and strategic framework for its promotion. FAO Sub-Regional Office for Central Asia (FAO-SEC) http://www.fao.org/docrep/017/aq278e/aq278e.pdf. Accessed 26 Jan 2018

FAO (2013) Conservation agriculture in Central Asia. Status, policy and institutional support and strategic framework for its promotion, FAO Sub-Regional Office for Central Asia (FAO-SEC) Ankara. http://www.fao.org/docrep/019/i3275e/i3275e.pdf. Accessed 22 Jan 2018

FAO (2010–2014) Fisheries global information system (FIGIS) [online database]. Available: www.fao.org/fishery/figis/en. Accessed 26 Jan 2018

FAO (2016) Save and grow: maize, rice, wheat. Rome. http://www.fao.org/conservation-agriculture/case-studies/kazakhstan/en/. Accessed 1 May 2020

FAO Food and Agriculture Organization of the United Nations (FAOSTAT). Available online: http://faostat3.fao.org/faostat-gateway/go/to/home/E. Accessed 25 Nov 2018

FAOSTAT Land use. http://www.fao.org/faostat/en/#data/RL/visualize. Accessed 30 Mar 2020

FAOSTAT Live Animals. http://www.fao.org/faostat/en/#data/QA/visualize. Accessed 30 Mar 2020

FAO The State of World Fisheries and Aquaculture (2016) Contributing to food security and nutrition for all, Rome, 200 p. http://www.fao.org/3/a-i5555e.pdf. Accessed 17 Jan 2018

Fitzherbert (2006) Country pasture/forage profiles Kyrgyzstan. FAO, Rome

Friedrich T, Derpsch R, Kassam A (2012) Overview of the global spread of conservation agriculture. Field actions science reports. Reconciling poverty eradication and protection of the environment. Special issue, 6. http://factsreports.revues.org/1941. Accessed 25 Feb 2018

Gomiero T, Giampietro M, Mayumi K (2006) Facing complexity on agro-ecosystems: a new approach to farming system analysis. Int J Agric Resour Govern Ecol 5(2–3):116–144

Gosudarstvennaya programma razvitya agropromyshlennogo kompleksa Respubliki Kazahstan na 2017–2021 gody (2017) [State program on the development of agroindustrial complex of Republik of Kazakhstan for 2017–2021] (2017) Astana. http://www.fao.org/fileadmin/user_upload/reu/europe/documents/compnew/Kaz_PDF5.pdf. Accessed 29 Mar 2020

Graham NA (2016) The Prospect for Regional Governance of Inland Fisheries in Central Eurasia. In: Freshwater, fish and the future proceedings of the global cross-sectoral conference. FAO of the United Nations/Michigan State University/American Fisheries Society, Rome/East Lansing/Bethesda, pp 333–338

Irrigatcya: Kyrgyzstan vyhodit na novyj vitok razvitya (2017) (Irrigation: Kyrgyzstan appears in the new round of the development) (2017). http://kabar.kg/news/irrigatciia-pravitel-stvo-kr-vykhodit-na-novyi- vitok-razvitiia/. Accessed 29 Mar 2020

Japan International Cooperation Agency (JICA) (2010) The study on regional development in Karakalpakstan in the Republic of Uzbekistan (Progress Report), vol 130. JICA, Chiyoda-Ku

Kazakhstan Agricultural Overview (2010) USDA Foreign Agricultural Service, Commodity Intelligence Report. https://ipad.fas.usda.gov/highlights/2010/01/kaz_19jan2010/. Accessed 26 Mar 2020

Kazakhstan: country fact sheet on food and agriculture policy trends (2017) FAO. http://www.fao.org/3/a-i7676e.pdf. Accessed 29 Mar 2020

Khalikov B (2010) New crop rotation systems and soil fertility. Nosirlik Yogdusi Publishing House, Tashkent

Khalikov B, Tillaev RS (2006) Practical recommendations on crop rotations in Uzbekistan. Uzbekistan Cotton Research Institute, Tashkent

Kienzler K (2010) Improving the nitrogen use efficiency and crop quality in the Khorezm Region, Uzbekistan. Ph.D. thesis, Rheinische Friedrich-Wilhelms-Universität Bonn, Bonn, Germany, 62 p

Kienzler KM, Lamers JPA, McDonald A, Mirzabaev A, Ibragimov N, Egamberdiev O, Ruzibaev E, Akramkhanov A (2012) Conservation agriculture in Central Asia – what do we know and where do we go from here? Field Crops Res 132:95–105

Klimov J (2010) Development of organic agriculture in Kazakhstan. Information on the Organic-Center. Conference proceedings, Astana, Kazakhstan, pp 31–33. www.conference.organiccenter.kz. Accessed 27 Jan 2018

Lerman Z (2007) Tajikistan: an overview of land and farm structure reforms. Discussion paper no. 2.08, The Hebrew University of Jerusalem. https://ageconsearch.umn.edu/bitstream/6050/2/dp080002.pdf. Accessed 25 Jan 2018

Lerman Z, Sedik D (2009a) Agrarian reform in Kyrgyzstan: achievements and the unfinished agenda. FAO Regional Office for Europe and Central Asia Policy Studies on Rural Transition No. 2009-1. http://www.fao.org/fileadmin/user_upload/Europe/documents/Publications/Policy_Stdies/Kyrgyzstan_en.pdf. Accessed 26 Jan 2018

Lerman Z, Sedik D (2009b) Sources of Agricultural Productivity Growth in Central Asia: the case of Tajikistan and Uzbekistan. FAO Regional Office for Europe and Central Asia Policy Studies on Rural Transition No. 2009-5. http://www.fao.org/3/a-aq337e.pdf. Accessed 25 Jan 2018

Lerman Z, Stanchin I (2005) Chapter 10: Agrarian reforms in Turkmenistan. In: Babu S, Djalalov S (eds) Policy Reforms and Agriculture Development in Central Asia. Springer. https://www.researchgate.net/publication/225927107_Agrarian_Reforms_in_Turkmenistan. Accessed 29 Apr 2020

Lerman Z, Pridhodko D, Punda I, Sedic D, Serova E, Swinnen J (2012) Turkmenistan. Agricultural sector review. FAO/EBRD, FAO, Rome

Lioubimtseva E, Henebry GM (2009) Climate and environmental change in arid Central Asia: impacts, vulnerability, and adaptations. J Arid Environ 73:963–977

Löw F, Prishchepov AV, Waldne F, Dubovyk O, Akramkhanov A, Biradar C, John PA, Lamers JPA (2018) Mapping cropland abandonment in the Aral Sea Basin with MODIS time series. Remote Sens 10:159. https://doi.org/10.3390/rs10020159

Mamilov A (2010) Determining risk agricultural practices of crop production in Central Asia – the first step towards quality and sustainability. Conference proceedings, Astana, Kazakhstan, pp 16–17. www.conference.organiccenter.kz. Accessed 26 Jan 2018

Meller S (2013) Silk and cotton– textiles from the Central Asia that was, Abrams, 336 p. https://books.google.lt/books?q=editions:ISBN1683355571&id=87vADwAAQBAJ&hl=lt. Accessed 27 Mar 2020

Mendikulova FJ (2008) Issyk-Kul – problem-ridden jewel of Central Asia. In: Sengupta M, Dalwani R (eds) Proceedings of Taal 2007: the 12th world lake conference: 2157–2160. http://www.Q-61.pdf. Accessed 10 Feb 2018

Mirzabaev A, Ahmed M, Werner J, Pender J, Louchaich M (2016) Rangelands of Central Asia: challenges and opportunities. J Arid Land 8(1):93–108

Mogilevskii R, Abdrazakova N, Bolotbekova A, Chalbasova S, Dzhumaeva Sh, Tilekeyev K (2017) The outcomes of 25 years of agricultural reforms in Kyrgyzstan. Leibniz Institute of Agriculture Development in Transition Economies. Germany. Discussion paper No 162

Novel Measurement and Assessment Tools for Monitoring and Management of Land and Water Resources in Agricultural Landscapes of Central Asia (2014). Lothar M, Abdulla S, Lischeid G (eds). Springer, Cham

Nurbekov A, Aksoy U, Muminjanov H, Shukurov A (2018) Organic agriculture in Uzbekistan: status, practices and prospects, FAO, Tashkent. http://www.fao.org/3/i8398en/I8398EN.pdf. Accessed 31 Mar 2020

Organic Centre of Kazakhstan. www.organiccenter.kz. Accessed 27 Jan 2018

Petr T, Mitrofanov VP (1998) The impact on fish stocks of river regulation in Central Asia and Kazakhstan. Lakes Reserv Res Manag 3:143–164. https://doi.org/10.1046/j.1440-1770.1998.00069.x

Plotnikov IS, Ermakhanov ZK, Aladin NV, Micklin P (2016) Modern state of the small (Northern) Aral Sea fauna. Lakes Reserv Res Manag 21:315–328. https://doi.org/10.1111/lre.12149

Pourkazemi M (2006) Caspian Sea sturgeon conservation and fisheries: past present and future. J Appl Ichthyol 22(Suppl. 1):12–16

Qadir M, Noble AD, Qureshi AS, Gupta RK, Yuldashev T, Karimov A (2009) Salt-induced land and water degradation in the Aral Sea basin: a challenge to sustainable agriculture in Central Asia. Nat Resour Forum 33:134–149

Qushimov B, Ganiev IM, Rustamova I, Haitov B, Islam KR (2007) Land degradation by Agricultural Activities in Central Asia. Climate Change and Terrestrial Carbon Sequestration in Central Asia, 137–146. https://www.researchgate.net/publication/279339126_Land_degradation_by_agricultural_activities_in_Central_Asia. Accessed 23 Feb 2018

Reddy JM, Jumaboev K, Matyakubov B, Eshmuratov D (2013) Evaluation of furrow irrigation practices in Fergana valley of Uzbekistan. Agric Water Manag 117:133–144

Royal Haskoning for GEF Agency of the IFAS (2001) Water and Environmental Management Project, Sub-component A1, National and Regional Water and Salt Management Plans, Regional Report No. 2, Phase III Report – Regional Needs and Constraints, Main Report (Draft) (WEMP)

Ruban GI, Khodorevskaya RP (2011) Caspian Sea sturgeon Fishery: a historic overview. J Appl Ichthyol 27(2):199–208

Satubaldina A (2019) Kazakh government to increase irrigated land area to 3.5 million hectares. Business, 3 January 2019. https://astanatimes.com/2019/01/kazakh-government-to-increase-irrigated-land-area-to-3- 5-million-hectares/. Accessed 27 Mar 2020

Savvaitova KA, Petr T (1995) Fish and fisheries in lake Issyk-Kul (Tien Shan), river Chu and Pamir lakes. http://www.fao.org/docrep/003/X2614E/x2614e10.htm#P22_1995. Accessed 11 Feb 2018

Sommer R, De Pauw E (2011) Organic carbon in soils of Central Asia–status quo and potentials for sequestration. Plant Soil 338:273–288. https://www.researchgate.net/publication/225342880_Organic_carbon_in_soils_of_Central_Asia-status_quo_and_potentials_for_sequestration/figures. Accessed 8 Aug 2017

Sommer R, Glazirina M, Yuldashev T (2012) Assessing the vulnerability of selected agro-ecosystems in Central Asia to threats resulting from climate change – production and productivity of wheat. Report of sub-component 3 of the ADB funded project, Adaptation to Climate Change in Central Asia and People's Republic of China. https://www.CA_CC Crop Modeling final report_ENG (1).pdf. Accessed 10 Nov 2018

Sommer R, Glazirina M, Yuldashev T, Otarov A, Ibraeva M, Martynova L, Bekenov M, Khalov B, Ibragimov N, Kobilov R, Karaev S, Sultonov M, Khasanova F, Esanbekov M, Mavlyanov D, Isaev S, Abdurahimov S, Ikramov R, Shezdyokova L, De Pauw E (2013) Impact of climate change on wheat productivity in Central Asia. Agric Ecosyst Environ 178: 78–99

Spoor M (2006) Land, markets and rural poverty in the CIS-7. Paper presented at the international conference 'Land, poverty, social justice and development', 9–14 January, Institute of Social Studies, The Hague

Spoor M (2007) Rural Poverty, Agrarian Reform and the Role of the State in Rural Growth and Poverty Reduction in Central Eurasia Ten Propositions regarding Agricultural Development and Rural Poverty Reduction. Technical consultation meeting, FAO Regional Office for Central Asia (FAO/SEC) Ankara, Turkey, 11–13 July 2007. http://www.fao.org/fileadmin/user_upload/Europe/documents/Publications/PaperEurasia_en.pdf. Accessed 26 Jan 2018

Status of the world soil resources (2015) FAO Main report, prepared by Intergovernmental Technical Panel on Soils (ITPS). http://www.fao.org/3/a-i5199e.pdf. Accessed 1 July 2017

Sultanova M (2010) Perspective of production and processing of bio-cotton in Tajikistan. In: Conference proceedings, Astana, Kazakhstan, pp 57–59. www.conference.organiccenter.kz. Accessed 26 Jan 2018

Swanson B, Meyer E, van Weperen W (2011) Strengthening the pluralistic extension and advisory system in Tajikistan. Report on the MEAS Rapid Scoping Mission. Second draft submitted to USAID Mission/Dushanbe, November 25, 2011

The State of Food Insecurity in the World (2008) FAO, Rome. http://www.fao.org/docrep/011/i0291e/i0291e00.htm. Accessed 1 July 2017

The United Nations Economic Commission for Europe (UNECE), Union Water Initiative (EUWI) MoAFIM (2015) Modern irrigation technologies and possibility of their application in Kyrgyzstan. National policy dialogue on integrated water resources management in Kyrgyzstan. Bishkek. https://www.oecd.org/env/outreach/KG_study_irrigation.pdf. Accessed 1 Apr 2020

Timirkhanov S, Chaikin B, Makhambetova Z, Thorpe A, van Anrooy R (2010) Fisheries and aquaculture in the Republic of Kazakhstan: a review. FAO Fisheries and Aquaculture Circular No. 1030/2. Rome, FAO. http://www.fao.org/3/i1596e/i1596e01.pdf. Accessed 1 Apr 2020

Turkmenistan agricultural sector review (2012) FAO Investment Centre. http://www.fao.org/3/a-i2911e.pdf. Accessed 26 Mar 2020

Tyan VS, Kosanov SO, Kogay LY (2005) Adapting vegetable crops to the extreme conditions of Priaralye. In: Kalb TJ, Mavlyanova RF (eds) Vegetable production in Central Asia: status and perspectives. Workshop proceedings 9–12 June 2003, Almaty, Kazakhstan, pp 50–52

Umarov Ch (2019) Sektor zhivitnovodstva v Tadzhikistane: problems ustojchivogo i sbalansirovannogo razvitya (Livestock sector in Tajikistan: Problems of sustainable development), Discussion paper # 190, Leibniz Institute of Agricultural Development in Transition Economies (IAMO). https://www.iamo.de/fileadmin/documents/dp190.pdf. Accessed 1 Apr 2020 (in Russian)

Umirsakov SI, Tautenov IA, Dschamantikov HD, Tochetova LA, Wilhelm MA, Schermagambetov K, Baibosinova SM, Abildajeva S (2011) Recommendations on conduction of spring field campaign in Kyzyl-Orda Oblast; KazAgro Innovazia: Astana, Kazakhstan (in Russian)

USDA FAS (2011) Uzbekistan Livestock report 2011 Date 27 April 2011. GAIN report UZ1101

Van Berkum S (2015) Agricultural potential and food security in Central Asia in the light of climate change 2 Issue, LEI Wageningen UR. http://edepot.wur.nl/351542. Accessed 25 Nov 2018

Vanderberg R (2018) Agriculture Minister Unveils New National Program for Livestock. GAIN Report, USDA Foreign Agricultural Service, KZ-06, 6/13/2018. Accessed 25 Nov 2018

Wolf SA, Allen TFH (1995) Recasting alternative agriculture as a management model: the value of adept scaling. Ecol Econ 12:5–12

World Bank (2001) Aral Sea Basin Program, Subcomponent A1, Report of the National Working Group of the Republic of Uzbekistan: Functional Relationship Between Salinity and Yields in Agriculture

Xenarios S, Laldjebaev M, Shenhav R (2019) Agricultural water and energy management in Tajikistan: a new opportunity. Int J Water Resour Dev. https://doi.org/10.1080/07900627.2019.1642185

6 Environmental Challenges in Globalization and Post-Modern Times

Abstract

Glaciers are the wealth of CA countries, and particularly of Tajikistan and Kyrgyzstan, where they are mostly located. Research on glacier mass balance confirms that most glaciers in the region are either receding or standing still. According to the researchers' forecast, by 2025 the territories of glaciers in Tajikistan and Kyrgyzstan will be reduced by 30–40%, and this will accelerate and make more frequent landslides, local floods, avalanches or mud flows, in the mountain and close to them areas. The rational use and development of water and hydro-energy resources in Amu Darya and Syr Darya basins with trans-boundary aspects and tasks have been and still are the most complex regional challenges before all CA countries. The climate warming which is primarily depended on the global atmosphere circulation changes brings new challenges for the natural and anthropogenic systems to understand their changes and find the most appropriate adaptation measures. The recent level of Caspian Sea rise seems to be caused by the combination of higher Volga discharges and lower evaporation from the Caspian Sea surface, in relation with the dominated atmosphere circulation patterns. The Aral Sea's Southern and Eastern parts, contrary, continue to desiccate with remained basin in the west and the recovering the Small Aral in the north. Concerning the future fate of this sea, different scenarios are discussed, from realistic to very pessimistic. Some regions like the Aral and Semipalatinsk regions of Kazakhstan are considered as the territories of ecological crisis, where the balance between natural processes and living components of ecosystems, including flora, fauna and the humans was affected mostly. In Kyrgyzstan, about 61% of all ecological accidents and disasters occur due to the natural processes, caused mainly by global climate. The CA countries are opened to the international partnership, while searching their paths towards higher region eco-sustainability.

Keywords

Glacier monitoring · Water resources management · Climate warming · Natural catastrophes · Eco-sustainability

6.1 Global Climate Warming and Glaciation; Glaciers as Indicators of Climate Warming

The glaciers not only retain water supply, permafrost, but also regulate river flows in CA region and the entire Aral Sea basin (Kayumov 2013). The reduction in glacier area and volume are considered to be the most reliable indicators of worldwide climate warming (IPCC 2013). Most of the researchers agree on the general trend of glacier mass loss, including acceleration since the 1970s (e.g. Sorg et al. 2012; Farinotti et al. 2015); however, as to seasonal changes, the studies disagree (Hoelzle et al. 2017). According to the Glacier inventory made in Soviet Union in the eighties of twentieth century and recent monitoring data, within the Tien Shan mountains, there are 15,953 glaciers with a total area of 15,416 km^2 and a total volume of 1048 km^3 (Katalog… 1973; World glacier… 2016). The four-fifths area of the Tien Shan glaciers is found in Kyrgyzstan and Kazakhstan (Fig. 6.1)

The largest glacial areas extend around Khan Tängiri and Victory peaks and the Eren Habirg Mountains. The largest among glaciers of the Tien Shan is Engil'chek (Inylchek) Glacier, which is approximately 60 km long (Bruk et al. 1999): it descends from the western slopes of the Khan Tängiri massif and branches into numerous tributaries (Fig. 6.2).

That glacier is included in the research program, conducted by U.S. Geological Survey (USGS) with aim to study the geochemical and isotopic content of precipitation, snow,

I. Jelen et al., *The Geography of Central Asia*, World Regional Geography Book Series,
https://doi.org/10.1007/978-3-030-61266-5_6

Fig. 6.1 Subdivision of the former Soviet Union (FSU) part of the Tien Shan (dotted lines) into: W—Western, I—Inner, N—Northern and C—Central Tien Shan, its main sub-ranges, and the study area of glaciers (black rectangle). (Source: Niederer et al. 2008)

Fig. 6.2 Glacier the Northern and Southern Inylchek. (Source: NASA http://earth.jsc.nasa.gov/scripts/sseop/QuickView.pl?directory=ESC&ID=ISS021-E-5654 Transferred from ru.wikipedia to Commons, Public Domain, https://commons.wikimedia.org/w/index.php?curid=12405170)

ice, and runoff samples from high-mountain glaciers in the middle latitudes of CA and the United States (Cecil et al. 2004; Water resources: state 2000).

Glacial action in the Tien Shan apparently is decreasing; most glaciers are either receding or standing still (Bruk et al. 1999). The glacier loss magnitude varies from one mountain and range location region to another.

The glaciers from the Zailiyskiy Alatau range of the northern Tien Shan mountains lost nearly 2 km^3 of ice a year between 1955 and 2000. One of the oldest glaciers monitoring in CA region and in the World began at the end of nineteenth century, in Kazakhstan, at the Tuyuksu glacier, 30 km south of Kazakhstan's largest city, Almaty. Since 1923 it has receded by nearly a kilometre, losing about 51 million cubic metres of ice (Aizen et al. 1995, 2007; Oledenenye gornyh territory…2015). Glacier degradation maximum was registered in the mid-1970s, decreased thereafter, and, starting from the mid-1990s, the annual mass balance became less negative (Severskiy et al. 2006, 2016). The water flow accumulated in glaciers is important during years of little precipitation, and at the end of summer when seasonal snow cover has mostly thawed. Thus, glaciers act as buffers, operating as flow regulators and providing security during periods of low flow (Kotlakov, Seversky 2006).

The research in the Balkhash-Alakol basin has shown that the rates of glacier reduction are in close correlation with changes in accumulation (winter mass balance) and summer temperature.

Pamir Mountains mainly stretch out in Tajikistan. According to Middleton and Thomas (2008), "pamir" is a geological term: it defines a flat plateau or U-shaped valley surrounded by mountains. A pamir forms when a glacier or ice field melts leaving a rocky plain and lasts until erosion forms soil and cuts down normal valleys. Glaciers of the Pamir high mountains supply water to over 60 million people in Tajikistan, Uzbekistan, Turkmenistan, Afghanistan and Xinjiang province of the People's Republic of China. By glacio-climatic regime and landscape, Pamir is divided for five major regions: Pamir-Alai, Western Pamir, Central Pamir, South-Eastern and Eastern Pamir (Aizen 2011). Western Pamir is receiving highest precipitation (up to 2000–2500 mm) but more than 60% is going to melt water. The dendrite type of glaciers is concentrated in the Central Pamir (Fig. 6.3).

Fedchenko Glacier which has been fluctuated in its volume and mass through the twentieth century, still remains the world's largest alpine glacier outside of the polar regions, 72 km long, 714 km^2 total area and 900–1000 m ice thickness at elevation over 5000 m (Aizen 2011).

Glaciers are the wealth of CA countries, and particularly in Tajikistan and Kyrgyzstan, where they are mostly located: they occupy more than 8400 km^2 in Tajikistan (about 6% of the country area) and 8169.4 km^2 in Kyrgyzstan (4.2% of the republic territory) (Tajikistan 2000; Kayumov 2013). Tajikistan has seen a seasonal decline in average rainfall levels in most regions, especially the west, north and east. In accordance with climate features, the projected trends of glaciers degradation rate in Tajikistan foresee that by 2050 the area of glaciers in country will decrease from 20% to 50%, at the same time many of small glaciers (<1 km^2) likely will be disappeared (Tajikistan..0.2003). The trend of climate warming leads to stable intensive reduction of glacier surfaces in Kyrgyzstan too. According to forecast, by 2025 the territories of glaciers will be reduced by 30–40% (Kyrgyzstan lost… 2017). At the same time other regions, which are neighbouring the largest mountain mass of CA, also experience seasonal impact of glacier melting process on climate change. For example, in the Aral Sea basin, where the local impact on microclimate is observed due to the drying of the sea and landscapes, and global warming impact. Glacial melt in the Pamir and Tien-Shan ranges is projected to increase, initially increasing flows in the Amu Darya, Syr Darya and Zeravshan systems for a few decades, followed by a severe reduction of the flow as the glaciers disappear (Glantz 2005). Field data indicate that significant changes in the seasonality of glacial flows have already occurred as a result of warming (Braithwaite 2002; Lioubimtseva & Henebry 2009). Later a risk towards the shortage of water, adversely impact ecosystems, food security and human health might occur (Kayumov 2013).

Many parts of CA are exposed to a variety of potential natural hazards related to climate warming. The accelerated glacier melting provokes more frequent landslides, local floods, avalanches or mud flows which may cause serious damages to the infrastructure and require considerable reconstruction and maintenance work particularly in remote mountain areas (Squires 2012). In general environmental problems in any world region, dependent or independent of climate change, lead the countries towards finding of the most suitable decisions for the situation to be improved. Human actions through many developing economic sectors also are in majority unsustainable. In CA region many management failures can be explained by continued governance models which were inherited from strongly centralized communistic-era planning system and legal weakness of poorly funded state institutions (Rahimon 2012).

6.2 Situation With Water Resources in CA Countries: General Overview and Forecast for Near Future

Water is an essential component of nearly all the mitigation and adaptation strategies—from carbon storage in terrestrial ecosystems to emerging clean energy technologies, to adapting to extreme weather events (White 2018).

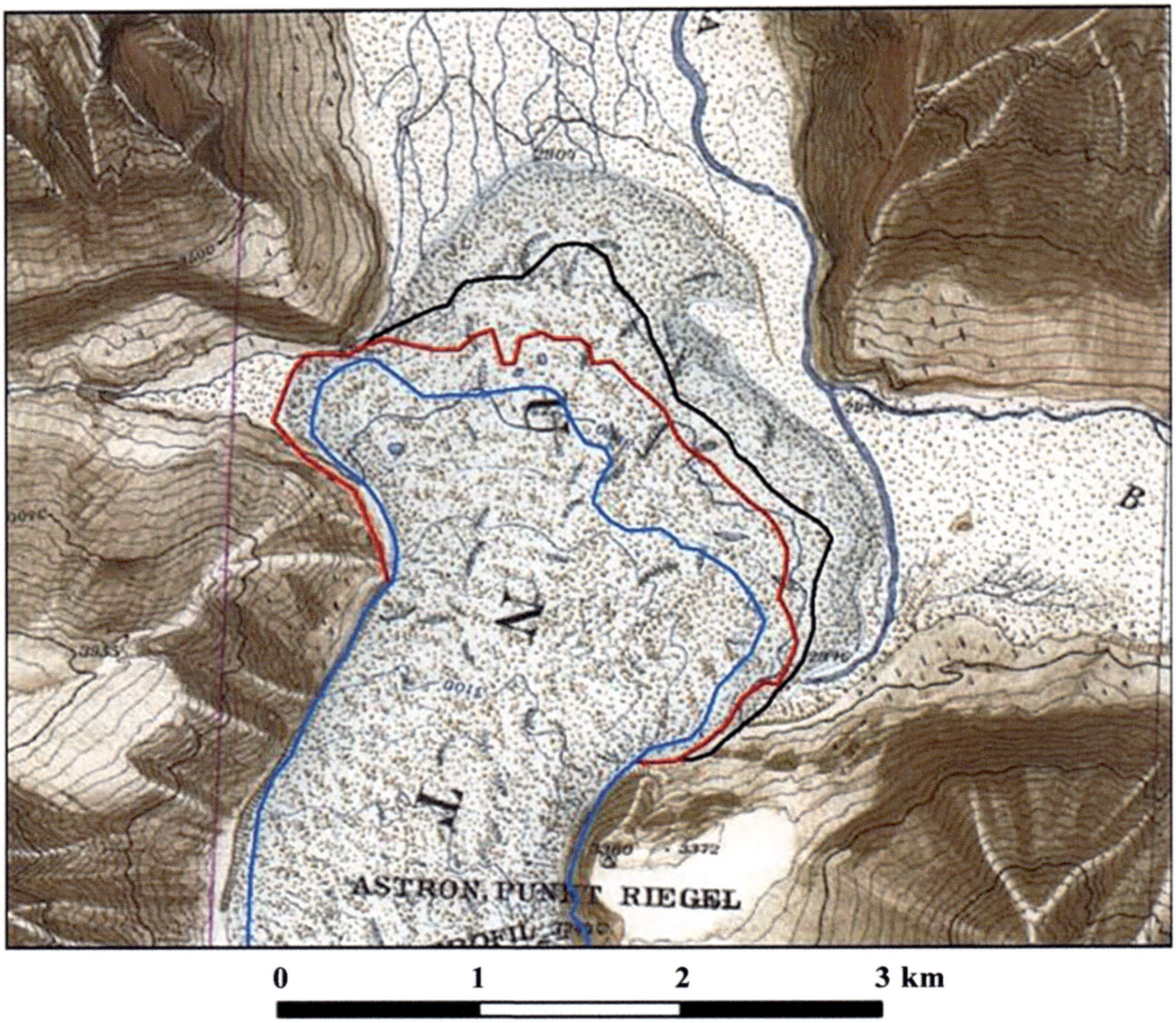

Fig. 6.3 Changes in Glacier Fedchenko area between 1928 (map in background) and 1958 (black line), 1980 (red line), and 2009 (blue line). (Source: Aizen (2011) https://www.researchgate.net/publication/233731981_Pamir_glaciers)

Humans during the twenty-first century continue to explore various water resources in different spheres of their activities. Water is identified as the number one priority for most of the Intended Nationally Determined Contributions (INDCs) adaptation actions to the climate change and is directly or indirectly related to all other priority areas (Fig. 6.4). The INDCs mean the measures, known as nationally determined contributions to mitigate and adapt to climate changes which are to be reviewed every 5 years according to the Paris Agreement (United Nations…2020).

World energy consumption in the water sector is expected to double through 2040, as a result of increasing desalination of seawater (IEA 2016), thus the more efficient and sustainable use of water resources remains as the most important task for the next decades.

6.2.1 Main Rivers and Their Basins in CA

CA has a long history of managing water resources because of their importance to the economic development, particularly of irrigated agriculture and hydro-energy. Previously economic development in all CA countries was governed by the USSR centralized planning, but since independence, countries have developed their own strategies. However, in the reality the access, usage and maintenance of surface water resources, including rivers, in CA are still conflicting,

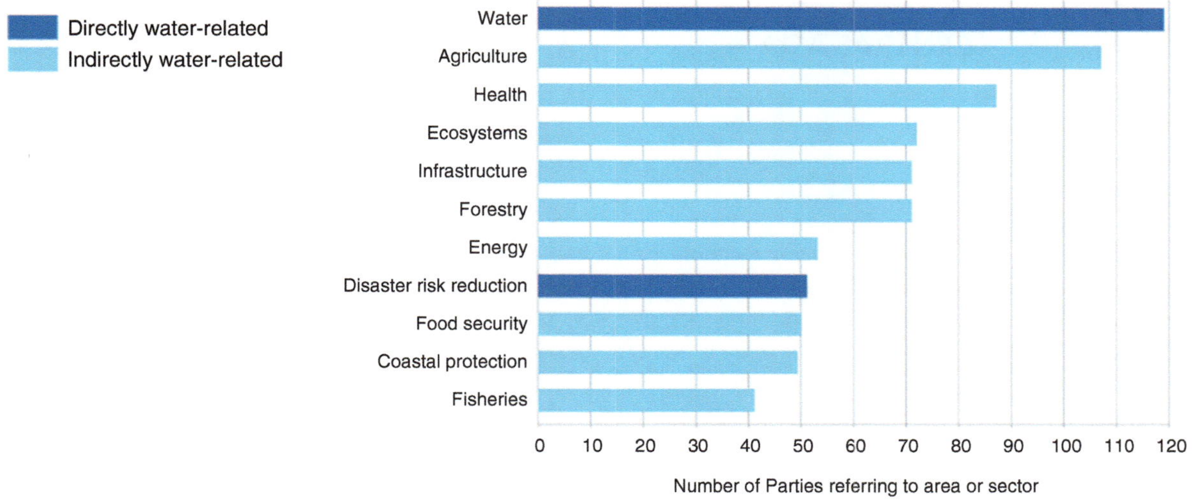

Fig. 6.4 Priority areas and sectors for adaptation actions identified in the adaptation component of the communicated INDCs; Party is considered as one of the UNFCCC (United Nations Framework Convention on Climate Change) countries. (Source: United Nations World Water Development Report 2020 https://unesdoc.unesco.org/ark:/48223/pf0000372985.locale=en)

Table 6.1 Main rivers and their basins in CA

Basin and sub-basins	Total area (km²)	Recipient	Riparian countries
Ob	2,972,493 from that KZ 734,543	Arctic Ocean (Kara Sea)	RU (MN, CN, KZ-via Irtysh)
Irtysh	1,643,000	Ob	RU, KZ, MN, CN
Tobol	426,000	Irtysh	RU, KZ
Ishim	176,000	Irtysh	RU, KZ
Amu Darya	n/a	Aral Sea	AF, KG, TJ, UZ, TM
Surkhan Darya	13,500	Amu Darya	TJ, UZ
Kafirnigan	11,590	Amu Darya	TJ, UZ
Pyanj	113,500	Amu Darya	AF, TJ
Vakhsh	39,100	Amu Darya	KG, TJ
Zeravshan	n/a	Desert sink	TJ, UZ
Syr Darya	n/a	Aral Sea	KZ, KG, TJ, UZ
Naryn	n/a	Syr Darya	KG, UZ
Kara Darya	28,630	Syr Darya	KG, UZ
Chirchik	14,240	Syr Darya	KZ, KG, UZ
Chu	62,500	Desert sink	KZ, KG
Talas	52,700	Desert sink	KZ, KG
Ili	413,000	Lake Balkhash	CN, KZ
Murgab	46,880	Desert sink	AF, TM
Tejen	70,260	Desert sink	AF, IR, TM

Source: Mueller et al. (2014)
Basic data and table from UNECE (2007), modified, n/a—not available; country abbreviations: AF—Afghanistan, CN—China, KZ—Kazakhstan, KG—Kyrgyzstan, MN—Mongolia, RU—Russia, TM—Turkmenia, TJ—Tajikistan, UZ—Uzbekistan

causing disputes both at the national/regional and international level. More than 6000 rivers (over 10 km long) originate in the mountains of CA region, including the Amu Darya and the Syr Darya Rivers (Integrated water resources…2014). As to the main rivers, they are mostly ending in the endorheic catchments except of the northern part of Kazakhstan, which is drained to the Arctic Ocean by Ob and its tributaries (Table 6.1 and Fig. 6.5).

The Ob River with its tributaries within Kazakhstan territory makes about 60% of the total country's outflow. The largest area of Kyrgyzstan belongs to the Syr Darya basin (62%), also there are smaller endorheic basins of the Chu, Talas and Assa, the large Issyk-Kul Lake and a group of small mountainous rivers drained to China. In Tajikistan according to the area distinguishes the Amu Darya basin (84%) with Syr Darya and the Zeravshan as followed. In Turkmenistan the Amu Darya basin is the largest (68%) and few others like Murgab and Tejen basins make about 32% of total area. In Uzbekistan which belongs to the Aral Sea basin, the Amu Darya sub-basin (86%) and Syr Darya sub-basin (14%) are the main (FAO 2012; Mueller et al. 2014; UNDP 2004).

The map shows that almost all territory of Uzbekistan and whole Turkmenistan and Kyrgyzstan with large areas in Kazakhstan and Tajikistan are marked in red as regions of extreme water stress. But climatic data prove that even the largest area of the CA region belongs to the region of extreme air temperature anomaly. With climate warming the droughts have become more frequent during the summer and autumn periods, especially downstream of the Amu Darya River and around the Aral Sea. Some of them can be extremely severe according to many impacts and consequences, like the drought of 2000–2001 (CACILM 2006).

There are many natural lakes in the mountainous areas and ravines of CA of various origins. The majority of large and deep lakes are of tectonic origin (Issyk-Kul, Song-Kel,

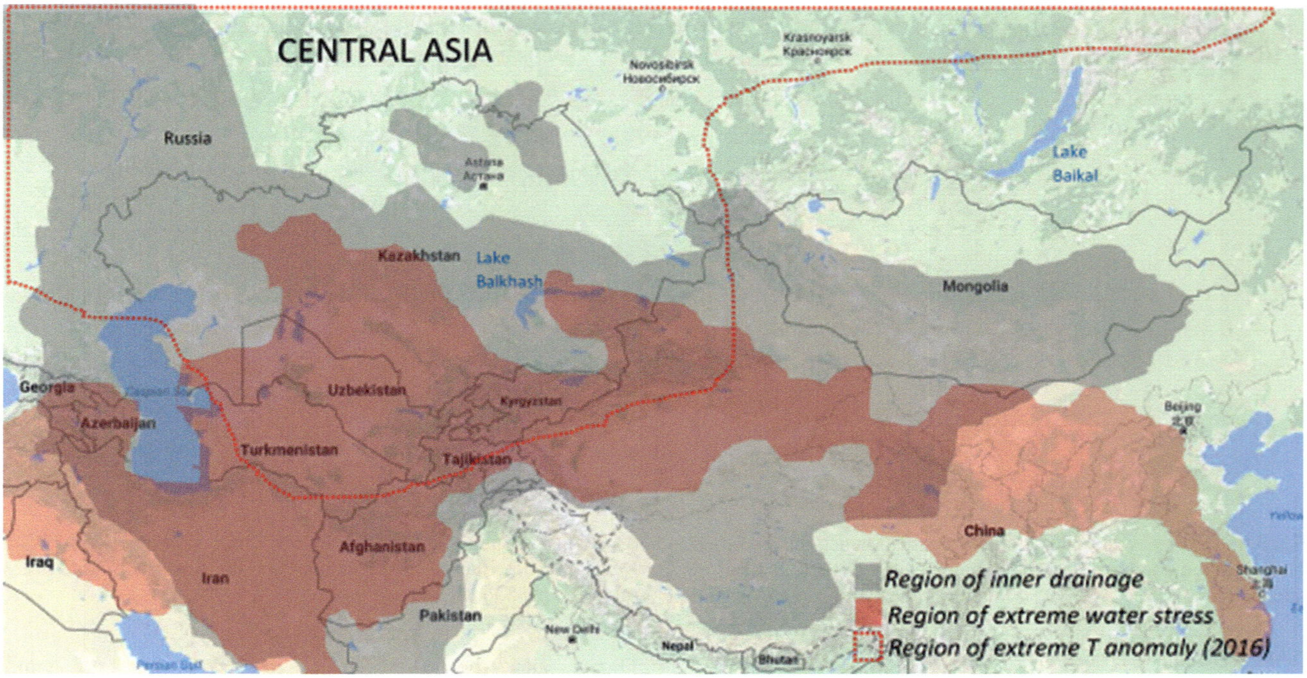

Fig. 6.5 Map illustrating the extensive regions of inner drainage (grey shade), regions of extreme water stress in CA, where water withdrawal is approximately equal to water availability [red shade. (Source: Karthe et al. 2017 after Smakhtin et al. 2004; WWAP 2012)] and regions of large temperature anomaly. Delineation by red-dotted line shows 2.5 °C deviation from pre-industrial temperatures in an example from January to June 2016 according to GISTEMP 2016

Chetir-Kel, Karakul and Sarichelek). Lakes originating from landslides, glacial and karst lakes are mostly distributed in the mountainous areas, which are usually freshwater or slightly saline, depending on the quality of in-flowing water. The majority of lakes located in the lowlands are shallow, often surrounded by saline soils on sandy background, with low shores and dense vegetation of reeds and rushes (CA WaterInfo 2011).

Thus, CA region remains as a vulnerable region in the light of global climate changes and the access and usage of water resources. Climate warming and the related water challenges are not only complex and interconnected, they are also intergenerational (United Nations..0.2020), that means the responsibility of recent generations and their decisions for the future generations coming.

6.2.2 Aral Sea Basin Water Resources

The water resources in the Aral Sea Basin consist of renewable surface and groundwater, as well as return water from anthropogenic use (wastewater and drainage water). The Sea water level started do decrease from 1960 due to the overuse the water for anthropogenic purposes, mainly irrigation, and due to the global climate warming, but from 2006 it stabilized in the northern Small Aral (Fig. 6.6).

The volume of sea water has dropped in all parts of Aral Sea from 1960, but mostly severe in the Big Aral from 1986, and in the Eastern Sea from 2006.

There are two major river sub-basins located in the Aral Sea Basin: the Syr Darya in the north, and the Amu Darya in the south with the Zeravshan River, and the Tejen-Murgab Rivers (http://www.cawater-info.net/aral/water_e.htm#lake). The total mean annual flow of all rivers in the Aral Sea basin is an estimated 118.43 km^3 (Fig. 6.7a), which varies in the large range between wet and dry years (The Aral Sea transboundary… 2013). The second highest annual flow is characteristic to the Amu-Darya, followed by the Syr Darya. As to the countries, the largest area belongs to the territory of Turkmenistan (almost 27% of the total basin area), followed by Uzbekistan (24.5%) and Kazakhstan (almost 20%) (Fig. 6.7b). Afghanistan is also included in this calculation.

The anthropogenic impact to water flow regime is not significant in the upstream zone of the main rivers, but in the downstream zone runoff regime is strongly influenced by large dams and artificial reservoirs. Because of significant withdrawal of water due to the agriculture and hydro-energy needs, also losses in the desert areas, the river runoff reaching the Aral Sea is significantly reduced under the impact of both anthropogenic and natural factors.

In the extreme dry years, the runoff corresponds to less than 10% for Amu Darya and less than 5% for the Syr Darya

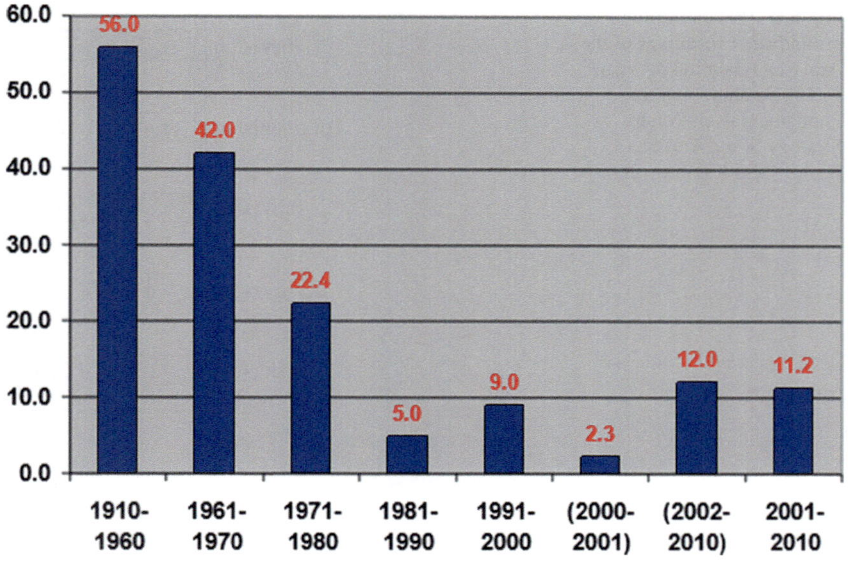

Fig. 6.6 Estimated average annual inflow to the Aral Sea 1910–2010 (in km³). (Source: Micklin 2016)

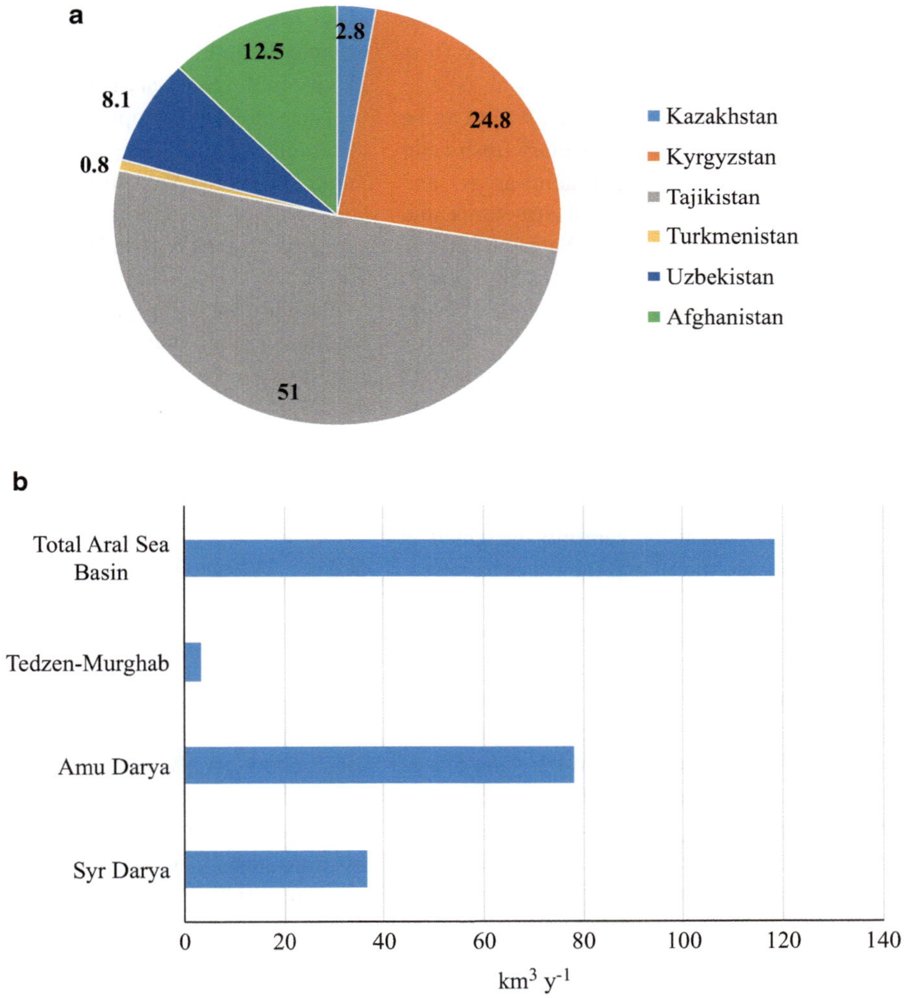

Fig. 6.7 Distribution of mean annual runoff in the Aral Sea Basin in km³/year (**a**) and in % by country (**b**). (Source: The Aral Sea transboundary… 2013)

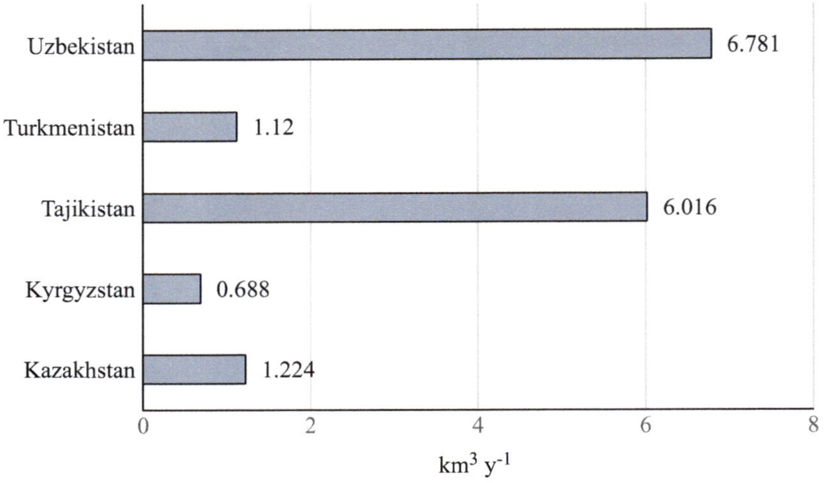

Fig. 6.8 Proven available groundwater resources of the Aral Sea Basin in km^3 year^{-1} by CA country. (Source: Fundamentals of Water Strategy of the Aral Sea Basin, 1996 Rahimov 2009)

(The Aral Sea transboundary… 2013). Observations show that climatic changes in regions where the essential part of river flow is at present formed from melt waters, the maximum flow values will move from spring to winter, whereas the majority of precipitation will fall there in watery form because of high air temperatures (Khamzayeva et al. 2009).

The groundwater resources of the Aral Sea basin can be divided into two parts: the natural flow or primary freshwater from the mountainous and water catchment areas, and groundwater filtrated from hydro-technical structures and irrigated land (secondary freshwater) (CA WaterInfo 2011). The proven available groundwater resources (in km^3 year^{-1}) in the Aral Sea basin according to the CA countries were estimated in 1990–1994 and showed in Fig. 6.8.

The largest groundwater resources were determined in Uzbekistan and Tajikistan, and the least in Kyrgyzstan. Estimated reserves of CA countries (Afghanistan not included) are about 31.1 km^3, of which 14.7 km^3 belong to the Amu Darya basin and 16.4 km^3 to the Syr Darya basin. The possible exploitation of groundwater has to be carefully weighted in order not to disturb the surface runoff and the water balance.

Return water constitutes a high proportion of water resources in the basin and are also a major source of pollution. In recent years, the annual mean values of return flows, consisting of drainage and wastewater from irrigation, industry and municipal users have varied between 28.0 and 33.5 km^3 with a larger amount coming from the Amu Darya basin. About 95% of the total return water is made by the drainage water and about 5% by untreated domestic and industrial wastewater. The large percentage of drainage water returned demonstrates the ineffective usage of water at irrigation. It might be re-used for irrigation, but the problem is not appropriate chemical composition (higher salinity, concentration of different agrochemicals and pesticides). Nowadays only about 15% of total return flows are re-used and more than 55% returned to rivers, often ending in natural depressions, from which the water evaporates.

More than 80 artificial water reservoirs in total were constructed in the Aral Sea Basin with the aim of better local water supply and storage. The aggregate capacity of these water reservoirs exceeds 60 km^3, of which about 44 km^3 is useable, including 17 km^3 in the Amu Darya river basin and 27 km^3 in the Syr Darya river basin.

There are 45 hydropower plants in the Aral Sea Basin of different capacity. The biggest plants are the Nurek (in Tajikistan on the Vakhsh River), with a capacity 2700 MW, and Toktogul (in Kyrgyzstan on the Naryn River), with a capacity 1200 MW (CA WaterInfo 2011; Khamzayeva et al. 2009).

Pressure on the region's water resources is expected to continue rising; thus, it is of the great importance of CA countries jointly managing their shared water resources, and developing techniques to use water more efficiently for different economic sectors, particularly for irrigated agriculture and hydro-energy (Russell 2018).

6.3 Water Resources by Country

6.3.1 Kazakhstan

Kazakhstan is the largest country by area in CA region, but water resources are distributed unevenly within its different parts: about 34.5% of resources are located in the eastern, 4.2%—in the northern, 2.6%—in the central, 24.1%—in the south-eastern, 21.2%—in the southern and 13.4%—in the western part. The total resources of fresh water make 524 km^3, 80 km^3 of these resources are concentrated in glaciers, 190 km^3—in lakes, 100.58 km^3—in rivers and from these 55.94 km^3 is formed within country, while the rest 44.64 km^3 forms within other countries. The total reserve of groundwater makes 15.6 km^3 which is partly used (5.6 km^3) for drinking and homeland needs, 0.8 km^3 is used for technical-farming-manufacturing purposes, 0.08 km^3 used

Table 6.2 River runoff in million cubic metre formed in the territory of Kazakhstan and in transboundary countries

Years	Total runoff	River runoff formed in the territory of Kazakhstan	River runoff formed in the transboundary countries
2010	143,600	77,200	66,400
2011	101,800	57,300	44,500
2012	92,700	49,200	43,500
2013	121,100	75,000	46,100
2014	109,800	63,500	46,300
2015	107,400	67,700	39,700

Source: http://doklad.ecogosfond.kz/водные_ресурсы

for farming and 0.09 km^3 is used for irrigation (Nacionalny doklad… 2016d).

The river runoff in the territory of Republic of Kazakhstan makes the larger part of the total runoff (Table 6.2).

Within Kazakhstan there are eight river basins, among them, the largest are Ertiskij, Balhash-Alakol, Aral-Syr Darya and Zhaiyk-Caspian (about 90% total water resources).

About 19.9 km^3 annually flows from China, 13.7 from Uzbekistan, 7.4 from Russia and 3.0 km^3 from Kyrgyzstan (Nacionalny doklad... 2016d).

The research on five rivers runoff of Zailiyskiy Alatau in Kazakhstan has shown that the total runoff depended on glaciers flow (Severskiy and Kogutenko 2016). The share of glacier flow due to ice melting in total runoff varied from 0.7 to 20.9%, depending on the time period in five rivers flowing from Zailiyskiy Alatau. The share of complete glacier flow in total runoff ranged from 2.3 till 69.8% depending on the time period in the same rivers. On average for 53 years the runoff due to year ice melting made 9.4%, and the complete glacier runoff accounted for 31.5% of the total annual runoff. The Talgar river total runoff was formed mostly by both ice melting and complete glacier flow.

There are about 48,000 of lakes in Kazakhstan, in total they occupy 4500 km^2. The majority of lakes are distributed in the forest-steppe zone and in the north of this zone. Prevailing lakes have no connection with inner seas.

6.3.2 Uzbekistan

Of the total 18,000 brooks, sais (mountain rivers) and rivers are there in Uzbekistan, more than 55% of them are in the Amu Darya basin and about 28% are in the basin of the Syr Darya. There are 525 different sized lakes in Uzbekistan, 53 reservoirs with a total useful volume of water of 15 km^3. In the basins of the upper reaches of the rivers Surkhan Darya, Kashka Darya and Chirchik there are 525 glaciers with a total area of 154.2 km^2. The estimated regional reserves of groundwater are the largest among CA countries. Most of groundwater deposits have a rather essential hydraulic interrelation with surface waters, thus it is estimated and approved which amount of groundwater reserve can be withdrawn from each aquifer. The dynamics of irrigated area in country from 1913 to 2010 has shown that after 2000 the area under the irrigation in Uzbekistan stabilized. A total length of the inter-farm and on-farm irrigation networks amount to 27,619.7 km and 167,378.8 km, respectively. 62% of inter-farm canals and 79.5% of on-farm canals have an earth channel. Pumping stations are lifting water for irrigation over 2.2 million ha in whole country (Water resources… 2011).

6.3.3 Turkmenistan

The main source of water for all human activities in Turkmenistan is the Amu Darya River, flowing from the Pamir and Tien-Shan Mountains along the entire length of the north-eastern border with Uzbekistan to the tragically dying Aral Sea (Kostianoy and Kosarev 2010; Zonn 2012). It brings about 84% of total fresh water resources in the country, which is equal about 36% of river's total runoff (Stanchin and Lerman 2008). Water intake from Amu Darya River is supplemented with surface runoff from three other rivers—Murgab, Tejen and Atrek, as well as from small rivers and springs. The total groundwater reserves are among the least in the total CA region (Kostianoy et al. 2011). For the economic development freshwater resources in Turkmenistan must be combined with other saline water resources.

The sediments of Caspian Sea record a complex pattern of transgressions and regressions. The fluctuation of sea level (Figs. 6.9 and 6.10) reflects on the deltas of inflowing rivers.

These fluctuations are not only independent of eustatic changes, but seem to be also grossly out of phase with them (Rychagov 1977; Varushchenko et al. 1987; Klige and Myagkov 1992; Ignatov et al. 1993; Kaplin and Selivanov 1995, cited in Kroonenberg et al. 1997). They are strongly related to the main changes in precipitation in the Volga basin, which supplies 80% of its inflow, as well as changes in evaporation at the sea surface (TED Caspia 1992; Naydyonov 1992; Rodionov 1994; Malinin 1994). Research has confirmed that within the last 70,000 years sea level has been as much as 75 m above (Rychagov, 1977; Svitoch, 1991) and possibly even 80 m below the present one. According to Kroonenberg et al. (1997) in a more distant geological past changes might have been even more extreme. Another record on the levelling during the twentieth century shows that the Volga delta area is subject to uplift up to 3 mm year^{-1} (Lilienberg 1985, 1993).

Thus, the recent sea-level rise seems to be due to a combination of higher Volga discharges and lower evaporation from the Caspian Sea surface, in relation with changes in the persistence of south-western circulation patterns (Klige and Myagkov 1992; Rodionov 1994). But the cause of this phe-

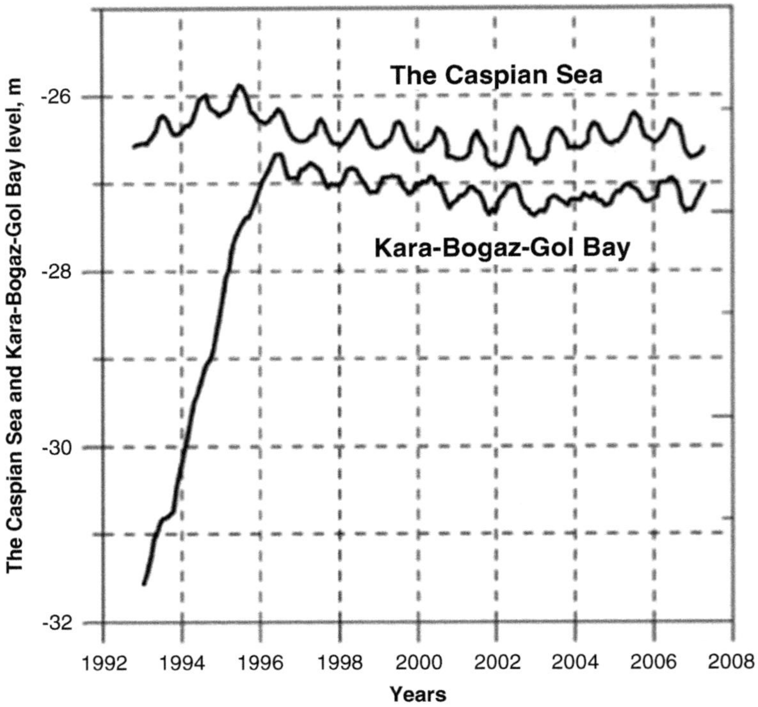

Fig. 6.9 Interannual and seasonal variability of the Caspian Sea (the upper line) in Kara-Bogaz-Gol Bay (lower line) level. (Sources: Kosarev et al. (2013) using the altimetric data of the TOPEX/Poseidon and Jason-1 satellites over the period from January 1993 to April 2007 according to the project "Surface water monitoring by satellite altimetry" LEGOS, France (http://www.legos.obs-mip.fr/soa/hydrologie/hydroweb/))

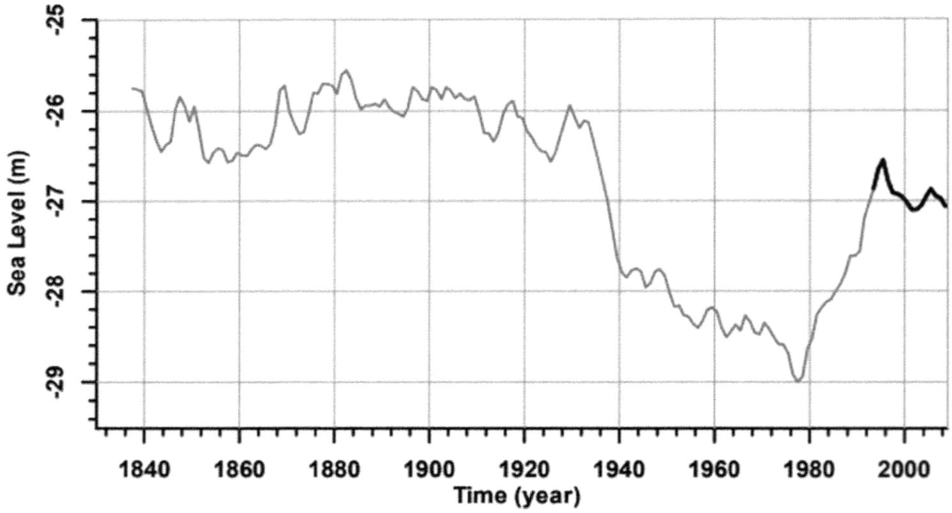

Fig. 6.10 Interannual variations of the Caspian Sea level measured by sea level gauges (grey line) and satellite altimetry (black line) from 1837 till 2009. (Source: Kostianoy et al. 2011)

nomenon is still a much-debated question (TED Caspia 1992; Naydyonov 1992).

According to Food and Agriculture Organization (FAO) calculations, Turkmenistan is the most wasteful user of water, not only in CA, but in the whole world, consuming six million litres per inhabitant per year. The Karakum Canal, which carries water to around half of the irrigated area in Turkmenistan, loses about one-third of its water due to the leak and poor management. Another example, the capital city Ashgabat consumes the same amount of water as Chicago, despite having just one-quarter of its population (Russell 2018). Still there are lots of space for changes in maintenance, exploration and saving of water resources.

6.3.4 Tajikistan

Tajikistan is the wettest country of the region, with 691 mm of average annual precipitation, from 100 mm in the south-east to 2400 mm at the Fedchenko glacier, in the middle of the country. About 947 rivers longer than 10 km, more than 1300 natural lakes together with more than 8000 km^2 of glaciers form about 51 billion cubic metre of water in the whole territory of the country (Russell 2018). But most of the surface runoffs (about 90%) are formed in the mountainous zone of Tajikistan, mainly within Amu Darya and Syr Darya river basins. The main runoffs come from Panji, Vaksh, Kafirnigan, Bartang and Zeravshan river basins as well. The irrigated agriculture with large cotton fields is the biggest consumer of surface water runoff. Domestic and industrial sectors consumed accordingly 3% and 4%. Industry and municipal services mainly use groundwater, but the current potential of groundwater resources use reached only 30%. However, due to intensive pollution of groundwater horizons because of infiltration of irrigational wastewater along the left bank of Syr Darya River there is the lack of drinking water in northern and south-western regions of Tajikistan. With further climate changes, particularly warming, the hydrological resources in Tajikistan might be shortened more frequently during the longer drought periods. Contrary rapid increase of air temperatures promotes in spring–summer season may lead towards formation of heavy rainfalls and flooding waters with dangerous mudflows and landslides (Tajikistan State... 2000). As to the hydropower potential, Tajikistan nowadays exploits only 5% of it, thus in near future, for example in 2028 with completion of Roghun Dam there might be arisen ecological problems in the CA since the large hydroelectric dams directly may disrupt water flow and affect downstream agricultural activities (Russell 2018).

6.3.5 Kyrgyzstan

Kyrgyzstan has considerable reserves of water resources. There are more than 2000 rivers longer than 10 km, which form about 50 km^3 of surface runoff. Most rivers are fed by glaciers and/or snow melt. Peak flows occur from April to July. Other water resources are the lakes and reservoirs with 1745 km^3 of water, and the glaciers with 650 km^3 of accumulated water reserves. The Syr Darya river basin covers 55.3% of the country: in Kyrgyzstan the river is called the Naryn River before it reaches the Fergana valley (Kyrgyzstan 2012…2013). The longest rivers are Naryn, Chatkar, Chu, Talas and Assa (Water resources: state 2000; Water resources and …2020). Most of surface runoff water is used under irrigated agriculture, other economic sectors use less of water resources. At present, Kyrgyzstan uses just 10% of its hydropower potential (Russell 2018). Still tensions exist between Kyrgyzstan and Uzbekistan in the Fergana valley. With climate warming, according to Dyurgerov (2010), the area of glaciation of the Tien Shan mountain system in 1960–2006 has decreased by almost 17%, the volume of glaciers in the same period decreased by 219 km^3. As a consequence future water runoff might be also vulnerable during the extremal seasons.

6.4 The Newest Research Data and Future Scenarios of Aral Sea Development

In the past (till 1960), the area of Aral Sea was 67,500 km^2 and it was the fourth largest continental water body worldwide (Micklin 2014). But to what will be the future of Aral Sea, there are at least two contradictious scenarios. Supporters of the first one argue that the sea will completely dry during the twenty-first century or perhaps, in the distant future. The other even argue that this forecast is simply wrong, since it disregards basic physical and hydrologic principles. Even in the highly unlikely event that inflows from the Amu Darya and Syr Darya Rivers were reduced to zero there would still be residual input of irrigation drainage water, groundwater and snow melt and rain that would probably maintain at least three lakes. The optimistic scenario for Aral Sea area in 2030, according to Micklin (2014, 2016), is shown in Fig. 6.11a.

In the summer of 2014 the Aral Sea reached near-complete desiccation, when its segregated water bodies totalled only about 10% of its historical area. If to interpolate the tendency of shrinking of Sea under more severe drought conditions, the 2020 rendering suggests shallow intermittent lakes, and potential avenues for restoration (McDermid & Winter 2017). Micklin (2014, 2016) calculated that for Small Aral, the inflow of 2.3 km^3 is all that is required to maintain stability at the 42 m level. The Kazakhstan Government is planning a second phase to the Small Aral restoration project in order to bring water to the Aralsk town back and improve ecological conditions to the fish and fishery as well as to sea transport. The main objection to this plan is concern over insufficient water available from the Syr Darya on an annual average basis to maintain this level (Micklin 2014).

It is doubtful the Aral could be restored to its former grandeur in the foreseeable future by reducing irrigation withdrawals in all neighbouring it countries. There is also a need to make grounded estimations on the possibility to use some groundwater for irrigation in the Aral Sea Basin. Though preliminary the total annual renewable groundwater resources in this basin may be as much as 44 km^3, the useable portion with low enough salinity and sufficiently near the surface has been estimated at around 16 km^3 or even less (Micklin 2000, 2016).

Looking for other alternatives, recently, two "mega-engineering" proponents have proposed a variant of the Siberian project, which would take water only from Lake Zaysan in Kazakhstan, which is the source of the Irtysh, and deliver it into the Syr Darya from where it would flow into

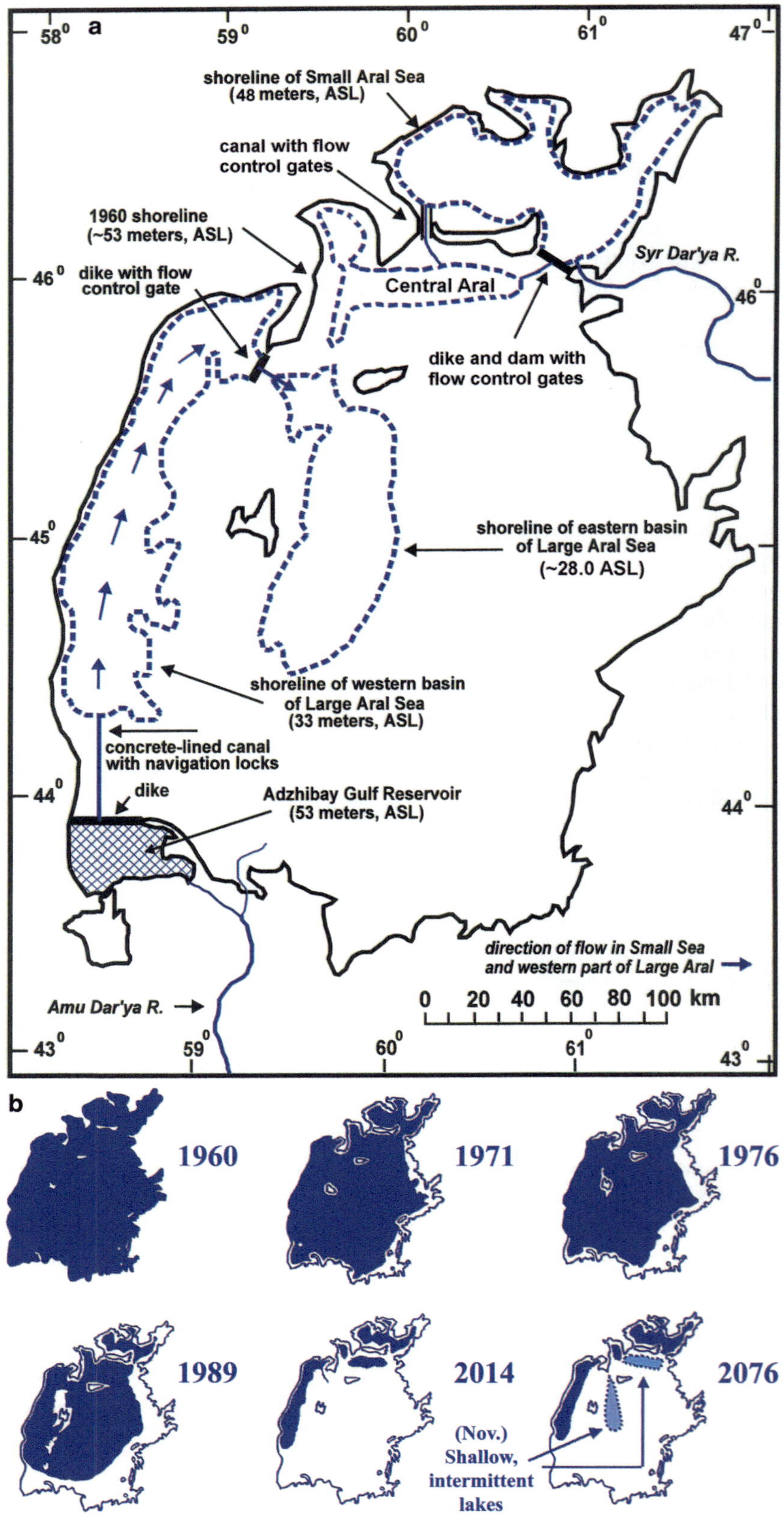

Fig. 6.11 Optimistic scenario of the future Aral Sea after 2030 (**a**); history of shrinking the Aral Sea from 1960 to 2020 (**b**). (Sources: Micklin 2014, 2016)

the Small Aral Sea (Badescu and Schuiling 2009). From one side the authors see a much lower cost for this project as the route would be considerably shorter and water would flow gravitationally to the Syr Darya River. But from another side it would require drilling a 100 km tunnel through a mountain range for which the costs are speculative. Many ecological and economic consequences of such a project are neglected or assessed weakly (Micklin 2014).

It is believed that some former functions of the Aral Sea might be restored on a smaller scale by creating and expanding the wetlands, which would serve as habitats for many fauna species, increasing landscape biodiversity, provision of environmental goods and services to the riparian zones, and buffering against desertification. As an example of the perspective research in this direction, the activities carried out in the Barsakelmes Biosphere Reserve near Aralsk have to be mentioned. The Reserve which was once established on the Barsakelmes island within the Aral Sea, nowadays has expanded within the 407,135 ha of terrestrial and marine water area. The territory of this Reserve constitutes as the world's only natural "laboratory" for the observation and conservation of different unique flora and fauna species of desert ecosystems under the climate changes. Among the main protected animal species are ungulates such as the saiga antelope (*Saiga tatarica*), the onager (*Equus hemionus*) and the goitered gazelle (*Gazella subgutturosa*) (Barsakelmes UNESCO…2016; Restoration of Barsakelmes…2018). Another example, the restoration of wetlands in surroundings of Lake Sudoche can be given. It is located in the Amu Darya's delta and is extremely important to the birds, lying in a vast desert in one of the most important paleoarctic flyways in Western Asia. Such a wetlands serve as a home of hundreds species of birds, several species of fish and mammals. The Government of Uzbekistan has created several lakes already, which are replacing the part of the fish catch of the former Aral Sea. A further wetlands restoration programme has been prepared with the assistance of the international donors and scientific assistance. It is hoped that it can become the Ramsar site in CA, directed to the restoration of the environmental functions of the wetlands and demonstrating these processes to the local people and the visitors from outside (Water-related vision… 2000).

6.5 Environmental Situation in the Largest CA Lakes and Their Basins

6.5.1 The Issyk-Kul Lake (Kyrgyzstan)

About 84% of lakes in Kyrgyzstan are located at heights 3000–4000 m in regions of tectonic origin (Water resources: state 2000). The country's largest lake Issyk-Kul (the area of water surface is 6236 km^2), located at an altitude of 1606 m, is the second largest mountain lake in the world (Fig. 6.12).

Issyk-Kul is 178 km long and 60.1 km wide, its shoreline is 669 km. The mean depth is 278.4 m, and maximum depth is 668 m (Kodyaev 1973). About 40% of its water area is covered by a depth of 0–10 m, but the lake water level is fluctuating, both within the same year and over longer periods. During the most of twentieth century there was a downward trend, but since 1986 it stopped decreasing and the lake surface area started to rise (Shnitnikov 1979; Konurbaev 1981; Savvaitova and Petr 1995).

The high mountain pastures to the south of Teskei Alatau and the slopes of the mountain ranges around the lake make up its drainage basin. Within the basin, 118 rivers and streams flow towards the lake, but only 49 of them, mostly in the eastern part, where precipitation is heavier, actually drain into it (Baetov 2006). Currently it is facing an array of problems that significantly endanger the lake. The research (Mendikulova 2008) has shown four categories of such problems: water pollution, infrastructure deficiency, endangerment of flora and fauna and lack of policy of enforcement.

The local industry and agriculture are the biggest pollution sources of the lake. An increasing content of ammonium, nitrite, zinc, copper, petroleum, organic waste and other harmful substances is observed in Tup, Djergalan, Djety-Ogyz, Cholpon-Ata and Chon-Aksuu Rivers—all of which discharges into Issyk-Kul, aggravating and already stressed ecosystem (Mendikulova 2008). Though all these changes can threaten the ecological system, the Lake Issyk-Kul is rich in phytoplankton, with 299 taxa identified. Blue-green algae (*Cyanophyceae*) dominate, but their standing crop is low. Among macrophytes Characeae are the most common (Savvaitova and Petr 1992). Zooplankton includes 117 taxa, but its distribution in the lake is uneven, with bays and shallows being richer than open water. Zoobenthos comprises 224 taxa. The original fish fauna comprised several endemic taxa, such as the common fish dace (*Leuciscus schmidti*), *Leuciscus bergi*, indigenous scaleless osman (*Dyptichus dybowskii*) (Savvaitova and Petr 1995). In September 2001, by decision of the coordinating committee of the UNESCO Man and Biosphere Programme (MAB), the Issyk-Kul Biosphere Reserve became No 411 on the list World Wide Network of Biospheres Reserves, it is also one of Ramsar sites (Baetov 2006).

Other ecological problems like contamination of lake water by radioactive materials (uranium-carbon) deposited from old uranium mines, and by precipitation, avalanches and water surface runoff reaching the lake have to be mentioned (Palesski et al. 2003).

6.5.2 The Balkhash Lake (Kazakhstan)

It is one of the largest lakes in Asia (Fig. 6.13) located in the south-eastern Kazakhstan. Balkhash Lake's basin is endorheic (closed) shared by Kazakhstan and China, with a

Fig. 6.12 The Issyk-Kul Lake's basin. (Source: https://commons.wikimedia.org/w/index.php?curid=292900)

Fig. 6.13 The Balkhash lake (upper left), eastern Kazakhstan satellite image. (Source: https://commons.wikimedia.org/wiki/File:Lakebalkhashbasinmap.png)

small portion in Kyrgyzstan. The lake lays in a land depression of tectonic origin.

The orography of lake's basin is very complex: it changes from high and low mountain systems to plains and sands. The climate is arid continental with annual precipitation less than 150–200 mm. The average depth of lake is about 5.8 m, and the max depth is 26 m. The area and water volume of the lake vary by 70–100% during the long-term periods. The eastern and western parts of the lake differ very much. The mean depth of the eastern part is 1.7 times more than of the western, and it has more saline water. The main reasons for such a difference are the huge inflow of the Ile or Ili River to the western part and the retarded exchange of water between the two parts (Lake Balkhash 2011; Imentai et al. 2015). The Ili is fed by precipitation, largely vernal snowmelt, from the mountains of China's Xinjiang region (Lineitseva 2011). Thus, with glacier degradation in the basin, the water level is somehow maintained at least in the western part. In the 1970 the Qapshaghay hydroelectric power station was built on the Ili River, and the diversion of water to fill the reservoir and to provide for irrigation reduced the flow of the by two-thirds and caused a decline in the lake's water level (Dostaj et al. 2012). In general, the lake like the Aral Sea is shrinking as a result of the diversion of water from rivers, it becomes more saline and polluted from the leakage of fuel-storage deposits and from copper mining and processing along its shores (Lake Balkhash… 2011a).

The groundwater level in the Ili Delta sank also, and due to this the riparian Tugay forests were degraded. Hydrogeologists have forecasted that underground, surface and salt runoffs into the lake basin will increase slightly (Zekster & Everett 2000). The most important current threats for the vegetation, fauna and biodiversity of the Ile Delta and southern Lake Balkhash are fires, logging of shrub vegetation, overfishing and grazing (Imentai et al. 2015).

6.5.3 The Karakul Lake (Tajikistan)

The 25-km diameter Karakul Lake laying at 3914 m above sea level (Fig. 6.14) is one of the world's largest high mountain lakes.

The lake was formerly known as Lake Victoria, named after Queen Victoria by the British cartographers/military geographers. It was marked as such by most maps of the area until the 1920s when a Kyrgyz name: Karakul, or the Black Lake, replaced the British name by the new Soviet cartographers. A peninsula projecting from the south shore and an island off the north shore divide the lake into two basins: a smaller, relatively shallow eastern one, between 13 and 19 m deep, and a larger western one, 221 and 230 m deep. It is endorheic and the water is brackish. The origin of the lake is unique: it was formed from a meteorite impact, leaving a crater with a rim diameter of 45 km. Islands formed from the central uplift can be seen in the northern and southern parts of the lake. The crater is relatively recent: its age is estimated variously as 25 Ma (million years from the present) (Taft et al. 2014; http://www.passc.net/EarthImpactDatabase/karakul.html; https://earthobservatory.nasa.gov/IOTD/view.php?id=4537), less than 23 Ma (Bouley et al. 2011) or probably from the Pliocene (5.3–2.6 Ma) (Gurov et al. 1993).

Recent studies have suggested that the basin is actually a north-western and south-eastern extensional rift. The most exciting preliminary result is the finding of shatter cones in metamorphosed sediments in the northern part of the peninsula and breccias on the central island as well as in the northern part of the rim. The site offers the possibility to investigate the possible interactions between impact cratering and tectonic activity (Bouley et al. 2011).

The basin and much of the Pamir mountain range fall within Tajik National Park, which in 2013 was named a UNESCO World Heritage Site (Hansen 2019). This area is also valuable for Tibetan avifauna: more than 90 species have been recorded near Karakul Lake (BirdLife International 2020).

6.5.4 The Project of Altyn Asyr Turkmen Lake (Turkmenistan)

One of the serious problems in Turkmenistan remains the depletion of water resources. The factors are both natural and anthropogenic. The last is mostly related with unsustainable practices of water resources management, particularly of irrigation and drainage water. Until now, almost all the collector-drainage water (CDW) without treatment is discharged into the Karakum Desert and partially to the rivers, thereby worsening water quality (Zonn & Kostianoy 2014). It was estimated that due to this CDW discharge in the Karakum Desert the enormous salinization and pollution of groundwater is taking place. In order to manage it, the Altyn Asyr (in Turkmenian "Golden Age") Turkmen Lake project was developed several years ago. The first project phase was completed in 2009 (Fig. 6.15).

This environmental project suggests that the shared network of drain manifolds will collect drainage water from irrigated lands through all the country and discharge them in the giant natural depression Karashor, located on the northwest of Turkmenistan. The area along the drain lines is planned to be planted with the salt tolerant agricultural crops, also pastures will be watered and fish farms will be built on the ponds with water from drain manifolds (Kostianoy et al. 2014; Zonn 2012).

Before the construction of the Altyn Asyr Turkmen Lake was launched, drainage and even waste water accumulated in the depressions of the Desert were coming not only from the

Fig. 6.14 The satellite image of Lake Karakul, Tajikistan. (Source: Created with NASA WorldWind by User:Vesta using Landsat 7 (Visible Colour) satellite image. Public Domain, https://commons.wikimedia.org/w/index.php?curid=501118)

Fig. 6.15 The Great Turkmen and Dashoguz collectors and the Karashor depression, future site of Golden Age Lake. (Source: Stone 2008)

local regions but also from Uzbekistan neighbouring areas. Thus, with the construction of the large artificial water body in the Karakum Desert there is a plan to improve the condition of irrigated land, raise agricultural productivity, reduce salinization processes and as a consequence, stop the soil degradation. With construction of this lake, not only environmental, but also economic and social problems can be solved in the region. There is an ambitious plan under the complex measures to decrease the pollution of the Amu Darya River (Turkmen Lake… 2015).

On the state level the Altyn Asyr Lake Development concept was approved for 2019–2025 period and the Action Plan started to be implemented. Upon completion, the lake will span 3500 km^2. The US$6 billion project is designed to drain swamps and combat the build-up of salt and other minerals that have degraded three-quarters of Turkmenistan's arable land and eroded archaeological monuments. According to government plans, it is intended to be filled by a 2650-km network of tributary canals. They are used for pumping run-off water from irrigated cotton fields towards the lake. But there are arguments for the negative consequences of this project. Some experts argue that the runoff will be insufficient to fill the lake, as the drainage water will evaporate or seep into the desert through unlined feeder canals. Others are sceptical of plans to purify the runoff, and finally the lake will become an artificial Dead Sea (Stone 2008).

This criticism has a reasonable ground: the areas along the Karakum canal, one of the longest irrigation canals in the world, nowadays are covered by numerous swamps and ponds, which exacerbated salinization.

6.6 Transboundary Water Management Problems and Steps for the Sound Solutions

The administration and management of water resources during the Soviet time in CA republics was centralized. Some districts were fed only with groundwater, others from large pumping stations with the surface runoff, while the rest used both the surface and groundwater for their needs (Thurman 2001).

Despite the first problems of land degradation and desertification under the intensive irrigation of that time, water was not rationally used in CA republics, and thus from 1960 all that has pushed towards the shrinking of Aral Sea. There were many efforts done to find different alternatives in order to increase the surface runoff to the Aral Sea, including so called "century project" to redistribute the runoff of rivers flowing in Siberia. But in August 1986, the Communist Party's Central Committee officially abandoned the "project of the century", a grandiose Scheme 15 years in the planning. It would have diverted the flow of rivers emptying into the Arctic and Pacific Oceans and pumped the water southward into the CA republics and Kazakhstan (Novikov 1986; World resources… 1988). The complicated multistage project, which would not have been completed until the middle of the twenty-first century, would have diverted a total of 120 km^3 of water a year—one-fifth of the Mississippi River's annual flow at its mouth (Micklin 1985). Why did the Soviets give up on their unprecedented water transfer plan? A major reason given by Alexander Yanshin, vice president of the USSR Academy of Scientists, is the reversal of conditions at the Caspian Sea. Prior to 1979, the Caspian had been drying up, but since then, the sea has been rising, undermining port facilities and threatening the Buzachi Peninsula's oil fields. Diversions from the north would have made matters worse (Why USSR… 1986; World resources … 1988).

National aspirations and the struggle for state-building following independence in 1991 shifted the concept of water security from a regional approach to country-focused policies. Still there are different points of view concerning role of water resources in the countries which are in the upstream of the main rivers, and those which are downstream. For the first group of countries as Kyrgyzstan and Tajikistan, water security is mainly interpreted as energy independence, reduction of disasters stemming from water, and sufficient agricultural water for irrigation and pastoralism. For the rest downstream countries, water security mostly means intensified agriculture, fisheries and pastoralism, and reduction of water scarcity incidents. Thus, without finding the consent attitude and compromises will be difficult for CA countries to achieve national water security and expected perspectives (Xenarios et al. 2018).

6.7 Recent Water Usage and Quality Problems

Over the past two decades, transboundary water resources management problems have emerged and revealed the weak points of recent water usage practices in CA. Geographical location and the transboundary nature of the main rivers determine the specific nature of water use in each country in the region (Yunussova & Mosiej 2016).

That can be illustrated by example, how is distributed the Amu Darya flow nowadays. The share of the Amu Darya flow allocated to Turkmenistan and Uzbekistan is 50% each of the actual river flow at the Kerki gauging station. This share is based on an agreement between Uzbekistan and Turkmenistan signed in January 1996, which supplemented the 1992 Agreement signed by the five Central Asian countries. But in reality, the annual renewable water resources in five CA countries are very different, even in normal as to the temperature and precipitation years. Though the actual surface water resources allocated to Turkmenistan and Uzbekistan are calculated every year, depending on the actual flow, but if the year is drier than many years average, situation with water resources is unsatisfactory in both the countries, which are downstream of the Amu Darya (The Aral Sea transboundary…2013). As a result water resources and particularly their quality problems accumulate and become more tense under the transboundary water usage.

6.7.1 Uzbekistan

The Uzbekistan's National report on water usage declares that water and energy management at transboundary level should be undertaken jointly by the countries and meet international norms. Moreover, it is necessary to develop a strategic plan and set clearly objectives of transboundary river water use that must contribute to safety in and sustainability of different water users and regions (http://www.cawater-info.net/5wwf/national_report_uzbekistan.htm). As the first step towards the better practical water usage in such transboundary areas can be presented the FAO-Turkey Partnership Programme (Uzbekistan and FAO 2016). Under this programme the implementation of irrigation water-saving technologies for increased water productivity in the Uzbek part of the Podshaota small transboundary river basin is planned. The region is developing intensive fruit and vegetable production. The objective of the project is to achieve a sustainable agricultural production in the basin by the means of water saving technologies and minimization of water deficits. The project is implemented hand in hand with farmer education and demonstration activities. The Government of Uzbekistan, with its development partners, made efforts include the introduction of integrated water resource management (IWRM) principles in selected basins and the transformation of administrative water management organizations to hydrographic basin irrigation system authorities. Only over the review period (2010–2016) 63 projects were identified as running or being running in the water sector of Uzbekistan. In terms of the number of projects per province, Ferghana province is covered the most frequently, while Surkhan Darya is the least. Fourteen out of 21 projects that cover Ferghana also cover other provinces. Only Karakalpakstan and Khorezm provinces are covered alone more frequently (>50%) rather than in combination with other provinces (Djumaboev et al. 2017).

6.7.2 Kazakhstan

The researchers Kurtov (2014), Sidorova (2008) and Poryadin (2014) point out that in terms of water scarcity, population growth and economic development of the countries in recent decades, the problem of rational use of transboundary rivers has worsened and becomes a mechanism of mutual economic and political pressure in the region. Thus, it was concluded that for transboundary water resources management there is a need for the assistance of UN organizations like United Nations Economic Commission for Europe (UNECE) and others. The intensive studies are carried out to quantify the transboundary pollution of rivers in Kazakhstan, mainly the Aral Sea basin (Burlibaev et al. 2012, 2013; Severskiy 2004; The future of the Aral Sea… 2014).

In the case of Kazakhstan, climate adaptation planning can be demonstrated as it is arranged in the Chu-Talas basin, shared with Kyrgyzstan (Fig. 6.16). This basin is one of the pilot basins in CA with a river basin organization, the Chu-Talas Water Commission, which initiated to work out the Adaptation to Climate Change and Long-term Programmes. Transboundary cooperation has been supported by the UNECE and other partners like United Kingdom, Sweden, Finland and Estonia (Climate-proofing…2018).

The Chu and the Talas Rivers are the major sources of water in irrigated agriculture, hydro-energy and support the livelihoods of more than three million people in Kazakhstan and Kyrgyzstan. The basin is highly vulnerable to climate change and the overall growth of aridity and the declining availability of water resources are the most likely and serious impacts of climate change in the basin. As a result of the forecasted climate change, glaciers in both basins may be fully exhausted by 2100. Regulation of flows of the Chu River between Kyrgyzstan and Kazakhstan is an issue already today (Transboundary cooperation... https://www.unece.org/env/water/centralasia/chutalas.html#c65768).

Initial climate change adaptation activities started in the basin in 2010 with modelling of climate change impacts and a vulnerability assessment, which was then elaborated into a set of climate change adaptation measures, covering issues from water quality to monitoring and education. Further the findings were assessed against their cost/effectiveness and integrated into a transboundary diagnostic analysis (TDA) and a Strategic Action Program (SAP). The SAP, when approved, will become the main document for transboundary management in the basin, facilitating cooperation, planning, funding and implementation (UNECE 2007). The assessment of adaptation costs, performed as part of the adaptation strategies and plans for selected basins (the Chu-Talas, the Dniester and the Neman river basins), revealed that the approximate costs for adaptation among the water-related sectors amounted to roughly €200 million each (DHI and UNEP 2016; United Nations…2020; The strategic Framework… 2015). These are the high costs, but necessary to be found from all the partners in the basin.

The next case on the situation with transboundary water basin issues was revealed after monitoring conducted during 2000–2011 in the Tobol-Torgay river basin, which extends in the border of Kazakhstan and Russia (Fig. 6.17).

The level of pollution and the quality of river water in Kazakhstan are controlled using both the state monitoring data and at transboundary checkpoints (Yunussova & Mosiej 2016). Only 9 rivers out of the 69 monitored in whole country in 2011 were under "pure" quality category (Kazgidromet 2012). For the monitoring in the Tobol-Torgay river basin, the chemical analysis of major ions, biogens, heavy metals, toxic substances including organic and organochlorine substances were conducted of the samples taken from the Tobol river and its tributaries Ayat, Ubagan and Togyzak during 12 years. The integrated assessment of anthropogenic load on the basin rivers was based on the DPSIR cause–effect relations model (Kristensen 2004). Research results have shown that over the research period, the maximum limited concentration (MLC) excess was recorded for groups of heavy metals and major ions like calcium, magnesium, sodium and potassium. The contamination with organic substances was the least in the row of pollutants. The hydrochemical background of natural waters of the Trans-Ural mountains, particularly in the Tobol basin, contain copper, zinc iron, manganese as well as fluorides. In some areas the quantities of heavy metals are higher, this is linked to the mining industry present in the Kostanay region. Particularly

Fig. 6.16 The location of the Chu-Talas transboundary watershed. (Source: Akbozova (2015) https://www.osce.org/pc/156466?download=true)

high levels of iron, copper and manganese were measured in 1995, where the iron levels were 3.9 times higher, the copper levels were 11.5 times higher and the manganese levels were 16 times higher than the average for the region. The calculations have shown that during high water and flooding periods more organic substances, chlorides, the sum of ions and heavy metals were added by the Ayat River than removed through the Togyzak and the Tobol Rivers, while during low-water periods, removal prevailed for all components in the water (Yunussova & Mosiej 2016).

It is to be accepted that different parties or countries in the transboundary basins often may have conflicting interests. Even within the same administrative unit conflicts between different users may arise. Three factors are necessary to be explored for the problem or conflict solving: knowledge, information and cooperation. The overall-appropriate solutions can be worked out only if different conflicting parties share data, discuss and understand the nature of water resources cycle and balance phenomena and know about the plans and interests of each partner in the basin (Water related vision…2000).

United Nations Environment Program (UNEP) promotes environmentally sound practices globally and recommendations concerning the risks which are projected to increase in the next 15–30 years, particularly for four hotspot regions: the Middle East, Central Asia, the Ganges-Brahmaputra-Meghna basin and the Orange and Limpopo basins in Southern Africa. Action should be taken now to reduce future costs and impacts (UNEP-DHI and UNEP 2016).

6.8 Extreme Ecological Situations, Technogenic Accidents and Natural Disasters

During the last 15–20 years due to the climate warming more and more extreme situations were registered in CA countries. According to the National environmental report from 2010 in Kazakhstan (Nacionalny doklad… 2011), there were registered 22,824 extreme situations and different accidents caused by natural and technogenic factors. During these disasters almost 6000 people have experienced material damage, and 1819 of them have died. The fires at home and at work place were the main reasons of the largest number of accidents. In general, the situation in Kazakhstan with technogenic accidents in 2010 was by 7.7% higher than in 2009.

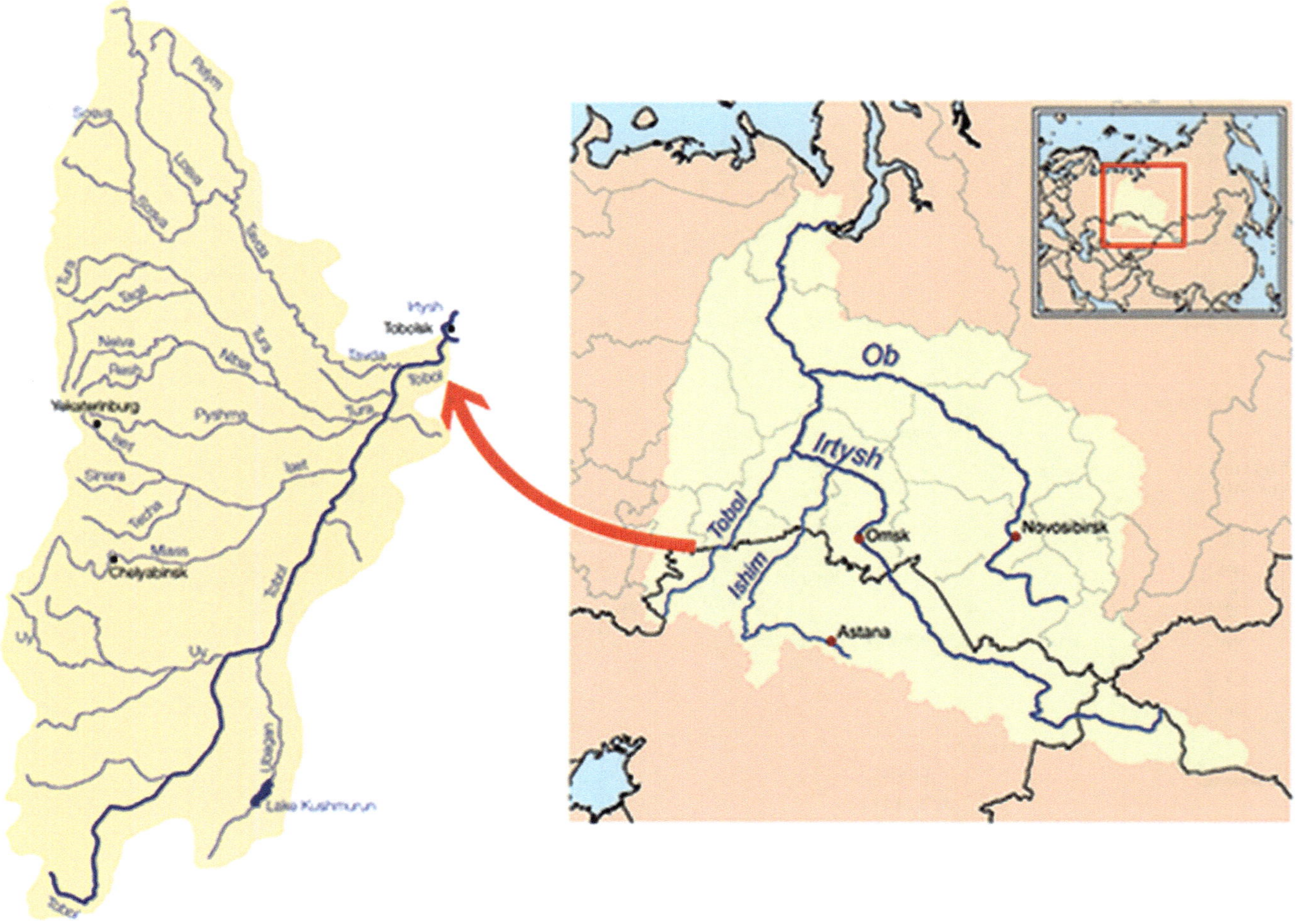

Fig. 6.17 Tobol-Torgay river basin, Russia–Kazakhstan. (Sources: Yunussova and Mosiej 2016; https://upload.wikimedia.org/wikipedia/commons/1/1b/Tobol_river_2_layers_en.svg)

The natural disasters have made about 18% of all accidents. Regarding the natural disasters, the most damage was made by heavy rains, floods in the mountain valleys, landslides, mud and stone flows etc. Mainly they were located in Ile-Alatau and Zhetysusky Alatau. Due to the extreme water runoff and atmosphere deposits, the dam of Ak-Eshke in Kyzylagash village was destroyed, and vast territory down the dam (with settlements) was flooded. It resulted the damage of 80% of buildings, bridge, part of railway, human and animal losses. About of 643 cases of forest fires were registered in 2010, most of them were caused by humans.

Due to infections, 12 people have died, and 2 cases of mass mortality of animals were registered in 2010: one of the highest death numbers of saiga took place in Zhanibeksky district, where the largest population of saiga used to live.

The Aral and Semipalatinsk regions of Kazakhstan are considered as the regions of ecological crisis, where the balance between natural processes and living components of ecosystems, including flora, fauna and the humans was affected mostly. Recently, the area of dried Aral Sea bottom occupies more than 50,000 km^2: the new solonchak-like desert has been formed there. The analogous situation takes place in the south of previous Aral Sea, in the territory of Uzbekistan. Dust and salts mixed with different agricultural chemicals are reaching by air not only the spaces in surrounding, but also are transported in the remote distances by wind and storms (Fig. 6.18).

Another spot of ecological crisis of Kazakhstan is located in Semipalatinsk. According to the research data from 2005 to 2007, the level of radioactive pollution in the environment of Semipalatinsk nuclear polygon is still higher than in other districts. The water is also contaminated with technogenic radionuclides like 137 Cs, 90 Sr. The pollution in this spot is not concentrated only within the polygon area, but also extends in the radius of hundred and more kilometres (Nacionalny doklad… 2011).

In Kyrgyzstan, about 61% of all ecological accidents and disasters occur due to the natural processes. For example, in the period from 2007 to 2011, the average number of such accidents was 288 per year. The majority of these disasters

Fig. 6.18 A massive dust and salt storm formed in the Aral Sea region, Kazakhstan on seventh May 2007. (Source: https://eoimages.gsfc.nasa.gov/images/imagerecords/18000/18344/centralasia_amo_2007127_lrg.jpg)

are caused by mud flows and floods (29% of all cases), landslides, avalanches etc. Of the period 2007–2011, the 2010 year was extreme in respect of these disasters, because the precipitation rate then was higher than average. Practically in almost whole territory of Kyrgyzstan there is a high risk of mud flows, but the highest occur in the basins of Chu, Talas, Naryn, Kara Darya and Lake Issyk-Kul. The landslides and avalanches cause 11% of all ecological accidents in Kyrgyzstan, and their number increases due to the seismicity, warming, heavy rain, groundwater dynamics, technogenic activities, which are made in the imbalance of physical processes on the slopes. Overall, the 7.5% of all country's territory is under the high risk of landslides, with the largest number at the Osh and Dzalal-Abad regions (Nacionalny doklad… 2012).

Tajikistan being the mountainous country and rich in landscapes and geographical components faces also with the physical disasters like mud flows, avalanches, landslides and floods, which with climate warming occur more often, till 25 cases per 10 years. In the wet 2010 year, the material damage made by the mud flows and floods for the country was estimated to more than US$600 million (Nacionalny obzor... 2012).

The processes of desertification can develop under the acting of natural and anthropogenic factors. They can be extremely severe when both factors act simultaneously. In Turkmenistan the anthropogenic factors are prevailing in the recent desertification processes. The aridization of climate in Turkmenistan can be seen in the more frequent draughts: they occur usually 7 years of 10. The extreme draught was registered in 1997–2001 years. Researchers of arid landscapes and climate in coordination with international experts have prepared the strategy of measures controlling the desertification intensity: among them the plantation/afforestation of moving sandy substrates, using technologies enriching the natural pastures, preventing soil salinization, applying the sun and wind energy (Nacionalny doklad... 2006).

In Uzbekistan the largest ecological problems occur in connection to land salinization and degradation, flooding, desertification, landslide, drainage system management, water pollution. Overirrigation of agricultural land has caused the deficit of drinking water in Karakalpakstan. The increase in stream runoff at lower reaches high mineralized drainage water caused the worsening of stream water quality. One of the methods of stabilization of ecological situation in Prearalie is the restoration of wetlands in the delta of the Amu Darya. It was determined that 1-year-old vegetation decreased the wind speed by 20.5%, the 2-year-old vegetation by 34.6%, and both grass vegetation together with trees and bushes have accumulated and stabilized the sand substrate. The project on the restoration and conservation of Tugay forests in delta of the Amu Darya in Karakalpakstan has been initiated to fulfil the above-mentioned ideas (Nacionalny doklad… 2013).

6.9 Topics for Further Reading and Discussions on the Content of Chapter 6

- The longest glacier monitoring in Kazakhstan at the Tuyuksu glaciological station.
- CA river water runoff in the extremely dry years.
- CA region places/areas which can be considered as the hot spots due to the arisen complex ecological problems: how they are in comparison to the hot spots in other Asian or European regions?

Recommended References

Aizen VB (2011) Pamir glaciers. In: Sigh VP, Singh P (eds) Encyclopedia of snow, ice and glaciers. Springer Publisher, U.K. Haritashya, p 1253. https://www.researchgate.net/publication/233731981_Pamir_glaciers. Accessed 6 May 2018

Farinotti D, Longuevergne L, Moholt G, Duethmann D, Mölg T, Bolch T, Vorogushyn S, Güntner A (2015a) Substantial glacier mass loss in the Tien Shan over the past 50 years. Nat Geosci 8:716–722. https://doi.org/10.1038/NGEO2513

Hoelzle M, Azisov E, Barandun M, Huss M, Farinotti D, Gafurov A, Hagg W, Kenzhebaev R, Kronenberg M, Machguth H, Merkushkin A, Moldobekov B, Petrov M, Saks T, Salzmann N, Schöne T, Tarasov Y, Usubaliev R, Vorogushyn S, Yakovlev A, Zemp M (2017a) Re-establishing glacier monitoring in Kyrgyzstan and Uzbekistan. Central Asia Geosci Instrum Method Data Syst 6:397–418. https://doi.org/10.5194/gi-6-397

Karthe D, Chalov S, Borchardt D (2014) Water resources and their management in Central Asia in the early twenty first century: status, challenges and future prospects. Environ Earth Sci 73(2). https://doi.org/10.1007/s12665-014-3789-1

Kazgidromet (2012a) Informacionnyj byulleten' o sostoyanii okruzhayushchej sredy Respubliki Kazahstan za 2011 god (Newsletter on the state of the environment of the Republic of Kazakhstan in 2011). Almaty, 29 p

Kostianoy AG, Kosarev AN (eds) (2010a) The Aral Sea environment. The handbook of environmental chemistry, vol 7. Springer-Verlag, Berlin\Heidelberg\New York, 332 p

Kotlyakov VM, Seversky IV (2006a) Glaciers of Central Asia: current state, changes, possible influence on water resources. Almaty

Lineitseva A (2011a) Changing water resources of rivers in Kazakhstan as a result of climate change impacts, 27–29 Sep, Porto, Portugal https://www.riob.org/fr/file/270333/download?token=9TbX4Emg

Lioubimtseva E, Henebry GM (2009a) Climate and environmental change in arid Central Asia: impacts, vulnerability, and adaptations. J Arid Environ 73:963–977

Micklin P (2014a) Chapter 15: Efforts to revive the Aral Sea. In: Micklin P et al (eds) The Aral Sea. Springer Earth System Sciences, pp 361–380

Nacionalny doklad o sostoyanii okruzhayuscey sredy i ob ispolzovanyi prirodnyh resursov za 2015 god. Vodnye resursy (2016a) (National report on the state of environment and on the use of natural resources for year 2015. Water resources) http://doklad.ecogosfond.kz/водные_ресурсы

Nacionalny doklad o sostoyanii okruzhayuscey sredy v Kyrgyzskoy Respublike za 2006–2011 gody (2012a) (National report on the state of environment in the Republik of Kyrgyzstan for 2006–2011) Bishkek http://www.cawater-info.net/green-growth/files/nature-kg.pdf

Nacionalny doklad o sostoyanii okruzhayuscey sredy v respublike Kazahstan v 2010 godu (2011a) (National report on the state of environment in the Republik of Kazakhstan for 2010). Almaty http://www.cawater-info.net/green-growth/files/nature-kz.pdf

Nacionalny doklad o sostoyanii okruzhajuscey sredy i ispolzovanyi prirodnyh resursov v respublike Uzbekistan (2008–2011 g.g.) (2013a) (National report on the state of environment and on the use of natural resources in the Republic of Uzbekistan for years 2008–2011) Tashkent, Izdatelstvo "CHINOR ENK". http://docplayer.ru/58059280-Gosudarstvennyy-komitet-respubliki-uzbekistan-po-ohrane-prirody-nacionalnyy-doklad.html

Nacionalny doklad. Vypolnenye konvencii OON po borbe s opustynivanyem v Turkmenistane (2006a) (National report. Execution of Convention UNO against desertification in Turkmenistan) Ashgabat http://www.cawater-info.net/bk/water_land_resources_use/russian_ver/pdf/turkmenistan-rus.pdf

Oledenenye gornyh territoryi Kazahstana (2015a) (The glaciation of mountaneous territory of Kazakhstan) Banner, Institut Geografii, Laboratorija Glaciologii). https://ingeo.kz/?page_id=2200

Russell M (2018a) Water in Central Asia – and increasingly scarce resource, EPRS | European Parliamentary Research Service https://www.europarl.europa.eu/RegData/etudes/BRIE/2018/625181/EPRS_BRI(2018)625181_EN.pdf

Severskiy I, Kogutenko L (2016a) Studying of cryosphere in Kazakhstan https://globalcryospherewatch.org/meetings/salekhard2016/presentations/3.1.3.c-L.Kogutenko-Kazakhstan%20cryospheric%20activities.pdf

Severskiy IV, Kokarev AL, Severskiy SI, Tokmagambetov TG, Shagarova LV, Shesterova IN (2006a) Contemporary and prognostic changes of glaciation in Balkhash Lake basin. Almaty

Severskiy IV (2004a) Water-related problems of Central Asia: some results of the (GIWA) international water assessment program. Ambio Royal Swedish Acad Sci 33(1–2):52–62. http://www.ambio.kva.se

Severskiy I, Vilesov E, Armstrong R, Kokarev A, Kogutenko L, Usmanova Z, Morozova V, Raup B (2016a) Changes in glaciation of the Balkhash–Alakol basin, Central Asia, over recent decades. Ann Glaciol 57(71):382–394

Tajikistan State of the Environment Report (2000a). http://enrin.grida.no/htmls/tadjik/soe2/eng/htm/water/state.htm

Tajikistan State of the Environment Report 2002 (2003a). http://enrin.grida.no/htmls/tadjik/soe2001/eng/htmls/climate/clim26.htm

Water resources and water management infrastructure of Kyrgyzstan (2020a) State Water Resources Agency under the Government of the Kyrgyz Republic. https://www.water.gov.kg/index.php?option=com_content&view=article&id=365&Itemid=1470&lang=en

References

Aizen VB, Aizen EM, Kuzmichonok VA (2007) Glaciers and hydrological changes in the Tien Shan: simulation and prediction. Environ Res Lett 2:1–10

Aizen VB, Aizen EM, Melack J (1995) Climate, snow cover, glaciers and runoff in the Tien. Shan Water Res Bull 31:1113–1129

Aizen VB (2011) Pamir glaciers. In: Sigh VP, Singh P (eds) Encyclopedia of snow, ice and glaciers. Springer Publisher, U.K. Haritashya, p 1253. https://www.researchgate.net/publication/233731981_Pamir_glaciers. Accessed 6 May 2018

Akbozova I (2015) Best practices on flood and drought response in the Chu-Talas basin and on effective governance approaches for addressing water-related disasters and climate change. Second Preparatory Meeting of the OSCE 23rd Economic and Environmental Forum Belgrade, 11–13 May 2015. The Commission of the Republic of Kazakhstan and the Kyrgyz Republic on the Use of Water Management Facilities of Intergovernmental Status on the Rivers Chu and Talas. https://www.osce.org/pc/156466?download=true. Accessed 14 April 2020

Akhmadov E (2008) Uzbekistan experiences serious water shortages. 05/28/2008 issue of the Central Asia Caucus Institute Analyst

Badescu V, Schuiling R (2009) Aral Sea, irretrievable loss or Irtysh imports ? Water Resour Manag 724 24(3):597–616. http://www.springerlink.com/content/0920-4741/?k¼Badescu. Accessed 10 Feb 2018

Baetov R (2006) Lake Issyk-Kul: experience and lessons learned brief, pp 193–203, http://www.worldlakes.org/uploads/13_Lake_Issyk-Kul_27February2006.pdf/. Accessed 11 Feb 2018

Barsakelmes UNESCO Biosphere Reserves (2016). http://www.unesco.org/new/en/natural-sciences/environment/ecological-sciences/biosphere-reserves/asia-and-the-pacific/kazakhstan/barsakelmes/

BirdLife International (2020) Important Bird Areas factsheet: Karakul lake and mountains http://www.birdlife.org. Accessed 09 Apr 2020

Bouley S, Baratoux D, Baratoux L, Colas F, Dauvergne J, Losiak A, Vaubaillon J, Bourdeille C, Jullien A, Ibadinov K (2011) Karakul: a young complex impact crater in the Pamir, Tajikistan. American Geophysical Union, Fall Meeting 2011, abstract id.P31A-1701, http://adsabs.harvard.edu/abs/2011AGUFM.P31A1701B/. Accessed 17 Feb 2018

Braithwaite RJ (2002) Glacier mass balance: the first 50 years of international monitoring. Prog Phys Geogr 26(1):76–95

Bruk SI, Rantsman YY, Allan NJR (1999) Tien Shan. In: Encyclopædia Britannica. https://www.britannica.com/place/Tien-Shan. Accessed 10 June 2017

Burlibaev MJ, Amirgaliev NA, Murtazin EZH, Shenberger IV, Perevalov AS, Burlibaeva DM (2013) Sovremennye sostoyaniya hidrohimicheskogo rezhima i toksikologicheskih parametrov v transgranichnyh rekah Balkash-Alakol'skogo bassejna (the current hydrochemical regime and toxicological parameters in transboundary rivers Balkhash-Alakol basin). Gidrometeorologiya i ekologiya 1:117–128

Burlibaev MJ, Murtazin EZH, Shenberger IV, Burlibaeva DM (2012) Sovremennoe sostoyanie zagryaznenie rek Central'noj Azii po transgranichnym stvoram v predelah territorii Kazahstana. V: Vodnomu sotrudnichestvu stran Central'noj Azii – 20 let: opyt proshlogo i zadachi budushchego (The current state of pollution of the rivers of Central Asia on the transboundary river stations within the territory of Kazakhstan. In: Water cooperation in Central Asia – 20 years: past experience and future challenges), Central'no- -aziatskaya mezhdunarodnaya nauchno-prakticheskaya konferenciya. Almaty, 20–21 sentyabrya 2012 g. Available at: www.libed.ru/konferencii-mehanika/464063-1-centralno-aziatskayamezhdunarodnaya-nauchno-prakticheskayakonferenciya-vodnomu-sotrudnichestvu-strancentralnoy-azii.php. Accessed 3 Apr 2018

CACILM (2006) CACILM multicountry partnership framework project document. Central Asian Countries Initiative for Land Management, Asian Development Bank, 70 pp.

CA WaterInfo (2011) The Aral Sea Basin http://www.cawater-info.net/aral/index_e.htm. Accessed 17 Feb 2018

Cecil LD, Green JR, Thompson LG (eds) (2004) Earth paleoenvironments—Records preserved in mid- and low latitude glaciers: Dordrecht, Kluwer Academic Publishers [now Springer], 250 p. Chapter 3 Drainage basin of the Aral Sea and other transboundary waters in Central Asia https://www.unece.org/fileadmin/DAM/env/water/publications/assessment/English/G_PartIV_Chapter3_En.pdf. s. Accessed 16 Apr 2020

Chembarisov EI, Lesnik TV (1995) To the preserve of surface waters of Central Asia. In: Denisov Yu M (ed) Issue of SANIGMI "fresh water". SANIGMI Press, Tashkent, pp 64–71. (In Russian)

Climate-proofing cooperation in the Chu and Talas river basins (2018) Support for integrating the climate dimension into the management of the Chu and Talas River Basins as part of the Enhancing Climate Resilience and Adaptive Capacity in the Transboundary Chu-Talas Basin project, funded by the Finnish Ministry for Foreign Affairs under the FinWaterWei II Initiative, Geneva https://www.unece.org/fileadmin/DAM/env/water/ChuTalas/EN_ClimateProofingChuTalas_web_10Dec2018.pdf. Accessed 14 Apr 2020

Djumaboev K, Anarbekov O, Holmatov B, Hamidov A (2017) Overview of water-related programs in Uzbekistan, Project report of the sustainable Management of Water Resources in rural areas in Uzbekistan. Component 1: national policy framework for water governance and integrated water resources management and supply part. International Water Management Institute (IWMI), Colombo, p 29. https://doi.org/10.5337/2017.211. https://www.researchgate.net/publication/321129067_Overview_of_Water_-_related_Programs_in_Uzbekistan/. Accessed 12 Feb 2018

Dostaj Z, Alimkulov S, Tursunova A, Myrzakhmetov A (2012) Modern hydrological status of the estuary of Ile River. Appl Water Sci 2:227–233. https://doi.org/10.1007/s13201-012-0034-5

Dyurgerov MB (2010) Reanalysis of glacier changes: from the IGY to the IPY, 1960-2008. Data Glaciol Stud 108:1–116

Farinotti D, Longuevergne L, Moholt G, Duethmann D, Mölg T, Bolch T, Vorogushyn S, Güntner A (2015b) Substantial glacier mass loss in the Tien Shan over the past 50 years. Nat Geosci 8:716–722. https://doi.org/10.1038/NGEO2513

FAO (2012) Conservation agriculture in Central Asia: status, policy, institutional support, and strategic framework for its promotion, FAO Sub-Regional Office for Central Asia (FAO-SEC). http://www.fao.org/docrep/017/aq278e/aq278e.pdf. Accessed 26 Jan 2018

Fundamentals of water strategy of the Aral Sea Basin (1996) Tashkent

Glantz MH (2005) Water, climate, and development issues in the Amu Darya basin. Mitig Adapt Strateg Glob Chang 10(1):23–50

Griniaev SN, Fomin AN (2009) Aktualjnye voprosy primenenya mechanizma birzhevoy torgovli dlia reshenya vodno-energeticheskych problem stran Centralnoy Azii. Analytichesky doklad, Moskva www.csef.ru (Actual questions of usage of stock exchange mechanism while solving water-energy problem for countries of Central Asia)

Groll M, Opp C, Aslanov I (2013) Spatial and temporal distribution of the dust deposition in Central Asia—results from a long term monitoring program. Aeolian Res 9:49–62. https://doi.org/10.1016/j.aeolia.2012.08.002

Gurov EP, Gurova HP, Rakitskaya RB, Yamnichenko AY (1993) The karakul depression in Pamirs - the first impact structure in Central Asia (abstract). Lunar Planet Sci XXIV:591–592

Hamidov A, Helming K, Balla D (2016) Impact of agricultural land use in Central Asia: a review. Agron Sustain Dev 36:6. https://doi.org/10.1007/s13593-015-0337-7

Hansen K (2019) Frozen Karakul lake, NASA Earth Observatory https://earthobservatory.nasa.gov/images/146006/frozen-karakul-lake. Accessed 09 Apr 2020

Hoelzle M, Azisov E, Barandun M, Huss M, Farinotti D, Gafurov A, Hagg W, Kenzhebaev R, Kronenberg M, Machguth H, Merkushkin A, Moldobekov B, Petrov M, Saks T, Salzmann N, Schöne T, Tarasov Y, Usubaliev R, Vorogushyn S, Yakovlev A, Zemp M (2017b) Re-establishing glacier monitoring in Kyrgyzstan and Uzbekistan. Central Asia Geosci Instrum Method Data Syst 6:397–418. https://doi.org/10.5194/gi-6-397. Accessed 3 Apr 2018

IEA (International Energy Agency) World Energy Outlook (2016.) Paris, OECD/IEA www.oecd-ilibrary.org/energy/world-energy-outlook-2016_weo-2016-en. Accessed 06 April 2020

Ignatov EL, Kaplin PA, Lukyanova SA, Solovyova GD (1993) Influence of the recent transgression of the Caspian Sea on its coastal dynamics. J Coast Res 9(1):104–111

Imentai A, Thevs N, Schmidt S, Nurtazin S, Salmurzauli R (2015) Vegetation, fauna, and biodiversity of the Ile Delta and southern Lake Balkhash – a review. J Great Lakes Res, Sebastian Schmidt Sabir Nurtazin. https://doi.org/10.1016/j.jglr.2015.04.002

Integrated water resources management in Central Asia: The challenges of managing large transboundary rivers (2014) Technical Focus Paper, Global Water Partnership. https://www.gwp.org/globalassets/global/toolbox/publications/technical-focus-papers/05-integrated-water-resources-management-in-central-asia.pdf. Accessed 04 April 2020

IPCC: Climate Change (2013) The physical science basis, contribution of working group I to the fifth assessment report of the intergovernmental panel of climate change. Cambridge University Press, Cambridge, 1535 p

Issanova G, Abuduwaili J (2017) Natural conditions of Central Asia and land-cover changes. In: Aeolian processes as dust storms in the deserts of Central Asia and Kazakhstan. Environmental science and engineering, Springer, Singapore, pp 29–49

Kaplin PA, Selivanov AO (1995) Recent coastal evolution of the Caspian Sea as a natural model for coastal responses to the possible acceleration of global sea-level rise. Mar Geol 124:161–175

Kara-Kul (2017) Earth impact database. University of New Brunswick http://www.passc.net/EarthImpactDatabase/karakul.html. Accessed 28 Aug 2017

Karthe D, Abdullaev I, Boldgiv B, Borchardt D, Chalov S, Jarsjo J, Li L, Nittrouer JA (2017) Water in Central Asia: an integrated assessment for science-based management, environ earth Sci 76:690 https://link.springer.com/content/pdf/10.1007%2Fs12665-017-6994-x.pdf accessed 6 may 2018

Katalog Lednikov SSSR (Glacier Inventory of the USSR) (1973) Centralnaya Aziaya (Central Asia), Tom 14 V 2, Kyrgyzstan Part 2. Hydrometeoizdat, Leningrad

Kayumov AK (2013) The first comprehensive international scientific expedition to study the state of the glaciers and the environmental situation in the upper reaches of Vakhsh and Panj rivers. Part 1. Glaciers and Hydrology. – Dushanbe: "Irfon", 154 p. https://www.wmo.int/pages/prog/www/OSY/Meetings/GCW-IM1/glaciers.pdf. Accessed 29 June 2017

Kazgidromet (2012b) Informacionnyj byulleten' o sostoyanii okruzhayushchej sredy Respubliki Kazahstan za 2011 god (Newsletter on the state of the environment of the Republic of Kazakhstan in 2011). Almaty, 29 p

Khamzayeva A, Rahimov S, Islamov U, Maksudov F, Maksudova D, Sakiev B (2009) Water resources management in Central Asia: Regional and international issues at stake, CIDOB 25, Barcelona. http://www.doc_asia_25.pdf. Accessed 16 Apr 2020

Kipshakbayev NK, Sokolov VI (2002) Water resources of the Aral Sea basin – formation, distribution, usage. Water Resources of Central Asia. ICWC Press Tashkent, In, pp 63–67. (In Russian)

Klige RK, Myagkov MS (1992) Changes in the water regime of the Caspian Sea. Geo J 27(3):299–307

Kodyaev GW (1973) Morphometric characteristics of Lake Issyk-kul. Izv Vsesoyuz Geogr Obschestva 105(4):362–365

Konurbaev AO (1981) Decrease in biological productivity of Lake Issyk-kul due to the declining water level. Biol. Osnovy Ryb. Khoz. Vodoemov Srednei Azii i Kazakhstana: 100–103, Ilim, Frunze

Kosarev N, Kostianoy AG, Zonn IS, Zhiltsov SS (2013) The Caspian Sea and Kara-Bogaz-Gol Bay. In: Zonn IS, Kostianoy AG (eds) The Turkmen Lake Altyn Asyr and Water Resources in Turkmenistan. Springer, pp 69–74

Kostianoy AG, Kosarev AN (eds) (2010b) The Aral Sea environment. The handbook of environmental chemistry, vol V7. Springer-Verlag, Berlin\Heidelberg\New York, 332 p

Kostianoy AG, Lebedev SA, Solovyov DM (2014) Satellite monitoring of the Caspian Sea, Kara-Bogaz-Gol Bay, Sarykamysh and Altyn Asyr Lakes, and Amu Darya River. In: Zonn IS, Kostianoy AG (eds) The Turkmen Lake Altyn Asyr and water resources in Turkmenistan, the handbook of environmental chemistry, vol 28. Springer-Verlag, Berlin\Heidelberg, pp 197–232. https://doi.org/10.1007/698_2013_237

Kostianoy AG, Lebedev SA, Solovyov DM (2011) Satellite monitoring of water resources in Turkmenistan. IWTJ I(1):4–15

Kotlyakov VM, Seversky IV (2006b) Glaciers of Central Asia: current state, changes, possible influence on water resources. Almaty

Kristensen P (2004) The DPSIR Framework. In: Detailed assessment of the vulnerability of water resources to environmental change in Africa using river basin approach [online]. Nairobi, Kenya. UNEP Headquarters. http://wwz.ifremer.fr/dce/content/download/69291/913220/file/DPSIR.pdf/. Accessed 11 Feb 2018

Kroonenberg SB, Rusakov GV, Svitoch AA (1997) The wandering of the Volga delta: a response to rapid Caspian Sea-level change. Sediment Geol 107:189–209

Kurtov A (2014) Vodnye resursy kak prichina konfliktov v Central'noj Azii (Water resources as a cause of conflict in Central Asia). Svobodnaya mysl'. No. 4. Available at: swom.info/entry/350-vodnye-resursy-kak-prichina-konfliktov-vcentralnj/. Accessed 29 May 2018

Kyrgyzstan lost 30 percent of glaciers and continues to lose (2017) Informacionnoe Agenstvo www.24.kg –Kyrgyzstan, Bishkek, https://24.kg/english/61077_Kyrgyzstan_lost_30_percent_of_glaciers_And_continues_to_lose/. Accessed 8 May 2018

Kyrgyzstan 2012 (2013) FAO Water Report 39. http://www.fao.org/nr/water/aquastat/countries_regions/kgz/print1.stm. Accessed 17 Apr 2020

Lake Balkhash (2011b) Encyclopaedia Britannica (contributors Darshana Dash & the editors of encyclopaedia Britannica) https://www.britannica.com/place/Lake-Balkhash. Accessed 16 Feb 2018

Lake Balkhash (2011a) International Lake Environment Committee. Promoting Sustainable Management of the World's Lakes and Reservoirs. World Lakes Database, https://web.archive.org/web/20110716085418/http://www.ilec.or.jp/database/asi/asi-54.html#K. Accessed 16 Feb 2018

Lilienberg DA (1993) Current trends in the endodynamics of the Caspian Sea and changes in its water level. Dokl Ross, Akad Nauk 33(6):745–750

Lilienberg DA (1985) Recent geodynamics of the alpine orogenic belt of southern Europe. Geomorphology 1985(4):16–29

Lineitseva A (2011b) Changing water resources of rivers in Kazakhstan as a result of climate change impacts, 27–29 Sep, Porto, Portugal https://www.riob.org/fr/file/270333/download?token=9TbX4Emg. Accessed 11 Feb 2018

Lioubimtseva E, Henebry GM (2009b) Climate and environmental change in arid Central Asia: impacts, vulnerability, and adaptations. J Arid Environ 73:963–977

Malinin VN (1994) The problem of forecasting the level of the Caspian Sea. Russian State Hydrometeorological Institute, Saint Petersburg, 160 p

McDermid SS, Winter J (2017) Anthropogenic forcings on the climate of the Aral Sea: A regional modeling perspective. Anthropocene https://doi.org/10.1016/j.ancene.2017.03.003. Accessed 12 Feb 2018

Mendikulova FJ (2008) Issyk-Kul – problem-ridden jewel of Central Asia. In: Sengupta M, Dalwani R (eds) Proceedings of Taal 2007: the 12th world Lake conference, pp 2157–2160. http://www.Q-61.pdf. Accessed 10 Feb 2018

Micklin P (2000) Managing Water in Central Asia. The Royal Institute of International Affairs, London.

Micklin P (2014b) Chapter 15: efforts to revive the Aral Sea. In: Micklin P et al (eds) The Aral Sea. Springer Earth System Sciences, pp 361–380

Micklin P (2016) The future Aral Sea: hope and despair. Environ Earth Sci 75:844. https://doi.org/10.1007/s12665-016-5614-5

Micklin PP (1985) The vast diversion of soviet Rivers. Environment 27(2):13 p

Mueller L, Suleimenov M, Karimov A, Qadir M, Saparov A, Balgabayev N, Helming K, Lischeid G (2014) Land and water resources of Central Asia, their utilization and ecological status. In: Mueller L, Saparov A, Lischeid G (eds) Novel measurement and assessment tools for monitoring and management of land and water resources in agricultural landscapes of Central Asia, vol XXIII. Springer, pp 3–60. http://www.springer.com/978-3-319-01016-8. Accessed 26 Aug 2017

Nacionalny doklad o sostoyanii okruzhayuscey sredy i ob ispolzovanyi prirodnyh resursov za 2015 god. Vodnye resursy (2016b) (National report on the state of environment and on the use of natural resources for year 2015. Water resources. http://doklad.ecogosfond.kz/водные_ресурсы. Accessed 3 Feb 2018

Nacionalny doklad o sostoyanii okruzhayuscey sredy v Kyrgyzskoy Respublike za 2006–2011 gody (2012b) (National report on the state of environment in the Republik of Kyrgyzstan for 2006–2011)

Bishkek http://www.cawater-info.net/green-growth/files/nature-kg.pdf. Accessed 07 May 2018

Nacionalny doklad o sostoyanii okruzhayuscey sredy v respublike Kazahstan v 2010 godu (2011b) (National report on the state of environment in the Republik of Kazakhstan for 2010). Almaty. http://www.cawater-info.net/green-growth/files/nature-kz.pdf. Accessed 7 May 2018

Nacionalny doklad o sostoyanii okruzhajuscey sredy i ispolzovanyi prirodnyh resursov v respublike Uzbekistan (2008–2011 g.g.) (2013b) (National report on the state of environment and on the use of natural resources in the Republic of Uzbekistan for years 2008–2011) Tashkent, Izdatelstvo “CHINOR ENK”, http://docplayer.ru/58059280-Gosudarstvennyy-komitet-respubliki-uzbekistan-po-ohrane-prirody-nacionalnyy-doklad.html. Accessed 7 May 2018

Nacionalny doklad. Vypolnenye konvencii OON po borbe s opustynivanyem v Turkmenistane (2006b) (National report. Execution of Convention UNO against desertification in Turkmenistan) Ashgabat. http://www.cawater-info.net/bk/water_land_resources_use/russian_ver/pdf/turkmenistan-rus.pdf. Accessed 7 May 2018

Nacionalny obzor na puti k „zelenoy“ ekonomike v Tadzhihistane, sostavlen v ramkah podgotovki k Mezhdunarodnoy konferencii OON po ustoychivomu razvitiju (RIO+20) (2012) (National overview on the road towards „green“ economics in Tajikistan, prepared under the frame of International UNO conference on Sustainable Development (RIO+20), UNDP Tajikistan, Dushanbe, http://www.cawater-info.net/green-growth/files/nation_plan_small_rus.pdf, Accessed 7 May 2018

Naydyonov VI (1992) Non-linear model of Caspian Sea-level fluctuations. Math Modell 4(6):50–56

Niederer P, Bilenko V, Ershova N, Hurni H, Yerokhin S, Maselli D (2008) Clim Chang 86:227–234. https://doi.org/10.1007/s10584-007-9288-6. Accessed 5 June 2018

Novikov Y (1986) The USSR Shelves a Water Diversion Plan. WorldPaper, Boston, November 1986, 6 p

Oledenenye gornyh territoryi Kazahstana (2015b) (The glaciation of mountaneous territory of Kazakhstan) Banner, Institut Geografii, Laboratorija Glaciologii https://ingeo.kz/?page_id=2200. Accessed 29 June 2017

Palesski SV, Nikolaeva IV, Saprykin AI, Gavshin VM (2003) Assessment of contamination of the Issyk-Kul valley natural waters with uranium mine wastes. J Phys IV 107:1013–1015

Poryadin VI (2014) Ekosistemnye resursy podzemnyh vod Kazahstana: metodologiya ocenki (Ecosystem groundwater resources of Kazakhstan: assessment methodology). Izvestiya NAN RK Seriya geologii i tehnicheskih nauk 5:47–57

Propastin P (2012) Patterns of Lake Balkhash water level changes and their climatic correlates during 1992–2010 period. Lakes & Reservoirs,https://onlinelibrary.wiley.com/doi/full/10.1111/j.1440--1770.2012.00508.x. Accessed 24 November 2018

Rahimon RM (2012) Evolution of land use in pastoral culture in Central Asia with special reference to Kyrgyzstan and Kazakhstan. In: Squires V (ed) Rangeland stewardship in Central Asia: balancing 31 improved livelihoods, biodiversity conservation and land protection, Springer Science+Business Media, Dordrecht. DOI https://doi.org/10.1007/978-94-007-5367-9_2. https://pasture.klink.asia/klink/9c1102/download?embed=true. Accessed 6 May 2018

Rahimov S (2009) Impacts of climate change on water resources in Central Asia. In: Water resources management in Central Asia: Regional and International issues at stake, CIDOB editions 25, Barcelona

Restoration of Barsakelmes Nature Reserve (2018) Kazakh TV YouTube. https://kazakh-tv.kz/en/view/society/page_194937_restoration-of-barsakelmes-nature-reserve. Accessed 17 April 2020

Rodionov S (1994) Global and regional climate interaction: the Caspian Sea experience. Kluwer, Dordrecht, 254 p

Russell M (2018b) Water in Central Asia – and increasingly scarce resource, EPRS | European Parliamentary Research Service https://www.europarl.europa.eu/RegData/etudes/BRIE/2018/625181/EPRS_BRI(2018)625181_EN.pdf. Accessed 16 Apr 2020

Rychagov GI (1993a) The level of the Caspian Sea in historic times. Bull Moscow State Univ Ser 5(4):42–49

Rychagov GI (1977) The Pleistocene history of the Caspian Sea. Unpublished D.Sc. thesis, Moscow State University, 252 p. Autoreferate 62 p

Rychagov GI (1993b) The sea-level regime of the Caspian Sea during the last 10,000 years. Bull Moscow State Univ Ser 5(2):3819

Savvaitova KA, Petr T (1995) Fish and fisheries in lake Issyk-Kul (Tien Shan), river Chu and Pamir lakes http://www.fao.org/docrep/003/X2614E/x2614e10.htm#P22_1995. Accessed 11 Feb 2018

Savvaitova KA, Petr T (1992) Lake Issyk-kul, Kirgizia. Int J Salt Lake Res 1(2):21–46

Severskiy I, Kogutenko L (2016b) Studying of cryosphere in Kazakhstan. https://globalcryospherewatch.org/meetings/sale-khard2016/presentations/3.1.3.c-L.Kogutenko-Kazakhstan%20cryospheric%20activities.pdf. Accessed 10 Feb 2018

Severskiy IV, Kokarev AL, Severskiy SI, Tokmagambetov TG, Shagarova LV, Shesterova IN (2006b) Contemporary and prognostic changes of glaciation in Balkhash Lake basin. Almaty

Severskiy IV (2004b) Water-related problems of Central Asia: some results of the (GIWA) international water assessment program. Ambio Royal Swedish Acad Sci 33(1–2):52–62. http://www.ambio.kva.se. Accessed 15 Oct 2017

Severskiy I, Vilesov E, Armstrong R, Kokarev A, Kogutenko L, Usmanova Z, Morozova V, Raup B (2016b) Changes in glaciation of the Balkhash–Alakol basin, Central Asia, over recent decades. Ann Glaciol 57(71):382–394

Shnitnikov AV (1979) Lake Issyk-Kul: nature, conservation and the prospects for lake utilization. Ilim, Frunze, 84 p

Sidorova L (2008) Gosudarstva Central'noj Azii: Problemy sovmestnogo ispol'zovaniya transgranichnyh vodnyh resursov (The states of Central Asia: the problems of sharing of transboundary water resources). Central'naya Aziya i Kavkaz 1(55):92–105

Smakhtin VU, Revenga C, Döll P (2004) Taking into account environmental water requirements in global-scale water resources assessments, Comprehensive Assessment of Water Management in Agriculture Research Report 2. Comprehensive Assessment Secretariat, Colombo

Sorg A, Bolch T, Stoffel M, Solomina O, Beniston M (2012) Climate change imapcts on glaciers and runoff in Central Asia. Nat Clim Chang 2:725–773. https://doi.org/10.1038/nclimate1592

Squires V (2012) Better land stewardship: an economic and environmental imperative, if there is to be sustainable development. In: Squires V (ed) Rangeland stewardship in Central Asia: balancing 31 improved livelihoods, biodiversity conservation and land protection. Springer Science+Business Media, Dordrecht. https://doi.org/10.1007/978-94-007-5367-9_2. https://pasture.klink.asia/klink/9c1102/download?embed=true. Accessed 6 May 2018

Stanchin I, Lerman Z (2008) Water in Turkmenistan. In: Spoor M, Arsel M (eds) The last drop: water, security, and sustainable development in Central Eurasia. Routledge, London. State department of Natural protected areas https://web.archive.org/web/20120331085424/http://www.tajikpark.tj/en/index.php?option=com_content&task=view&id=1&Itemid=2. Accessed 1 Feb 2018

Stone R (2008) A new great lake—or dead sea? Science Magazine http://aquadoc.typepad.com/waterwired/files/turkmenistan_lake.pdf. Accessed 5 Feb 2018

Sustainability and Sustainable Development (2008) Circular Ecology http://www.circularecology.com/sustainability-and-sustainable-development.html#.W_rZxej7TIV. Accessed 25 Nov 2018

Svitoch AA (1991) Fluctuations of the Caspian sea-level in the Pleistocene (Classification and systematic description). In: The

Caspian Sea. Paleogeography and Geomorphology of the Caspian Area in the Pleistocene. Nauka, Moscow, pp 3–100

Taft L, Mischke S, Wiechert U, Leipe C, Rajabov I, Riedel F (2014) Sclerochronological oxygen and carbon isotope ratios in Radix (Gastropoda) shells indicate changes of glacial meltwater flux and temperature since 4,200 cal yr BP at Lake Karakul, eastern Pamirs (Tajikistan). J Paleolimnol 52:27–41

Tajikistan State of the Environment Report (2000b). http://enrin.grida.no/htmls/tadjik/soe2/eng/htm/water/state.htm. Accessed 9 Feb 2018

Tajikistan State of the Environment Report 2002 (2003b). http://enrin.grida.no/htmls/tadjik/soe2001/eng/htmls/climate/clim26.htm. Accessed 20 Jan 2018

TED Caspia (1992) Main conclusions of the technical economical report. Moscow, 48 p

The Aral Sea transboundary river basin (2013) FAO regional report, http://www.fao.org/nr/water/aquastat/basins/aral-sea/index.stm. Accessed 17 Feb 2018

The future of the Aral Sea lies in transboundary co–operation (2014) Thematic focus: environmental governance, ecosystem management, climate change, UNEP environment for development https://na.unep.net/geas/getUNEPPageWithArticleIDScript.php?article_id=108. Accessed 8 Apr 2018

The Strategic Framework for Adaptation to Climate Change in the Neman River Basin (2015) In: Korneev VN, Volchak AA et al. (expert group) United Nations Development Programme in Belarus and United Nations Economic Commission for Europe. Brest, p 64

Thurman M (2001) Agriculture in Uzbekistan: Private, Dehqan and Shirkat Farms in the Pilot Districts of the Rural Enterprise Support Project. December 2001

Transboundary cooperation in Chu and Talas river basin (2020) UNECE – ASIA – environmental policy. https://www.unece.org/env/water/centralasia/chutalas.html#c65768. Accessed 14 Apr 2020

Turkmen Lake: water conservation – a key priority of the environmental policy of Turkmenistan (2015) State News Agency of Turkmenistan http://www.turkmenistan.gov.tm/_eng/?id=4376. Accessed 5 Feb 2018

UNDP (2004) Water resources of Kazakhstan in the new Milenium. Aseries of UNDP publication in Kazakhstan, Almaty. http://waterwiki.net/images/a/ad/KazakhstanWater.pdf. Accessed 27 Aug 2017

UNECE (2007) United Nations, Economic Commission for Europe, convention on the protection and use of transboundary watercourses and international lakes. 1st assessment of Transboundary rivers, lakes and ground waters. United Nations, New York\Geneva. http://www.unece.org/fileadmin/DAM/env/water/blanks/assessment/assessment_full.pdf. Accessed 27 Aug 2017

UNEP-DHI and UNEP (2016) Transboundary river basins: status and trends, summary for policy makers. United Nations Environment Programme (UNEP), Nairobi

United Nations World Water Development Report (2020) Water and climate change. UNESCO, Paris. https://unesdoc.unesco.org/ark:/48223/pf0000372985.locale=en. Accessed 05 April 2020

Uzbekistan and FAO (2016) Partnering for sustainable production intensification and food security, FAO 4 http://www.fao.org/3/a--ax466e.pdf. Accessed 12 Feb 2018

Varushchenko SI, Varushchenko AN, Klige RK (1987) Changes in the regime of the Caspian Sea and closed basins in time. Nauka, Moscow, 240 p

Wandemberg JC (2015) Sustainable by design. Amazon, p 122. ISBN 978-1516901784

Water related vision for the Aral sea basin for the year 2025 (2000) UNESCO, France, 237 p. http://unesdoc.unesco.org/images/0012/001262/126259mo.pdf. Accessed 27 Aug 2017

Water resources and water management infrastructure of Kyrgyzstan (2020b) State water resources agency under the government of the Kyrgyz Republic https://www.water.gov.kg/index.php?option=com_content&view=article&id=365&Itemid=1470&lang=en Accessed 17 April 2020

Water resources management in Uzbekistan (2011) Preparation and publishing of the booklet was supported by the Global Water Partnership of Central Asia and Caucasus (GWP CACENA), Tashkent http://www.gwp.org/globalassets/global/gwp-cacena_images/news/water_resources_management_in_uzbekistan.pdf. Accessed 3 Feb 2018

Water resources: state (2000) Environment state of Kyrgyz Republik, National electronic report http://enrin.grida.no/htmls/kyrghiz/soe2/english/waters.htm. Accessed 10 Feb 2018

What Is Sustainability and Why Is It important (2018). https://www.environmentalscience.org/sustainability. Accessed 25 Nov 2018

White M (2018) Watering the Paris Agreement at COP24. Stockholm International Water Institute (SIWI) blog www.siwi.org/lates/watering-the-paris-agreement-at-cop24/

Why USSR Abandoned Diversion Scheme (1986) World Water 9(10):44–45

World resources 1988–89 (1988) An assessment of the resource base that supports the global economy with data tables for 146 countries. World Resources Institute. http://www.wri.org/sites/default/files/pdf/worldresources1988-89_bw.pdf. Accessed 9 Feb 2018

WWAP (World Water Assessment Programme) (2012) The United Nations World Water Development Report 4: Managing Water under Uncertainty and Risk. Paris, UNESCO. http://www.unesco.org/new/en/natural-sciences/environment/water/wwap/wwdr/wwdr4-2012/

Xenarios S, Shenhav R, Abdullaev I, Mastellari A (2018) Current and future challenges of water security in Central Asia. In: World Water Council (ed) Global water security: lessons learnt and long-term implications. Springer, pp 117–142

Yunussova G, Mosiej J (2016) Transboundary water management priorities in Central Asia countries – Tobol River case study in Kazakhstan. J Water Land Dev 13(X–XII):157–167. http://www.itp.edu.pl/wydawnictwo/journal; http://www.degruyter.com/view/j/jwld. Accessed 11 Feb 2018

Zekster IS, Everett LG (2000) Groundwater and the environment – applications for the global community. CRC Press, pp 76–78 https://books.google.lt/books?id=NueJf_WhSw4C&pg=PA76&dq=Balkhash&redir_esc=y#v=onepage&q=Balkhash&f=false. Accessed 16 Feb 2018

Zonn IS, Kostianoy AG (2014) The Turkmen Lake Altyn Asyr. In: Zonn I, Kostianoy A (eds) The Turkmen Lake Altyn Asyr and water resources in Turkmenistan. The handbook of environmental chemistry, vol 28. Springer, Berlin, Heidelberg, pp 159–176

Zonn IS (2012) Water resources of Turkmenistan. In: Zonn I, Kostianoy A (eds) The Turkmen Lake Altyn Asyr and water resources in Turkmenistan. The handbook of environmental chemistry, vol 28. Springer, Berlin\Heidelberg, pp 59–68

7 A Historical Periodization: From Nature to Early Stages of Human Settlement, to Classic Age

Abstract

In spite of its spatial extension and the absence of transport routes, the CA area played a central role—actually a crossroad—between the different Eurasian civilizations. Migration movements of nomadic and sedentary populations affected this region and they represented the inner core of the Silk Road, the most important commercial route between China and the West. During the centuries of the Ancient and Middle Age, two particular trends characterized the Central Asian political and socio-economic dimension: the presence and the co-existence of the nomadic and sedentary lifestyle. The process of settlement was not irreversible in this area where different types of ways of living could cyclically change. Furthermore, settled and nomadic population interacted and influenced each other in cultural and social terms and they occasionally established symbiotic relations in different timeframes of history. Due to its geographical location, Central Asia assumed a role of economic and cultural link between Asiatic cultures, especially Chinese and Indian, and Europe. The transit of most important trade routes in the world through CA was an important catalyst to regional economic development. Moreover, it represented the place where technological innovations were transmitted from east to west and vice versa. Finally, it was the meeting point of different religions and traditions, which imprinted an indelible mark on the regional culture. It is interesting to analyse the history of Central Asia since the ancient times, particularly focusing on socio-economic elements, which characterized the region throughout the centuries.

Keywords

Historical periodization · Early human settlements · Past main geopolitical configurations · Nomad and oasis cultures · Civilization sequences · Timurid classic imperial age

7.1 Early Stages and Ancient Cultures

During the Neolithic Age, a large number of evidences of human settlements and cultures were registered in Central Asia: they can be generally distinguished between communities of the southern CA, mostly linked to cultures of Middle East, and cultures of northern CA which moved to the south area from the steppes. The first categories were characterized by an agriculture economy, for example, Dzheitun culture located in nowadays Turkmenistan,[1] while the second ones based their activities on farming and animal husbandry, such as Kelteminar culture situated in Khorezm region.[2]

This different development proceeded in the Bronze Age. In the south zone, large settlements were erected by Oasis

[1] More precisely the settlements were discovered in the northern part of the Kopet Dag which is predominantly characterized by arid and dry climate. The soviet archaeologist Sarianidi argued that there is a high possibility that the members of this culture had developed primitive irrigation techniques to farm that zone: "Special palaeographical studies have proved that in prehistoric times this arid zone could not have been farmed without irrigation, most probably by what is called the catchment technique, which uses simple banks to retain flood-water, or channels to divert it as required", Sarianidi V., 1992:112; nevertheless Margereta Tenberg claimed that it is difficult to be certain of Sarianidi's hipothesis: "Les analyses archéobotaniques de Djeïtoun ne permettent pas de determiner avec certitude si la culture des céréales était pratiquée grâce au seul support desprécipitations (agriculture sèche), vraisemblablement quelque peu plus élevées au Néolithique qu'aujourd'hui, ou si elle nécessitait un complément en eau apporté d'irrigation à partir de la rivière, comme cela avait été suggéré précédemment", Tengberg Margareta, 2013.

[2] "The archaic aspect of the Kelteminar culture is in all respect that of a prehistoric society still at stage of the commandeering economy based on hunting, fishing and, to a lesser extent, gathering. (...) The complete absence both of cultivated cereals and domestic animal bones, other than those of dogs, proves that Kelteminar tribes had not yet advanced to a production-based economy", Sarianidi V., 1992.

I. Jelen et al., *The Geography of Central Asia*, World Regional Geography Book Series,
https://doi.org/10.1007/978-3-030-61266-5_7

cultures, which used first forms of irrigation technologies[3] and bronze metallurgy; the most known example was the Oxos civilization located in the proximity of Oxus (Greek name for Amu Darya). In the steppes region, the Andronovo culture, composed by Indo-Iranian subjects, stood out for its ability in horse domestication and for use of chariots for warfare. Their sustenance in the Kazakh steppes was mainly based on farming, pasture and also on agriculture. They turned out to be progenitors of several Iranian nomad population, including Scyths, Massagetes and Sakas. The steppes and the Oasis cultures got in touch and it seems that part of Andronovo population adopted a more sedentary lifestyle.

At the beginning of the Iron Age, both civilizations faced crucial developments: Oxos culture expanded in several oases of southern CA and it founded more complex urban settlements supported by more advanced infrastructures, such as irrigation and drainage (ID) channels, streets and fortified walls. Organized political structures were developed in order to better rule these sedentary realities. In the steppes zones, domestication of horses and the refining in metal production techniques lead to the phenomenon of Nomadism. The main reason of this development could probably be identified with the necessities of livestock, which was the core of steppes people economy.

A nomadic lifestyle based on seasonal migrations could have better supported the research of pastures in order to feed increasing cattle, which they needed to sustain growing population. Furthermore, nomadic way of life permitted to people living in the steppes to improve their military skills, thanks to horses riding abilities and use of mounted archers. The population of the steppes started to be split into tribes and clans in order to guarantee a better protection and safety to pastures and clan's members. The Scyths was one of the major nomadic cultures of CA, they moved to a large area between Dzungaria and Syr Darya basin and they started migrating to the southern regions. Around the eighth century BCE, they raided southern CA and Middle East sedentary settlements; local population failed in containing Scythian population because of their organization and military strength. Nomadic incursions continued until the sixth century BCE, when the Achaemenid civilization expanded towards Central Asia.

7.2 The Achaemenid (Persian) Empire (550–330 BCE)

The conquest of CA by the Achaemenid Empire represented the first influence established in the region by an organized sedentary state (Fig. 7.2).

Under the reign of Cyrus II, the Persian military expansion reached the southern bank of Yaxartes (Greek name for Syr Darya). During the conquest some nomad population were subdued, but other tribes continued to threat Persian power: Cyrus II died during a military campaign against the Massagetes, an Iranian nomadic people who inhabited the zone located in the northeast of the Caspian Sea. Part of the nomadic Sakas was instead integrated in the Persian state and they were enrolled in the military campaigns of the Empire. The conquered area of Central Asia was divided into four satrapies (provinces): Sogdiana, Bactriana, Margiana and Khorezm. In order to defend the acquired territories, Persian kings erected new fortresses along the borders, including Marcanda (Samarqand) and Cyropolis (Istaravshan).

Nevertheless, the central authorities demanded high taxes to the provinces (paid in gold, silver, grain but also in lazurite), the local administration was left to indigenous rulers. Likewise, Achaemenids did not try to impose Zoroastrian worship, which was the religion of the royal court, but they generally tolerated indigenous religions. Relevant infrastructural works had been built in the region, such as long-distance road and innovative irrigation systems.

These systems, called Qanat, allowed transferring water supply from mountainous zones and rivers to arid regions of Iranian plateau. They were based on a series of vertical shafts connected to underground canals, which drew from groundwater; the main channel was inclined in order to exploit gravity redirecting flows from the water source to the point of use. These irrigation tools made possible the cultivation of new categories of plants in the region, such as peaches, apricots and melons and significantly boosted demographical index of the area (Fig. 7.1).[4]

[3]As several scholars have pointed out, the development of irrigation systems during the Bronze Age could be influenced also by climate change: "L'aridification climatique de la fin du troisième millénaire est un fait attesté par les reconstitutions paléoclimatiques pour le Proche et le Moyen Orient comme pour l'Asie centrale. Ce qui pose question ce sont les conséquences de cette aridification climatique pour les sociétés humaines et la manière dont elles s'y sont adaptées. L'étude géomorphologique des sites de Sabzevar au Nord-Est du plateau iranien et de Bam situé au Sud-Est du plateau iranien montre qu'entre le Chalcholitique et l'Âge du Fer les resources en eau diminuent mais que grâce à des contextes tectoniques favorables au piégeage de la nappe phréatique les societies humaines s'adaptent en changeant de localisation et en adoptant de nouvelles techniques d'irrigation, en l'occurrence, la technique des qanâts", Fouache Éric, Francfort Henri-Paul, Cosandey Claude, Adle Chahryar, Bendezu-Sarmiento Julio et Vahdati Ali A., 2013.

[4]The importance of Qanat in the development of Persian Civilization is pointed out in the abstract of Alemohammad Seyed Hamed, Gharari Shervan, 2010: "(...) the success of the Achaemenid Empire is somehow related to the Iranian's knowledge of water. Considering that many ancient big empires were formed near big rivers such as: Nile, Tigris, Euphrates and Yangtze, Iran was the only country who become the largest empire in the history without being near a big river. During the Achaemenid Empire, if someone would have cultivated an arid land by the use of Qanat their tax would have been waved for 5 years", Alemohammad Seyed Hamed, Gharari Shervan, 2010; see as well Burri Ezio, Del Bon Andrea, Ferrari Angelo, Ragni Pietro, 2018.

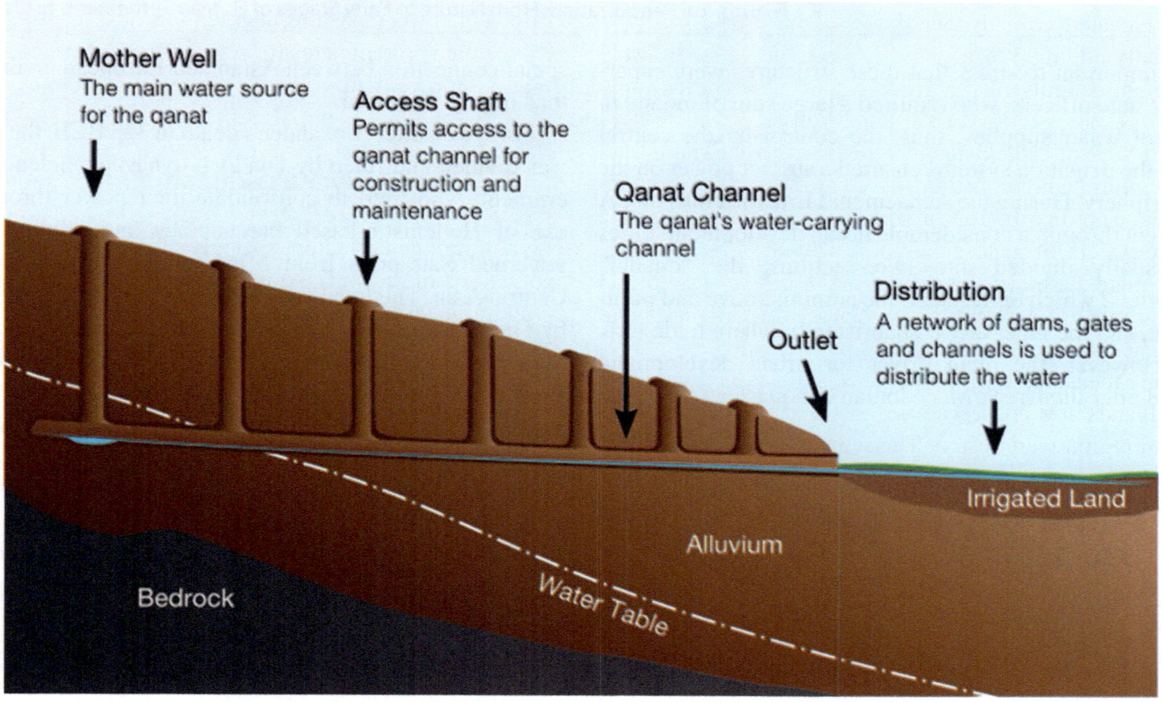

Fig. 7.1 General Schematic of a "qanat" system. (Source: Samuel Bailey, https://commons.wikimedia.org/wiki/File:Qanat_cross_section.svg?uselang=it, accessed 5/7/18)

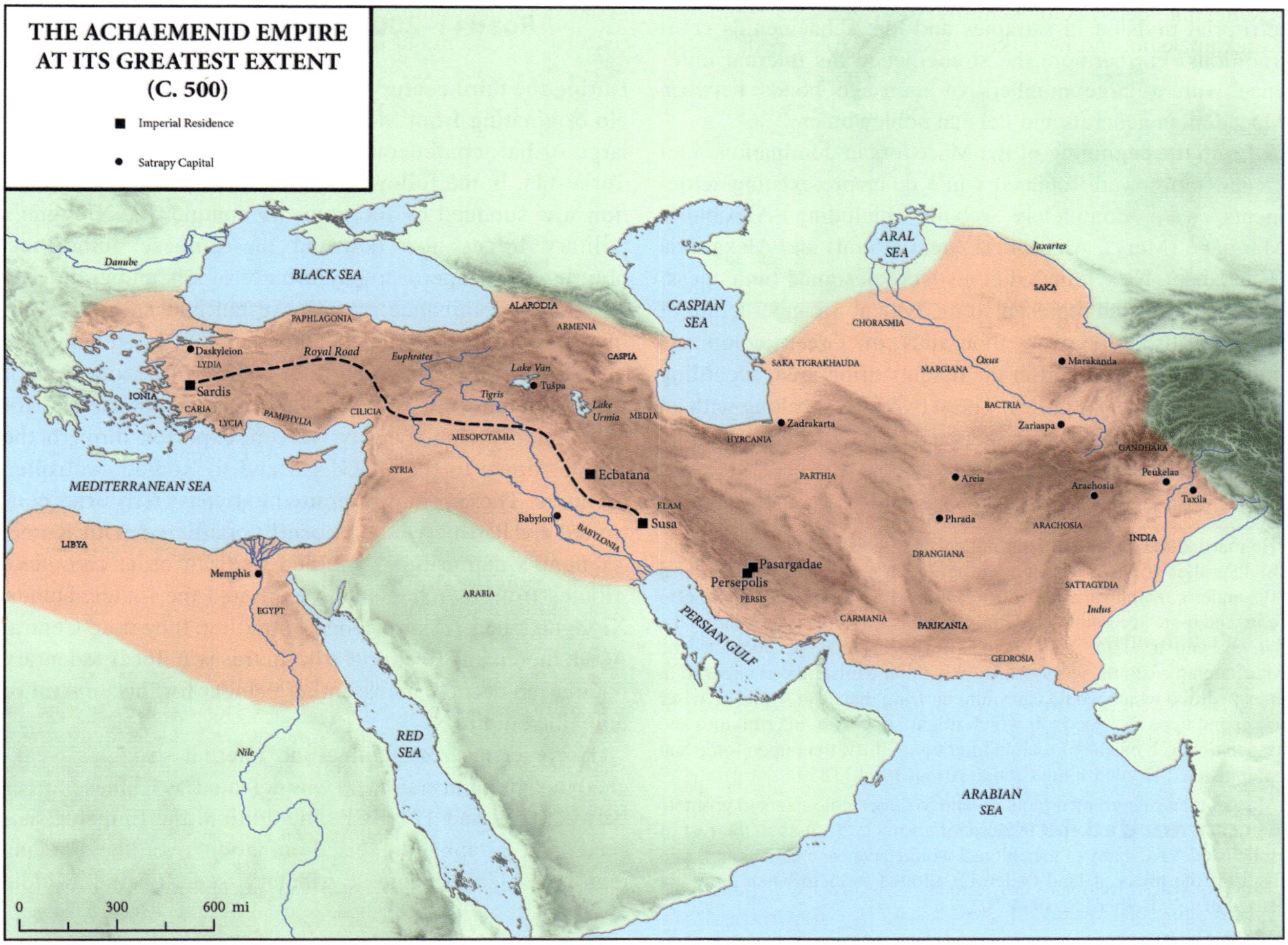

Fig. 7.2 Achaemenid Empire map. (Source: https://commons.wikimedia.org/wiki/File:Achaemenid_Empire_at_its_greatest_extent_according_to_Oxford_Atlas_of_World_History_2002.jpg. Original creator: Mossmaps Corrections according to Oxford Atlas of World History 2002, The Times Atlas of World History (1989), Philip's Atlas of World History (1999) by पाटलिपुत्र (talk) – This file was derived from: The Achaemenid Empire at its Greatest Extent.jpg. Licensed under the Creative Commons Attribution-Share Alike 4.0 International (CC BY-SA 4.0) license)

It is important to stress that these structures were supervised by state officers, who required a large sum of money in change of water supplies; thus, the control by the central state on the irrigation systems ensured a stricter power on the rural periphery. During the Achaemenid Era, the cities of CA oases went through a considerable urban development: cities were usually divided into two sections, the "citadel" ("kuhandiz") which represented the administrative and political core, and the inner city ("shahristan"), where trade took place. However, the main driver for urban development occurred after the Greek/Macedonian conquest.

7.3 Alexander's Empire and the Hellenistic States (330–150 BCE)

Alexander's conquest did not involve all the territories of the Achaemenids; most of Sogdiana, Bactria and the Fergana valley were annexed, but Khwarazm and the territories between the mouths of Oxus and Yaxartes remained under the rule of Sakas tribes. In order to keep the social and political cohesion of the Persian Empire, Alexander preserved the territorial division in satrapies and the Achaemenids court traditions. Furthermore, he strengthened his internal influence with a large numbers of marriage bonds between Macedonian generals and Persian noblewomen.[5]

From the beginning of the Macedonian domination, a lot of new cities were founded while other pre-existing settlements were completely rebuilt, including Alexandria Margiane (Merv), Alexandria Areia (Herat) and Alexandria (Kandahar). New founded cities were designed on a chessboard plan, which represented the typical urban structure of Greek polis. Moreover, thousands of Greek colonizers moved to the cities and in the neighbouring area, according to the governmental policy. These two factors, along with an expansive monetary[6] policy adopted by the Macedonian rule, stimulated economic growth and trades. The economic and social connection between Asian and European civilizations thus highly increased.

Few years after Alexander's death in 323 BCE, the Empire was divided and ruled by Diadoch dynasties. Seleucid government, who tried to consolidate their power through the use of Hellenistic-based bureaucracy and administration, governed Satrapries from Minor Asia, Mesopotamia and Central Asia. The local economy continued to be influenced by Greek traditions, such as craftsmanship and manufacturing techniques. Moreover, Seleucid's rule further improved the circulation of money due to the exploitation of silver and gold mines in Bactria. This factor allowed the expansion of international trade and the establishment of new routes with India and China. Nevertheless, the growing pressure of nomadic tribes and the ascending of new rival powers (Rome and Parthia) weakened Seleucid's power. Around 250 BCE, Parthian population started to occupy Seleucid's territories and to replace Hellenistic authority (Fig. 7.3).

7.4 The Imperial Nomadism of the Xiongnu (Around 200–54 BCE), the Han Dynasty and the First Silk Road (1–200 CE)

During the third century BCE, the Xiongnu, nomadic people originating from Mongolia, started to be organized in large tribal confederations and moving towards East Turkestan. In the following years, local sedentary population was subdued by the nomadic conquerors. Xiongnu's military forces also defeated the Chinese authorities, obliging the Emperor to pay tributes of silk and grain. The Xiongnu's dominance over Turkestan had an unforeseen impact on the economics dynamics of the region; the nomadic control permitted the opening of new routes for silk in direction of western markets. Silk goods used for payments by Han dynasty were re-exported through the steppes down to the Black sea and to Roman controlled harbours. These routes permitted to bypass territories controlled by Parths, who frequently conflicted with Rome. Xiongnu's authorities promoted trade in order to collect tributes from merchants who exploited the routes. Due to Xiongnu's lack of knowledge in trading, they used Central Asian merchants to export silk; in this way local sedentary population became an essential element for the survival of nomadic empire.

However, a great famine and internal struggles weakened the Xiongnu state who was defeated by Chinese forces between 119 and 121 BCE. In addition, the Emperor succeeded in establishing his domination over the Fergana region, which became a tributary state of the Middle

[5] He married the daughter of an influenced Sogdian nobleman, Roxana. In "Parallel Lives", ("Βίοι Παράλληλοι"), Plutarch mentioned Alexander's maintenance of local customs and traditions: "Now, also more and more accommodated himself in his way of living to that of the natives, and tried to bring them, also, as near as he could to the Macedonian customs, wisely considering that whilst he was engaged in an expedition which would carry him far from thence, it would be wiser to depend upon the goodwill which might arise from intermixture and association as a means of maintaining tranquillity, than upon force and compulsion"; Dryden John, Clough Arthur Hugh, 1859.

[6] Macedonian rule granted plots of land to Greek colonizers to promote migration. As it is reported in Bernard abstract: "(…) as in the rest of Hellenized Asia, many of the colonist would have been landowners who lived off the tracts of land ("kleroi") allotted to them when they first settled there", Bernard P., 1994:102.

Fig. 7.3 Seleucid Empire map. (Source: Thomas Lessman, https://commons.wikimedia.org/wiki/File:Seleucid-Empire_200bc.jpg, licensed under the Creative Commons Attribution-Share Alike 3.0 Unported (CC BY-SA 3.0) license) (Accessed 7/5/18)

Kingdom. Later on the Chinese Empire established diplomatic relations with the Kushan, a nomadic population who moved to the southern areas under the pressure of Xiongnu invading Bactria and Sogdiana, leading to new possible shipping routes to reach Roman harbours. Furthermore, Han authorities succeeded in reaching an agreement with Parthian Empire on the development of silk trade through their territories, opening a third way to Roman markets. Definitely defeated, the Xiongnu accepted in 54 BCE to surrender and to become vassals of the Emperor, leaving to Chinese Empire the control of East Turkestan trading routes.

During the first two centuries CE, the political stability reached by the principal economic actors (Middle Kingdom and the Roman Empire) and by trading intermediaries (the Kushan and the Parthian Empire) permitted the huge development of trade flows, in particular the silk export. Silk manufacturing know-how was an exclusive domain of Chinese Empire, where the world production was broadly centred. It was considered a valuable luxury good and it was used in China and in CA as medium of exchange and effective currency.[7]

The main customer of Chinese goods was the Roman Empire, which imported tonnes of silk in exchange of precious metals and glassware. Central Asian inhabitants actively participated on trade processes not only as commercial intermediaries but also by exporting valuable products; in particular, Fergana's horses were extremely appreciated in China as well as particular foodstuffs (as apricots and peaches). From the regions that were under Kushan Empire's control, Sogdiana and Bactria, the main exportations consisted of metal raw materials, such as gold, silver and lazurite, and agricultural products including cotton, melons and wines. The Parthian state exported textiles and carpets, but they mostly participated to the trading process thanks to their preferred geo-economic positions: they could claim a large amount of taxes in form of custom duties on the traded goods in both directions.

As mentioned earlier, three major routes linked the Chinese core silk production to Roman final customers:

- The "Steppes route": from Marcanda and Merv it headed up the Oxus river through the Khozerm region until the northern shore of the Caspian sea; it moved forward in direction of Volga's path and it reached Azov sea. From there, silk stocks continued to travel by sea. It represented the less used route because of its length and because it was not considered as a safe route.

[7]Xinriu Liu focused his studies on the use of silk products as a settlement currency: "Except for gold, silk was the most popular currencies; envoys sometimes had to pay for their food in silk. As the influence of the Han Empire increased the oasis states had to provide free food and lodging to the Chinese envoys and to send tribute, including delicacies such as raisins, as well as precious stones, furs and other Central Asian specialities. But the Han emperors did reward them with silk and other Chinese goods as gifts", Liu X., 1988:14.

- The "Sea route": from the Chinese territories it crossed the Tarim Basin[8] to Kashgar; then it went through Pamir (or it moved across Hindu Kush and Khyber Pass) along the Indus valley to Barbaricon harbour on the Indian Ocean.
- The "Parthian route": from Kashgar it continued to Merv passing through the Bactria or Fergana region, then it went ahead in Parthia to the Euphrates valley; it reached Roman provinces and commodities could be delivered by ship from Tyre or Antiochia. It represented the safest route, but its access depended on the political situation between Rome and Parthia, which was characterized by several conflicts and periods of peace.

At the beginning of the third century, trading flows across CA started to fade because of two main reasons: first, the political and economic internal problem of principal actors. After violent internal struggles, the Han dynasty disappeared and the Chinese Empire split up into three different kingdoms. After the collapse of the Imperial domination, the Tarim Basin fragmented into a large number of weak city-states, which represented an easy target for nomadic raids.[9]

At the same time, the Roman Empire started to decline because of its over-extension that made it vulnerable to external military pressure, and the crisis of its economy principally based on conquest and exploitation. The situation radically changed also in CA after the Sassanian invasion of Parthia and their substitution to Parthian at the head of the Persian state in 226 CE. Second another invasion from steppes of inner Asia took place at the beginning of the century; the nomadic people of Hephthalites (or White Huns), native of Mongolia, migrated toward the southern regions and raided cities of Central Asia, particularly in the Kushan Empire. Weakened by continuous tribe assaults, the Kushan state was conquered by Sassanians forces and absorbed into their Empire in 230 CE. Between the fourth and the fifth centuries, the Hephthalites settled in a large confederation and moved towards the south area, first conquering the Kazakh steppes and the zone near the Caspian sea, then entered in Bactria, Sogdiana and the Tarim Basin. Therefore, the presence of nomadic tribes definitively prevented all the economic and trading activities from developing in CA.

7.5 The Turk Domination (550–630 CE)

The period between the fourth and the fifth centuries was characterized by political instability and economic stagnation within the region. The collapse of the main sedentary states and the domination of the White Huns increased insecurity and turmoil within the area causing the demolition of important infrastructures such as of several irrigation systems and the loss of the economic know-how including precious metalworking techniques, skills and knowledge on trade and specialized manufacturing. Nomadic rulers established new settlements and locations, whereas historical cities as Samarqand fell into ruin. The sole exception was represented by the activities of Sogdians merchants who dealt with Huns's commercial and monetary businesses, thus increasing their influence within the nomadic state.

Around the 550 CE, another nomadic civilization emerged in the region, the Turks; they were originated from Altai Mountains and they were able to use heavy armoured cavalry. Thanks to their military advanced resources, Turks rapidly expanded from the zone of Gobi desert to the heart of Central Asia, particularly in the region latter named Turkestan. Together with Sassanian forces, they defeated Hephthalites and established a new nomadic empire from Manchuria to the Black Sea. The Khaganate consisted on a confederation of tribes ruled by nomadic aristocracy, economically based on animal husbandry. Although Turks

[8]This endorheic basin played a pivotal role throughout the centuries due to its geopolitical and geoeconomic relevance since it was crossed by the main route linking China to Central Asia. As Sir Aurel Stein pointed it out: "Geography in other respects, too, seems to have singularly prepared the Tarim Basin for its chief historical *rôle*. It was serve as the channel through which the ancient civilizations of China on the one and of Persia and India on the other, both the latter already stimulated by Hellenistic influences, were first brought into prolonged contact", Stein Aurel, 1925.

[9]Nomads frequently used to plunder Tarim Basin oases and trading caravans by exploiting the mountain passes of the Tien Shan: "Another point to be noted here has a direct bearing upon the historical past of the Tarim Basin and of the old road which, stretching along the southern foot of the Tien Shan, connects it with the westernmost marches of China. I mean the opportunities which the Tien Shan range, notwithstanding the continuity of its rampart, offers to nomadic neighbours on the north for plundering inroads upon the oases and trade routes in the south. These opportunities are due to the mountain rampart being pierced at intervals by passes practicable during a considerable portion of the year for mounted men and transports". Nevertheless, it is necessary to underline that nomadic invaders traditionally aimed at raiding and collecting tributes from oases' settlements rather than permanently conquering them. This behaviour is probably "fully accounted for by the extreme aridity of the climate, itself a result of the geographical position of the basin, which makes the local rain or snow fall a perfectly negligible factor as regards agriculture. (...) The same deficiency of atmospheric moisture restricts grazing to the narrow belts of riverine jungle. (...) This point is of distinct historical importance; for it explains why the great migrating tribes of Wusun, Sakas, Yüehchih, Huns, Turks, Mongols and the rest whom, as we now from Chinese historical records, the last 2000 years saw in successive possession of the northern slopes of the Tien Shan, were always ready to raid or to make tributary the oases of the Tarim Basin, but never crossed the range permanently to occupy it. (...) nature, by dening grazing-grounds to the vast basin between Kunlun and Tien Shan, had protected it against ever becoming the scene of great migratory movements and of such upheavals as are bound to accompany them", Stein Aurel, 1925:391, 403.

imposed tributes to the defeated population, they allowed self-governance to local authorities and freedom of worship.

The rise of the Turkic Empire boosted the economy of the region and it had a positive effect on transcontinental trade, especially on silk exports. The territorial scale of the empire and the military presence of Turks forces strengthened the security of the steppes route, which permitted to bypass Sassanian territories and to directly reach Byzantium, main customer of silk products. Moreover, the Chinese political situation stabilized in the second half of the sixth century when the Sui dynasty succeeded in re-unifying the divided kingdoms, permitting the restart of trade flows. During the period of nomadic migrations, the Middle Kingdom lost the monopoly of silk manufacturing in favour of Byzantine and Asiatic producers, although Chinese production remained the highest in term of quality and convenience. Turks sent diplomatic embassies in China and Constantinople and they arranged to obtain trading agreements to develop a direct route for the commercial flows.

Such as during the Huns domination, Sogdians were in charge of economic activities within the nomadic Turkic Empire. This complementarity of roles between Nomad and Sedentary societies represented a regular phenomenon in Central Asia. Thanks to their expertise, Sogdians carried out trading business, acting as commercial intermediaries among China, Persia, Byzantium, the Baltic area and Indian kingdoms. Furthermore, they performed banking functions, lent credits[10] and exchanged currencies, in particular Chinese silk and Persian silver. Part of Turkic population settled down and they partly absorbed uses and the culture of sedentary Iranian people, giving birth to a fruitfulness cultural melting pot.

After a period of internal tension and disputes, in the year 588 the Khaganate split up into two parts, the Eastern and the Western Khaganate; the Eastern state included Mongolia and Gobi's area, while the Western one approximately covered the region of Turkestan. At the beginning of the seventh century, Chinese Imperial government took advantage of the inner uprisings of the Khaganate state to invade the East part and to turn it into a vassal. At the end of the seventh century, the Middle Kingdom succeeded in extending its influence also on Western Khaganate, including Transoxiana and Fergana cities such as Samarqand, Bukhara and Tashkent. These events paved the way to a further deployment of trade networks and to a general economic development of the region. Nevertheless, a new political and religious power was rising from the West and it would have radically changed the history of Central Asia: the Arab Caliphate and the Islamic religion.

7.6 The Arab Caliphate Era (630–875 CE)

Arab forces took advantage of the weakened Persian and Byzantine Empire, consumed by decades of war between each other, and they rapidly advanced in Syria and Persia, absorbing the remain provinces of Sassanian state in 650 CE.

The second phase of conquest started at the beginning of the eighth century when Arabs took Samarqand, Khiva and Bukhara. In few years, the Caliphate extended its dominance over the totality of Khozerm, Transoxiana and Fergana; finally, Tashkent fell in the 750 CE. Thus, the rule of Western Khaganate ended and the Chinese Empire, defeated in the battle of Talas by the Caliphate army in 751, lost most of its influence on West Turkestan.

In the very first phase after the conquest, Arabs initiatives towards conquered populations and territories were very cruel and harsh. In many cases, the cities were crushed and looted and many people were put to death or enslaved, while irrigation infrastructures, fruit trees and cultivated lands were seriously damaged. Moreover, a lot of properties were seized in favour of Arab aristocracy and Islamic religious establishment, and new taxes were imposed to local population; in particular non-Islamic people had to pay "jizya", a heavy tax applied on persons who did not convert to Islam. The required amount was so high that this tax promoted "en masse" converting and it played an important role on the eradication of traditional worships. Landowners had to pay in addition "kharaj" or land tax, while trading and manufacturing activities were charged of tributes on sales.

Nevertheless, after the first period of chaos, Caliphate authorities significantly worked in order to rebuilt infrastructures and to boost regional economy. Street and irrigation systems were repaired and the cultivation of typical products was sustained. Furthermore, Arabs introduced the *tiraz* system, textiles factories owned and controlled by the public authorities, in order to improve the quality and the competitiveness of textile local industries in the silk trade production. Arabs rule re-launched the urbanization of the historical centres of Central Asia; the expansion process developed new urban centres alongside former city's core and introduced the *rabadas*. It was a district designed for economic activities where bazaars, craft store, manufacturing and industrial factories and trading centres were placed. The mosques, religious and social heart of Islam towns, were built near the "sharistan".

The Islam religion radically changed the social and cultural structure of Central Asian societies also through the

[10]The central role of Sogdian merchants is underlined by De la Vaissière who reports from the annals of Tang dynasty: "Ils excellent au commerce et aiment le gain; dès qu'un homme a vingt ans, il s'en va dans les royaumes voisins; partout où on peut gagner, ils sont allés". Furthermore De la Vaissière mentions references by the Armenian geographer Anania Skiraktasi (610–685 CE): "Les Sogdiens sont des marchands riches et entreprenants", De la Vaissière Étienne, 2004.

introduction of socio-political entities, such as "waqf",[11] which consisted in properties entrusted to religious authorities. They could consist in cultivated lands and productive economic structures, including shops and factories, which benefited of tax relief and inalienability. The ownership of this properties provided economic wealth and political power to Muslim clergy. Another important institution was the "iqta'" system: it consisted in granting lands of public property to members of military and civilian entourage: the new landlords were in charge of collecting "kharaj" taxes on behalf of the central state in return.

This system was commonly used to reward vassals and to entrust the rule of provinces to loyal subjects. In all conquered territories the Islamic legal system, "Sharia", was applied. Caliphate authorities reformed the local administration and the body of law regulating trade rules and measures systems to protect merchants and to stimulate trade and economic activities. Finally, new currencies system was introduced and it was based on golden ("dinars") and silver coins ("dirhams")[12] and Islamic businessmen adopted credit and payment techniques, including check and letter of credits, in order to simplify transaction and to decrease the risk. In this way, the Arab government achieved to create a profitable common market which extension covered territories from Atlantic shores of North Africa to Central Asia oases.

The progresses introduced by Arabs with regard to trade activities and the political control of two sedentary states on the commercial routes of CA lead to the improvement of long distance commercial activities. Although Tang dynasty and the Caliphate crashed for geopolitical influence of West Turkestan, both states developed a structured commercial network, which re-launched the economic activities within the area. Commercial flows enormously increased in terms of quantity and variety of exchanged goods: silk, tea, ceramics and paper from China; carpets, textiles and silver from Persia; horses, glass and jade from Central Asia regions; spices and medical herbs from India; finally, gold from the Sahara region. Flourishing of commercial exchanges had a remarkable and positive impact on productive and manufacturing activities of Fergana, Sogdiana and Bactria, and Sogdians merchant elite regained the fundamental role of trade intermediaries and bankers.

This second large-scale development of trades along Central Asia covered three different periods of time between the seventh and the ninth century: the first phase (from 675 to 745 CE) was the period of major economic expansion and it was characterized by political stability and the absence of major conflicts between regional actors. During the second period (from 745 to 800 CE) several events undermined the political and economic balance of the region. The Eastern Turkic Khaganate, which was a "de facto" Chinese protectorate, was weakened by internal disputes and was invaded by the Uighurs, another nomadic population. The Middle Kingdom fixed a complementary relationship with the new nomadic neighbours: Chinese authorities paid a very high tribute in silk supplies to Uighurs, the Khanate[13] guaranteed protection of caravans on the routes, horses supplies and military support to Tang dynasty in return.

This agreement weakened Chinese control on the main commercial routes and increased the cost of silk commodities, therefore reducing the volume of trade. Moreover, the defeat in the battle of Talas (751)[14] and domestic turmoil caused An Lushan rebellion (755–763 CE) weakened the internal power of Tang dynasty and put an end to Chinese influence in CA. Also in the central Asian provinces controlled by the Caliphate, the Arab ruler had to face several uprisings led by local landlords. In the end central authorities granted, the lowering of land taxes and accorded to indigenous aristocracies major autonomy of power; this leads to the conversion to Islam of the local aristocracy as payback. In addition, Tibet forces tried to take advantage of Tang weakness and occupied a major part of Tarim Basin in order to obtain control over the trade routes. All these factors lead to the fragmentation of commercial routes and to the substantial decrease of trade flows until the beginning of the ninth century.

During the third stage (from 800 to 875 CE) the volume of trade partly re-increased as a result of the emergence of a new branch of the Silk Road. After Tang and Uighurs armies succeeded in recapturing Dzungaria and the northern part of the Tarim Basin, merchants and traders travelled on a route which bypassed positions controlled by Tibet forces. This part of the road passed northern of Tien Shan through Dzungarian territories and arrived in Tashkent and

[11] Epkenhaus Tim, 2013.

[12] It is important to underline that the introduction of the new currency system followed a progressive approach: "The Arab conquest of Central Asia in the first quarter of the eight century produced social and political changes that resulted in an extensive penetration of the caliphate coinage into Central Asia but no immediate destruction of local mintage and money circulation. Between A.D. 720 and 760, local coins were employed along with Arabi dirhams and fulus. This period of coexistence came to an end after the 760s, when coins of pre-Islamic type were no longer minted and quickly went out of the circulation", Zeimal' E., 1994.

[13] The Khanate is a political entity typical of Nomadic Turkic tribes.

[14] The battle of Talas also represented a cultural watershed for Central Asia and the Western world, because it was a turning point in the spreading of paper industry: "(...) Arab sources reported that their paper industry started at Samarkand in the middle of the eight century when some Chinese prisoners of war were taken there after the battle of Talas river in 751 AD. (...) With the introduction of paper-making to the Muslim world and its spread during the eighth and ninth century a revolution took place in the industry. Writing material was freed from monopoly and paper became a very inexpensive product", Salam Muhammad Abdus, 1994:161.

Samarcanda. This period represented the apex of Uighur khanate in terms of territorial expansion and wealth. They managed to exploit economic benefits derived from the passage of Dzungarian routes in their territories and they could require a high amount of tributes China in terms of silk supplies. Textiles products were later re-exported towards Arab territories and exchanged with silver or gold coins, increasing Khanate's income.

However, after a period of civil war, which eroded their power on the region, Uighurs were military defeated in 840 CE by Kyrgyz, a Turchik-Mongol nomadic population,[15] who invaded the territories of the Khanate. Part of Uighurs tribes achieved to migrate to the East of Turkestan where they took control of two kingdoms inhabited by people of Indo-European origin, Qocho and Ganzhou. In few decades, they abandoned nomadic practices and embraced a sedentary and urban lifestyle. They developed remarkable trade activities and they kept on acting as commercial intermediaries between the Caliphate and China, exchanging silk to horses and jade. Furthermore, they improved large agriculture and manufacturing skills, especially in the production of cotton.

The absence of Uighur military support and the breakdown of trade flows undermined the authority of Tang dynasty, which was already weakened by autonomy tendencies from the outlying provinces. After a long period of internal conflict, the Tang dynasty ended in 906 CE and the political and economic core of the Empire moved to southern regions. The provinces of CA within the Caliphate started to gain more autonomy. Iranian local elite occupied the upper level of the administration and the provinces were ruled by indigenous dynasties, such as in Transoxiana and Fergana. Because of the decrease of trading with China and the East Turkestan, these regions of CA reoriented their economy and trade network towards the Caliphate inner market. During this period, Central Asian manufacturing and craftsmanship considerably developed and its products were traded in the whole Arab state. Samarqand first became famous for the production of glass, and second for paper, after Chinese know-how was transmitted to Arabs.

The introduction of *tiraz* permitted the implementation of a large-scale production leads to the trade of textiles; in particular Bukhara became an important manufacturing centre of cotton and silk. Islamic merchants[16] played the role of intermediaries among the major world market and they established commercial hubs in Europe, Africa, India and China. Nevertheless, in the middle of the ninth century internal struggles between Arab dynasties destabilized the Caliphate and deeply damaged trade and economic activities. Furthermore, external pressures from nomadic Turkic tribes kept on eroding the stability of CA provinces. In the end of the ninth century, the Abbasid Caliphate lost political control on the main part of its former territories, although its formal authority was still recognized (Fig. 7.4).

7.7 The Samanid Empire (875–999 CE)

The Samanids, an Iranian-Persian dynasty had already ruled Transoxiana under the Arab state. After the break-up of the Caliphate, they gained independence "de facto" and they expanded their rule on the main part of CA, in particular on Khorassan, Khwarazm, Afghanistan and on the Fergana valley. The Samanid emir, the absolute monarch of the Empire, strengthened its authority with an efficient central administration and the setting up of a massive military force, mostly composed by Turkic slave soldiers called "ghulams". The construction of fortresses and defensive fortifications along the borders was improved in order to contain Nomadic invasion from the steppes (Fig. 7.5).

During the Samanid domination Central Asia re-gained economic prosperity and cultural flourishing. Bukhara, the capital city of the Empire, became the cultural and religious centre of the region[17] and the heart of Persian civilization. Samarqand represented the economic core of the Samanid state thanks to its famous paper and glass production and its slave's market. During this period, the mining sector was significantly implemented, in particular with regard to the extraction of silver and gold from the area close to Tashkent and to the exploitation of rich mines of iron and cooper of Fergana. Textiles and specialized agriculture remained high values activities for the whole region.

The importance of Turkic presence in the Samanid army gradually increased and Turkic commanders reached to cover high positions in the Empire's hierarchy. The emir distributed land and properties to Turkic leaders through the "iqta" system, making them powerful landlords. The influence of military elite grew until they succeed to establish an autonomous state against the will of Samanids in the Eastern part of Afghanistan, the Ghaznavid Emirate, and they ceased the payment of tributes and taxes to Bukhara. This event started a period of internal struggles, which weakened the

[15]"The nomadic Kyrgyz were a powerful Turko-Mongol tribal union centred in the uppermost Yenisey region". Barisitz Stephan, 2017:69.

[16]They usually belonged to Sogdian/Persian but also Jewish ethnic groups. Numerous Jews lived and practised trade businesses within the Caliphate especially in Baghdad, where the largest Jewish community of the region had settled down, Kizami N.A., 1998:370.

[17]"(…) the Samanid era marked a turning point in the process of Central Asia Islamization. The Samanid era witnessed, for the first time, the establishment of one of the greatest culture, political and military centres of the Islamic world in Central Asia itself, on the border of the eastern non-Islamic lands. Transoxiana was transformed from a remote provincial backwater into the most powerful and important centre of the Islamic Empire, in virtually every field of endeavour", Tor D.G., 2009:287.

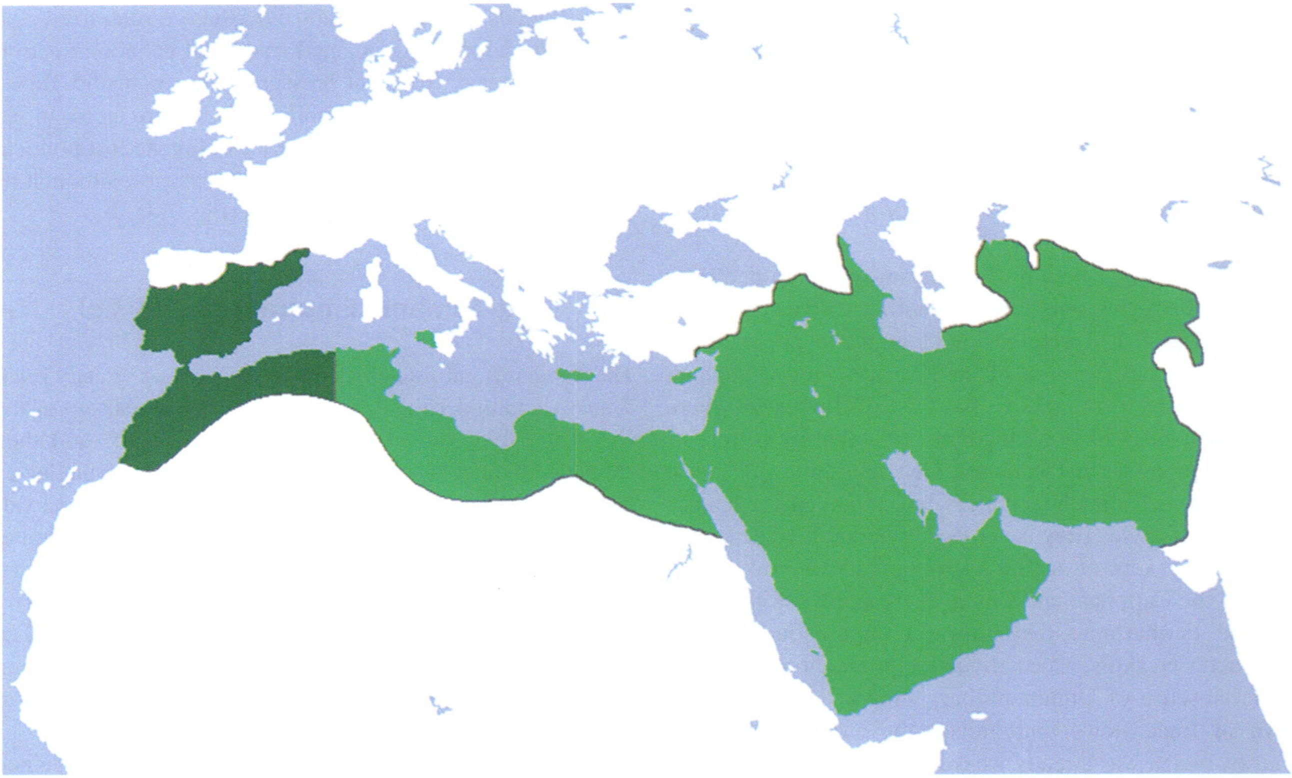

Fig. 7.4 Abbasid Caliphate Expansion. (Source: Gabagool, https://commons.wikimedia.org/wiki/File:Abbasid_Caliphate_most_extant.png, accessed 5/7/18)

state structures and it was left undefended to nomadic invasions. In the 999 CE, the Karakhanids, a nomadic population with Altaic origins invaded the Samanid Empire in cooperation with the Ghaznavids the conquered territories were split between both invaders.[18]

7.8 The Nomadic Dynasties Domination (999–1206 CE)

During the eleventh and twelfth centuries, political splintering and instability characterized the social frame of the area. The change in power of different nomadic populations remarkably undermined the performance of local administration inherited by sedentary states and considerably damaged the economy of the area.

At the beginning of the eleventh century, the Karakhanid Khanate included Zhetysu steppes regions, part of East Turkestan and the recently conquered regions of Fergana and Tashkent. Their expansion in the former Samanid territories reached the Amu Darya, which was the border to the Ghaznavid state. During the first period of their domination, the nomadic conquerors overturned the social framework of the area; they suppressed the Sassanid administrative institutions and they carried out a massive redistribution of lands among the members of Turkic aristocracy, excluding "wafq" properties.[19] These changes caused the political fragmentation of the region into small entities,

[18]"The Ghaznavids were also Turks, with their roots among the nomads of the Eurasian grasslands (…). The Ghaznavids came to power as former slave soldiers of Samanids (…). (…) The Ghaznavids display the phenomenon of the rapid transformation of a line of "barbarian," originally Turkic, slaves into monarchs within the Islamic tradition", Barisitz Stephan, 2017:89.

[19]Karakhanids converted to Islam during the tenth century, before invading the Sassanid state. Their Kingdom "was the first Muslim country among the Turkic nations. (…) we can affirm that the Islamic penetration into East Turkistan did come from the Karakhanid Kingdom. The advent of Islam among the Karakhanids is as unclear as the early history of their Kingdom. Arab and Persian sources from the 11th to the 13th century pointed to a figure called Satuq Bughra Khan who was the first convert of Islam among the Turkic Khans", Li Tang, 2005; the importance of Satuq Bughra in the conversion of Karakhanids is also stressed in the work of Jürgen Paul where he mentioned the evidence of Jamal Qarsi: "Satuq Bughra Khan (…) fut le premier des souverains (qaghan) turks de Kachgar et de Ferghana à se covertir à l'islam, c'était pendant que al-Muti'lillah était calife, pendant le règne de l'émir 'Abd al-Malik b. Nuh, le Samanide", Jürgen Paul, 2001, paragraph 22, from the site http://journals.openedition.org/asiecentrale/614, accessed 26/6/2018.

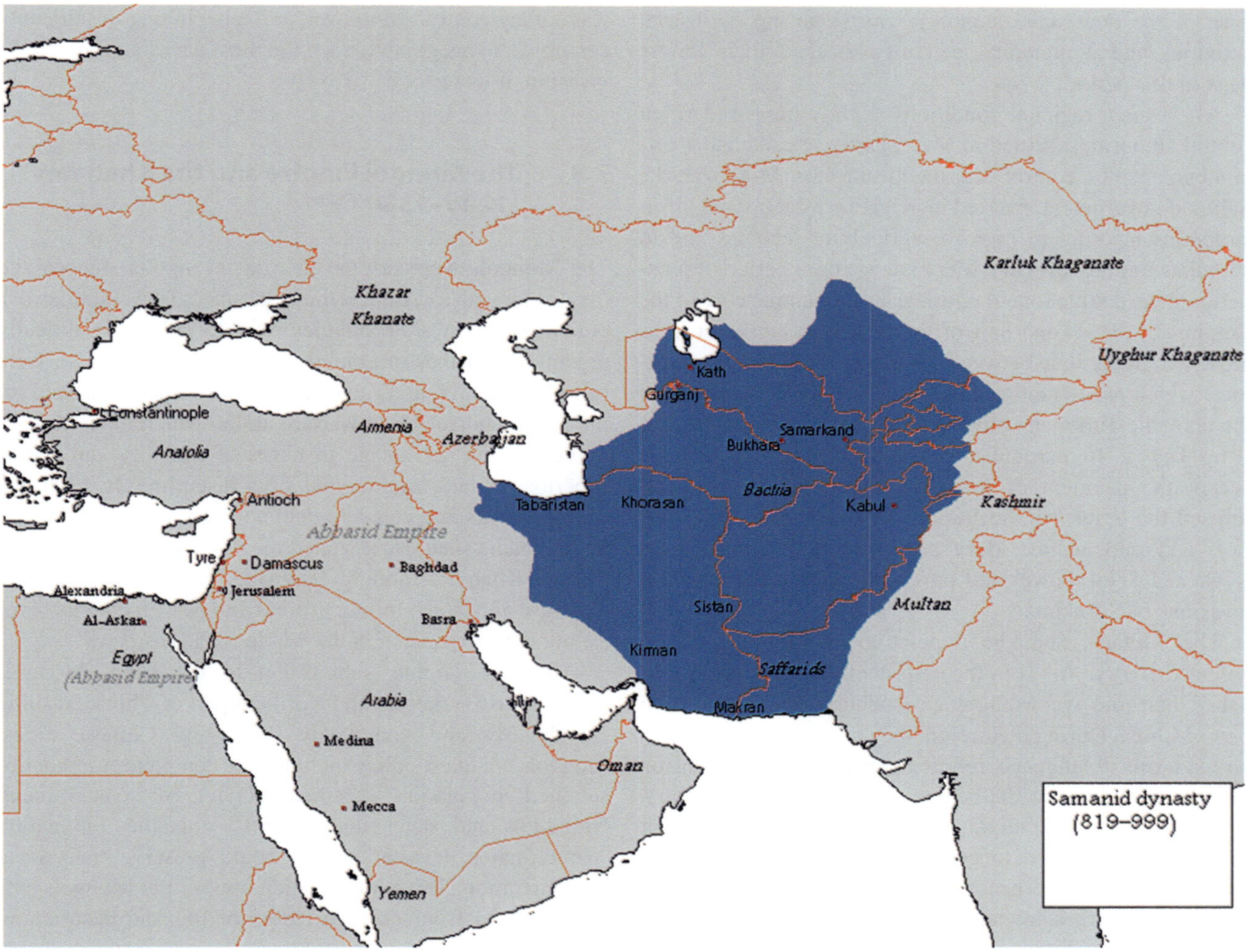

Fig. 7.5 Samanid Empire map. (Source: Arab League, English Wikipedia, https://commons.wikimedia.org/wiki/File:Samanid_dynasty_(819–999).GIF, accessed 5/7/18)

which were autonomously ruled by local nobles, who were allowed to coin their own money. After the partial settlement of the Karkhanid population and their turning to a sedentary lifestyle, the Khanate strengthened its rule re-introducing part of the former administrative institutions and regulating the monetary policy. Also the economic situation of the region re-gained stability, and local activities recovered.

The Ghaznavid conquest included Khorassan, the entire Afghanistan, Khozerm and all the territories of CA, south of Amu Darya basin. During their domination, Ghaznavid forces regularly launched several raids against Indian kingdoms in order to loot the rich territories of Punjab and Hindustan. Plunder represented a core activity for Ghaznavid economy. Taxation and tributes were the other main sources of income; high tax pressure burdened the population, in particular sedentary inhabitants of Oases, while trade and manufacturing activities considerably decreased. The nomadic rule represented in this case a period of despotic domination and economic stagnation.

In 1035, the Seljukid, native Turkmen tribes from Oghuz state, located in Aral Sea area, invaded the Ghaznavid state damaging agriculture and all the economic activities. In 1040, Seljukid forces achieved to definitely defeat Ghaznavids and later they forced the Karakhanids to accept their domination. Turkmen tribes managed to establish their government over a large area in a very short timeframe; the Seljukid Empire included the main part of CA and Middle East, including Margiana, Bactria, Fergana, Khorassan but also Iran, Iraq, Azerbaijan and part of Anatolia. The rule of the provinces was entrusted to higher members of the sultan family who re-distributed properties of lands among Turkic aristocracies and military leaders (indeed something similar to the European feudal system of the same time). Farmers and peasants, that were sedentary and generally of Iranian origins, subdued to Turkic noblemen and represented the base of the new state economy; craftsmanship, industrial activities and commercial activities reached their lowest peak in this period.

Afterwards, regional conditions slightly improved when central authorities, settled in Merv, partly re-establish a certain bureaucratic control over the local rulers. Moreover, the Seljukid government invested in irrigation systems and infrastructures in order to improve agriculture incomes and to stimulate trading upswing. Merv knew a brief period of prosperity in the double role of political and economic core of the Empire. In the second half of the twelfth century, tensions arose within the state between Seljuk semi-sedentary authorities, which partly embraced Islam and Persian culture, and the nomadic Turkmens tribes placed in the northern bank of Amu Darya. In particular, nomadic populations did not accept the payment of taxes to central government and refused the Sejuk administration; they gave rise to several riots and raid against Merv authorities that considerably weakened Sejuk power and paved the way to the nomadic invasions of Karakhitay.

The Khitans, nomadic population of Mongol origins, defeated Kyrgyz tribes in the early tenth century and established their rule over Mongolia, Manchuria and the Northern part of China. During their reign, they absorbed Chinese culture in terms of language, religion and script. The expansion of Jurchen[20], dated at the beginning of the twelfth century, forced the Khitans to migrate west and to invade neighbour states. They managed to conquer East Turkestan, forcing Qocho's Kingdom to become a tribute state, and the main part of West Turkestan, defeating Sejuk forces. The Kithans, now called Karakhitay, subdued the Karakhanids and Turkic autonomous provinces of Central Asia, such as Khozerm shahs, and founded the Western Liao Empire.

The new nomadic state did not organize bureaucratic institutions and did not centralize the administration, preferring an indirect form of control. The core of Karakhitay rule was located in the steppes regions of Zhetysu and Dzungaria, while the territories placed south of Amu Darya were vassals of Western Liao authority. Karakhitay authorities allowed local inhabitants to autonomously rule their territories in return of tributes and they adopted a religious tolerance policy. Furthermore, Liao government tried to re-establish commercial flows through CA and to improve intercontinental trade, implementing connections on the SR and providing security for traders and caravans. During the Karakhitay, domination trade exchanges along the steppes route and the southern routes moderately re-increased. Nevertheless, local population unwillingly accepted to pay high tributes to Liao's rule and then they triggered uprisings and armed rebellions, such as in the case of Uighurs, undermining Karakhitay control. Moreover, at the beginning of thirteenth century CA was going to face the most devastating nomadic invasion of its history.

7.9 The Mongol Empire and the Khanates (1206–1380 CE)

The Mongols were originated from Mongolia steppes and they belonged to Altaic ethnic group.[21] They embraced a pure nomadic lifestyle and they were socially and politically organized in tribes or confederations of tribes, which were ruled by the khan. Their military strength consisted of impressive horse riding warfare skills, which gave Mongol hordes a huge advantage in terms of mobility and speed, superior military training and strict discipline. In 1206, the Mongol tribes appointed Temuejin as leader of the new unified nomadic state. He assumed the role of supreme chief, or Gengis Khan in Mongol language. He re-organized the structure of Mongol military forces and he triggered the largest nomadic expansion in the whole world history.[22]

After the Uighur submission to the khan, Mongols invaded East Turkestan and Northern part of China, crashing Jin dynasty and completely destroying Chinese cities. Mongols conquest policy included the complete surrender of subdued population or their slaying or enslavement. Kingdoms and states that accepted nomadic domination spared major devastation and could preserve their local administration. The entities which opposed to Mongols suffered instead from massive losses of life and destruction.

[20]They were a Tunguz population who later founded the Chinese Jin dynasty.

[21]In his abstract, Louis Hambis mentioned several sources from the "Sin-T'ang-chou" (History of Tang Dinasty) which are considered as the first written mentions on Mongols: "Droit au nord des Che-Wei, il y a ce que l'on appelle la branche septentrionale de la tribu Na (?). Au nord de celle-ci il y a les Grandes Montagnes: Au delà des montagnes se trouvent les Ta Che-Wei qui habitent le long du fleuve Che-kien, lequel sort du lac Kiu-louen et se détourne ensuite à l'est. Au sud du fleuve vit la tribu des Mong-wa", Hambis Louis, 1970:126.

[22]Several scientific researches argued the existence of a direct link between the Mongol invasion and the climate upheaval which took place around the beginning of the 13th century: "(…) the expansion of the Mongols under Chinggis Khan occurred in a consistently wet and warm environment not witnessed during any other time in the last 1,112 y. The warm and consistently wet conditions of the early 13th century would have led to high grassland productivity and allowed for increases in domesticated livestock, including horses. (…) our tree-ring evidence now shows that rapid expansion of the Mongols after their unification is correlated with favourable climate conditions, which were conducive not just to increased pastoral production but to the political centralization and military mobilization that would make conquest possible. The successful campaigns of the Mongols between 1206 and 1225 against the Tangut, Jurchen, and central Asian regimes enabled the construction of a solid and sophisticated politico-military state, which in a more advanced state of the conquest could support itself not just with local resources but also with the exploitation of conquered region", Pederson Neil, Amy E. Hessl, Nachin Baatarbileg, Kevin J. Anchukaitis, and Nicola Di Cosmo, 2014:4376–4377.

Western Liao Empire deliberately submitted to Mongols in 1218, while Khorezm state resisted and was later invaded in 1219. Thereafter, Mongol conquest spread into whole West Turkestan with dramatic consequences; in 1220 Samarqand and Bukhara were torn down together with the totality of irrigation infrastructures and fields of the region, while the most part of the population were slaughtered or enslaved; the oasis of Tashkent was razed and the cities of Fergana suffered from the same fate. In 1221 Herat, Merv and Kabul fell and were afterwards destroyed. After Gengis Khan death in 1227, the Mongol expansion moved on towards Middle East and Anatolia, while in the north it absorbed Russian principalities and reached Eastern and Central Europe.

The earliest period of Mongols domination was characterized by instability and distress. The conquest of CA largely damaged also economic activities, specialized manufacturers were enslaved and sent to the steppes and many cultivated areas were converted to grazing in order to supply nomadic livestock. Moreover, the nomadic authorities not only maintained previous taxes system, but also charged new in-kind tributes to local sedentary population, turning them into slavery defaulting debtors. There were several differences with the former rule: for example, the ethnic element was not a discriminatory factor to advance in the social hierarchy of Mongol state. Furthermore, Mongol tribes did not impose their body of law, the Yassa, to subdued people but they left local elite free to retain their previous rules. In addition, the Mongols were quite tolerant with other worships. As other nomadic people, Mongols started to recruit functionaries among literate and skilled vassal populations, such as Uighurs, Chinese and Iranians, to manage the administration of the state (Fig. 7.6).

Around 1250 CE, the social situation in the conquered territories of CA was stabilized and the economic conditions better improved. Mongol authorities blocked the conversion of cultivated lands into grazing and allowed sedentary people to farm them. Moreover, Khan's rule reformed the tax system in order to reduce the fiscal pressure in charge of local inhabitants, in particular fees on trade activities decreased. Mongols began to realize the importance of trade flows within the area and the geo-economic relevance of the region, thus trying to stimulate trade with the introduction of important institutions. They developed a structured and efficient network of post stations along all the main routes in order to improve communications and transports. In addition, they introduced a system of safe passages that permitted to travel along the Imperial routes and to establish the importance of the message/travellers according to the materials of their passport.[23]

To re-launch trade activities and economic growth, nomadic imperial authorities encouraged the improvement of coinage activity, which had been abandoned after the conquest causing a high shortage of currencies. Nevertheless, Mongols did not adopt an official currency but they kept on using local coins, such as silver from Persia and paper money from China. In the second half of the thirteenth century, trade on the Silk Road rapidly reached the highest peak since the Caliphate-Tang dynasty Era.[24]

The reasons of this regained prosperity could be explained with the high level of security on the route that was guaranteed by Mongol authorities from the Black Sea to China, the efficient communications network and the lack of custom fees along the intercontinental routes. Due to these factors, prices on commercial commodities significantly decreased and trade exchanges between China and Europe flourished, in particular with Italian cities. Cities of Central Asia, placed in the core of new trade blossom, were re-built and re-populated and regained the functions of intermediary markets. After the death of Ogoday Khan, heir and son of Ghengis Khan, the Mongol Empire was split up in several sub-states or khanates that formally acknowledged, the superior authority of Great Khan:

- The Great Khanate: it included the North part of China, Mongolia and Tibet. It was the most important part of the Empire from a political point of view because the Great Khan directly ruled it. Over the years, the Khanate was deeply influenced by Chinese civilization and the court gradually absorbed some typical traditions of the Middle Kingdoms. During the reign of Kubilay Khan,[25] the capital was moved from Karakorum in Mongolia to Khanbalik in Northern China. In addition, the Khan assumed the title of Emperor and adopted the name Yuan for his dynasty. The Khanate expanded in Southern part of China and conquered the reign of Song.
- Chagatay Khanate: it consisted of East Turkestan, Central Siberia, part of West Turkestan, including Tashkent area, Fergana and Bukhara, and Northern Afghanistan (Kabul). After the first period of instability, central authorities ordered the rebuilding of the irrigation systems and

[23] This unique system was called "yam" and it was described in "Historia Mongolorum" by Friar Giovanni Da Pian del Carpine: "The Emperor sends whomever he wishes as a messenger, and he sends as many of them as he wishes wherever he wishes. One must give these messengers the horses and expenses they commander without delay. No matter they come, whether as a tribute bearers or messengers one must likewise give them horses, carts and money", Da Pian del Carpine Giovanni, 1996:65.

[24] For these reasons this period was named "Pax Mongolica"; the political stability and the economic prosperity are resumed by a common figure of speech from the period: "a maiden bearing a nugget of gold on her head could wander safely throughout the realm".

[25] During his reign, Marco Polo travelled along the Silk Road and visited the Great Khanate, describing this experience in his masterpiece "Il Milione".

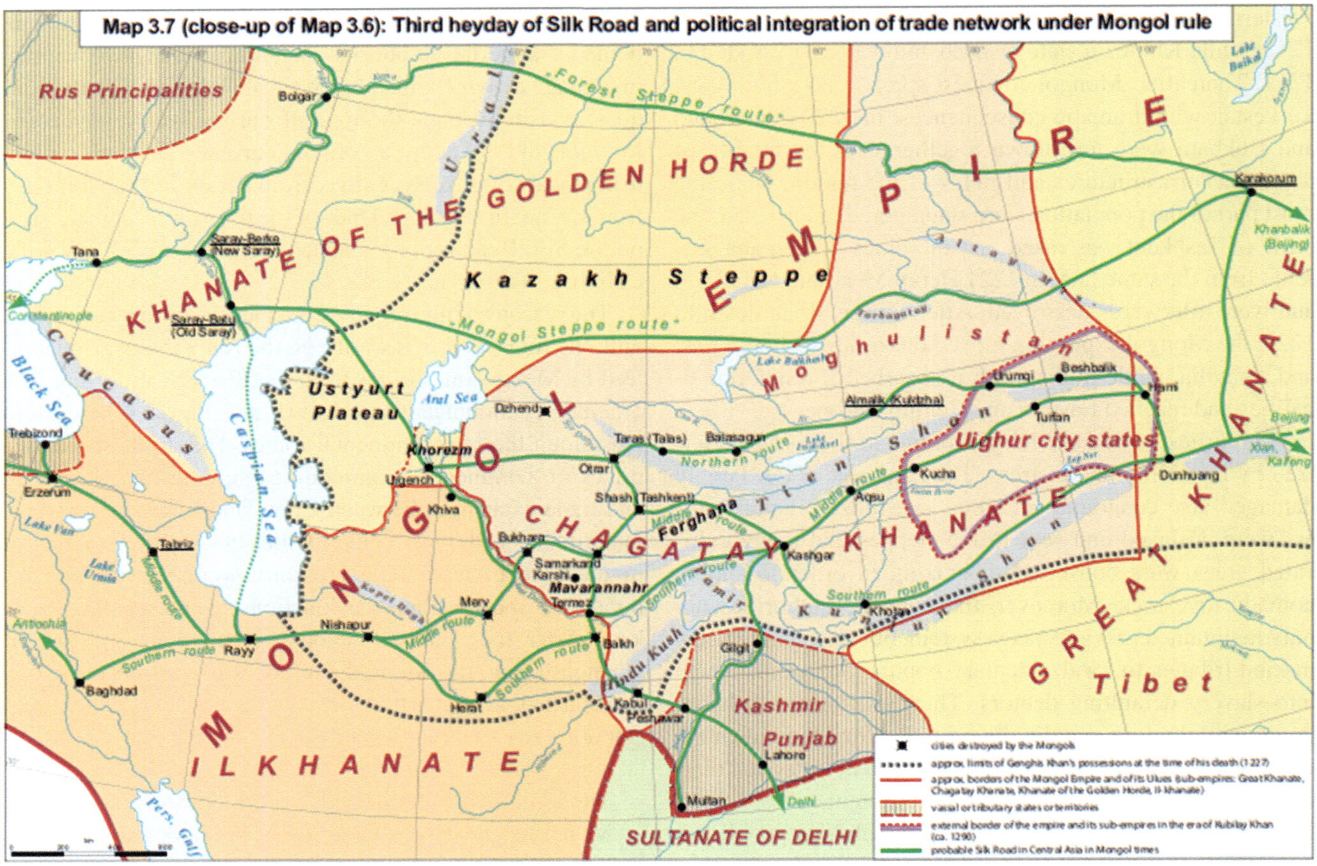

Fig. 7.6 Silk Road and political integration of trade network under Mongol Rule. (Source: Courtesy Barisitz Stephan (2017):105)

entrusted local rule to the Islamic elite. Mongols rule tried not only to re-launch trade and markets of CA but they also improved industries promoting the local production of silk. Moreover, nomadic authorities introduced a new local currency system in order to match increasing monetary demand originating from the development of regional economy.

- Golden Horde Khanate: it grouped together the territories of Khwarazm, Kazakh steppes, Western Siberia, Rus states and the main part of nowadays Ukraine. The main routes, which linked China to Europe, crossed the steppes territories that belonged to the Golden Horde. Furthermore, the Khanate played a role of commercial crossroad also between Baltic and East Europe and India. Through the steppes route, Chinese silk and spices were exchanged with silver, luxury commodities such as perfumes and jewellery, woollen cloths and velvets coming from Europe, mainly deriving from Italy. In order to better improve trade flows, nomadic authorities allowed Italian city-states, especially Genoa and Venice, to establish commercial colonies in the Black Sea. With the intermediary function of Italian merchants, the intercontinental market was significantly re-launched. In this zone, Mongol domination started a massive process of enslaving of native specialized labours and craftsmen, in particular from Eastern Europe and Russian territories.
- Il-Khanate: it was formed by Persia, Iraq, zone of Herat and Kandahar, Transcaucasia and part of Anatolia Peninsula. Even though their military success, Mongols did not achieve a stable control of Eastern Mediterranean harbours so they could not develop a significant trade network within this area. Mongol domination did not provide any economic advantage in the area; the destruction of the main part of the irrigation systems and the conversion of cultivated lands into grazing caused the desertification of large zones, especially in the Persian provinces. Moreover, the large-scale migrations of nomadic population from the steppes triggered a new process of distribution of land among the Mongol aristocracies, who introduced an

economy of exploitation levying high tributes to Iranian inhabitants.

The process of settlement of nomadic population varied from different regions. It was stronger in the Great and Il-Khanates, while population of Chagatay and Golden horde's khanate remained close to nomadic traditions. Partial shift of population to sedentary lifestyle eased the cultural melting pot and the absorbing of Turkic traditions of the conquerors. Moreover, Mongols who inhabited CA, Middle East and steppes regions embraced Muslim religion, in particular variants inspired by Sufi movement.[26]

The khanates were de facto autonomous but they recognized the superior authority of the Great Khan and they kept peaceful relations among each other until 1280. The lack of clear rules on dynastical succession, the struggles derived by geopolitical and geo-economic factors and the increasing difference among Mongols in terms of culture, religion and traditions triggered several conflicts among the khanates. Most of these struggles concerned the control over the Asiatic trading routes.

These clashes, in particular between semi-sedentary Mongols and pure nomadic ones, damaged Central Asian cities and populations that had not completely recovered. After 20 years of wars, the political situation was stabilized at the beginning of the fourteenth century, and trade and the economic prosperity of CA re-flourished for 40 years on. Nevertheless, the Black Death, which was originated from Southern part of China,[27] spread into Central Asia around 1338 carried by trading caravans along the SR. The population suffered from another demographic collapse; trade and the main economic activities almost completely ceased. The plague expanded on the steppes route and reached the Genoese colony of Kaffa on the Black Sea, from which it spread into Europe. The Khanates, weakened by the plague and the conflicts among each other, were not able to oppose to new political powers, which were rising in the area. In the second half of the fourteenth century, Tamerlane, a Turk-Mongol noble, achieved to conquer Transoxiana, thereafter he was able to expand his rule on the rest of CA.

7.10 Timurid Empire (1380–1500 CE)

Timur (or Tamerlane) obtained the control of Transoxiana in 1370 thanks to the support of Mongol tribes but also of Persian local merchants and the Muslim elite. His military success derived from a new military tactic, which combined the use of Mongol cavalry and Iranian heavy infantry. After receiving the title of Emir, Tamerlane imposed his rule on Khorassan, Khwarazm and Tashkent. He re-introduced the Mongol conquest policy against his enemies; cities that resisted the invaders were razed, as happened to Herat and Urgench. In the last decade of the fourteenth century, Tamerlane's conquest absorbed Fergana, part of East Turkestan, Persia, Mesopotamia, Afghanistan and Transcaucasia.

In the same period, Turkic-Mongol forces reached the Punjab plundered and destroyed the city of Delhi. At the beginning of the fifteenth century, Timur's Empire expanded in Syria, Anatolia and Asia Minor. Thereafter, Tamerlane started a military campaign against the Golden Horde Khanate for the monopoly on the trade routes. He sacked and destroyed the cities near Volga and the Black Sea, seeking to interrupt commercial flows on the steppes and to redirect flows in direction of the southern routes, controlled by his forces.[28]

After the first wave of destruction, likewise the Mongol invasion, a period of structural and economic re-balancing followed. Although Timur's military campaign caused a huge number of casualties and destruction over Central Asia, the Timurid domination represented a period of development from the economic and cultural point of view. The cities of Central Asia knew a remarkable demographic and architectural growth, including Tashkent and Samarqand. In particular, Samarqand, which was the capital of the Empire, reached its highest peak of prosperity; impressive mosques and madrasas were built within the inner town, and the cities became the centre of renaissance of the Islam culture (with as principal representative and symbol Ibn-Sina, Avicenna, Fig. 7.7).

During this period, Samarqand represented also the industrial and economic core of the Timurid state, although the main part of its wealth derived from looting and high taxes imposed on the other regions of the Empire (Fig. 7.8).

Tamerlane's army sacked and seized a large number of properties and resources, which were then divided among Turkic-Mongol aristocracy based in Transoxiana. Moreover,

[26] "The conversion started in the 1260s in the Golden Horde, when Khan Berke converted to Muslim faith, spread to the Il-Khanate, where Ghazan Khan adopted Islam in 1295, and to the Chagatay Khanate, with Khan Tarmashirin converting another generation later, in 1330". Fragner 2008, pp. 48, mentioned in Barisitz Stephan, 2017, pp. 116.

[27] Although the disease originated in China, several studies located the main plague outbreak in the northern part of Karakorum mountain range. Furthermore, these studies found a direct link between the climate change, which occurred in Central Asia in the 14th century, and the spread of this disease: "Our findings support a scenario where climate fluctuations that positively affect tree-ring growth in the juniper trees in the Karakorum mountains range also affect climate in a larger region in a way that can promote and synchronize plague outbreaks among the rodent populations of Central Asia. When the climate subsequently becomes less favourable, it facilitates the collapse of plague-infected rodent populations, forcing their fleas to find alternative hosts", Schmid, B. V., Büntgen, U., Easterday, W. R., Ginzler, C., Walløe, L., Bramanti, B., & Stenseth, N. C., 2015.

[28] Barisitz Stephan, 2017, p.139–140.

Fig. 7.7 Tajikistan, Turzunzoda, 2017, Ibn-Sina, scientist founder of medicine, is widely celebrated today as Tajik glory with giant posters. (Source: Photo: Igor Jelen, August 2017)

Fig. 7.8 Samarqand, Uzbekistan, 1994, Registan square, world famous Central Asian Islamic symbol. (Source: Photo: Igor Jelen)

numerous craftsmen and specialized workers were enslaved and transferred in the region from all the provinces of the Empire, impoverishing the peripheral areas of the state. The Emir's rule introduced a deep centralization of political power towards state institutions and started the erection of numerous public works, such as irrigation systems, schools, hospitals and new infrastructures, which supported the flourishing of commercial activities.

The Timurid authorities codified a new body of law, which combined elements of *Yassa*, Mongol rules, and *Sharia*, Islamic code of law. The code included a definitive adjustment on properties of lands, which was organized into four categories: the *khalisa*, land on direct control of the Emir, *mulk*, which represented private properties, *wafq* and *soyurghal*, lands given as reward to loyal subjects. Timur's domination secured the trade routes and allowed commercial flows to re-start. Furthermore, the central authorities restored the Mongol efficient communication system based on post stations in order to assist the streaming of travellers and merchants. Under the Timurid era, commercial exchanges increased but they did not reach the high level of the previous Mongol domination.

After the death of Tamerlane in 1405, the rule of the emirate passed to his heirs; they did not succeed in preserving the political control over the Empire and they were forced to withdraw from a large number of territories. In particular, they suffered from pressures from Uzbeks, a nomadic population located in the steppes. Furthermore, internal struggles between local Turkic-Mongol noblemen and Central authorities forced the Emir to allow more autonomy to his vassals, thus decentralizing the decision process.

Nevertheless, the fifteenth century was named in CA the "Timurid Renaissance" era, due to the impressive cultural blossoming of the region. Unlike Tamerlane who was mainly interested in the urban and architectural development,[29] his offspring boosted the progress of humanistic disciplines and scientific discoveries. During the rule of Timur's grandson Mirza Muhammad Taraghay bin Shahrukh, called Ulugh Beg or the "Great Ruler", the most important cities in Central Asia turned into centres of academic and scientific knowledge; Herat, Samarqand, Merv and Bukhara became famous for their mathematicians, astronomers, poets, literates, medicines and philosophers. Ulugh Beg built the most important "madrassa" of the Empire in Samarqand, which was specialized in mathematics and algebraic studies; Al-Khashi, the most famous Persian mathematician, studied indeed in this institute. Moreover, the Khan oversaw the building of an impressive astronomical observatory where he pursued astronomic studies by himself.[30]

After a phase characterized by political turmoil, the reign of Husayn Bayqara represented another period of cultural and art development. The new ruler was a strong supporter of figurative arts and literature and he rounded up in his court the best painters and literates from the whole region.

In this period, a positive economic trend appeared; the lands property reform boosted the agricultural production, while an impressive anti-inflationary monetary policy[31] was able to strengthen trade activities and to reduce the general price index. After a first period characterized by troubled and hostiles relations, Timurid Empire established with the Chinese Ming dynasty a profitable trade network. Typical Chinese commodities, such as silk, paper and porcelain continued to be exchanged with silver, velvet, horses and luxury goods from CA.

Despite the positive economic and cultural results, the political stability of the Empire was weakened by internal turmoil and external pressure. In the second half of the fifteenth century, several domestic conflicts took place, especially between Naqshabandia, a branch of Sufi movement, which claimed a more progressive idea of Islam and more radical religious factions. Moreover, the armed raids carried out by Uzbek tribes undermined the unstable Emir's authority. Finally, at the beginning of the sixteenth century Uzbeks army achieve to conquer Bukhara and Samarqand and to conquer Transoxiana and Fergana, putting an end to the Timurid Empire and to the Central Asia golden age.

[29] Evidences of the fundamental role of Timurid architecture also in its political feature: "Timurid architecture, as part of the country's "golden heritage" ("oltin meros"), is used to boost the Uzbek population's sense of belonging and pride through the construction of an ethno-national identity," Paskaleva Elena, 2015, abstract.

[30] In "The Lost Enlightment: Central Asia Golden Age from the Arab conquest to Tamerlane" the author stresses on the central role of Ulugh Beg studies: "Il suo compendio dei dati ottenuti, chiamato la raccolta La raccolta di tavole astronomiche del khan, fu chiaramente un lavoro di collaborazione che aveva coinvolto soprattutto Al-Khashi. (…) Il catalogo stellare incluso nell'opera era più completo di qualsiasi precedente elenco, e molto più accurato; anzi, è considerato la più autorevole guida del firmamento compilata tra Claudio Tolomeo nel II secolo e Tycho Brahe (1546–1601) alla fine del Cinquecento". Starr S. Frederik, Einaudi 2017:591.

[31] The inflationary trend was provoked by the decentralization of minting which caused an unsupervised and excessive supply of low value coins. In 1428 the central authority introduced a new monetary reform to contain inflation; it included the ban of low weight cooper coins and their exchange with new heavier currency. As Davidovich points out: "The main goal of the reform was to provide the market with a constant and stable cooper coin for retail trading and, at the same time, to increase revenue from the minting of such coinage", Davidovich E. A. and A. H. Dani, 1998:414.

References

Alemohammad Seyed Hamed, Gharari Shervan (2010) Qanat: an ancient invention for water management in Iran, Water History Conference, Delft, Netherlands. http://hamed.mit.edu/sites/default/files/Qanat_WHC_2010.pdf. Accessed 26 June 2018

Barisitz Stephan (2017) Central Asia and the Silk Road, economic rise and decline over several millennia. Springer, Cham

Bernard P (1994) The Greeks Kingdoms of Central Asia. In: History of civilizations of Central Asia, volume II, the development of sedentary and nomadic civilizations: 700 B.C. to A.D. 250. UNESCO Publishing, Paris

Burri Ezio, Del Bon Andrea, Ferrari Angelo, Ragni Pietro (2018) Analisi ambientali e valorizzazione culturale dei qanat di Yazd, patrimonio dell'Umanità (UNESCO), Asiac annual conference 2018, 5–7 December 2018, University of Trieste – Gorizia campus (manuscript)

Da Pian del Carpine Giovanni (1996) Historia Mongolorum quos nos Tartaros appellamus. Branden Publishing, Boston

Davidovich EA, Dani AH (1998) Coinage and monetary system. In: History of civilzations of Central Asia volume IV, the age of achievement: A.D. 750 to the end of fifteenth century, part one: The historical social and economic setting. Unesco Publishing, Paris

De la Vaissière Étienne (2004) Les Sogdiens, un peuple de commerçants au coeur de l'Asie, from the site: https://www.clio.fr/BIBLIOTHEQUE/les_sogdiens_un_peuple_de_commercants_au_cOe_ur_de_l_asie.asp. Accessed 26 June 2018

Dryden John, Clough Arthur Hugh (1859) https://en.wikisource.org/wiki/Plutarch%27s_Lives_(Clough)/Alexander. Accessed 26 June 2018, citing Plutarch's Parallel lives

Epkenhaus Tim (2013) Defining normative Islam: some remarks on contemporary Islamic thought in Tajikistan – Hoji Akbar Turajonzoda's *Sharia and Society*. In: Heathershaw John, Herzig Edmund, 2013, the transformation of Tajikistan: the sources of statehood. Routledge, London and New York, pp 95–110

Fouache Éric, Francfort Henri-Paul, Cosandey Claude, Adle Chahryar, Bendezu-Sarmiento Julio et Vahdati Ali A (2013) Les régions de Bam et de Sabzevar (Iran) : une évolution dans l'implantation des sites archéologiques et dans la gestion des ressources en eau compatible avec l'hypothèse d'une aridification croissante du climat entre 2500-1900 BC, In: Cahiers d'Asie centrale [En ligne], 21/22 | 2013:2, mis en ligne le 30 septembre 2014, consulté le 21 novembre 2018. URL: http://asiecentrale.revues.org/2081

Fragner B (2008) Die "Khanate": Eine zentralasiatische Kulturlandschaft vom 15. Bis zum 19., mentioned in Barisitz Stephan 2017

Hambis Louis (1970), L'Histoire des Mongols avant Gengis-Khan d'apres les sources chinoises et mongoles, et la documentation conservee par Rasidu-D'-Din, in "Central Asiatic Journal: international periodical for the languages, literature and history and archeology of Central Asia", 14(1/3)

Kizami NA (1998) Popular movements, religious trends and Sufi influence on the masses in the post Abbassid Period. In: History of civilzations of Central Asia volume IV, the age of achievement: A.D. 750 to the end of fifteenth century, Part one: the historical social and economic setting. Unesco Publishing, Paris

Li Tang (2005) History of Uighur religious conversions (5th – 16th centuries), Asia Research Institute, National University of Singapore, 2005, downloaded from http://www.ari.nus.edu.sg/wps/wps05_044.pdf. Accessed 26 June 2018

Liu X (1988) Ancient India and ancient China: trade and religious exchanges, AD 1–600. Oxford University Press, Delhi

Neil Pederson, Amy E. Hessl, Nachin Baatarbileg, Kevin J. Anchukaitis, Nicola Di Cosmo (2014) Pluvials, droughts, the Mongol Empire, and modern Mongolia. PNAS 111(12):4376–4377. From the site https://doi.org/10.1073/pnas.1318677111, Accessed 24 Nov 2018

Paskaleva Elena (2015) Ideology in brick and tile: Timurid architecture of the 21st century. Cent Asian Surv 34(4):418–439. https://doi.org/10.1080/02634937.2015.1118207

Paul Jürgen (2001) Nouvelles pistes pour la recherche sur l'histoire de l'Asie centrale à l'epoque karakhanide (10e–début 13e siècle", In: Cahiers D'Asie Centrale: Études Karakhanides, 9, 2001, http://journals.openedition.org/asiecentrale/614. Accessed 26 June 2018

Salam Muhammad Abdus (1994) Reinassance of science in Islamic countries. World Scientific Publishing Co. Pte. Ltd., Singapore

Sarianidi V (1992) Food-producing and other neolithic communities in Khorasan and Transoxiana: Eastern Iran, Soviet Central Asia and Afghanistan. In: History of civilizations of Central Asia, volume I, the dawn of civilization: earliest times to 700 B.C. Unesco Publishing, Paris

Schmid BV, Büntgen U, Easterday WR, Ginzler C, Walløe L, Bramanti B, Stenseth NC (2015) Climate-driven introduction of the Black Death and successive plague reintroductions into Europe. Proc Natl Acad Sci U S A 112(10):3020–3025

Seyed Hamed, Alemohammad, Shervan Gharari (2010) Qanat: an ancient invention for water management in Iran, Massachusetts Institute of Technology 2010, downloaded from the site: http://hamed.mit.edu/sites/default/files/Qanat_WHC_2010.pdf. Accessed 26 June 2018

Starr S. Frederik (2017) The lost enlightment: Central Asia's golden age from Arab conquest to Tamerlane. Paperback Einaudi, Milano

Stein Aurel (1925) Innermost Asia: its geography as a factor in history. Geogr J 65(5):377–403. https://doi.org/10.2307/1782547

Tengberg Margareta (2013) Économies végétales et environnements en Asie centrale du Néolithique à l'époque sassanide: la contribution des disciplines archéobotaniques, in: Cahiers d'Asie centrale [En ligne], 21/22 | 2013, mis en ligne le 30 septembre 2014, consulté le 20 novembre 2018. URL: http://asiecentrale.revues.org/2057

Tor DG (2009) The Islamization of Central Asia in the Samanid Era and the reshaping of the Muslim World. Bull Sch Orient Afr Stud Univ Lond 72(2):279

Zeimal' E (1994) The circulation of coins in Central Asia during the Early Medieval Period (Fifth–Eighth Centuries A.D.). Bull Asia Inst 8, 245(new series) http://www.jstor.org/stable/24048777. Accessed 22 Nov 2018

8 Modern Era and Modernization Processes Until the Soviet Collapse

Abstract

Sometimes, human processes have a tendency to accelerate, a fact which induces unavoidable asymmetries, since the changes usually occur on different scales, in different moments and in different spaces: it is the situation which typically characterizes modernity and which, originally induced by a set of practical and technological changes, would soon exert further effects on many dimensions of the geographical reality. Colonial expansionism, and further modernist changes, would represent the major aspects of such processes, involving and affecting almost the whole world. Russian colonialism and then Soviet domination represented the main actors of these epochal changes in the CA. However, after three-quarters of a century, the Soviet empire also collapsed, leaving local populations and institutions without any reference for territorial and social organization or for cultural elaborations either. In fact, Russian and Soviet domination penetrated deep into the local cultures, so any impact, either material or immaterial, from the collapse would continue to exert an effect for an extended period.

Keywords

Modernity changes · Colonization · Colonial frontier in the steppe · Institutional territorialization · Territorial state expansion · Soviet revolution in Central Asia · Post-Soviet passage · Formation of new geopolitical units · Independency of local societies · Transition time migrations · Post-independence adaptations · 1989 conflicts

8.1 The Epochal Change

8.1.1 Early Travellers, Missionaries and Merchants

The movement of merchants, explorers, pilgrims and of other travellers along the SR began in ancient times, and from that time forward, it was never completely interrupted, even during times of war and invasion, even in the most critical of moments.

For centuries and millennia, travellers on this path left few or no documents. Usually, merchants and transporters did not travel the entire road, from west to east, form the north to the south, or the reverse, but just a segment of it, because the merchandise could be better loaded and delivered by local carriers, using specific vehicles—chariots, animals or caravans etc.—in different environments (mountain passes, disorienting steppe and desert spaces).

Nevertheless, starting from late Medieval period (in the "pax mongolica" epoch) it is possible to observe the cases of several travellers completing the whole itinerary as there is information and written evidence (diaries, notes, descriptions, letters etc.) of such travels.

This prefigured the beginning of a new type of literature, which would have a strong impact over the following centuries, with western traveller-writers representing new attitudes and a new expansionist epoch.

It includes a number of missionaries (often from the Franciscan order) undertaking long journeys, departing from European cities; they were mainly friars (namely the "intellectuals", who had the occasion to write and preserve their observations), inspired by spiritual medieval culture, but already demonstrating a kind of modern curiosity.

The missionaries had a different approach to all other travellers who had covered such routes; they had a habit of writing and of describing. This was an operation that, considering the times, was demanding. They took notes in difficult circumstances, conserving papers and memories until

I. Jelen et al., *The Geography of Central Asia*, World Regional Geography Book Series,
https://doi.org/10.1007/978-3-030-61266-5_8

they came back home—after years of long travel. Sometimes, they were sent on diplomatic, rather than idealistic missions, with the duty of describing and reporting for the domestic institutions which had sent them—a friars' order, a brotherhood or directly the Pope, who was the head of western Christianity at that time; they had the possibility to accumulate information and to organize further proselytist actions (typical of the Catholic traditions), bringing the "happy news" to whom, they imagined, was without God.

These would be followed—as it is possible to imagine—by further adventurers, venture captains, members of confraternities, or simply people in search of opportunities or who were escaping some circumstances; these movements became more consistent in the late Medieval period (after the Crusades), with the European society slowly but firmly opening to the outside world, demographically and economically expanding.

Then they were merchants in particular, who for different reasons (to those of the missionaries), were not used to writing about the countries they visited or to render public any business news. This is with the notable exception of the Venetian Marco Polo. He dictated his memories, namely his descriptions about the marvellous countries he had the chance to visit in the Far East, and along the road while crossing the CA countries.

He did this towards at the end of his career, when he was already in prison, arrested by the Genoese rivals of Venice; he described mountain ridges and strange animals, oases and fabulous cities, sophisticated cultures and impressive rituals in a book that has not lost any of its fascination over the years and which is still a popular read today, even considering some approximations in his narrative and as regards his memory.[1]

Soon it became the "best seller" of the period, circulating in an extraordinary manner, soon being translated into the popular languages of the period (from the vulgar French to elitist Latin and vernacular Italian), becoming the basic element of knowledge about the "orient" in western countries for centuries. So, Marco Polo became famous not just because he travelled these paths, but because he wrote (actually dictated) his memories, propagating information, as well as stereotypes, approximate descriptions and images of those countries and populations, consequently stimulating the curiosity of generations of his countrymen (among them even Columbus).

[1]Lentz W., 1933.

8.1.2 From Explorations to the Beginning of Colonization

With modernity, the situation changed; with the progressive affirmation of a new political culture, institutions and various organizations started to develop exploration systematically, which had essentially been pursued individually until that moment, by people with precise motivations. This period started in the seventeenth century, possibly promoted by individuals who had previously read the Jesuits' narratives—such as those of Matteo Ricci and Bento de Goes.

This movement contributed to overcoming the medieval religious monopoly of information concerning these regions (that was based on the interpretation of the Holy Books), to the benefit of a pluralistic idea of culture, that had the possibility to develop freely in diverse circumstances. These explorations showed a special interest in developing knowledge about any aspect of the newly discovered places and populations, and of their picturesque way of life.[2]

Indeed, the Jesuit travellers were already characterized by a new awareness, namely by the awareness of belonging to an organization (the counter-reformation church, namely the religious hierarchy), which wanted to systematically develop knowledge and possibly influence over the new territories—among them China particularly and in general the "orient".[3] This new narrative, composed systematically on the basis of instructions, relied on direct observation, as an attempt—made by "intellectuals" at the time—to find an explanation for the diversity of cultures, places and populations, which had been made evident by the circulation of such descriptions and that consequently needed a rational explanation.

The next step coincided with eighteenth century Enlightenment (to some extent, a consequence of the same seventeenth century Jesuits' findings regarding the diversities of the cultures, reinterpreted as a new cultural paradigm); this would signify the start of a wider development process, prompting industrial development and colonial capitalism, with the definitive expansion of the western civilization, usually, at the expense of the native populations.

[2]Marvin Harris, 1971; the geographical knowledge demonstrated soon to be an instrument for political expansion.

[3]Also for looking to roots of their own religion, intending the Christian religion as originally an oriental one; then, possibly looking for ancient legendary Christian communities, like the Nestorian ones, lost in the Orient since time immemorial; see the legend of Prete Gianni, "presbyter Johannes"; in the centuries, pilgrims, missionaries and Enlighteners were followed by colonial armies and representatives of organized nineteenth century states, preparing the road for masses of migrants arriving with mechanized transport (rail, ship); Marvin Harris, 1971; Wessels G., 1992; Meloni Alberto, 2017; Jelen Igor, 2002.

This was a period in which it is possible to observe a dramatic upset (overturning) of the power relations on a wider scale, with the progressive affirmation of centralized institutions, accumulating an unprecedented quantity of power, using new political and cultural weapons.

This also concerned further territorial exploration, which would be organized by companies and institutions, and then directly by governments, in a political way. They would increasingly sponsor exploration enterprises through academies, private or public "colonial companies", establishing "ministries of new territories" appropriately equipped and deliberately founded with this intention in mind.

8.1.3 Geographic Discoveries and the Beginning of Modernity

The modernist turn induced a planetary-scale change, upsetting a millenary continental organization, relegating, after the discovery of the oceanic maritime routes, the core of Eurasian landmass to the periphery of the new world order. In this scenario, the CA territories became an enormous outback, and eventually the space and resources for the incumbent colonial expansion. This process would soon be facilitated by a sequence of further changes with which western countries would generally acquire superiority compared to the rest of the world.

At first, the western states invented and learned to use some technological weapons (lethal fire weapons, with organized artillery); secondly the same powers, as true territorial-state organizations, had the ability to consolidate themselves and their new technological equipment in centralized bureaucracies and permanent armies. In this way, they could further develop innovative knowledge, constructing long range infrastructures, starting large-scale public works and pursuing an efficient schema of statehood.

Then, they started a process that would last for centuries and that, from the European core, would expand into most of the oecumene, spreading more intensive forms of resource exploitation, connected increasingly with industrial economics. In the case of CA areas, this assumed a character of occupation, with settlers expropriating territories to the detriment of both nomadic groups and ancient oasis civilizations.

Such patterns particularly characterized the described steppe "corridor", the narrow—considered on this scale—strip between the Taiga forests (approximately, on the north side), and the CA deserts and mountains, crossing Eurasia transversally. Indeed, the advancement of new states (representing structured and sedentary civilizations) was concentric, proceeding from all the directions towards the "empty" continental "heartland"; but from the north-western side, it became more intensive, mainly coinciding with Russian imperial expansion.

In fact, the Russian colonizers, in search of cultivable soil, and freedom from their domestic aristocratic regime, started a process of appropriation of such territories, while applying agrarian methods: in this sense, colonial expansion assumed a form of revenge of the sedentary populations towards the steppe tribes, considered as barbarians, who had ruled and ultimately enslaved the populations of this side of Europe in the previous centuries (until XIV), expanding just beyond the same steppe-taiga line, assumed as a kind of natural frontier between the nomads and sedentary populations.[4]

In an early period, the expansion was led by the Cossacks, originally peasants mainly escaping serfdom or possibly, even earlier, nomads conforming to Orthodox Christianity, settling in the "no man's land" between cultivated land and the steppe; they conserved some original habits, and like the nomads, a highly mobile way of life, practising cattle breeding combined with agricultural activities; then—once sedentary power acquired military superiority—they started to expand eastwards, beyond the frontier, rapidly exploring and occupying wider areas.

Soon this movement would be a state organized movement, with the czarist power taking control of expansion, organizing trade routes, supply chains and military outposts, eventually carrying out a war of conquest. The conventional start of this new era—on this side of Eurasia—is assumed to be the conquest by the armies of Ivan Grozny ("the terrible") of Volga khanates, Kazan (1552) and Astrakhan (1556); it signified the change in nomad-sedentary relations, starting the expansion of what can already be considered a modern state along the trade routes, in the direction of various Asian destinations.

In this way, the organized (western) state would advance into apparently empty spaces, following the merchant caravans and the Cossack avant-gardes, who would accomplish further extraordinary enterprises, crossing the entire Siberian space reaching the Pacific, exploring the Arctic area, up to the occupation of the west coasts of the American continent.

However, in the other direction, southwards, early Russian attempts were unsuccessful, delaying expansion. This is probably also due to the relatively low interest of settlers in conquering the environment of the arid steppes (currently central Kazakhstan), in principle superfluous to farmers of European origin.

In fact, early expeditions into this area, in the early XVIII century, at the time of Peter the Great, resulted in a sequence of military disasters: the first Russian armies were sent to the

[4]The Taiga southern limits approximately correspond to current Russian-Kazak-Mongolian border, which may be considered a natural but also cultural discontinuity between nomads and sedentary.

area with the aim of extending control over a further segment of the SR, as well as of the liberation of Russian slaves captured by Turkmens—supposedly a sensitive motivation, used for mobilizing popular sentiment—and the destruction of the tribal raiders, but they were bloodily defeated.

The situation changed after the Russian–Persian wars in 1828, when the Russians acquired a definitive advantage in the area, with the Caspian sea becoming a kind of Russian lake, opening the route to the entire area; the east coast port of Krasnovodsk (Turkmenbashi today) would then become the starting point for the occupation of the road to Ashgabat, predisposing the further occupation of the whole Amu Darya valley.

8.1.4 Becoming an Object, Instead of Being a Subject, of World Geography

The schema for colonial expansion was more or less similar. The early pioneers were soon followed by merchants, adventurers and fur hunters, and then by groups of field diggers, farmers seeking free land to settle and other organized colonizers.

Then, after the political institutions started to acquire control of such movements during the seventeenth century, state-owned companies, public institutions and organized corporations also settled in the area. For each group these spaces—as for any new colonial acquisitions—represented different things: a kind of promised land, a way of escaping from serfdom, a unique opportunity or simply a resource to be exploited.

It also became a kind of empirical penitentiary, namely the destination for czarist opponents, criminals or simply members of persecuted associations; they were sent to these remote regions with the intention, besides that of punishing them, to contribute to settlement purposes.[5]

Further movements in the centuries of ideological changes and of religious reforms, regarded confessional discriminates, outcast groups and sects, possibly belonging to ethnic minorities, who intended to refound their communities in remote villages, far from any imposition. All these represented, not always intentionally, a "resource" for colonization, and confrontations with local indigenous populations soon began.

In this context, a special role was played, specifically in CA, by groups of Muslim Tatars; they originally belonged to the communities of the Khanates in the Volga region, which, after being expropriated by the Russians, had started a migration in this direction.

They usually came from the communities of traders (culturally already integrated into the empire's structures), which would be encouraged by czarina Caterina in the eighteenth century to settle these places at the edge of the empire with the purpose of diffusing literacy, religion and any kind of "modernist" culture: an "enlightened" way of extending control over the populations of the steppe, apparently intractable to any integration.

They did indeed play a decisive role. The Volga Tatars were capable of communicating with the local CA population, contributing to the organization and integration of the tribes; they were also teachers and missionaries, starting what would later become the movement of the Jadid—literally "new method", which represented the precedent for any Muslim reformist movement in the area from that time forward.[6]

The "enlightened" czars and their governments soon realized that the continental outback represented a reserve of resources, and then a base for further expansion; they understood the importance of the human element of integrating the native populations into the imperial context.

Therefore, they started to encourage the settlement of populations arriving from the European side of the empire, prospecting the increase in demographic pressure and political tensions, at the expense of the populations of the steppe.

The Russian authorities started the exploitation of natural resources, like minerals, soil and wood, and of agro-pastoral productions, of which the new territories were immensely rich, and that would soon become necessary to feed the growing metropolitan industry.

Beside this, the czar had a broader need to reinforce control of long-range trade routes, towards China and the Far East, southern and parallel to, the Tea Road (stretching through the Russian Siberian cities to Mongolia and China) and the Zibeline Road (penetrating the Siberian forest), north of the SR, looking for a more direct way to reach the Orient.

Furthermore, the Russians also had to prevent the expansion of other empires in the same area: therefore, the aim was both of a strategic and economic character, continuing the war of conquest, preparing permanent garrisons, for generals in search of glory, but also to the benefit of soldier–peasants migrating in search of land.

For all them, this enormous area appeared as an opportunity, simply an available space, where they could not find any trace of organized life, and where they could simply settle down, founding villages, building towns and roads, connect-

[5]In some cases, deportation in such areas would become, rather than a punishment, the concession of a further possibility, what in Russian literature may be interpreted as a "redemption"; see also Rossi Marina, 1997.

[6]Buttino Marco, 2003; Barisitz Stephan, 2017; Freeman Jacqui, 2013:32 fs; Moxley Alissa, 2013:64; Adeeb Khalid, 1998.

ing outposts with a fortified lines of "ostrog" and "forts"[7]; then building long range supply chains, equipped with communications systems and, later, telegraphs and railroads.

But this occupation represented a lethal threat for the native population. In fact, with their simple movements, the colonizers—not always intentionally—fragmented the area of the local nomads; this consequently led to the construction of new networks of borders, infrastructures and towns and they could simply no longer accomplish their survival activities.

Thus, this "mobile" frontier in the steppe became the typical colonization scenario: geographers, military, as well as journalists, illustrators, merchants, missionaries and spies—the modern western travellers—could observe such situations, taking notes and writing their diaries, describing places, cultures and peculiarities. In this way, the colonial narrative became an expedient to encourage colonial migrations: such remote lands must have appeared as somewhat charming to western readers, who could observe images of people without homes, perennially on the back of a horse or camel, carrying everything on their "kibitka" (the typical nomads' home on wheels); they were breeding herds and cattle in the endless space of the steppe, without (apparently) a destination, challenging extreme environmental conditions. In this confrontation, the Western colonizers (mainly arriving from Russian empire) matured the idea that the more technological developed civilization would become dominant, not just politically and economically, but culturally as well, finally legitimizing a policy of exploitation.

8.1.5 Sequence of Explorations

A forerunner of colonial exploration is the geographer Von Humboldt; he had been sent by the czar to CA countries, with the objective of producing descriptions of these territories, but keeping silent, as per contract, about sensitive political and human aspects.[8] This, with further explorations would open the period of the classic CA orientalism—namely of the "western" knowledge of the Orient especially by the Russians, with explorers and geographers, geologists and ethnographers, each of whom would have the chance to found a specific branch of the new science.

They were often enrolled as soldiers, adventuring in this area, motivated by a passion purely for exploration or just looking for conquest and glory; in other cases, they were true scientists motivated by the discovery of new cultures and territories, applying modernist study methods which had just been codified by Enlightenment philosophy. Sometimes, they were non-Russian (originally Germans like Radlov, Polish like Prževal'skij, Holland, Baltics, and even a steppe insider, like the Kazakh Valikhanov), but were called from all over the empire to the court of the czar to undertake the systematic study of the new territories.

All of them may be defined as the pioneers of oriental studies, obviously from a western point of view, nevertheless signifying a fundamental step in the history of knowledge: new territories were mapped and inventories of resources drawn; languages codified, oral tales and epics transcribed; local idioms registered in dictionaries and grammar books; habits, beliefs and values registered in ethnographic collections.

And so on with the interpretation and collection of data, measurements, descriptions and any further scientific activity in order to accumulate and render these new territories functional to the new state organization and to the new economy.

Almost anyone who undertook explorations in this area, in this period, was able to reach some worldwide glorious success. Geographers like Pëtr Petrovič Semënov (1827–1914, pupil of Von Humboldt), or like General Nikolaj Prževal'skij explored the remote inaccessible mountain regions: the former was later called Tjan-Šanskij in his honour; the latter gave his name to the famous horse, which today is almost extinct.

The geologist Fedčenko discovered what was considered the largest glacier in the world in those times, the poles not having yet been discovered.

The czarist official and spy Cokhan Valikhanov, explored the frontier of the empire, becoming interested in local cultures; he started the study and the transcription of what would be defined as the longest epic tale that had ever been composed—since it was recited orally for hundred thousands of lines of verses—namely the Kyrgyz poem Manas.

The philologist Radlov discovered and studied the mysterious inscriptions of the ancient Altaic populations, representing the starting point of any further studies in this sector, while the historian Barthold started to systemize the CA culture in the wider context.

Besides what the western geographers and scientists claimed to have discovered, such circumstances were also the occasion for an encounter between cultures, which sometimes took on unexpected significance. Among them it is necessary to cite the deported Dostoevsky, who spent many years confined in Omsk, in the openness of the steppe, and his comrade, the Kazakh prince and scientist who has already been cited, Valikhanov,[9] a genial figure, pioneering the dialogue between colonialists and colonized (but playing an ambiguous role).

[7] Namely the Russian 'linija', Barisitz Stephan, 2017.

[8] Von Humboldt A., 1975.

[9] Valikhanov Č. Č, 1985; Valikhanov capt., Veniukof M., et alii, 1865.

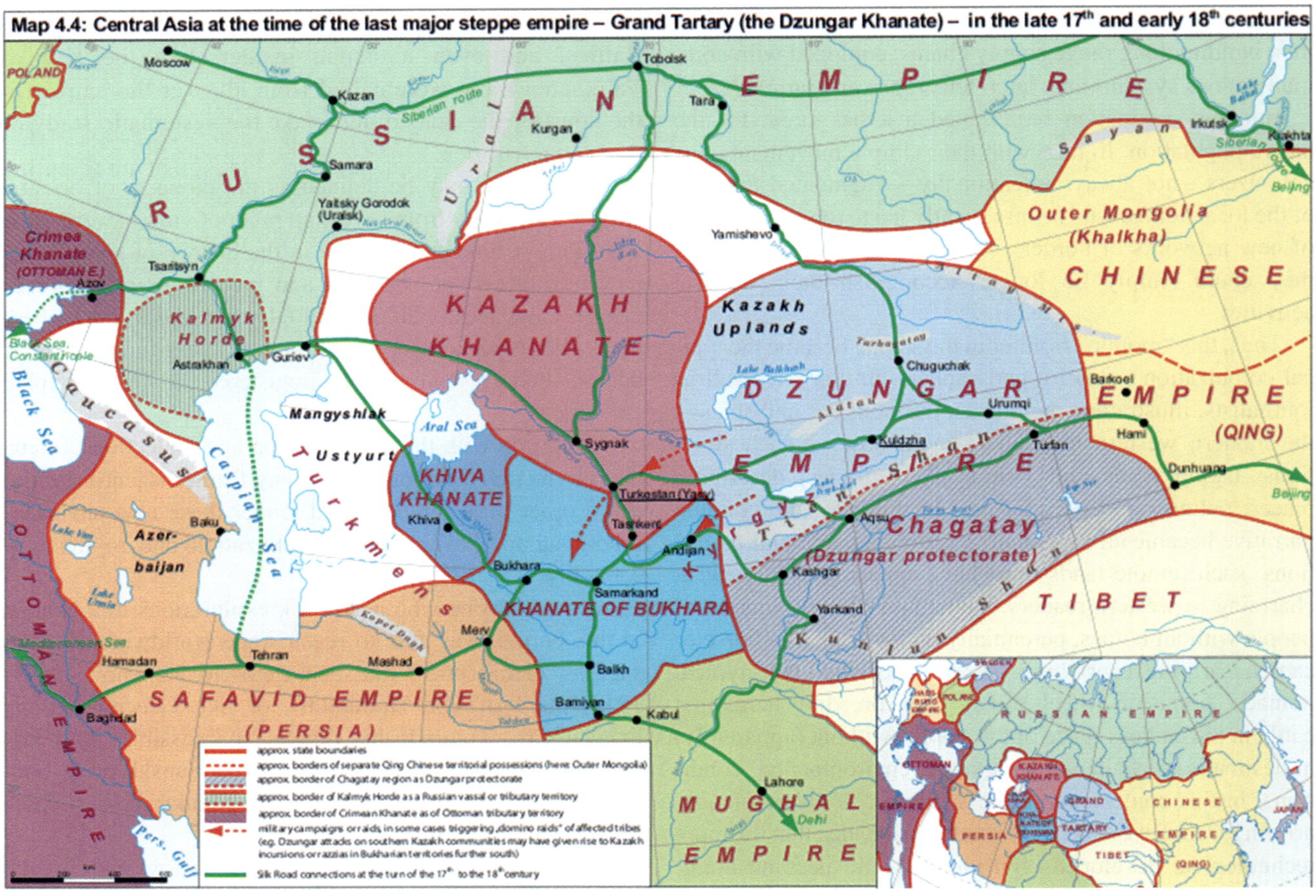

Fig. 8.1 Central Asia in the time of the last major steppe empire, the Dzungar Khanate. (Source: Courtesy of Barisitz Stephan, 2017:173)

Then, other European colonial powers also tried to intrude into this space, sending envoys and expeditions, sometimes secretly: a space indeed, as was soon evident, which would be claimed by the Russian Empire in competition with the neighbouring empires; these were respectively expanding from the south (the British one), or resisting on the eastern side (under China Qing dynasty, which in this period reoccupied the Tarim and Dzungarian area, denominating them as "western provinces", Xinxiang), and on the south-western lines (the post-Savafid Iran). All this happened "over the head" of the local emirates, Bukhara and Kokand, which would rapidly be incorporated by the Russians (Figs. 8.1 and 8.2).[10]

Not only empires and nations participated in this "race" of exploration and missions, but so too did a number of organizations not directly interested in the area and not particularly interested in territorial expansion, companies, trade organizations and simple individuals. This is the case of independent explorers—an activity that in this period became a professional practice—businessmen, geologists, cartographers, reporters and illustrators, who described and reproduced images, circulating them within the new media (journals, magazines, books and encyclopaedia printed figures), greatly impressing their metropolitan readers.

Some of them became very popular, publishing books and newspaper articles about their adventures, looking for the "roof of the world" (Pamir), for a possible Shangri-La in the Himalayas, for further mythical representations, or simply for archaeological treasures or agricultural possibilities; sometimes they just became the "stars" of the new popular exotica imagination.

Among them the Hungarian born, British naturalized citizen, Sir Aurel Stein, archaeologist and alpinist, and the Swedish Swen Hedin, Richthofen's pupil, who explored the area, looking for the "wandering lake" (the Lop Nor), continuously appearing and disappearing in the vastness of the Taklimakan basin, which was considered a crucial resource for making such arid extensions cultivable; he drew the first systemic Central Asia Atlas, published posthumously, definitively describing the geography of this—until that moment—mysterious region.[11]

[10] As said, after a history of military non-success which had occurred in the previous centuries.

[11] Hedin S., 1913; Nazaroff P., 1933.

Fig. 8.2 Central Asia in the Late Nineteenth Century. (Source: Courtesy of Barisitz Stephan 2017:243)

8.1.6 The Confrontation Scenario

Often such movements just became a metaphor for a military campaign with its leaders in search of glory, for merchants in search of easy gain, for archaeologists in search of vanished cities and for any kind of adventurers in an (apparently) lawless territory. All this contributed to the creation of a typical colonial scenario, characterized by technical and military–political supremacy, in which the modern state could plan any kind of conquest.

It resulted rather in confusion, characterized by intentional and non-intentional confrontation with local populations and ancient cultures; it immediately highlighted a gap, with transformation processes accelerating as consequence of the application of new technologies (military, industry, railroad and telecommunications), realized consequently in the context of modern states. The confrontation evidenced discrepancies, with colonial apparatus affirming definitive supremacy, yet appearing—to the eyes of the local populations—as almost anything metaphysical and equipped with monstrous capability. It was immediately followed by the spread of a corresponding ideology, in order to justify the conquest, indeed inspired by the same ideologies which tried to explain such differences; the local "genre de vie" was defined as something archaic, destined soon to die out (even when, indeed, it was not much different from the economic survival characteristics of preindustrial rural Europe).

Such confrontation was based on a typical modernist idea of progress, which would also produce many ambiguities over time. It accelerated the impact on local societies, characterized by their "medieval" features, and defined by the western rhetoric, as despotic- and feudal-aristocratic societies based on serfdom in the oasis, and—in the open spaces of deserts and steppes—by tribal habits and "warrior democracy". Such representations were created in a purposefully simplistic way by XIX positivist geographers, applying western theories to the local situation.[12]

Indeed, the supremacy scenario was the result of a set of processes, of discoveries and of technical innovations, which

[12] Khazanov A. M., 1994; an ideological manoeuvre, ethically and scientifically not tolerable, as it is possible to state today retrospectively.

made western politics much more efficient than any other country in the rest of the world: a supremacy that would last for half a millennium, which would redefine the non-western culture as something barbarian or inefficient, and depending on points of view, sophisticated, but baroque and self-referential, not really compatible with any idea of progress.

Besides these considerations, the geo-political element possibly occasioned the decisive one: the simple contact with the modern idea of a centralistic state, capable of systematically using artillery and fire power to maintain permanent armies and colonial apparatus and which proved to be disastrous for the local populations, and this continued until the same modern state definitively conquered the entire territory available, and until, in the twentieth century, it would provoke the definitive assimilation and the cultural (and sometime genocidal) destruction of indigenous populations.[13]

At the same time and in a complementary manner, the political process induced the migration of millions of farmers arriving from the overpopulated European countries (that indeed were characterized by centralization processes, meaning expropriation of peripheral and rural areas, sometimes devastated by civil wars, induced by the same processes of construction of the modern state): entire generations were just fleeing from "modern" European despotisms, religious discrimination, political and, increasingly, racial persecution, looking for new opportunities and for the chance to create an ideal state.

This happened in different periods, from times of early exploration to mass migration characterizing the second half of the nineteenth century (they were the liberated peasants in particular, after the abolishment of serfdom in 1861); finally to the twentieth century establishment of a totalitarian regime (with the SU), and to late modern policies in the 1950s and 1960s.

In these terms, the colonialism era appears difficult to define univocally, since victims and persecutors are often interchangeable, with "effects and causes" chains sometimes inverting; sometimes the colonial conquests were planned, sometimes they happened spontaneously, sometimes originally expressing an ideal intention (i.e. to bring the modern civilization in these remote areas), sometimes realizing a brutal predatory attitude from the very beginning.

8.2 Colonial Times—The Consolidation

8.2.1 The Fulfilment of Colonial Expansion

After an early phase of "casual" expansion, the later waves of colonizers brought with them a new perception of territory, pursuing a systemic exploitation scheme. At first, on their arrival, the colonizers founded settlements, creating continuity with supply lines, linking local resources with metropolitan areas, setting up infrastructures and occupying the whole available territory.[14] They usually either founded new outposts and forts in the open spaces or occupied the rich oasis towns on the biggest CA rivers, sites of ancient and sophisticated civilizations; in a second moment, they completed the occupation of the wide peripheral territories used by nomadic shepherds as pasturage areas.

This movement alone was the cause of a structural change: the simple construction of a new road, of a fortress or of a farmers' hamlet, the reclaiming of a forest or of a swampy area, interrupted the itineraries of the herders, and of the system of transhumance, based on a fragile ecological calendar in such a difficult environment.[15] This fact could provoke the destruction of the entire "genre de vie", intentionally or not. To the eyes of the colonizers, the life of the steppe was not usually visible, or only at certain moments of the year: nomads intentionally would not leave any traces on the territory they were occupying, and they exploited the land seasonally, in a non-destructive manner, since they expected to use the same resources in the following years in the context of a long-term cycle. In this way, those groups could adapt and survive in the "longue durée" in such extreme conditions, elaborating indeed a very smart strategy.

On the contrary, the western-sedentary colonizer intended to impact the local territory deliberately, following a precise plan in order to appropriate those resources, demonstrating the signs of the new sovereignty.

8.2.2 The Stages of the Conquest

Sometimes, the colonial armies anticipated the civil settlers, at others they followed them, continuing expansion until they reached the margins of the CA "continent", and the territories occupied by further empires beyond the frontier line.

As the epilogue of the race to gain the "empty spaces" was the establishment of the "Pamirski post" (currently Murgab) in 1893, on the side of the mountains (in the south-east); so too for the Geok Tepe oasis final battle against the Turkmens (1881) on the desert side (in the south): these events signified the end of a phase, but the start of a further period as well, anticipating expansion beyond those natural barriers—the Amu Darya river and the southern mountain ridges.

Indeed, from this moment forward, aware that the expansion spaces were close to exhaustion, imperial policies began to change. Instead of planning the conquest of new territo-

[13] Pianciola Niccolò, 2001, 2004; Marzhan Thomas, 2015.

[14] Barisitz Stephan, 2017.

[15] Jelen Igor, 2012.

ries, the imperial authorities began to consolidate internally, using the available ones more intensively. From that moment on the priority would be consolidation of the conquest and the organization of the confrontation with the new contender on the other side of the frontier—namely another European power expanding in the opposite direction, in this case the British Empire advancing from the south, from the Indian sub-continent.

The 1907 Russian-British Entente evidences the preoccupation with elaborating such conservative geopolitics, planning buffer areas (like the Wakhan corridor and indeed like much of the desert and remote mountain areas) and consolidating the frontiers; in this moment, a "game" of influences and manoeuvres was started whose real goal was the avoidance of a direct war between these powers, who could consider themselves, at that point, as satisfied with their colonial acquisitions. A period that would be defined as the "Great Game", specifically in corresponding Russian narrative, as the Shadow Tournament (Турниры теней).

Nevertheless, even though they reached a kind of "modus vivendi", they left some uncertainness along some segments of the borderline in the areas where, due to physical reasons, it was difficult to mark perimeters and definitive lines; this is the case of deserts, high mountains and glacier areas.

While the southern border was precisely delimited by natural barriers (the Nebit Dag mountain with the Persian empire and the Amu Darya with Afghanistan), the problem of marking the border was more evident in the frontier with China; here several borders could not be defined at all, and this occurred for a long period, until recent times; it concerned the boundary between the current Kazakh, Kyrgyz and Tajik republics with China (see later).[16]

In the same period a project of consolidation of all the areas of new acquisitions began, no longer to be considered as a "mobile" expansion belt; it meant the construction of military outposts, of infrastructures, but of settlements as well, which would slowly change from garrison winter quarters to true colonial towns.

These were quickly structured in networks, with facilities, linkages and administration, which would have further consequences "inside" the territories and populations, becoming areas of assimilation and "tout court" of destruction of the indigenous cultures, totally disoriented by such changes; rarely such encountering and melting represented the chance for reciprocal advantage.

Finally, the construction of settlements contributed to the formation of a vivacious colonial society, made up of the military, businessmen, officials, and ambitious adventurers and their families, arranging an acceptable "western" style life around them. These "new towns" were distinct from the pre-existing ones, defined as "old towns", demonstrating the difficult integration between the colonial and local society (a fact that would soon have tragic consequences) (see Fig. 8.3 for the case of Samarkand).

8.2.3 The Railroad

Lastly, starting in the second half of nineteenth century, the construction of the main colonial railway included the CA periphery in the metropolitan system structurally, opening new possibilities for colonial exploitation and, to some extent, development.

The construction of the railroads articulated the advancement of colonial expansion which occurred on three axes, the Trans-Caspian, the Trans-Aral, and in the planned (but constructed later, in the 1930s of the twentieth century) Turk-Sib ones.

The Trans-Caspian (also called the Central Asian Railway, Russian: Среднеазиатская железная дорога) constructed from 1879 onwards would soon connect Krasnovodsk on the Caspian coast with the Turkmen oasis and Ashgabat, then crossing the Amu Darya, connecting to Zarafzhan valley, Bukhara and Samarqand (reached in 1888), reaching Tashkent and Andijan (1898); since then, it has represented the main instrument of the Russian authorities' dominance as well as the major instigator of modernization for local society and economics.

The second axis, the Trans-Aral Railway, ran across steppes and deserts, connecting with the main branch of the Trans-Siberian railways, east of the Aral sea, from Orenburg, to Tashkent, and was subsequently completed in 1906 and rapidly became the major rail terminal in colonial CA.[17] The same railway network alone induced a new geography of settlements, creating a set of cities that immediately assumed a new importance to the detriment of the old political precolonial centres.

Finally—and already in Soviet times—the last big railway, the Turk-Sib, operating since 1928 from Barnaul to Tashkent (actually a side branch of the Trans-Siberian railway) which completed the CA network; it connected the central steppe area of Kazakhstan, the Zhetysu cities, Almaty and Bishkek, with the major network and with transcontinental routes (it is noteworthy to say that, since 1990 a track to China from the junction Aktogay is open, through the Dzugarian gate completing a new trans-continental itinerary).[18]

In different periods, the colonial approach changed, and this greatly depended on the efficiency of the railway connections. In the first phase, the Russian occupation had left some authority in the frame of an indirect governance

[16] Polat Necati, 2002; Megoran Nick, 2017.

[17] Barisitz Stephan, 2017.

[18] Reid Patryk, 2017; Heathershaw John, Herzig Edmund, 2013:21.

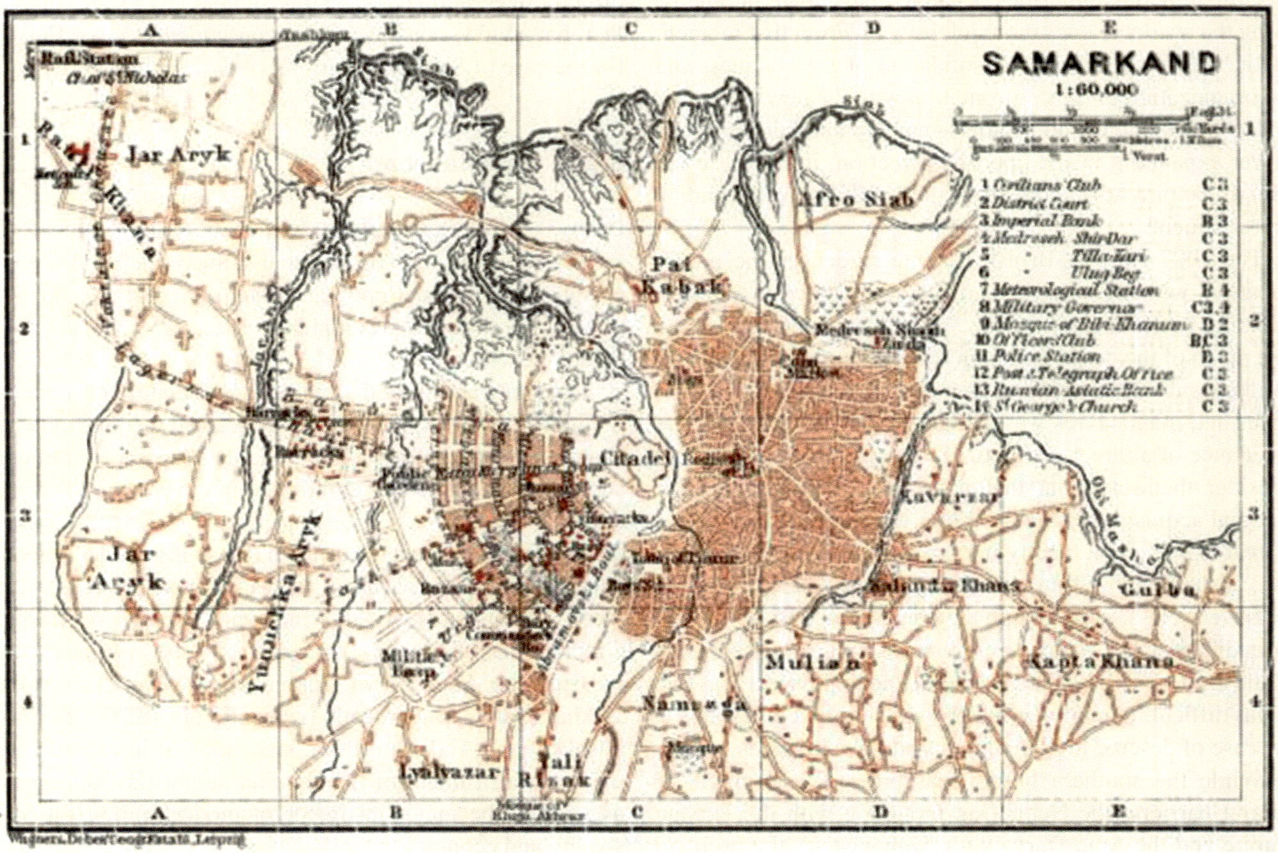

Fig. 8.3 Samarkand 1910 from Baedecker K., Russia with Teheran, Port Arthur and Peking: handbook for traveler, Leipzig, 1914; it is evident the diverse organization of the new town, left side, from the traditional town, separated by the citadel, right side. (Source: https://commons.wikipedia.org, cited in Tonini Carla, 2017:43)

through vassal states to local traditional powers, namely the old emirate of Bukhara and Kokand (and Khiva), which had presumably the "know how" to govern the local territory, as well the peripheral and sparsely populated areas of these newly conquered regions.[19]

With time this approach changed, and as soon as the czarist power realized it was capable of extending its direct rule—possibly thanks to the setting up of infrastructures and the use of modern political technology, it did not hesitate to suppress the precolonial emirates, imposing a new order.

Thus, the Russian conquest had diversified impacts, forming a new set of urban centres, some barely apparent, as in the case of remote peripheries (of Tien Shan and Pamir valleys), desert and steppe areas, where the colonialist influence remained limited, leaving the traditional institutions (clan, tribes and villages) the possibility for survival.

[19] see the figure of the "pristan" the Russian authorities adopted to stabilize their influence in the wide borderless territory of the steppe, Sultangalieva Gulmira, 2014:63.

8.2.4 Change in Economy

On such extended surfaces, the railway proved to be an essential instrument in order to ensure the continuity of supply over such great distances (military, industrial and civil), connecting long distances on a regular time schedule: it would be the main instrument for organizing these wide surfaces, maintaining surveillance and control over areas which were sparsely populated, but rich in resources and in which to spread the ideology of the empire.

It immediately induced a change in the whole local structure, and in land-use patterns, which were reorganized even more intensively with the progressive expansion of modern economies and with the acceleration of immigration—especially with reforms which were occurring on the European side of the Empire, above all the abolishment of serfdom.

These facts determined the development of a typical colonial economy, and to a kind of production functional to industrial needs that completely converted the CA periphery, from that time forward becoming dependent on European centres. Originally, local economics were organized on a domestic scale—considering the fact that agrarian products cannot be easily transported with traditional technology—and mainly for subsistence purposes; such produce could be exchanged at oasis-city bazaars, visited by surrounding pastoral populations, and, for a limited fraction, long range as silk-route trade.

It relied on traditional produce like rice, grain and other cereals, vegetables and fruit, and on such precious harvests like grapes, melons, citrus fruits and pomegranates, and secondarily, for limited produce such as manufacturing crops like cotton or mulberry (for the production of silkworm).[20]

The original distinction was between the river oasis culture in the south, with irrigation possibilities (for rice and cotton, especially in the Fergana valley and in the oases along the major rivers), and the extended arid surfaces, usually run by transhumance tribes; the prairies in the northern-eastern regions (Zhetysu, and the regions crossed by Irtysh), also represent another category and proved to be suitable for the cultivation of grain and other middle-latitude cereals, even if it had usually been used for extensive pasturage in pre-modern times.[21]

The colonial economy soon initiated a change in production patterns, and consequently of soil and resource organization, transforming survival subsistence agriculture into specialized production, functional to the needs of modern development and suitable to be commercialized through modern transport means.

8.2.5 Change in Local Governance

Soon after the conquest, the Russian colonizers realized what their interest could be in the local economy: initially they pursued the strategic goal of consolidating borders, establishing control and surveillance outposts on long range caravan routes, and in general ensuring security in the newly acquired territories. These both for defence from outside and for internal purposes, especially considering the turbulent nomad populations—and especially against some indomitable rebel raiders, soon defined as "outlaws" and "criminals", who would continue to oppose colonial authorities for decades.

Soon, the colonizers set up policies to increase local production privileging the most efficient ones, then, considered local environmental conditions, climate and water (in areas relatively close to glaciers and surrounded by mountains, enjoying regular and abundant water flows), essentially for cotton. This in a context of increasing of industrial demand, and of market pressure, for such a commodity, that became even more requested in some contingencies, especially in the 1860s, during the American Civil War, which provoked the interruption of deliveries of such raw materials to Europe.

This development was further stimulated by the construction of the railroads, which gave impulse to the rapid increase of any kind of trade, in both directions; this with regard to import of commodities typically requested by colonial economies from the metropolitan regions; it is the case of construction material, fuel, utensils and equipment, manufactured products and others.[22]

It meant a switch from the local self-consumption production, to a kind of ancillary economics, with the authorities consequently planning the extension of their monopoly on water sources, reorganizing irrigation and further functions (transport, work organization etc.) in a manner exclusively functional to the interest of metropolitan industry.[23]

This change was relatively easy to carry out considering that local economics was completely dependent on water and that water appliances were easy to control in a centralistic way: colonial authorities soon began with the reconstruction of canals and ID facilities, in order to acquire more power over the local society; this operation was combined with the contemporary abandoning of the old-style locally operated "qanat" system,[24] that needed complex maintenance work, that would soon become completely useless.

It signified a programme for the construction of infrastructures, excavation of canals, building of dams and aqueducts, applying modern technology, using modern construction materials, machinery and irrigation pumps. In this way, it was possible to transform the whole of local society, making it dependent on new technology—monopolized by the colonial power—completing a structural change of the whole area in just few years.

The change affected all features of local life, namely the regime of land ownership, local rules and regulations, inducing the progressive extension of control over peripheral nomadic populations, who would be induced and also forced to change their condition from that moment forward. In general, it had consequences for the entire human and natural system, destroying the equilibrium that had been built up over millennia and that would have irreversible consequences.

8.2.6 Nineteenth Century Occupation of Spaces

In this period, control of the colonial territories became more intensive. With the progressive occupation of lands, the tra-

[20] Barisitz Stephan, 2017.

[21] Significatively, Zhetysu in colonial time was named with the Russian calques Semirechye, "land of the 7 rivers".

[22] Reid Patryk, 2017.

[23] Barisitz Stephan, 2017.

[24] Burri Ezio, Del Bon Andrea, Ferrari Angelo, Ragni Pietro, 2018.

ditional groups became isolated from territorial, water and agro food resources; furthermore, they were often cut-off from their usual supply network, from bazaars and from the towns on the river oases, like the Fergana and the Zarafzhan oases; it was a vital question for nomadic groups, such oases being their supply centres—especially in crisis and famine periods.[25]

In principle, the political impact did not concern culture or religious matters in colonial times. However, the material changes soon started to induce effects on these dimensions of local life as well; often, due to expropriation and to limitation in their use of water, or just because of the construction of infrastructures, tribes were separated from their usual itineraries, from pasturages as well as from cultural and religious centres (sanctuaries, pilgrimage itineraries, holy places); sometimes, they just disaggregated, losing their identity markers—both cultural and territorial.

These were actually important because, around them, the nomads could elaborate identitarian elements, essential for maintaining compactness and communication among continuously moving groups; they were used to living separately on the transhumance itinerary, even when they belonged to the same clan or tribe, then reunifying seasonally for particular events, organizing rituals and meetings.

So, the pastoral communities were progressively pushed into marginal areas, their vital space becoming more and more limited, with scarce possibility to reorganize a life which was based on cattle breeding over broad swathes of land surfaces and no longer compatible with the intensive economics carried out by sedentary farmers. They were made to search for new pastures, more and more distant, continuously reorganizing their itineraries, even in trans-frontier areas, crossing the mountain passes towards China and, in the South, over the Pamirs (from the Great to the Little Pamir), in the Wakhan corridor and the Panshir valley, in the direction of Afghan territories, or elsewhere.

In fact, the usual traditional range of shepherd groups—essentially Kyrgyz in Tien Shan and Pamir area, Kazakh in the steppe, Turkmen in the Kara Kum and Karakalpak in the Kizil Kum desert area—was very broad, with the whole tribe moving freely but following consuetudinary itineraries; in the new situation they were compelled to explore new territories in the highlands or deep into the desert.

But the option of migration became less and less practicable. With the progressive extension into the fringes, the czarist empire came into contact with neighbouring ones—the Chinese and British colonial borders, progressively limiting such residual spaces. They finally established hermetically closed borders towards the outside, often militarized, and functioning as a check point and control for the internal part (namely on populations resident in the border area).[26]

Such situations represented an unnatural situation for traditional mobility; the border of the sovereign state was no longer just a buffer, or a "tribal area" out of state control as it had been in the early colonial times.

On the contrary, it was a precisely drawn line, marked and regulated by a treaty signed by the diplomatic bureaucracies in a faraway capital of the empire, completely unknown to the indigenous populations; it would progressively be more intensively controlled by garrisons, customs officials and frontier infrastructures.

In such borderlands the nomads, which had previously been used to migrating indifferently beyond the same borders, had either to adapt to new circumstances, changing their style of life or to rebel.

So, they found themselves ironically in territories with increasing strategic interest for the new regime, as they were located at the frontier of huge colonial empires (in the crucial moment of a planetary struggle), where the political pressure became even more intensive.[27]

In such circumstances, it was not conceivable that the nomadic groups should compromise—organized in "aul", mobile clan communities, identifying themselves in some wide pasturages area—surrendering their periodical migration, or their nomadic culture, and it was a matter of life and death. Adaptation to the new lifestyle was simply impossible; for nomadic shepherds the work of the settled farmer, and in general heavy work, in a condition of "constraint" to a certain land, was considered not much more than working as a slave for aristocrats; it was simply the premise for a rebellion.

8.2.7 WWI, "Basmaci" Rebellion, Revolution and Civil War

Yet for a while—until the Soviet revolution—the situation of the small groups nomadizing these areas, at the limits of oecumene, remained tolerable. In these times the main instigators of colonization—pursued by farmers of European origin, coming from all the parts of the Russian empire—gave priority to the most favourable territories, considering accessibility, soil, water, climate, and in general agrarian and settlement potential. It was the case of the Zhetysu and of northern-

[25] The prohibition to visit the bazaars was an especially severe instrument used eventually the by local Kahn, in precolonial times, towards the surrounding nomad populations.

[26] Shahrani M. N. M., 1979.

[27] During the "basmaci" rebellion many Kyrgyz groups moved to China, following trails that they had always been accustomed to running; but then they found themselves divided by borders that in the new century would became hermetic closed frontiers; Shahrani M. N. M., 1979; see as well Ferrando Olivier, 2013:46, note 7; Sodiqov Alexander, 2013:49.

eastern Kazakhstan, until that moment scarcely inhabited, not yet settled by sedentary populations.[28]

However, soon the colonists expected to be able to settle the peripheral areas as well, with the exploitation of the land becoming more intensive, correspondingly to demand—for an increase in agricultural production—arriving from the main imperial centres.

The end point of the colonization process coincided comprehensibly with the arising of further tensions characterizing the beginning of the new century, which would soon manifest on all sides, even in the same Russian population. Once started, it would further aliment contrapositions, finally resulting in revolution and civil war, namely, locally in the rebellion of the "basmaci".

The rebellion started because of tensions arising internally to the local populations; it initially needed the intervention of the Russian troops, as required by Kokand khanate, to restore order during the uprisings of local nomads against the same khanate. Then, the rebels attacked Russian authorities in Khujand, in Fergana valley in 1916, in what became a general revolt, spreading later over the whole area.[29]

Evidently the situation in colonial CA was already compromised, and exacerbated by tensions induced by the engagement of the Russians in WWI, for whom the rebellion would signify a kind of internal front; the "casus belli" of the rebellion was the reaction towards the requisitions made by the Russians in war time, and occasionally the way in which such requisitions were made, eventually using words defining cattle requisitions, that sounded intolerably outrageously for local populations, inducing violent reactions.

This both by the urban Muslim population and by the nomadic Turkmen, Kazakh and Kyrgyz, then by virtually all the Turkish peoples, with a rebellion that soon assumed further proportions, eventually ethnic- and anti-colonialist ones, especially among the populations with nomadic and warrior traditions, living at the edges of the empire.

Even when the peace order on Khujand was achieved in the Fegana valley, clashes continued and expanded everywhere in CA until 1920, as well as later in more remote regions. Consequently, such revolts meant many indigenous populations abandoned the region—following a traditional attitude, reacting with the solution of migration when faced with a certain crisis—moving on this or on the other side of the political (and perceived as artificial) border.

This happened for Kyrgyz and Kazakhs, and for other nomadic populations, that either rebelled or migrated[30]; and this until the new Soviet rule was able to establish a new order, defining what would be a modern state border from that moment forward, marked by state officials and certified by international treaties: a line on the territory that would also define a new concept of sovereignty (and of state-ideology), that was to be considered indisputable, with no resistance or opposition, or simply no trespassing, all of which would have been considered as subversive.

Such tensions soon degenerated into a widespread civil war, melting into further conflicts, provoking large scale devastation (especially considering the large use of modern weapons, and in general of modern technologies).

It was to be increasingly characterized (as usual for civil wars) by uncertainties, and by the sudden changing of alignments, overcoming pre-existing forms of identification. The modern ideologies proved to exert an incredibly persuasive capability, spreading out transversally (through ethno-tribal situations and language, religion and local cults, and correspondingly nomadic or sedentary status, tribal ascendancy, loyalty to the ancient regime, to a dynasty, to a community or to a brotherhood).

In fact, it is not possible to understand the changes characterizing that epoch without considering the power that modern ideas would exert as mass persuasion devices—communism, above all, then nationalism, industrial-capitalism and "tout court" modernism as innovative social forces. This especially considering the corresponding vast technological improvement, which, beside political appearances, induced a "de facto" change in power relations: a minority of "modern" workers could assume strategic power over the whole system.

This was the case of the crucial role assumed by the railway workers, then an organized minority of modernized "proletarians", leveraging further organized "classes" of industrial workers and military; they had the possibility of using the train, probably the most important device for the control of the whole area; based at Tashkent and Ashgabat, the hubs of the local railway network, they could immediately establish "Soviets" connected with the main Russian centres of the revolution. Already on 13 November 1917, the Tashkent Soviets had gained power in the entire region, defending the revolution from contra-revolutionaries.

8.3 Soviet Times

8.3.1 Early Times, After the Bolshevik Revolution

Soon after—indeed contemporary to—the power grab on the European side of the empire, the Soviet rule appeared in the peripheral CA areas, stretching along the connection lines delineated by the railway.[31]

[28] Barisitz Stephan, 2017.

[29] But different version and descriptions; Buttino Marco, 2003.

[30] Buttino Marco, 2003.

[31] as said, the main protagonist of the same revolution was a minority of modernized proletarians, that had effective control of the key industrial technique: a political mistake indeed made by the elite at the time as they left control of such vast power in the hands of this minority; indeed none at that time could have had any idea of the power the new technology would exert.

The new rule imposed itself completely on all territories, fighting and finally defeating the factions of "whites", loyalists and "basmaci".

So too, for any further resistance, or for any other kind of movement, perhaps inspired by diverse motivations, which tried to take advantage of such a situation of disorder.

Among them, the "pan-turchik" movements, trying to raise the local population of supposed ethnic affiliation in neo-national terms; the most relevant one, which was represented by Enver Pascià, a former official of the Osman army (dismantled after WWI), leveraged sentiment against Russian (colonial or Soviet) rule, trying to establish a Turkish-identity state; he joined the "basmaci", trying to bring them to the cause, but was irremediably defeated in 1921 by the Tashkent Soviet.[32]

Likewise for neo-Islamic movements, more similar to anti-colonial uprisings than to the self-defined "jihad", as happened in Andijan in 1898, which set a precedent for religious resistance (even when the "holy war" remained an "exception" in this phase)[33]; so too for further movements mixing local and ideological motivations, which spread in periods of change and of vacuums of power.

The Red Army finally prevailed, under the energetic and merciless guide of Commandant Frunze, as well as other revolutionary leaders who were inspired by ideals and who were of extraordinary persuasive capability, mixing a number of anti-colonialist motivations, and who eventually managed to mobilize local populations against the traditional authorities who were presented as horribly despotic. In general, the communist revolution presented itself as somewhat messianic, overcoming the usual contrapositions characterizing the ancient regime (serfs and masters, poor and rich, nomadic and settled farmers), transversal to the usual religious and clan-national divides.

But the imposition of communist power did not happen linearly, and sometimes came about in a contradictory manner (as will soon be evident); sometimes, it reached opportunistic compromises with local progressive powers, which had spread in the meanwhile and which were inspired by reformist movements.

This is the case of the Alash orda—the "red orda"—that imposed its rule in Kazakh towns and tribes, promoting an apparently revolutionary intention, according to the Bolshevik party, but formed mainly of local Kazakh young nationalists. They had success in establishing their power in Verny (then renamed Alma Ata, and since 1991 Almaty), in the area of the steppe, taking advantage of being isolated for some time both from the Tashkent Soviet and Moscow. This situation—possibly prefiguring an adaptation of the absolutistic attitude of the revolutionary elite—lasted just a few months, finally being suppressed by the central Soviet power in 1920.[34]

The new rule soon showed its intrinsic character, of an uncompromising power, with the intention of realizing a completely new order, without exception (also in the colonial regions). In principle, it represented the populations and accordingly the new power had to emancipate from the ancient regime. In this sense, the new power evidenced a contradiction, pursuing the reconquest of a colonial territory as demonstration of the victory of a proletarian revolution, which was originally based in the industrialized area of European Russia: the Soviet rule had to find new justification for its control, since in principle it meant liberation from both the colonial empire and the traditional Khan despotism.[35]

This would be found in the program of renovation of the archaic reality, namely in the combination of a modernistic inspiration and a Soviet ideology (as per Lenin's slogan indeed, "the revolution means Soviet plus electricity").[36]

8.3.2 Modernism as a Power Consolidation Device

The revolution—as for any revolution, in principle—had to appear from the outset as something irreversible, otherwise it would appear as useless, without meaning: the pre-existing social situation had been immediately redefined as "medieval" (applying western schemas), characterized by the "warrior democracy" of the nomads and the feudal aristocrats enslaving farmers in the oases; to some extent such systems did coexist with the capitalistic colonial regime, and it was impossible to renew from all points of view.

The Soviet rule imposed itself using a combination of ideological and practical motivations, combined—by the agit-prop, the secret police infiltrated into society, the political commissaries as well as by mass organizations—with modernistic motivations which were to bring the benefits of the new era to the populations.[37]

In fact, the propaganda proclaimed that the revolution would soon bring to both exploited colonial and European

[32] then pan-nationalistic ideologies echoing the nineteenth century "spring of peoples" movements, Silvestri Tommaso, 2015–2016:26.

[33] Hambly G. (a cura di), 1970:200; Robinson 1989:124; Bennigsen 1989; Bacon 1990:112–113; Hisao Komatsu, 2009.

[34] Hambly G. (a cura di), 1970:197.

[35] Buttino Marco, 2003.

[36] Hughes Thomas P., 2004, quoted by, p.299: "Lenin said 'Communism is the Soviet power plus the electrification of the whole country'", citing Vladimir Lenin, Collected works, XXV:490–91, in Russian; civil war 1917–1920; see as well Freeman Jacqui, 2013:35.

[37] Even when not foreseen originally by the theory, since Marx predicted that communism would realize in an industrialized advanced country, not really in a feudal-age colony.

populations the benefits it had already brought to the industrially advanced regions.

It meant the combination of tactical manoeuvres (the need to strengthen power), of scientific modernism (faith in technology, bringing progress, economic and territorial planning, substituting self-regulating markets) and of persuasive ideology (since it would also signify the liberation from millennial serfdom in the archaic CA society).

Practically, it had the effect of elaborating and imposing centralistic politics, without alternatives, with additional and a much more intensive transformation of local economics. This occurred in order to re-establish a dependency chain among the industrial centres of the region, and the build-up of centralistic planning practices.

Such enormous exertions had consequences on all levels of reality, changing culture and society permanently, imposing, with violence, a new style of life, pursuing the destruction of the entire ancient regime; it was the case, among the others, of religious clerics, of "bei" and "bi", traditional aristocrats, of clan leaders and traditional shamans; and it was the case of the new class of "kulaks", then enriched farmers, accused of speculating on the general misery.[38]

Over time, the whole material reality changed, restructuring both urban and rural territories: a process which developed without any control, finally destroying entire social categories, and weakening entire ecosystems—simply because the new radical economic paradigm was applied indiscriminately.

In fact, the establishment of a centralistic organized economy, made possible by scientific planning organization, induced the reorganization of economic practices and of the whole agro-industrial production in order to develop the maximum profitability for any resource and for all activities.[39]

8.3.3 The New Planning Practice

From that moment forward any element of reality (human, economic and territorial) would depend on centralistic planning, a utopia of programming any variable concerning development and human life (therefore a scientific view which can be traced to nineteenth century positivism). This planning system was applied to supply chains, to structures and infra-structures—oriented radially, then in the direction of centre-peripheries— to technology and industrial development, sometimes applying methods of social- and ethno-engineering, including deportation and internal migration, forced and induced sedentarization. This scheme with its mixed policies and with the combined use of material force and propaganda demonstrated the superiority of the Soviet organization combine through the transformation of technological miracles.

The segmentation of elements of human and territorial life was the premise for the realization of a system suitable to being perfectly planned (a kind of human-economic Fordism), as well as for the realization of a sort of "divide et impera" politics. For this reason, it was necessary to proceed with both the restructuring of economic cycles and with the reconfiguration of administrative regions (and the setting up of borderlines).

Finally, the new rule was able to gain further legitimization, because, as a final result of such an operation, it should have been evident that only such centralistic power was able to rule and regulate the whole system.

In this way, the new power would realize a condition of pure arbitrariness, much more than simple colonial exploitation: the regime used its totalitarian power in order to decompose and recompose reality in a number of elements (human, social, economic etc.) which could easily be manipulated, and then finally controlled. In this context, any possible source of opposition would be completely annihilated, and also deprived of any legitimization, appearing as just an "enemy of the people" to be merciless destroyed.

8.3.4 The Reconstruction of the Geographical Units

This program meant the need to completely reconstruct reality, namely of the geography of urban centres and facilities, pursuing the change in lifestyle of indigenous peoples, namely the sedentarization and irreversible destruction of traditional cultures, beliefs, social modes, religions and values.

The same restructuring of geographical–political units had to demonstrate the capability for power: the setting up of new administrative categories was the evidence that the new power would be able to mould the entire reality, reorganizing mobility and territorial benchmarks. And that it was able to change the references the population usually applied to explain identity—starting from an individual intimate dimension—tribal and clan organization, mahalla (traditional city quarters) and village community, supply patterns but also religion traditions, ethno-nation divides and language use.

And so too for other references, for territorial borders, for identity and functional relations; this through the reorganization of economic connections, used in order to create dependency as well as through such operations as the naming of a new city capital or of new regional hierarchies (upsetting the pre-existing ones), through the destruction or reconstruction of landscape references, both in urban (e.g. modern residen-

[38] Jelen Igor 2002, interviews.

[39] Hughes Thomas P., 2004; a kind of state-capitalism indeed.

Fig. 8.4 Kazakhstan, Nursultan (Soviet side city), 2017, Soviet time "chruščëvka". (Photo: Igor Jelen)

tial compounds, Fig. 8.4) and rural areas (pilgrimage itineraries, traditional architecture, irrigation infrastructures).

This meant the formation of a new geopolitical setting, with borderlines redefining ethnics and nationalities (just like political-social constructions) and redesigning any territorial relations (for accessibility, supply, social and culture relations). This all occurred in the context of the application of a modern method, with border and territorial sequences designed in order to configure a set of concentric administrative units.

These were consequently classified in bureaucratic categories ("sovetskaja socialističeskaja respublika", SSR; "avtonomnaja sovetskaja socialističeskaja respublika", ASSR; "avtonomnaja oblast'", AO; and then "oblast", "rayon" and "micro-rayon"), and in production units (sovchoz, kolkhoz, leschoz), and inserted into a hierarchy of dependencies.[40]

This policy is neither easy to reconstruct nor easy to evaluate with the documents and information currently available; generally speaking, it is possible to say that the internal reconfiguration of CA territories followed certain ethno-language criteria[41]: delimitation lines, the territories of major linguistic communities, with some ethnic groups defined as "titular"; a fact that, beside any rhetoric, ensured some rights, and in particular that of learning and teaching the mother tongue.

However, the manipulatory intention seems to be predominant; in fact, Soviet power maintained the attitude of using identity elements as tactical instruments, in order to produce or to manipulate loyalties and to influence the local communities, either settled in remote regions or in ancient urban centres.

Above all, autonomy was to be conceded to entities not representing any real danger, in order to marginalize the elites which had survived and were a residual of the pre-revolutionary era; it is especially the case of the ancient urban centres, of the religious traditions of Samarqand and Bukhara, and of the Fergana cities, traditionally representing respectively the Tajik-Persian- and the Turkish-Uzbek cultural elites (especially from a cultural and confessional point of view).

This happened in particular in the case of the prestigious urban centre of Samarqand, prevalently Farsi (Tajik) speaking, that may have represented an obstacle for the new power: the city had been separated from its Tajik hinterland in 1928, becoming for a while the capital of Uzbek SSR—mainly Turkish speaking— then in 1930 it was replaced as capital city by the more "Soviet" Tashkent.

So too in other cases, it was an overriding practice throughout the Asiatic Russian provinces; it is the case of the Kazakh situation, assimilation that took place by decree in 1936 when the Kazakh territory, in its entire extension was only made up of 25% of a Kazakh speaking population, so weak as not to represent a possible threat to the power.

In such conditions, similarly as in other colonial contexts, it was considered impossible for the natives to maintain a distinct identity, as the population was close to complete assimilation, and "tout court" to extinction (nevertheless, it is to say that this event is considered, by the same Kazakhs,

[40]Dagiev Dagikhudo, 2014:16.

[41]Cosentino Italo, 2017.

even today, the beginning of their reconstruction as an ethno-national unit in the current configuration).[42]

8.3.5 The Organization of the Soviet Administration

For a while, the new Soviet administration was characterized by a general confusion, with uninterrupted processes of rectifications and adaptations; this involving all internal policies, including the elaboration of new circumscriptions, often resulting in operations of pure geopolitics.[43]

Sometimes such changes signified the struggles of some lobbies, claiming some section of territory, sometimes they were the result of underground manoeuvres in order to sideline some still influential authority.[44]

However, the main interest was in the power of dividing territories and populations, confusing natural and functional borders, applying the same method the new rule had adopted in society, so preventing the formation of any stable reference (which could potentially be used for organizing resistance).

After the revolution in 1921, in the territory of the Bukhara, Kokand and Khiva khanates, the ASSR of Turkestan was established, this being the general name for such regions, meaning the "land of the Turks", with capital Tashkent, already the most important CA town since becoming the regional main railway terminal.

From the territory broadly encompassing the southern oasis belt, the Uzbekistan SSR was created in 1924 (since, 1928, as said, including Samarqand),[45] and the Turkmen SSR, Tajikistan, originally an ASSR included in Uzbek territory, was recognized as a full republic in 1929.

On the northern steppe side, in same period (in 1920) the Kyrgyz ASSR was established, originally in the frame of the Russian Federative SSR, then renamed as "Kazakh", roughly coinciding with the territory of historical Kazakh khanate; from it in 1925, the Kara-Kyrgyz AO was established, on the south-eastern side of this, including part of Zhetysu (Semyreč'e) and its mountain outback in the Tien Shan, it was named as Kyrgyz ASSR.

Such ways of naming these spaces was possibly a reminiscence of the colonial ethnographic attitudes; other names used for Kara-Kyrgyz, among the others, were "Dikie Kamni Kirgizii", the "wild mountain" Kyrgyz (a picturesque definition in order to distinguish the mountain settlements from the steppe pastors, defined as "kazakh"), and "Burut"; but these were just a few of the different names the different clans used in different situations, sometime with the intention of obscuring their identity (as typical for nomadic tribes).

Finally, the Kazakh and Kyrgyz territories would both be recognized as RSS, but mixing respective names, with possibly the idea of deliberately establishing new divides, namely territorial borders (but which for nomadic and clan tradition did not have such great significance).

In all of the considerations, the geopolitical factor always played a relevant role, often representing the principle motivation. It is the case of special status regions in some marginal areas (a kind of "buffer") or in areas contiguous to foreign borders, with the intention of creating better conditions of control. It is the case of Pamir AR (Gorno Badakhshan) in relation to Tajikistan, and previously of the same Tajikistan, originally a ASSR of Uzbek SSR; it is the case of Kyrgyzstan in relation to the Kazakh SSR; a further case is that of Karakalpak AR in western Uzbekistan, in the Khorazm area, on the mouth of Amu Darya in the Aral sea, that was settled by a pastoral population more similar to neighbouring Kazakh and Turkmen than to those of the Uzbek urban culture.

8.3.6 Internal Geopolitics and Functional Intersections

The operation had the primary intention of making docile instruments of the internal geopolitics of the territorial units. However, it also had a further motivation in an economic and social sense.

Soviet territorial organization was an important system used to include any resource (agrarian, mineral, energetic etc.) into the new industrially planned system, extending over the whole territory and the whole population, as well as to nomadic and scattered mountain groups, which were always trying to remove themselves from centralistic control. The new power, engaged in an epochal effort for industrialization, needed to mobilize all resources, human or natural, central or peripheral, integrating them into the same wider industrial circuit.

Such marginal territories were no longer to be considered "res nullius"—and then spaces in which counter-revolutionists and "basmaci" could survive—but elements of a top down network of commands, reflecting an appropriate segmentation (and specialization) of economics, becoming disciplined gears of the centralistically dominated "machine". They had to be adapted to the requirements of the new production units (sovchoz and kolkhoz, sometimes very extended), and to organize large scale public works (as the Karakum canal and the Great Fergana Canal, completed

---

42 Ertz Simon, 2005; Pianciola Niccolò, 2001, 2004; Marzhan Thomas, 2015; Akyildiz Sevket, Carlson Richard, edited by, 2013:16.

43 Reid Patryk, 2017.

44 Bond A. R., 1991; such practice continued for the entire soviet period, at different levels, as an effective geopolitical instrument; see as well Akyildiz Sevket, Carlson Richard, edited by, 2013:5.

45 Paskaleva Elena, 2015:420.

respectively in the '30s and '40s), which were to be planned transversally to administrative borders.[46]

This was necessary in order to create spaces suitable for an efficient flow of commands, to carry on exploitation with the maximum of efficiency and to make modalities of scale economics and industrialized agriculture possible, with application of machinery and chemicals. It was also necessary to predispose wide surfaces, which could be easily irrigated and worked for industrial agriculture (cotton), and industrial agro-food (grain and further cereals), namely productions prescribed by the SU planned economics (rather than for local needs).[47]

8.3.7 Borders Functional to Economic and Political Planning

The Soviet rule demonstrates the tendency to use such territorial discontinuities as an "instrumentum regni", signifying omnipotence of power, rather than for administrative usefulness. Such configurations represented and have to be perceived as artificial units, far from representing references of ideals; they were often functionally unsustainable, with peripheries overlapping centres, borders segmenting transport routes, and functional units transversally extending over natural divisions (barriers, configurations, and circumscriptions).

The establishment of sovchoz, kolkhoz, kombinats and similar large-scale production units often overlapped (transversally) the administrative units, "de facto" carrying out public duties, like organizing supply, promoting industrialization and local development.

This occurred particularly in areas suitable to being organized with modern techniques, and which, until that moment, had been barely inhabited, like the reclaimed steppe, the irrigated plains in Fergana or in southern Tajikistan (in which a mass immigration and the urbanization of large areas was expected).[48]

The resulting administrative figure evidences all kinds of confusion and discontinuities, with the mixing of functions and urban networks, with exclaves and enclaves appearing here and there on the map.

The regional capital towns (Tashkent, Almaty, Bishkek, Dushanbe, Ashgabat, that became the capitals of the new independent states) were localized close to foreign borders (namely not in central positions), possibly signifying their low (or inexistent) strategic significance. The same "republic" configuration was evidence of the absence of any centre-periphery schema, without any radial-organized infrastructures (concentric as usual for administrative units), evidencing, on the contrary, the confusion of functional lines for energy, food, services and production, for transport of water and further commodities.[49]

Sometimes such elaborate borderlines intersected river flows, mountain ridges, valleys, watersheds and further natural elements usually assumed as natural boundaries; in other cases they evidenced geometric lines, crossing peoples and territories as well as trade and transport itineraries indifferently, separating the cities with their Christallerian (functional) tributary area; obviously they never took pre-existing mobility schema into consideration, such as seasonal migrations, transhumance itineraries and bazaar accessibility.

Such situations were particularly evident in some areas and were critical for the whole order of things because of their economic and political importance.

This is the case of Fergana in particular, a kind of "puzzle", that seems to have been deliberately organized in order to establish interdependence among several parts; the same for borderline sections, for the Turkmen-Uzbekistan border on the Amu Darya, for the line dividing Kyrgyz and Kazakh, and Kazakh and Russia, posing a problem that is still evident in current politics (once such borders assumed a true national-political character).[50]

8.3.8 Language Geopolitics

Linguistic (and communication) issues were of critical significance from the beginning of the Soviet era, considering both the original extremely complicated linguistic geography of the SU as well as the ideological necessity: it was essential to apply, alongside the new regime, a functional communication device, in order to make the commands that the new modernist power would use understandable, even to populations sparsely settled in remote areas.

Evidently, it was essential for the new power to elaborate a kind of linguistic geopolitics, in order to make it possible for the citizens to participate in Soviet-modernist programmes: they had to participate in social rituals, to be able to read and learn instructions, usually written, and which were also to be distributed by telephone, cinema and radio, and had to be able to use other such innovative instruments.

The issue was elaborated concretely in different policies, which provoked an already complex multi-lingual and multi-alphabet area in CA, sometimes with paradoxical effects.

It is possible to recognize several steps in this policy. In the early post-revolution times, the communist idealist expe-

[46] Reid Patryk, 2017:22, 31; Kassymbekova Botakoz, 2011; Teichmann Christian, 2007, 2016.

[47] Reid Patryk, 2017.

[48] As in Vaksh River Valley in South Tajikistan agro industrial complex, Reid Patryk, 2017.

[49] Dagiev Dagikhudo, 2014; PWC 2011; Brill Olcott Martha, 2012.

[50] Megoran Nick, 2017; Edgar Adrienne Lynn, 2004.

rienced various linguistic "uniformation" programs that coincided with a process of Latinization ("latinizazija") of Russian and of CA local languages. It was intended as an instrument for pursuing a certain degree of communicative uniformation, but also possibly for pursuing a program of export of the revolution into Western Europe and elsewhere (the program of Trockij).

In general, the Soviet policy elaborated an ambitious process of "jezikovno stroytelsvo" (linguistic reconstruction) in order to adapt and eventually reform the 172 languages spoken in the Soviet territory, with 52 alphabets (of them effectively about 16 would be reformed).[51]

A few years after, in 1923–1934 the politics evolved, to some extent contradicting the early stages, elaborating a policy defined as "korenizacija",[52] literally "rooting"; it changed the basic concept, signifying a new effort for indigenization and territorialization of the ethnics; it would mean the combination of different languages with the described formation of a centre-periphery geo-administrative hierarchy.

Soon after with Stalin, the policy changed again, possibly evidencing a program compatible with the new principle of "socialism in one country", then the affirmation of a neo-Russian nationalism, even when mitigated by the persistence of bi- or pluri-lingual practices (necessary for functional purposes, rather than for identitarian ones). It introduced a progressive russification; even when not openly imposed, but "de facto" anyone in the SU had to know Russian—also considering the increased political pressure and the bureaucratization affecting the Soviet apparatus.[53]

It is possible to observe a further change, in the later Soviet era, when the authorities, Chruščëv and Brežnev the Communist Party secretaries, started to talk in public to a undifferentiated "Soviet population"; this possibly defining a "slijanje" (слияние) policy, which promoted the merging of a "national content in a Soviet unique form",[54] signifying the achievement of a further revolutionary goal.

A further phase can be traced roughly to the same time, and characterized—on the contrary—by a kind of recovery of local culture, mainly for foreign politics purposes, indeed as a manifestation of affinity with developing decolonizing Asiatic countries; it signified the development of a world scale ideological instrument in the frame of a kind of "domestic Third-Worldism".[55]

8.3.9 Restructuring Alphabets

Sometimes, such evolutions combined with changes in the alphabet system: after the originally used Arab alphabet (used by a minority of clerics and literate people), a modified form of Arab followed, then Latin and finally Cyrillic in 1936.

In fact, the population of these countries had to shift through four writing codes in just a few years: changes that produced strong disorientation especially for the cultural elites (a deliberately pursued effect of such internal geopolitics).[56]

The—by some authors defined—"forced" alphabetization was thought to be an instrument for pursuing political integration, a program to emancipate the people from presocialist "obscurantism"; but also, to pursue the fulfilment of totalitarian power. Before the revolution the great majority of people could not read or write in any language, while the narrative function among the populations was realized through orally transmitted and declaimed tales—of which the Kyrgyz Manas poem and the Kazakh "batyr" mythologies represent the most extraordinary examples.[57]

These measures were intended to be the starting point of a mass alphabetization program, which was pursued on an extraordinary scale. It affected the basic education of any demographic class, notwithstanding gender, age and place; it was meant, more than a basic education program, as a truly ideological instrument used to pursue the cultural mobilization of the masses, which had to adapt to a new standard of continuing education on all levels; "learn, learn, learn" was much more than a school program, it was an ideological incentive.

8.3.10 The Realization of Totalitarian Power

Soviet power lasted 3 or 4 generations; during its long history this power manifested in different degrees of intensity, sometimes and for long periods reaching the intensity of true totalitarian power, in other phases mitigating its impacts: phases that it is possible to define in terms of assessment and of central consolidation (considering opportunistically

[51]Cosentino Italo, 2017; local language and dialect differentiation is to be considered essentially as a relic of pre-modern "longue durée" times, previous to national standardization.

[52]Dagiev Dagikhudo, 2014:16.

[53]It is to consider the fact that Russian also became the language of the international communist movement in those years, with many ideologists arriving in SU, also as refugees from the domestic persecutions, undertaking a political apprentice.

[54]Kaiser R. J., 1994:113; Monteil V., 1957:85ss; Werth N., 1993:215; Akyildiz Sevket, Carlson Richard, edited by, 2013:8.

[55]Even when not explicitly declared, it was manifest in many aspects of political practices; the Second Asian-African Writers' Conference was held in October 1958 in Tashkent in Central Asia, programming anticolonialist initiatives; see as well Kanet Roger E., 2006.

[56]Cosentino Italo, 2017.

[57]Yilmaz Harun, 2013.

potential opposition especially in such remote regions), and also phases characterized by contradictory attempts at renewal, with the power appearing more "human" and motivated in its search for consensus.

In fact, just after the revolution, the priority was the repression of any pre-existing identity, which could at any moment become a reference for resistance: Soviet power had to demonstrate that the revolution was something definitive and irreversible, both ideologically and materially. These instruments were combined with further sedentarization policies, as well as with expropriation of territory and resources managed in a pre-modern way (transhumance itineraries, seasonal pasturages, and especially water and agrarian soil), shaping even more dramatic consequences.[58]

Nomadism, in particular, and any related social and economic form of it (e.g. the loyalty to a clan, a 'zhuz'), was to be considered as not compatible with the standard of the revolution (then ironically reproducing a kind of colonialist superiority complex); this meant the annihilation of the subsistence base of the local populations, provoking starvation, and destruction of a genocidal nature, that characterized the '20s and '30s especially for the Kazakh, Kyrgyz and Turkmen populations.

So similarly for latifundium and old style agriculture practiced in the oases, which would often exploit local populations with serfdom-like conditions: here the authorities combined expropriation with collectivization, and often with the mass displacement of populations transferred from mountain regions; this especially in Tajikistan an Kyrgyzstan, with populations from remote valleys resettled in cotton and intensive agriculture areas on the plains, especially in southern Vaksh, in Glissar and in Fegana.[59]

These policies were combined with further manoeuvres aimed at eradicating identity traditions; among them atheist propaganda, destroying religious landscapes and ridiculing religious representatives; so similarly for further symbols and representatives of the traditional order like clans, brotherhoods, tribes and village authorities, breaking down the sense of belonging to a communitarian culture, and in short to a village and to a "home".

Sedentarization proved to be tragic, simultaneously depriving the local populations of their cultural roots and economic means and it triggered fierce resistance: shepherds preferred flight than to change their life, and to slaughter their cattle and horses rather than deliver them to the new collective farms.

In this period, steppe populations (and in general the population of the remote areas) declined dramatically by one/third, with the Kazakh alone reduced in number by one million, due to starvation and mass emigration, with the entire intellectual, religious and tribal elites, being destroyed.[60]

At the time of Stalin this indiscriminate repression assumed a systemic character, degenerating finally into the "great purge": in this period, entire generations of local politicians (indistinctively targeting local officials, clan-leaders or Muslim clerics), intellectuals, writers (which had a great role in Russian and consequently in colonial CA society), and even harmless citizens were exterminated.

8.3.11 Displacements and Deportations

This program was realized on such an extended scale that it paradoxically created a planning instrument as well: in fact, once the CA republics were integrated, planned migration and deportation, combined often with sedentarization, were used as a pseudo-colonization instrument, taking peoples to remote areas (as indeed was a typical instrument in times of colonization, not just for the czarist empire).

In this period, CA and Kazakhstan (as well as Siberia and northern Russia) became a land of confinement; millions of prisoners were simply arrested and exiled to those areas in order to settle them in deportation camps (like that prototypical of the Karaganda lager, Kar-Lag, Fig. 8.5).

This policy interested individuals (sometimes deported with families that had to settle in separate camps), but also entire communities and ethnic groups.

With WWII, the deportation plan was accelerated, with entire populations suspected of conspiracy with the German Nazi invader who were made to resettle; it included about one million people, among them about 400,000 Volgadeutsche, Koreans from the Far East, Crimean Tatars and several North Caucasus populations, especially Chechen, Ingush, Kabardins, Balkars, and South Caucasus; among the latter the Georgia (Stalin-native republic) Meshketian Turks, inhabiting a region bordering Turkey, suspected of not being loyal to SU, which was as usual just a pretext as in several other such cases.

Further populations that were deported were originally from central and eastern European countries, previously invaded by Soviet troops, like in the Baltics and the Polish, often as political opposition or just suspected of being that.[61] Finally, also foreign communist activists, that had previously

[58]Pianciola Niccolò, 2001, 2004; as usual indeed for any absolute power, the resulting effect was that the "power" used them, finally, when it lost much of its original legitimization, as an instrument simply for conserving itself.

[59]Reid Patryk, 2017; Ginzburg N.S., 1986; Jelen Igor, 2002, interviews.

[60]Whether it was directly and deliberately perpetrated by the authorities (often with the significant participation of insiders, since the revolution ideal persuaded a number of natives as well), or it was the side effect of a politics pursuing collectivization, is a matter of discussion; however, it meant a catastrophe for the local population, that in many cases was close to total destruction.

[61]Pianciola Niccolò, 2001, 2004; then, repression and induced famine, in order to exterminate the pre-revolutionary social strata, and forced collectivization, planned in late '20s and '30s.

Fig. 8.5 Kazakhstan, Dolinka, Karaganda region, august 2017, Kar-lag Mamočkino children cemetery. (Photo: Igor Jelen)

found refuge in SU, were then deported because they had matured a critical attitude or for other reasons, in the meanwhile.

This plan of mass-scale deportation was a geopolitical manoeuvre which was instigated in order to realize a systematic "mixing" of the Soviet society; this to avoid the formation of resistance by nationally inspired opposition, and also in order to foster the repopulation of the regions at the edges of the empire (even considering the Russian preoccupation, evident in any time of its history, of settling the scarcely populated territory, bringing under centralistic control such huge extensions).

8.4 Post WWII Period

8.4.1 The Transition from Stalin to Chruščëv Era

As for any revolutionary regime, based on idealistic impulses, the Bolshevik power also progressively lost much of its original push; in this sense, and at the right point in time, the victory in WWII offered a unique opportunity for the renewal of its legitimization, and also a big gain in terms of prestige on the new international scene.

In fact, soon after the grab for power, the new rule started to look for further and self-renovating incentives.[62] Productivity decreased, demonstrating that the top down initiative was insufficient for pursuing growth, and that in the post-Stalin era it was impossible to move forward with expansion by simply encouraging obedience, solely through political commands in the usual totalitarian frame, manipulating the consensus and diffusing a sense of fear.

Such policies proved to be possible (at least to some extent) in periods characterized by the relative stability of the economic paradigm, as in the time of steel and coal, mechanic-based technology, suitable to be governed in a Fordist way, namely with sequences of standardized and controllable operations.[63] But this only lasted for a while, then the profitability curves show—as usual at the end of a cycle, the paradigm becoming mature and then obsolete—an assessment, reduced almost to zero; this without even considering the further effects, and especially the impact on society, territory and environment.

Soon after, the regime had to invent some other remedial devices in order to drive the economic process, and some other goals to motivate the new generations of Soviet citizens (different from the consumerist freedom typical for western countries). It can be seen in the competition with the USA for world supremacy, the arms race and technological military innovations, of the rhetoric of cosmonautics and of the conquest of the moon.

It was also a more practical question of real economics, of agricultural production and infrastructural efficiency, pros-

[62] what in western economies derived eventually from middle class consumerism, defining welfare and prosperity, both individual and public, as goals; and then from the characteristic dynamics originated by open society and by widespread social initiative (beside any possible manipulation).

[63] Hughes Thomas P., 2004; the soviets soon learned the functioning of the Fordist principles, extending such policies to the territory, in the whole of society; however, such forced planning was often rather inefficient, above all in CA peripheries, leaving the space for improvisation; see also Reid Patryk, 2017:20.

pecting a new objective for the Soviet disaffected population. It is the case of the imagination of a new frontier, to be found in a renewed internal colonial expansion, based on reclamation of new territories in the empty spaces of the steppe and elsewhere inside the Soviet "continent", as an extreme attempt to render agricultural work productive—at least for internal needs, and for basic products.

Such politics were extremely successful prompting a new wave of investments, mainly in the military-industrial complex, once the push stimulated by the WWII victory had been exhausted, through the '50s and into the early '60s; this was further alimented by post-war recovery and "baby boom", and possibly by a certain relaxation which occurred after the death of Stalin, followed by the prospect of relative liberalization: these elements prefigured a new optimism inducing the same illusion once more; that the system would be capable of renewal, finding in itself the momentum to keep on with its development, both in metropolitan areas and in Eurasiatic peripheries.

8.4.2 "Virgin Lands" Campaigns

The "virgin lands" campaign in Kazakhstan (and elsewhere) was one of such initiatives, with the intention of transforming the scarcely productive extension of the steppe—in particular those more usable for agriculture in north-eastern Kazakhstan—in the "breadbasket" of the Soviet empire. Such projects intended to transform the steppe soil into a fruitful agricultural region; it was followed by a, more or less, planned migration with people attracted by the idea of colonizing a new land, organizing facilities and settlements, and creating new possibilities for the generations of the post-war Soviet nation.

It represented what is possible to define as a semi-ideology, a goal to give purpose to the peoples, to direct their energy and their imagination: a usual practice for authoritarian regimes and that the Soviet rule in particular applied with continuity over its entire history (and also characterizing the post-Soviet times, see later).[64]

The "campaign" had mixed results; it had been applied on the fragile steppe soils, not really suitable for such intensive exploitation; productivity increased for a while, but the environment soon showed up negative consequences for the eco-systemic. Among them erosion, loss of humus, and further side-effects; further negative consequences deriving from the abuse of fertilizers and chemicals contaminating the natural resources, with intensive cultivation damaging aquifers, vegetation and further resources (evidencing paradoxically a scarce planning capability, an attitude that was lacking in terms of measuring externalities and environmental impacts).

For some years, it had considerable success and attracted two million (of mainly) Russians from the European provinces to Kazakhstan between 1954 and 1962, who moved voluntarily to the "virgin lands" (the "tseline"); such migrations were popularly considered as a possibility to create better living conditions. This could have been the case for young couples in a demographic baby boom time, hoping to have an individual family home, better services and possibly taking away a little bit of pressure from the authorities (this especially after Stalin's death).

However, comprehensibly it had contradictory consequences for the natives. In fact, this migration had the effect of further marginalizing the local populations, indirectly or directly causing, among the further tragic effects, their demographic weakening; in particular the Kazakhs decreased to 30% of the Kazakhstan SSR total population (with the ethnic Russians up to 43% in 1959), risking being culturally either assimilated or just destroyed (with their "extinction" as an ethno-cultural entity considered almost unavoidable, as in a typical colonial scenario).[65]

This was just one of the manoeuvres of "internal geopolitics" carried out in this period, in order to re-equilibrate the resource/population ratio in the newly claimed areas, in the frame of wider social-economic planning.

Further migrations regarded a regional or state level; this occurred in the Tajik and Kyrgyz republics, with mountain populations more or less forced to move towards the southern Tajik reclaimed steppe, Zhetysu and Fergana valley respectively. In these areas the Soviet planning set up new towns, infrastructures, and large works in order to settle these populations; in fact, the entire CA settlement pattern resulted as modified in just a few decades.

8.4.3 Domestic "Third-Worldism"

For all these reasons, the "campaigns" could not appear but contradictory; however, the regime gained a new lease of life due to its legitimization.

Furthermore, the same circumstance represented the basis for a new set of policies, and in fact the regime was also able to use these results in order to give purpose to a new role on its Asiatic side, acquiring a certain recognition among emerging Third World populations arriving right at the moment of post-colonization. The socialist way often appeared as an alternative to the capitalistic free market's "rapaciousness", and in the '60s it represented a realistic alternative, removing any accusation of acting out a neo-colonial attitude in its non-European peripheries, which was not compatible with the Marxist principles. On the contrary, the ideologists could

[64] Kudaibergenova Diana T., 2013.

[65] Marzhan Thomas, 2015:460.

Fig. 8.6 Uzbekistan, 1984, images from the illustrated book published by local government "Uzbekistan" "Uzbekistan", Izdatelstvo planeta Moskva, p.344

present the Soviet portion of Asia as a model for Third World movements (Fig. 8.6).

This new idea was represented by the Tashkent sequence of conferences, in a period when the Soviet universities—for example, the Moscow Patrice Lumumba University—were engaged in a programme to attract students from the developing countries: the intention was to spread such principles as well as to shape a new post-colonial elite (instructed in spreading "liberation" movements), demonstrating the new international role of the SU, expecting to represent the developing decolonizing countries.[66]

In this new context, the attitude towards CA indigenous culture changed significantly, with the Soviet communism propaganding the achievements of the revolution in questions regarding development—namely for a problem the western system demonstrated as unable to offer a solution for.

Indeed, some figures were encouraging, evidencing some of the achievements of the socialist states; it is the case concerning statistics about infant mortality, alphabetization, diffusion of the welfare state, demonstrating that centralistic planning could be a solution for such pressing situations.

[66]Kanet Roger E., 2006; http://www.blackpast.org/gah/afro-asian-writers-conferences-1958-1979, accessed 22.4.2018; The Second Asian-African Writers' Conference was held in October 1958 in Tashkent in Central Asia.

8.4.4 The Challenge to Nature—The Apotheosis of Soviet Modernism

In this period, Soviet communism manifested some of its most characteristic features, using modern technology, with modernist ideology which became the means to demonstrate the capability of the regime. Among them, especially, the capability of solving the most ancestral anxiety the Russian (Soviet) citizen had been forced to face, namely the confrontation with an overwhelming idea of nature.

In fact, it is not possible to fully understand Soviet philosophy without taking into consideration the context in which Russian culture was originally formed, namely the relations with the typical Russian environment, in all its dimensions and characterized by wide extensions, inhuman distances, impenetrable forests, frozen soil and long winters; in general with elements that seem to be impossible to manage, if not under the guide of a strong central power. In this context Russian communism had always represented the semblance of a modernist challenge towards the mastery of nature.[67]

In an early stage, such power was represented by the czarist empire, eventually elaborating the Great Russia messianic ideology; the consequent evolution, and the elaboration of the modern techno-politics, would offer communism a further chance in this sense, to represent a regime capable of controlling such spaces, to emancipate the human condition from this oppressive natural burden. It meant centralistically

[67]Defined in Stalin times with the slogan of the "Great Plan for the Transformation of Nature"

Fig. 8.7 Russian Federation, South Siberia, pipeline, 2013. (Photo: Igor Jelen)

developed networks for the needs of any citizen, such as electric light, heat water, food in standard portions (in Soviet times with the access to a "people's canteen" and to public supply-shopping centres).

So similarly, for fuel, transport and communications, for welfare and social solidarity as well as for railroads and pipelines, for networks connecting each single residency and any remote settlement (Fig. 8.7): all these would offer a visual idea of connections to the "centre", that could ensure the satisfaction of any basic need. And so for any further social circumstance: the new technology of construction would prepare covered and protected (from weather, stormy wind, snow etc.) spaces for social events (sport, theatres etc.) at any time of the year; and then accessible health and education centres on any level, from kindergarten to universities.[68]

Finally, this also occurred in the case of the colonization of huge surfaces, potentially fertile but which had just been ignored until that time, land to be made cultivable with the help of modern technology, with tractors, harvesters and agro-machines.[69] Endless steppe surfaces would become orderly rows of fields for potatoes, cereals and cotton; so similarly for milk and meat from the Pamir and Tien Shan highland pasturages daily picked up by 4-wheel automotive transport and conveyed to food-processing factories and then delivered anywhere.

And so forth for any production and for any environment, materially representing the idea of integration in the Soviet "fatherland": trucks and amphibious vehicles were crossing mountain ridges, swampy areas and frozen soils in any season, the railway was running continental distances, helicopters and airplanes were flying over the mountain ridges, reaching any remote place.

All these images were re-elaborated by propaganda, presented as evidence of the ability of the Soviet power to reach any place, finally to "domesticate" nature: a task that arguably none had managed with such success before. Such projects were connected with further projects on an even bigger scale, with the aim of making natural forces simple elements in an engineering game; it is the case of river water diversion, of the excavation of long-distance canals, bringing the water into the desert, of the construction of dams and hydroelectric plants, of a general urbanization and industrialization of the whole area.

This push, in the intention of the ideologists, had to be progressed beyond any limit; the space program and the Baikhonur cosmodrome—in the middle of the Kazakh steppe—was the demonstration of the possibility of the conquest of extra-atmospheric space, and the Semipalatinsk (Semey) nuclear testing area was the demonstration of the capabilities of dealing with powerful, almost metaphysic forces, evidencing a definitively reached capability.

However, beyond a certain point, the program encountered a threshold; finally the industrial paradigm came to a conclusion in a fundamentally useless gigantomaniac attitude, producing uncontrollable externalities (pollution, soil and resources consumption, rise of maintenance costs).[70]

8.4.5 Last Stage, Convention and Adaptation

In such situations, Soviet power lost much of its residual capability for persuasion; the '70s, after Chruščëv's relative

[68] Akyildiz Sevket, Carlson Richard, edited by, 2013:19.

[69] Hughes Thomas P., 2004.

[70] Klüter H., 1992:20–38; Hughes Thomas P., 2004:264.

liberalization, were the years of abandonment of any ambition for renewal—beyond that of an image of conformist obedience to the regime. This new situation had consequences both in the international arena and internally.

In international relations, it signified the search for appeasement (in the Helsinki treaty era), meaning the elaboration of a new rhetoric of the "konvencija", (purposing the maintenance of "status quo"), promoting "de facto" a "compromise" between SU and the USA.[71]

From an internal point of view, it signified that the local "nomenklatura" took the place of the "party", becoming something "conservative", signifying the (more or less intentional) abandonment of any further ideological ambition of changing reality. It signified, in particular in CA, a new policy of appeasement, with the Soviet power actually renouncing its revolutionary spirit. The central power started to negotiate a kind of "tacit social contract"[72] with local informal powers, popularly defined as "mafias" (namely the spontaneous forming of clan and lobbies), covering illegal or semi legal activities spreading simply due to the inefficiency of the official economics.

In fact, the whole reality—social and economic—was changing, adapting, in times of stagnation after the Chruščëv's illusory initiatives, into an inefficient system characterized by opaque and submerged economies, far from what the Soviet theory was used to prescribing.

This occurred with informal bazaars, often organized at official workplaces, for instance the "black bourse", contraband and corruption on all scales, from public services, to social activities or small trade (to be taken into consideration the grotesque scarcity afflicting public supply centres). And so too for activities of any kind, in the frame of the emergence of informal practices, which masqueraded or were just exhibited as rituals of obedience.

In fact, the whole area of economics had been left in the hands of local clans (namely hidden power structures, defined as "klanovost'", tribalism),[73] formed in a parallel way, or de facto covered by local power, diffusing corruption—a kind of parallel salary system—mis-accountability, falsifying data about production, covering up thefts and robbery, manipulating centralistically programmed data.[74]

This evolution had a corresponding development on a macro-economic scale; the next step, after the Chruščëv campaigns, relied on the discovery of large hydrocarbon (HC) quantities, in the '70s, also inducing pragmatic adaptation.[75] In this period, the Soviet "nomenklatura" realized that they could survive by becoming the suppliers for western countries of gas and oil, inaugurating a new phase that would also characterize the post-Soviet economic paradigm (starting the build-up of a network of east-west oriented pipeline networks).

It was, in fact, fundamentally a stagnating economic system, combined with political disappointment, which had degenerated into a kind of game of roles, a system in which nobody believed anymore, trying simply to adapt and to gain time. In this context, Perestroika and Glasnost in the late '80s were not perceived as true reforms (even when considering that, in Russian history, reforms were usually imposed top down, not maturing bottom up); namely they were not considered as sincere and realistic attempts pursued by the top of the regime, but simply as the umpteenth case of centralistic manipulation and this especially in the CA periphery of the Soviet empire, where such reforms had few or no consequences at all.

8.5 Collapse and Post-Soviet Passage to Current Times: The Unexpected Break Up

8.5.1 Premises and Causes of Collapse

Whatever can be considered as the cause of the SU collapse, it is evident that the system had lost any efficiency and any capability some time before, "surviving on its own". For this reason, it is also interesting to observe that the collapse had not effectively been foreseen by anyone.[76]

Reasoning about the causes of the collapse is not just an academic game, but it may help in analysing the different evolutions which have occurred.

It is possible to identify some of the different causes: pressure from outside (from western capitalist powers) and

[71] reflected as well in the narrative of CA writers like the Kyrgyz Cingis Ajtmatov; writers as in the Russian tradition, had in CA, in Soviet times, a disproportionate role.

[72] Black C. E. et alii, 1991:290.

[73] Ro'i Yaacov, 1991.

[74] The obsession for central planning facilitated and induced such side effects as the proliferation of a shadow economy, small, and then large scale corruption and bribery; systematic stealing at workplaces in part as reaction or adaptation to the inefficiency of the system, in part for alimenting small corruption practices, namely the paying for public duties was a common practice in soviet times; such effects indeed were unavoidable considering the extremely long command "chains" of the public organization in Soviet territory, and the overall bureaucratization; Marzhan Thomas, 2015:469; Rasanayagam Johan, 2014:11; Jelen Igor, 2002; this was the obvious consequence of the absence of an internal suitable organization of controls based on the separation of duties, but also the consequence of a conceptual uncertainty in the definition of property, the limit between "private" and "public" categories being "de facto" non-existent.

[75] Krasnov G.A., 1987.

[76] With the exception of few French sociologists, Emanuel Todd and Carrère D'Encausse, whose hypotheses were based substantially on demographic projections, rather than on structural or political interpretations; Carrère D'Encausse H., 1978; and of and some narrative writers like Le Carré; see as well Shabad T., 1980.

functional weakening from inside—then a kind of implosion; an incapability of going beyond the mechanic-coal-steel paradigm (suitable of being fordistically managed), to an electronic-digital based one in which authoritarian control would be less efficient in principle.

Further reasons to take into consideration are a kind of self-extinguishment due to widespread inefficiencies or to some peculiar decisions, or potentially due to mistakes made by politicians who promoted the Perestroika policy as a consciously extreme attempt to modernise the system.

For some authors, the main reason can be traced to the apparent risk of losing supremacy by the Russian majority: almost a social-Darwinist scenario deriving from a situation of increase in Muslim Asiatic populations and corresponding Russian–European demographic regression, in times of CA demographic booming.

In this context, the elite probably matured the idea it was impossible to conserve power, unless engaged in demanding retro-guard wars (like that following the Afghanistan 1979 invasion, or like the "basmaci" rebellion, still vivid examples for Russian politics); so, a sequence of decolonization conflicts was to be expected.[77]

8.5.2 The Situation After the Collapse

Comprehensibly, the collapse of the totalitarian regime had consequences in all circumstances of the life of the citizens. Rarely the population (as well as the institutions) reacted rationally, for instance by increasing their participation in political decisions and maturing a sense of responsibility; sometimes they merely reacted with confused protestations and street rebellions.

More often than not they just adapted to circumstances in a passive manner with apathy, remaining under the threshold of political visibility, as was usual for the Soviet citizen, recovering the "parallel" economic circuit, which they had already practiced previously. So, they learned to survive, setting up spontaneous bazaars, organizing self-production, arranging transport, for example, car sharing or collective taxi and "private" "maršrutka", and when possible, getting closer to natural and human resources (then returning to villages). In this way, they could limit the impact of the breakdown of the system, yet at the same time, they also perpetuated the same deviations characteristic of that lifestyle.

In other circumstances, and for different reasons, when the local communities were completely dependent from the central power, the situation soon became disastrous. This happened for large scale production units, with kolkhoz and sovkhoz closing down, remaining without raw materials and means of any kind; so for public facilities for transport, education or health, for centralistically delivered commodities, like medicines, spare parts for machines, energy, fuel; this also occurred for security and further typically political functions.

This fact caused a shockwave in many places of the periphery, both urban and rural, greatly exposed to such scarcity, and especially in CA. Whenever it proved to be possible, the local community regressed to a pre-Soviet modality, namely to a pre-modern style of life. In such circumstances, the people, desperately looking for a means of survival just went back to rural life, in search of resources (arable land, pasturages, water, self-made food), and to the villages ("kishlak" and "aul"), from which they had migrated many years previously. They often left the Soviet industrial "new towns", which were either completely abandoned or sometimes involved in post-Soviet conflicts, going back to the community (both in urban mahalla or in rural aul), in search of the references they had lost during the different periods of sedentarization or migration.[78]

8.5.3 Regression in All Senses

The catastrophic situation, provoked by the collapse of the highly integrated system, was evident everywhere; functionality of the system and activities were simply 'emptied' (hospitals without medicines, transport means without fuel, industries without raw materials), while the Soviet-style infrastructures in concrete, asphalt or metal, often just collapsed like a "a castle of cards". (Figs. 8.8 and 8.9).

The halt in centralistically organized maintenance, which was already in an obsolete condition accelerated this degradation; this must also be taken into consideration as regards the Soviet monolithic way of organizing any function, per large scale unities, usually hyper-dimensioned and simply difficult to maintain.

The collapse was even more dramatic since the revolution had erased the previous strata of civilization, its traditional techniques and social-solidarity markers (what people need for survival, organizing mutual help, communication and practical knowledge). Often the people had no alternative than regression to an elementary style of life. (Fig. 8.10).

It did not mean—by any means—the return to an idyllic village situation; the disappearance of the overwhelming, even when despotic, idea of state provoked the proliferation of local illegitimate powers (sometime the former kolkhoz "brigadir" vexing the employees, sometimes just improvised gangs or criminals), exposing the population, above all the

[77] Indeed, the successive evolutions – as will emerge later - will refute such scenery, with the demographic rate by the native population, which boomed in the '70ies, then rather soon decelerating.

[78] Jelen Igor, 2000 and 2002.

Fig. 8.8 Tajikistan, Khujand, 1997, gas refuel. (Photo: Igor Jelen)

weak social strata and the minorities, to all kinds of arbitrariness.

This aggravated the already existing fractures between society and power, with the emergence of new divisions in the territory, with the arising of new categories of new poor (often older people who had retired with fixed pension in devaluated currency), the new rich and new "oligarchs"—often nomenclature friends, seeking rent positions, degenerating into underground clan politics.

It meant the return to pre-modern production and technique, for example, reconstructing yurts or traditional homes, "kibitke" and "palatke", and recovering transhumance itineraries; in other situations—if they could not remember that technique—they just prepared some slums (bidonvilles) with scrap metals, used pneumatics and further waste materials recovered from the post-Soviet ruins.

But many had literally no place to go, and had no alternative than to desperately lie down, ask for help and assistance, becoming the victim of some criminal organization or—finally—just rebelling and starting some conflict (as happened in many emptied kolkhozes in Tajikistan, in Fergana valley, and more or less anywhere).

Such movement interested, not just the native populations (sedentarized or expropriated), but also in general, the descendants or directly the surviving individuals belonging to populations that had once been deported, migrating to the nations where they or their ancestors had come from (Germany, Greece, Poland, Israel, South and North Caucasus, Eastern Europe, supposedly, in a far flung past); this also for millions of CA Russians and Russianized citizens who started to migrate back, often choosing the metropolitan areas on the European side of the new Russian republic (the s.c. "vozvraščentsy", see later). A mass out-migration—that characterized the entire transition time—that was partly

Fig. 8.9 Kyrgyzstan, Fergana valley, 1994, broken concrete aqueduct. (Photo: Igor Jelen)

Fig. 8.10 Kyrgyzstan, Pamir Alaj, July 1994, on the summer pasturages. (Photo: Igor Jelen)

compensated by the "repatriation" of the co-nationals who had once migrated to other SU republics.[79]

It is the case of the population in Kazakhstan, which declined in this period by 30%, but that was compensated elsewhere by about half million ethnic Kazakh "returnees" from Mongolia, China and elsewhere from FSU, coming back "home", attracted by renationalization policies and by favourable economic contingencies.[80] Similar processes characterized the other republics, where the economic situation was less favourable, provoking further migrations, sometimes temporarily, from the poorest regions in Tajikistan and Kyrgyzstan especially, to the further CSI republics (especially Russia and Kazakhstan).

8.5.4 1989–1991 Tensions and Clashes

The first case of post-Soviet (transition) tension in CA, may be considered as the 1986 Alma-Ata unrest, in the period of Perestroika, well before the Soviet collapse: it started as a spontaneous street protest against the substitution of the Kazakh first secretary of the Party with an ethnic Russian "aparatčik", sent directly from Moscow; it occurred possibly as a kind of localistic reaction against top down imposed reforms.

The riot, later called as Jeltosqan ("December") uprising, also spread to other Kazakh cities, being bloodily suppressed, provoking approximately 200 civilian deaths, and many rioters were imprisoned. It represented a precedent, but the attitude towards (and the narrative about) such an event even today sounds contradictory: it had a popular character, but occurred as a street revolt (then contrary to the conservative attitude of the post-Soviet and current power); it had undisputedly a neo-nationalist Kazakh character, but contrasted what was considered at that time as a principle of stability, rather than a top down commanded "reform"; in general it mixed anxiety and expectations, identitarian claims and anti-colonialist motivations.

The most important conflicts emerged just after, or actually contemporary to the SU collapse and CIS foundation (what most observers consider a kind of inconsistent surrogate of the disappeared SU), in the Fergana valley, and then in several important cities like Dushanbe, Novyi Uzen', Ashgabat, Osh and elsewhere.

They were caused by confusion,[81] more or less characterized by the same dramatic screenplay with the rioting of a disoriented population, abandoned in the emptied kolkhozes, literally panicking about remaining without food, fuel, a home and safety (because of the disappearance of planned supplies); even more dramatic was the situation of food and the lack of basic need supplies for some peripheral populations, as in some mountain valleys of Tajikistan, where this was possibly one of the main motivations for the sparking of civil war.[82]

In fact, the riots usually occurred in a spontaneous manner, exploding and immediately spiralling out of control,

[79] Ro'i Yaacov, 1991.

[80] Marzhan Thomas, 2015:460.

[81] Ro'i Yaacov, 1991; Elebayeva A. B., 1992.

[82] Then starting a kind of "war of the kolkhozes"; Heathershaw John, Herzig Edmund, 2013, p.8, citing Roy 2000; the mountain remote populations were possibly more exposed to risks of scarcity, being more dependent on the public deliveries because of environmentally extreme conditions.

characterized by outbursts of ferocity and escalating into home-to-home violence. In such circumstances the people gathered on the streets, at the crowded bus stations or at the bazaars, finding themselves in improvised clashes, in which the only identity they would instinctively recognize was that of ethno-linguistic character (also evidenced by exterior semblances).[83]

Such outbursts soon degenerated, with social groups (organized, ethnically, regionally, by religion or language etc.) fighting against each other; sometimes occurring in trans-frontier contexts such in Fergana (even when the whole CA space may be considered a trans-frontier area, considering the proximity and the configuration of the national borders), becoming even more dangerous.

Starting from Perestroika times, in 1989, the records registered pogrom-like assaults against minorities like Armenian and Meskhetian Turks in Novy Uzen', provoking 150 victims and the consequent flight of 150.000 Meskhetian Turks from the Fergana Valley.[84]

Further aggressions regarded the Jewish population in Andijan, as well as diffused hostile actions against Russian as well as locally concentrated minorities.

The assaults were often perpetrated by local gangs made up mainly of titular nationality members, usually youngsters, suspected of being covered by local national polices; they mainly had as their goal the intimidation of minorities, inducing them to leave, then accelerating the migration of "vozvraščentsy" and of further populations, at this point to be defined as "minorities" in the new nation.

The target of these aggressions was usually minorities that proved to be more vulnerable, possibly considering the international situation. This is the case of minorities without a nation-state that could potentially represent them, like the Meshketian Turks and other Caucasians (Lezgins, Checens, Ingush, Ossetian and other remnants of the nationalities deported into CA during WWII), becoming a kind of scapegoat for the general situation of tensions. Likewise, social minorities such as impoverished or elderly Russians or russianized peoples who remained in the CA cities suffered particularly after the change of economic regime (e.g. persons just became "foreign citizens", in a kind of statelessness, remaining without pensions and any assistance service).

The clashes in Osh, South Kyrgyzstan, June 1990, between the Uzbek minority, but locally the majority, and the Kyrgyz population, were provoked ("casus belli") by the social housing programs; in fact, the Uzbeks condemned the public programs which they believed would favour the national Kyrgyz, so apparently discriminating on an ethnic base. These clashes soon degenerated, they lasted weeks and provoked a number of victims, with country authorities arranging curfew and police repression; they happened close to the borders, then became even more dangerous, and risked degenerating into an inter-national war.[85]

All this time the role of MDV Minister of the Interior troops (special units, until obviously the SU was in force) was ambiguous, even when it was evident that "the nationalities would prefer the domination of Moscow [...] to the anarchy that would necessarily result from their independence".[86] A further question is the spread of drugs and alcohol among the rioters, probably distributed instrumentally by local criminal organizations; so also for the "excessive prevalence of firearms among the population", used initially as a means of self-defence by groups which soon became organized militias and could be politically manipulated.[87]

The most dramatic case is that of Tajik civil war that occurred between 1992 and 1997, with tensions lasting for some time after (see later); it originated from tensions typical of the de-Sovietization process, then alimented by a sequence of—what it is possible to define as—"errors". These "errors" were generally linked to the incapability demonstrated by the different sides of organizing a dialogue and of being unable to avoid misunderstandings and manipulations, as usually occurs in such circumstances (this especially by the established authority).

8.5.5 Further Period Clashes

The tensions did not end, and periodically emerged again in different circumstances, for the whole transition period; however, sometimes they changed from spontaneous unpredictable street rioting to more organized movements, or also into social-economic claims (housing, welfare, jobs etc.).

In some circumstances, protest assumed the form of confessional-radical contraposition (with eventual terrorist degeneration), with corresponding disproportionate state reactions often creating a kind of vicious circle. Indeed, many of the tensions that exploded during the immediate post-Soviet period were exhausted soon after their break out, but a few continued to manifest, recodifying into a political struggle, becoming structural (generally characterizing the opposition to a new despotism)[88]; this especially in the case of Tajikistan and, to a lesser degree, Uzbekistan.

It is possible to list in this sequence, among other events, the Batken "invasion" in 1998 (in Kyrgyzstan, perpetrated by religious extremists), and the terrorist attacks in Tashkent (2000), with auto-bombing targeting president Karimov

[83]Rotar Igor, 2006; Megoran Nick, 2017.

[84]Rotar Igor, 2006; Ro'i Yaacov, 1991.

[85]Elebayeva A. B., 1992; Megoran Nick, 2017:208.

[86]Ro'i Yaacov, 1991.

[87]Ro'i Yaacov, 1991.

[88]Then further alimented by factors originated "inside" the conflicts, Jelen I, 2012:551.

unsuccessfully; both events had geopolitical consequences, with the reciprocal closing of borders (especially on the sensitive Fergana side), and reciprocal recriminations and accusations between governments (especially between Karimov and the Turkmen president Nyazov); both evidenced the arising of religious extremism, and the scarce readiness of national law enforcement apparatuses (often so inexpert that they themselves further alimented the tensions).

More recently, uprisings and their consequent bloody repression occurred in Andijan in 2005, confirming the change in conflict phenomenology (from ethno-political to a religious one), demonstrating a tendency for radicalization; the same city, localized in the mentioned Fergana valley, had become the heart of a religious fundamentalist movement and of resistance against President Karimov.[89]

Periodical ethnic clashes have occurred in south-west Kyrgyzstan, in the Osh area, until recently; this is the case of the 11 June 2010 events in particular. Indeed, in a period of Kyrgyz vacuum of power[90]; such clashes happened suddenly and again with similar pretexts—the supposed privileges attributed to nationals to the detriment of minorities. Observers usually reported that local gangs entered into rivalry, possibly representing opposite ethnic factions, in this case Uzbek and Kyrgyz, with the latter eventually supported unofficially by the state police, who nevertheless denied this accusation. The situation rapidly escalated into home-to-home fighting, resulting in hundreds killed and thousands injured.

Undeniably, it is difficult to understand the dynamics of the 2010 events, whether (and to what extent) they started in a spontaneous or planned way, eventually mixing with local tensions or criminal activity, or becoming a cover for trafficking or for other activities. Fortunately, the situation did not escalate further, with the Uzbekistan government deciding not to intervene across the border for the protection of its co-nationals.[91]

However, such clashes had a consequent after-effect of reciprocal accusations on a governmental level and the spread of a climate of mistrust between, not just governments, but also populations. Some of the sides evoked, as in a typical CA screenplay, the intervention by the Russian army, justified as humanitarian aid. Possibly the factor which finally de-escalated the situation—even when considering the impressive number of victims—was the new role exerted by the SCO (see later).

8.6 Migrations

8.6.1 Late Soviet Time Migration

The out-migration of Russians, or of people not originally from (then once migrated in) CA countries, had already begun in the late Soviet period. It possibly consisted of people who had recently immigrated into Kazakhstan during the "virgin land campaign" times, or in analogous circumstances.[92]

In the same period, it essential to mention a further movement essentially regarding national minorities and possibly facilitated by the Helsinki treaty evolutions (1975). Something similar had also happened before, after Stalin's death, a repatriation of individuals and populations who had once been deported; it essentially regarded Caucasian people, as well as Crimea Tatars.

A major wave of migration in the Perestroika period is represented by the Jewish origin population (or supposedly since many possibly pretended such identity in order to have the permit to leave); it started in the late '80s, in correspondence to neo-nationalistic tensions, possibly threatening the minority, and of some agreements between SU with Israel.[93]

At that moment almost the entire Jewish community of CA (and of all SU) was allowed to flee to Israel (and in limited numbers to other western nations). Among the migrants were also those belonging to the historical settlements of Tashkent, Bukhara and Samarqand, about 10,000 (of the almost one million from all over the SU), and to small groups of Persian speaking Jews elsewhere in Tajikistan and Kyrgyzstan: communities limited in number, but of extraordinary cultural significance, that signified a definitive loss in cultural and broadly human terms; they had not just immigrated during the different times of colonial or Soviet rule,

[89] McMahon Robert, 2005; indeed a "chain reaction", with un-stabilities feeding in on themselves, Rotar Igor, 2006; World Report 2015: Uzbekistan, Events of 2014, https://www.hrw.org/world-report/2015/country-chapters/uzbekistan, accessed 23.4.2018.

[90] In 2010, another "revolution" emerged ("April revolution"), originally similar to a street protest (possibly carried out by an organized minority), but soon reaching the political threshold. In the same years, further tragic riots spread in Fergana, Kyrgyz side, with clashes between majority and ethnical Uzbeks, and further minorities, with refugees estimated in the hundreds of thousands between internally displaced and refugees seeking to flee crossing the border to Uzbekistan. In this case, the riots were connected to the context of both, nation- and local rivalries, of retaliations and provocations that characterize this ethnically mixed area, divided by incoherent borderlines – indeed a legacy of the soviet past. The unrest was, as informed independent UNO observers, "orchestrated, targeted and well-planned", Najibullah Farangis, 2010; indeed, such conflicts evidence a preoccupying potential for diffusion (in a kind of imitation effect), feeding (alimenting) indefinitely, and also multiplying the tensions, spreading into politically sensitive border areas.

[91] Possibly because Uzbekistan had just stepped out of the SCO, the geostrategic "umbrella" especially established to regulate such situations.

[92] Klüter Helmut, 1993; Ro'I Yaacov 1991.

[93] Klüter Helmut, 1993.

but were of the original populations, with a history dating back to the second Ct. B.C.

8.6.2 Transition Times Migrations

The first comprehensible effect of liberalization, after independency, was the start of a sequence of migrations, from what was considered for centuries as a deportation destination, rather than a "promised land", as usual for colonies. It took on different meanings and forms, developing in a more or less organized manner; in general, it can be interpreted as a kind of movement reciprocal to that represented by forced migrations (or deportation, sedentarization et similia) occurring much time before, in the different SU epochs. This especially considering the nationalities involved, whether or not they may refer to a titular foreign nation-state eventually supporting such a "return" (like Germany, Ukraine, Israel, Korea, Baltic or South Caucasus states), signifying a kind of reparation for a historical injustice.

It is necessary to consider the populations that did not have a country of reference, and that are more or less a kind of SU "orphans"; the same for the populations of mixed character—possibly the majority, about 40%—that really did not have a definite post-Soviet identity, and then a definite place to go back to, namely a nation state capable of hosting them.[94]

Further migration trends happened on a local scale. Among them internal migration to the new independent states must be considered, that re-established a centre-periphery scheme, to which an internal migration flow also corresponds, in terms of demographic and workforce re-equilibration. This particularly occurred due to the rapid growth of urban centres and new city capitals as well as to replace the departed populations. In early transition times, the movement interested a kind of reflux to rural areas (the geography of regression, as described); then, once the economic situation improved, it once more changed direction, developing from rural and mountains areas to the new centres of the NIS. It can be said that such movements were sometimes ephemeral, sometimes temporary, and therefore not properly registered by statistics in this chaotic period.

Besides this, the continuation of principal migration tendencies can be registered, representative of titular nationality internal repatriating from other parts of CIS and elsewhere (e.g. Kazakh and Kyrgyz from Russia, Mongolia and China to Kazakhstan and Kyrgyzstan, the "oralmans").[95]

It depended on the apparent situation and obviously on the economic situation offering work as an opportunity for family reunification, the gaining of citizenship, and further benefits. Furthermore, the movements induced by new tensions, especially in the border areas with mixed linguistic tendencies must be remembered; for example, Uzbek, Kyrgyz and Tajik populations expelled or voluntary crossing in mass over the border with the neighbouring Fergana countries, because of recent disastrous clashes (2005 in Andijan and 2010 in Osh).

Finally, new emigration patterns from the poorest CA countries must also be considered—notably Tajikistan and Kyrgyzstan to other rapidly developing countries, especially to Russia and Kazakhstan, and other CIS countries, for which they do not need a visa; from these destinations they could send remittances that would represent primary economic resources of the states they migrated from for decades (but this movement rate has declined over the last few years because of the oil price crisis affecting the host countries, mainly based on HC rents).

8.6.3 Change in Migration Significance: A Kind of Trans-Continental Commuting

Having said this, it is not altogether easy to talk about minorities as per modern times (it is no longer a "taken for granted" variable); populations evidence a tendency to mix and to move—not just in territorial, but also in identitarian terms - much more so than in the past, and often in reversible terms. This change provokes further questions, like those about dual citizenship, visa regulations, reciprocal acknowledgment of rights, the status of the minority languages (this in particular with concern to use in official circumstances of the Russian idiom in the new nations-states).

In fact, mobility shows some unexpected evolutions over time; in several cases, it is possible to observe a subsequent reversal movement of returnees going back to CA republics—even in limited numbers and for limited periods—namely to the places those people originally migrated from. This is the case of individuals that obtained the citizenship of the destination (namely original) nation in the meanwhile, and who once the tensions characterizing the transition were exhausted, would return to CA, possibly due to business or private reasons (e.g. family reunification, to visit the places where they spent their childhood).[96]

This regularly happened for Russian "vozvraščentsy" ("Возвращенцы", see later), but also for a number of Jewish, Armenians, Koreans, Europeans and other nationalities[97]; in these circumstances, this movement occasionally configures

[94] Klüter Helmut, 1993; Ferrando Olivier, 2013:35 fs.; Sodiqov Alexander, 2013:50 fs.

[95] Esenova Saulesh, 1998; Kenzhebekovna Kalshabaeva, Akbota Senbayevna Seisenbayeva, 2013.

[96] Klüter Helmut, 1993; Ro'I Yaacov 1991.

[97] Indeed, it seems the obtainment of a dual citizenship or at least of a permanent visa is often a major motivation of such migrations' application; see as well Klüter Helmut, 1993.

a kind of commuting (a circular movement), taking on further significances. Thanks also to new favourable conditions (e.g. the low-cost travel tariffs, visa liberalization, pacification and relaxation of attitudes towards minorities), people often move temporarily, for example, because of study, looking for jobs or visiting relatives, then maintaining relations and travelling periodically all over the CIS space.

In this way, the "vozvraščentsy" contributed to re-establishing and to maintaining a common cultural post-Soviet area, with a multiplication in exchanges and relations in all shapes and sizes, representing an important element in general normalization (confusing in a movement that is to be considered typical for unavoidable globalization).

8.7 After Independency

8.7.1 The Gaining of Independency: General Questions

Notwithstanding a decade of Perestroika reforms, none could imagine such an incumbent change, and in fact, independence happened in CA almost in an unexpected way, highlighting a total lack of preparation by local institutions and populations. Paradoxically, when it happened, the same nomenklatura did not believe in this change, to the point that they even organized a sequence of referenda to reject or delay the idea of "independence" (as happened in Kazakhstan), possibly leveraging on the sense of incredulity and of the widespread sense of uncertainty.[98]

It was simply impossible to think that Soviet power had just disappeared (as had often happened in history in hypertrophic empires, surviving to the evolution of the wider context, maintaining a veil of prestige combined with the fear they may dissipate, or just because they have no alternative).

In fact, the elite and populations appeared disoriented, at least initially. Far from representing a chance for progress, the gaining of independency was perceived as a risk, because of the appearance of a vacuum of power that local lobbies tried to avoid at any cost (since it would put the continuity of leadership and the power system that was framing it into discussion).[99]

Then (after initial incredulity, which created a situation of impasse) the first question was the consolidation of power, ensuring the continuity of the legitimization in front of internal actors, of neighbours (inside the post-Soviet fragmentation process, that had a continental dimension) and of the wider international community (IC).

All this in a context of potentially destabilizing problems; among such problems, socio-economic fragility (due to welfare collapse), political claims, mass migration, tensions deriving from rising demographic asymmetries—especially considering the wave of young generations, claiming for jobs, housing, freedom etc.—and others.[100]

This also occurred because, soon after independency the local elites, namely the old regime recollocated in the new frame, suddenly realized that they had enormous power which involved managing a huge treasure of natural and economic resources, among others.

A major concern lay in the forms of opposition, potentially overlying any social contraposition, or organization mode (regional and tribal, ethnic and language, civil and economic); and this especially as regards religious resurgence, taking encouragement from neighbouring situations.

This can be especially seen in the case of the Khomeini revolution in 1979, that originally started as "street rioting", as a spontaneous meeting of the population, in occasion of the Friday prayer at mosque square.[101]

And it is the case of the years-long heroic resistance against the Soviet invasion in contiguous Afghanistan in the '80s, which shaped a brand-new generation of "partisans" who were often religiously inspired and supported by several foreign, even western countries. These individuals had little occasion to reintegrate into civil life after the dismantling of the resistance (often literally no families, no job and no home to go back to, often having grown up in some refugee camp, recruited by some militia at a young age).

8.7.2 New Geographical–Political Units

The NISs, emerging from the fragmentation of the Soviet map, assumed per default—as is usual in these circumstances—the pre-existing administrative borders among the different SSRs, which with all its imperfections and critical

[98] The path towards independence in these confused years was not so linear; the referendum in March 1991 saw 88.7% of voters approving a proposal of preserving the Soviet federation, even when in a "renewed form"; in December of the same year, after the independence proclamation, the referendum had given as result the 99% of votes favourable for independence; for a while – and to some extent even today -, the major element of uncertainness for CA systems was to trace the situation in Russia; this especially in occasion of the attempted "August putsch" (1991), that contributed decisively to the ending of the SU, with the consequent foundation of the CIS and the acquisition of independency by the side of CA republics.

[99] Ro'i Yaacov, 1991; Babajanian Babken, 2015:514.

[100] As said, in Carrère D'Encausse H., 1978.

[101] Brill Olcott Martha, 2012:40 fs.; even when the Iranian revolution can be interpreted as the long term "choc de retour" generated by excessive and rapid modernization, considering the dramatic gaps arising between the rural and urban, secularism and religion and others such contrapositions.

points, represented the only practicable option at that moment in time.[102]

Indeed, they corresponded roughly to an ethnic and language majority (defined as "eponym" or "titular"), with a prevalent religion, or religious variant: a fact that made it possible to imagine a reasonable settlement of the eventual contrapositions, preventing any escalations and further desegregation.

On the contrary, from a geo-strategic point of view, the new map left a sequence of questions open, evidencing a potentially scarce territorial compactness and a sequence of contentiousness that will probably arise in the near future. It is the case of infrastructures' overlapping, of mutual border acknowledgments and demarcation, of the appearance of enclaves, of the formation of trans-border minorities, and of further similar questions.

8.7.3 From Destatalization to Liberalization

Since independence, local leaderships have realized the possibility of managing natural assets independently, a kind of "treasure" that in the previous regime, in conditions of difficult accessibility to the international market, and of a centralistic political monopoly, they could not benefit from.[103] This was considered the true priority of politics, signifying not just the material exploitation of such resources, but the need to organize a whole brand new apparatus capable of carrying this out as well.

This is the case of off-shore oilfields and of mountain mines, of water reservoirs, ID and hydroelectric facilities, impossible to be managed with the obsolete Soviet know-how (also considering that the previous management had often returned to Russia as "vozvraščentsy"), and of agro-food and agro-industrial production.

It was also a question of political functioning, for example, making difficult decisions between alternative strategies, concretely defining new policies; among them the decision between privatization and neo-statalization, between the establishment of a completely free market and the conservation of some centralistic control, between integration within the international community and isolationism, and so on.

8.7.4 Privatizing the Post-Soviet Economics

In an initial phase, denationalization, combined with privatization, appeared as the best slogan, meaning the attempt to restore the damage done (the expropriations) provoked a long time before, to some extent (the exact reversal of the communist revolution). In this way, the new government could also gain in legitimization, promoting the formation of a free market and of a new class of citizens, participating in political and economic life.

But, such intentions proved to be difficult to realize, due to many issues, both structural and cultural; among others it is possible to list the weakness of the new institutions, the absence of an evolved social base (namely of a middle-class supporting privatization), of private initiatives and capitals, and of market self-regulating rules.[104]

Furthermore, some resources proved to be impossible to privatize because they were non-disposable, in practice. This is the case of territorial and of natural resources (in general perceived as "common"), of industries processing such resources (most of which still operating industries); then of assets which are difficult to account for (and then to be evaluated with a reliable price). This also considering the characteristics of the industry (tracing them back to Soviet tradition), mainly heavy and obsolete, incapable of dialoguing with the market (since they were intended to be instruments for centralistic planning).

Activities, furthermore, characterized by relevant externalities, above all on a social and environmental scale. It proved to be difficult or even impossible to privatize industries and mines, infrastructures and facilities, which even when underperforming, represented the territorial frame for entire regions: they were conceived as integrated "company-towns" providing administration services and territorial organization.

So too, for primary activities (agriculture, natural resources and cattle breeding), which were organized in units characterized by heavy social structuring: kolkhoz, sovchoz and leschoz (in peripheral and remote areas, "lesnoe chozjastvo"), they were to be considered as basic social-administrative, rather than mere productive units.

In fact, few of the existing companies proved to be suitable for privatization, since they were providing (even when in poor conditions), a set of services like hospitals, sanitation, housing, public transport or education. Such services obviously proved to be impossible to be measured with standard accountability, therefore impossible to be directly

[102]This process has usually been defined as legitimate, following the assumption that, in the absence of further elements, the administrative borders – in the frame of a federal state - can be adapted as political borders, consequently to the change of the power qualification, from the administrative level to sovereign-political one. Consequently, with a change of scale, the administrative local elite become the government of the successor state; Necati Polat, 2002.

[103]This situation would last for a while; yet in in 2011, 2/3 of Turkmen gas was delivered through the Russian state-owned Gazprom pipelines network, Anceschi Luca, 2017; see as well http://news.bbc.co.uk/2/hi/business/5314940.stm, 5 September 2006, accessed 22.4.2018,"Russia reaches Turkmen gas deal"; but today this figure is rapidly becoming obsolete, since the "stans" could diversify on different markets, geopolitically leveraging eventually on China, or other contiguous countries.

[104]Haghayeghi Mehrdad, 1997.

removed (also impossible to be managed in a truly corporate manner).

8.7.5 Activities Surviving to Transition

In such situations (a difficult passage to an open market), the new sovereign power changed its attitude; just after having consolidated its position, it soon realized the difficulties—and the risks—of privatization politics, and initiated a new strategy, beginning with a new (paradoxical) renationalization of the same precious assets.

These resources, in fact, proved to be easily manipulable, politically useful, and then suitable to being concentrated in a centralistic way. They could be "used" both on international markets, in order to get hard currency (or as a kind of diplomatically convertible "money"), and internally for social control purposes (eventually for "buying" the opposition or to remunerate the oligarchs).

It is the case of the long list of natural resources, above all obviously HC (including coal, largely used in energy-industrial and civil activities) in Kazakhstan, Turkmenistan and Uzbekistan. Then, it is the case of other major heavy industry plants, making easy profits through energetic resources, or for raw material preliminary processing (aluminium, uranium, gold, nickel); the same for some agro-industrial activities, like cotton and some other cultivations.

It is the main reason why the "new" (actually old) elite would soon build new monopolies, heralding what would be defined as a "patrimonial–autocratic" rule (actually the re-edition of the classic concept of "oriental despotism").[105] It interested entire sectors to the point that some authors talk about a process of renationalization, namely of a "fictive transition".

But this involution highlighted a limit; besides renationalization purposes, these assets urgently needed foreign intervention in order to upgrade production modalities (hardware, software and know-how). Therefore, local governments had to negotiate such interventions from outside, unavoidably conditioning internal politics.

This is the case of cited resources- or energy-exploiting industries, of mines, energy- and agro-industries, that attracted foreign investors, mainly MNCs specialized in those sectors that, taking advantage of weak local transition regimes, were seeking favourable (or biased) contracts. Such heavy industries were able to survive, maintaining profitability, also thanks to high entry barriers (which usually characterize such sectors), namely the high investment necessary to start up such kinds of industry; this also considering the high externalities (social and environmental, above all) that may discourage the foundation of new such impactful plants elsewhere.[106]

Many of such contracts would soon be contested and adjusted, or interested by corporate-like contentions; among them it is possible to list the Kumtor gold mining (Canada MNC Centerra Gold), Turzunzoda aluminium (with regard to some functions, the ownership having remained public, also because of the long-lasting civil war),[107] and the metallurgic Karaganda plant branch (Mittal Indian investments).

Among the biggest contracts (for public work) are those regarding the dam on the Vaksh river in Tajikistan; among them, the Roghun dam project under construction since Soviet times (the highest in the world, once finished) is currently managed by an international consortium guided by the Italian Salini Impregilo (see later).

[105] Babajanian Babken, 2015:514; current literature insists on such definitions as patrimonialism, sultanism and similars, however geography scholars had already defined such situations with categories as "organicism", "oriental despotism" or "geo-determinism", applying such concepts in different occasions; see Heathershaw John, Herzig Edmund, 2013:31.

References

Adeeb Khalid (1998) The politics of Muslim cultural reform. Jadidism in Central Asia, Berkeley

Anceschi L (2017) Turkmenistan and the virtual politics of Eurasian energy: the case of the TAPI pipeline project. Central Asian Sur 36(4):409–429. https://doi.org/10.1080/02634937.2017.1391747

Babajanian B (2015) Promoting empowerment? The World Bank's village investment project in Kyrgyzstan. Central Asian Sur 34(4):499–515. https://doi.org/10.1080/02634937.2015.1095967

Bacon EE (1990) Central Asians under Russian rule,. Cornell University Press, Ithaca, Or Ed 1966

Barney W (2011) Geographies of global Internet censorship. GeoJournal 76:1–23. https://doi.org/10.1007/s10708-010-9393-3

Barisitz Stephan (2017) Central Asia and the silk road, economic rise and decline over several millennia. Springer, Cham

Belorusskoye Bogatyri - Byelorussian giants, Moskva, Panorama (1991)

Bennigsen A (1989) Islam in Retrospect. Central Asian Sur 8-1(1989):89–109

Bennigsen A Lemercier Quelquejay C, a cura di Fasana E (1990) L'Islam parallelo. Marietti, Genova

Black CE et al (1991) The modernization of inner Asia. M.E. Sharpe, Armonk

Bond AR (1989) Soviet official admit past policy of deliberate map distortion; calls for increased access to large-scale maps and images mount. Soviet Geogr: 76–80

Bond AR (1991) Oblast reorganization in Kirghiz SSR. Soviet Geogr: 133–136

Brill Olcott Martha (2012) Tajikistan's difficult development path, Brookings Institution Press, 27 nov 2012

Buttino Marco (2003) La rivoluzione capovolta. L'ancora, Napoli

Carla T (2017) Samarcanda, Bukhara, Khiva. Città dell'Asia Centrale, pp.29–59, in: Città dell'Asia, Marzia Marchi a cura di, Bononia University Press, Bologna., 2017

[106] Then the consumptions of such industries in terms of environmental-natural and energetic resources, including soil and landscape, that in densely settled areas would probably mean much higher costs.

[107] Brill Olcott Martha, 2012.

Carrère D'Encausse H (1978) L'empire éclaté. Flammarion, Paris

Cosentino Italo (2017) Lingua, toponomastica e struttura urbana: il caso del russo nello spazio sovietico e post-sovietico, ASIAC annual conference, November 6-7th, 2017, Dipartimento di Culture, Politica e Società dell'Università di Torino (oral presentaiton)

Dagiev Dagikhudo (2014) Regime transition in Central Asia. Routledge, New York

Edgar Adrienne Lynn (2004) Tribal nation: the making of Soviet Turkmenistan. Princeton University Press, Princeton

Elebayeva AB (1992) The Osh incident: problems for research. Post-Soviet Geogr: 78–88

Ertz Simon (2005) The Kazakh Catastrophe and Stalin's order of priorities, 1929–1933. Evidences from Soviet Archives. Stanford's Student Journal of Russian, East European, and Eurasian Studies, 1, pp 1–12, https://www.scribd.com/document/149928998/Kazakhstan-2. Accessed 5 July 2018

Esenova S (1998) 'Tribalism' and identity in contemporary circumstances: the case of Kazakhstan. Central Asia Sur 17(3):443–462

Ezio B, Andrea DB, Angelo F, Pietro R (2018) Analisi ambientali e valorizzazione culturale dei qanat di Yazd, patrimonio dell'Umanità (UNESCO), Asiac annual conference 2018, 5–7 December 2018, University of Trieste – Gorizia campus (manuscript)

Farangis N (2010) https://www.rferl.org/a/UN_Cites_Evidence_Kyrgyz_Clashes_Orchestrated_While_OSCE_Official_Fears_Ethnic_Cleansing/2072710.html, accessed 23.4.2018, UN Urges Kyrgyz Government To Act Decisively Amid Turmoil, June 16, 2010, RFE/RL Kyrgyz Service, accessed 5.7.2018

Ferrando Olivier (2013) Soviet population transfer and interethnic relations in Tajikistan: assessing the concept of ethnicity. In: Heathershaw John, Herzig Edmund 2013:35–48

Freeman Jacqui (2013), The emancipation of women in soviet Central Asia from 1917 to 1940: strategies, successes and failures. In: Sevket Akyildiz, Richard Carlson, 2013, social and cultural change in Central Asia: the soviet legacy. Routledge, pp 32–44

Ginzburg NS (1986) A microgeography of settlement in the Pamir highlands. Soviet Geogr: 398–435

Haghayeghi M (1997) Politics of privatization in Kazakstan. Central Asian Sur 16(3):321–338. https://doi.org/10.1080/02634939708400994

Hambly G (a cura di) (1970) Asia Centrale. Storia Universale Feltrinelli, Milan

Harris CD (1994) Ethnic tensions in the successor republics in 1993 and early 1994. Post-Soviet Geogr: 185–203

Harun Y (2013) The soviet construction of Kazakh batyrs. In: Sevket A, Richard C (eds) Social and cultural change in Central Asia. The soviet legacy. Routledge, London and New York, pp 45–62

Heathershaw John, Herzig Edmund (2013) The transformation of Tajikistan: the sources of statehood. Routledge, London and New York, pp 95–110

Hedin S (1913) Durch Asien Wüsten: drei Jahre auf neuen Wegen in Pamir, lop-nor, Tibet und China. F.A. Brockhaus, Leipzig

Hisao Komatsu H (2009) From holy war to autonomy: Dār al-Islām Imagined by Turkestani Muslim Intellectuals. Cahiers d'Asie centrale, pp 449–475, 17/18/2009, Accessed 22 Apr 2018, http://journals.openedition.org/asiecentrale/1286

Hughes TP (2004) American genesis: a century of invention and technological enthusiasm 1870–1970, 2nd edn. The University of Chicago Press

Hughes K (2017) From the Achaemenids to Somoni: national identity and iconicity in the landscape of Dushanbe's capitol complex. Central Asian Sur 36(4):511–533

Jelen Igor (2000) Repubbliche ex sovietiche dell'Asia Centrale. Utet libreria, Torino

Jelen Igor (2002) Tra i kirghizi del Pamir Alaj. Forum, Udine

Jelen Igor (2012) Appunti di geografia politica ed economica. Aracne, Roma

Kaiser RJ (1994) The geography of nationalism in Russia and the USSR. Princeton University Press, Princeton

Kaiser RJ (1995) Nationalizing the work force: ethnic restratification in the newly indipendent states. Post-Soviet Geogr 36(2):87–111

Kalshabaeva K, Seisenbayeva AS (2013) Some problems of repatriation and adaptation of representatives of the Kazakh Diaspora of Central Asia in the Historic Homeland Bibiziya. Middle-East J Sci Res 15(1):20–26. https://doi.org/10.5829/idosi.mejsr.2013.15.1.11054

Kanet RE (2006) The superpower quest for empire: the cold war and soviet support for 'Wars of National Liberation'. Cold War His 6(3):331–352. https://doi.org/10.1080/14682740600795469

Kassymbekova B (2011) Humans as territory: forced resettlement and the making of Soviet Tajikistan, 1920–38. Central Asian Sur 30(3–4):349–370. https://doi.org/10.1080/02634937.2011.607916

Khazanov AM (1994) Nomads and the outside world, second edition, The University of Wisconsin Press, or. ed. in Russian 1983

Klüter H (1992) Russland und die Auflösung der Sovietunion. Geographische Zeitschrift Jahr 80, Heft 1

Klüter Helmut (1993) People of German descent in CIS States – areas of settlement, territorial autonomy and emigration. Geojournal 314:419–434

Krasnov GA (1987) Il commercio tra paesi a diverso sistema: tendenze e prospettive. In: Parboni R. e Wallerstein I., a cura di, 1987, L'Europa e l'economia politica del sistema-mondo. F.Angeli, Milano, pp 63–79

Kudaibergenova Diana T (2013) National identity formation in post-Soviet Central Asia: the Soviet legacy, primordialism, and patterns of ideological development since 1991. In: Akyildiz Sevket, Carlson Richard, edited by, 2013, Social and cultural change in Central Asia. The soviet legacy, Routledge, London and New York, pp 188–200

Lentz W (1933) War Marco Polo auf dem Pamir? Zeitschrift der Deutschen Morgenländischen Gesellschaft 86, NF11:1–32

Marvin Harris (1971) L'evoluzione del pensiero antropologico, Il. Mulino, Bologna

Marzhan Thomas (2015) Social, environmental and economic sustainability of Kazakhstan: a long-term perspective. Central Asian Sur 34(4):456–483. https://doi.org/10.1080/02634937.2015.1119552

McMahon Robert (2005) Uzbekistan: report cites evidence of government 'Massacre' In Andijon, Radio Free Europe/Radio Liberty, 7 June 2005, https://www.rferl.org/a/1059147.html. Accessed 5 July 2018

Meloni Alberto (2017), Quando i gesuiti erano i francescani. La Repubblica 20 febbraio 2017

Megoran Nick (2017) Nationalism in Central Asia: a biography of the Uzbekistan-Kyrgyzstan, boundary, Central Eurasia in Context, 2017, University of Pittsburgh Press. https://doi.org/10.2307/j.ctt1vjqrk6

Monteil V (1957) Les musulmans soviétiques. In: Collections esprit Frontiére ouverte. Seuil, Paris, Editions du

Nazaroff P (1933) Hunted through Central Asia. Oxford University Press, Oxford, NY. 1. edition 1932

Özcan GB (2015) Introduction: market adaptations, interventions and daily experience. Central Asian Sur 34(4):409–417. https://doi.org/10.1080/02634937.2015.1103580

Paskaleva E (2015) Ideology in brick and tile: Timurid architecture of the 21st century. Central Asian Sur 34(4):418–439. https://doi.org/10.1080/02634937.2015.1118207

Pianciola Niccolò (2001) The collectivization famine in Kazakhstan, 1931–1933. Harvard Ukrainian Stud 25(3–4):237–251

Pianciola Niccolò (2004) Famine in the Steppe. The collectivization of agrocolture and the Kazakh herdsmen, 1928–1934. Cahiers du monde russe 45:137–192

Polat N (2002) Boundary issues in Central Asia. Transnational Publishers, Ardsley

Pryde PR Bradley DJ (1994) The geography of radioactive contamination in the former USSR. Post-Soviet Geogr: 557–593
PWC (2011) Doing business guide Tajikistan 2012–2013, http://www.eurasianbusinessbriefing.com/wp-content/uploads/2015/04/pwc-doing-Business-in-tajikistan.pdf. Accessed 20 May 2018
PWC (2016) Guide to do business and investing in Uzbekistan 2016 edition, https://www.pwc.com/uz/en/assets/pdf/dbg_2016.pdf. Accessed 20 May 2018
Rasanayagam J (2014) The politics of culture and the space for Islam: Soviet and post-Soviet imaginaries in Uzbekistan. Central Asian Sur 33(1):1–14. https://doi.org/10.1080/02634937.2014.882619
Reid P (2017) 'Tajikistan's Turksib': infrastructure and improvisation in economic growth of the Vakhsh River valley. Central Asian Sur 36(1):19–36. https://doi.org/10.1080/02634937.2016.1204533
Robinson F (1989) Atlante del Mondo Islamico. De Agostini, Novara
Rossi Marina (1997) I prigionieri dello zar. Ugo Mursia Editore, Milano
Rotar Igor (2006) Resurgence of Islamic Radicalism in Tajikistan's Ferghana Valley, the Jamestown Foundation, Terrorism Focus, 3/15, April 20, 2006, https://jamestown.org/program/resurgence-of-islamic-radicalism-in-tajikistans-ferghana-valley/. Accessed 22 Apr 2018
Ro'i Yaacov (1991) Central Asia riots and disturbances, 1989–1990: Causes and Context. Central Asia Sur 3:21–54
Roy O (1992) L'échec de l'Islam politique. Éditions du Seuil, Paris
Roy O (1997) La nouvelle Asie Centrale. Éditions du Seuil, Paris
Rumer BZ (1989) Soviet Central Asia. A tragic experiment. Unwin Hyman, Boston
Saidazimova Gulnoza (2008) Central Asia: child labor alive and thriving, Radio Free Europe/Radio Liberty, http://www.rferl.org/content/article/1144612.html. Accessed 22 Apr 2018
Sevket Akyildiz, Richard Carlson (2013) Social and cultural change in Central Asia. The soviet legacy. Routledge, London and New York, pp 188–200
Shabad T (1980) The 1979 Census and some demographic trends in the Soviet Union. GeoJournal 4/1:84–90
Shahrani MNM (1979) The Kirghiz and Wakhi of Afghanistan: adaptation to closed frontiers. University of Washington Press, Seattle
Sodiqov Alexander (2013) From resettlement to conflict: development-induced involuntary displacement and violence in Tajikistan. In: Heathershaw John, Herzig Edmund 2013:49–66
Sultangalieva G (2014) The role of the pristavstvo institution in the context of Russian imperial policies in the Kazakh Steppe in the nineteenth century. Central Asian Sur 33(1):62–79. https://doi.org/10.1080/02634937.2014.885746
Silvestri Tommaso (2015–2016) Le strategie della federazione russa nel teatro del mar Caspio dopo la caduta dell'Unione Sovietica. Master Thesis, University of Trieste, academic year 2015–2016
Teichmnan Christian (2007) Canals, cotton, and the limits of decolonization in Soviet Uzbekistan, 1924–1941. Central Asian Sur 26(4):499–519. https://doi.org/10.1080/02634930802018240
Teichmnan Christian (2016) Wildscapes in Ballyhooland. Shock construction, Soviet colonization, and Stalinist governance. Cahiers du Monde Russe 57, 2016(/1)
Valikhanov ČČ (1985) Sobranje sočinij v pjatu tomach. Glavnaja Redakcija Kazachskoj Sovetskoj enciklopedii, Alma-Ata
Valikhanov capt., Veniukof M et al (1865) The Russians in Central Asia. Translated by J. and R. Mitchell, Edward Stanford, London
Von Humboldt A (1975) a cura di Milanesi M. – Visconti Viansson A., La geografia, i viaggi. F.Angeli, Milano
Warf B (2016) Global geographies of corruption. GeoJournal 81:657–669. https://doi.org/10.1007/s10708-015-9656-0
Werth N (1993) Storia dell'Unione Sovietica. Il Mulino, Bologna
Wessels G (1992) Early Jesuit Travellers in Central Asia, 1603–1721, Asian Educational Services, New Delhi - Madras 1992. Ed. or. 1924
Wheeler G (1962) Racial problems in Soviet Muslim Asia. In: Seconda edizione. Institute of Race Relations. Oxford University Press, London
World Report (2015) Uzbekistan, Events of 2014

9 The Geographical Mosaic

Abstract

Local cultures and populations finally had the chance to recover their independence and their identity, even if in a slightly ambiguous manner: the current figures of the five "stans" evidence a mix of aspects and opportunities but risks as well. The heritage of the Soviet system did indeed prove to be easier to overcome than expected (at least for aspects concerning any material impact), and the local independent states have finally been able undertake a journey of stabilization and progress.

Keywords

Presentation of Kazakhstan · Kyrgyzstan · Tajikistan · Uzbekistan · Turkmenistan

9.1 General Description and Particular Definitions

Depending on the point of view, CA geography can be defined in a single macro-region, rather than a set of individual countries, nations and sovereign states: the basic data, as well as the list of resources and cultural features evidence an inextricable knot of elements unifying the populations of the region. However, the current economic-political situation is not the result of any unilinear evolution delineating the progressive formation of coherent self-containing territories over time. On the contrary, it is often the result of a sequence of shocks and of the arbitrary overlapping of new borders and of new powers.

The best way to present the current situation is probably to shift between different ways of describing it, alternating individual and common descriptions.

For some aspects, it is necessary to analyse the political unit that represented the prevalent social format in the different epochs. In the past, it was the case of "aul", the clan and tribe and at the other extreme of the scale, of city-states on the trans-continental SR and of universal steppe-empires. It is currently the case of what is defined as the nation-state, repurposed in CA in post-Soviet times (possibly becoming obsolete, replaced in many functions by emerging super- and trans-national configurations).

Such "zooming" is necessary so that it is possible to observe the phenomena on different scales (even if it is necessary to avoid the risk of any excessive simplification of the scenario at the same time). So too, for any function (production, mobility organization, identification pattern, communication mode) and for any human activity.

Basically, in CA, it is possible to distinguish between different specialization belts, each one corresponding to a different cultural adaptation: a grain cultivated area in the north-eastern steppe, a livestock and extensive farming area in the central steppe region, a river-oasis irrigated cultivation belt in the southern areas with cotton and rice prevailing and, furthermore, a livestock and seasonal mountain agriculture area in the south-eastern mountain region.

This differentiation can also be applied to further general characteristics: for settlement patterns, with the distinction between urbanized and sparsely populated areas, that overlay (and intertwine) the traditional schema based, respectively, on nomadic and sedentary populations. So too for the establishment of industrialized agriculture in modern times, "urbanizing" the rural areas with kolkhoz and sovchoz, shaping entire landscapes and so also in remote areas, for the continuation of subsistence living (for local community subsistence) and self-sufficient communities.

This is also the case for the different phases in recent history in terms of modernization, from industrialization to the current HC-driven economics, to possibly the next step in a post-modern and post-carbon passage. All these aspects have developed in the background of the geographical constraints characterizing the different areas.

I. Jelen et al., *The Geography of Central Asia*, World Regional Geography Book Series,
https://doi.org/10.1007/978-3-030-61266-5_9

This is the case of the landlocked condition affecting all CA countries and of their related environmental characteristics, like endorheic conditions and climatic continentalism, determining impacts on all scales, on human and natural situations. In particular, the imbalance in accessibility to resources, first and foremost water of whatever origin — underground, meteoric, surface and flowing from the mountain glacier — and agricultural soil; and in general, determining a basic fragility on ecosystems (and of the same human organizations).

Such material constraints affect politics and economics in different ways, and has inspired the current attempt to unlock CA territories to the outside while planning impressive infrastructure corridors; it is possible to find such projects on the agenda of each local government (with projects occasionally combining with the purposed new versions of the SR, represented through different variants and slogans).

Generally speaking, such policies underline the need to escape certain ideas of continentalism that represent the most obvious feature of CA countries, provoking ecological fragility, economic dependency and other critical aspects. But which—at least to some extent—seem to be a matter of perception, rather than a true material constraint, since current political and technological evolution has proved to be capable of disrupting the territorial "trap" (namely, landlockedness in any form) that has conditioned local human experiences (at least as intended until few decades ago) in its entirety.

9.2 Kazakhstan: The Fatherland of the Steppe

9.2.1 Basic Description

As the largest country, Kazakhstan represents all the distinctive features of the CA geography, evidencing huge economic potential and a self-evident impact in international geopolitics. Indeed, traditional technology shows most of its surface as unproductive, appearing even today virtually as desert-like or sparsely populated. The same name evokes such characteristics, since the name "Kazakh"derives from an ancient Turkic word meaning "wandering", echoing the traditional "genre de vie", for people continuously moving in search of resources.

The "fatherland of the steppe" refers to the most authentic tradition of CA populations. It has a surface of 2.7 million sq.km, and a population of 18.5 million inhabitants, ranking ninth in the world for surface area; it is the largest among the landlocked countries and one of the least inhabited, about 7 inh. sq.km;[1] it is also the richest in terms of resources of any kind, agrarian, mineral and energetic, mainly hydrocarbon; this sector of Kazakh economics represents about 60% of the whole CA GDP, a figure that evidently represents a tendency for the highest potential but also as a risk of becoming dependent on this resource.

From many points of view, today it appears as the dominant country in CA, even when it is the most exposed and geopolitically weak at the same time, being involved per default in the most important situations in the area. It has borders in common with all other CA countries (with the exception of Tajikistan), as well as with the external great powers (Russia, for 6.800 km, and China, about 1.782 km), for which much of the CA's destiny historically depends, and on the western side with the Caspian Sea. On the northwestern side, it borders beyond the Ural river onto the European continent for a surface representing 6% of the whole Kazakh surface (more than just a geographical curiosity).

Continentalism is the obvious characteristic of the country: Kazakhstan is the symbol of Eurasia and includes, especially in the north-east, large surfaces of potentially useful land (that can be rendered cultivable after some treatment), and in general huge natural resources whose exploitation is to be considered at an initial stage.

The country has an extraordinary variety of landscapes, climates and environmental conditions, from steppe, arid or semi-arid areas, to mountain slopes, flatlands and plains, as well as the taiga; from picturesque eroded rocky canyons, to marine deltas and coastal lagunas, glaciers and high-altitude plateaus, swampy depressions and salty lakes. Its territory also has abundant water resources, even if distributed in a rather uneven way, consisting mainly of rivers flowing from the south-east mountains in a north-west direction. The biggest of them, the Irtysh, arrives from China and flows further on to southern Siberia, into Ob'; the Ili also arrives from Chinese territory, presenting some pollution problems; others—the Chu and its tributaries—from the Zhetysu endorheic basin, flow into (as does the Ili) Lake Balkhash.

It is typical transition country, having been crossed by Eurasiatic roads since ancient times, connecting the Far East with the extreme edge of Europe, as well as connecting the northern part of the Asian continent with urban civilizations on the SR; it is also a transition country from a cultural and ethnic point of view, and it is the country where Russian domination—coinciding with the modern epoch—had left its most persistent and important traces.

The same ethnic structure represents a legacy of history, even now including about 1/3 of populations of Russian and further European origin; these reside mostly in Zhetysu, in urban area (especially Almaty), and in the north-eastern steppe belt, which after reclamation and colonization waves both in czarist and in Soviet times, is suitable for cultivation without any artificial irrigation. In this area, the Russian presence is—from all points of view—a structural compo-

[1] https://www.cia.gov/library/publications/the-world-factbook/geos/kz.html, accessed 22.4.2018; Yilmaz Harun, 2013:53.

nent of the Kazakh society and is not perceived as a minority; this is also the case because the Kazakh society tended to assimilate further any European or non-European minorities that had remained without a role or status in the new national Kazakh-led country.

This is the case when the Russian population in Kazakhstan decreased by 38% in 1989, when it had been the majority in 7 of the 20 regions of the Kazakh republic, to the present day 23%. The ethnic Kazakh make up about 63% today and have also increased in percentage and in absolute numbers, thanks to migration of co-nationals (the "oralmans") mainly arriving from other CIS republics, Mongolia and China (mostly in transition times, between 1991 and 2005).[2] Among the other minorities, there are Tatars (1.3%), Ukrainians (2.1%), Uzbeks (2.8%), Belarusians, Uighurs, Azerbaijanis, Poles, Lithuanians, Germans and Koreans;[3] they are few in numbers, and possibly decreasing, but still maintain a significant role.

The current ethno map is the evident result of a troubled history of spontaneous and planned movements, but especially of the deportation of minorities which occurred in Soviet times, sometimes simply as "preventive" punishments. It is the case of the Volga Germans, Ukrainians and further middle-eastern Europeans; it is the case of North Caucasians, and to a lesser extent of South Caucasians (especially Kurds and Meshketian Turks from Georgia, among the populations punished by Stalin, for diverse and instrumental reasons); and then of Far Eastern Koreans and from other parts of Asiatic Russia; it is the obvious case of political opponents having been deported for other reasons than nationality (but above all Russians) in the different stages of Soviet rule and to some extent also in colonial times.[4]

After the transition tensions, today the attitude towards minorities has changed completely, since they are no longer perceived as a risk for the unity of the country (with the possible exception of an ambiguous sentiment characterizing Russian communities on the borders). This is also considering the Kazakhstan trend in economics that is tendentially opening up and is based on international trade. It is comprehensible as it is based mainly on HC export and also because of a general cultural attitude of openness, differentiating the country from other FSU states (possibly to be considered as a cultural legacy of the nomadic attitude).

In this frame, their presence and in general the ethnic plurality of the country are officially shown and considered as an advantage, if not for any other reason than to make "cultural mediation" possible in the new global environment.

[2]Marzhan Thomas, 2015:460; Kenzhebekovna Kalshabaeva, Akbota Senbayevna Seisenbayeva, 2013.

[3]Source Wikipedia, Kazakhstan, accessed 23.4.2018.

[4]Rossi Marina, 1997.

9.2.2 From the Dramatic Past to the Present

While natural resources are generous and abundant, demography represents a critical point of the Kazakh country: after the tremendous shocks suffered in Soviet and colonial times, population numbers (in the late Soviet times) stabilized; then in transition times, they again weakened because of outmigration and persistence of unfavourable bio-demographic factors, and possibly because of the worsening of welfare services (healthcare, education, mobility).

Today, the demography is recovering, with the population reaching about 20 million (considering immigrants), even though the population density is still very low. As mentioned, it is among the lowest in the world, and irregularly distributed over rural and urban areas, and in the different parts of the country.[5] Currently, the government sustains the demographic policy with fiscal, welfare and urbanistic programmes (social housing, care and welfare facilities planning), but with mixed results.

The territory of Kazakhstan is settled in an asymmetrical manner, with the population scattered over huge surfaces, barely fixed in one particular place. Traditionally, the country was inhabited by pastural tribes, practicing long-range transhumance, often organized on the base of some geographic axis, e.g. north-south, plain-mountain, arid and wet steppe (prairies) around the lake Balkhash, or along some rivers, in order to take advantage of periodical changes in availability of resources.[6]

They were not "perfect" nomads, since they probably ran repetitive itineraries—marked in some manner — and were accustomed to spending the winter in permanent structures, possibly celebrating festivals and identity rituals in particular moments of the year, in particular places, in proximity of inter-changing centres; from here they had the possibility to trade or organize their movement, on the SR or on one of its variants, that crossed and practically covered the entire Kazakh territory.

The change of such basic schema started in early modernity when the entire area covered by Kazakh migrations began to stabilize into the perimeter currently defined by the CA steppe, between Lake Balkhash and the Caspian Sea, between the taiga tree-line in the north and the oasis khanates in the south; in these new circumstances, the Kazakh population began to divide into three groups, occupying and controlling specific territories, even when maintaining a semi-nomadic lifestyle.

Then the tribes started to acquire a more definite identity, possibly as a cascade effect of further movements and particularly due to the formation of structured states by neighbouring populations (especially considering the formation of the Uzbek

[5]Marzhan Thomas, 2015:460.

[6]Barisitz Stephan, 2017.

khanate in the seventeenth century in the south). This combined with further elements incentivizing a process of renewed ethnogenesis, subsequent to the invasion of people of Mongolian origin, pressing from east and west, wandering over the same Eurasiatic steppe corridor and sharing similar economic resources—but difficult to assimilate because of their different cultural, Buddhist religious and tribal backgrounds.

These circumstances represented a reference for the formation of a distinct identity: the Kazakh "nation" emerged "in opposition" to that of the Mongolian Dzungars—defined as the "last nomadic empire in the steppe" — and Kalmyk (from the western side, Volga region) invaders.[7] It is likely that they just needed a foe in order to define themselves: a moment presented today in the narrative as the "great disaster", when the Kazakh were defeated by further steppe people (which from that time forward would become the "traditional" enemy for the Turkish Muslims).[8]

In the following centuries the political life of Kazakhstan would be closely connected to Russian and Soviet state evolution, finally becoming the Kazakh SSR in 1936 (after a terrible repression, assuming possibly a genocidal character,[9] not much different to other colonial and totalitarian situations), in the frame of the SU. However, this moment, beside its tragic impact, is considered by the same current Kazakh elite as the milestone for the formation of a modern idea of the Kazakh nation.

The current situation—as an independent state—is characterized by the gigantic task of keeping the sparsely settled people together, reconstructing an entire new national landscape: new towns, facilities and skyscrapers appear everywhere in new urban environments, financed by national programmes—alimented by generous natural capital—inducing further investments cycles. The traditional obstacles, represented by environmentally extreme conditions, that have always impeded stabile settlement, can be overcome using new technologies, and new capabilities, that continuously create new opportunities, are possibly changing the perception the Kazakhs have about their own territory.

9.2.3 Renovated Ethnogenesis: Establishing of Nation State but Revival of the "Klanovost"

In recent modern times, traditional identity—namely the element structuring the identity of the people of the steppe—entered into contradiction with territorialized institutions, inducing a change in the social units. This happened to some extent intentionally, being provoked by the simply encounter of the Kazakh groups with colonial institutions rather early in the eighteenth century. Indeed, such cultural processes, developing over very long periods, have some impact even today.

The pressing in from outside forces—from Russians, but also from the Chinese and Mongol sides, respectively—did not render the tribal order more compact; the rivalries continued, to some extent spontaneously, as a consequence of the same continuation of the nomadic "genre de vie" in the undifferentiated space of the steppe, and possibly increased as a consequence of the space availability reduction. Finally, the weakened tribes were obliged to seek negotiation, ironically asking protection to the Russian empire which had appeared in that period as the new leading power on this side of the Asiatic continent (and without any particular tradition of conflict with the Kazakh tribes, because they had only just appeared in the area).

Previously, the life of the tribes took place in a more or less flat territory, in which they could identify a repetitive list of elements—water courses, deserts, mountains — as reflected indeed by the same nomadic toponomies, always reproducing the same names and the same geographical schema (Ak Suu and Kara Suu, Kizil Kum and Kara Kum, etc.). Russian colonization impacted this "genre de vie", inducing a deep transformation in the Kazakh cultural body, but without completely eradicating the nomadic fierce sensitivity, that (more or less imagined and narrated) would continue to represent an element of their identity.

To some extent, it seems that even today this original form of identification, based on the group[10] and clan, maintains some effectiveness; it is based on the presumption of a common ancestry, often celebrated in semi-sacred terms (a heroic horse rider, a warrior fighting the "Mongols" and saving the tribe),[11] usually combined with other cults and reminiscences.

Such representations are based on segmented schemas, suitable to the life of groups continuously moving over wide areas and proving to be conveniently flexible in order to face the variability of pastural life in such extreme environmental conditions. They are deliberately invisible to the eyes of foreigners and celebrated in particular moments, in circumstances connected with the transhumance calendar, when the identity of the group has to be represented and consolidated.

Besides the pastoral routine, such organizations were flexible enough to allow the groups the sudden formation of larger aggregations (horde- or nomadic-empire like), potentially for defence purposes; in other cases, they aggregated in

[7] Heathershaw John, Herzig Edmund, 2013:11; then a possible case of identity formed by contrast, as a reaction to the negation of that same identity, as typical in a colonial or subalternity circumstance; Korostelina K., 2011; Jelen Igor, 2011; for a general description of these times see Bussagli M., 1970.

[8] As presented in Muslim Turkish nomad traditional narrative.

[9] Pianciola Niccolò, 2001, 2004.

[10] Collins Kathleen, 2003.

[11] Yilmaz Harun, 2013:53; such image has been eventually actualized with the Skyft origin "golden man", Kudaibergenova Diana T., 2013:164.

proximity to commercial routes crossing this area, to organize trade, caravans or riders (plunder), to start a war or a migration, depending on circumstances. Such ways of thinking — although losing much of their original meaning, and combining with further elements of the traditional culture, sometimes just in folkloristic terms — even today represents a reference. In fact, although its significance as an effective identity marker is questionable, the sense of belonging to a clan is still felt, prefiguring what some observers defined as a sense of "klanovost" ("tribalism"), namely as something tendentially sectarian, in opposition to that of the "state" and of the "law".[12]

9.2.4 Zhetysu and Almaty

The south-eastern part of the country is the most suitable for settlement, and for development, lying in a mild climate prealpine area, where colonial settlement originally formed at the border with China; it coincides with the historical region of Zhetysu or Semyreč'e (in Russian "7 rivers", implicitly meaning a place with an abundance of water).

It is the area where the most important towns can be found and of the former capital town Alma Ata, currently Almaty, "city of the apples". In this case, it was also a name expressing a significant element of attraction and motivation for colonial migrants travelling for weeks and months, crossing arid steppe and deserts to get there. In fact, the town enjoys a pleasant climate and a nice alpine surrounding landscape, despite the infamy it acquired as one of the favourite places for deportation in Soviet (and czarist) times.

The city was founded with the original name of Vierny, as a border fortress on the uncertain boundary between the Czarist and the Chinese Empire provinces, on the path stretching between the mountain ridge and Lake Balkhash.

Lying in the middle of a fertile area but not permanently settled, it became the target destination of migrants, both free and forced, from all over the Russian Empire. Its growth accelerated when it became the terminal of the Turk-Sib railway, completed in 1930 (planned in colonial times; its works were suspended during the turmoil following revolution). Then, it became the most important urban centre in the area and a multinational city right from the outset, characterized by its vivacious society whose members were colonial officials, businessmen as well as deported politicians and intellectuals and their descendants.

The early independence period was particularly favourable for the city which became a sovereign state capital (for a while), rapidly acquiring the character of a globalized town, made up of self-defined "diasporas" (the Baltic, the Caucasian, the Korean "diaspora") which re-emerged in those times of liberation, acquiring a new role: a cosmopolite and tendentially liberal character challenging the nationalistic revival of the new Kazakh politics. Such true international culture is still current and indeed testified to by its inhabitants on many occasions.

9.2.5 Regional Articulation

The central area of the country is characterized by its borderless succession of steppes and deserts, by extremely harsh climate conditions, almost uninhabited, and it is the site of the most important Soviet deportation camp, symbol of the totalitarian regime, Karaganda, the destination of hundreds and thousands of deported people. The Kar-Lag coincides with the extended open-air coal mines, one of the biggest in the world. The mines represented the wealth but also the damnation of this country, which for generations were considered the only relevant resource of the country.[13]

This area in the middle of the continent—exposed to stormy weather, cold in winter, hot in summer—has represented a kind of impregnable fortress of CA since ancient times in which the nomads used to have to search for refuge: in such an environment their only possibility was that of moving continuously, in search of wellsprings and seasonal pasturages, on itineraries that only they would know, following signs that for others are usually invisible. This place, which has been defined as the mackinderesque heartland, was chosen by the Kazakh new leadership for the construction of the fantastic new capital town, Nursultan (formerly Astana), with a project designed and realized in just a few years with the help of many "archi-stars" such as Norman Foster, Renzo Piano and others (see Fig. 9.1). These projects would completely change the territory, and the whole nation's urban structure, changing the perception of the local population regarding their own territory as well.

This new landscape, with its architectural artificial beauty, can be considered as a challenge to the adverse natural conditions, demonstrating that the new Kazakh fatherland would be capable of ideally and practically challenging its steppe destiny, giving continuity to the modernist intention; mainly investing the revenues of the Kazakh treasure, namely from the HC industry.

In fact, the recent chapter of Kazakh history is characterized by the discovery of immense oil and gas resources, mainly in the western part of the country on the Caspian coast, with oil fields concentrated in coastal and offshore marine deposits.

This is possibly the most promising side of the country, with a rapidly growing industry which is nevertheless characterized by some problematic incompatibilities. This is due

[12] Collins Kathleen, 2003; Ro'i Yacoov, 1991; Esenova Saulesh, 1998.

[13] Rossi M., 1997.

Fig. 9.1 Kazakhstan, Nursultan skyline, 2017. (Photo: Igor Jelen)

to the fact that such activities—concentrated in coastal areas—can be in contrast with further types of development, for tourism and amenities, navigation activities and the fishing industry (especially for the precious and endangered sturgeon, for producing caviar).

The north-eastern part of the country is also relatively suitable for settlement and agricultural use, and in fact it was the preferred settlement area for farmers arriving from European Russian provinces, who immigrated for generations into the area. Today, the area is characterized by the most important minority residing in the tendentially neo-nationalist Kazakhstan, namely the Russian one (a majority in its own settlement area) along the Russian middle-Siberian borderline.

The Russian (and Soviet) colonizers settled here in different waves, following the construction of railways; then especially the construction of Trans-Siberian route (running just north of the Kazakh-Russian border), of the subsequent Trans-Aral Railway, completed in 1906, and finally of the Turk-Sib (from Barnaul to Almaty and Tashkent), connecting the whole CA to the Soviet developing eastern Siberian side.

The immigration was incentivized and supervised by a special Governmental Agency (Migration Department, Переселенческое Управление), fostering a typical colonial expansionist policy in order to pursue the numerical increase of colonists on this side of the frontier of the empire. The numbers account for 400,000 Russians had who arrived in the area before the revolution times—this signified the end of classic colonization, but not of mass migration used as a political-planning device. Comprehensibly such immigration, pursuing the methods of sedentary farmers, incompatible with the almost invisible (because seasonally moving) nomad groups, triggered the adverse reaction of the local population, which resisted in different situations.

Even today relationships between the majority of Kazakhs and the minority of Russians (or indeed of non-Kazakhs) represent an element of tension—even when not overtly evident. But the pluri-cultural attitude (possibly a relic of the exogamic attitude of the steppe traditions) seem to be prevalent:[14] society resulting from centuries of mixing cannot just be compressed into a neo-nationalistic schema, with minorities that may find themselves embarrassed by a situation of dual loyalty—in this case the Russians or Kazakh ones, intended as mutually exclusive options.

The relationships between the two communities (to whom the other minorities, either Slavonic or Turk, or from other origin, would adhere) will probably be crucial for the long-term development of the country. Among the different questions, there is the linguistic and alphabetic one, considering that the Russian language is the majority locally, even in the north-east, and still prevalent in institutional and public life; it has the constitutional status of a language of interethnic communication, which means the consequence that the non-Kazakh-speaking have the possibility not to learn the Kazakh language, and not the reverse. In fact, this highlights a low level of skill in the Kazakh language by non-ethnic Kazakhs.[15]

The further critical element is of an ethno-demographic character; after independence, the out-migration of Russians and of further nationalities soon started, who perceived their

[14] Barthold W., 1968:464; Shahrani M. N. M., 1979:159.

[15] Kudaibergenova Diana T., 2015:442.

status as uncertain and usually moved to the European provinces of Russia (in the case of Russians), or to the respective nation-states, as happened for the German, Ukrainian, Jewish, Armenian and Korean populations.[16] This fact could destabilize the equilibrium of a scarcely populated country; therefore the Kazakh government gave priority to the elaboration of a policy in order to pursue stable development, harmonizing the ethno-linguistic and geographical diversities. These last evolutions, in society and economics, seem to prefigure the maturation of a kind of bilingual equilibrium, namely the attitude to consider a multi-cultural condition as normal one (as indeed accepted in many countries of the world).

9.2.6 Resources and Activities

Kazakhstan is the richest country of CA in terms of resources; they comprehend any resource you can possibly imagine, especially mineral resources, some of them may be considered of strategic importance, like lead, chromium, zinc, uranium, coal, manganese, copper and iron; Kazakhstan exports gold as well, for which it is among the first ten as regards the largest reserves in the world, and diamonds.

The main wealth of Kazakhstan derives from a gigantic availability in terms of HC and especially in oil: the country reaches the capacity of three million barrels per day, being among the first 10 largest producers in the world, 11th largest for reserves of both petroleum and natural gas, continuously opening and exploiting new fields, among which there are some classified as "supergiants", of which the estimation can only be approximate.

It should be remembered that such resources are relatively difficult to manage, with oil extraction being rather expensive, since most of oilfields are located on off-shore deposits in the Caspian Sea, far from destination markets; so Kazakhstan has to plan and organize its principal activity carefully in order to make it profitable enough, especially today and possibly in the middle term future, with international oil quotation relatively low.

Besides oil and gas, coal is still also much used, especially for internal purposes, mainly for energy production, by older energy power plants—considering that 70% of such plants in Kazakhstan use coal, creating indeed several ecologic problems.[17] These are especially concentrated in Karaganda and Pavlodar (in the northeast) regions, lying close to the major coal fields of the country, coinciding with the oldest area of Soviet industrialization; further ecological concern regards the rapidly growing western industry, pivoting HC investments on the Caspian coast, where they cannot but produce externalities on the fragile ecosystem represented by a closed basin.

Indeed sensitivity towards ecological impact seems not yet to be sufficiently developed; this is possibly due to the perceived urgency of structuring new national economics (considered as a priority compared to ecological recovery) and to further contextual elements, such as the relatively low demographic density, possibly diluting the environmental impact over the wider territory.

The authorities are increasingly giving such questions attention which is motivated, besides the obvious questions of public health, also by increasing maintenance and operative costs. In fact, the internal values chain—namely the capability of generating new activities for self-alimenting further growth — is still limited and "ballasted" by the same coal, and in general by the HC industry, considered as technologically mature.

Today, the HC industry contributes to about 80% of the Kazakh GDP (and for 60% of the whole CA economy), rendering a fundamental image of a rentier state, with revenues from this commodity—mainly exported—signifying a huge opportunity but also a risk. This is connected to the dependency induced by the same wealth, with the attitude to passively exploiting natural resource, without devoting much attention to innovation (as indeed typical for rentier states).

A further risk derives from the relatively low capability of the oil industry for generating new activities, namely on the relatively poor value-creating effect—those resources being for the main part exported as non-processed raw material. Kazakh authorities seem to be aware of this; they have already begun with economic diversification programmes, as also evidenced by Expo 2017 organized in Nursultan (Astana), which concentrated on renewable and no-fossil energy sources (see Fig. 9.2).

Further activities are mainly connected with the HC industry, and are mainly financed with the revenues of the relative export. This is the case of the construction and infrastructures industry, which has registered booming growth rates over two decades, and that comprehends the impressive construction of new towns (notably Nursultan and Aktau city projects, in western Kazakhstan) and a whole network of new facilities.[18]

Something similar has also happened for tertiary activities, such as trade, transport, public and private services, culture-related economics, education, health and welfare, which are without exception in a condition of rapid development. In fact, besides the natural resources, the most precious aspects are to be considered those concerning the political quality of the system, which are attracting foreign

[16] as said, a tendency concerning above all the Russian community in the "near abroad", and as well probably the further non-European Russian provinces; see as well "Limes" 2014.

[17] Kudaibergenova Diana T., 2015; Marzhan Thomas, 2015:476.

[18] Kudaibergenova Diana T., 2015; Zhumabayeva Kamila, 2016.

Fig. 9.2 Kazakhstan, Nursultan / Astana Expo, August 2017, main entrance. (Photo: Igor Jelen)

investments both for private and public companies, and usually rely on long-term "returns" expectation (contributing, in a circular manner, to further stabilization).

Recently, the Kazakh economy seems to be ready for a further step, incentivizing advanced social-financial services, such as life insurance, pension funds, health and education programmes; these would consolidate a typical middle-class attitude in life-cycle investments, programmed in the long term (meaning the exit from a condition of system precariousness).[19]

Further investments with similar significance concern high added-value functions such as medical and scientific research, as well as high education and technological innovation (concentred in Almaty and Nursultan). All such functions are still quite dependent on HC revenues that are comprehensibly used by the authorities as a "foundation stone", in order to leverage the development of more advanced activities consequently.

9.2.7 Agriculture and Agriculture-Related Activities

Having said this, agriculture maintains its extraordinary potential, even when it represents just 5% of the whole GDP; it is a critical activity considering both the production chain (e.g. for the agro-food industry development) and the local territorial impact (namely its role in the consolidation of the local society, beyond the current mere "commandeering" phase).

It consists of cereals (grain, corn), potatoes and vegetables in the extensively cultivated steppe in the north; of vegetables, fruits, cotton and rice, prevalently in the south, in the intensively irrigated Syr Darya oasis cultures; and of livestock representing the overwhelming economic activity, concerning about 70% of the agricultural classified land, consisting in endless steppe prairies, eventually used as pasturage or hay production.

Cattle breeding, important as a basic activity, proves suitable to be modernized and integrated into wider economic cycles (as well in agro-food industry, milk, meat, leather, wool processing), demonstrating development possibilities and considering other sometimes neglected productions.

This is particularly the case for high-quality and special-niche productions (considering the current market expectations), thanks to the availability of uncontaminated environmental resources, e.g. for organic farming and bio-production, and thanks to a rich artisan tradition; they represent a promising market, e.g. for products like kurut, ayran, yogurt, kefir, and kümmys in the different variants and from the different environments suitable to be commercialized with the appropriate marketing and packaging methods.

9.2.8 Infrastructural Programmes, Promises and Semi-Ideologies

Beside this, the government is in search of new drives, as those based on HC and on the building industry, in neo-Keynesian terms, possibly losing efficiency in the near

[19] The Gazzette of Central Asia, 23 January 2013, http://gca.satrapia.com/+unified-pension-fund-recommended-in-kazakhstan+, accessed 22.4.2018.

future. This seems to be particularly interesting in terms of the investments in both internal traffic and long-range routes, namely the SC revival of the SR development axis.[20]

Such intentions take the concrete form of the "Nury Zhol"(Bright Path) project, as it was named recently in 2014 by the former President Nazarbayev in a public discourse. It consists in a giant infrastructure plan, based on massive public investments, essentially in the realization of the Kazakhstan segment of the new (actually "old") idea of SR as purposed recently by the Chinese government.

Such a programme would dramatically improve the connection between the different sides of Eurasia, developing the role of Kazakhstan as a "bridge" between the continents; it consists of a set of lanes equipped with advanced technology for high-speed and high-capacity trains, roads and highways, hub and intermodal centres, as well as other facilities.[21]

But the project means much more; it is the way to diversify accessibility to the international market for all kinds of goods, including raw material and energy, finally liberating the CA economies from the perception of the landlockedness, and from a situation of dependency (at first from the Russian infrastructures), which at the moment of independence was possibly the first priority for the local regimes.[22]

Furthermore, it can also attract traffic that otherwise would develop on other paths (Trans-Siberian and Arctic North route, and further routes, terrestrial or maritime, south of Caspian, through Iran and the Middle East and many other possible variants), incentivizing development along the road, namely on its different intermediate stations. It is not just to be intended as a direct line between two distant points, but a kind of economic area (the BRI, "belt and road initiative"), to be combined with further SR variants.[23]

Such plans would strengthen collaboration with the Far East's rapidly growing markets, at the same time offering the Kazakh government the possibility of maintaining centralistic control over its economics (the building and transport industry mainly public and state-owned). In fact, it consists of a kind of plan, offering the domestic system (corporations, institutions, economic investors, and to the society as a whole), an opportunity, a new development driver.

Possibly its final objective is the shifting from the HC-driven economics to a trade-infrastructural driven-one, giving further impulse to growth. It is in fact also a device to avoid any stagnation effect (a kind of flattening of profitability curves, which periodically affects the top-down economics, as patrimonial-authoritarian Kazakh economics is to be considered).

Such policies express the worry, typical for centralistically guided countries, that the economy might remain without any push-pull factors, configuring a saturation effect and finally a crisis.[24]

Indeed, beside its undisputable economic value, it has been defined critically as a kind of "semi-ideology" supposedly characterizing the semi-democracies, whose societies are not yet capable of orienting themselves, but need a political guide. It seems to be a device for pursuing different goals, for leaderships seeking legitimization, and in general for mobilizing resources, justifying a policy, impeding the (parasitic) bureaucratization of the system, and similar.[25]

9.3 Kyrgyzstan: Horsemen, Mountains and Glaciers

9.3.1 A Basic Description

The Kyrgyz people present some cultural and historical affinities to the Kazakhs, from whom they were separated during Soviet times: Kyrgyzstan became an SSR in the

[20] Marzhan Thomas, 2015:467.

[21] https://en.tengrinews.kz/politics_sub/Nurly-Zhol-infrastructure-project-to-strengthen-EEU-258109/, 25.12.0.2014, accessed 22.4.2018; Babajanian Babken, 2015:514; Kudaibergenova Diana T., 2015:452; for "Kazakhstan 2030" see on The Strategy for development of the Republic of Kazakhstan until the year 2030, Official Site of the President of the Republic of Kazakhstan, http://www.akorda.kz/en/official_documents/strategies_and_programs, accessed 22.4.2018.

[22] Silvestri Tommaso, 2015–2016:97; it is possible to say, at the starting moment of its independent history, the priority of Kazakhstan was that to access to international market without dis-quieting Russia (and further neighbours), but in the same time eventually out-flanking Russian monopolies.

[23] especially with the "XXI Ct sea silk road", inducing other investments and growth programmes; in the same context, the Chinese authorities decided to establish an Asiatic Infrastructure Investment Bank with a capital of 100 bl dollars, including China, as a main partner representing a capital of 29,7 bl dollars, and further Eurasiatic countries such as India and Russia; Page Jeremy, 2014; Borzatta Paolo, 2015; indeed this "new" SR is just one of the periodically purposed variants on the west-east axis, in competition with each other, in order to anticipate some other possibility, and in order to re-orient investments and to catalyse initiatives; the mere operative aspect of trans-continental travel is not decisive; in fact it is to consider that even today, with the existing steppe-train, using rather obsolete tracks, it is possible to cover this distance in about 15–18 days from Far East to main Europe and Mediterranean cities (a time-distance sufficient to satisfy a great deal of commercial needs); there are already some operative block-trains from Western Europe to Far East; that from Mortara freightage hub, close to Pavia, Italy, to Shanghai, run along the SR — as traditionally defined -, using existing infrastructures, it takes about 18 days for the journey, Lanzetti Eleonora, 2017.

[24] While in further contexts, based on pluralistic economics, the initiative pushing development is generated by a bulk of operators, public and privates, independently contributing to a game of demand and offer.

[25] Kudaibergenova Diana T., 2015; Babajanian Babken, 2015:514; it resembles the soviet "plurennial planning", rather than neo-Keynesian policies, that has to be realized in a context of a prevalent middle-class society.

1930s, as the result of a geopolitical operation, probably intended to be a buffer on the Chinese border, creating a special administrative area. In fact, it represents the southern mountainous appendix of its larger neighbouring country of the steppe, with 40% of its territory at over 3000 m. altitude.

Its geography is characterized by the sequence of high mountain ridges developing around the main Then Shan orogenetic axis, extending transversally from south-west to north-east, then promulgating into Chinese territory; the higher peaks are Jengish Chokusu (Pik Pobedy), 7439 m. at the borders with China, and the spectacular north face of the Khan Tengri (7010 m.), whose peak signposts the three corner border with Kazakhstan and China.

Basically, the Kyrgyz territory consists of a sequence of deep valleys, oriented east-west and not very coherent with the existing infrastructure lines (mainly constructed in Soviet times, assuming a much wider scale of connections, then transversally to the current national one); these are sequentially the valleys formed by the river Naryn, that flows from the Tien Shan glaciers in direction NE-SW, to the Fergana valley, where it inflows in the Kara Darya and then in Syr Darya; and by the Kyzyl Suu river, flowing parallel southern to Naryn, then being renamed as Vaksh river in Tajik territory, a tributary of Amu Darya.

The north-eastern side of the country is occupied by the depression, in Khan Tengri area, with Lake Issyk kol, and the surrounding area, namely the Kyrgyz side of the Zhetysu, formed by the rivers Chu and Talas; in this area the capital city Bishkek can be found, known in colonial times from 1862 until 1926 as Pishpek and from 1926 to 1991 as Frunze, renamed after the Georgian-origin Red Army commandant in CA.

Such schema also define the two most important settlement areas that are divided by the same Ala-Tau ridge, the prolongation of the Tien Shan in the south-western side of the country: the north-eastern side with the capital city, and the south-west, namely the Kyrgyz side of Fergana valley, with the second-largest city of the country, Osh. These two regions both occupy one-seventh of the country's surface and are the most important areas, where the largest part of the population is concentrated, as well as the most important economic and cultural assets of the country; this schema additionally prefigures a kind of dualism, that has also proved to be politically relevant in the recent past.

Comprehensibly the country is rich in rivers and lakes, and of artificial reservoirs and dams, alimenting the most important industries of the country (hydroelectric power plants); among the many lakes, the most important is that of the crystalline-waters of Lake Issyk Kool, south of Bishkek and a favourite alpine-like recreational centre ("kurort" and "sanatorium"), with alpine "turbaza" equipped for trekking and excursion activities during the Soviet time and is nowadays recovering its tourist attractiveness (with about one million visitors a year, mostly in the summer period).

In general, the country configures a set of assembled territories, often valleys cut off from the respective outback (cul-de-sac like) by transversal borderlines. They are connected with inefficient infrastructures that must cross high passes, linking remote settlements (namely, settlements not connected through a main downstream valley or a plain, where the major centres usually lie); one must consider that the passes are usually blocked in the winter season because of snow and avalanches, "de facto" dividing the country geographically.

This occurs in the north-east and south-west sides of the country in particular, between which the only link coincides with the extra-territorial circular railway, connecting the southern Osh oblast, through Tashkent, with Almaty and Bishkek, crossing three foreign countries (Tajikistan, Uzbekistan and Kazakhstan), passing around the Ala-Tau ridge. For this same reason, different parts of the country occasionally gravitate in opposite directions, the first towards Kazakhstan and Almaty conurbation, the second attracted by the Fergana valley: this is just one of the many aspects that highlight the need of these states to collaborate in sensitive trans-border questions.[26]

So similarly, for the Naryn and Ak Su valleys upper sides, on the south-eastern side of the country, close to the China and Tajik borders which are almost inaccessible in some periods of the year.

9.3.2 Landscape, Settlement Structures, Internal Organization

The same physical configuration of the country also overlaps other variables, ethno-national, economic and political ones in particular. This is particularly the case for the southern part that is characterized by the massive presence of ethnic minorities especially Uzbek and Tajik, and by exposure to the influence of the respective neighbouring countries. This area has different characteristics from the rest of the country, considering the over-populated cities of the valley, the conservative and prestigious religious tradition of its centres, the ethnic mix and the territorial organization, based on irrigated agriculture.[27]

The north-eastern side (with Bishkek and the southern side of Zhetysu), on the contrary, is prevalently Kyrgyz national, with the Russian presence especially noticeable; Bishkek is quite a modern city, open to outside influences, characterized by a rather vivacious civil society (recognized as the most advanced in the whole area), which proved to be

[26] Megoran Nick, 2017.

[27] Megoran Nick, 2017.

the pillar of a pluralistic regime, possibly an exception in the CA scenario.

The main aspect of the country—the "mountainess"—is combined with the condition of "rurality", with no more than one-third of the population living in urban areas (mainly Bishkek and the Fergana valley cities); this situation represents a challenge for state organization, namely the setup of programmes for the integration of peripheral areas. Much of the population lives in remote settlements, carrying on traditional economics, having recovered after the collapse of the Soviet organization.

In the moment of the regime change, the former kolkhoz employees (mainly located in the plains) re-discovered the highland pasturages and the ancient caravan routes crossing the mountains, bordering China and Tajik Pamir. In such regression, the populations recovered their traditional organization, based on communitarian activities, and suitable to survival economics.

In the same circumstances, in these areas far from state control, the spread of illegal activities such as contraband and any kind of trafficking was also observed.[28]

Comprehensibly the most urgent political intervention, and the most relevant investments are oriented towards solving the problem of accessibility of these parts of the country, practically the whole mountain wilderness.

The agenda of the post-Soviet government is full of such resolutions especially concerning infrastructures; among them and thanks to financing by the Asian Development bank, the road connecting the north and the southwest side has been completed and improved, from Osh to Bishkek, from the Chu valley to the Fergana valley. And so, for a long sequence of projects (tunnels, bridges, by-passes, new roads and highway, railway and urban transport facilities, international and regional airports).[29]

9.3.3 The Mountain Outback

The central side of Kyrgyzstan has had a rather important role since colonial times, because of its strategic character on its borders with China, with which it is connected along some high mountain passes. Projects foresee the construction of railways and transport infrastructures on the Torugart pass,[30] and, furthermore, the restoration of a road from Andijan to Osh, and further on to China. These are to be considered key projects, matters for negotiation with the Uzbek government, which is especially interested in the same infrastructure having no border in common with that country.[31]

The Kyrgyz mountains are rich in minerals, especially gold (Kumtor gold mining), uranium, coal, antimony and "rare land" (earth metals) of increasingly strategic importance. On the contrary, it has little hydrocarbon, satisfying energetic needs with hydroelectric resources and with the Toktogul dam on the Naryn river. The dam was built in the 1970s, providing a great deal of the country's electricity needs and exporting such commodities to neighbouring countries (Uzbekistan at first).

It is possible to say that the most important resource in this area derives from perennial snow deposits representing a reserve of pure water usable in all seasons — a fact that at such latitudes is not obvious, and that is a matter of concern due to climate change. This is also considering the current revival of community-based economics: water flowing down from mountains is accessible for foothill villages, directly by the populations through local canal networks, arriving at any home and into any courtyard (where it is possible to practice a valuable kind of horticulture).

In fact, the Kyrgyz state surface includes the sources of the major rivers of the area (Naryn for the Syr Darya, Kyzyl Suu-Vaksh and Zarafzhan tributaries of Amu Darya basin, and the Issyk Kol basin, and the Zhetysu rivers), namely glaciers and the snow-covered mountain slopes, that the country shares with the nearby Tajikistan.

The economy of the country is based on primary activities, even if severely limited because of its topography and other environmental conditions. It comprises mechanized (scale-production) agriculture in the northern plain (cereals, potatoes, vegetable and fruits); cotton and rice in the irrigated plantations in the Fergana valley; and cattle-breeding, the main activity practiced on the highlands, rather productive considering the abundance of water and pasturages.

These activities are relatively easy to organize by the local populations, since they can directly access environmental and social resources without the mediation of bureaucracies or expensive technology. This is the case of summer pasturage, of forests for wood, of wildernesses for wild animals, for fishing and hunting, for gathering wild fruits, herbs or roots (like ginseng, "zolotoj koren") which are rather abundant in such places and useful for a number of preparations — alimentary, pharmaceutical, etc.

In these circumstances, it is possible to start up some activities, recovering appropriate traditional techniques, based on the ecological calendar and environmental circumstances. This is particularly the case when considering the use of animals in many activities, especially for cultivation,

[28] Botoeva Gulzat, 2015; such activities indeed can represent a bias for the beginning for a certain development path.

[29] It is to consider that Kyrgyzstan is located on the main axis of the SR and on the further transcontinental routes, and that such accessibility could open the remote regions of the country to the wider economy.

[30] Mashrab Fozil, 2015; Jelen Igor, 2002.

[31] Megoran Nick, 2017:52.

mobility and transport, which had a considerable effect in improving the quality of both environment and human needs (in comparison with late Soviet industrial agro-food production).

Indeed, it is a kind of activity that is much more strenuous (less productive in terms of human work), but sufficient to ensure survival for small communities. Above all, it is carried out in an independent way from any (potentially despotic) power, enjoying communitarian solidarity (as usual in such remote peripheries, where the public infrastructure may provide such services with difficulty, as well as rehabilitation and maintenance of infrastructures themselves).

It is hardly surprising that local agriculture has developed rapidly since the 1990s, with private or privatized, family-run activities representing about one half of the whole agricultural production, and already in 2002 representing more than 1/3 of the GDP, and about half of employment.[32] The most important primary output (milk and milk by-products, meat, leather, wool) shows the rather important potential in these kind of productions.

Apparently labels relating to CA culture and territories are symbols for quality and excellence in textiles and fashion marketing, evidencing value for expensive products like kakarol, markhor (falconeri wild goat), kashmir, astrakhan and similar; possibly an underestimated resource of which the local shepherds do not have much awareness (Fig. 9.3).

At the moment, such activity represents basic survival economics which is difficult to integrate into the rest of the economics; it is difficult to account for, and to adapt for commercial purposes (as said, such activity being prevalently domestic and family run, not producing usually significant surplus).

However, it has relevant potential, since traditional agriculture is able to implement new appreciable market segments, such as bio-organic farming (considering market tendencies, a kind of production to be exploited), eventually in synergy with tourism, handicraft and other activities which in Soviet times were practically unknown.

As said, the northern Tien Shan side of the country was always one of the favourite settlement places in the whole CA because of its abundance of water, amenity landscapes, mountain valleys and plateaus with a favourable climate, and at the same time it is easily accessible. This particularly so in late-Soviet times, when it became a popular resort attraction area but was then almost abandoned with the collapse of SU, in transition times. Such facilities are nowadays recovering with difficulty, opening up for foreign tourists interested in the exceptional beauties of such alpine landscapes, recovering and developing new forms of tourism, limited in numbers but potentially rich for local development (Fig. 9.4, especially in spectacular Pamir Alaj mountains).

Fig. 9.3 Kyrgyzstan, Pamir Alaj, 1997, Lyailiak valley, Ozgoruš village, textiles domestic manufacturing. (Photo: Igor Jelen)

The "community-based" tourism that is relatively easy to develop, could possibly represent the first stage of amenity economics, highlighting synergies with the local traditional economics.[33]

9.3.4 VIP Projects: An Updated Idea of Development

The more recent initiatives in Kyrgyzstan are inspired by a new approach, changing modernist development theories and changing the perspective; they are based on the recovery of rural life as a resource (inverting the modernist-productivist scale economies), concretely meaning the development of projects in the context of community-driven development; sometimes, they are financed by international investors and donors who intend to address the poverty situation directly, involving the rural population and out-flanking the local elites (possibly representing lobbies intercepting subsidiza-

[32] Such activities are often submerged and family run-operated, therefore such number are probably under estimated.

[33] Botoeva Gulzat, 2015; see as well Lonely Planet 2014.

Fig. 9.4 Kyrgyzstan, Pamir Alaj, august 1994, Parus (sail) peak north face, m.5400. (Photo: Igor Jelen)

tions, the major problems of the top-down development programme had to face).

This is the case of the "village investment project" (VIP), purposed recently by the World Bank with the aim of orienting international investments in order to give incentive to the communitarian way, developing local-scale projects, especially small and social-use infrastructures. This concept seems to be especially adequate for remote and scattered population regions like the Kyrgyz mountain outback, focusing on spreading micro-projects, based on principles of empowerment of rural "communities".[34]

The programme consists, rather than in few great projects (as the described Chinese and Kazakh purpose of a "new" SR) in numerous small-scale investments, based on a social way of making accountability (rather than in exclusively monetary terms), namely trying to measure an effective return in social terms (concretely improving the local life, fostering "inclusiveness").

The "community base tourism", combined with further local-scale activities, means a bottom up way of doing economics with local operators (or just private homes) offering hospitality, accommodation and services: an activity that has proved suitable to circumstances, avoiding the intrusive impact of mass tourism. It is also important as a social strengthening element, enforcing territorial identity and eventually contributing to the prevention of indiscriminate mass-migration (flight from rural peripheries), that represents a major risk in such situations of encounters between modern life and remote culture.[35]

[34] Babajanian Babken, 2015:514.

[35] Botoeva Gulzat, 2015; see as well Lonely Planet 2014.

9.3.5 Remoteness as a Chance

Kyrgyzstan is occasionally referred to as "the Switzerland of Central Asia"; indeed it is more than a naïve metaphor, since it is, like Switzerland, poor in natural resources and mainly mountain community base organized; and this is not just because of its mountain topography but because of a peculiar cultural/political situation as well, as it is a country lying in the middle of a complicated scenario: it borders with regional powers, and it is characterized by a multinational society (rather than a stratified society, composed of majorities and minorities), and by deep internal diversities; finally it is characterized by what it is already possible to define—after 25 years of independence — a tradition of neutrality, and by a kind of mountain and geopolitical "remoteness".[36]

In fact, remoteness represents a problem, but in some circumstances, it can become a resource. This happened in times of totalitarian regimes, when the populations considered life on the mountain as a refuge from the pressures of power; and it has also happened in times of recent crisis, with people escaping the consequences of the Soviet system changes.

Today the situation is changing again, with people sometimes returning to the main urban centres (that are acquiring new role as capitals of new sovereign states). However, it is important that the rural population does not perceive such a situation as a handicap, in the context of a possible new marginalization (a kind of inferiority complex).

Indeed, assuming the recent technical and organizational development (ICT, as well as transport revolution, improvement of worldwide mobility) remoteness can be considered

[36] Megoran Nick, 2017:99.

rather as a matter of perception, than as a material disadvantage.[37] In fact, experience teaches that such situations represent handicaps but can also bring important advantages, considering the availability and accessibility of vital resources, as well the possibility of staying out of geopolitical troubles (that affected the wider CA spaces in the last decades).

Finally, such remoteness represents a chance for the conservation of natural areas as well as of an extraordinary traditional culture, that is presumably better preserved in these peripheries: mountain valleys can become folkloristic parks or "open air" museums, nature and ethnographic "experiences" are seen in a new vision, making them a kind of precious attraction (at the same time maintaining authenticity and a role as a cultural reference).

As a further main aspect of this resource, the Manas, the "steppe Iliad", must be considered. It has been defined as the most extended epic poem that has ever been composed, particularly when orally orated, by specialized traditional singers, the "manasci". It includes hundreds of thousands of verse lines, that describe the different phases of the pastoral communities, representing a precious asset—the cultural product of a unique but disappearing nomadic life — that in this new context, postmodern and globalized, can be exploited in many ways.[38]

9.4 Uzbekistan: Ancient Civilization and Mythical Cities

9.4.1 Basic Description

Considering the geographical position, its prestigious past, as well as basic data, Uzbekistan is competing with Kazakhstan as the dominant and most representative country in the CA region: it represents the further prevalent identity of CA populations, the oasis glorious civilizations, located exactly on the central segment of the SR itinerary; it is the area in which the ancient Sogdian society developed, representing the most evolved merchant community for centuries, and it is still now the site of the most important urban and industrial areas of the region.[39]

It is central from any point of view; it borders with all the other four CA countries, a fact that, also considering the common border with Afghanistan, in the south, leads one to imagine the exposure on all scenarios characterizing the region. This is also considering the precarious aspect of such borders, still to be stabilized from many points of view, for trans-frontier relations, minorities protection, traffic regulation, military-strategic questions; and that the country has no common borders with the two "great powers" incumbent on the region (China and Russia), that signifies the possibility of conducting relatively autonomous politics.

9.4.2 The Cities of Uzbekistan: Tashkent

Differently from the other "stans", Uzbekistan is relatively highly urbanized and densely settled, especially in the strip along the river oasis on the Zarafzhan, and in much of the Fergana valley that belongs to the country.

The capital, Tashkent with 2.3 million inhabitants, the metropolis of CA, lies in the Circhiq and Keles (both Syr Daria tributaries) river oasis, at the centre of several ancient caravan routes; it has a millennial history, but it became a regional centre relatively recently in colonial times. Then, it accelerated its growth after becoming the terminal of the Trans-Caspian railway (1889), and of Trans-Aral railway from Orenburg, completed in 1906.[40]

In WWII times, it was chosen as the site for many industries, especially military ones, that had to move from the European provinces of the SU in order to escape from the Nazi invasion. Then, after the war, they stayed there. The city became the centre for further strategic productions—aeronautics, aero-spatial, satellite industries also thanks to the favourable climate and meteorological conditions, necessary for such industries.

From that time forward, it became the hub for the whole CA and also for further Asiatic Russian colonies, conserving this role in the Soviet century, a kind of symbol for the non-European side of a communist empire: it signified in the popular imagination something exotic, but also the role of a punitive destination, e.g. for public clerks and officials, or of a deportation area.

At the same time, it became an example of the Soviet modernization of Asiatic towns. This happened in particular when, in the 1960s and 1970s, the SU opened up to a further stage of its history, undertaking global politics (eventually taking advantage of the contraposition arising in that period between western-colonialists and de-colonizing countries).[41]

[37]Özcan Gül Berna, 2015:414; Megoran Nick, 2017.

[38]Megoran Nick, 2017.

[39]Sogd world famous merchants, running the SR, have been for centuries a kind of myth in Florence and Venice as well in any side of Eurasia; possibly they had invented the double accountability, the key element for the development of the economics in modern sense, that Medioevo Italian merchants would later diffuse anywhere.

[40]Barisitz, Stephan, 2017:249.

[41]Purposing itself as a possible alternative to capitalism; see Uzbekistan, 1984, a kind of propaganda book government "white paper", typical for soviet times; then soviet Asia became a model pretending to represent the "proletarian" Third World engaged in de-colonization struggles.

In these periods, the city became the seat of international "third-worldist" conferences, of universities specialized in the "communist" way for developing countries, and of further similar initiatives, attracting students from the "south" of the world, possibly a new generation of African, South American and Asian leaders.

After independence, it was re-dimensioned to a nation state scale, losing part of its importance, and losing much of this top-down induced richness, namely its role in the industrialization programme pursued during Soviet times, both civil and military.

At the same time, it acquired a national capital status, gaining new capabilities; it was chosen as headquarters by many multinational companies, NGOs, media networks, private associations and international organizations active in the whole CA region.

Thus, it became the reference for the locally growing IC, also leveraging on local ethno-linguistic minorities—a residual aspect of its history — that could now exert a role of a cultural "bridge" between the local society and the wider international environment (e.g. Korean or German local communities, and furthermore, Israeli, Baltics, Armenian and others). De facto, it is the only true global city of the area (competing with Almaty, which lost some functions because of the transfer of the capital-town status to Nursultan).[42]

This is the case in the transition period even if suffering due to the same tensions that characterized other parts of the country (the Fergana valley and the southern Uzbek border with Afghanistan), namely for instability induced by terrorist opposition, mainly represented by religious and political radicals. It was rocked by repetitive terrorist attacks in 1999, with the authoritarian President Karimov becoming a major target for religious extremists.[43]

9.4.3 Samarqand and Bukhara

These are world-recognized cultural capitals, namely extraordinary examples of Islamic art and culture, and in general of the CA humanity heritage. They survived the programmes of large-scale Soviet destruction, on the contrary, becoming the object of important monument conservation programmes, signed by Lenin immediately after the revolution.[44]

Such programmes intended to "save" the old "material culture" monuments while at the same time, "decontextualizing" them, namely restructuring the whole surrounding area, supposedly depriving them of religious and social symbolism.

At the same time, the same Soviet urbanist decided to build up brand-new towns instead of destroying the old traditional ones. The new town being what they defined as an ideal "socialist town", close to the old one. The probable intention was to represent an example of the new civilization, saving the traditional architecture as a paradoxical negative example: the idea was (presumably) to demonstrate the efficiency of the new organization compared to the narrow streets of the old city quarter.

This occurred in particular for modern services, for water and food supply, heating system and transport (public bus, tramway), easily organized in the linear and wide Soviet schema of "perspektive" and "bulevar". Furthermore, for tree-planted parks, public facilities, government buildings and districts, and in general for the new modernist and functional planning.[45]

But the Soviet all-day praxis meant bad maintenance with an accumulation of inefficiencies also at urbanistic scale as well (this especially because of the weakness of internal functioning of the state apparatus, beyond officially demonstrated intentions).

Indeed, after a period of decadence, and of insufficient conservation during the late-Soviet and transition time, now the old town cores have been recovered a great deal. These cities, together with further Uzbek sites, Khiva, Shakhrisabz and others, today may use such cultural assets to develop international functions. They have become the seat of numerous institutions, foundations and universities, attracting an increasing number of foreign visitors, developing a promising tourist economy.

9.4.4 Fergana Valley

The valley is the symbol of the oldest tradition of CA, a wide water-rich oasis fertilized by close surrounding high-mountain rivers. It has been the site of important cities, manufacturing centres and bazaars since ancient times, among them being the regional capital, then the capital of the khanate of Kokand, and Khujand, the town possibly founded by Alexander the Great in ancient times.

It is characterized by sophisticated cultures based on urban development, intensive land use (especially thanks, as said, to abundant water), manufacture and trade, contrasting with the surrounding mountain and nomadic outback. It is currently divided between three countries, the biggest part in Uzbekistan, by a line that does not follow any ethnic, language or geo-strategic criterion (that indeed, due to the com-

[42] As usual, the global cities, as defined by Saksia Sassen, acquire in importance, then possibly diverting their role from that of representing their own nation-state outback.

[43] Jonson Lena, 2006.

[44] Paskaleva Elena, 2015:420.

[45] As said, obviously at first for control purposes, giving the impression of visibility and accessibility.

plexity of the area, would be impossible to be applied coherently). It is the result of millennial mixing, today representing an interweaving in all senses, of minorities, traditions and economies either contrasting or combining, depending on the circumstances. The valley is highly urbanized — the only place in CA with such characteristics — and is characterized by a long history of rivalries, with political and religious identities continuing to confront each other.

It is periodically shaken by tensions, being the cultural and political hub (the "cradle") for the whole CA space; this was the case in particular in late Soviet times, and then in the transition period (actually until present times), when it was demographically booming; a complicated situation that has been aggravated because of both natural causes and immigration consequent to the recent super-imposition of new borders which dramatically divided its organic unit (creating an inextricable knot).

A border that represents a challenge for the local population; the post-Soviet crisis periodically provokes tensions, with people crossing the border in search of asylum into the neighbouring country perceived as their own fatherland. Such movement sometimes assumes mass dimensions which may leave one to imagine a kind of tacit pact in the exchange of populations in order to render their own territory ethnic, nationally more homogeneous. Currently, the Uzbek side of the valley is hosting hundreds of thousands of refugees, citizens of Uzbek origin, who arrived from the nearby countries.[46]

Its original importance is due to the flourishing of the ancient urban tradition, as well as its basically rich agriculture and to commercial activities connecting local manufacture and long-range transport; it is also the centre of prestigious religious schools, spreading their influence — a conservative Islam tendency — over the whole area, a place of pilgrimages and holy destinations.

Among them is Osh, the SC CA "Mecca", site of the "Suleiman throne", lying on the Kyrgyz side but predominantly Uzbek inhabited, which is positioned very highly as one of the Muslim holy destinations.[47]

All this makes Fergana a critical area (a kind of political "barometer" for the whole CA), where tensions and innovations have continuously mixed, sometimes starting reform movements, occasionally degenerating into contrapositions.

The worst situation occurred (as said) in 1989—90, just before the collapse of Soviet Empire, when it seemed as if the entire local system was close to exploding, continuing around Osh, on the Kyrgyz-Uzbek border into the 1990s and with periodical recurrence until recently.[48]

Further large-scale clashes happened recently in Andijan in 2005, with hundreds of deaths, massacres followed by mutual accusations: indeed today, the clashes are treated as a kind of taboo subject since no-one wants to speak of them, but just to forget and remove[49] what occurred (the city has also been a centre of religious, eventually anti-Russian, resistance in the past, dating back in history to 1898, when a "holy war" was declared against the colonial rule)[50]; then further conflicts occurred in the Osh area in 2010.[51]

Besides local tensions, the valley is problematic from a geopolitical point of view; in the last few decades it has become the intersection of three points, namely the area where three new independent countries meet—countries without much political experience, which have to put into practice aspects of common life into several dimensions, from diplomacy to border surveillance as well as economic and social relations. This is the case of transit controls, of visa permits, of commercial regulations—in such specialized bazaar areas—subject to restrictions depending on political-level relations which are crucial in a trans-frontier area that is characterized by minorities and border economies, in a territory that has only recently been divided.

The transition period registered the entire repertoire of such tensions, with reactions and retaliations, showing the whole range of political aggressive initiatives, without considering local sensitivity, or the true needs of the population.

This is the case of when border passes are periodically closed, or the setting up of check points, militarization and the fortification of borderlines, or any kind of provocation, the arresting and rejection of commuter workers, treated as extremists and terrorists, often in a contentious manner. It is — in a strategic military sense — the place for a characteristic "flexing of muscles", where the respective governments occasionally position barbwire fences, opposing "no man's lands" and buffer areas, with reciprocal denial for the recognition of some border marking—considering that borderlines in the valley have never been properly marked because they have had no effective character.

There have been reciprocal accusations, threats of invasion (and of latent "ethnic cleansing"), and the whole rubric of border reactions and tensions that "young" nation-states, which have not yet been fully consolidated, usually apply.

[46] Mclellan David, 27 Jul 2017; Pan Philip P., June 18, 2010.

[47] Since, it is said, 4 visits to the town would be equivalent 1 (the "hajj") to Mecca, signifying a sign of respect to the latter, but also a merchants' astute manoeuvre in order to make the visits to this town more frequent; interview, Osh, 1994.

[48] Elebayeva A. B., 1992; Megoran Nick, 2017.

[49] Lonely Planet, 2014; World Report 2015: Uzbekistan, Events of 2014, https://www.hrw.org/world-report/2015/country-chapters/uzbekistan, accessed 23.4.2018; McMahon Robert, 2005.

[50] Robinson 1989:124; Hambly (a cura di) 1970:200; Bennigsen 1989; Bacon 1990:112–113.

[51] Megoran Nick, 2017.

For all these reasons the conflict potential, as well as the extraordinary cultural and economic wealth and correspondent formation of differentials, is much higher here than in other areas.

Recently, with the new President Mizoyoyev, the country seems to be experiencing a period of possible political relaxation; but true normalization needs time (actually, it is questionable whether such normalization can be managed and purposed by someone from the same elite that administered the country over the last few decades).

9.4.5 Oasis Agriculture: The Basic Resource

Thanks to the extraordinary combination of climate and resources, the Uzbek agriculture was able to develop precious water-consuming cultivations, like cotton, fruit, silkworm breeding, vegetables, rice and all kinds of crops. Cotton, in particular, had already been cultivated here in ancient times, but in modernity, with opening up to the wider market and the improvement of transport, it became the most important and sometimes exclusive production of the area.

This was due to the reconversion which occurred in colonial and Soviet times, transforming the area from a survival and local bazaar-exchange agriculture, to an industrial one: the reconversion relied on hard infrastructures like ID networks, reservoirs and artificially excavated channels, mill and pump aqueducts, and provoked a general transformation of society.

From that time onwards, the local economy lost its legendary diversified richness (deriving from both its commercial position and agriculture productivity), and specialized in functions connected to industry, either locally or for export to European textile factories. It is still a major asset and currently the most important export-oriented production, carefully managed by the government in order to get convertible currency.

This transformation occurred to the detriment of local economics and land use patterns: pre-colonial oasis agriculture was based on a wide range of products, on a biodiversely rich environment (especially considering the tugai along the fluvial forests). Agriculture became ancillary to industrial growth (based on scale economics and technological standardization) at the expense of the original diversified cultivations, expropriating local resources and annulling traditional knowledge.

This occurred until the system reached a critical threshold, when the need for regulation of the use of waters flowing mainly from neighbouring areas (today foreign countries) that control the up-stream sources became evident. This critical point was reached sometime in the 1970s, namely at the moment when (as in the rest of the world), the damage of indiscriminate industrialization became irrefutably evident. It is the macroscopic case of the drying up of the Aral sea, which occurred unexpectedly, with neither of the rivers that had alimented the sea (Syr Darya and Amu Darya) reaching it, but drying up before the mouth (it is to consider that Aral sea, the only large water body of the country, represented a closed basin, therefore much more important and ecologically fragile).[52]

There have been further consequences, namely a sequence of disasters, due to soil erosion, salinization, sterilization and similar. These appeared mainly in late-Soviet times, after a long-term silent accumulation of inefficiencies, pollution and further externalities, demonstrating the basic fragility of the modernization model.[53]

9.4.6 Cotton as Political Monopoly

The current agricultural organization seems rather functional for political control purposes; in fact such resources can be concentrated monopolistically, with the state exerting a latifund-like role (considering that much arable land is state-owned, and managed in a centralistic manner), impeding any diversification, either of society or economics.

Such organization, even when profitable from a strictly financial point of view, has resulted as inefficient over time, producing externalities that must be considered. Above all, it is characterized by high labour consumption and by obsolete work modalities, inducing scarce social benefits as well as inhibiting any innovation.

Sometimes—as referred to by many sources—the authorities employ forced or self-defined "voluntary" work, indeed work provided by categories such as prisoners, the unemployed, students on their vacation period, or even children (as it was in Soviet times, originally for social-education rather than for mere exploitation purposes; but obviously a controversial and sensitive question). It seems, rather than an economic activity, an activity suitable for pursuing social control purposes (Figs. 9.5 and 9.6).[54]

[52]In fact, from a general point of view, applying efficient criteria in use, accumulation and redistribution (this even considering the climate changes, difficult to estimate, possibly irreversible), water is not to be considered a limited resources, but a renewable one; a disputable issue but above all a question of efficient use.

[53]Rumer B. Z., 1989; above all, these cases demonstrated that the environmental dynamics do not occur in a linear way, but in a catastrophically unpredictable manner, fundamentally uncontrollable by any regime.

[54]"at cotton-harvest time, all students and teachers are still mobilized as unpaid labour to help in the fields", including children, a fact that has provoked the boycott of several international textile brands, according to some sources; International Business Publication, 2015:109; see as well World Report 2015: Uzbekistan, Events of 2014, https://www.hrw.org/world-report/2015/country-chapters/uzbekistan, accessed 23.4.2018.

Fig. 9.5 Proud Uzbek woman showing Soviet honours medals in Fergana valley, 1994. (Photo: Igor Jelen)

Fig. 9.6 Kyrgyzstan, Katran village, common voluntary services at village school, August 1994. (Photo: Igor Jelen)

Not surprisingly, today employees in agriculture account for 26% of the total labour force but contribute to 18% of Uzbek GDP (as per 2012). It was the premise for the establishment of monoculture economics, oriented at profitable international markets, but substantially alimenting a new kind of "nomenklatura" (oligarchy).[55]

9.4.7 Industry and Manufacturing Traditions

Uzbekistan is characterized by a relatively old industrialization, with textile industries which had already started to develop in the nineteenth century, in colonial times, and previously by an artisan and pre-modern industry. After the revolution, the country, notably Tashkent area, became the centre of an early industrialization programme. This industrialization accelerated further during WWII, when much of the Soviet European cities' industrial equipment was dismantled, and then transferred and re-assembled here.

It developed further in later Soviet times especially in advanced sectors (namely in those times concerning the "industrial-military complex", with the SU engaged in a world competition as a superpower), requiring the immigration of skilled technicians and engineers from other Soviet provinces.

The Tashkent TAPOich (Chaklov), currently "Tashkent Mechanical Plant" (TMZ), aero-manufacturing plant is the symbol of this industrialization period; it was transferred

[55] Brill Olcott Martha, 2012.

here from Moscow (where it was founded in the 1930s) in 1941; it remained one of the most important aerospace industrial plants in SU until the end of the 1980s, as a typical hard capital-intensive industry, top-down planned and managed.

The factory experienced its heyday before Perestroika, with about 30,000 employees and relied on expertise which had been attracted and concentrated here from all over the SU. After the collapse of the SU, it fell into a crisis period, reducing and slowing production, with many qualified workers and engineers leaving the country, many workers being made redundant, often out-migrating as "vozvraščentsy", and production becoming soon obsolete (today it keeps production at a minimal standard).[56]

So that, during transition, it would be unthinkable to maintain the plant at the usual level of productivity; the government started policies and interventions in order to adapt the factory to the down-scaling of the market (from SU transcontinental scale to Uzbek national-state).

Indeed, after independence, its marketing activities (and access to the market) was made more difficult by the same governmental neo-mercantilist attitude; this meant above all an alternate (and fundamentally ambiguous) relationship with Russia (as well with other foreign economies), not ideal to ensure the necessary continuity for an industrial sector of this kind (that needs to be managed in a corporate manner).

In the later stage of its existence, it was to see a possible re-launch of production in a number of different ways; China ordered 38 airplanes in 2005—possibly dissimulating an acquisition attempt in this way, or a "technology grabbing" operation, considering the circumstances—but without results; then it declared bankruptcy in 2014. Recurrently, there are rumours of a revival of production; this especially with reference to the recovery of the airplane market, comprehensibly important for CA countries which need good regional-scale connections, developing national airplane production.[57]

Besides this, further market-oriented sectors developed soon after independence in the frame of a new economy, tendentially based on the growing, somewhat urbanized, middle-class needs, rather than on the military-industrial complex as in Soviet times (when the light industry consumerism-target production was not a priority, or mostly just ignored). The car and mobility-automotive-industry developed rather rapidly, starting collaborations with foreign investors, bringing know-how and updated technology into Uzbekistan.

Further collaborations regard joint ventures with General Motors (GM Uzbekistan), with the American-Korean GM Daewoo, a joint company of which the government maintains substantial control. Further agreements have been signed with Japanese and Turkish auto companies for the production of buses, lorries and other automobiles.

Furthermore, for railway and for investments connected with infrastructures, contracts have been drawn up by Spanish railway constructors for an Afrosiyob high-speed train connecting Tashkent with Samarqand (and Bukhara), the major cities of the country (possibly contributing to overcoming the dualisms between the two cities that disturbs the state politics as well).

Further major contractors are dealing with public and private constructions, communication, ICT, transport and administrative appliances. Tashkent and its urban area are usually chosen as a hub for such productions, considering its skilled workforce and industrial traditions, and the potential domestic market (in the automotive and transport industry, for large consumer and commodities production). It represents a good opportunity for international manufacturing and for multinational corporations which are always competing with each other to cover such emerging markets, looking for the "pole position" in starting some development cycle.

Further economic investments leverage the country's natural wealth, which is considerable—even though not as affluent as its neighbours — from several points of view. The availability of minerals like gold (fourth largest in the world), copper, silver and uranium is particularly attractive for foreign investors; and then of coal and HC, available in significant amounts, enough for internal consumption in order to ensure energetic self-sufficiency. It is also a relatively rich and net exporter off gas (11th in the world in natural gas production).

9.4.8 Internal Fragility and Exposure to International Market

The data evidences that Uzbekistan has been a fast-growing country for two decades (similarly to other economies of the area): it is currently rapidly recovering from the decline which occurred in the transition period and nowadays shows an accelerated growth of about 7%—8% GDP per year. This growth evidences the potential of the country, but it also may hide some elements of weakness that will probably emerge in the middle term, once the current favourable wave is exhausted (as it is typical indeed for any transition period).

[56] Fazendeiro Bernardo Teles, 2015: 485, 490.

[57] considering that regional air-transportation is usually managed on a national scale-monopoly, then excluded from the international competition, then possibly representing an opportunity for a protected investment environment; see as well Özcan Gül Berna, 2015:413ss; PWC 2016.

In fact, Uzbek economics is still producing heavy externalization, which sooner or later will have consequences for the monetary balance sheet. This is the case of the agricultural production (with cotton, rice, watermelons and exotic fruits of which Uzbekistan is one of the major world producers and exporters), but where the universal situation is not cost considered (as said, due to intensive and highly industrial water- and soil-consumption, largely polluting, due to massive use of chemicals, that also continues in post-Soviet times).

Furthermore, such specialization puts small-local farmer activities at risk (that are of key importance for social and territorial stability); this agricultural management proves to be convenient for trade purposes, but it makes local production vulnerable to international market shocks.

The country has apparently declared that it is to pursue a reformist strategy, but observers remain unconvinced. This is because growth is based mainly on natural capital exploitation, managed by a kind of new nomenclature that is used to having control of those resources to the detriment of bottom-up initiatives and social demand (to the detriment of social inclusiveness). It means a fundamentally biased growth, with an increasingly bureaucratized apparatus, exposed to corruption (see corruption index, whenever based on perceptions, then difficult to be interpreted),[58] with a "nomenklatura" (clan-based elite) progressively "closing" itself off and monopolizing the resources.[59]

Over time, it may be possible to understand if these are just temporary "growth problems", considering the rapidity of the changes, or if they will become something of a more structural nature, permanently hindering the country's economics.

9.4.9 True or Fictitious Development?

The country has demonstrated that it is capable of overcoming the uncertainties that characterized the transition period; it was mainly due to political unrest, the spread of radical movements that originated mainly from situations of impoverishment and of system inefficiency.

The recovery (as for other CIS economies indeed) started when the country reached a situation of sufficient stability in geopolitical terms (after 2000); analysts currently think this has happened thanks both to the reforms realized in that period (just after independence, possibly following the liberation climate of those times), as a kind of delayed effect and due to contingent international market conditions (favourable commodities prices). However, considering that the starting level of the growth cycle was so low (because of the de-structuration consequent to de-Sovietization), it was to be expected that the country would enjoy a phase of accelerated growth sooner or later.

Statistics consider the starting point of the current economic cycle a very low one—positioned in the early 1990s, when the nominal GDP failed due to de-structuration occurring in transition times — when the new independent states accomplished some structural reforms, economically indispensable, but very "expensive" from a social point of view.

The fact that the production gross index in 2008 was double than in 1995 (the minimum after Soviet collapse) when measured in constant prices (per IMF) must be considered.[60] But, it is a matter of discussion (as usual for any emerging and fast-developing country), whether it is to be considered true self-alimenting growth, or just growth obtained thanks to the possibility of covering up some externalities (especially social and environmental costs). In general, it is possible to say that Uzbekistan has precious natural and human resources, useful for making its economy resilient and socially inclusive, but it has to adjust the fundamental aspects of its functioning.

Finally, it is to be considered that the country still relies significantly on remittances from migrant workers, possibly the first element of internal economics (as usual for CA countries, except HC rentier states Kazakhstan and Turkmenistan), remaining highly vulnerable.

9.5 Turkmenistan: A Land Between Deserts

9.5.1 Geography, Environment and Landscape

Before 1967, when the artificial Karakum canal (one of the longest ever excavated, about 1100 km long) was completed, Turkmenistan could not really be considered as a "territorial" entity; populations were sparsely settled and affiliated with tribal groups on the base of an extensive archaic economy, based on long-range transhumance.[61]

The only territories suitable to be inhabited are the thin strips of oases at the foothills of the Kopet-dag mountains in the south of the country, where the only urban centres like Ashgabat can be found, and the coastal bank of the rivers Amu Darya, Morghab and Tejen.

[58] Warf Barney, 2016; Lamy Frederick, 2013:146.

[59] the EIU define the Uzbek government as "hostile to allowing the development of an independent private sector, over which it would have no control", Wikipedia, Uzbekistan, accessed 29.3.2018, citing the Economist Intelligence Unit

[60] http://www.imf.org/en/Publications/WEO/Issues/2018/01/11/world-economic-outlook-update-january-2018, accessed 23.4.2018, cited by Wikipedia.

[61] Edgar Adrienne Lynn, 2004.

The Kara Kum canal changed this situation, with the abundance of water carried by the canal contributing to the development of settlements, urban centres and economic activities (in particular internal navigation and irrigated agriculture). An enormous enterprise—bringing the water to the desert—that changed the destiny of an entire population, but which also created a sequence of side-effects, among which the decreased carrying capacity of water of the cited rivers, especially of the Amu Darya in its lower course, and a sequence of environmental and climatic changes at different scales (connected to the drying up of the Aral sea).

Consequently, the outcome of such a big enterprise is to be considered contradictory; the water-use technique applied—based on open air water flow—means an intolerable water loss due to evaporation and filtering down, combined by misuse and poor maintenance of pipes and canals, which besides the obvious benefits to agriculture, can be considered disastrous. And this was in a land characterized by continentalism that proved to be highly vulnerable to ecological stress, as highlighted by the sequence of archaeological ruins, possibly the demonstration of changes that had occurred since ancient past times: Turkmenistan is the land of ancient civilizations ruined in the dust, representing the evidence of the power of environmental changes (Nissa, Merv, Köneürgenç).

Indeed, the Turkmen territory still offers a wide variety of landscapes, but the situation is critical, even when slowly recovering from Soviet times (when the ecological catastrophe was "de facto" ignored). The forest is especially endangered on the middle and lower flow of Amu Darya, the "tugai", subject to significant ongoing reduction, also needs immediate intervention. The current authority has started diverse programmes to restore natural environments, counteract desertification and apply re-forestation policies, but their effects have to be measured in the long term (as obvious for any re-naturalization processes).[62]

9.5.2 Traditional Life and Adaptation to the Arid Environment

Turkmenistan is what the international reader would imagine as CA, with its untouched beautiful sandy deserts, arid mountains ridges, and archaeological ruins, precious testimonials of abandoned and forgotten civilizations; further images are those of camels and caravans, and of the fierce expressions of the nomads, which is still possible to see today in the faces of Turkmen population. A further symbol of the country was—and still is—the "akkal tekke" horse, a special race, bred here since ancient times, with the calcareous water of this Karst region reinforcing their bones; it has always been a symbol of the powerful instrument of the Turkmen raiders.

The current population still maintains some codes of the ancient tribes — the last ones to be forced to live a sedentary life among the CA people. They demonstrate this sentiment in different ways, e.g. setting up a yurta close to the modern home, keeping on with the breeding of camels and horses (mainly for cultural or folkloristic purposes), preserving rituals and peculiar religious habits, even when forbidden by the diverse authorities at various times.

Sometimes, it seems as if they consider the desert as their true original home. The Kara Kum represents the possibility for pasture in some months and, as the nomads well known, in winter it is covered by a layer of green grass, thin—indeed almost invisible for people not accustomed to this environment—but consistent enough to ensure the survival of cattle. A desert for the rest of the year intolerably arid and hot, impossible for any kind of organized life—even today—with people forced to move continuously for survival.

In pre-modern times, the tribes were used to nomadizing the areas surrounding the sedentary settlements, alternatively and depending on circumstances, visiting or attacking such cities, spoiling bazaars, invading the oases that were situated at the edge of the desert. With such periodical movements, they demarcated a commuting path between the northern urbanized belt on the Zarafzhan river, especially Bukhara, and the southern Persian towns, where they sometimes carried out raids, escaping any retaliation by moving rapidly back to the desert where neither the traditional nor the early modern armies could follow them.

As usual, their culture was apt to a mobile "genre de vie", based on production and use of light instruments (weapons, equipment, goods), excluding any stable or fixed installations.

Such constraints influenced all aspects of their life, forcing them to develop a particular material culture, made up of high-value portable objects, but of small dimensions. These were sophisticated artisan works, artistic jewellery, carpets and textile products. They were made from the raw material the nomads could obtain from their own work, like leather, wool and horn, and occasionally from plunders and raids (like gold or silver soon melted and re-shaped into new artefacts).

The artisan skills of the desert wanderers were much appreciated, so they could always supply the bazaar merchants or caravans crossing the desert with such highly

[62] "Three million trees planted each year in and around Ashgabat" in the context of a major reforestation project, counteracting desertification in the whole country, Turkmenistan, that has been known as the "land of orchards" since the ancient times, United Nations Environment Programme 2008; from a certain point of view protecting and restoring the ecosystem is to be considered the most profitable investment; but indeed it is necessary to evaluate, beside the apparent advantages, whether such programmes are sustainable and self-renewable, bringing a concrete benefit to the environment in the long term.

requested products (among them, the famous carpet labelled "Bukhara", typical of Turkmen production).

9.5.3 The Canal

With Russian expansion, the Turkmen residual population was marginalized; the new potentate applied new technology in order to consolidate its rule, and to conquer the desert where none had managed to (or had the interest) dominate before.

Industrialized agriculture, irrigation capabilities, infrastructural networks and further modern innovations, were evidence to the eyes of the poor shepherds—who had never before experienced such technologies—of the thaumaturgic force of communism. The new power demonstrated itself capable of transforming the desert into fruitful fields, tireless caravan travel into modern transport, using railway and machines, and so on, finding a solution for many aspects of traditional life, organizing resources in planned economics.

In the 1930s, the new politics planned the complete reorganization of agriculture, with forced enrolment) on big infra-structural works and especially in canal excavation campaigns, with local Turkmen definitively forced to abandon their style of life, erasing from their culture what eventually remained of their nomadic freedom. Ultimately, they were assimilated, often forced to carry out manual labour, that for the shepherd and warrior was the most unworthy and humiliating kind of work, a kind of "slave" labour on the "ground" (not on horseback), namely working in agriculture or on the canal excavation.

With Soviet rule, it was not just a question of exploiting a remote periphery inhabited by a hostile population rebellious to any purpose of "civilization", but it was a question—as proclaimed by propaganda—of liberating people "imprisoned" in feudal habits, to be integrated into the new order. Furthermore, it was a question of "domesticating" such a primordially extreme environment which was impossible to settle, using a modernistic plan while at the same time preventing (thanks to wise planning) an intolerable waste of material and human resources.

All these intentions coincided with the project of the Kara Kum canal, created using forced labourers, employed in excavation work who were definitively "integrated" in the new order through their obligation to carry out this heavy manual work. The "Turkmenistan Nile", the way it was called (like indeed a Soviet pharaonic "donation"), diverted the waters of the Amu Darya (flowing down from sources situated in what are nowadays foreign countries) to the Turkmen desert. Around this infrastructure, the whole system of new settlements was constructed, of cotton and rice fields, new villages networks, roads and bridges, a whole new country. The canal, with works started in the 1930s, was finally completed in 1967, excavated by "brigades" of thousands of workers, "de facto" bound to a new form of serfdom.

9.5.4 The Treasure Under the Sand

The central Kara Kum, the traditional refuge for the nomads, has proved to be the place of most important evolution, since rich as a HC, especially in gas deposits, meaning the recent evolution of the country's economics. Presently, Turkmenistan is one of the most important producers and exporters of gas, credited as having the world's fourth-biggest reserves of natural gas. The discovery of new fields has continued until recently, with the discovery of what has been defined as the second-biggest gas deposit in the world in 2011, the Galkynysh field, with estimated reserves of around 21.2 trillion cubic metres.[63]

This wealth does not immediately lead to authentic and durable development: the HC abundance produces some incentives for local economy, but it also represents the risks of "easy money" which may often occur in such circumstances (supposedly at the beginning of a development cycle). Above all, it is the instrument with which the state extended its influence on society. In fact, this richness proves to be double-edged, inclining towards a centralistic monopoly to the detriment of economic diversification, disseminating unfavourable side-effects.

Among others, it signifies the risk for indiscriminate consumption and for the spread of attitudes that are not particularly compatible with the standards of sustainable development and to be pursued in the frame of paternalistic politics. Turkmenistan can be considered a paradisiacal place for car drivers, electricity users, water and natural gas consumers, and in general for consumers of usually expensive goods that there have been directly provided by the government—since 1993—freely or heavily subsidized: in the form of electricity, gas, water and fuel, as well as salt which is very important for herding sheep and goats.[64]

In general, the government used to provide generously welfare services, goods and facilities of any kind to its citizens, in a kind of implicit trade-off between basic needs and obedience (but renouncing some basic civil rights).

Such politics characterized the rule of Saparmurad Niyazov, the former nomenclature leader who, after independence, was appointed as president for life with progressive

[63]Anceschi Luca, 2017:414.

[64]Decree of the People's council, 14 August 2003, for commodities to be subsidized until 2013; further benefits regard gasoline, that is currently free; recently such policies have been changed since defined as "infective", as Nyazov's successor President Berdimuhamedov said, "Turkmenistan leader wants to end free power, gas, and water", Deutsche Welle, 2018; PWC 2010:5.

constitutional amendments until his death in 2006. His politics gave continuity to a kind of isolationistic dictatorship, which soon degenerated into a grotesque post-Soviet personality cult (finally being compared to that of North Korea and of other "rogue states").[65] The new President Berdimuhamedov, the former minister for health (then insider to government circuit) has continued this government style, even if currently showing signs of relaxation, finally prospecting a possible turn in politics.[66]

9.5.5 From Nomadic Genre de Vie to Hydrocarbon Abundance: True Wealth?

The recent history of Turkmenistan is characterized by its dramatic changes from a traditional way of life to the Soviet (modernist) constrictions, to the current independence and HC abundance, namely the formation of a rentier state. It is also the history of the sudden passage from the sand of the desert—as a natural environment in which to organize survival economics—to "white gold", namely to water abundance, after the excavation of the canal, and then to "blue gold" (natural gas). However, this has not meant any automatic and consequently true development.[67]

The early discoveries in gas deposits happened some time ago, in the last decades of the Soviet rule, when the SU became a major supplier of HC, mainly from its Asiatic provinces, to western European countries.[68] Explorers and researchers had the chance to discover the huge extension of oil and gas fields, but they did not have enough time to exploit these resources. With independence, the question was re-purposed, making it the fortune of the new independent Turkmenistan.

Such richness gives the country the typical appearance of an "oil emirate"—as it is called journalistically, by international media—an (almost) empty sand container, but full of precious HC.

The question was, at that point, how the country could benefit from this richness considering that the main routes were blocked northwards by Russian monopolies (recently in 2016 Gazprom announced publicly the immediate cessation of purchasing gas from Türkmengaz, with "de facto" closure of the "Russian route" because of some disputes)[69]; southwards by Iran and Afghanistan territories which still represented a critical situation and furthermore, on the western side, by the Caspian maritime borderline disputes.

It must also be considered that the access to deposits was not easy technologically speaking and there was the real risk the HC richness would become just a missed chance over time.

9.5.6 Accessibility to International Markets

For all these reasons, improving the connection capability can be considered quite urgent. The Turkmen government has tried to set up alternative infrastructural routes, especially for HC from the beginning of its independent life, but this operation has not proved to be easy, considering the isolationist obsession of Niyazov, then the "impermeability" of Turkmenistan to foreign investments.

In 1997, Turkmenistan opened a little pipeline to Iran (Korpedzhe—Kord Kuy). Then in 1999, it started to work on the eastward Turkmenistan-China pipeline (TCP, with CNPC partnership),[70] across Uzbekistan and Kazakhstan, through Dzungarian gate, that became operative in 2009 (projects financed by EBRD and IDB)[71]; this last facility substantially improved the country's capability. Today, China purchases 73% of Turkmen gas, the lion share of the country's export, evidencing a shift from Russian to Chinese acquisitions (an important change considering that China, differently from Russia and Iran, is an HC net consumer, then a client, not a competitor of Turkmenistan economics).[72]

The southern way proves still to be impracticable, at least until recent changes, and Iran's return on the international scenario, after decades of "sanctions" and market marginalization; this also considering the natural barrier of the eastern mountain bordering region (eventually blocking transition in the direction of such energy-hungry countries like India and Pakistan).[73]

[65] Warf Barney, 2011:5; see also the Freedom House index of political freedom, www.freedomhouse.org, accessed 23.4.2018.

[66] It is still considered by independent observers and NGOs as a repressive government, scarcely inclined to regional collaboration; such politics could be considered both, as a consequence of a self-perception as a weak country, surrounded by powerful and hostile politics, or an expedient, namely an instrumental kind of neutralism, functional to the interest of local elite, that in this way can pursue politics based on arbitrariness.

[67] It is the typical history of the encounter (eventually clash) between nomads and sedentary life, with a sequence of wars, without any possibility of compromise; Barisitz, Stephan, 2017.

[68] Krasnov G.A., 1987.

[69] Anceschi Luca, 2017; PWC 2010.

[70] PWC 2010; even when the principal reason of such impermeability is possibly the preoccupation that such intrusion would influence and also upset the local clan system of power.

[71] Anceschi Luca, 2017:417.

[72] Anceschi Luca, 2017; Silvestri Tommaso, 2015–2016.

[73] the possibility of delivering the liquefied gas via sea has spread rather recently; it depends obviously on navigation possibilities, considering that the Volga-Don channel is controlled by Russia, and the ports on the Caspian and Black seas are controlled by competitor countries; however, considering the constraints evidenced by the geographical map, and the reciprocal necessity to gain access to international market, it is evident that a basic free movement condition would be in the interest of everyone.

Besides this, Turkmenistan (as usual for all emerging countries) urgently needs to implement international exchanges, not just for selling its natural resources, but also to acquire expertise for the construction of oil and oil-derivative refineries, pumping stations and pipelines, organizing stock and transports, and further mineral explorations. It must acquire technology that only advanced countries, which often are also the final customers of these same products, can provide.

This is a necessity that can be considered as urgent since pipeline construction programmes develop in the long term (with an even longer time estimated for return on investments), risking that they become periodically obsolete (considering that in the meanwhile everything can change, the type of energy used and technology of distribution, as well as the same kind of energy source prevalently used).[74]

9.5.7 Pipelines as Geopolitical Influence Tool

Besides the east, and the China relationships, a further project must be mentioned. It was scheduled for 2012 (TAPI, connecting Turkmenistan, through Afghanistan, to Pakistan and India, Fig. 11.9), which makes it possible to imagine further potential presumably in the shape of additional important export routes in the future, relying on investments of $3.3 billion or more, attracting the interest of many international operators.

Indeed, this evidences some critical points: at the moment the project is not much more than an "infrastructural promise", likely purposed with speculative motivations (configuring a kind of renewed CA "great game" played with pipelines and infrastructural corridor projects).

There are several reasons for this, in particular the Afghan southern way represents a kind of "cap" for such projects (and for the whole CA accessibility). After the TAPI "groundbreaking" ceremony, nothing was done, neither the pipeline route determined, nor the operative steps defined. The ceremony happened with a spectacular exhibition of "friendship", with an ostentatious signature of agreements, prospecting the setup of an "energomost" ("energy-bridge") connecting the continents, supposedly solving all problems.[75]

Such projects have been combined with further works, in order to improve the internal transport capacity, especially on the west-east axis: President Berdimuhamedov proclaimed in 2010 the project of a new pipelines connecting the major deposit in eastern Turkmenistan to the Caspian coast, considerably improving the capacity of the country (a huge project of a pipeline of about 1000 km, possibly a resurrection of the Karakum canal myth, re-purposing HC instead of water).[76]

9.5.8 The Question of Investing and Spending this Money

As said, the question of which way to spend this money is not a secondary one (even if this sounds paradoxical): notwithstanding the problem of access to the international market, the HC revenues have been (and will be presumably for a while) rather high, configuring a huge amount of extra-profits.

The government continuously has to take decisions regarding which way to spend this money, in order to pursue growth, and has to control such spending, in order to prevent the side-effects such activities may induce, even if not intentionally.

Such operations—and the realization of such potential value—require wide capabilities in a country that has just a few million inhabitants, characterized by its relatively poor economics and infrastructure, and by a public apparatus lacking in expertise (necessary to support such spending without encountering deviations and mismanagement which are usual for such rapidly growing countries). It must be

[74]The realization of a pipeline is a complicated matter, not just for planning the itinerary, but also considering the whole cycle of the infrastructure's existence, namely royalties, surveillance, maintenance, accounting on quantity delivered; actually it requires a high level of collaboration, planned in the long term; Anceschi Luca, 2017; it is to consider, besides the economically demanding scale of intervention, that the extremely articulated structure of such an industry (pipelines thousands km long, extended extraction fields, huge accumulation deposits and stock), makes such a network vulnerable to geopolitical instabilities (e.g. considering the risk about terrorist attacks, or eventually foreign powers manoeuvres and speculations, in general the relations with neighbours).

[75]Indeed, such a celebration was possibly a reminiscence of the soviet habits of fictitious public work inaugurations.

[76]D'Amato Giuseppe, 27.10.2010; Indeed, the IC is increasingly becoming sceptical about such giant scale projects; both, in the case of scenario evolution, energy-source diversification and ongoing discoveries of new deposits which may change expectations, and the correspondent HC quotations, as well the calculation concerning return on such investments, usually planned for the middle-long term. The HC sector proves to be rather unstable, besides the listed reasons, also for further variables, such as marketing and technological reasons, for stocking, safety, shipping and logistics, that may be subject to relevant and unpredictable changes; it is the case of modalities of shipping gas and oil, as well as coal, that may change depending on the use of tankers of liquefied gas or of opening of new pipelines, changing conditions between long term contracts or spot (then liberated and short term), among sovereign states or independent brokers, and depending on many more of such variables; such evolution can render useless any attempt to gain a fixed rent from this position (also considering the construction times characteristic for such infrastructures, that are commeasured essentially on the long term, while prices and other commercial conditions can change suddenly).

taken into consideration that the local society is not (yet) able to represent an autonomous entrepreneurship (also because the politics of the country seems to impede the formation of spontaneous private economics, similarly to other CA and CIS countries).

At the moment, the only possibility is to rely on a top-down economics, based on centralistic investments in the sectors that the public apparatus can develop (maintaining them, at the same time, under its control).[77] Besides the obvious extraction industry, infrastructures and building industries organized per big contracts, are currently booming, being the better way to re-invest HC revenues.

The better example of such polices in CA (besides the already described Nursultan and the western Kazakhstan Caspian cities) is Ashgabat. Turkmen authorities, since the time of Niyazov, have set up an extraordinary programme of (re)constructions, with an impressive use of precious materials and luxury architecture, making it a kind of marble-city, with new urban landscapes in entire neighbourhoods, official and government, residential and social buildings, "arenas" and auditoriums, and stadiums, museums and hotels.[78]

Similar investments have interested religious buildings; among them are the huge Saparmurat Hajji Mosque in Geok Tepe, considered the biggest in CA and suitable for up to 10,000 worshippers; the Turkmenbashi Ruhy Mosque; and a further list of newly constructed spectacular holy buildings (possibly pursuing a political manoeuvre on the religious practice).[79]

More recently, programmes regarding the coastal development of Avaza have been proposed,[80] as well as further projects concerning environmental and amenity improvements (Fig. 9.7).

9.5.9 The Recent Turn in Turkmen Policy

All of this has happened in the recent context of a tendency towards the opening up of the territory, for instance, the organization of the 2017 Asian games, indoor and martial arts disciplines, in Ashgabat. They were organized under the patronage of President Berdimukhamedov, who seemed to want to inaugurate a change in the political tradition of the country in this way. It seems he wants to capitalize on the improved stability, possibly induced by a sequence of other evolutions regarding the area (the opening up of their neighbour Iran in its post-sanctions era, and especially the effects induced by the SCO agreement signed between the major regional power incumbent on CA, see later).[81]

It proved to be a good way to spend a great deal of money in a reasonably economic way, achieving a set of further effects, like self-celebration (see the grandiose opening cer-

Fig. 9.7 Expo Astana (Nursultan), August 2017, Avaza resort advertisement, Turkmen pavilion. (Photo: Igor Jelen)

[77]Obviously, any activity considering environmental resources, like HC extraction but also agricultural land, construction and urbanistic projects, depend on the possibility of obtaining public permits, considering they rely on bureaucratic procedure and licences.

[78]Gaynor Kelly Lee, 2017.

[79]Corley Felix, 2005a, b; but constructed usually at some distance from political buildings — see later;

[80]PWC 2011:12.

[81]Parenti Fabio Massimo, Adda Iacopo, 2017:347.

emony in occasion of the event orchestrated by the Italian designer Balich).

Possibly such evolution would be supported (and alimented) by some moderate tourist development, leveraging on the natural and cultural beauties of the country; for instance, the Karst caves with underground lakes, artificial gas flames issuing from a burning crater (Darvaza), as well as leveraging on further local traditional resources, like Akal tekke horses' carousel and folkloristic nomadic traditions, artisanal handicrafts, desert and sea-side climatic attractions.

9.6 Tajikistan: Peace After War

9.6.1 Basic Geographical Description

Tajikistan has always been considered as the weak "ring of the chain", and the poorest among the CA countries. It was already an emigration country in Soviet times—considering that the Tajik community is even now one of the largest communities of immigrants in Moscow—where its poverty had been already stereotyped.

This is not always easy to understand; among the possible reasons, there is its difficult accessibility—the position at the extreme periphery of the "empire", difficult to integrate from any point of view, both in czarist and Soviet times, for infrastructures, economies and planning policies.

A further reason can be traced to the linguistic-ethnic diversity, the Tajiks being an Iranian-speaking people, while their neighbours are prevalently of Turk origin—a fact that might represent an element of marginalization.[82]

A further material reason is the remote character of its mountain settlements, especially of the Pamir area, represented by its archaic origin populations (probably, simply to be considered autochthonous), as well its topographic and geographic condition, unfavourable for any production purposes (at least in modernist terms).

A further element impeding the "normal" development of the country was (until a few years ago), the not yet (at that time) recognized czarist/Soviet, and currently Tajik border with China: it included Pamir highlands, with high-altitude plateaus, peaks and glaciers, where borderlines were objectively difficult to mark, considering the technology of those times. In fact, this question has been just recently settled, in 2011, with the acknowledgments over a small portion of territory from Tajikistan to China. It was made possible thanks to the general improvement of international relations: a "good will" treaty following others in the context of a geostrategic agreement which Russia and China confirmed through the establishment of the SCO (see later).[83]

The area has always represented a critical point in the relations between SU and China, occasionally crossed by nomadic shepherds, then during Soviet times, hermetically closed and watched. It is characterized by a history of contentiousness which can be traced back to colonial and Soviet times (the "basmaci" rebels), with refugees escaping over the high mountain passes which are difficult to monitor. At the same time, the country appears, made up of inaccessible valleys, vulnerable to foreign and neighbouring influence, above all from the Afghanistan border which is difficult to control and always presents the risk of infiltrations.[84]

This is also considering that the borderline is extremely articulated, exposing the country to continuous outside influences. In fact, crossed by international roads, this region is often characterized by the spread of illegal economies (of special concern today is the emergence of opium poppies and hashish trade), with all the consequences such problems can induce. It is the case of the corruption of the local police and of the formation of criminal networks which easily transform in such remote borderlands, into political contraposition, "guerrilla" or terrorist movements.[85]

It is enough to take a glance at the map in order to realize the reasons for the weakness of this disarticulated state. Dushanbe is an ancient town on the route between Kabul and Bukhara, but it is displaced compared to the actual state territory and not able to exert its functional and political influence on its far-flung peripheries. The country indeed appears as the typical artificial Soviet construction, resulting in a set of places, cultures and resources combined with a "divide et impera" style. It always represented an issue for the Soviet planners, who applied different policies in order to consolidate their rule.

In fact, they soon started to move the remote populations towards the "new towns" of the plains, namely in regions suitable to be industrialized, especially for the production of cotton (applying agro-industrial and mechanized techniques in these plains); the southern Vaksh valley registered a booming demography in the 1970s and 1980s (that had tragic posthumous consequences, being among the major premises for the successive civil war, together with the failed project of ethno-social engineering). Furthermore, they combined such politics with the already described big work policy, like canal

[82]This especially considering the problematic question of relations with the Uzbek in neighbouring countries, characterized by mixed and bilingual populations, influencing the formation of a national identity; to some extent for the presence of a consistent Tajik minority in neighbouring Afghanistan; Dagiev Dagikhudo, 2014.

[83]"Limes" n.8/2014.

[84]Necati Polat, 2002; Shahrani M. N. M., 1979; Brill Olcott Martha, 2012:101:76ss; the Badakhshan represents a kind of outgoing flow into Rasht valley, dividing "de facto" two Tajik sides, with a border line difficult to check, exposing it to such infiltrations from Afghanistan.

[85]Megoran Nick, 2017:44 fs; De Danieli Filippo, 2013:143.

Fig. 9.8 Tajik Aluminum Company at Turzunzoda, August 2017, surrounded by cotton and rice fields. (Photo: Igor Jelen)

excavation, and the setting up of dams, industries and further centralistically managed facilities.[86]

9.6.2 Economy and Resources

It is not surprising that, considering such circumstances, migrants remittance is the first economic source, prefiguring a kind of dependency from outside economies, and this especially from Russian economics—at least until the recent crisis, with the downturn of HC quotations in 2012–2014—and from further CA rapidly growing countries. A further relevant source is industry and in particular the gigantic plant in Turzunzoda (one of the major aluminium smelters in the world), producing exclusively for export for the wider world market, the Tajikistan Aluminium Company (TALCO).

It has been built up on the basis of a top-down programme, "bringing industry" in the middle of the idyllic Glissar valley, including the construction of a nearby company town (Turzunzoda), surrounded by rice and cotton fields, and cattle breeding prairies, with the landscape of glaciers and snow peaks as a backdrop (Fig. 9.8).

Constructed in the mid 1970s, it has undergone a series of upgradings; it is alimented by Nurek dam electricity and by imported Uzbek gas; it is state-owned, representing the main Tajik asset, indeed a critical asset. It appears as disproportionate and difficult to manage in the narrow nation-state perimeter, resulting after SU fragmentation.

It requires relevant expertise from many points of view: international market accessibility, raw material and energy supply, management and technological upgrading; the question of environmental externalities is especially critical, with polluted emissions impacting cross-border areas. In general, it seems to represent an object much larger than the current Tajikistan economy can manage.[87]

Further assets regard the agro-industrial production in the south-western Amu Darya valley, especially cotton, for which the same consideration as for Uzbekistan must be made: agriculture which has spread to the detriment of original agriculture activities, modernistically organized on the basis of scale economics, requiring huge consumption of natural and water resources, exerting high environmental impact. In fact, cotton is—after aluminium production—the main country asset, representing about the 60% of the agricultural production, using about half of the irrigated agricultural soil, leaving one to imagine a tendency to monocultural dependency.

Besides this, the true richness of Tajikistan is simply the abundant pure water—the "white gold"—flowing down from Pamir plateau glaciers: the rivers, actually the upper flow of the biggest CA rivers, Amu Darya and Syr Darya (considering its tributaries Kara Darya and Naryn, flowing from Kyrgyz mountain) design the topography of the country, flowing in the direction of the Aral sea endorheic basin, comprehending practically the whole CA area.

The rapid upper flow of these rivers is not easily accessible; they are used by mountain farmers for an efficient horticulture, thanks to side channels, excavated upon moraine

[86]Shabad T., 1980:87.

[87]The Gazzette of Central Asia, 2012, Tajik Aluminium Output Falls 11.6%; Laruelle Marlene, Asian development bank, 2016; Brill Olcott Martha, 2012:85.

terraces, which can reach villages and mountain fields using simply inclined pipes that are self-made and easy to be maintained. It is indeed a precarious situation since such streams easily transform seasonally into floods, sometimes provoking disastrous events (mud-slides, debris-slide, etc.).[88]

On the lower side of the valleys, water is also relatively easy to access; but the facilities (aqueducts, concrete pipes, motor pumps and irrigation systems) of Soviet origin are in poor condition, forcing the farmers to recover, whenever possible, their original ancient ID know-how.

Water is becoming of even more strategic importance in post-Soviet times consequent to the segmentation of local sovereign states; it occurred that the sources (and the upper course) of the major rivers (and of their tributaries) were separated from the main places of consumption (irrigated oases, urban-industrial centres and large cities in Fergana, Uzbekistan and Kazakhstan); this fact can obviously induce tensions, namely the arising of contrasting interests, in conservation, accumulation and use of the same resource over the different stages of the river course.[89]

9.6.3 Resources for a New Development Pattern

Much of the Tajik territory is uncontaminated mountain terrain; about a half of the country's surface is over 3000 m asl., comprehending the Trans-Alay mountains, a prolongation of the Tien Shan on the south-western side of the country, and the Pamir mountain system, occupying the central-eastern part of the country. It comprehends some of the biggest alpine glaciers of the world, the Fedčenko, and some of the highest peaks in CA—Ismoil Somoni Peak (Communism peak) 7495 m. and Ibn Sina (Avicenna) Peak (Lenin peak) 7134 m.; they have been currently renamed with traditional or neo-nationalistic names, although the Soviet and other local names are still in use.

The extensions of arable soil are limited, mostly to the oasis plains or to even more limited surfaces on mountain slope terraces; but the good combination of solar radiation, climate conditions and water availability make both high productivity and high-quality agriculture possible. So do in the valleys and river plains suitable for being settled or cultivated, on the Tajik side of Fergana, in the central-southern side of the country, in the Glissar valley, around Dushanbe, and in the southern Vaksh valley in the Qurghonteppa and Kulab areas; these account for just one-tenth of the total surface of the country, where the major part of the history and of the economy of the country is understandably concentrated.

The country is rich in other natural and cultural resources, which in the current transition (to a post-industrial economy) are of great interest, potentially inducing some diversification in economics. Such resources would make a kind of bottom-up development possible (the water resources usually being accessible without the mediation of particular technical or political power), starting from family and locally run agricultural activities.

The remoteness of most parts of the country, the diversified altitude, climate and topography, means the possibility for the preservation of high biodiversity, of its soil and landscape characteristics and of its interesting and unique naturalistic attractions.

The same for cultural resources, in terms of ethnography and folklore, linguistic and religious particularities, where each of them has some cultural value to be reconverted into economic opportunities. This is the case of tourism and of any kind of amenity and quality economy, of natural biosphere reserves, simply of pure water, air and soil, that represent a chance to start up smaller niche economics but which may be significant for a small country. It is not just a project on paper, as can be demonstrated by the increasing number of Pamir, Piotra Velikogo, Alaj and Fan mountain visitors coming from all parts of the world, with Dushanbe international airport becoming a hub for activities such as alpinism, biking, trekking and similar. The peripheral valleys have often proved to be an opportunity for salvation in the past (for refugees and people resisting some power) and possibly a resource for future development.[90]

9.6.4 Migration from Mountain to Plain

In modern times, with an idea of Fordist productivity prevailing, the mountain area was considered mostly non-productive (as it was impossible to use standardized methods or industrialized ones, e.g. to be cultivated with agro-machinery), and eventually was abandoned (Fig. 9.9). This also occurred because of the fragmentation of settlement, with villages scattered in deep valleys, particularly inaccessible in winter as well as in other seasons; simply, the modernist planner could not imagine any kind of life in these isolated areas. This resulted in the predisposition of migration plans, forced or not, of these populations to downhill and urban areas (mainly in centre-south of the country).

[88] Brill Olcott Martha, 2012:112.

[89] ID processes and, in general, the cycle of water is difficult to be measured and managed and requires the contribution of the whole population and of the entire society; the risk of waste and mismanagement is always real.

[90] It is important that the local population understands this, overcoming any self-perception of marginalization, undertaking development occasions offered in remote areas by innovations arising in times of globalization and ICT.

Fig. 9.9 Tajikistan, Rasht valley, August 2017, in the background Piotra Velikogo Mountains. (Photo: Igor Jelen)

It is the case of the population of the idyllic Rasht valley (historical name Karategin, with the small town of Garm, once a "basmaci" resistance centre, involved in 1990s civil war), of the north-western mountains ancient minorities (among them the ancient Yagnobi populations, descendants of the Sogdian civilization), and of the Pamir plateau.[91]

Such people migrated—actually were displaced—in those times to the foothill industrial areas, mostly settling in "modern" dwellings and planned "new towns", losing the connection with their home villages. The movement especially concerned south Tajikistan, where large complexes, exploiting new irrigated fields, suitable for scale-production agriculture required a large workforce for the booming cotton agroindustry.[92]

Indeed, such resettlements had different results, with the new populations not really being assimilated—besides a superficial integration—leading to the arising of tensions and to a basic contraposition between the mountain and plain, which emerged in its whole intensity at the time of Soviet collapse. In fact, the immigrated and local populations always demonstrated a difficulty in dialogue, where the mountain population which had settled in the oasis or in the new towns always maintained language and habits, and even clan-like organization, and solidarity.[93]

9.6.5 The War

Independence in 1991 happened—as usual for the Soviet CA "provinces"—not just as a conscious acquisition, but rather as an accident, as a consequence of the sudden collapse of the recognized single authority; then an escalation effect started immediately. In fact, with the disappearance of Soviet power, which had been considered as definitive until that moment, the different social and previously artificially

[91] Panaino A., Gariboldi A., Ognibene P., 2013.

[92] They were displaced not just for ideological or economic reasons, but because of the conditions of the original settlements, because of the possibility to exert modernized functions in these difficult topographies (for production, accessibility and surveillance). It is to say that these populations have always been considered with suspect by the authorities, and that these areas are characterized by a tradition of rebellions. The resettlement policy brought people from different origin and culture together in the new soviet towns, creating occasion for a difficult coexistence; these people, sovietized by force, and mixed with all the other immigrated populations, preserved nevertheless much of their original culture; it is possible to say, the resettlement followed a general plan, predisposed by geographers and engineers, motivated by functional-ideological reasons, in the form of a project of constructing an integrated, communist-inspired new society, notwithstanding the cultural particularities, considered secondary compared to the material changes, finally leading to the formation of uniform soviet population everywhere in the country; Ginzburg N.S., 1986; Kassymbekova Botakoz, 2011; Mostowlansky Till, 2014; as said the Gissar and Vaksh valley in southern Tajikistan evidenced the most rapid growth in population due especially to in-migration and to cotton agro-industry development, Shabad T., 1980:87; Kassymbekova Botakoz, 2011, it was reported that between 1925 and 1940, 48,000 households moved from the mountains to southern Tajikistan because to secure the border as they could not rely on non-local population; see also Reid Patryk, 2017:20; Ferrando Olivier, 2013:37; Sodiqov Alexander, 2013:49.

[93] clan and further traditional community formats had new relevance "in the alienating urban environment", where the populations were settled; Esenova Saulesh, 1998:448; Heathershaw John, Herzig Edmund, 2013:58.

aggregated, components, taken by surprise, found themselves in a condition of mutual mistrust.

The war started as a street protest in Dushanbe in 1991, initially as a manifestation carried out by religious and liberal opposition—considered in those circumstances moderate indeed—which wished to start a confrontation with the old-style government, expecting a change.[94]

Then this movement clashed with the power, quickly radicalizing (as usual in a situation of conflict) interlacing with further tensions and contaminated from contiguous conflict areas (in Afghanistan).

Indeed the "casus belli" is not particularly evident, and it is not realistic to define the conflict just as an explosion of social "nervousness", or just as the consequence of the shock of liberation.[95]

At the moment of SU collapse, the old power did not show any intention of changing and it continued, as if nothing had happened. In particular, it kept on controlling the economic assets of the country monopolistically (cotton production in irrigated plains, energy and hydropower) and other still working industries (especially the aluminium smelter in Turzunzoda), but it also continued to control survival basic resources like agro-food production, insisting on the autocratic method.

Consequently, the confrontation degenerated into manipulation, creating a greater effect of confusion. It is possible to classify different phases of the conflict; after initial and spontaneous "street rioting", the SC "war of the kolkhozes" followed, with populations grouping in respective villages, defending food and land, and their general means of survival (water, machineries, etc.), then a kind of panic ensued.[96]

The next phases registered a change, with the formation of contrapositions based either on ethno-linguistic divisions, with Kulab and Khujand regional groups opposing the "mountain" immigrated populations (as previously considered, especially in the south of the country), mainly Garmi and Pamiri.

The subsequent escalation and the general confusion made any classification complicated, with clashes spiralling out of control. Sometimes, they degenerated in pogrom-like actions, home-to-home fighting in such mixed societies, with frequent cases of ethnic cleansing against the immigrated populations, indeed often even against the second generation from previous migration. The main characteristic was the diffused formation of spontaneous self-defence militias, often just gangs controlling neighbourhoods or settlements arbitrarily using violence.

It was a "typical" civil war, without an evident frontline to make the opposed factions recognizable: a fact that multiplied the sense of fear, triggering a logic of pre-emptive attacks (in order to avoid retaliation), carried by self-organized militias with factions occasionally forming and overlapping pre-existent social aggregations.

Among them, it is possible to list brotherhood or clan loyalty, ethno-linguistic differences, regional belonging, political and religious groups, with elements of "warlordism", manipulation of the industrial plants (like Turzunzoda) and cotton farms (beside aluminium, the main country assets), with managers siding with one group or the other, in a general confusion, that was feeding in on itself.[97]

The numbers regarding the war represent the dimension of the disaster that occurred to such a small country of just a few million inhabitants, with up to 100,000 estimated dead, and around 1.2 million "displaced people", of them 500,000 migrating outside the CA (mainly to Russia, to former Soviet republics or Western countries) during wartime, because of impoverishment, persecution or other economic and political reasons.[98]

9.6.6 War "inside Effects"

The war soon "produced" a typical "environment", with a collection of "liberated areas" and guerrilla "sanctuaries", with criminals liberated from prison (immediately appointed as militia commanders) and local "warlords" and professional fortune soldiers, deserters and privatized militias simply seeking for plunder and spoils, and finally "war entrepreneurs" and "war experience tourists", and what would be soon defined as "foreign fighters".

In a further step, such scenarios would induce the formation of gangs active in civil areas, not just on the frontline, using terrorist methods (car bombs and similar to spread panic, kidnapping for ransom, human shields or suicide bombers manipulated with drugs), indefinitely feeding fear and mistrust. Usually, such tensions are fed by outside factors, like trafficking and smuggling, and tensions originated

[94] It is important to say that the Islamists still in 1990s were usually non-radical, often deriving from the Afghan partisan movement formed during war against soviet invasion in 1979, at that time supported by western countries and IC in general.

[95] Sodiqov Alexander, 2013:57; De Danieli Filippo, 2013:143.

[96] Heathershaw John, Herzig Edmund, 2013:8, see also Ferrando Olivier, 2013:37; Sodiqov Alexander, 2013:49; Brill Olcott Martha, 2012:147ffss.

[97] Heathershaw John, Herzig Edmund, 2013; Hryniuk Olga, 15 January, 2018; Brill Olcott Martha, 2012:183; Megoran Nick, 2017:44; Sodiqov Alexander, 2013:60; Akyildiz Sevket, Carlson Richard, edited by, 2013:8; Straub David, 2013:174; it is a typical case of conflict feeding itself, confusing original causes with induced ones, namely the so called "inside causes", with effects triggered by the same war explosion, making evident differences that until that moment were considered fictive or obsolete, like clan and brotherhood, confessional or ethno-linguistic nuances; Jelen Igor, 2012; Straub David, 2013:181; Dagiev Dagikhudo, 2014:189 fs.

[98] Rashid Ahmed, 2002; Rotar Igor 2006.

in contiguous scenarios—in this case in Afghanistan, as well as in Pakistan, Uzbekistan, Caucasus and elsewhere; they risked destabilizing sensitive border areas like the one with China (already characterized by the Xinxiang problems, namely by Uighur minority), finally converging into "globalized" Islamist and terrorist organizations.[99]

For several years, Dushanbe has been under curfew and military rule with the city criss-crossed by self-proclaimed militias, engaged in what it seemed to be a low-intensity war or just fighting among gangs with retaliations and rakings, with civil entrenching in compounds sometimes escalating into massacres, sides and the frontline changing in an unpredictable manner.

Over time, what it is to be considered as the main cause of rivalry emerged (the SC Islamic resurgence), with the recovery of religious codes as a "mask" (identity marker).[100] Having being repressed during Soviet times, the religious tradition re-emerged asking for acknowledgment and respect, claiming a political role that, being a theocratic culture intrinsically not compromise-prone, would immediately be understood as a revolutionary pretension. Thus, exactly at that moment, the Islamic Revival Party (IRP), the political arm of the Muslim community, gained momentum when the atheist system collapsed, and the ruling communist party, actually the "nomenklatura", appeared to no longer be politically legitimized, but was simply trying to maintain itself and its privileges.

9.6.7 Pacification Experiments

The war lasted about 5 years, possibly ending when it had exhausted its original motivations and thanks to a set of favourable contingencies; in fact, pacification seems to have been effective. This took place with the help of the IC, which supported the 1994 elected President Rahmon, once Rahmonov (that time defeating the contender Abdullaionov, representing the religious party), with 58% of the votes, in elections carried out in a situation of confusion, in a devastated country, but which had been widely recognized.[101]

Another favourable circumstance is the one represented by occasional quiet international circumstances; in fact, the neighbouring "great powers" appeared not to be very interested in feeding these tensions, with the IC reacting in a relatively efficient manner. Russia organized and led a peace-keeping operation at CIS level in 1993, including various former Soviet countries, with 25,000 soldiers. It had the special intention of ensuring the surveillance of the southern border with Afghanistan, indeed a vulnerable point for the whole CIS area.[102]

The figure of Rahmon, until that moment a middle-rank representative of the nomenklatura, was decisive in playing a role of intermediation; the parts finally reached a peace treaty in 1997, when the government and the opposition signed an agreement under the mediation of the IC and the UN special representative; the agreement prescribed a procedure based on the cessation of violent actions and the recognition of the opposition—consolidated in a quota of 30% of "ministerial positions" reserved for them in a kind of "great coalition", or "peace alliance".[103]

This first agreement, combined with the more detailed one of 1998, prescribed the pacification path, defining the reconstruction of the social basis, including the return of refugees, a general amnesty and the possible integration of the opposition into a future political game. From that moment forward the tensions begin to diminish, with the war eventually transforming into isolated attacks, local instabilities or occasional rebellions, but losing its wider political character.

As a further step, the government started a stabilization programme; it tried to give some impulse to the economy in order to contribute to an initial reconstruction and development effect.

For a while the peace treaty was considered with scepticism; this especially in the peripheral areas, where the war overlay further contrary situations (centre-periphery, mountain—plain, or rural-urban). It set out a sequence of steps that ought—in its intentions—to induce pacification progressively on all scales, from military to social questions, aiming for the integration of the opposition, possibly the most difficult task, requiring much more time to take place, usually the passage of a generation.

Further elections in 1999 (98% of the vote for Rahmon) and 2006 (79%) confirmed the presidency with success close to plebiscite, delineating an involution (to a more autocratic situation); however, pacification did move ahead, even when boycotted by several opposition parties and even when it was criticized both by internal and international observers.

20 years on from the end of the war, it is possible to say that the internal tensions must still not be understated; tensions and

[99]Fortunately, the international situation in those years was revealed as favourable, with — as said — the local power showing an interest in not interfering, and to de-escalating the conflict. Evidently also geopolitical aspects played a decisive role; perhaps it is to consider the role of neighbouring China and its sensitiveness concerning the risk for a SU-like disintegration, namely the fear of any escalating in peripheral "colonial" regions of Xinxiang, and even in Tibet and Inner Mongolia.

[100]as an anti-modernist rebellion, not, as it could be expected, a contraposition between tyranny and liberals.

[101]https://www.wrmea.org/1995-june/tajikistan-after-the-elections-post-soviet-dictatorship.html, accessed 22.3.18; De Danieli Filippo, 2013:144.

[102]Silvestri Tommaso, 2015–2016:115.

[103]Nichol Jim, 2010.

aggressions, even when occasional, characterize accounts of recent years. In early 2000, a group of self-declared "1000" irredentists (mutualizing a patriotic symbolism) sparked an insurrection in Tajik side of Fergana valley and attempted a kind of "coupe d'état" (or possibly a mutiny). Their claims represented typical transition and pacification problems, e.g. the return and re-integration of army officials and troops to civilian life, the return of refugees, the re-establishment of the state of the law (in fact implicitly signifying a kind of normalization, dismantling the war economics).[104]

9.6.8 Structural Elements of Instability—Conflicts and Tensions Periodically Erupting

After the peace treaty, tensions did not completely disappear but periodically assumed unexpected semblances, combining with other impetuses (foreign interests, interferences, terrorism, trafficking); then they re-emerged periodically as a true proxy war, that tends to manifest in "weak areas" which lack an effective power to control the territory. But incidents, mutinies, protestations, as well as terrorist attacks which are difficult to classify, did not impede the continuation of the pacification processes.

In 2010 some officials of the dismantled opposition army (active during the civil war) escaped from prison and carried out some attacks in the Rasht valley (as said, a sensitive area and the anti-governmental headquarters during the war, also due to the proximity of the border with Afghanistan) killing about 30 soldiers; again such limited scale attacks sprung up in 2012.[105] Further fights have been registered, with ambushes on the national army in recent years. Recently, high-ranking officials have deserted and defected to the "Islamic state", signifying local tensions combining with the role of ISIS-like organizations. So too, for other case of extremists, criminals and smugglers, in diverse cases of trafficking on the road to Afghanistan.[106]

A negative sign is Russia sending new troops into Tajikistan in 2015, most likely meaning a change in attitude of the Russians towards this country, usually considered a "near abroad" country, namely an exclusive area of influence (even when under the SCO geostrategic umbrella, see later).

Another negative sign is the persistence of impoverishment that seems to highlight a structural issue as well as a form of imbalance in society, between different regions, the western and eastern parts, as well as clan and cultural/regional rivalries—Kulab in the south, Khujand in Fergana, and so forth.

Many of these poor migrants represent a mass group which is easy to manipulate. Even now remittances from migrated workers (mainly in Russia, since they did not need visa for CIS countries) represent the primary economic source, estimated for about half of Tajik GPP, making the country the world's most dependent on remittance by emigrants.[107]

Assistance from international programmes remains crucial to ensuring the continuation of pacification process, as it is necessary to support rehabilitation and reintegration programmes of former civil war figures into civilian roles. This is especially relevant in the remote valleys, for minorities considered as split from the centralistic role of Dushanbe (Fig. 9.10).

So, after several years, it is note-worthy that places such as the Rasht valley, as well other previously war-torn areas, have been placed into Baedeker tourist books as amenities and places to sightsee. They are indeed a true paradise for alpinists and nature friendly visitors, who have never had the chance to visit Pamir and Tien Shan mountains (since they had been closed since Soviet times, and then because of war). The next step is the most important for moving forward with stabilization policies. It consists in programmes for supporting the remote populations even after pacification and with projects regarding home and infrastructure reconstruction, natural hazard preventions, irrigation and agricultural improvements, namely tangible actions, destined for the improvement of living conditions and to overcome the perception of marginalization.[108]

A probable issue might be represented by the recent ban (2015, see later) of the IRP, who signed the Tajikistan Peace Agreement of 1997; such government decisions could result in some questions for the future of the pacification process.[109]

9.6.9 Dushanbe Global City

Pacification has had spectacular effects, with major testimony in its urban reconstruction with the building industry booming, and in general economic recovery. Rebuilding is usually the first step of post-war recovery considering the obvious necessity of restoring the destroyed structures as well as the need for jobs and the "abundance" of a non-qualified workforce (as the former fighters usually are, after laying down

[104] Khayrullo Fayz, 24 July 2012.

[105] https://www.rferl.org/a/Tajikistan_Says_Kills_Three_Suspected_Islamist_Militants/2193377.html, accessed 22.3.18; http://www.bbc.com/news/world-asia-pacific-11565443, accessed 22.3.18

[106] Deutsche Welle 2019; these events demonstrate that conflict can reproduce itself indefinitely, remodifying and transforming, becoming anything suitable to being instrumentalized for any purposes.

[107] https://www.economist.com/news/asia/21584999-russia-attempts-draw-tajikistan-and-kyrgyzstan-back-its-orbit-remittance-man, accessed 22.3.18.

[108] Brill Olcott Martha, 2012.

[109] Deutsche Welle 2019.

Fig. 9.10 Tajikistan, Rasht valley, August 2017, headquarter of ONG working in the valley. (Photo: Igor Jelen)

Fig. 9.11 Tajikistan, Dushanbe, August 2017, government building of the 1940s, representing Athenian classic style. (Photo: Igor Jelen)

their weapons, looking for jobs in the new circumstances); it is a crucial phase in the pacification period when it is also necessary to represent peace as "convenient".

Today, the inviting capital city is the centre of a country growing in all senses: this pleasant place at 800 m asl., had already been completely reconstructed after the earthquake of the 1940s, based on the typical socialist plan. At that time, it must have represented an "ideal town", and possibly attractive for settlers arriving from northern parts of the SU, resulting in an intriguing mix of European and Asian.

It was a typically planned town with wide spaces and gentrified views, with Soviet landscape engineering of the 1940s imitating the classic Athenian style as an appearance of democracy (evidencing a sense for the paradox, for manipulation purposes); it is decorated with parks and fountains, and shadowed by lines of poplars and plane trees, mitigating temperature in summer hot months, developing on a grid of "boulevards", an orthogonal rational urbanistic design (Fig. 9.11).

Dushanbe "survived" the civil war, demonstrating an ability to rapidly overcome tensions after the end of the conflict, possibly deriving from its characteristic as a multi-cultural post-Soviet capital. Its challenge today is that of representing the different parts of the country, building new confidence, unifying the clans from the south and from the north as well as the Fergana cultural area, the southern cotton-based plains and the mountains in a single political procedure. In general, unifying the centre with the peripheries, the small provincial towns and the vivacious global city and mediating between the country's diverse identities and the international world (Dushanbe is rapidly re-discovering its ancient vocation).

In this new context the cultural resource of this city may represent an authentic advantage: it soon became a site of a vivacious community, in which ethno and religious minorities, either local or immigrated, may exert a privileged role, if for no other reason than to experience and practice multinational culture and linguistic knowledge which can be considered as a source of opportunities.

Economic development has been accelerated by further steps in pacification; at the moment it is rather building-industry driven, thanks to a massive programme of public, and to lesser extent private investments; this consists in the construction of the impressive Dushanbe government district and other public buildings, a typical operation for a country that wishes to consolidate itself politically.

So, similarly for the programme of public infrastructures like dams, highways and mountains roads, tunnels, ID networks restoration, improvement for transport companies, wealth, education and welfare facilities; this economic push should induce, as is the intention of the government, an additional step to be able to achieve a kind of self-sustainable development.

Possibly such growth, even when artificially induced, will attract many out-migrants back home, as well as tourists and visitors, amenity or "nostalgia" migrants (many from Russia and further CIS countries) visiting what was once upon a time, the southern exotic side of the Soviet empire.

References

Anceschi L (2017) Turkmenistan and the virtual politics of Eurasian energy: the case of the TAPI pipeline project. Central Asian Survey 36(4):409–429. https://doi.org/10.1080/02634937.2017.1391747

Asian Development Bank (2016) Tajikistan promoting export diversification and growth, country diagnostic study, Metro Manila

Babajanian B (2015) Promoting empowerment? The World Bank's village investment project in Kyrgyzstan. Central Asian Survey 34(4):499–515. https://doi.org/10.1080/02634937.2015.1095967

Bacon EE (1990) Central Asians under Russian rule. Cornell University Press, Ithaca. Or Ed 1966

Barisitz S (2017) Central Asia and the silk road, economic rise and decline over several millennia. Springer, Cham

Barthold W (1968) Turkestan down to the Mongol invasion. In: E.J.W. Gibb memorial. Luzac, London

Bennigsen A (1989) Islam in retrospect. Central Asian Survey 8(1):89–109

Borzatta P (2015) Ultima chiamata per l'Europa, 02 aprile 2015, https://it.wikipedia.org/wiki/Nuova_via_della_seta, accessed 5 July 2018

Botoeva G (2015) The monetization of social celebrations in rural Kyrgyzstan: on the uses of hashish money. Central Asian Survey 34(4):531–548. https://doi.org/10.1080/02634937.2015.1092742

Brill Olcott Martha (2012) Tajikistan's Difficult Development Path, Brookings Institution Press, 27 nov 2012

Bussagli M (1970) Asia Centrale e mondo dei nomadi. In: Storia universale dei popoli e delle civiltà. Asia Centrale e Giappone. UTET, Torino

Collins K (2003) The political role of clans in Central Asia. Comp Polit 35(2):171–190. https://doi.org/10.2307/4150150

Corley F (2005a) Turkmenistan: President's personality cult imposed on religious communities, Forum 18, http://www.forum18.org/archive.php?article_id=522, 1 Mar 2005, accessed 11 April 2018

Corley F (2005b) Turkmenistan: 2004, the year of demolished mosques, Forum 18, http://www.forum18.org/archive.php?article_id=481. Accessed 23 April 2018, 4 Jan 2005

D'Amato G (2010 Oct 27) Turkmenistan. Diversifying export routes. Ashgabat's strategy, europarussia.com, http://www.europarussia.com/posts/1748. Accessed 23 April 2018

Dagiev D (2014) Regime transition in Central Asia. Routledge, New York

De Danieli F (2013) Counter-narcotics policies in Tajikistan and their impact on state building. In: Heathershaw J, Herzig E (eds) The transformation of Tajikistan: the sources of statehood. Routledge, London/New York, pp 143–160

Deutsche Welle (2018). http://www.dw.com/en/turkmenistan-leader-wants-to-end-free-power-gas-and-water/a-39152012. Accessed 23 April 2018

Deutsche Welle (2019). https://www.dw.com/en/tajikistan-17-killed-in-border-outpost-attack/a-51129060. Accessed Dec 2019

Edgar Adrienne Lynn (2004) Tribal Nation: The Making of Soviet Turkmenistan, Princeton: Princeton University Press

Elebayeva AB (1992) The Osh incident: problems for research. Post-Sov Geogr 33(2):78–88

Esenova S (1998) 'Tribalism' and identity in contemporary circumstances: the case of Kazakhstan. Cent Asian Surv 17(3):443–462

Fazendeiro BT (2015) Uzbekistan's 'spirit' of self-reliance and the logic of appropriateness: TAPOich and interaction with Russia. Central Asian Survey 34(4):484–498. https://doi.org/10.1080/02634937.2015.1114780

Freedom House index of political freedom. www.freedomhouse.org. Accessed 23 April 2018

Ferrando Olivier (2013) Soviet population transfer and interethnic relations in Tajikistan: assessing the concept of ethnicity, in: Heathershaw John, Herzig Edmund, 2013, pp. 35–48

Gaynor Kelly Lee (2017) Transformations in Turkmen higher education: current opportunities and challenges at a new university, Central Asian Survey 36(4): 473–492. https://doi.org/10.1080/02634937.2017.1367645

Ginzburg NS (1986) A microgeography of settlement in the Pamir highlands. Sov Geogr 27(6):398–435

Hambly G. (a cura di) (1970) Asia Centrale. Storia Universale Feltrinelli, Milan

Harun Y (2013) The soviet construction of Kazakh batyrs. In: Sevket A, Richard C (eds) Social and cultural change in Central Asia. The soviet legacy. Routledge, London and New York, pp 45–62

Heathershaw J, Herzig E (2013) The transformation of Tajikistan: the sources of statehood. Routledge, London and New York

Hryniuk O (2018 Jan 15) The enduring saga of Tajikistan's TALCO dispute, http://www.cisarbitration.com/2018/01/15/the-enduring-saga-of-tajikistans-talco-dispute/, accessed 23 April 2018 http://www.imf.org/en/Publications/WEO/Issues/2018/01/11/world-economic-outlook-update-january-2018, accessed 23 April 2018, cited by Wikipedia

https://en.tengrinews.kz/politics_sub/Nurly-Zhol-infrastructure-project-to-strengthen-EEU-258109/, 25 Dec 2014, accessed 22 April 2018

https://www.cia.gov/library/publications/the-world-factbook/geos/kz.html, accessed 22 April 2018

https://www.economist.com/news/asia/21584999-russia-attempts-draw-tajikistan-and-kyrgyzstan-back-its-orbit-remittance-man, accessed 22 Mar 2018

https://www.rferl.org/a/Tajikistan_Says_Kills_Three_Suspected_Islamist_Militants/2193377.html, accessed 22 Mar 2018; http://www.bbc.com/news/world-asia-pacific-11565443, accessed 22 Mar 2018

https://www.wrmea.org/1995-june/tajikistan-after-the-elections-post-soviet-dictatorship.html, accessed 22 Mar 2018

International Business Publication (2015) Uzbekistan country study guide, vol 1. Strategic Information and Developments, Washington DC

Jelen Igor (2002) Tra i kirghizi del Pamir Alaj, Forum, Udine

Jelen Igor (2011) Forming new national identities: a retrospective. Natl Ident 13(4):379–388

Jelen Igor (2012) Appunti di geografia politica ed economica. Aracne, Roma

Jonson L (2006) Tajikistan in the new Central Asia: geopolitics, great power rivalry and radical Islam. I.B.Tauris, London

Kassymbekova B (2011) Humans as territory: forced resettlement and the making of Soviet Tajikistan, 1920–1938. Central Asian Survey 30(3–4):349–370. https://doi.org/10.1080/02634937.2011.607916

Kenzhebekovna Kalshabaeva, Akbota Senbayevna Seisenbayeva (2013) Some problems of repatriation and adaptation of representatives of the Kazakh Diaspora of Central Asia in the Historic Homeland Bibiziya. Middle-East J Sci Res 15(1):20–26. https://doi.org/10.5829/idosi.mejsr.2013.15.1.11054

Khayrullo F (2012 July 24) Tajikistan clashes: 'Many dead' in Gorno-Badakhshan, http://www.bbc.com/news/world-asia-18965366, accessed 10 April 2018

Korostelina K (2011) Shaping unpredictable past: national identity and history education in Ukraine. Natl Ident 13(1):1–16

Krasnov GA (1987) Il commercio tra paesi a diverso sistema: tendenze e prospettive, pp. 63–79, in Parboni R. e Wallerstein I., a cura di, 1987, L'Europa e l'economia politica del sistema-mondo, F. Angeli, Milano

Kudaibergenova DT (2013) National identity formation in post-Soviet Central Asia: the Soviet legacy, primordialism, and patterns of ideological development since 1991. In: Akyildiz Sevket, Carlson Richard, edited by, 2013, Social and cultural change in Central Asia. The soviet legacy, Routledge, London and New York, pp 188–200

Kudaibergenova DT (2015) The ideology of development and legitimation: beyond 'Kazakhstan 2030'. Central Asian Survey 34(4):440–455. https://doi.org/10.1080/02634937.2015.1115275

Lamy Frederick (2013) Social networking practices: Continuity or rupture with the Soviet past? In: Akyildiz Sevket, Carlson Richard, edited by, 2013, pp. 145–159

Lanzetti E (2017) La nuova «Via della Seta». Dall'Italia alla Cina in treno partendo da Mortara, "Corriere della Sera", 25.11.17

Laruelle M, Peyrouse S (2013) Globalizing Central Asia: geopolitics and the challenges of development. M. E. Sharpe, Armonk

Limes (n.8/2014) Cina, Russia, Germania unite da Obama

Lonely Planet (2014) Asia Centrale, EDT Torino

Mashrab F (2015) Bishkek Puts Brakes on China–Kyrgyzstan–Uzbekistan Railway, Eurasia Daily Monitor 12(199) November 3, 2015, https://jamestown.org/program/bishkek-puts-brakes-on-china-kyrgyzstan-uzbekistan-railway/, accessed 22 April 2018

McMahon Robert (2005) Uzbekistan: Report Cites Evidence of Government 'Massacre' In Andijon, Radio Free Europe/Radio Liberty, 7 June 2005, https://www.rferl.org/a/1059147.html, accessed 5.7.2018

Mclellan D (2017 July 27) 10 Facts about Uzbek Refugees, https://borgenproject.org/10-facts-uzbek-refugees/, accessed 22 April 2018

Megoran Nick (2017) Nationalism in Central Asia: A Biography of the Uzbekistan-Kyrgyzstan, Boundary, Central Eurasia in Context, 2017, University of Pittsburgh Press. https://doi.org/10.2307/j.ctt1vjqrk6

Mostowlansky T (2014) The road not taken: enabling and limiting mobility in the Eastern Pamirs. Internationales Asienforum 45(1–2):153–170. ABI, Arnold-Bergstraesser-Institut

Necati Polat (2002) Boundary issues in Central Asia. Transnational Publishers, Ardsley

Nichol J (2010) Central Asia's security: issues and implications for U.S. interests. Congressional Research Service, Washington DC

Özcan GB (2015) Introduction: market adaptations, interventions and daily experience. Central Asian Survey 34(4):409–417. https://doi.org/10.1080/02634937.2015.1103580

Page J (2014) China to contribute $40 billion to Silk Road Fund new trade-development push gathers momentum, in http://www.wsj.com/articles/china-to-contribute-40-billion-to-silk-road-fund-1415454995, 8 Nov 2014, accessed 9 May 2018

Pan Philip P (2010 June 18) U.N. doubles estimate of Uzbek refugees as crisis grows in Kyrgyzstan, http://www.washingtonpost.com/wp-dyn/content/article/2010/06/17/AR2010061705954_2.html?sid=ST2010061802267, accessed 22 April 2018

Panaino A, Gariboldi A, Ognibene P (2013) Yaghnobi Studies I. Papers from the Italian Missions in Tajikistan, Indo-Iranica et Orientalia, Series Lazur, Milano, Edizioni Mimesis

Parenti FM, Adda I (2017) Are we going to live in a post-NATO world? A critical perspective on trends, obstacles and possibilities. GeoJournal 82:345–354. https://doi.org/10.1007/s10708-015-9693-8

Paskaleva E (2015) Ideology in brick and tile: Timurid architecture of the 21st century. Central Asian Survey 34(4):418–439. https://doi.org/10.1080/02634937.2015.1118207

Pianciola N (2001) The collectivization famine in Kazakhstan, 1931–1933. Harv Ukr Stud 25(3–4):237–251

Pianciola N (2004) Famine in the Steppe. The collectivization of agriculture and the Kazakh herdsmen, 1928–1934. Cahiers du monde russe 45:137–192

PWC (2010) Doing business in Turkmenistan 2012–2013, https://www.pwc.com/uz/en/assets/pdf/doing-business-guide-in-turkmenistan-2012-2013.pdf, accessed 20 May 2018

PWC (2011) Doing business guide Tajikistan 2012–2013, http://www.eurasianbusinessbriefing.com/wp-content/uploads/2015/04/pwc-doing-Business-in-tajikistan.pdf, accessed 20 May 2018

PWC (2016) Guide to do business and investing in Uzbekistan 2016 edition, https://www.pwc.com/uz/en/assets/pdf/dbg_2016.pdf, accessed 20.5.2018

Rashid A (2002) Jihad: the rise of militant Islam in Central Asia. Yale University Press, London

Reid P (2017) 'Tajikistan's Turksib': infrastructure and improvisation in economic growth of the Vakhsh River valley. Central Asian Survey 36(1):19–36. https://doi.org/10.1080/02634937.2016.1204533

Rotar Igor (2006) Resurgence of Islamic Radicalism in Tajikistan's Ferghana Valley, the Jamestown Foundation, Terrorism Focus, 3/15, April 20, 2006, https://jamestown.org/program/resurgence-of-islamic-islamic-radicalism-in-tajikistans-ferghana-valley/. Accessed 22 Apr

Ro'i Y (1991) Central Asia riots and disturbances, 1989–1990: causes and context. Cent Asian Surv (3):21–54

Rossi Marina (1997) I prigionieri dello zar, Ugo Mursia Editore, Milano

Robinson F (1989) Atlante del Mondo Islamico. De Agostini, Novara

Rotar I (2006) Resurgence of Islamic Radicalism in Tajikistan's Ferghana Valley, the Jamestown Foundation, Terrorism Focus, 3/15, April 20, 2006, https://jamestown.org/program/resurgence-of-islamic-radicalism-in-tajikistans-ferghana-valley/, accessed 22 April 2018

Rumer BZ (1989) Soviet Central Asia. A tragic experiment. Unwin Hyman, Boston

Sevket Akyildiz, Richard Carlson (2013) Social and cultural change in Central Asia: The Soviet Legacy, Routledge.

Shabad T (1980) The 1979 Census and some demographic trends in the Soviet Union. GeoJournal 4(1):84–90

Shahrani MNM (1979) The Kirghiz and Wakhi of Afghanistan: adaptation to closed frontiers. University of Washington Press, Seattle

Silvestri T (2015–2016) Le strategie della federazione russa nel teatro del mar Caspio dopo la caduta dell'Unione Sovietica, Master Thesis, University of Trieste, academic year 2015–2016

Sodiqov Alexander (2013) From resettlement to conflict: development-induced involuntary displacement and violence in Tajikistan, in: Heathershaw John, Herzig Edmund, 2013, pp. 49–66

Straub David (2013) Deconstructing communal violence during the civil war in Tajikistan: the case of the Pamiris, in: Akyildiz Sevket, Carlson Richard, edited by, 2013, pp. 174–187

The Gazzette of Central Asia (2012) Tajik aluminium output falls 11.6%, http://gca.satrapia.com/+tajik-aluminium-output-falls-116+, accessed 6 July 2018

The Gazzette of Central Asia (2013 Jan 23) Unified pension fund recommended in Kazakhstan, http://gca.satrapia.com/+unified-pension-fund-recommended-in-kazakhstan+, accessed 22 April 2018

The Strategy for development of the Republic of Kazakhstan until the year 2030, Official Site of the President of the Republic of Kazakhstan, http://www.akorda.kz/en/official_documents/strategies_and_programs, accessed 22 April 2018

Marzhan Thomas (2015) Social, environmental and economic sustainability of Kazakhstan: a long-term perspective. Central Asian Survey 34(4):456–483. https://doi.org/10.1080/02634937.2015.1119552

United Nations Environment Programme (2008) Annual Report, UNEP Division for Communications and Public Information, Nairobi

Uzbekistan (1984) Izdatelstvo planeta, Moskva

Warf B (2011) Geographies of global internet censorship. GeoJournal 76:1–23. https://doi.org/10.1007/s10708-010-9393-3

Warf B (2016) Global geographies of corruption. GeoJournal 81:657–669. https://doi.org/10.1007/s10708-015-9656-0

Wikipedia, Kazakhstan, https://en.wikipedia.org/wiki/Kazakhstan, accessed 23 April 2018

Wikipedia, Uzbekistan, accessed 29 Mar 2018, citing the Economist Intelligence Unit, in: "Uzbekistan: Economic Overview", eurasiacenter.org (page not found)

World Report (2015) Uzbekistan, events of 2014, https://www.hrw.org/world-report/2015/country-chapters/uzbekistan, accessed 23 April 2018

Zhumabayeva K (2016) Kazakhstan's caspian coastline offers various vacation options, Kazakhstan tourism, 10 July 2016, https://astanatimes.com/2016/07/kazakhstans-caspian-coastline-offers-various-vacation-options/, accessed 22 April 2018

10 From Culture to Material Aspects

Abstract

CA reality maintains a characteristic of high mobility in all senses, from a human and material point of view as well as from a cultural point of view: values, communication modes, habits, expectations, cults and religious rituals represent the basic elements of a society, but they also represent the more difficult variables that must be studied, possibly because they assume, by definition, an immaterial dimension. The basic characteristic of the current cultural pattern is characterized by two opposite tendencies, one of opening up (mainly induced by information and communications technology [ICT], international commerce and globalization) and the other is completely the opposite, of politically locking-in, prefiguring a kind of neo-national ideology—but not without its contradictions.

Keywords

Central Asian cultural traditions · Identity markers · Cultural soviet legacy · Religion · Linguistic and alphabetic changes · Ethno-linguistic map · Social changes · Popular culture

10.1 Culture as a Common and Unifying Element

10.1.1 Culture: Material and Immaterial Elements

The CA countries may be represented as a fundamental unit, even if their history has diverged in some cases and sometimes in quite a dramatic way. This occurred particularly in critical periods, as per invasions, wars and revolutions, when the cultural geography of the area underwent processes of fragmentation (as in other circumstances in history, the reversal of reunification), with the formation of new units, and disaggregation of the older ones.

Sometimes, geopolitical units develop indifferently and transversally to ethno-cultural units; this happened in the past with the hordes integrating into nomadic empires, with bazaars connecting long-range caravan routes, or merchant communities forming city-states. This happened in times of post-Soviet transitions well, when the individual countries gave the impression of the desire to start out on an individual path,[1] but not without contradictions.

In fact, the NIS overlapped the once federative republics (Soviet Socialist Republic [SSR]) in which the national code appeared preordained due to its geopolitical convenience: the current form of the nation-state (that came about very "late" in comparison to other experiences) cannot but result as anything but "reconstructed", if not fictitious. The new 1991 border—as any new border—separated new national units in a highly imperfect way. Therefore, it was (and still is) to be expected that innumerable trans-national elements would still keep the populations of these countries periodically together (beyond the national–political intentions).

It is the case of religious and cultural roots, of the presence of minorities and languages "on the other side" of the new border, that the "new" national borders cannot divide, but in a violent way. It is evidently the case of material–infrastructural issues, of ecological processes and energy supply chains, that occur on a typical trans-frontier scale, requiring reciprocity and dialogue, in general, on geographical–political questions and also on functional and cultural questions.

Simply, the "stans" have too much in common to be able to undertake such a divergent path. They are purely united by the same geography, with distances that result as being too close, borders that intersect each other and territorial configurations that are continuously being confused. Beyond their intentions, these neighbouring societies cannot but rep-

[1] Akyildiz Sevket, Carlson Richard, 2013; Urbanc M, Boesch M, Jelen I. 2008.

I. Jelen et al., *The Geography of Central Asia*, World Regional Geography Book Series, https://doi.org/10.1007/978-3-030-61266-5_10

resent a kind of "mirror" for each other in which they observe and eventually imitate each other, recovering a fundamental perception of unity.

10.1.2 Internal Re-nationalization Tendencies

After independence, as soon as things became clear enough, a general tendency towards a political re-nationalization started. It was carried out almost instinctively by the new governments in order to increase internal national compactness, as well as their representation towards the outside. This tendency soon appeared to be problematic, considering both the fragmentation of the CA space and its basic multi-cultural foundations.

As it is possible to observe nowadays, there is still generic pressure for national uniformity, sometimes even regarding changes in culture and ethnic composition—for example, obtained through assimilation, induced migrations or cultural standardization policies; this either privileging the titular nation or re-nationalizing other populations, some belonging to the class of Soviet "orphans" (see later), without any clear identity.

Official statistics evidence the results of such politics. For instance, in Kyrgyzstan, the ethnic Kyrgyz increased from around 50% in 1979 to over 70% in 2013, while other groups decreased, presumably because of migrations, assimilation, mixed marriages or similar circumstances.[2] Russians, Ukrainians, Germans and Tatars decreased from 35% to about 7%—much of them having left for their country of origin, essentially Russians and Germans.

Similarly in Kazakhstan, where after having reached maximum numbers during the "Virgin Lands" campaign, ethnic Russians started to decrease in numbers, as well as other non-Kazakh populations, possibly those descending from immigrated and deported people in the different phases of Soviet history.[3] Such tendencies, known as "vozvraščentsy" migration, had already started in the 1970s, in an almost imperceptible manner.[4] In the period of transition, this reduction accelerated and Russians decreased from 38% in 1989, within 7 of the 20 regions of the Kazakh Republic, to the current 23%, usually settled and concentrated in the north and eastern parts, in the Zhetysu and Almaty urban areas.

In other CA countries such trends occurred at an even higher scale, such that Russians and Russianized people today seem to be in a small minority. This situation can be observed in Tajikistan, Turkmenistan and Uzbekistan, where numbers were determined respectively, and to different extents, by out- or in-migrations, centralistic pressure or economic contingencies, or safety issues (especially in Tajikistan because of war, even when Russians were not usually the target of hostile attitudes).

In these circumstances culture and cultural pressure have often been used as a political instrument; it is evident that the neo-nationalistic one is today the most widespread ideology, even if compensated by (opposite) globalization tendencies to some extent, assuming a political significance. In fact, the outside actors (not just political but also economic and cultural actors) soon started to interfere in CA, being perceived in transition times as a vacuum to be filled with a soft power, namely with any kind of influence (commercial, cultural, religious etc.).[5]

10.1.3 Cultural Change Connected with Post-Soviet Transformation

Considering the long-lasting Soviet rule and the depth of its penetration into the local society, it is to be expected that its culture affected the dynamics of the inter-generational transmission of behavioural patterns and beliefs continuing to impact societies (delineating possible incongruences in the new structural context).

Nevertheless, it is difficult to estimate such impacts because, as said, they are obviously immaterial and difficult to measure or identify.

Furthermore, manipulation effects between the different dimensions of Soviet modernization must be taken into consideration, in which the assimilation of technology and of other cultural elements has been imposed in a context of high pressure, sometimes even with violence. This is with regard to functional aspects (e.g. expropriation, impositions, new laws), but also to culture (identity processes, ethnic and linguistic acknowledgement, even basic alphabetization). Therefore, and for this reason, at the moment of independence, a reaction by the side of the destroyed (repressed) identities was to be expected, or at least the persistence of a diffused diffidence towards the new authorities.

Such sentiment manifested in different ways. This was the case of economic regression and of the return to rural areas; it was the case of the revival of traditional powers,

[2] http://www.irinnews.org/feature/2005/04/19/focus-post-akayev-russian-exodus (accessed 22 March 2018) Spoorenberg Thomas, 2015; Spoorenberg Thomas, 2013.

[3] Klüter Helmut, 1993.

[4] Klüter Helmut, 1993; Ro'i Yaacov, 1991.

[5] Such situations often happened in CA, in the past as today; local cultures seldom had the power to maintain their originality—with the exceptions of the remote populations or of vanishing nomadic groups; they often had to face outside influences that manifested either with wars and invasions or with subtle cultural manoeuvres, like religious interpretations, literature and alphabet changes, and further—today ICT inspired—communication code diffusion; it can be said that even today the surrounding powers look at CA as an expansion area either for culture, soft-power or economics.

even when to some extent fictitious, of the "resurgence" of religion (that, as usual in such circumstances happens in its most conservative variation, is often radicalizing), of "klanovost" (clan identity revival) or other non-modern codes.

All these processes impacted the already weakened society that was put under pressure by new autocratic powers. The same for "new" or "renewed" values, spread through global dissemination, for social rituals and consumerist habits, considering the capability of new technologies, inclining towards democratization or, in other circumstances, to passive consumerism (indeed prefiguring the risk for renewed alienation).

This happened on different scales, from levels of high politics (institutions, ideologies) to intimate individual dimensions, and in "popular culture", namely daily practices (food or clothing habits, micro-social rituals, family and vicinity-neighbourhood experience) that have a decisive meaning for human systems. In fact, they may appear connected with stereotypes, perceived in a superficial way, imposing an "extraneous" scenario.[6]

The question concerning the "excess of democratization" is a sensitive one and exposure to global tendencies, that consequently means the disappearance of usual sets of references, may trigger counter-reactions (as has possibly happened recently with Islamist radicalization, coming from populist neo-nationalism or even neo-communist attitudes).

10.1.4 Much More than an Identity Marker

The formation of the new states meant the affirmation of new identity patterns, of renewed perceptions of loyalty and of social organization. These are processes that do not occur in a linear manner, remaining politically ineffective.[7]

This happened for all cultural variables and especially for the most relevant ones—religion, traditional and tribal values, as well as additional social organization modes; for instance, in religion they tend to articulate spontaneously in local cults, sects, brotherhoods, faith and ritual variants, with the consequent fragmentation of the clerical organization, with religious hierarchies, ulama and mullah sometimes in competition with one another, and continuously dealing with a typically CA "popular" version of Islam.[8]

[6]It would signify the justification of a new standard represented by multinational, by social- and mass-media, controlled by the "big business", possibly "bringing into all homes the American culture"; Warf Barney, 2011.

[7]The maturation of such identity "markers"—such as nationality, class or civil society, intended as the maturation of the consciousness of the common interest —usually takes time and in many circumstances has proved to be ephemeral.

[8]Muminov Ashirbek, 2014; Hanks Reuel R., 2016.

It must be considered that these passages occur in a multicultural environment (that had already formed in colonial and Soviet times) that can make such changes simply appear as anachronistic.

There is little space for reconstruction manoeuvres as is the case for the attempt to manipulate religion in a neonationalistic sense in order to get a wider popular consensus; the same with the corresponding new version of consumerism, visible in the major urban landscapes (characterized by media, ubiquitous global brands, advertisement sounds, images, music and "jingles"), with their many signs of modernism (e.g. the spread of communication electronic devices, of sport and national popular entertainment events).

Having said this, the path to achieving a self-conscious civil society, which is to some degree autonomous from the power structure, appears still to be long and arduous. In such post-totalitarian (and neo-authoritarian) environment, the most relevant sentiment is still that of a fundamental fear (eventually dissimulated) towards politics (whatever it may be), with power structures considered popularly as something arbitrary, from which it is necessary to defend oneself.

10.2 Religion and Religious Variants

10.2.1 Original Islam Spread in a Diversified Way

Whatever one may assume as the original definition,[9] religion is the basic element of cultures, modelling entire civilization cycles; it has demonstrated to be very powerful as it is connected with education and cultural transmission modes and with intimate elements of identity, usually assimilated in childhood, inspiring societies and political systems.

Some aspects of religion may consolidate in institutions, forms and procedures (cult, rituals); sometimes they occur in a standard representation for all populations, which however maintain a tendency for the spontaneous elaboration of cults and liturgical forms (besides the institutional prescriptions).

In CA it is possible to recognize different strata, namely different waves of religious standardization, sometimes coinciding with some political events, such as invasions, colonizations and revolutions or any other kind of material or cultural innovation. The original Islamization in CA had different intensities and happened at different times. It can be traced to early Arab conquests, in some other cases it occurred much later and even now—some authors say—the

[9]Religion as an independent elaboration of the human being or as a socially constructed one, possibly induced by the experience humans have at any moment of their life with the dimension of "impossibility"; some authors, like Appadurai, define religion as the "format" capable of monopolizing (or expressing) the imaginative energy of individuals, therefore, essentially, a political question.

process of assimilation of such a monotheistic religion cannot be considered complete.

From that time, Sunni Islam has been the predominant religion in the area. After the Arab victory in the eighth century, it spread to the southern urbanized side, along the line of oasis cities, then slowly spread out towards peripheral territories, where pastoral populations conserved practices and traditions of pre-Islamic times until recent centuries.[10]

It seems that in these areas the standard religion was assimilated in a rather superficial way, leaving space for a more differentiated set of religious variants.[11]

In particular, Islam spread more intensively on occasion of the spread of imperial power, when such power had the chance to consolidate itself, structuring political and military hierarchies over extended territories, for longer periods. This happened especially after the Arab conquest and dominion, at the time of the Samanid (862–999) and Timurid dynasties (fourteenth and fifteenth centuries), as well as with the Golden Orda.

Then it furthermore consolidated—to some extent in a paradoxical manner—in times of early Russian colonial expansion in CA, when czarina Caterina sent Tatar missionaries to the steppe. This was possibly to contrast tribal particularism, in those times expressed with pre-Islamic cults (shamanist or similar), spread by groups scattered over wide surfaces, refractory to unitary rule. The Tatars (at the same time teachers and merchants, representatives of empire culture, but capable of communicating with the steppe populations with whom they shared the same Turkic origin) intended to organize these dispersed shamanist-prone hordes in order to make it easier to integrate them into the expanding empire.

Then, later, in the nineteenth century, with the acceleration of colonial processes, the situation changed again; increased outside pressure induced a further wave of Islamization but in this case with reverse motivations, namely as a reaction against Christian colonialists by the natives. The reference to religion essentially became a way to maintain an identity (an ideological proxy), in order to render the groups more internally compact. In the CA case it often happened in times of colonial invasion.

In general, the recovery of the religious code became an identitarian instrument to oppose any kind of invasion and change, or modernization in any form, sometimes reflecting the impossibility of adapting to new situations. It is important to consider that such waves, even when going deep into the past, still have consequences in daily practices today. The religious culture of CA populations appears stratified in many circumstances, with elements of different origins continuously mixing and combining.

10.2.2 Religious Specificities in CA Society

Currently, CA Islam, and religion in general, is characterized by different phenomena: on the one hand, there is spread of a spontaneous evolution, namely by a popular expression of spirituality and on the other the recovery of the official religious institutions. Both such effects are to be considered as long-term consequences of Soviet atheism and religious repression that failed in the struggle against official religion (probably as it made the same fundamental mistake as many western cultures, assuming Islam to be monolithic, addressing its repression mostly against the institutional side of the local religion).

On the contrary, even in Soviet times, "parallel" Islam was widely practised in a hidden or dissimulated manner, organizing pilgrimages to holy places, mausoleums and graves (the most important of them was the one in Osh, at the Suleiman mountain), as well as praying (*dhikr*) and devotion rituals; they sometimes comprehend Sufi inspired or further traditional, shamanic or clan-tribal variants. There are many beliefs, local variants and personal interpretations inspiring and mixing such elements, as well as pre-Islamic cult reminiscences (even if not recognized, since they would delegitimize Islam as the exclusive religion of CA to some extent).[12]

Beyond this, it is important to say that syncretism in rituals is rather diffused in CA. It is possible to find traces of Buddhist, Zoroastrian, Hindu, Jewish and ancient Christian influences (tracing back to Nestorian beliefs); these presences have been well documented since ancient times and find some cultural continuity into the present day. The same concept, but in different circumstances, happened with other religions following recent immigrations such as in the Orthodox church and the Ashkenazi Jewish religion and is possible for other churches established in CA in the last few centuries.

All this occurred in the new context of the post-Soviet states in which they represent new minorities, practising religion outside the state control but enjoying secularized constitutional rights, generally being tolerated.

It is possible to evidence two main tendencies; the most predominant is that of the new nation-state that tries to make the official religion ancillary to established power, fearing

[10] Esenova Saulesh, 1998; Muminov Ashirbek, 2014; Epkenhaus Tim, 2013.

[11] Alberti and Naserdinova, 1993; Bennigsen A.- Lemercier Quelquejay C., a cura di Fasana E., 1990; Radford David, 2014:23–27; see also https://web.archive.org/web/20070702212941/http://asia.msu.edu/centralasia/Kyrgyzstan/religion.html (accessed 23 April 2018); it is difficult to say to what extent the weakening of standard religious practising depends on secularization processes, as happens in any much part of the modernized-globalized world.

[12] Muminov Ashirbek, 2014; Bennigsen A.- Lemercier Quelquejay C., a cura di Fasana E., 1990.

radical degeneration, an approach that proved often to exaggerate such risks, using it to justify a despotic repression.

The second is that of elaborating a normalized form of Islam, coexisting with other recognized faiths in a secularized reality, configuring a commonly accepted ethical standard, assumed as (moderation) a mitigation element (therefore eventually encouraged by governments).[13]

In fact, the current CA landscape barely has any religious symbols, especially in urban areas, consequent to the spread of a secularized iconography that seems to drive the cultural transformation from all points of view. This fact evidences the question of the relationship between the Islamic institutions and post-Soviet society, especially for leading classes, resident in new global cities, exposed to international flows and influence.[14]

In these circumstances it is possible to verify that religion is more adaptable than expected, "de facto" assuming the typically discrete and moderate character religion usually assumes in advanced contexts (indeed as an expression of free choice, not as an imposition). This has occurred even when the Islamic tradition evidences strong activism, possibly due to the presence of centres of ancient prestigious importance (especially Bukhara, Samarqand and the Fergana cities, possibly competing with each other for the moral–religious primacy), where ideas are continuously spreading, either in a reformist or in a radical or conservative sense.

From this point of view, another important element in CA tradition is the one represented by the (already described) Jadid, "new method", a precedent in this matter that can be presented as a failed attempt to adapt the local Islamic tradition to modernist change.[15]

10.2.3 Religious Stratification

As said, the predominant part of the population is Sunni, with the exception of a significant Shi'a minority; further minorities are rather scattered following some ethnic-territorial distribution, or because of being historically rooted in some area. This is the case of the Ismaili Shi'ites in Badakhshan, of the Imamites who immigrated from Iran to Samarqand, and of Orthodox Christians in Russian-speaking areas.[16]

The persistent portion of atheists is possibly the heritage of Soviet times, while the question of incumbent globalization and consequent diffusion of cosmopolite attitudes is yet to be appropriately studied and it contributes to the spread of ecumenical and syncretic, as well as of agnostic, tendencies.

Official statistics are of relative help. This is due to difficulty in classification and the modality of data collection. It must be considered that the self-declaration modality, in matters of religion, can be influenced and then distorted due to the recent past of official atheistic propaganda and religious repression, which may also have been violent. Furthermore, the effect of progressive confusion in all senses must also be taken into account as it usually signifies religious overlapping, for example, in mixed families or in further social mixed contexts.

Globally in CA there are 54% self-declared "non-denominational" Muslims, of them Sunni self-declared 18% and 1% Shi'a ("non-denominational Muslim" being a generic formula used by Muslim believers who do not identify with a specific Islamic denomination); in Uzbekistan, where Islam is more rooted, comprehending, as said, the most prestigious religious centres, 88% of the population self-declare as Muslim, mostly Sunni, with a 5% Shi'a minority.

Colonial times were characterized by the arrival of Christians, Orthodox and other beliefs, depending on immigration (German Protestant, Polish Catholics, Caucasus Christian and Islam variants). In colonization and Sovietization times, the newly arrived eastern European Jews (Ashkenazi) mixed with Sephardic communities that date back to ancient times with their main centre of aggregation in Bukhara; they represented a pillar of local history until their mass migration to Israel in Perestroika times. Similar discourses can be made in the case of other minorities that had migrated or been deported several times in CA, representing special religious traditions (like Koreans, Mongolians and Dungans as well as further populations).

Another difference concerns the basic divide in the CA, highlighting a late Islamization of peripheral populations (Kazakhstan, Kyrgyzstan and Turkmenistan essentially), who maintained local practices, differently from early Islamized populations in the urban areas, whose practices are closer to original systems (therefore possibly exposed to the risk of radicalization today).

In the steppe and mountain areas, far from power centres, signs of previous religions survive as well as residuals of spontaneous cults, like Sufi or tribal shamanism.[17] This is

[13]Considered as a tool in order to recover order, discipline, morally to protect as stated by Kyrgyz former president's daughter Bermet Akayeva in 2007: "Islam is increasingly taking root across the nation [...] many mosques have recently been built and the Kyrgyz are increasingly devoting themselves to Islam, [...] not a bad thing in itself. It keeps our society more moral, cleaner". Eurasianet.org. 17 July 2007, cited by Wikipedia, accessed 23.4.2018; see as well Hanks Reuel R., 2016.

[14]Radford David, 2014; Muminov Ashirbek, 2014.

[15]Adeeb Khalid, 1998.

[16]Muminov Ashirbek, 2014.

[17]Hays Jeffrey, 2008; Minority Rights Group International, 2018.

especially evident in Kyrgyz practices that possibly preserve legacies from their original Siberian places (like visits to holy rocks, waterfalls or fountains, the practices of prayer trees and similar).[18]

So similarly, for Kazakh, Turkmen and Karakalpak cultures, where liturgies often reflect clan loyalties or traditional codes in the tribal dialectics.[19]

The most important autochthonous minority is that of the Pamiri Ismaili faith, that results from an original adaptation of a Shi'a variant, representing an identitarian element in that remote region, mixing ethnic and religious characteristics. The same for other remote areas' minorities (Yagnobi, Garmi and others, depending on classification), where peculiarities survive that represent extraordinary testimonials.[20]

10.2.4 Islamic Rediscovery and "Resurgence"

Transition times are characterized by the rediscovery of neglected identities, also in religious terms. This also occurred when in Soviet times Islamic practice never completely disappeared, at least in a private dimension, if not in public.

In general, the attitude of the Soviets towards religion is to be considered ambiguous. Their official politics considered religious practices as residuals of the medieval period, alongside popular idolatries and superstitions. Sometimes a real ideological hatred towards traditional religion developed commanding the destruction of material signs, of cult places, with mosques and madrasas being closed, often destroyed, as well as cemeteries, sanctuaries, pilgrimage sites, prayer cult places, with believers usually being derided and persecuted and sacred symbols ridiculed.

Especially in post-revolution times, the Soviets tried to eradicate religious behaviour and beliefs that might divert the attention of the population from ideological targets (to some extent similarly exerting a metaphysical character), representing an alternative to absolute obedience to the new power.

In other circumstances, the power structures demonstrated a more pragmatic attitude, seeking a kind of compromise, even trying to manipulate religious institutions. This often took place in periods of crisis when the Soviet rule understood that it would be difficult or not particularly convenient to completely eradicate traditional practices.

This double standard is evident in some aspects of the CA reality; for example, while monuments and high religious expression are approached with respect (e.g. for Samarqand and Bukhara mosques and monuments), the attitude towards daily religious practice is completely diverse. Repression and systemic destruction were applied to religious symbols and places (village mosques, holy itineraries, sanctuaries) massively attended by the population.[21]

It is not surprising that, in post-Soviet times, religion appeared as a powerful element of motivation, signifying a new wave of Islamization, legitimizing itself due to the fact that traditional religion had been repressed (or possibly reoriented to a kind of political metaphysics, covered by official atheism) for generations. The "resurgence" is indeed an expected reaction (as a direct consequence of the post-Soviet liberation). It was consolidated by the sudden contemporary disappearance of other social references, with religion re-emerging as a possible identity marker, substituting the secular institutions.

For all these reasons post-Soviet society evidences a critical question concerning relations between religion (symbols, rituals, cleric's social role etc.) and the new power, with "young" and tendentially secularized governments (namely the old nomenclature, giving continuity to Soviet political culture) starting confrontations with religious institutions, sometimes seeking legitimacy from religious sides.

This confrontation is made more complicated because Islam is not characterized by structured religious clerics and hierarchies (at least Sunni, prevalent in CA), capable of elaborating a uniform attitude. Often the new power searches for a compromise, in order to avoid conflicts, but in some cases, extremism has triggered tensions.

Sometimes it would seem to be a modern "investiture controversy" (as had happened in the past in Europe and elsewhere), with material and spiritual powers looking for balance in their relations (that for current times take on a key significance), seeming indeed far from being stabilized with continuative overlapping. The secular government usually interferes in religious institutions, eventually using their symbols and codes, while the "presidents" often invent themselves as the leaders of the believers.[22] They build mosques, madrasas and maktabs, and apparently facilitate

[18] Muminov Ashirbek, 2014.

[19] Traditionally, these populations prefer to follow the consuetudinary tribal "adat", rather than the sharia rule; Buttino Marco 2003.

[20] Panaino A., Gariboldi A., Ognibene P., 2013.

[21] The concern about Islamic monuments in historical cities already started in colonial times, after the conquest, and it continued into Soviet times, giving continuity to the precious tradition of Russian orientalist studies; just after the revolution, the Soviet authority promulgated a decree concerning the "importance of conservation of monuments and cultural artefacts, considering the scientific importance of them", including the establishment of a specialized school of restoration in Samarqand in 1937" (Paskaleva Elena, 2015:420); indeed such policies combined into a kind of double standard, since at the same time the authorities planned the destruction of many mosques and other religious places all over the Soviet space; Tonini Carla, 2017.

[22] Epkenhaus Tim, 2013; Brill Olcott Martha, 2012:39; mullas and clerics in CA seem to be increasingly under the control of political authorities, for example, they have to do examinations, to be registered and so on.

the spread of religious culture and education, potentially with social-moralization intentions; further, they complete "hajj" (the mandatory pilgrimage to Mecca), charity (zakat), shahada, salah and sawm, carefully giving their example as good believers, respecting Islam's "five pillars".

At the same time, presidents are afraid of the new religious wave, as a possible element triggering the mobilization of peoples, possibly in the forms of uncontrollable street revolutions (as demonstrated by the Iran revolution, 1979, the more recent "Arab Springs" and finally the ISIS—Islamic State of Iraq and Syria—movement), with waves of rebels suddenly meeting and rebelling. These are facts that evidence that it is necessary not to underestimate—as the Soviet rule did—religion as a political element.[23]

Rarely, indeed, as post-Soviet history demonstrated, has the political power structure exaggerated, deliberately manipulating religion symbols. The most evident case is that of Turkmenistan, with the previous Turkmen President Niyazov who imposed the teaching of Islamic principles in schools in the frame of a political cult of personality. He published his own written text, in separate volumes, from 2001 to 2004, named the "Ruhnama" (a title echoing Persian literary tradition), to be given the same status as the Quran, exhibited and mandatorily learned in mosques and madrasas;[24] then he imposed a special formula at the Friday prayer, self-purposing himself as a religious leader.[25]

But mostly new presidents—inheriting a "discrete" attitude from the Soviet tradition—are keen to demonstrate subtle behaviour (applying a kind of diversion method) and do not invent themselves as divinities, or as a caliph (namely as successor of the Prophet, as has happened in other circumstances), remaining a "step outside" religious disputes.[26]

10.2.5 Radicalization of Religious Parties

The political use of religion prefigures the risk of manipulation, namely the application of religious codes to secular practices (an intrinsic contradiction). In the new neo-nationalistic environment, the situation is even more complicated, with some leaders believing that religious hierarchies can represent a challenge for the unity of the new state that is urgently seeking social compactness.[27] So too, for the presence of minorities, representing a motive for possible disaggregation (especially when concentrated in some regions).

This occurred mainly in Tajikistan and Uzbekistan, areas characterized by a millennial tradition of Muslim elites, with prestigious schools and mosques, where the Soviet repression was perceived as even more violent; these same counties are also more exposed to external tensions (especially infiltration and contamination from Afghan "sanctuaries").

The formation of parties and movements presenting a political programme had already occurred in Perestroika times and spread massively in early independence times. These "labels" often assumed diverse consistence, often difficult to interpret accurately (as is usual for any political movement at the beginning of its history), mixing national revival, liberal and religious elements.

This is the case of large popular movements in early transition times, like Birlik in Uzbekistan and Rastokhez (Revival) in Tajikistan. Soon after the war broke out in the early 1990s, movements merged into the United Tajik Opposition (UTO). Consequently, the spaces of mediation tended to be more limited, inducing clerics and militants to radicalize (as usually happens in a context of conflict).

Further cases are those of the Hizb-ut-Tahrir, a local branch of the organization, of a rather sectarian character, founded in the Middle East in the 1950s. Typical religion-based parties are the Islamic Movement of Uzbekistan (IMU) and Islam Renaissance Party (IRP), going down different paths but usually radicalizing.

The Tajik Civil War was the tragic consequence of a failure in this dialogue, rather than any manipulation from the inside or outside, while the consequent pacification represented, for opposite reasons, a "normalization", namely an attempt for integration of the Islamic political role. The same pacification process recognized this role reserving 30% of "ministerial positions" to the religious opposition parties, besides other inclusive measures and guarantees.[28]

Such integration programmes rejected the extremist side at the same time, namely the militant Hizb-ut-Tahrir, whose declared programme relied on the realization in Tajikistan of the Islamic state (Caliphate) (even when, as they originally declared, without imposition, "through conducting ideological work").[29]

In Uzbekistan, even if not degenerating into war, the tensions have heavily conditioned the whole transition time through to the current day. The security apparatuses tend to react disproportionately applying mass prosecution measures, possibly triggering a sense of injustice: the religious

[23]The current radical (actually radicalized) factions take inspiration—paradoxically—from the local tradition in Islam reformism like that represented by Jadid, possibly originated by the early encounter with colonialist forces in CA, that induced an early maturation of a national-religious consciousness (Buttino Marco, 2003).

[24]Corley Felix, 2005b.

[25]Corley Felix, 2005a, b; the newly built holy place is a matter of argument and quarrel since its walls represent writings both from the Quran and the Niyazov Ruhnama, which is perceived as outrageous by Muslim clerics.

[26]Rasanayagam Johan, 2014:7, 12.

[27]http://www.forum18.org/archive.php?article_id=190, accessed 11.4.2018.

[28]Nichol Jim, 2010.

[29]Muminov Ashirbek, 2014; Karagiannis Emmanuel, 2009.

party is considered illegal and its members are indiscriminately arrested and charged, often without evidence; there are reports about torture and further inhumane treatment. In general, some aspects of the religious practices are strictly under state control; many mosques have been limited for Friday prayer, since this represents the risk of forming a crowd and potentially an uncontrollable popular meeting.[30]

10.2.6 Re-Islamization and Re-nationalization

Currently, religious extremism seems to have lost much of its capability and can hardly figure as a driving political motivation;[31] rather it seems to be mainly inspired by revanchist-restoration impulses (indeed a weaker motivation). In general, it is possible to say that religion maintains its typical coordination capability, playing an essential role in order to avoid—as in other phases of history—the desegregation of basic social ties.[32] The new risk is, indeed, that it can become a further object of re-nationalization, with new governments making it manipulable, and potentially an instrument to better organize and strengthen national loyalty.

In Soviet times the mufti seat was in Tashkent, where it was established in 1941 in war times, when the power structures were looking to recover popular consensus, and it was responsible for "religious affairs" for all CA republics.[33] Then, in times of independence, it was reorganized and subdivided into national "Muslim boards",[34] essentially subject to national authority—even if the Tashkent mufti maintained a recognized influence.

The new national authorities, that continued the tradition of a secular state, have not contributed much to the reconstruction of religious institutions. In many cases reconstruction was materially supported, at the time of transition, by international investments (mainly from Gulf countries, like Saudi Arabia and Kuwait, eventually through private foundations) for madrasas, mosques and further religious monument reconstruction, promoting teaching in the Arabian Quranic language and possibly lobbying the local religious parties.

But soon such interventions were considered with a sentiment of diffidence by the consolidating post-Soviet states. The political approach changed, with local religious organizations being increasingly "nationalized", with their foreign (mainly Saudi Arabian) presence often being contrasted as they were suspected of diffusing political influences (often just labelled as "Wahhabism", defined as subversive ideology).[35]

The question of political risk is still open; the dialogue with the multi-faceted aspects of local Islam, which has never been completely under the control of a clerical hierarchy or even the political institutions, is fluid. It is culturally seminal but also exposed to the risk of new radicalization. This is the case of local mystics, or of self-defined Sufi traditions, easily manipulable since hardly institutionalized; and it is the case of radicalization risks, considering that religion can easily become a proxy for defining a subversive attitude.

10.2.7 Currently a Device for Strengthening Power

The situation in CA has appeared to normalize, integrating the predominant religion in the frame of a secularized society; it seems to be the only option available to maintain a pluralist attitude considering the actual situation, the presence of minorities and in general the diversities induced by economic and cultural opening and similar processes. But such opening up, when occurring too rapidly, as in the case of globalization and secularization, risks exposing the local society to new and continually consequent shocks.

Recently, politics has developed new regulations in order to restrict the spread of new faiths that tend to proselytize, abusing the weakness of some social groups. The new regulations limit their activity and recognition, prescribing registration and further obstructive procedures.[36]

This is the case of groups of evangelical Protestants, like Adventist Baptists and Jehovah witnesses who represented an element of preoccupation for local authorities for a while, spreading to some precarious social contexts, among impoverished people in particular.[37]

[30]Rasanayagam Johan, 2014:9; Brill Olcott Martha, 2012:40ss.

[31]As indeed also happens for other ideological cycles in modern times, with new ideologies forging entire generations, offering the elite an extraordinary cultural weapon: simply they seem progressively to lose influence, "de facto" sometimes disappearing as political elements, possibly opening up the time for new ideological (post-ideological) elaborations.

[32]Akayeva Bermet, 2007.

[33]Muminov Ashirbek, 2014,

[34]Paskaleva Elena, 2015:426; Muminov Ashirbek, 2014.

[35]Rasanayagam Johan, 2014:11; Muminov Ashirbek, 2014.

[36]One must remember the tendency to increase restriction for religious proselytism, concerning, for example, minimum numbers of members, mandatory registration procedure and the like; a law about such limitations was signed in 2009 in Kyrgyzstan by President Bakyev.

[37]Rotar Igor, 2003; see https://www.hrw.org/reports/2003/uzbekistan0903/5.htm, accessed 9.5.2018, about roles and functions of mahalla committees; see also Moxley Alissa, 2013:66; Brill Olcott Martha, 2012:40fs; Sultanova Razia, 2011.

In particular in times of transition, the traditional faiths were likely to perceive themselves as threatened compared to such new well-funded (but neither recognized nor historically rooted) churches.[38]

But in the confrontation with established religions, governments have demonstrated a prudent attitude; on the one hand, they exhibit respect for symbols, for example, promoting construction of mosques and of religious practices. On the other hand, they carefully predispose celebrations in order to avoid, also in daily practice, politics and religion from overlapping. This occurs in planning the construction of the new important mosques in de-centralized areas for instance, far from political urban centres, pursuing separation and control of education programmes, and some kind of control over social rituals and hierarchies that potentially interfere in the nomination of new clerics.

10.2.8 Ecumenism, Syncretism and Secularism Tendencies

It is possible to observe a further tendency, complementary to the one described (of the normalization of religious practices), that seems to find diffusion in the current situation. It is interesting to observe that new messages—also official ones—seem to support a kind of syncretism among the religions that are historically predominant in CA countries.

It consists in the underlining of common aspects among Islam and Christianity (and even Judaism, Buddhism and others), possibly predisposing a kind of ecumenical (non-divisive) role of religion. It is the case of official iconography that uses some symbols that are relevant both for religion and secular institutions; it is the case of a certain representation of the "book", to be considered as the "constitution" as well as a sacred representation (as prescribed by the religion of the "book"; see Fig. 13.2), deliberately mixing meanings in an evident search for legitimization.

So similarly, with a kind of "sacralization" of elements from tradition (such as Manas, the Kazakh batyrs or historical figures such as the Samanid and Timurid kings), popular attitudes are already considered as semi-sacred.

In general, this practice tries to exclude aspects that represent any motivation for "division" among pluri-religious populations, "relativizing" the political-conflictual potential.[39]

Furthermore, such approaches (can eventually be considered as an attitude of understatement for religious and political significances) can also signify an expedient to prevent the establishment of new cultural exclusiveness (a "unique" religion), avoiding the risk of new conflicts.

[38]Radford David, 2014; it is the case popularly ascribed to Christian western confessional sects, considered rich in material means, exerting a persuasion potential and inclined to promote conversion—assumed in general as something negative.

[39]As happens especially in Uzbekistan, Paskaleva Elena, 2015:426.

10.3 Linguistic Situation and Communication Codes

10.3.1 The Language Variable

Language, similarly to religion, is a powerful identity marker, usually assimilated in childhood (the "mother tongue", possibly the first sound any child may hear in its life), almost unchangeable for life, connected with all other identity manifestations. Having said this, in premodern times, language as factor of identity was not so important, daily life usually being made up of more indicative (concrete and self-understandable) behaviours. In fact, it is very difficult to trace conflicts in history due to language diversities, or even language incomprehension or for similar reasons.

This issue is particularly relevant in CA: There has always been a tradition of bi- and pluri-lingual attitudes (that were once probably more diffused than in modern times, characterized by nationalization policies). Dialects and local variants must also be taken into account. It was the case of ancient capital cities (such as the ones in Kokand, Bukhara and Khiva khanates), and of the Silk Road (SR) centres (such as Samarqand), themselves tendentially pluri-cultural. It is also the case for remote tribal areas, characterized not only by linguistic and ethnic particularism, but also by exogamic traditions (culturally open to exchanges).

Indeed, the situation in CA was difficult to define: the groups were (and still are) continuously differentiating from each other, using different codes both for identitarian and functional purposes, to evidence territorial collocation, for practising cults and rituals and for pursuing economic needs. This occurred in all circumstances, for itinerant communities of merchants, artisans or pilgrims as well as for pastoral and warrior tribes (but in the latter cases possibly not in a written manner).

With modernity the situation changed, since the language question increasingly became a political–national issue, connected with new technologies enabling communication on a mass scale and with the possibility of spreading ideological–functional instructions (typical for modern apparatuses, usually written ones using a standardized language). Language classification became an element of affiliation, similar in intensity to few others, such as religious, ethno-tribal or ideological (or economic) divisions, as it is possible to recognize even now.

10.3.2 Situation—Uzbekistan

The most used languages belong to Turkic family, diffusing in CA in different waves, mainly after the fifteenth-century invasion (or migrations) and overlapping the original Persian language (or of other origin), which was the original linguistic strata. Then a process of segmentation of such tribal groups started in what would later be considered the formation of national idioms (in this case of Uzbek language), sometimes coexisting with the previous languages in mixed ethnic populations (namely the Persian-speaking population resident in main urban centres).

Considering the current constitutional set-up, Uzbek is the only state language, written since 1992 in the Latin alphabet and spoken by 85% of the population. In Samarqand and Bukhara, Tajik is widespread (defined historically as "Sart"), spoken bilingually mainly by ethnic Tajiks; the same for other places in the country in Fergana, in Syr Darya oblast, as well in Shakhrisabz, for approximately 10–15% of the Uzbek population (official figures about 4%).[40]

These figures are difficult to interpret as they represent different identity aspects, such as language used, ethnic sense of belonging and citizenship, and do not necessarily coincide (namely Russian- or Tajik-speaking citizens may not identify themselves immediately with Russian or Tajik identities). This occurs especially in urban and bazaar areas, where identitarian elements are continuously confused with official languages and dialects used interchangeably.

Among the other minorities, one must consider the Karakalpak ("black kalpaks") language, spoken in the autonomous republic on the western side of the country on the Amu Darya mouth on the Aral Sea (which has today almost disappeared), where it has an official status. It is closer to the Kazakh language, with which it is mutually understandable; Russian is spoken by 14% of the population as a first language and by more (about a half) as second language, even when the Russian population only accounts for about 5%.

Russian is progressively diminishing in use, being replaced by the nominal language in official circumstances. It is an issue that directly impacts the situation of Russian and Russian-speaking minorities. It depends on many factors, among them the officially recognized status or the contingent relations between the governments—the "high politics"—and others that do not depend on the actual political intention. The current language law defines the use of Russian with a transitory formula, as the language of higher education, but of difficult interpretation and application.[41] In this case, the authorities possibly decided to leave the language to spontaneously evolve; the question is whether Russian will still have a certain official role in the country in the future.

[40] Foltz Richard (1996).

[41] Maurizio Massimo and Tomelleri Springfield Vittorio (2018); Megoran Nick, 2017; possibly the Russian influence will persist in CA even when the Russian community numerically decreases.

10.3.3 Turkmenistan

The official language is Turkmen, as defined by the 1992 constitution, with Russian largely spoken or understood in urban areas; respectively Turkmen is the first language for 72% of the population, Russian for 12% and Uzbek for 9%. The language law of 1990 defined Russian as the "language of communication among the nationalities",[42] intending it for internal use among the different non-Turkmen-speaking minorities—a formula frequently used in CA, meaning a guarantee for Russian speakers, but it supposedly signifies a transitional character.[43] The Uzbek-speaking minority is rather territorialized in the Amu Darya valley, close to the Uzbekistan border; other languages account for 7% and represent all the nationalities of the Soviet empire and the contiguous countries in minimal numbers.[44]

10.3.4 Kazakhstan

Kazakhstan is officially a bilingual country, with Kazakh slowly substituting Russian as the more widely used language; Russian is used in official circumstances, by nearly all people participating in public life, even if not as their first language (as their mother tongue); in fact, just 64.4% native speak Kazakh; Russian still is, and possibly will remain, a diffused common language. The Constitution, in Article 7, describes Kazakh with the status as the official language and Russian is ascribed a peer (equal) status in public offices and in administration. It is written in Cyrillic, but it is foreseen that the Latin alphabet will replace Cyrillic "by 2025".[45]

10.3.5 Kyrgyzstan

Similarly to Kazakhstan, in Kyrgyzstan Kyrgyz is the official language, but Russian is used predominantly in public spheres and also in administration—even though there are also signs that the national language is gradually expanding

[42] Cosentino Italo, 2017.

[43] Gaynor Kelly Lee, 2017, p.475.

[44] Some sources report that the populations have decreased, Wikipedia "Turkmenistan", accessed 11.4.2018; https://www.cia.gov/library/publications/the-world-factbook/geos/tx.html, accessed 11.4.2018; Flynn Moya, 2004.

[45] Megoran Nick, 2017.

into the public communication sphere. Russian has a co-official status, although the law about the "promotion of the Kyrgyz language" prescribes a transition, namely it defines such co-official status as temporary. As said, Uzbek and Tajik are spoken in enclaves (the former notably in the Osh area) and locally may represent relative majorities.

10.3.6 Tajikistan

Tajikistan's official language is Tajik, spoken by 84.3% of the population; it is a Persian vernacular with ancient literary traditions and wide speaking areas, even beyond the current political border, in Fergana and the Zarafshan valley, in some Afghanistan provinces and in contiguous Badakhshan. Uzbek is the principal minority language spoken mainly in Fergana, by 13.8%. As in other CA countries, the Tajik constitution of 1994 recognizes Russian with a status of "inter-ethnic communication", namely a language used among the territorial ethnics, "de facto" sparing non-Tajiks (Russians also of other Soviet ascendency) the learning of Tajik.[46]

Russian is usually spoken as a second language, as a business or administration language, but is less rooted because the Russian ethnic population in the country is decreasing—this with the exception of Dushanbe, the "global city" and the areas interested in the past by a relevant immigration (such as the new towns in the south), where the "Soviet" Russian-speaking generation is still rather numerous, even if not well represented politically. In the remote areas diverse languages are diffused; they are considered officially as Tajik dialects but are often rather distant from the standard.

The mountain outback is an area of several minorities, scattered over wide surfaces. Among them are those of Yagnobi, in the upper Zarafshan valley, spoken by about 25,000 people, possibly a relic of ancient populations dating back to Sogdian merchant communities. They were forced to migrate from their original remote valley in the twentieth century.[47] Similarly for the Gorno Badakhshan (Pamiri) peoples who, as legend says—rediscovered in current times, possibly to increase the interest among visitors—date back to Alexander the Great's soldiers who settled in the area. They are considered Tajik, but depending on criteria, their languages are classified as eastern Iranian dialects.[48] They practise a particular Ismaili faith (cult), to be classified as a Shi'a branch, and, thanks to isolation in the Pamir highlands, they have preserved very ancient traditions, folklore and habits.

10.3.7 Language Official Status and Current Use

In post-Soviet times, it is possible to observe further changes to the language, combining with those that originated from Soviet "slijanje" (слияние, "melting", "fusion") policies in the 1970s.[49] Depending on different points of view, it represented an extreme attempt at russification or the opportunistic acknowledgement of some local majorities (respectively, in the eponymous republic SSR or in autonomous province AO), in opposition to other local minorities.

The situation is in evolution. Russian is recognized by several countries as a "language among the nationalities" ("inter-nacionalni jazik"), an official language and the main language (being spoken transversally by nearly everyone, the language is spoken by local nationals, not always by minorities); it is used currently by elite groups (and also by elderly generations, without exception, since they learned it in military service, in school or in other circumstances), so it represents a "closed group" that is probably destined to decrease in number.

Therefore, its status in the long term will probably depend on the conservation of territorially defined minorities (as it is the case of north-eastern Kazakhstan, in Zhetysu, in northern Kyrgyzstan, and in general in urban areas), and by connected geopolitical issues.

Current evolution seems rather to recognize Russian as a lingua franca, possibly reaching some form of equilibrium, with minority languages' claims—after decades of neo-nationalist policies—losing political significance and governments accepting and realizing the advantages of a bi- or pluri-lingual status.

So, regarding the question of the language diversity, the fact that small communities continue to use their own language is no longer seen as a risk; on the contrary, in some cases, it is seen as a territorial characteristic, to be promoted, where local idioms are endangered on the red list, making it a matter of protection.[50]

10.3.8 Linguistic—Political Implications

The role of language, as well as of any kind of communication, as a factor that produces identity is undisputable, yet its

[46] Cosentino Italo, 2017; for adults the learning of a language is a difficult task, and this even more for a language not considered particularly useful, since not spoken widely, even when representing a national official language.

[47] Ginzburg N.S., 1986; Panaino A., Gariboldi A., Ognibene P., 2013.

[48] Ognibene Paolo, 2018.

[49] Marchi Marzia, a cura di, 2017.

[50] As considered e.g. in the UNESCO Atlas of the World's Languages in Danger.

effectiveness as an identity marker changes depending on opportunities and situations.

It is a question related to the fact of whether a certain language minority is territorialized or not, and obviously to the fact that language has a nominal status in cross-bordering regions (prospecting the risk for territorial claims). It is related to a question of power relations, as usually minorities are subject to particular self-perception (also often adopting elusive strategies, because of historical experiences, of subalternity or dominance traditions). This is the issue in particular in Kazakhstan, with its Russian minority especially along the border with Russia and the question of Tajiks in Uzbek cities.

A further question is represented by the belonging of a language to a wider super-national group, potentially evoking ancestral affiliations; this is the case of all the Turkic languages (like Uzbek, Kazakh, Kyrgyz and Turkmen), and also of minority languages such as Uighur, Karakalpak, Azeri, Meshketian Turk and others that belong to the same language family. The same for Slavonic languages (Russian and Ukrainian and eastern European languages), and for Persian, namely Tajik spoken outside Tajikistan, as well as for several Iranian dialects spoken by small minorities.

The fact that the respective populations do or do not understand each other (considering the differentiation of the languages), namely whether they are eventually interchangeable (to be considered then as dialects of the same main idiom), seems not to be very relevant. However, this may represent a political discussion, for example, to underline or, on the contrary, deny some affinity. This is the case of Karakalpak, which is similar to the Kazakh language, or to the Pamiri dialects, that are classified or not, depending on circumstances, as Tajik (Eastern Iranian or just autochthonous languages).[51] A similar question also comes up between Russian and other Slavonic languages. This occurred in times of resurgence of national cultures, with neo-national academies and other similar institutions underlining (in an artificial way) the differences between some languages (which until that moment had not been noticed at all) and corresponding groups.

In some circumstances it is possible to note a surreptitious changing of status for the benefit of the local original language (in a visual language landscape, bureaucracy and public communication, school and education). The most recent trend—as said—is that of accepting bi- or multi-linguicism as a normal situation (in Kyrgyzstan and Kazakhstan, notably), with mutual benefits to all parts of the nation. This evolution should lead to a kind of normalization of the multi-lingual and multi-alphabet question, de-potentiating its political significance over time.[52]

[51] Ognibene Paolo, 2018.

[52] Such an issue is possibly destined with the passage of time to decrease, mitigating some incompatibilities spontaneously, considering that usually new generations easily assimilate new languages, and that in general the political context (and the attitude towards a certain minority) may change (and also due to cultural–technical devices' improvement, in a new ICT world, in which the political relevance of the linguistic variable should decrease).

10.3.9 Alphabet Change Question

During the twentieth century, CA populations were subject to mass alphabetization programmes and had to shift in their language use between four codes over just a few years. It is possibly a unique case in world history, characterizing a period of dramatic structural changes.[53]

In fact, the use of the alphabet is to be considered a powerful device and is connected with both the use and spread of certain languages, and with the plans of some new power structures. It means a question of political recognition and the change of the entire generation's memory. Before the Russian conquest, and before the revolution, the Arab language and Arab-Persian alphabet were used by local elites and by emirates' courts, by clerics, writers and poets, by Tatar merchants and in general by the cultural and economic elite.[54]

After the revolution, the Latin (rather than the Cyrillic) alphabet took the place of the Quran-Arabic, imitating the example of the Kemalist Anatolian alphabet change after WWI (in order to secularize the culture and to assimilate economics, previously monopolized by non-Turchik minorities, mostly in that context Greek and Armenians, using their own alphabet). The idea was to cut off from old (and promote the new) cultural roots; furthermore, it was congenial to the early Bolshevik utopias where the formation of a universal alphabet was imagined (applying linguistic science to the worldwide revolution communist programme).

In correspondence to a further turn in politics, the Cyrillic alphabet was introduced, with more or less the same intentions, possibly pursuing an assimilation with Russian, which was to be considered as definitive.

10.3.10 Alphabet Current Situation

The latest tendency is to change the alphabet again, inspired by globalization, with the Turkic-speaking countries inclined to turn to Latin, which could facilitate the inclusion of CA into the world economy. Only Uzbekistan and Turkmenistan

[53] Maurizio Massimo and Tomelleri Springfield Vittorio (2018); Cosentino Italo, 2017; Schwartz Christopher, 2013:196.

[54] As usual in pre-modern contexts, literacy was considered a sacralised skill (Adeeb Khalid, 1998); in times of czarist empire, Latin script was used by elites, as well as by immigrants and minorities; French was diffused in pre-revolutionary Russia, as well as German or Polish, and, with the respective alphabets, Greek, Yiddish, Armenian, Tatar and other linguistic minorities.

have effectively changed to Latin while the other republics did not or have not yet.[55]

Kazakhstan is officially a bilingual state (even when Kazakh is slowly replacing Russian for internal purposes). The government has repeatedly announced its intention to change its alphabet. This should have taken place by 2025 (updating a previous deadline placed at 2015).

The possible alternative is—in the circumstances of increasing opening, possibly avoiding a further alphabet change—to use a third language for public communication. This would be English, for example, for street and urban announcements, public information, as well as international contracts, instructions or exchanges; this to the benefit of foreign visitors, for trade and business, and for international communication in general.

The example of this solution was the Astana (Nursultan) Universiade and Expo 2017, as well as further large events with the three languages systematically used at all levels of communication (written and oral instructions, visual landscape, notices). With the generational passage, it will be evident which tendency will prevail, maintaining Russian, privileging Kazakh or using English as an intermediate element.

In Kyrgyzstan, similarly to Kazakhstan, Russian maintains the status of the official state language, possibly prefiguring a bilingual situation (it is to be considered that Russians historically settled in Zhetysu and are still prevalent in Bishkek). Kyrgyz, as for the other Turkic languages in the area, was written in Arabic alphabet in the past, then in the 1920s (1928) it turned to Latin, and then in 1941 to Cyrillic.

In pre-Soviet times the Uzbek elites used Chagatai Turk with adapted Arab characters, which was replaced in 1926 by Latin, and in 1940 by Cyrillic; after independence, in 1993, Uzbekistan went back to a Latin script, which was gradually diffused in school, public communication and official writings.

Turkmenistan has changed to Latin characters recently, Russian being little spoken, if only by cultural elites, and Uzbek—spoken by Amu Darya oasis minorities—has already shifted to Latin script.

In Tajikistan for a while, in the 1990s transition, there was an intention to adopt a Persian-Arabic alphabet used in Iran. The Constitution of 1991 mentioned Russian as the "language for inter-ethnic communication", and a later amendment decided the role of the Russian language for official roles only, especially for legislative ones or for similar purposes. A further amendment tried to restore the Russian status in administrative and political communication and law making, alongside Tajik; the situation remains ambiguous and possibly reflects the contingent relations with Russia and the refractoriness of the non-titular nationalities of using the Tajik language standard. Actually, it seems that the use of Latin characters is spreading, following tendencies in information and communications technology (ICT) and an increase in international relations on all scales. The uncertainness represents a critical question above all for non-titular nationalities (always tempted by the idea of leaving the country).[56]

However, generally speaking, the alphabet question today, with the widening of computer writing, with application of automatic transliteration and further capabilities (e.g. using digital screen adapting writing in public communication, open or public spaces, for instructions, functional and geographical indications), is probably a less relevant question.

10.3.11 Toponomastic: Recoding of Places and Names

The question of toponyms (of any kind and in any place) is especially significant in CA, since it represents an element of officiality, of status recognition, in a highly dynamic cultural landscape. It was a particularly sensitive question in transition times (that is usually also a transition between communication codes).

They affect public communication dimensions, producing a visual and immediately perceivable impact restructuring the environment in which people are carrying out their daily life. It is a question connected to that of identity and with ideological roots, for example, in the case of change of names of streets and squares (deleting names such "Lenin", "Stalin" or "Beria", namely the same memory of the revolution).

This instrument is commonly used to "communicate" to the populations some element of ideology, inducing the individuals insistently to repeat such denominations, then day by day assimilating some ideological elements (e.g. the name of a new national leader, the celebration of a "proletarian" holiday).[57] A similar issue concerns its impact on the private dimension, for personal and family names, for documents and registrations, for local place names and for any kind of signs, on any scale.

The practice of renaming is commonly used by politics, especially on occasion of revolutionary changes—as had happened in the Soviet past.

It is also happening today, prefiguring the exactly reverse operation of the one realized in Soviet times, both recovering original names and inventing completely new ones. Often this transformation is the occasion for further distortion, sometimes reflecting some ideological influence (traditional, religious or neo-nationalistic).

[55] Schwartz Christopher, 2013:196.

[56] Maurizio Massimo and Tomelleri Springfield Vittorio, a cura di, 2018.

[57] Farangis Najibullah, 2016.

It is the case of public places, streets and squares, monuments, naming of public institutions and schools, denomination of holidays and other events; so too for landscape elements, natural environments and sightseeing, mountains, as well as cities and places, private and family names; all these are renamed with their old pre-Soviet names, or however with names with some special impact. It is a question connected to memories but also to legitimization patterns, on all levels. In particular, the change of once Russianized (and Sovietized) individual names can be considered today a good indicator of Russian philia or phobia; similarly, to rediscover pre-revolutionary names, sometimes supposedly of sacred or religious, of ancestral Turkic or Tajik origin, sometimes indeed inventing brand-new names.[58]

10.3.12 The Global Evolution of Communication Modes

It can be considered that the new ICTs are rapidly changing the scenario, with the continuous formation of new classes and the claiming of new political status.[59] The spread of new communication habits discloses a wide series of possibilities, both for citizens and institutions. Social or mass media, cellular- or smart-phones and further innovative devices, occasionally appearing on the market, represent instruments widely used for social organization purposes ("flash mob") or for other similar situations, which may be perceived as politically dangerous.

Furthermore, the diffusion of new communication devices to some extent also induces new legitimization patterns. They tendentially permit a free flow of information that is difficult to control by the power structures who cannot just prohibit them, and that may react (to such loss of influence) by elaborating even more sophisticated instruments of control.

The power structure has to adapt to such situations to some extent; it cannot "tout court" impose a totally invented "reality" in a hermetically locked society as happened in totalitarian nineteenth-century regimes.[60]

The use of cell phones in CA has quickly become universal—as expected—with about 22 subscribers in 2015 in Uzbekistan, and the number of internet users exceeding 12 million.[61] These numbers have possibly increased as of today thanks to the more recent diffusion of further innovative devices that are even more invasive to social and individual behaviour.

Such numbers are not easy to estimate, since anyone may use such instruments that are now completely user-friendly making the question of digital alphabetization a relative one (indeed it is not necessary to know how read and write to use such devices like tablets or smartphones, at least for a basic use). The ICT virtually covers the whole population, configuring a sensitive question for the established powers, who—as usual in a situation of semi-democracies—have tried to extend their control over digital networks with censorship or manipulation at diverse levels.

This recently occurred after the "Arab Spring" unrest (that demonstrated the application possibilities of such devices), possibly representing the evolution of the post-Islamist tensions (a current nightmare for authorities, since it represents a civic motivation, particularly difficult to de-legitimize).[62]

10.4 Ethno-National Map

10.4.1 Description, Data, Demo: Ethno and Urban Map

For some populations—steppe, desert and mountain populations that sedentarized relatively late—the current national classification does not make much sense. It is a typical modernist "construction" and is rather perceived as a political instrument when for them the true map is still based on ethno-clanic divisions (even if overlaid by cosmopolite and globalization-induced lifestyle today).[63]

The same can be said of the urban population, located along the SR cities, that has a long tradition of linguistic and ethnic promiscuity, for whom the neo-nationalist classification is to be considered as fictitious (considering that in cities such as Samarqand and Bukhara they were traditionally diffused with bi- or pluri-lingual capabilities). So too, for popu-

[58] Obviously following also diverse axes, e.g. depurating a supposed Persian original expression from a Turkic superposition or the reversal; Megoran Nick, 2017:36fs; the most obvious case is that of the Tajik President Rahmonov, who changed his name to Rahmon.

[59] Warf Barney, 2011.

[60] It is necessary to say that the diffusion of such user-friendly devices does not necessarily mean the improvement in the quality of culture, in education and, in general with that, in the "social capital".

[61] PWC (2016:4).

[62] In Uzbekistan there are "signs of relaxation" but which possibly should be considered in the context of the "honey-moon" period of the new president; http://www.bbc.com/news/world-asia-16218808 (accessed 23 April 2018), Uzbekistan profile – Media, 1 November 2017; in the last decade censorship has increased possibly due to the "Arab Spring" and other similar events, with government blocking access to proxy servers; the Uzbek government has been defined as an "enemy of internet" by NGOs like "Reporters Without Borders", Uzbekistan toughens Internet censorship, 11.10.2012, http://www.uznews.net/news_single.php?lng=en&cid=30&nid=20980 (accessed 23 April 2018); but it is difficult for governments to achieve total control of such a way of communication. Some authors assume that the new digital culture would be intrinsically pluralistic, since it is impossible to completely manipulate, and due to its "filtering" through any barrier, even when there is always the possibility of abuse and of the instrumentalization of such technologies.

[63] Then "klanovost" Ro'i Yaacov, 1991; see also Esenova Saulesh, 1998.

lations beyond any manipulation that were characterized by general ethnical mixing both in rural and in urban spaces in Soviet times.

The new boundaries have overlapped the ancient territorial categories, inducing further incongruences. This is the case even when the denominations of CA republics (of the previous SSR and AO) arguably follow a coherent ethno-language criteria:[64] the Soviet regional delimitations cover, sometimes crudely, the territory of a particular ethno-linguistic community, defined as a titular ethnic group (even when it must be considered that this was usually in a very weak condition; so for instance for the SSR of Kazakhstan, established in 1936, when it had just 25% of titular populations: the Kazakh language was acknowledged officially even when it was at the edge of extinction as a cultural identity, if not as a human community).[65]

Besides these interpretations, it is evident that the current ethnic map of CA is the heritage of the past, but also of the continuing mixing that has occurred up to the present time—considering the high mobility that even today characterizes the resident population. It is questionable to what extent it represents the actual situation; it is made even more confused by the presence of pluri-linguicism or of a situation in which it is objectively difficult to ascertain an identity.[66]

10.4.2 List of Nationalities

The actual ethno map results from a long-term stratification, with the confusion of ancient autochthonous populations, new or old immigrants and deported populations, concentrated in cities or dispersed in small communities, between ancient towns—"Sovietized", but evidencing an Islamic heritage or peculiar colonial culture (e.g. bazaar city quarter or "самостоятельные дома" blockbau-style wooden homes).

Besides the major populations, there are small communities that are difficult to estimate numerically and to some extent preserve their compactness. For instance, the case of Koreans, Pontian Greeks, Tatars, Roma, Volga-Germans (mostly descendant of the Volgadeutsche, deported in CA during WWII), as well as of other eastern Europeans and South and North Caucasians (Fig. 10.1 a, b and c).

Bukharan Jews, descendants of ancient communities, migrated to Israel and elsewhere after the collapse of the

a

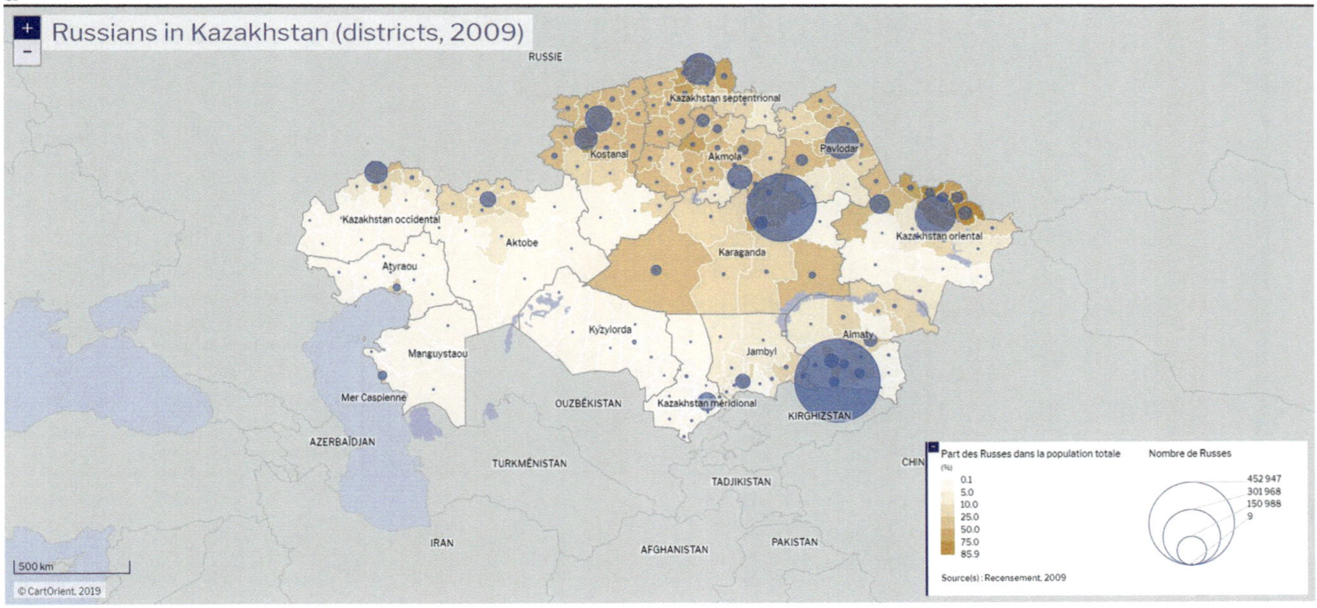

Fig. 10.1 (**a**) Russians in Kazakhstan, 2009. (**b**) Uzbeks in Kazakhstan, 2009. (**c**) Germans in Kazakhstan, 2009. (Source courtesy Thorez Julien, www.cartorient.cnrs.fr)

[64] Marzhan Thomas, 2015.

[65] It was considered modernistically and deterministically destined for extinction, as usual in a colonial scenario; in Kazakh SSR in 1959 the Kazakh population was about the 30% of the entire population, with ethnic Russian 43%, after the "Virgin Lands" campaign; nevertheless, it is to be considered that this acknowledgement is generally thought of, by many Kazakhs, as the founding moment of the Kazakh ethnicity in a modern sense.

[66] The ethno-national classification assumes critical importance in modern times, but it is a question which is possibly losing importance today due to de-territorialization processes and the application of powerful communication techniques incentivizing mobility on all scales.

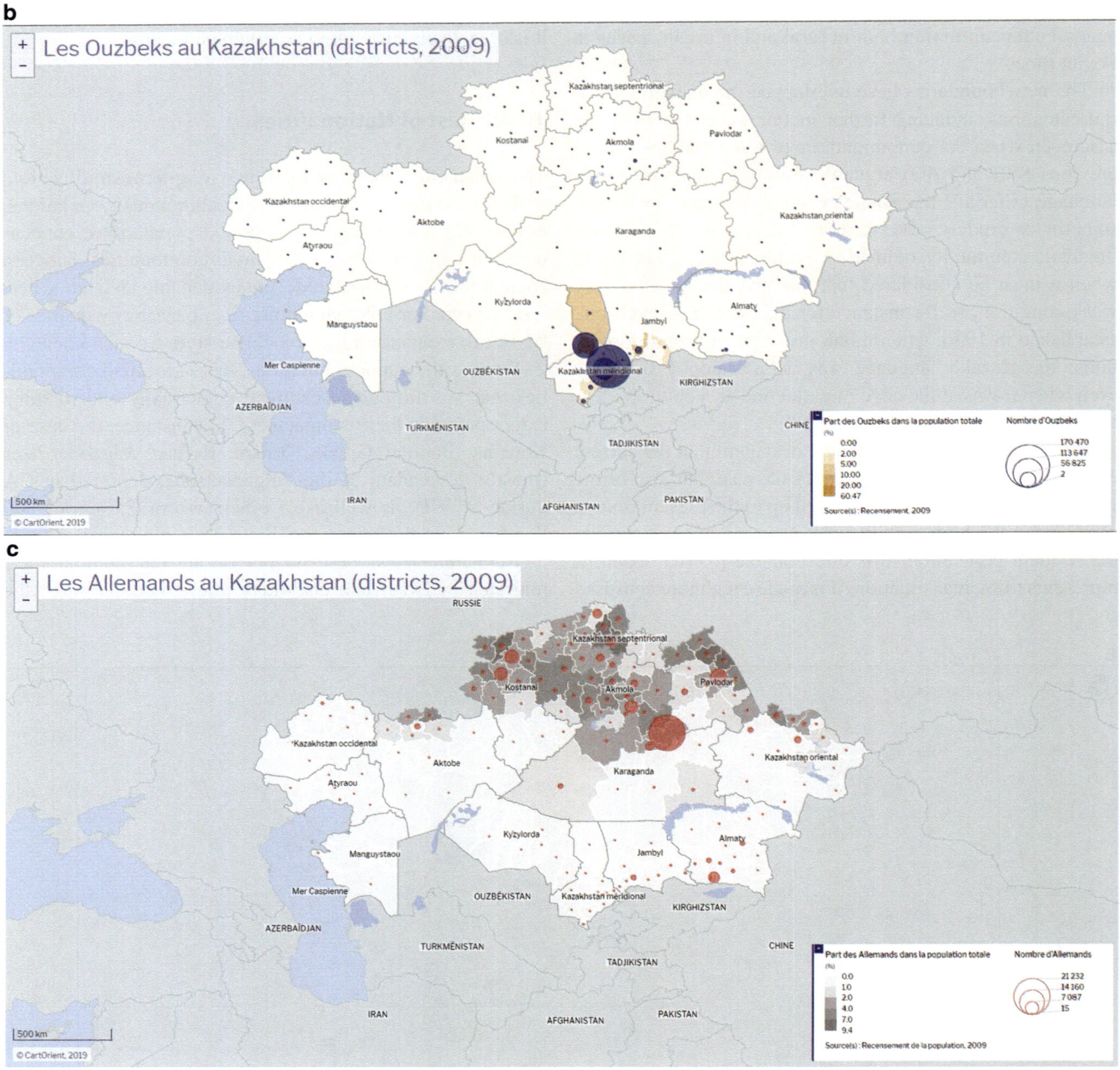

Fig. 10.1 (continued)

Soviet Union (SU); other Ashkenazi Jews who arrived from eastern Europe during WWII are still scattered in small communities or mixed with other communities. Further distinctions should be made properly between Russians and Cossacks (obviously to be distinguished from Kazakhs), who arrived from south Russia and western Ukraine in early colonial times and who are today mainly descendants of the pioneer colonizers, and also between Russians belonging to different in-migration waves since early colonization times until the late Soviet "Virgin Lands" campaign in the 1950s.[67]

[67] The mandatory inscription of belonging to a minority was an aspect of the Soviet ethnographic policy: the list of the Soviet nationalities is an extraordinary testimonial of humanity, and at the same time, paradoxically, the pre-condition of their persecution but also of their conservation; Bregantini L. (a cura di), 1997.

The most evident aspect to consider is whether the national population lives in concentrated or in scattered forms and whether they have an inclination (propensity) to mix and to maintain geo-ethnic compactness beyond the generational passage. Russians have been concentrated in northern Kazakhstan (even when combined with Ukrainians, Belarusians and further Europeans) since colonial times and again since the further waves of immigration until the Chruščëv period (engaged in "Virgin Lands" campaign).

Another problematic question is the one represented by non-eponym minorities concentrated in definite areas (Tajiks in Samarqand and Bukhara, Uzbek in Osh and in Kyrgyz Fergana area), who have become a "local majority", often characterized by prestigious cultural remnants, residing in areas bordering with nations, where that nationality is titular (as common case in this area, indeed).[68] This is the case of Russia for Russian Kazakhs, of Tajikistan for Afghan Tajiks, of Uzbek republic for Kyrgyzstan and Tajikistan Uzbeks, residing on the other side of the border and so on.

The situation is made more serious because of the configuration of the borders themselves, as in many cases they seem to be designed to deliberately extenuate tensions (not the contrary) with minorities located just across the border in exclaves that are difficult to manage from many points of view (communications, trade, cultural ties, security and surveillance). In fact, the ethnic map presents some critical situations (also considering that national belonging is in many cases difficult to demonstrate, the current population being the descendants of groups who in-migrated centuries ago). It was probably much more dangerous in the immediate time after independence, when combined with other issues (as in 1989–1991 clashes, see later).[69]

Practically, all CA countries present some territorially critical questions that in principle seem to be without a definitive solution and that will probably be the object of indefinite negotiation, with periodical reciprocal consultations (as indeed it seems to be a widespread habit in such situations, also in other parts of the world, with sovereign states accepting some "modus vivendi", delaying a kind of definitive solution).[70]

10.4.3 Territorialized and Non-territorialized Nationalities

Kazakhstan Republic figures report (2010) 63.6% Kazakhs, 23.3% Russians, 2.9% Uzbeks, 2.0% Ukrainians, 1.4% Uighurs, 1.2% Tatars, 1.1% Germans and 4.5% "others", with significant numbers for Azerbaijanis, Poles and Baltics (Fig. 10.1 a, b and c). We must consider that some groups were deported as whole "national communities" by Stalin in proximity to WWII, namely Volga Germans, Koreans, Meshketian Turks and other Caucasian small groups (Chechens, Ingushes and Kabardino-Balkarians). The trend seems to be to the advantage of the indigenous Kazakh population considering the "vozvraščentsy" out-migration, as well as that of other peoples who have a fatherland capable of hosting them (or eventually departing for another country, for which they could obtain a visa).[71]

Uzbekistan is CA's most populous country (31.5 million, about half of the total CA population), and represents the most populous national element in the whole CA space. Titular nationals are about 80% of the total population, there are about 5.5% of Russians (further estimations refer about 8%), and Tajiks make up 5% (officially), Kazakhs (3%), Karakalpak (2.5%), Tatars (1.5%) and others (1996 estimates). As said, the most important minority in Uzbekistan is that of Tajik origin, that accounts for, depending on estimation criteria, a figure of 5% to 10–15%, or even up to 30%. However, it is practically impossible to ascertain in which measure the Uzbek-speaking Tajik may be considered as Uzbek nationals, or the contrary; this especially in Samarqand (where they represent the majority), Bukhara and few other centres, close to the Tajik border (assuming easily a political significance).

Turkmenistan's population in majority is defined as "ethnic Turkmen"; significant minorities are Uzbeks and Russians (this was made by "vozvraščentsy" and dropped to a minimal number), with smaller numbers of Kazakhs, Tatars, Ukrainians, Kurds (native to the Kopet Dagh mountains a significant minority for a political question, to be found anywhere in the Middle East, but without their own national country), Armenians, Azeri, Baloch and Pashtuns. Some sources report tensions still reflecting traditional "invisible" clan divisions inside Turkmen nationality.

Official figures recount that in Tajikistan there are 84.3% Tajiks, 13.8% Uzbeks, 0.8% Kyrgyz, 1.1% others; peculiar to Tajikistan (possibly one of the major cause of the 1990s conflict) is internal displacement from remote regions (from Gorno Badakhshan and other mountain areas), for many reasons (political surveillance above all, in border areas, but as well accessibility and infrastructure planning, water exploitation policies, with possible resettlement of local communities).

The situation in Kyrgyzstan is described by Table 10.1.

The new states periodically organize a census and are to some extent, as the Soviet state was, obsessed by the necessity of measuring all the variables in order to predispose

[68] Foltz Richard, 1996.

[69] Ro'i Yaacov, 1991; Megoran Nick, 2017.

[70] A typical post-modern situation, assuming the national-state as an imperfect construction, that has to face and accept its own imperfections looking for international cooperation; see later.

[71] Klüter Helmut, 1993; Brill Olcott Martha, 2012:92.

Table 10.1 Kyrgyzstan: languages used

Language name	Native speakers	Second-language speakers	Total speakers
Kyrgyz	3,830,556	271,187	4,121,743
Russian	482,243	2,109,393	2,591,636
Uzbek	772,561	97,753	870,314
English		28,416	28,416
French		641	641
German		10	10
Other	277,433	31,411	

Source: Wikipedia, Kyrgyzstan (https://en.wikipedia.org/wiki/Kyrgyzstan, accessed 4 July 2018)

planning. But the reliability of such measurements is questionable, not just because of the high mobility of social components, but in general because of the same difficulty in measuring the different minorities in epochs of high cultural-ethnic mixing.

This also regarding macro-data; for example, some sources reported that in 2012 the population of Turkmenistan had decreased to less than the estimated number of 5 million, and it is difficult to understand why this happened—possibly because of the introduction of a dual passport regime, or possibly because of migration or because of natural demographic phenomena.[72]

In such circumstances, it cannot be otherwise. The data account for a long list of small minorities (Koreans, Baltics, Greeks, Germans and others) that are probably descendants of second- or third-generation immigrants (or deported), possibly deriving from intermarriages, or just assimilated, due to cultural spontaneous mixing, to some other group; they rarely still speak their original language and many of them have some kind of "dual citizenship", if permitted or prescribed by some legal system.[73]

Such variables (just as for many others, arising in globalization times) are difficult to measure. This is due to accelerated demographic dynamics and to the new evolution regarding residency—considering multiple dwelling—and migration, which often evolves into a kind of commuting or temporary migration even over large distances; temporary movements are usual for the current style of life, which is taking advantage of new ICT and of further mobility facilitators ("low cost", "high speed" and similar modalities). In these terms, migration is often intended as something reversible, in the context of a progressive de-territorialization (especially considering countries like Kyrgyzstan and Tajikistan heavily affected by temporary migration).

10.4.4 "Orphans" of the Soviet Empire

Since the late Soviet "slijanje" policy, prefiguring a mix of the ethnics (with "different nationalities shaping into a unique proletarian content"),[74] it is possible to observe the formation of a kind of particular group, culturally homogeneous, suddenly becoming a kind of orphan of the deceased Soviet "fatherland".

In transition times, they have literally no place to go and no institution representing their rights. In fact, they represent nationalities without a state (like several south and north Caucasians, Volga- and Crimean Tatars, and others) or just populations of mixed character, potentially non-titular, or displaced, in their own republic (like Pamiri and Garmi populations in South Tajikistan). Some groups, like the Dungans,[75] and Uighurs, do not enjoy a favourable status because of geopolitical questions. Similarly, for east European Slavonic and non-Slavonic minorities, that tend to assimilate with Russian groups (Ukrainians and Belarusians notably).

Such populations are (and were) used to communicating with other nationalities and with the official institutions—and also sometimes among themselves—in Russian. For a while, in transition times, they tended not to learn the titular language (Kazakh, Kyrgyz, Uzbek, Turkmen and Tajik), possibly because they could not accept the idea that they were living in a new country. In fact, after three generations of Soviet rule, they still represent a kind of post-Soviet "ethnic group", without a definite collocation.

In early transition times, there were about 10 million Russians in CA, self-declared, or "ethnic Russian",[76] that means about 30% of the population. It is necessary to add them to millions more eastern Europeans of Slavonic origin (mostly Ukrainian), or other groups culturally Russian assimilated.

Another significant percentage consists in individuals or groups that cannot easily self-identify with a definite nationality due to intermarriages or to spontaneous assimilation, simply due to the fact they had lost—with a generational passage—their original language or other elements of national

[72] The question about the reliability of these statistics is open; usually Soviet citizens had to declare a single nationality, a habit continued mostly today; but it is to consider that about 40% of the Soviet population is (was) possibly mixed, and that many citizens cannot speak the original national language, neither do they conserve the knowledge or practise the national culture: "each soviet passport [...] includes an extra line concerning the nationality of the person. It is unequivocal and irrevocable. In the case of mixed marriages only one parent can pass on his/her nationality to the child"; Klüter Helmut, 1993:419

[73] As said, 40% of the Soviet population was estimated as ethnically mixed; Klüter Helmut, 1993.

[74] Kaiser R. J., 1995:113; Monteil V., 1957:85ss; Werth N., 1993:215; Peyrouse Sebastien, 2008.

[75] Ferrando Olivier, 2013:37; Sodiqov Alexander, 2013:49.

[76] Dunlop J. B., 1994:208.

affiliation; so that now they cannot declare themselves as anything but Russian.

In fact, the knowledge of Russian (a language that usually maintains a constitutional or "de facto" status of "internationalities" language) represents an advantage for economic and social activities. Minorities, who have forgotten their own mother tongue two or three generations on from the original migration, have no alternative,[77] and they insist on speaking Russian.

With the generational passage, as indeed for many such ethno-identitarian incongruences, the question will be probably settled, finding a natural solution (since younger generations learn easily the local idiom, integrating in the new societies).

10.4.5 Potential Conflictual Situations

In CA the concept of "nationality" is difficult to define since it has only emerged recently in times of neo-nationalist disputes. In fact, the current groups are not easy to classify on the bases of objective elements such as religion, ethnicity, idiom and dialect, territorial origin or descendance (even if in Soviet times nationality was assumed as exactly determined in the context of a mandatory ethnic accountability).[78]

This especially in mixed areas, like the Fergana and Zarafshan cities (Samarqand and Bukhara), where the population is characterized by pluri-linguistic habits.

In some circumstances, the ethno-linguistic differentiation corresponds to differences in status or socio-economic conditions (as among Kyrgyz and Uzbek populations in Osh area), reflecting further contrapositions such as urban/rural or mountain/plain.[79] In these cases the differentiation maintains a dangerous character and can periodically explode and is used as an excuse for other "invisible" oppositions.

The situation of the Pamiri populations is quite particular. They were originally from the Gorno Badakhshan province and have been disaggregated on different occasions and resettled—together with other mountain Tajik populations—in South Tajikistan in the Vakhsh Valley, in an area of new cotton- and rice-fields kolkhoz needing manpower, and occasionally elsewhere.[80]

Sometimes they managed to go back home, as happened after the SU collapse in particular, but often this option was impossible in practice. This occurred above all for the second- or third-generation descendants, who had integrated into the new cities in the meanwhile and did not have a great deal of motivation to go back to their original village, which was often just abandoned or no longer existed.

These populations were forced to migrate for different reasons; their remoteness always represented a problem for Soviet politicians and an obstacle for planning practices. Sometimes they were also suspected of scarce loyalty, occupying remote territories that had been the headquarters of "basmaci" rebellions in the past. Evidently, such valleys on the border of the "empire" had even higher strategic importance because they represented an external sensitive boundary.

In the new destinations they maintained folklore and traditions and also clan solidarity attitudes. In fact, such a conservative character prevented true integration, and this is a major reason for recent civil war (it should be observed that the Pamir area was not involved in war). Today, several programmes for stabilization, recovering traditions, supported by Ismaili Aga Khan and further foundations, international and local, have attempted to maintain the compactness of this population.[81]

10.4.6 North Kazakhstan

The most characteristic situation is in Kazakhstan, and partly in the urban province of Kyrgyzstan, where the Russian settlement is to be considered strongly rooted, representing a further titular nationality.

In fact, the Russian presence and penetration in Kazakh culture is deep, anticipating complete assimilation in some periods (not necessarily just of the native populations to the Russian standard, but also the reversal, as a result of a kind of "steppe charm" that fascinated generations of Russians and other Europeans). Therefore, it occurred that the Kazakh population became a minority in its "own" country.[82]

In Kazakhstan today nearly everyone has spoken Russian for generations, on all levels, in the family, in private relations as well as in public circumstances; even the youngest generations do not always learn Kazakh appropriately as it is

[77] Kudaibergenova Diana T., 2015:442.

[78] Klüter Helmut, 1993; Megoran Nick, 2017.

[79] Megoran Nick, 2017.

[80] Ginzburg N.S., 1986; Reid Patryk, 2017; Panaino A., Gariboldi A., Ognibene P., 2013; besides political purposes, Garmi, Yagnobi and other populations in remote valleys were displaced, or induced to migrate, from their high mountain settlement, for functional reasons, as it was a typical policy in modern times everywhere, to integrate such populations into the state institutions.

[81] See http://www.akdn.org/where-we-work/central-asia/tajikistan/agriculture-and-food-security-tajikistan (accessed 23 April 2018).

[82] Even when Russian descendants of the old settlers argue that it is actually their "own": a quarrel about the right of primogeniture for a country that, as for many others, has been inhabited since primordial times by different populations; indeed, such definitions, and in general definitions taking into account "primitive" elements (like standard language, ethnic-cultural characteristic and religion), or even "primogeniture" rights (autochthony, ancient character of the settlement) are even more controversial in CA; see also Steinicke Ernst et al. 2016; an obsolete quarrel, but in CA very diffused as usual for young nation-states (Suyarkulova Dokhira, 2013).

perceived as an original but limited purpose language, and they learn it as a second language preferably to English or other languages used in international contexts.

A situation that necessitates a compromise must be found. From here, a set of institutional acrobatics, among them—and the most important—at a constitutional level, defines Russian as a "language of communication among the nationalities" (or "inter-ethnic" and similar, namely using slightly different formulas, adopted by other constitutional rank laws). In this way, anyone who wants to have an active social role in society (e.g. undertaking some career or looking for a public job) has to know Russian, belonging "de facto" to a population of newly defined "Kazakhstani", comprehensive of all, majorities and minorities alike.[83]

10.4.7 Identity Markers

The current national classification (Tajik, Uzbek etc.) emerged in modern times as a result of the disputes concerning the necessity to construct new geopolitical units. Therefore, it does not make much sense to determine primogeniture in CA, whether Persian/Iranian, Mongol or Turkic, or of any other origin: the CA culture can be traced back much deeper, into pre-historical times. The descent of the local population is confirmed by recent genetic analyses. It derives from different populations, which either migrated or were autochthonous, from mixed Siberian, Altaic, Mongol, Iranian or simply unknown origins, arriving in practice from all directions.[84]

Mainstream opinion considers that the Soviet politics has never been openly nationalist, and even less racist: such an opinion may sound strange, in particular to the deported and persecuted populations. In fact, Soviet culture had been inspired originally, since revolutionary times (from the beginning), by an anti-nationalist culture, the same revolution leaders often being representatives of minorities, with nationalism being considered a decadent "bourgeois" or even czarist ideology (against which socialism originally formed).[85]

Such attitudes had repercussions for the whole Soviet period and even today it is difficult to register deliberate discrimination policies against minorities. In the ethnically mixed CA, it represents a kind of heritage of the Soviet culture, but it is also something to be considered as unavoidable (in fact society being originally so mixed) in order to preserve peace on all levels.

The ethno-national variables inspired the clashes that occurred in the early 1990s in Fergana, periodically, and that have often been boiling over until current times; but it is disputed whether such opposition represented anything spontaneous, such as the consequences of a true ethnical hate or if they were simply manoeuvred operations. Sometimes they seem to have had the intention of establishing a climate of tension against non-national groups, pressing them to migrate. This is particularly the case of Russians or of Russian-assimilated people in a period when the Russian state could not exert much influence in these situations; so too for other minorities that do not have a nation-state capable of politically representing them.[86]

Evidence of politics hostile towards minorities have occurred in some circumstances, for example, in Niyazov's Turkmenistan; in those times, indeed, such politics was combined with isolationistic (rather than nationalistic) purposes, xenophobe attitudes, limitations for travelling abroad and restriction on communication and on any kind of freedom.[87]

Then it was possibly confused with a general repressive practice, with systematic abuse of civil rights, not just of those concerning language or ethnic characteristics. It is the case of some discrimination against former non-Turkmen citizens, Russian and Uzbek mainly,[88] as well as against local minorities (namely the Baloch Iranian minority, and the Uzbek nationals in Amu Darya Valley, close to Uzbek border).[89] Possibly the post-Niyazov policy will relax such authoritarian attitudes.

But it is questionable whether such an attitude does reflect a genuinely popular sentiment; on the contrary, it is common to observe evidence of a popular pluri-cultural attitude: CA citizens are proudly witnessing that "Almaty medunarodny gorod" and "Tashkent intertnacional'no mesto".[90]

[83] A question gradually becoming obsolete, with the government opting for a bilingual status, as is considered quite normal by many nation-states in the world.

[84] "The ethnic and historical continuity between the Kirgiz and the people living today under the same name in the USSR is supposed but not proved" (The Encyclopedia of Islam, 1980, "kirgiz"); see also Shahrani 1979:49–50 and Wheeler 1966:21–22; the distinction between Kyrgyz and Kazakh emerged in the middle of the eighteenth century, when the settlement process of some steppe tribes in the mountain slopes of Zhetysu (Semirečye) began; so similarly for further populations; see also Spuler 1970:476ss and Hambly 1970:147ss.

[85] Maurizio Massimo and Tomelleri Springfield Vittorio 2018.

[86] Ro'i Yaacov, 1991; Megoran Nick, 2017; Dagiev Dagikhudo, 2014.

[87] Russians "flee" Turkmenistan, http://news.bbc.co.uk/2/hi/asia-pacific/3007598.stm, Friday, 20 June 2003 (accessed 23 April 2018).

[88] Regarding Russian students' enrolment in universities, see "Turkmenistan: Russian Students Targeted". Institute for War and Peace Reporting, 21 February 2005.

[89] The news reports periodically of discrimination cases; it is to consider that the sources about Turkmenistan are especially uncertain (Anceschi Luca, 2017).

[90] Jelen Igor, 2002, interviews.

Fig. 10.2 Kazakhstan, Astana (Nursultan), former "Virgin Lands campaign" hub (Soviet city side), August 2017, "chruščëvka" neighbourhood. (Photo: Igor Jelen)

10.4.8 Migrations and Mobility

The population of the CA has always been characterized by a high degree of mobility, in all senses. It obviously happened because of migrations, forced or induced, consequent development of cultural ties, and mutual assimilation; it also possibly happened because of residual transhumance habits, commercial activities or religious traditions—pilgrimages and visits to sanctuaries.

This particularly so in Soviet times, with populations maintaining rather mobile habits even when controlled and subject to a "passport" rule. They moved frequently apparently for official missions or work commitments; so too for military service, that was usually practised in different places and often very distant from usual residency, representing an instrument to pursue a further degree of inter-ethnic assimilation in times of "slijanje" politics.[91] Thus, for study and education purposes, young students and also children sometimes enrolled in "internaty" and colleges far from home (but often all these reasons were simply pretextual and mobility was simply an expedient in order to escape the pressure of totalitarian power).

The same planning (pseudo-colonial) practices persuaded the people to move, following commands and sometimes opportunities (the opening of new factories, organization of "campaigns" like that of "tseline", construction of new industrial towns or of "chruščëvka" neighbourhoods); keeping the population mobile can sometimes be considered a government method in itself (Fig. 10.2).

Such lifestyles continue even today and are fostered by the easiness of travelling over CIS space, where a visa is not required, and presumably by the presence of local "colonies" of co-nationals (e.g. of Tajik and Kyrgyz in Moscow), helping in the travel organization. Today, seasonally around 10% of Uzbekistan's labour forces, and 20% of the Kyrgyz and Tajik ones, are abroad, mostly in Russia. Nursultan, Atyrau and other Kazakh towns and industries (considering the hydrocarbon and construction industries) are offering work, but this percentage may change unexpectedly considering the rapidity of economics and international processes (e.g. today due to crises consequent to the lowering oil prices especially affecting Russian economics).

Today such mobility has different motivation that can be considered either cultural or demographic, with travel also depending on life cycles, family traditions, gender and age; for example, young people and students usually look for a chance to transfer to some other countries as a life experience, in search of better opportunities; often the intention of migration is related to safety perception, or to a similar question affecting the scarce confidence of citizens concerning their own country's future.

Economic motivation is still prevalent, with individuals followed by families and entire groups, undertaking a certain itinerary while trying to escape misery and precariousness; in some areas it assumes a mass dimension, provoking a destructive impact through entire societies (in general the more the original community is closed, the higher is its

[91]Jelen Igor, 2002, interview n.14; for example, many Kyrgyz and Tajiks were sent to, or applied for, maritime destinations; it is to be considered that in SU much of the usually civil functions were militarized with ordinary citizens subject to military obligations even in daily activities, often just "commanded" to move anywhere due to service duties—therefore "komandirovka".

demographic–social fragility, as, for example, mountain valleys or some urban neighbourhoods).

The out-migration of youngsters and of young women represents a particularly sensitive and popularly perceived problem (especially in some poor regions, during and after the civil war);[92] these in particular demonstrate as inclined to move to rich countries because of combined marriages or other kinds of relations, potentially incurring the risk of exploitation.

A problem which is not visible but exerts a high emotional impact is perceived in terms of the need to protect the national young generation (mixing a sentiment of diffidence against foreigners). It is a non-secondary element of political involution, with people clamouring for a more authoritarian government capable of stopping such intolerable speculations.

10.4.9 Demographic Trends

Local society experienced a baby boom, namely a demographic rapid increase in the 1970s (a decade later than the Russian European population that increased a great deal in the post-WWII period). This represented a non-secondary reason for the emergence of a conviction, that Muslim CA society would overcome the Russian (European) power in CA and that the SU—and its Russian component especially—would have no alternative but to abandon the CA space (unless engaged in a series of disastrous de-colonization wars). Carrère D'Encausse described this situation in terms of a strategic risk arising in those times, prefiguring the "explosion" of the empire ("empire éclaté").[93]

Indeed the demographic situation changed rather quickly (demonstrating that those predictions were exaggerated); it stabilized and soon started to worsen for the local Asiatic and Muslim components too, especially for the numbers reporting the average number of natural born children per woman (as expected indeed, considering the general trend in the rest of the world).

Today the demography still represents a weak aspect of local politics, with neo-nationalistic authorities pursuing policies promoting increase, either by supporting demographic natural development (e.g. improving health and motherhood care, predisposing fiscal exemptions for numerous families) or by incentivizing immigration (especially by nationals residing abroad), or internal migration (changing the migration towards a characteristic of transition, namely the return to the village, a kind of post-Soviet counter-urbanization).

Official data report that demography in all the countries of CA is increasing steadily (Tables 10.2 and 10.7); but there are contradictory interpretations, considering natural changes, displacements and conflicts, temporary migration and commuting, family reunification and other elements making any estimation sometimes difficult (sometime the deliberate distortion of such numbers due to different reasons has also been hypothesized).

In some cases, demographic figures remain critical, continuing to increase dramatically especially in already high-density areas. This is the case of Fergana and in other metropolitan areas especially in Uzbekistan; this country has the only urban and densely populated areas in the region, with a high proportion of young people making up the population, namely 34.1% aged up to 14 (in 2008), signifying one of the youngest countries in the world. Considering that Uzbekistan accounts for about one-half of the total CA population, this figure will surely have some impact.

The recent economic growth happened over a rather territorial and uneven manner and configured new migration flows inside the CA republics, mitigating the demographic surplus over time (at least in terms of employment deficit); some regions, such as Kazakhstan, mainly function as poles of attraction for internal migration, rapidly becoming extended urbanized areas (thanks to HC-driven industry); it is another element evidencing the necessity of an inter-state coordination.[94]

It would generally seem that a certain disequilibrium between urban and sparsely populated peripheries is arising, starting mass migration without truly justified motivations (with people simply abandoning certain places perceived as marginalized, starting an emulation snowball effect). It is to say that until now such migrations did not provoke excessive problems for metropolitan areas that were the destinations of the migrations, possibly as planning practice appears still effective, maintaining an acceptable standard in welfare ser-

[92] Usually "young males" is the category most inclined to migrate, but the migration of girls and young women signifies, rather than the preoccupation of a patriarchal sense of pride, a kind of sense of menace concerning the biological future of the nation; it characterizes many developing countries, perceiving themselves as exposed to the risk of human exploitation, with the most active cohorts of the population attracted to powerful and richest countries; this law prescribes a ban for women under the age of 23 from travelling abroad without a parent or guardian, with the purpose of "increased morality and preservation of the gene pool" passed in the Kyrgyz parliament in June 2013; Trilling David, 12 June 2013, https://eurasianet.org/s/kyrgyzstan-passes-controversial-girl-travel-ban (accessed 23 April 2018); Kyrgyzstan Passes Controversial Girl Travel Ban.

[93] Carrère D'Encausse, 1978; Spoorenberg Thomas, 2015:115–133; Shabad, 1980.

[94] See about the recent accident involving migrants, https://eurasianet.org/s/behind-uzbekistans-bus-tragedy-a-story-of-grinding-despair (accessed 23 April 2018): Behind Uzbekistan's Bus Tragedy, a Story of Grinding Despair. Only five people onboard were able to jump to safety out of the vehicle, leaving 52 migrant laborers dead, 19 January 2018.

Table 10.2 Central Asia: total population

	2012	2013	2014	2015	2016	2017	2018
Kazakhstan	16,792,089	17,035,550	17,288,285	17,542,806	17,794,055	18,037,776	18,276,499
Turkmenistan	5,267,900	5,366,375	5,466,328	5,565,284	5,662,372	5,757,669	5,850,908
Uzbekistan	29,774,500	30,243,200	30,757,700	31,298,900	31,847,900	32,388,600	32,955,400
Tajikistan	7,874,835	8,059,769	8,252,833	8,454,028	8,663,579	8,880,268	9,100,837
Kyrgyz Republic	5,607,200	5,719,600	5,835,500	5,956,900	6,079,500	6,198,200	6,315,800

Data source: Data Bank, World Development Indicator, The World Bank; https://databank.worldbank.org/reports.aspx?source=2&series=SP.POP.TOTL&country=#. (Accessed 19 February 2020)

Table 10.3 Central Asia: improved water source (% of population with access)

	2012	2013	2014	2015	2016
Kazakhstan	93.1	93.0	92.9	92.9	–
Turkmenistan	–	–	–	–	–
Uzbekistan	87.3	–	–	–	–
Tajikistan	71.7	72.7	73.7	73.8	–
Kyrgyz Republic	87.6	88.4	89.2	90.0	–

Data source: WHO/UNICEF Joint Monitoring Programme (JMP) for Water Supply and Sanitation, https://data.worldbank.org/indicator?tab=all (Accessed 24 October 2017)

Table 10.5 Central Asia: CO_2 emissions (metric tonnes per capita)

	2008	2009	2010	2011	2012	2013	2014
Kazakhstan	14.8	13.2	15.2	15.6	14.5	15.4	14.3
Turkmenistan	11.5	10.0	11.2	12.0	12.3	12.4	12.5
Uzbekistan	4.5	3.8	3.6	3.8	3.8	3.4	3.4
Tajikistan	0.4	0.3	0.3	0.3	0.3	0.4	0.6
Kyrgyz Republic	1.4	1.2	1.1	1.3	1.8	1.7	1.6

Data source: Data Bank, World Development Indicator, The World Bank, https://databank.worldbank.org/reports.aspx?source=2&series=EN.ATM.CO2E.PC&country=# (Accessed 20 February 2020)

Table 10.4 Central Asia: access to electricity (% of the population)

	2012	2013	2014	2015	2016	2017
Kazakhstan	99.8	99.9	100.0	100.0	100	100.0
Turkmenistan	99.9	100.0	100.0	100.0	100	100.0
Uzbekistan	98.8	99.9	100.0	99.9	100	100.0
Tajikistan	99.1	99.2	100.0	99.6	99.8	99.3
Kyrgyz Republic	99.8	99.8	99.8	99.9	99.9	100.0

Data source: World Bank, Sustainable Energy for All (SE4ALL) database from the SE4ALL Global Tracking Framework led jointly by the World Bank, International Energy Agency, and the Energy Sector Management Assistance Programme, https://databank.worldbank.org/reports.aspx?source=2&series=EG.ELC.ACCS.ZS&country=# (Accessed 19 February 2020)

vices (consider general life condition as evidenced in Tables 10.3, 10.4, 10.5 and 10.6).

The situation today is far from satisfying (as considered by neo-nationalistic authorities), but it is stabilizing, with basic indicators steadily improving (Tables 10.8 and 10.9); the migration of "vozvraščentsy", typical of transition times, has stopped (and in some cases reversed), eventually "normalizing" as multi-purpose travel (or temporary transfer), occurring periodically in individuals' life cycles (not induced by particular necessity).[95]

[95] Such movements seem to transform into a kind of continental-wide commuting, with groups of Tajiks and Kyrgyz settling down in Moscow, where they represent a significant minority (considering that, media refer, Russia became for a while a major immigration target in the world, actually the second after the United States), but maintaining a rather mobile attitude, without deciding to move there definitively, periodically returning home.

10.5 Society in Evolution: Rapidly Changing

10.5.1 Social Stratification and Mobility, After Transition

In transition times, society structure was subject to high stress (whenever it did not collapse as expected by the most pessimistic scenarios). The changes occurred under the control of state apparatus that demonstrated an autocratic attitude. This fact prefigured an incongruence between a highly mobile society, rapidly changing, and rather "lazy" and conservative institutions.

In this context new references are emerging, exerting a potentially revolutionary character: transition to new power sometimes occurs with violence, sometimes in a Soviet-like "frozen" environment, with grey-dressed "aparatčiki" trying to appear as the guarantors of continuity, making decisions in the backstage of the official politics (although apparently following democratic procedures).

The imposition of new local national majorities has changed the social composition profoundly with the affirmation of a new titular eponymous national society (even when more or less ethnically mixed). Almost everywhere Russian has remained the "lingua franca", but it has suffered because of the spread of other idioms, diffused internationally and regionally, especially for those born into the digital generation, who do not show any "nostalgia" towards the Soviet past.

Table 10.6 Central Asia: forest area (sq. km)

	2011	2012	2013	2014	2015	2016
Kazakhstan	33,090.0	33,090.0	33,090.0	33,090.0	33,090.0	33,090.0
Turkmenistan	41,270.0	41,270.0	41,270.0	41,270.0	41,270.0	41,270.0
Uzbekistan	32,643.8	32,532.6	32,421.4	32,310.2	32,199.0	32,087.8
Tajikistan	4104.0	4108.0	4112.0	4116.0	4120.0	4124.0
Kyrgyz Republic	6690.0	6610.0	6530.0	6450.0	6370.0	6290.0

Data source: Data Bank, World Development Indicator, The World Bank, https://databank.worldbank.org/reports.aspx?source=2&series=AG.LND.FRST.K2&country=# (Accessed 20 February 2020)

Table 10.7 Central Asia: population growth (annual %)

	2012	2013	2014	2015	2016	2017
Kazakhstan	1.4	1.4	1.4	1.5	1.4	1.3
Turkmenistan	1.7	1.8	1.8	1.7	1.7	1.6
Uzbekistan	1.4	1.5	1.6	1.7	1.7	1.7
Tajikistan	2.2	2.3	2.3	2.4	2.4	2.4
Kyrgyz Republic	1.6	1.9	2.0	2.0	2.0	1.8

Data source: Data Bank, World Development Indicator, The World Bank, https://databank.worldbank.org/reports.aspx?source=2&series=SP.POP.GROW&country=# (Accessed 20 February 2020) 25/11/2018

Table 10.8 Central Asia: life expectancy at birth, total (years)

	2011	2012	2013	2014	2015	2016
Kazakhstan	68.9	69.6	70.4	71.6	72.0	72.3
Turkmenistan	66.9	67.1	67.3	67.5	67.7	67.8
Uzbekistan	69.9	70.1	70.4	70.6	70.9	71.1
Tajikistan	69.0	69.3	69.6	69.8	70.1	70.3
Kyrgyz Republic	69.6	70.0	70.2	70.4	70.6	70.9

Data source: Data Bank, World Development Indicator, The World Bank, https://databank.worldbank.org/reports.aspx?source=2&series=SP.DYN.LE00.IN&country=# (Accessed 20 February 2020)

10.5.2 Social Life as a Game of Dissimulations

One must consider that the Soviet rule often did not destroy the social "forms", but it just overlaid them; sometimes it applied a kind of social "calques", with old meanings (contents) superimposing on them new forms, generating several adaptations.

It is the case of several peripheral institutions and even of party memberships, with each clan or family unit usually "enrolling" their own component in order to have a (supposedly) friendly "insider" in the institution. It is the case of production units (kolkhoz or leschoz) and organization charts, that superimposed the pre-existing social forms (like brotherhoods, extended families, village communities), therefore reproducing them in the new order as a traditional schema.[96]

This phenomenon affects all moments and all segments of individual and social life, shaping a conformist attitude that was a strategy to survive in Soviet times and as also occurs today in the renewed authoritarian context to some extent.

Examples of such adaptation, based on widespread dissimulation, are official missions, the "komandirovka", arranged as private travel or business, or also as family visits or as pilgrimages, or private business arranged illegally in public offices. Further cases are those of "aksakal" and mullah, the traditional authority in the villages becoming "brigadiers" and organizing the leschoz work. So too, for brotherhoods becoming the reference for corporative organizations, for example, of taxi drivers or bazaar traders (this last example indeed signifies belonging to a clan organization, a form that seems to pervade the post-Soviet society transversally).[97]

Further examples of such confusion can be considered in the "cult" for monuments and sacred places, meaning confusion between meanings, possibly between party and religious rituals, "hajj" and "komandirovka" (prototypically the mission to Moscow), new technologies and the supposed "miracles" as propagandized by the power. In fact, the manipulation of symbols for producing persuasion has produced several paradoxical effects.

Such situations are not necessarily in contradiction with each other. The traditional pre-existing forms can survive, re-emerging in diverse contexts, assuming different meanings. In fact, the new power uses the same strategy to some extent, manipulating symbols and meanings in order to get consensus.

For instance, in rural areas the re-emerging tribal values are not always in contradiction with those of the modern state's values, where the assimilation of such archaic institutions is absorbed into the new regulations; for example, the elderly persons councils in the villages (the "aksakal") are assumed by the Kyrgyz Constitution as officials and local judges. The same for mahalla in Uzbek cities, for "chruščëvka" dwellings, for rural communities and new apartment buildings, for professional organizations, housing city-quarters organizations, based on traditional hierarchies:

[96] Esenova Saulesh, 1998.

[97] Bennigsen A.- Lemercier Quelquejay C., a cura di Fasana E., 1990; Esenova Saulesh, 1998.

Table 10.9 Central Asia: mortality rate, under five years (per 1000 live births)

	2011	2012	2013	2014	2015	2016	2017	2018
Kazakhstan	18.2	16.2	14.5	13.1	11.9	11.0	10.4	9.9
Turkmenistan	57.9	56.1	54.2	52.4	50.7	49	47.4	45.8
Uzbekistan	34.1	31.9	29.8	27.8	25.9	24.2	22.7	21.4
Tajikistan	41.9	40.7	39.7	38.7	37.8	36.9	35.9	34.8
Kyrgyz Republic	27.7	26.1	24.7	23.5	22.3	21.1	20	18.9

Data source: Data Bank, World Development Indicator, The World Bank, https://databank.worldbank.org/reports.aspx?source=2&series=SH.DYN.MORT&country=# (Accessed 20 February 2020)

the traditional forms, representing self-organization (spontaneous) attitudes, are not necessarily in contradiction with modern changes and adaptations.[98]

10.5.3 Re-emergence of Traditional Values and Practices

The collapse of the all-encompassing Soviet state had an especially dramatic impact on daily practice, often spreading a kind of panic (finally provoking clashes like the ones in 1989–1991). It seemed that the same idea of organized life would disappear, leaving an intolerable vacuum. In these circumstances people instinctively looked for new forms of coordination, or solidarity groups, in order to provide for basic needs (safety, wealth, food, housing, education, mobility etc.).[99]

This is the case of extended families and village communities, as well of tribal hierarchies (which had never really completely disappeared), of aksakal and patriarchal (masculine and gerontocratic) authority, and then especially of traditional-religious references. This consequently meant the recovery and application of such "traditional" forms in current situations (a kind of regression, with obvious consequences such as nepotism, familyism, small-scale corruption, possibly a residual of traditional communitarian gift economics).[100]

Usually such informal procedures remain invisible to the eyes of outsiders (foreigners), or of the authorities. They were embodied in community rituals, whose affiliates would never properly recognize them as illegal simply because they are perfectly embedded into the community values and considered as usual (and eventually denied as prescribed by "takkya" practice, namely by the right to lie, in order to defend themselves and their organization in case of threat; see later).

The same for mahalla in urban Uzbekistan and other Fergana cities, in a "chruščëvka" or in new apartment buildings, overlapping pre-existing solidarity forms or simply developing a mutual help system in communitarian complicity, among neighbours (therefore they sometimes do not want to abandon such Soviet-time quarters, even when they are in poor state of conservation, because the "invisible" landscape made up of social ties still works, especially for solidarity and assistance functions) (Fig. 10.2).

So too for religion in CA, in particular those practised in popular forms, signifying diversification often outside of the recognized Islamic standard. In fact, it relies on rituals often inspired by Sufi practice, often considered by official Islam as unacceptable, or even heretical. It means the organization of cults and pilgrimages based on mausoleums and tombs of saints, possibly leaning towards sectarianism, namely the attending and the enrolment into brotherhoods (potentially secret ones), based on a specific cult or on a ritual oath. Often, they entail practices of non-Islamic rituals, connected with elements such as sacred rocks, holy trees, prayer flags, fortune-teller sanctuaries and other devotional formats.[101]

These are possibly traces of other religions, perhaps reflecting tribal cults, sometimes regenerating old cults or just inventing new ones (in the frame of a popular spiritualism, spontaneously forming, above all, in times of changes)[102] (Fig. 10.3). This is particularly the example of Sufi brotherhoods and for the so-called parallel and non-official forms of Islam, which are difficult to define since Islam is not characterized by a very structured hierarchy.[103]

The rural areas, especially in Kyrgyzstan, retain self-defined shamans (possibly echoing in the latter case the Siberian origin of this population), who assume different functions, for example, of the traditional medical doctor, or more simply herb and medicinal experts, to be consulted in a wider sense (e.g. organizing marriages or mediating con-

[98] Haghayeghi Mehrdad, 1997:338; Cossutta, 1996a, b; https://www.hrw.org/reports/2003/uzbekistan0903/5.htm (accessed 9 May 2018): Role and functions of mahalla committees; particularly significant is the case of the Kazakhstan established traditional council of elderly peoples as supervisory institution in a privatization operation regarding the city; see as well Jelen Igor 2002, interview n.6.

[99] Roy O., 1997; Esenova Saulesh, 1998; Megoran Nick, 2017:78fs.

[100] Lamy Frederick, 2013:146.

[101] Rasanayagam Johan, 2014:9.

[102] Muminov Ashirbek, 2014; Hanks Reuel, 2016; Radford David, 2014:17ss.

[103] Echoing the Soviet-time "red mullah", eventually considered a "false mullah", often reciprocally accusing of being heretical.

Fig. 10.3 Fergana, 1997: prayer tree close to Sulayman Mountain, Osh. (Photo: Igor Jelen)

flicts), or of tradition experts (even when sometimes just as performers, eventually for some non-religious purpose, today for tourists and visitors).[104]

Such attitudes extend to other environments, becoming a social practice, losing all forms of sacredness. This affects neighbourhoods and villages, associations and corporations that are blessed and celebrated with semi-sacred rituals and cults, superimposing the corresponding civil procedures. It gets confused with residual clan categories and with the attitude of assuming a "tribal" identity (the attitude of thinking in a sectarist-factional way, possibly contradicting such concepts as rule of law and "state" as common interest).[105]

Such attitudes have weighty consequences on all levels and especially in politics. For instance, it means that instinctively presidents would hire from their original clan and then get structurally linked to the collaborators, like praetorians, military and public officials, ambassadors, political representatives and high-rank managers, to whom he would delegate parts of his personal power.

In this way the political game assumes a segmentary form, sometimes taking inspiration from traditional pre-existing models (even clan or hordes, as well as brotherhoods or ethnic affinities), then further developing forms of non-transparent internal struggles. Sometimes, this would re-emerge in the forms of regional affiliations; the map of such affiliations is extremely differentiated, and comprehensibly of extreme interest for the decision makers (e.g. it comprehends Tashkent, Samarqand or Fergana clan in Uzbekistan, Kuljaby and Khujandy, Garmi and Pamir in Tajikistan).

Sometimes it derives from other unpredictable forms of aggregation of interests formed following the privatization of assets after independence, in transition times (e.g. those representing "hard powers", financial-corporate groups, HC lobbies, some strategic industrial sectors or some parallel or hidden powers, as reported by the media defining them picturesquely as "satrapia" or "mafia").[106]

Today all these merge into a general schema of patronal-clientele politics, based on arbitrary and personalist attributions, possibly shaping future political problems.

10.5.4 Brotherhoods, Sects, Charismatic Movements and Social Segmentation

The brotherhood is typical of the Sufi tradition, diffused in CA, where it is perceived as domestic. It re-emerged after the Soviet era, when it was functioning to some extent as an invisible hierarchical organization, even when often in the frame of a "role play", evidencing a certain attitude to elusive social behaviours.[107]

It is possibly a legacy of the previous regime's repression that induced a kind of fear to exhibit affiliations, and it still persists today. It may also be a legacy of the ancestral nomad organization (namely the difficulty in assuming a definitive collocation, also in identitarian terms); or also a legacy of the scarcely structured Islamic culture, different from modernized apparatuses, heavily territorialized and structured (and also a consequence of the fear induced by the twentieth-century totalitarian experience).

[104] Radford David, 2014:17.

[105] For example, the "trade mafia" is considered the cause of the cited conflicts that occurred in the Osh oblast, in 1990 (Elebayeva, 1992:80; Roy, 1997:199; Rasanayagam Johan, 2014:11).

[106] Esenova Saulesh, 1998; Kudaibergenova Diana, 2015; such social formats function usually in an invisible way, not necessarily overlapping the forms of official politics (parties, trade unions, corporations, associations); in fact the "klanovost" contributes to reproducing the concept of sectarian ways of thinking, definitively biased by some egoistic interest (a synonym of "anti-social" or anti-state, incapable of contributing to the public interests). This way of operating developed and expanded also on a purely political level, namely at top positions, consequently feeding some backstage struggles for the power, the predominant political form in current scenario.

[107] Even the secret police, KGB or similar, was not capable of penetrating it; thanks to the use of takkya by their members, namely making large use of dissimulation, they could defend themselves against the despotic power (Bennigsen A.- Lemercier Quelquejay C., a cura di Fasana E., 1990).

In principle, they are informal and flexible organizations, and this has made it possible for the same organizations to survive in different situations, overlying any kind of structured organization. During the Soviet epoch as well as today, the "sheik" (the brotherhood leader), or similarly other local leaders ("bay", the "bi", or whatever defined hierarchical figure), could be a leschoz manager or a Party official.

Sometimes the character of the brotherhood is so vague that the same followers do not even know exactly that they take part in the same brotherhood.[108] They simply respect rituals and habits or take part in some communities as common daily practices, without knowing exactly what this signifies.

In general, the brotherhoods do not represent a political risk; however, they show an evident tendency for sectarianism, with small groups under the influence of a charismatic "sheik", which may also degenerate into some hidden organization, inclined to radicalize.

10.5.5 Culture, Education, Leisure and Art as a Welfare System

The current idea of culture still shows a certain influence from the Soviet legacy. It is hierarchically organized and strictly controlled by the power structure. In those times there was no true concept of free time and leisure (considered as a bourgeois or individual attitude), and any entertainment was inspired by an idea of (communist) education and of social improvement. Much attention was devoted to the organization of great events, intended as mass-celebrations, that in time would lose their true significance, becoming mere liturgies.[109]

In the background there was a general all-encompassing idea of the culture as a socially useful function, to be incentivized but strongly regularized for groups and individuals that had the possibility (and the duty) to practice any kind of physical or artistic activity, integrated into the frame of the official institutions (school, university, sports association).

It involved other civic services, such as education, holidays and free-time (post-work) organization, spreading the practices of sports, open-air and nature activities (e.g. trekking and alpinism), performative and figurative arts. This was to be considered in all contexts, for traditional or modern activities, for example, classical ballet and music, traditional folklore and local ethnographic production.

This idea of culture is—and was—rather structured, and carefully considered. Even during transition, in the worse times of crisis, culture remained a priority, with structures that were always open and perfectly functioning (theatre, opera, ballet, but also museums and local activities), with organizations and staff remaining operative, even when relying on scarce resources and on personnel literally without salaries (also because, obviously, neither the new national currency was recognized yet, nor did the new state have an operative budget).

10.5.6 Education: Soviet Heritage, Obsolescence and Transition

Similar to culture, the same can be said of education that still evidences much of the same stereotypes as well as the advantages of the Soviet way.

In the past education was materialistically oriented, aimed at creating perfect executors; it was specialized in hard sciences, while the humanities were considered rather as a method for indoctrinating society, concerning disciplines like "istoricheskij i dialekticheskij materializm", actually a socialist version of literature, history and geography.

In fact that education system—as for any social function in Soviet times—was incapable of any self-reform and soon became characterized by its obsolescence: a caricature of the communist regime, as can be expected for such an isolated system and a further demonstration of the fracture that occurred between institutions and reality.[110]

This was especially evident in college and university programmes that did not include disciplines such as business, marketing and corporate administration, which are typically "open economics" subjects; similarly for foreign languages, literature (with exception of socialist writers) and international studies (when not at an elitarian level, for diplomacy); geography was limited to physical matters with no mention of sensitive political issues (like ecological impacts or socio-territorial inequalities) where teachers and academics carefully tried to avoid privileging more neutral matters.[111]

This gap would become particularly evident in the context of the cross over to digital electronic and a post-Fordist paradigm. Students continued to study theoretical, and politically neutral, sciences in a repetitive and notionistic-mechanical manner paradoxically becoming "overqualified" in these matters yet losing any connection with reality.

Nonetheless, such institutionalized ways of studying were without alternatives and were slowly becoming the symbol of a decadence (of a ritualized self-referential education system) of the apparatus.[112] Study programmes were oriented to the necessity of the central apparatus (evidently refractory to

[108] Rashid, 1994; Bennigsen A.- Lemercier Quelquejay C., a cura di Fasana E., 1990.

[109] Moxley Alissa, 2013:65; in Soviet times they were defined as "culture workers", echoing a materialistic approach.

[110] Akyildiz Sevket, Carlson Richard, 2013:27.

[111] Hughes Thomas P., 2004.

[112] Gaynor Kelly Lee, 2017; author's interview.

Fig. 10.4 Kyrgyzstan, Katran Village, village school, August 1994. (Photo: Igor Jelen)

market economics), specialized in maths and physics, engineering for the heavy Fordist industry, warfare, aerospace and similar kinds (also these rather in a manner soon becoming obsolete).

This was also the case of schools that were considered the basic institution where Soviet citizens would be shaped, signifying a vehicle to be filled with instructions imposed by the power structures. Towards this institution, students and families are (and were) used to demonstrating an extraordinary sense of respect, bordering on sacredness. Possibly going back to school in September is even today (as it was in the previous regime) one of the most important civil social events.

Rural and urban families—also from the remote steppe and mountain villages—even when very poor, prepared clothes and materials for months in advance (pencils, notebooks etc.) for their children so that they could attend the village school in a dignified manner.

In transition times, the institutional collapse did not involve the education network much, which continued to function at local levels; consequent reforms were rather slow, and, at least for primary levels, school remains not much different from how it was in Soviet times. Entering into a school classroom, even until recently, one had the impression of entering into a gallery of the previous regime's idols. There was a list of cosmonauts and scientists (usually Russian), with the Mendeleev (Russian scientist) table periodical elements hanging everywhere (Fig. 10.4). Sport, music and art facilities invite the practice of all kinds of the same activities especially in the context of a school based on severe selection processes and exams.

Such a Soviet-style education system, whenever capillary and efficient, was destined (especially in later times, and especially for higher school levels) to go off track: all students were studying basic hard sciences with programmes that had not been updated for possibly over half a century, excluding any subject useful to interpret any evolution in knowledge up to current times.

The accelerated changes created "de facto" an entire generation characterized by obsolete education standards. This for a population not just made up of "illiterates", but by a majority of individuals simply incapable of adapting to the new reality or of participating in internationalization processes and also incapable of solving concrete daily problems, tendentially being prone to obedience, expecting the "power" to find a solution for everything.

However, the previous system also had some advantages. Education was free and easily accessible at all levels with a system of scholarships (sometimes compulsory), making it possible for anyone (even for people from remote villages) to attend university, possibly choosing to go to college to enrol by themselves, quite far from their original residence. This was also a method used to facilitate uprooting and to incentivize mobility inside the "empire", similar to military service and other civil duties.[113]

The current apparent high level of literacy is the legacy of that organization, when "alphabetization" was perceived rather as a mandatory duty and it was also used as a device to induce obedience. In fact—wherever it is measured—literacy is very high even in poor countries, such as Tajikistan devastated by war and mass migration.[114]

[113] Jelen Igor, 2002, interviews.

[114] According to the statistic register in 2011, there were 170 "physicians" per 100,000 people in Tajikistan; Educational System in Tajikistan, 2018; UNDP, 2016; Human Development Report, accessed 6.7.2018; PWC 2010:6.

Some data are encouraging, but the situation is different depending on the context, if it is rural or urban, and depending on the organizational capability of the new state. The major risk is the one represented by the worsening of state-organization, with regression in public school attendance. The risk of economic or gender bias due to conflicts, worsening of safety or of economic situations is particularly evident. Some sources regarding Tajikistan schools refer that a percentage, between 4.6% and 19.4%, of children do not attend or are excluded from primary education (possibly because engaged in work, or because they attend the parallel madrasa school), and of them an even higher percentage of girls, due to the gender factor.[115]

10.5.7 Education at Different Levels

The education system suffered in particular during transition (as for all other welfare services) due to shortages. This was especially the case for the high school and university system (in Soviet times it was completely subsidized), functioning almost without resources. On the contrary, at a ground level, kindergarten and primary schools could be locally self-organized to some extent.

The new authorities started to carry out reforms rather late trying to update programmes but then encountering severe difficulties because of economic crises, cuts in social spending and teacher's low wages. The consequences of such situations (sources report continuously) are the spread of corruption, of threats and other similar problems among teachers who were (as usual in the public system in a transition phase) often marginalized (also because they represent the institutional culture, considered as symbols of the ancient regime, with difficult adaptation to the new situation).

In general, such institutions proved particularly difficult to reform. The quality of a welfare system cannot be simply planned with decrees and liberalization measures, but it needs the gradual accumulation of new "know-how" as well as cultural and organizational capabilities (internal control, evaluation system, teachers' experience, school reputations).

The school system suffered especially because of the crisis of centralistic planning; it often has had to challenge the emergence, in the kishlaks, of a parallel religious education system (essentially in the village maktab or madrasa) and in the cities, of well-financed private middle-high class schools and colleges destined for wealthier and privileged families (as usually happens in situations of public system crisis).

Besides this, lately it has been possible to observe a recovery and even the flourishing of new high schools and universities and courses. This is to be considered in part as a reaction to the difficulty of maintaining a high degree of quality in public Soviet-style universities, in part as the expression of a new entrepreneurship (with private individuals investing in higher education facilities), in part as a consequence of the inflow of foreign institutions. Several examples of such institutions have changed the panorama of the CA education system rapidly, at least at a university level. The current offer of courses is astonishing, making it possible to imagine that a true improvement in the knowledge system has taken place (and which is possibly in contradiction with the involution of the political reality).

Besides the standard public universities, that continue to recover and improve, in just a few years many new higher education institutions have been set up; among them the University of Central Asia with several campuses in Kyrgyzstan, Tajikistan and Kazakhstan as well as universities mainly dedicated to minority cultures like the Deutsch-Kasachische Universität (DKU) in Almaty (attended not just by a German minority, about 1.1% of the population, but in general by pupils learning German), and the "Slavonic universities", devoted mainly to the Slavonic minorities (meaning not just Russian, but also Ukraine, Belarus and for further Slavonic languages).

Tashkent is the site of the local branches of the Moscow State University and of other top institutions such as the Singapore Institute for Management, Westminster International University, Turin Polytechnic, Russian University for Oil and Gas, and INHA University.[116]

Furthermore, it is possible to observe a multiplication of international seminars, summer schools and other educational events above all in culturally prestigious centres such as Samarqand, especially for learning humanities, languages, cultures, and in other main cities—Tashkent, Dushanbe and Almaty—for business, economic geography and foreign literature. Further institutions—such as the Eurasian Studies department at Nazarbayev University in Nursultan—organize courses in foreign language (besides Russian and Kazakh).

A number of new high standard management-oriented schools have been founded, some with Russian and English as the teaching language, like KIMEP and KazGU ("gosudarstvennyj universitet") in Almaty and in Nursultan, meaning the possibility to access world cultural institutions characterized by numerous international agreements.

Some centres specialize in education functions (humanities, business, medicine) possibly developing specific expertise, also attended by foreigners, specializing in local languages, cultures, archaeology or ethnography (offering comprehensibly extraordinary opportunity for fieldwork).

[115]Educational System in Tajikistan, 2018; UNDP, 2016; Human Development Report, accessed 6.7.2018; Akyildiz Sevket and Carlson Richard 2013:26.

[116]PWC 2016:3; Gaynor Kelly Lee, 2017.

All this has meant a multilingualism trend (as in the whole world) capable of offering a wide spectrum of subjects and high-quality education to new generations. Most of such programmes are heavily sponsored by the governments—especially HC "rentier states" that have invested greatly in higher education and in connected research institutions, just as the Kazakh case demonstrates.[117]

The new university has become the key element (as usual freedom in culture and quality in education) for the formation of a new generation of responsible and aware citizens—capable of participating in social life, while spontaneously elaborating a set of values. In fact, it is the definition of a civil society capable of administering public and private wealth, becoming autonomous to some extent (not capable of being blackmailed) from the power structure. It is not easy to achieve such a development (it is not a linear evolution), since it usually takes generations and it depends much on the attitude of the power, which should be aware of this, supporting such attitudes.[118]

10.6 Popular Culture and Social Organization

10.6.1 Entertainment, Sport, Hobbies, Social Rituals, New Style of Life

As said, the popular idea of culture still highlights the Soviet legacy, with culture intended as an organized activity, not really comprehending the true idea of leisure and entertainment.

In the new era, societies have started to change their attitudes, assuming the idea of leisure as a free individual choice, developing and imitating a set of habits typical of evolved societies. At the same time governments have understood why they had the chance to use such expectations (developing a consumerist culture) in order to condition the wider public opinion and fundamentally to gain consensus.

In fact, post-Soviet power rediscovered the potential of free time activities—suitable for being planned to some extent—as socio-political aesthetics. This can be seen in the case of sport, entertainment, and of popular culture, big events, organized and diffused with ICT, internet, TV, cinema and other devices in general.[119]

[117]Prefiguring a risk for political bias (Gaynor Kelly Lee, 2017); experience proves that it is difficult to invest in such activities, like planning the functioning of a top research or university centre; in fact the quality standard of such institutions may be achieved over time, accumulating reputation, developing a set of further quality functions (above all the formation of a new class of expert and independent professors).

[118]Boboc Cristina, 2017a, b.

[119]Warf Barney, 2011.

It is the category in which the conservation issues (namely the interest of the power) appear complementary to a kind of individual freedom. This can especially be taken as the case for the practice of sports, which was very common in Soviet times, when it was one of the few elements of continuity with western societies and potentially the only way for individuals to emerge without being subject to heavy political conditioning.

Today, local societies have applied a similar concept, developing rediscovered traditional activities alongside the conventional ones, in terms of sport and leisure and eventually in touristic and folkloristic terms—considering their potential ethnographic interest. This is the case of wrestling, horse riding, archery, falcon hunting and many other practices simulating aspects and values of traditional societies; these are actualized like "ulak tartish" and "kyrz guu", repurposed as athletic or martial arts, Greek-roman wrestling, "polo", usually in the format of spectacular shows (Figs. 10.6 and 10.7).

Further activities proved easy to combine with traditional practices; it is the case of trekking, outdoor and alpinist sports, and of other traditional activities that were soon modernized and opened up to wider use—tourist attractions, sports events and national celebrations. Other activities included those ideally suitable to environmental conditions, like ice hockey in Kyrgyzstan, which was successful and which soon became one of the preferred activities for thousands of children in the village streets.

10.6.2 Social Events, Holidays, Traditional Calendar, Traditions

Often such activities intersect within local and traditional recurrences. The new society has elaborated a partially new calendar, namely a time segmentation based on events, social rituals, behaviours and cultures—that maintains the same function of coordinating society.

It is traditionally a powerful instrument with which to regulate societies, merging significances: a subtle surrogate in order to attain a "religious" (from Latin word "religare", "keeping together") legitimating potential.

Today, the resulting calendar shows an interesting syncretic character, a mix of local and global, Soviet legacy and Islamic celebrations, intertwining religious holy days like Kurman Ait (the day of sacrifice; Fig. 10.8), Orozo Ait and traditional Navruz (Fig. 10.5), finding continuity in the new context. Furthermore, it evidences typically neo-nationalist holidays, like "Independence" or "Fatherland" days, underlining a combination of Soviet and post-Soviet (like "pobeda", proletarian work "heroj", WWII resistant) celebrations that reflect the combination of identities recoded (as

Fig. 10.5 Tajikistan: poster celebrating Navruz 2017. (Photo: Igor Jelen)

Fig. 10.6 Uzbekistan, Fergana Village, 1997: wrestling training. (Photo: Igor Jelen)

in the case of the "Fatherland Defendant", 7 May) into a curious list of inter-references.

In Kazakhstan, Kyrgyzstan and Uzbekistan, predominantly Islamic countries, orthodox Christmas is officially celebrated, as well as the religious last day of Hajj; in all republics they celebrate, as for the whole world, civic holidays like 1 January, 8 March and 1 May. Transversally, they celebrate, even when renamed in different ways, possibly signifying diverse political nuances of the date 9 May, "The Great Patriotic War" or "Against Fascism Victory Day", or as recoded in Uzbekistan "Remembrance Day"; in Kyrgyzstan, 7 November survives, the "Day of the Great October Socialist Revolution", generally no longer celebrated (in Russia either); even on occasion of the 2017 centenary of the revolution, that was not celebrated at all.

The biggest feast transversally celebrated (Uzbekistan, Kyrgyzstan and Kazakhstan) is the 21–23 March Nauryz Meyramy (in 2013, depending on the moon calendar), an archaic spring holiday popular over the whole CA territory.

Then there are the political national feast days: in Kazakhstan, 7 May as "Defender of the Fatherland Day", 6 July as "Capital City Day", 30 August as "Constitution Day", 1 December as "First President Day" and 16–17 December as "Independence Day"; in Kyrgyzstan, 23 February as "Fatherland Defender's Day", 7 April as "Day of National Revolution", 5 May as "Constitution Day", 8 May as

Fig. 10.7 Kyrgyzstan, Bishkek,1997: ulak tartish playing. (Photo: Igor Jelen)

Fig. 10.8 Kyrgyzstan, Ozgoruš Village: Kurman Ait (Sacrifice Day celebration), spring 1997. (Photo: Igor Jelen)

"Remembrance Day" and 31 August as "Independence Day"; in Uzbekistan, 14 January as "Day of Defenders of the Motherland", namely "Vatan Himoyachilari kuni", 1 September as "Independence Day" or "Mustaqillik kuni" and 8 December as "Constitution Day" or "Konstitutsiya kuni". Another important seasonal date is the initial school day, celebrated everywhere on 1 October as Teacher's Day.

Further mixing is induced by the "celebration days" prescribed by the IC, especially by UN (the day of the "children work", environment, landscape, "against the mafias"), motivated by the necessity of consolidating of civic values transversally to local societies.

10.6.3 Lifestyle and Other Popular Cultural Elements

The institutional culture (education, art, performative or figurative, the so-called high culture) reflects (and is reflected in) innumerable kinds of local and spontaneous practices that have cultural significances as well as the popular and daily practised "low" culture. It is the case of individual and domestic traditions (regarding dress, everyday organization, home and private space organization), of cuisine and food habits, social customs, lifestyle, popular beliefs and rituals that represent the concrete daily aspect of culture (that in principle have to be preserved from any standardization).

Such elements segment daily life, classifying time into the categories of work and no-work, home and no-home, leisure and rest, periodically alternating with religion (prayer, pilgrimages, charity) well classified in certain moments and places (changing indeed for the typical Soviet-organized free time, to the individualistic western style vacation), as well as other social activities. Such elements are not just important in terms of their curiosity or ethnographic characteristics, but they also represent, in the context of the new postmodern reality, resources and references, both for social organization and economic development (as many examples worldwide may demonstrate).

This is the case of gastronomic traditions that are the expression of the agro-food economics, that manifest both in identity references and also as incentives for economic production—for internal consumption, but also for trade, tourism and export. The Uzbek and Tajik (and Uighur) culinary traditions represent an extraordinary variety in particular, as is usual for places with a great variety of agricultural production, which in this case is in the irrigated oasis. It is based on a rich repertoire of cereals, rice and grains, of fruit, herbs, meat preparations, either for immediate consumption or for conservation, as well as for ritual and cultural preparations (noodles "beshbarmak", and similar).

On the contrary the nomad diet was traditionally based mainly on meat and milk, and the products that derivate from them (ayran, kaymak, kurut, kümmis), namely on simple resources, but nevertheless precious regarding the modality of production and conservation in such uncontaminated and extreme environments (that rely on sophisticated techniques). The same can be said of other materials and social aspects (handicraft, popular medicine, traditional literature, social games and "passage rituals") that survived the decades of Soviet standardization. These possibly may be recovered today, potentially to incentivize and improve local production for business purposes, showing its importance for the strengthening of identity.

Such new approaches emerge in a new economy based on a responsible concept of consumerism, on "happiness" and "amenity" functions (tourism and export, above all), that could be factors that drive entire systems and that have simply been underestimated so far. So too, in a parallel way for the nomadic traditions, which manifest a precious patrimony of culture, handicraft and artisan skills, like jewellery, ceramics, gold and precious stones, textiles, carpets and rugs, felt and wool manufacture, leather and textile handicraft, wood furniture, utensils and instruments for horse riding, mobility, yurta construction kits (typical for people traditionally continuously moving); regarding tribal lifestyle, not to mention aspects of their humanistic and spiritual culture.

Such products are not just souvenirs for tourists, but they signify the extraordinary capability of local peoples to adapt to extreme conditions, today maintaining their exotic charm; and this in all circumstances, in remote mountain valleys, in steppe "auls", in mahalla and "chruščëvka" city-quarters—some of them extremely well decorated and maintained—organizing street feasts, celebrations and holidays, reported and transmitted even by the media. Some of these are collected in the ubiquitous museums (ethnographic, archaeological or whatever) of Soviet origin, repurposed in current times as "national museums", representing a unique cultural patrimony.

10.6.4 Traditions' Revival: Changes, Risks and Opportunities

Popular culture represents the basic element in understanding how effective changes have affected a certain human system; this occurs especially in times of opening up (to globalization), both in structural and political terms. Sometimes the populations demonstrate a high capability of adapting to new circumstances; for example, girls and women learn to have two kinds of dress codes, the traditional and the modern ones, adapting their wardrobe to different spaces and moments; so too for men, who learn to play modern sports, but continue to practice traditional activities and games (simulating traditional activities such as cattle breeding and horse riding), depending on circumstances. Similar expedients can be prepared for any such situation, specializing places, significances and times, representing an acceptable compromise in order to safeguard and preserve local traditions, without hindering development and opening up to the wide world.

Indeed, after the unveiling of modernity, the recovery of local and traditional cultures and landscapes has proved to represent a great potential in terms of social inspirations representing a reference to the "internal" boosting of culture. It is the case of traditional arts, of extraordinary architecture and construction styles, of material and immaterial cultures, and of a general sense of pride in unique traditions.

It is the case of Uzbek and Tajik literature, representing continuity with the Persian narrative then repurposing a millenary tradition in contrast with radical religious and "realistic" Soviet literature; just think of the fabulous "1000 nights" and of the tales of Sheherazade.

So as well for the extraordinary steppe nomadic epics—the Kyrgyz Manas and the Kazakh "batyr"—and for several others.[120] All these always represent a new motivation for cultural renewal, motivations for which, after the "greyness" of Soviet times, the whole culture of CA must try to reconvert the usual stereotypes in order to establish new identities.

10.6.5 Recent Evolution Connected to Globalization

Besides the "internal" side, namely the enhancement of neglected cultural elements, the question today is the opening up of CA space to the influence from the outside world but avoiding the exposure of the local society to indiscriminate external influences.

Post-Soviet society was not yet ready for this change, since it had been influenced and shaped by decades of

[120]Yilmaz Harun, 2013:53.

anti-foreigner propaganda. Today it is a problem, since the changes also impact all populations, affecting them on all scales: it means opening up to and accepting outside influences without perceiving this in a distorted manner (biased by some inferiority or superiority complex), and eventually managing this as a cultural and economic opportunity.

The current landscapes evidence such incongruences, for example, old bazaars, with informal relations and characteristic crowds, coexisting with crystal-sparkling shopping malls; old style habits, either religious or Soviet-culture inspired, suffering from the intrusion of westernized lifestyles; so too for the world famous Uzbek and Uighur cuisine banalized by international standardized cooking, fast food restaurants—proliferating all over the place—versus "šašlik" kiosks; and further, traditional "čaikana" conversation versus TV talk shows, horse riding and carousels versus football stadiums: indeed coexistence should be considered as possible but the experience of rapid modernization could trigger reactions and regression.

10.6.6 Organization of Big Events and Inclusion in International Network of Events

With this in mind, the recent tendency towards a new politics of big events usually organized for propaganda purposes is important to demonstrate stability and reliability, but which exerts an obvious effect on local society. It represents a tendency to opening up, improving the transparency of social procedures and also offering economic opportunities.

This is the case of the 2011 Asian Winter Games (the first time Kazakhstan hosted such events since its independence; Fig. 10.9), the 2017 Universiade Games and the 2017 Expo in Nursultan (Astana).

Fig. 10.9 VII Asian Winter Games logo, Nursultan (Astana) and Almaty, Kazakhstan

In general, Kazakhstan has demonstrated it is capable of organizing such events, on a large scale, as well as participating in a number of international moments, promoting sport and its international role, with its professional bicycle and football teams participating in competitions in western countries.

A further example is the Astana (Nursultan) economic forum,[121] as well as a long sequence of fairs held in the major CA cities, integrating them into the international trade network; Nursultan's (Astana) recent role in Syrian negotiations can be remarked on, even if under the patronage of Russia, accrediting it a prestigious diplomatic role for the Kazakh capital city.

The Kazakh Republic will possibly be imitated by the other CA countries. Further examples of such intentions are the Ashgabat Asian Indoor Games, and martial arts and other applications for such events have been put forward by other cities (Tashkent, Dushanbe, Bishkek).

Among other things, the nomination of Samarqand as the city of culture by UNESCO in the year 2020 must be underlined and that the same Samarqand, with Khiva, Bukhara and Shakhrisabz, have been included in the UNESCO World Heritage List, increasingly driving cultural tourism economics.[122] It is possibly the definitive evidence of acquired stability, making such organizational efforts possible, integrating themselves into the network of international events.

10.6.7 Communication, Mass and Social Media

ICT—diffusion and infrastructure—statistics are volatile and rapidly changing; some data report that these countries, after post-Soviet stagnation, started rapid growth, becoming kinds of "new tigers"—in terms of ICT "consumption" and use. But the situation evidences several asymmetries; the World Economic Forum's Network Readiness Index (NRI), a cited indicator concerning the updating of a country's ICT level, is high for Kazakhstan, even when still very low for Kyrgyzstan and Tajikistan.[123]

This is not necessarily a big disadvantage, at least considering the risk of over or hyper-technologization (namely disproportionate use of technology, "jumping" some adaptation steps, eventually to the detriment of social or traditional relations).[124]

This is the case of "jumping" from the phase of the traditional market places to that (to some extent fictitious) of the

[121] Kudaibergenova Diana T., 2015:443.

[122] Paskaleva Elena, 2015:427, 436; UNESCO, World Heritage Centres, accessed 23.4.2018.

[123] World Economic Forum's Network Readiness Index (NRI) (n.d.); see also Warf Barney, 2011; http://reports.weforum.org/global-information-technology-report-2016/, accessed 23.4.2018

[124] As it happened in diverse occasions in similar rapid growth situations.

modernized mall or of internet marketing, without passing through the modernization of the old picturesque bazaar and so ruining the social setting represented by that economic organization; so too, in similar circumstances for many of the described cases of "popular culture".

For social media and similar devices, the situation is different but fundamentally ambiguous. The power structures perceive them as dangerous, since their use proves to be uncontrollable and unpredictable (actually they were used in recent street rebellions, especially during "Arab Springs", and as well as by subversive organizations). On the contrary the power structures have nearly total control of mass media, press and newspaper, internet (increasingly subject to control and filters) and TV channels, which they can use and also manipulate.[125]

Television and radio in fact maintain a predominant role as source of information (and of entertainment); it results indeed in a flow of standardized information, diffused by state channels, and by the omnipresent Russian "pervyi kanal", representing the basic cultural reference for the majority of the population – even decades after of disappearing of FSU. The authorities place particular importance on movie production; some governments usually invest in movie colossal-style production (movies like Mongol, Nomad The Warriors in 2006, realized by a French-Kazakh cooperation), which are important as self-representation devices, for the re-elaboration of identity, to be standardized and diffused top down.[126]

References

Akayeva B (2007) Kyrgyzstan: time to ponder a federal system, http://www.thefullwiki.org/Bermet_Akayeva, accessed 11 April 2018. In: Eurasianet

Alberti A, Naserdinova B (1993) Più sciamano che mullah. Credenze preislamiche presso i Kirghisi. Marietti, Milan

Alexander S (2013) In: John H, Edmund H (eds) From resettlement to conflict: development-induced involuntary displacement and violence in Tajikistan, pp 49–66

Alissa M (2013) The concept of traditional music in Central Asia: from the RevolutIon to the independence. In: Sevket A, Richard C (eds) Social and cultural change in Central Asia. The soviet legacy. Routledge, London and New York, pp 63–71

Anceschi L (2017) Turkmenistan and the virtual politics of Eurasian energy: the case of the TAPI pipeline project. Cent Asian Surv 36(4):409–429. https://doi.org/10.1080/02634937.2017.1391747

Bennigsen A, Lemercier Quelquejay C, a cura di Fasana E (1990) L'Islam parallelo. Marietti, Genova

[125] Having said this, the "pervy canal" of the Russian TV seems to be still fundamental; it is watched per default wherever in CA, especially in Kyrgyzstan and Kazakhstan, and it is considered as a primary and authoritative source of information (author's interview).

[126] https://theculturetrip.com/asia/kazakhstan/articles/10-amazing-kazakh-films-you-should-see/, accessed 23.4.2018; Akyildiz Sevket and Richard Carlson, 2013; Mowell Barry, 2013:78.

Boboc C (2017a May 30) 'Middle-class, limited-edition'? Middle class subjectivities in urban Azerbaijan. Caucasus Analytic Digest 95:11–14

Boboc C (2017b) Middle class subjectivities in urban Azerbaijan, ASIAC annual conference, November 6th–7th, 2017, Dipartimento di Culture, Politica e Società dell'Università di Torino (oral presentaiton)

Bregantini L. (a cura di) (1997) I numeri e i luoghi delle minoranze etniche dall'Atlantico al Pacifico. Istituto di Sociologia Internazionale, Gorizia

Carrère D'Encausse H. (1978) L'empire éclaté. Flammarion, Paris

Christopher S (2013) In: Sevket A, Richard C (eds) The relics of 1991: memories and phenomenology of the post-soviet generation, pp 188–200

Corley F (2005a) Turkmenistan: President's personality cult imposed on religious communities, Forum 18, http://www.forum18.org/archive.php?article_id=522, 1 Mar 2005, accessed 11 April 2018

Corley F (2005b) Turkmenistan: 2004, the year of demolished mosques, Forum 18, http://www.forum18.org/archive.php?article_id=481, accessed 23 April 2018, 4 Jan 2005

Cosentino I (2017) Lingua, toponomastica e struttura urbana: il caso del russo nello spazio sovietico e post-sovietico, ASIAC annual conference, November 6th–7th, 2017, Dipartimento di Culture, Politica e Società dell'Università di Torino (oral presentaiton)

Cossutta M (1996a) Diritti e libertà dell'uomo nelle repubbliche ex sovietiche, "Rivista di Diritto Pubblico e Scienze Politiche", anno VI, n. 2/1996, pp. 409–423

Cossutta M (1996b) Diritti fondamentali negli stati ex sovietici asiatici. Le costituzioni del Kirghizistan, Uzbekistan e Turkmenistan, "Rivista di Diritto Pubblico e Scienze Politiche", anno VI, n. 1/1996, pp. 129–140

Dagiev D (2014) Regime transition in Central Asia. Routledge, New York

Dokhira S (2013) Statehood as dialogue: conflicting historical narratives of Tajikistan and Uzbekistan. In: John H, Edmund H (eds) The transformation of Tajikistan: the sources of statehood. Routledge, London and New York, pp 161–176

Dunlop JB (1994) Will the Russian return from the near abroad? Post-Sov Geogr 35(4):204–215

Educational System in Tajikistan (2018). https://www.classbase.com/Countries/Tajikistan/Education-System, accessed 23 April 2018

Elebayeva AB (1992) The Osh incident: problems for research. Post-Sov Geogr 33:78–88

Esenova S (1998) 'Tribalism' and identity in contemporary circumstances: the case of Kazakhstan. Centr Asia Surv 17(3):443–462

Farangis N (2016) Tajik name-change craze targets Turkic words, https://wwwrferlorg/a/tajikistan-name-change-turkic-words-rahmon/27529687html, RFE/RL Kyrgyz Service, February 03, 2016, accessed 11 April 2018

Flynn M (2004) Migrant resettlement in the Russian federation: reconstructing 'homes' and 'homelands'. Anthem Press, London

Foltz R (1996) The Tajiks of Uzbekistan. Cent Asian Surv 15(2):213–216. https://doi.org/10.1080/02634939608400946

Frederick L (2013) In: Sevket A, Richard C (eds) Social networking practices: continuity or rupture with the soviet past? vol 201, pp 145–159

Gaynor KL (2017) Transformations in Turkmen higher education: current opportunities and challenges at a new university. Central Asian Survey 36(4):473–492. https://doi.org/10.1080/02634937.2017.1367645

Ginzburg NS (1986) A microgeography of settlement in the Pamir highlands. Sov Geogr 27(6):398–435

Haghayeghi M (1997) Politics of privatization in Kazakstan. Cent Asian Surv 16(3):321–338. https://doi.org/10.1080/02634939708400994

Hambly G. (a cura di) (1970) Asia Centrale. Storia Universale Feltrinelli, Milan

Hanks RR (2016) Narratives of Islam in Uzbekistan: authoritarian myths abd the Janus-state syndrome. Cent Asian Surv 35(4):501–513

Hays J (2008) Minorities and minority issues in Tajikistan, http://factsanddetails.com/central-asia/Tajikistan/sub8_6b/entry-4874.html, accessed 23 April 2018

http://reports.weforum.org/global-information-technology-report-2016/, accessed 23 April 2018

http://www.forum18.org/archive.php?article_id=190, accessed 11 April 2018

http://www.irinnews.org/feature/2005/04/19/focus-post-akayev-russian-exodus, accessed 22 Mar 2018

https://eurasianet.org/s/behind-uzbekistans-bus-tragedy-a-story-of-grinding-despair, accessed 23 April 2018, Behind Uzbekistan's bus tragedy, a story of grinding despair. Only five people onboard were able to jump for safety out of the vehicle, leaving 52 migrant laborers dead. Jan 19, 2018

https://theculturetrip.com/asia/kazakhstan/articles/10-amazing-kazakh-films-you-should-see/, accessed 23 April 2018

https://web.archive.org/web/20070702212941/http://asia.msu.edu/centralasia/Kyrgyzstan/religion.html, accessed 23 April 2018

https://www.cia.gov/library/publications/the-world-factbook/geos/tx.html, accessed 11 April 2018

https://www.hrw.org/reports/2003/uzbekistan0903/5.htm, accessed 9 May 2018, Role and functions of mahalla committees, accessed 6 July 2018

Harun Y (2013) The soviet construction of Kazakh batyrs. In: Sevket A, Richard C (eds) Social and cultural change in Central Asia. The soviet legacy. Routledge, London and New York, pp 45–62

Human Development Report., http://www.hdr.undp.org/en/data, UN Development programme, accessed 6.7.2018

Institute for War and Peace Reporting (2005 February 21) Turkmenistan: Russian students targeted, https://iwpr.net/?p=rca&s=f&o=175615&apc_state=henirca2003, accessed 8.11.2020

Jelen I (2002) Tra i kirghizi del Pamir Alaj. Forum, Udine

Kaiser RJ (1995) Nationalizing the work force: ethnic restratification in the newly independent states. Post-Sov Geogr 36(2):87–111

Karagiannis E (2009) Political Islam in Central Asia: the challenge of Hizb-ut-Tahrir. Routledge, New York

Khalid A (1998) The politics of Muslim cultural refom. Jadidism in Central Asia, Berkeley

Klüter H (1993) People of German descent in CIS states – areas of settlement, territorial autonomy and emigration. Geojournal 31(4):419–434

Kudaibergenova DT (2015) The ideology of development and legitimation: beyond 'Kazakhstan 2030'. Cent Asian Surv 34(4):440–455. https://doi.org/10.1080/02634937.2015.1115275

Martha BO (2012) Tajikistan's difficult development path. Brookings Institution Press. 27 nov 2012

Marchi Marzia, a cura di (2017), Città dell'Asia, Bononia University Press, Bologna, 2017

Marco B (2003) La rivoluzione capovolta. L'ancora, Napoli

Maurizio Massimo e Tomelleri Springfield Vittorio, a cura di, 2018, Rivoluzione visiva attraverso visioni rivoluzionarie: alfabeti, cinema e letteratura in URSS, Dipartimento di Lingue e Letterature Straniere e Culture Moderne, Universita di Torino, Torino., ISBN 978-88-7590-134-9

Minority Rights Group International (2018) World directory of minorities and indigenous peoples, http://minorityrights.org/country/tajikistan/, accessed 23 April 2018

Monteil V (1957) Les musulmans soviétiques. Collections esprit Frontiére ouverte. Paris, Editions du Seuil

Muminov A (2014) Traditional and modern religious-theological schools in Central Asia, CA&CC Press AB Sweden, CA&CC Press AB Publishing House (Sweden), http://www.ca-c.org/dataeng/09.muminov.shtml, accessed 5 July 2018

Nichol J (2010) Central Asia's security: issues and implications for U.S. interests. Congressional Research Service, Washington, DC

Nick M (2017) Nationalism in Central Asia: A Biography of the Uzbekistan-Kyrgyzstan, Boundary, Central Eurasia in Context, 2017, University of Pittsburgh Press, https://doi.org/10.2307/j.ctt1vjqrk6

Ognibene P (2018) Gli alfabeti per le lingue iraniche orientali del Tagikistan. In: Maurizio Massimo e Tomelleri Springfield Vittorio, a cura di, 2018, pp. 85–94

Olivier F (2013) Soviet population transfer and interethnic relations in Tajikistan: assessing the concept of ethnicity. Heathershaw John, Herzig Edmund., 2013:35–48

Panaino A, Gariboldi A, Ognibene P (2013) Yaghnobi studies I. papers from the Italian missions in Tajikistan, Indo-Iranica et Orientalia, Series Lazur, Milano, Edizioni Mimesis

Paskaleva E (2015) Ideology in brick and tile: Timurid architecture of the 21st century. Cent Asian Surv 34(4):418–439. https://doi.org/10.1080/02634937.2015.1118207

Peyrouse S (2008) The Russian minority in Central Asia: migration, politics, and language. Kennan Institute, Washington, DC

PWC (2010) Doing business in Turkmenistan 2012–2013, https://www.pwc.com/uz/en/assets/pdf/doing-business-guide-in-turkmenistan-2012-2013.pdf, accessed 20 May 2018

PWC (2016) Guide to do business and investing in Uzbekistan 2016 edition, https://www.pwc.com/uz/en/assets/pdf/dbg_2016.pdf, accessed 20 May 2018

Radford D (2014) Contesting and negotiating religion and ethnic identity in post-Soviet Kyrgyzstan. Cent Asian Surv 33(1):15–28. https://doi.org/10.1080/02634937.2013.871831

Rasanayagam J (2014) The politics of culture and the space for Islam: Soviet and post-Soviet imaginaries in Uzbekistan. Cent Asian Surv 33(1):1–14. https://doi.org/10.1080/02634937.2014.882619

Rashid A (1994) The resurgence of Central Asia, Islam or nationalism? Oxford University Press, Zed Books, London

Reid P (2017) 'Tajikistan's Turksib': infrastructure and improvisation in economic growth of the Vakhsh River valley. Cent Asian Surv 36(1):19–36. https://doi.org/10.1080/02634937.2016.1204533

Reporters Without Borders, Uzbekistan toughens Internet censorship, 11 Oct 2012., http://www.uznews.net/news_single.php?lng=en&cid=30&nid=20980, accessed 23 April 2018

Ro'i Y (1991) Central Asia riots and disturbances, 1989–1990: causes and context. Centr Asia Surv 3:21–54

Rotar I (2003) Tajikistan: religious freedom survey. Forum 18 News Service, Oslo

Roy O (1997) La nouvelle Asie Centrale. Éditions du Seuil, Paris

Sevket A, Richard C (2013) Social and cultural change in Central Asia: the soviet legacy. Routledge, London

Shabad T (1980) The 1979 Census and some demographic trends in the Soviet Union. GeoJournal 4(1):84–90

Shahrani MNM (1979) The Kirghiz and Wakhi of Afghanistan: adaptation to closed frontiers. University of Washington Press, Seattle

Spoorenberg T (2013) Fertility changes in Central Asia since 1980. Asian Popul Stud 9(1):50–77

Spuler B (1970) Central Asia from the sixteenth century to the Russian conquests. In: The Cambridge history of Islam, vol 1A. University Press, Cambridge

Steinicke E, Jelen I, Lieb GK, Löffler R, Čede P (2016) Slovenes in Italy: a fragmented minority. Eu Countrys 1:49–66

Sultanova R (2011) From shamanism to Sufism: women, Islam and culture in Central Asia. Tauris, London

The Encyclopaedia of Islam (1980) Bosworth C.E., Van Donzel E., Lewis B., Pellat CH., edited by, E. J Brill, Leiden

Tim E (2013) Defining normative Islam: some remarks on contemporary Islamic thought in Tajikistan – Hoji Akbar Turajonzoda's *Sharia and Society*. In: John H, Edmund H (eds) The transformation of Tzajikistan: the sources of statehood. Routledge, London and New York, pp 95–110

Marzhan Thomas (2015) Social, environmental and economic sustainability of Kazakhstan: a long-term perspective. Cent Asian Surv 34(4):456–483. https://doi.org/10.1080/02634937.2015.1119552

Tonini C (2017) Samarcanda, Bukhara, Khiva. Città dell'Asia Centrale. In: Città dell'Asia, Marzia Marchi a cura di, Bononia. University Press, Bologna, pp 29–59

Trilling D (2013 June 12) Kyrgyzstan passes controversial girl travel ban, https://eurasianet.org/s/kyrgyzstan-passes-controversial-girl-travel-ban, accessed 23 April 2018

UNDP (2016) Education in Tajikistan, accessed 6 July 2018; http://hdr.undp.org/sites/all/themes/hdr_theme/country-notes/TJK.pdf, Human Development Report 2016 Human Development for Everyone Briefing note for countries on the 2016 Human Development Report Tajikistan

UNESCO, World Heritage Centres, http://whc.unesco.org/eng/list/603/documents/, accessed 23.4.2018

Urbanc M, Boesch M, Jelen I (2008) Kultura in regionalna politika: Kultura kot dejavnik regionalnega razvoja Alp. Geografski Vestnik, 79/1:39–48

Warf B (2011) Geographies of global Internet censorship. GeoJournal 76:1–23. https://doi.org/10.1007/s10708-010-9393-3

Werth N (1993) Storia dell'Unione Sovietica. Il Mulino, Bologna

Wheeler G (1966) The peoples of soviet Central Asia. The Bodley Head, London

Wikipedia "Turkmenistan", accessed 11 April 2018

World Bank, Tagikistan, https://data.worldbank.org/country/tajikistan, accessed 27 March 2018

World Economic Forum's Network Readiness Index, https://networkreadinessindex.org/ (n.d.)

The Material "Container": Structural and Infrastructural Aspects

11

Abstract

The CA territorial outline consists in structures and infrastructures, facilities and regulations, all developing on different scales, from local and national to cross-border and trans-continental ones. It also relies on a definite practical culture, supporting and influencing any kind of development and territorial management. In CA spaces, this element is even more important considering its landlockedness and the location of the NIS, depending on international commerce and supply chains. It is evident that national individual planning inevitably involves neighbouring systems, creating an inseparable network of connections. It must also be considered that a territorial network of this kind is a permanent one in practice, since the hardware of a country cannot be changed in the short term, but only through careful programming in the middle long term. Therefore, it is comprehensible that the new territorial system that developed from the one of Soviet origin, still appears today as fragmented because of the overlapping of the new geopolitical fragmentation. On this basic schema, the NIS has recently started an impressive programme of public works of particular importance.

Keywords

Administrative geography · Soviet legacy in planning · Structures and infrastructures · Planning on urban and regional scale

11.1 Territorial and Urbanistic Aspects

11.1.1 Geo-Deterministic Approach

The Soviet ideology was based on a geo-deterministic principle, and through its entire history it attempted to organize a certain geography in order to influence human life at any moment in time. This on the basis of the assumption, that practicing, repetitively and obsessively, some behaviour, then ritualizing spaces and times, would induce the interiorization of some beliefs (as indeed prescribed by the materialist principle characteristic of Marxist scientism).

Such attitudes continued into the post-Soviet era to a certain extent, with the new governments demonstrating their intention to use territorial patterns as a kind of "material" device for inducing a sense of loyalty and indeed for all aspects of societal organization.

This is evident on all scales, with governments starting impressive reconstruction programmes. It is the case of new public spaces, with shopping centres, international CBDs, stadiums and new sports and entertainment facilities, "arenas" and "malls". It is the case of the direct construction of new neighbourhoods, of new apartment buildings and "tout court" of totally new cities—then the fascinating evergreen dream of the "ideal city" as an instrument for forging a new society. Then, on regional and national scales, it is the case of infrastructures for re-equilibrating the country space, adopting a radial centre-periphery model in order to obtain a "nationally" integrated territory; and furthermore, on an international scale to connect the local territories to the continent-wide network for any kind of transport and mobility (pipelines, commodity supply, export–import etc.).

I. Jelen et al., *The Geography of Central Asia*, World Regional Geography Book Series,
https://doi.org/10.1007/978-3-030-61266-5_11

11.1.2 Territorial and Urbanistic Planning

In order to consider the "hardware" of the local system, it is necessary at first, to distinguish between urban and extra urban, regional and national, between internal and wider dimensions, linking the CA countries with the outside world. This must be done in order to classify mobility and in general organization modes (connections, mutual integration, discontinuities, and capacities), to be able to estimate their efficiency.

A glance at the current map is sufficient to evidence what the first priority is: re-scaling the infrastructures inside the new national perimeters which were drawn up after the Soviet collapse. In fact, the infrastructural networks (rail, roads, pipelines and energy transmission grid) were originally dimensioned on a much wider scale; today they have to be adapted to the new borders, configuring new territorial "containers".[1]

Such networks have to be restructured in order to predispose homogeneous nation units for all functions (economics, transport, surveillance of border, commuting, supply): from the new capital cities (functioning also as international gateways) to peripheries and eventually the peripheral regions transversally (contrasting centrifugal attitudes which must always be considered a risk in such disarticulated spaces).[2] At the same time, the possibility of a new risk is evident that this type of structuring of new national systems, with corresponding new borders, would create the effect of reciprocally blocking the NIS from each other, segmenting the whole territory on a continental scale.

11.1.3 Urbanistic—A Question of Soviet Cultural Heritage

It is possible to say that the current territorial situation evidences a set of elements that are the legacies of the previous regime and that will presumably condition the "stans" in the long term. Some of them can be considered negative while others may eventually take on a positive meaning.

Among the worst heritages of the Soviet time is the poor quality of urban "hardware" (buildings, infrastructures, construction materials), and above all a scarce ecological standard.[3] Among those to be considered as positive, there is a substantial culture in planning (territorial, urban and in any sense): regional spaces were and still are coordinated in a relatively good way, with inter-modal facilities, even when mostly equipped in an obsolete manner, organized on the basis of effective zoning. This regarding both industrial and urban areas, as well as the regional and rural territory in a wider sense.

In particular, the metropolitan areas evidence good standards in land use and functioning of facilities, with the authorities managing to maintain acceptable governance even in times of rapid changes, making sure that externalities neither accumulate nor annihilate growth benefits. This is the case of environmental impact (waste management, drinkable water and commodity procurement), real estate "bubble" and soil over-consumption, individual traffic proliferation and mobility inefficiencies.

The current urban space in CA appears as an ordered sequence of homogeneous areas (then of "bloki", well maintained parks, residential neighbourhoods, decorated and spacious squares and linear connections), public compounds and housing units.[4] This is due to all the urbanistic government parameters, for example, for facilitating the organization of services and the start-up of new initiatives, ensuring efficiency in functions like accessibility, mobility, public transport, warehousing, shopping and supply accessibility as well as for the organization of an efficient ranking and integration of multilayer traffic means.

All this reduces the risk of implications typically for big cities in transition like urban sprawl, improvised housing (e.g. abandoned industries, with the appearance of post-Soviet "ghost towns"), health and security problems. This especially considers the risk for urban–social ghettoization for new under classes that may eventually perceive themselves as excluded from the wider context, then induced to rioting and potentially illegal activities and crime—to some extent a consequence of those urban inefficiencies.[5]

The Christallerian settlement structure seems to be effective at any level. The major cities do seem to be ready to open up towards the outside while maintaining a role in regional integration at the same time (a role of intermediation between the wider space and the local community). This represents the current major challenge, namely the necessity of modernizing, and opening, without losing their own authenticity (namely in local representativeness for local society in times of changes).

This necessity has an immediate effect on an urban scale, with city spaces maintaining old connotations, but developing new activities; with appropriate zoning, the conservation

[1]Brill Olcott Martha 2012; particular problematic are the cases of Tajikistan and Kyrgyzstan.

[2]Secchi Roberto and Spita Leone, a cura di 2018; Scharr and Steinicke 2012; Thorez Julien 2018.

[3]Beside residents' health, the environmental rehabilitation for tourism purposes is especially important; media has recently reported cases of foreign tourists organized in group being "surprised by salt-storm, get seek" in Khiva, L. P. 2018.

[4]Marchi Marzia, a cura di 2017; Tonini Carla 2017; Marchi M., Tonini C., a cura di, 2009.

[5]Currently in CA towns, there is not the overwhelming confusion that is mainly characteristic of the urban metropolitan space of many contiguous non-Soviet spaces, also considering the relevant refugee and migration waves occurring in the recent past, and no particular problems.

of the old "zelenny bazaar" would not be in contrast with the setting up of the new city malls; so too for the conservation of the old picturesque city district which can be maintained while at the same time developing new CBDs, making the coexistence of the new and the old possible and overcoming possible incongruences.

11.1.4 Metropolitan and Regional Planning: Mobility, Infrastructures et Similia

The urban schema, inherited from the Soviet and to some extent colonial eras, is still to be considered efficient and usable for development purposes. The basic design highlights the search for symmetry, the most salient characteristic of the Soviet city, which was originally conceived mainly for surveillance purposes, ensuring accessibility and visibility on an orthogonal grid of the urban "chessboard". It proves to be useful today for other purposes (mobility, services organization, social inclusion), and in general for a flexible use of spaces (as necessary in a rapidly changing environment).[6]

The same spaces had to represent a political–ideological gallery of icons and therefore they were decorated with political symbols, with the statue of Lenin (and Stalin) at the centre of major square, to be visible from far away—so that ideally the "gaze" of power would follow the citizen in all moments of their daily life. But they also used to be pleasant places, in order to attract Soviet "colonizers" especially to CA who would have been accustomed to other climates and landscapes. Therefore, they were decorated and planted with trees lining the streets (mitigating temperature, light and sound) and metropolitan parks (resembling the northern forest, quiet and silence) with fountains and pavilions, botanic gardens and zoos in order to assure that the Soviet citizen, accustomed to a Russian-like landscape, would have a tolerable life in such hot regions.

This situation has made complete restructuring possible by following the criteria and using the symbols of the ideal city. This especially in some periods, when the planning practice had the chance to rebuild from scratch entire new urbanistic stratifications. This is the case of the 1940s and 1950s where the landscape evidenced a peculiar construction style, with public buildings resembling (ironically) classic Greek architecture, imitating the early form of democracy (presumably as a legitimization device) (Figs. 11.3 and 9.11).

It must be said that such restructuring, on such a scale, was also made possible by a sequence of earthquakes which periodically devastated CA cities, especially Dushanbe (in 1930), Ashgabat (1948) and Tashkent (1966): in this way it was possible to realize a schema characterized by a certain sense of volumes and distances, interlacing monumental squares, urban and social facilities and so obtaining an impressive effect.

11.1.5 The Building of New Urban Landscapes

Such organization presents further advantages; even when considered overdue in its symbolism, the chess-board city proves to be easily convertible for further development purposes, with reserves of space usable for any kind of adaptation. This is the case of new transport systems, of the enlargement of building and facilities, of the organization of intermodal hubs.

This also regards the construction of new city quarters and sometimes of entirely new towns, of leisure, shopping centres and malls, representing a challenge for the old organization of the territory with new infrastructures possibly substituting bazaars and a "station" for collective taxies (maršrutka), almost spontaneously appeared in the transition period (as an example of the described parallel mode of organization).[7]

Special consideration must be made when considering the construction or the rehabilitation of religious monuments; the impressive marble-white Friday Prayer mosques, usually located in the non-central areas, so as not to appear incumbent, like a political challenge, to the government buildings.

This is the case of the new splendid mosques in Dushanbe and Nursultan, and of those in Turkmenistan, some of the largest in CA, the Ärtogul Gazy mosque in Ashgabat and the Türkmenbashy Ruhy mosque, which have been built far from the city. It is a particularly sensitive question; these mosques usually have appropriate dimensions, capable of hosting a certain number of believers, and are carefully monitored by the authorities. They are planned considering the danger which might derive from amassment of people, potentially when they are lectured to by a charismatic leader, and which might create the premise for uncontrolled opposition and "street revolutions".[8]

So too for other sensitive urban areas, such as bazars, bus-stations, social events facilities (auditoriums, stadiums), intended as spontaneous social gathering places, discouraging situations that in the past had been potential areas for the start of mass demonstrations or similar phenomena.

[6] In its intentions, the totalitarian town evidences a symmetrical planning, reflecting a sense of metaphysic order, but also the functional use of making street manifestations more difficult, or of inducing in the citizens a perception of being dispersed and impotent in front of the power; Marchi M, Tonini C., a cura di 2009.

[7] A kind of double organization, with planned services, sometime ineffective, coexisting with the spontaneous ones, self-organized in terms of "grey economics", which characterized also transition times, and reproduced also today in the new circumstances.

[8] Like the one in Iran 1979, an example which the CA elites are always aware of; so too for the much closer example of Arab springs, demonstrating that no power can be considered effectively "safe" from spontaneous city uprising; Rasanayagam Johan, 2014:9

11.1.6 A Gallery of Icons

The reconstruction of new urban and regional landscapes is evidence of the capabilities of such rentier states, pursuing mass-scale investments and constructions. The major evidence of this cash surplus can be seen in the sparkling crystal and marble presidential buildings, surrounded by superb government structures, obelisks, parks, fountains and artificial lakes.[9]

This scenography is intended to represent the new powers, depending on the circumstances, combining the revival of local cultures, opportunely revisited in neo-nationalistic terms, religious elements—instead of the rarely surviving Lenin statue—and a long list of new symbols. These may represent the founding myth of the country, equestrian statues of the original heroic kings, either with turbans or with crowns, with swords or sables (meaning in popular imagination a cultural distinction between orient and occident), and in various additional representations. This is the case of Manas in the Bishkek Ala-Too square, of the Somonids in Dushanbe, and of Timur in Tashkent; other representations celebrate ancient poets, scientist, or writers, signifying national glorious roots (Figs. 11.1 and 11.2).

Such urban iconography is usually associated with other means like posters, "murales" paintings, and other elements of communication combined with names of streets, city squares and "rayon", demonstrating the intention, respectively, for de-Sovietization or re-Islamization, for Turchization and Arabization, or alternative tendencies, sometimes newly invented.

11.1.7 Preservation of Ancient Urban Centres in Soviet Times

The current structure of CA towns usually appears as a multi-layered one—thanks to the relative abundance of space—with the new districts being built close to the oldest, without destroying them; so, it is possible to observe cases of interesting multi-nuclei cities. The Russian colonial city can be found close to the modern one, the Soviet city close to the Islamic old town without overlapping each other.

Sometimes traditional signs and monuments persist even when at the edge of the newly built up neighbourhoods; this is the case of "zelenny bazars" in Dushanbe, of beautiful but poorly maintained Almaty wooden-house district, and of the "chruščëvka" neighbourhood in Nursultan, possibly disappearing in the near future.

In the most important cases, the preservation happened, thanks to early post-revolutionary "material culture" protection acts, to some extent contradictory to the revolutionary spirit, usually prescribing material restructuring or even the destruction of the previous landscape.[10] Possibly this is due to the fact that the planners had the intention of maintaining a visible example of the narrow street Islamic-mediaeval town, with feudal palaces, fortresses, prisons and repression techniques used by pre-Russian despots (as in Bukhara especially).

Fig. 11.1 Kyrgyzstan, Manas poster representation, 1997. (Photo: Igor Jelen)

[9]Gaynor Kelly Lee 2017; Scharr and Steinicke 2012.

[10]A decree signed personally by Lenin, Paskaleva Elena 2015; but besides monuments classified as "ethnic" or "material culture", the authorities planned destruction and "de-contextualization", Tonini Carla 2017.

Fig. 11.2 Kyrgyzstan, Bishkek, Manas park, 1997. (Photo: Igor Jelen)

Some of them were redeveloped as museums so that the new Bolshevik rulers could achieve a legitimization effect, representing themselves as the liberators from that inhuman regime (a fact that would paradoxically represent an advantage, precisely, preventing destruction).

In fact, the marvellously unique old towns of Bukhara and Samarqand were spared the revolution's destruction fury and re-construction plans,[11] so that today they represent a spectacular cultural celebration as well as an asset for a promising tourism industry.

11.1.8 Post-soviet City

The post-Soviet CA city is rapidly developing; all the major towns are practically unrecognizable for anyone visiting them a few years later. The most impressive new element is the government district, monumental and disproportionate, evidencing the anxiety of the new government for self-representation. They may also appear as a cement buffer (like a geopolitical-urban expedient) against their own societies, consisting in institutional buildings, wide perspectives, monumental compositions and other elements, evidencing the relative inaccessibility of such districts (Figs. 11.3 and 11.4).

Another element arising in this urban landscape is the new globalized CBD, as in Tashkent, Almaty, Dushanbe, and obviously in Nursultan, the whole city in itself a new CBD. It is made up of administration centres, international hotels, museums, business and shopping centres. In a surrounding concentric belt, there are usually malls, skyscraper-like apartment buildings, stadiums and big event facilities (arenas, auditoriums), as well as the university, research centres and offices. They cannot be considered as "closed" modern enclaves (as in Soviet or transition times), but as spaces mediating with the outside world. It is the environment for a "nouveau riche" local society, for foreign residents and for an increasing number of tourists—not very numerous yet, but who represent an important new presence, promoting cultural adaptations.

In fact, in post-Soviet times, the major cities had the chance to open to outside relations, sometimes becoming a national-city capital, developing new functions, sometimes booming and rapidly changing role and its own social composition.

It is the place for a kind of post-Soviet "melting pot", with an ongoing confusion of new signs, western commercial symbols, oriental electronics and high-tech appliances (Korean, Chinese or Japanese), in short, a globalized style of life. All this occurs in a monumental town that maintains the basic schema of the previous system, avoiding slum city-quarters and confusion, with well-functioning services, in an ordered sequence of "bloki" and "chruščëvki", crossed by tramways, trolleybuses and—in Almaty and Tashkent—an underground.

Some bazaars were revamped already in Soviet times, following building standards, using massive steel and reinforced concrete, reorganizing standardized spaces for traders which meant that they lost some of their main characteristics.

This occurred for the famous Chorsu bazaar in Tashkent, modernized in the 1980s, consequent to the earthquakes in 1966. It is a kind of "concrete" open-air market, characterized by standard spaces for different functions, as a paradoxical experiment to induce a kind of "order" into the intrinsic "dis-

[11] As well as by over-building and speculation, as may happen in conditions of real estate free market that is in principle damaging to landscape and monument protection.

Fig. 11.3 Tajikistan Dushanbe, 2017, new government district. (Photo: Igor Jelen)

Fig. 11.4 Kazakhstan, Almaty, 1997, government presidential district. (Photo: Igor Jelen)

order" of the bazar: an attempt to conciliate "modern" functions (organization, accessibility, sanitary vigilance) with control purposes.

In particular, the public transport system is very efficient, both in big metropolitan areas and in provincial towns. The Tashkent underground transport system is possibly the best example of this, a three-line underground, planned and built in 1977, renovated and extended in 2001.

These infrastructures and especially the underground stations—as usual in Soviet tradition—are very highly decorated and adorned, with frescos, murals and icons of Soviet propaganda, like cosmonauts, revolutionary heroes and "great patriotic war" celebrations (possibly the metro tunnels were also equipped as nuclear war shelters, in the context of the territorial defence programme, following a programme of social militarization, in the framework of the Cold War period).

Almaty also has an underground system that appears today almost a legacy of the SU. It is clean, efficient, rapid, well decorated and museum like (surprisingly not Nursultan, the new sparkling Kazakh capital which seems even more strange if we consider that Nursultan was completely rebuilt on "virgin ground" with a totally new design, which would have theoretically made the organization of such efficient mobility routes even easier).

Further lines of public buses and tramways operate in an excellent manner, combined with an ubiquitous set of taxis, registered and not, collective or individual long range connections to a "last mile" scale; the same widespread habit of practicing a kind of hitchhiking (a further habit inherited from Soviet time), a further example of the parallel organization of services, contributing to general standards of accessibility.

11.2 The Key Factor: Mobility and Infrastructures

11.2.1 Tendencies for Opening up

Infrastructures are the basic elements in order to deal with the initial problem in CA, namely its landlocked-ness and general accessibility problems.[12] A situation currently far from being satisfactorily amended and that has been further aggravated because of the overlapping of new political borders, defining today what are to be considered as new sovereign states.

There is a widespread awareness that mobility improvement is necessary to support general tendencies for opening up to give incentive to commercial ties, setting up a network of efficient connections. The symbolical moment in this sense can be considered as the Expo Astana (Nursultan) 2017 in Kazakhstan, as well as other such international events that express this area's desire to be integrated into the wider world (and also the neo-nationalistic purpose of exhibiting itself to the IC).

The same for other such initiatives (direct and indirect foreign investments—FI, trade, tourism, technical innovations), for culture and sport, as well as for occasions of exchange among universities and students, agreements with foreign institutions, museums, academies and a kind of cultural exchange.

In this way, the big metropolis cities like Tashkent, Almaty and Nursultan (in a kind of "win win" dualism inside the Kazakh Republic), Dushanbe, Bishkek have rapidly become international hubs while Samarqand and Bukhara have an ever increasing number of visitors; at the same time large areas in the mountain (Pamir, Tien Shan, Fan mountains and Alaj unique landscapes), the seaside or areas of particular natural interest have become areas of great touristic attraction.

The persistence of a multi-cultural society and multicultural attitude in the same towns, and indeed everywhere may represent a further advantage in this sense, namely, consolidating the same tendencies for opening up.

[12]Brill Olcott Martha 2012; PWC 2010, 2011, 2012–2013.

11.2.2 Transport, Mobility and Infrastructures: Intra-national

The NIS have to face a huge challenge, namely to re-integrate and to create the intra-national system: the networks constructed in Soviet times obviously did not consider the current border segmentation. Today, the national lines often trespass the borders, indicating a critical issue for the continuity of any kind of connection. These networks, furthermore, at the moment of independence presented many problems of obsolescence, both technologically and economically and they need to be urgently adapted. The question is particularly urgent in remote areas in Tajikistan and Kyrgyzstan, where accessibility is to be considered a priority, but also in other countries, considering the extended distances and the poor efficiency of existing regional transport.[13]

In fact, the geographical constraints (difficult topography, huge distances, geological fragility) condition all aspects of social life; this especially so in mountain districts, with deep valleys and passes of more than 3000 m, tunnels and serpentines.

The Tien Shan and Pamir high mountain roads—among the most picturesque—are also inadequate, presenting ongoing risks of avalanches, snow in winter, mud slides and debris flow in all seasons, due to extreme meteorological events and, furthermore, to hydro-geological instability.

In such circumstances, regression, not just in remote areas and not just for local purposes, to pre-modern practices like horses or donkeys and caravans is not surprising. They are easy and inexpensive to manage in such conditions, usable without the support of a network of services and energy supply, and often, in those circumstances, even more efficient as modern tools (e.g. practicing shorter transversal high mountain mule-treks, instead of down-valley infrastructures).

Besides transport for people and freight, the question concerns any kind of flow (information, energy), the reconstruction of highways, ID systems, pipelines, energy and information networks; the reconstruction of an entire economic system in a space characterized by dramatic disequilibria.

11.2.3 Transport, Mobility and Infrastructures/ Transcontinental

It is not surprising that, after independence, governments soon started to elaborate giant programmes of infrastructure (re)construction, often supported by international institutions, both corporate or national, or sometimes "tout court"

[13]PWC 2011, 2012–2013.

financed by foreign governments—especially China, that has a direct interest in developing such infrastructures.[14]

Such programme, if realized at such large scale, would result in a general change in CA geography and possibly in the populations' perception of their own territory (much more "open" than it used to be). Evidently, this new internal transport grid has to find an extension in a framework of international systems in order to facilitate communications between countries that depend on each other for a wide set of functions.

It is a question of roads and railways, namely of routes and of junctions, to be reorganized as a gateway for entire regions—that otherwise risk paralysis in development programmes—hubs for traffic on all scales in airports, river and lake ports, in order to reorganize local transport on diverse ways—from inter-continental to the "last mile", in intermodal settings, water, rubber-tyre, iron etc.—overcoming discontinuities both natural and political.

The main goal is not just to ensure economic circulation, but in general to render each country accessible to the open market, avoiding the possibility of using some infrastructures in a distorted and manipulated way (eventually geographically "blackmailing" the neighbours); otherwise it would probably become useless sooner or later.

11.2.4 Rails and Roads

The formation of new borders has definitively upset the schema of traffic modes so that it is necessary to restore functional hierarchy, considering any mode and frequency today. This evidences a sequence of further questions concerning regulations, surveillance, duties and customs, and obviously visa concessions: such tasks cannot be carried out in a short time, it takes a period of common work, relying on neighbouring reciprocal confidence, through agreements—possibly multilateral, considering the accidental configuration of CA borders—as well as the accumulation of habits and the sharing of experiences.

In fact, as a result of such fragmentation, the main infrastructures appeared as useless, this especially so in peripheral areas; Kyrgyzstan installed two unconnected railroads, the southern and the northern ones; similarly in Tajikistan with recent inaugurations, with three railroads, mainly west-east oriented, but non-connected.[15] And so forth for many other kinds of connection, transport and linkages.

At the same time, such a situation also presents a unique opportunity for each of the CA countries, the chance for a new start, rebuilding a new system and designing new functional areas from scratch (in Christallerian sense). It represents the opportunity to modernize the whole system, in its entireness, making it possible for the governments to elaborate some strategies.

It is the case of the decision to invest either in public or in private transport, in the collective or individual, sustainable or spontaneous mobility: the recovery of a public idea of transport may be essential in order to give systemic impulses to the entire network (then fostering intermodal integration), but the individual ones would enforce the sense of freedom and of free enterprise.[16] And so for any further activity.

Comprehensibly, the NIS devoted particular attention to rail infrastructures, of utmost importance for integrating the new national territories; they have started projects for high speed and high capacity trains, that would be of great utility considering topography as well (especially in steppe and deserts, rather flat, non-problematic, then making construction and depreciation cost not very expensive), and distances[17]

It is the case of Uzbekistan, where a network of high speed electric trains was constructed after independence; the first of such programmes has connected Tashkent with Samarqand and Bukhara since 2011, with an Afrosiyob train, constructed by Soanu, Talgo Patentes Talgo S.L. (Spain); similarly in Kazakhstan between Almaty and Nursultan, and between well connected major cities, so that high-speed trains connect Almaty with Petropavlovsk (on the country's north-south axis) in 18 h, and Nursultan and Almaty in 7 h (competing with regional airplanes on middle-long ranges). Both of these high-speed trains can represent a definitive solution for major connection problems for both countries, linking the opposite sides of their national territory.

Rail improvement is important to serve the commuter distance better (improving the whole mobility system), considering the rather good potential customer basin in some areas, such as larger cities, populous regions and tourist centres; it is also a good chance for bulk transport. This is also the case

[14]And that is becoming a major creditor of CA countries; Indeo Fabio 2018, 137–138.

[15]PWC 2012–2013:9; PWC 2011:4; but as said recently the new Dushanbe–Qurghonteppa–Kulob railroad has been inaugurated, see later, http://www.president.tj/en/node/12789#hisor, Inauguration of Dushanbe–Qurghonteppa–Kulob railroad, 24.08.2016, accessed 5.7.2018.

[16]Such options would soon configure a crucial trade off with a critical impact on the long term; the same considering further aspects, namely the pay-off about such distances and the use of transport means, respectively, for collective and individual purposes, considering such parameters as sustainability (namely environmental costs, insurance, "carbon tax", etc.) and efficiency.

[17]The technological evolution can suddenly change the cost curves of infrastructures, namely the breakeven point for the different carriers concerning freight, distances and mode: any distance scale, and any kind of transport, is characterized by specific cost, that ensure the better connectivity, but it can change; e.g. the airplane was considered the better transport on long-medium ranges until a decade ago, connecting and integrating peripheral parts of countries to centres, until the recent "high speed–high capacity" train program proved to be a good alternative over these distances.

Fig. 11.5 Tajikistan, Rasht valley, August 2017, works for Roghun dam. (Photo: Igor Jelen)

over the wider areas and long distances, both national and inter-national.

11.2.5 Trans-Continental Corridors

The opening up of the already cited China-Europe road through the Dzungaria gate, through central Kazakhstan and southern Russia seems of particular importance, possibly configuring an effective trans-continental "land bridge", integrated in a re-launched SR, sometimes re-defined as OBOR ("one belt, one road") programme[18]: "[t]he new Eurasian land bridge economic corridor will involve Russia by linking China with Europe via Kazakhstan and Russia. There are other sub-corridors, or spurs, of the main projects that are particularly relevant: the Khorgos–Aktau railway, linking the Kazakh port in the Caspian Sea to the main BRI trade gateway of Khorgos; the China–Kyrgyzstan–Uzbekistan Railway, linking the western Xinjiang cities of Kashgar/Kashi, via the southern Kyrgyz city of Osh, to Andijan in eastern Uzbekistan; China–Central Asia gas pipeline, which at present is mainly fueled with Turkmen gas but, following the realization of Line D, will involve all five Central Asian countries as both suppliers and transit countries."[19]

In this sense, the country could represent a central segment of a new trans-continental axis, in particular considering the role of the Chinese-Kazakh border terminal: the passenger trains from Ürümqi, initially stopping in Yining, then in 2013 arrived to Khorgos "dry port"—the central point on this route. In the meanwhile, in 2011, it completed a 293-km railway from the Khorgos border passage to Almaty Zhetygen terminal; in December 2, 2012, finally the Chinese and Kazakhstan railways have been connected.[20]

Trains for such traffic result as being the privileged option, being sustainable and efficient; but they need a higher level of organization capability, for example, coordinating time schedules and other traffic facilities, setting up intermodal services and connection nodes (needless to say, the basic train organization of flows, especially at the initial stage of a development path, induce a need for the organization of the whole economic and social system, which it is to be considered as a strategic issue).

The open question is the recovery of the side branches, residual branches of the Soviet network, for bulk traffic; they currently often serve obsolete and abandoned Soviet factories or mines (as in Karaganda areas, Fig. 11.6); the situation of passenger transport is even worse, and needs a threshold critical mass in order to be feasible.

Among the major connections currently under construction is a railway from Kazakhstan–Turkmenistan–Iran, on a north-south axis; it is the case of Zauzen–Bereket route, connecting the western Caspian coast with Iran (a key country in this scenario), a way that would contribute to develop traffic on the southern side, towards sea routes and ports.

Another example is in Turkmenistan–Afghanistan–Tajikistan (TAT),[21] with the recent completion of the 88 km railroad connecting Atamurat–Ymamnazar–Akina, follow-

[18] https://edition.cnn.com/travel/article/trans-siberian-road/index.html, accessed 29.3.18.

[19] Indeo 2018.

[20] Indeo 2018.

[21] http://gca.satrapia.com/+adb-suspends-its-participation-in-supporting-construction-of-the-tat-railway+, accessed 29.3.18.

Fig. 11.6 Kazakhstan, Karaganda region, industrial zone. (Photo: Igor Jelen)

ing the Memorandum of Understanding (MoU) signed by the countries' presidents in Ashgabat on March 20, 2013: The project has been developed over a distance of 400 km and will link Afghanistan, the town of Akina–Andkhoy to Atamurat–Ymamnazar in Turkmenistan and Panj in Tajikistan. Such projects are essential to be able to "adjust" this tremendously complicated geography. It will be completed in 2020, and it will substantially improve the accessibility of a peripheral area, by-passing the Uzbek part of that itinerary, obsolete and dating back to the Soviet era, adding a 35 km link to the northern Afghan town of Andkhoy to this.

Such programmes are especially important for Tajikistan, considering the articulated geography of the country. Recently it has opened roads and railways towards important sides of the country, the centre and the south, the capital city area with Qurghonteppa and Kulab; the most important is the completion of the tunnels on the road Dushanbe–Kulab (Fig. 11.7) connecting the two sides of the country in a few hours, on a comfortable and convenient road. On the contrary, the peripheral mountain areas (Pamir, Trans-Alai, Piotra Velikogo Mountains, with especial regard to Rasht valley) are not yet connected with acceptable roads, neither with public transport—a rather urgent question indeed, also considering its extension over the Kyrgyz–Chinese border that would improve connections on a wider scale.[22]

The current list of work in progress is impressive and it makes it possible to imagine to what extent local economics can take advantage of connection improvements, sometimes rehabilitating obsolete Soviet infrastructures, sometimes constructing brand new paths.

An updated list includes a series of works in progress or which are just being completed. They include the rehabilitation of the route from Dushanbe to Uzbek border, in the direction of Samarqand (Chanak border post), Dushanbe–Kulma (Chinese border), Qurghonteppa–Nizhny Pyanj, to the Afghan border. It includes several viaducts and tunnels—high engineering projects with particular impact on mountain country—usually financed by international donors or financial institutions; those concerning the mountain passes of Anzob, Shakhristan, Shar-Shar and Chormazak, under construction or recently opened.[23]

In particular, the works for the Kulob-Kalaikhumb highway have recently been started, supported by IDB, with the main aim of the project being to "support social and economic development of the south-east region of Tajikistan by means of providing a year-round ground-based transportation service between the western part of Tajikistan and the eastern region of the Gorno Badakhshan Autonomous Region (GBAO)".[24]

[22]http://news.tj/en/news/chormaghzak-tunnel-renamed-khatlon-tunnel-and-shar-shar-tunnel-renamed-ozodi-tunnel, accessed 29.3.18; https://blogs.fco.gov.uk/leighturner/2013/05/07/trade-tunnels-transit-and-training-in-mountainous-tajikistan/, accessed 29.3.18.

[23]Another basic question is the capability of organizing regular public transport from centres to peripheries; in Tajikistan, currently, there is no public transport scheduled from the capital city to the peripheries especially Garm, Pamir and the south.

[24]Recently IDB has provided a US$20 million loan to Tajikistan for the construction of Kulob-Kalaikhumb highway, May 17, 2017, https://news.tj/en/news/tajikistan/economic/20170517/239927, accessed 5.7.2018.

Fig. 11.7 Tajikistan, 2017, new tunnel on the main road Dushanbe-Kulab. (Photo: Igor Jelen)

11.2.6 The Southern Way

The most impactful and imaginative projects concern the southern way, namely the access to open seas. The bridge constructed in 2004 on the border between Afghanistan and Tajikistan, thanks to a special investment made by the United States—also for strategy purposes supposedly—has facilitated access to the southern part of the continent.

This recent arrival has been well exploited and used politically by the local governments. This also considers the side effects induced by projecting and constructing intra-national infrastructures: the pacification in Tajikistan, hopefully stabilization in Afghanistan and signs of the opening up of Iran, which prospect a brand-new gateway.

In 2009, a convention for the building of a 1300 km highway and rail system from Tajikistan through Pamir and Gorno Badakhshan was signed, to Pakistan sea ports[25]; furthermore in 2012, the presidents of Tajikistan, Afghanistan and Iran signed an agreement about planning a new network of roads, railways and all kinds of pipelines (oil, gas and also for water) to connect the three countries. They recently inaugurated the Dushanbe–Qurghonteppa–Kulob railroad, which is an important "link" in this chain.[26]

Turkmenistan opened a new railway to Mashhad as well as a pipeline in the same direction; it is noteworthy that the former President Niyazov had already opened a small pipeline connected to Mashad with the marginalized Iran in 1996, which has recently been widened[27]; above all the recently opened pipeline to China of great strategic significance for the whole CA space must be mentioned.

11.2.7 The General Trend

The basic intention is that of rendering accessible the international market for local natural resources (especially HC), emancipating it from any kind of geopolitical monopoly. Therefore, the projects consider in principle all the directions; but as a priority they are interested in the eastern side and improving relations with China, then the south (the north and the west being controlled mainly by Russian state-owned corporations to some extent).

The situation, after a quarter of a century of transition shows some real improvement. In particular, for the major HC exporters Kazakhstan and Turkmenistan (besides the TAPI "infrastructural promise", Fig. 11.9), who have realized their plan to diversify infrastructural capabilities, overcoming any obstruction as it is sometimes heavily opposed by neighbouring super powers (as said, the priority of independent Kazakhstan was to gain access to inter-

[25]http://www.tajikistanmission.ch/business-and-investment/infrastructure.html, accessed 29.3.2018.

[26]"If you look at the economic importance and advantages of the project, we can say that the length of the former railroad, Dushanbe-Qurghonteppa, which crossed the border, was 432 km, and, as a result of the construction of tunnels and bridges, it was possible to reduce the route length of 313 km, and the length of the current road is only 119 km," http://www.president.tj/en/node/12789#hisor, Inauguration of Dushanbe-Qurghonteppa-Kulob railroad, 24.08.2016, accessed 5.7.2018.

[27]Then the Tejen–Serakhs–Mashad railway, Wikipedia, "transports in Turkmenistan", accessed 29.3.18; https://www.economist.com/node/10181134, accessed 29.3.18.

Fig. 11.8 Kyrgyzstan, road from Bishkek to Osh, April 1997. (Photo: Igor Jelen)

national markets without dis-quieting its northern neighbour).[28]

In Kyrgyzstan, the improvement of the Osh-Bishkek road, not always open in winter, represents the most important public work (Fig. 11.8); the government has further plans to give continuity to this road to China with a 3500 m pass into the Talas Valley in the northwest, and then constructing a straight road to the border, over the Torugart pass, from Osh to the Chinese border, taking on an evident political significance.[29]

The CA countries have the intention, in particular, of improving access to Chinese markets, re-purposing the fascinating SR tradition. Some projects are financed by oriental institutions (Chinese, Japanese, Korean) in order to make transport and communication over the CA mountain ridge easier. This is a gigantic operation, difficult to finance especially considering the costs and the continuing changes in technologies in transport and communications (that impedes mobilization of large capital in what could become obsolete very early).

However, transcontinental corridor maps and projects do not represent a dream, but a concrete possibility considering the complementarity of CA with the surrounding economies and the availability of capital to be invested; this especially with east and south, since any exchange with China and India seems to have the most relevant potential, prefiguring investments with high profitability, exactly what international finance institutions—usually moving in stagnant economic situations—are looking for (Fig. 11.9).

11.2.8 The Central Asian Soviet Railway—Historical and Strategic Significance

The old Soviet railway remains a basic network, even if extraordinarily obsolete, and practically useless in modern terms, connecting CA with Russian and Siberian provinces.[30] It is almost a piece of history, with symbolical impact of what the major transport has meant since the times of colonization and the Soviet revolution. The train to Moscow was the symbol of any "komandirovka" for any "aparatčik", or anyone who was taken this train, and was of principle importance.

Comprehensibly, the early railway hubs became with the time major cities and references for the entire CA transport system and having being constructed with an extraordinary effort, it became the instrument for ensuring the long-term Russian geopolitical control over the entire area as well as the major penetration axis for modernization, spreading out over all the CA spaces, making even remote peripheries accessible.

The first one, the Trans-Caspian, ran from Krasnovodsk to Ashgabat in 1879, then to the Zarafzhan valley, Fergana and Tashkent; the Trans-Aral railway, completed in 1906, ran from Orenburg to Tashkent, while the Turk-Sib, a Trans-Siberian southern branch, started to operate in the 1930s, even if planned in late colonial times.[31] This connected the

[28] Silvestri Tommaso 2015–2016:97.

[29] "China realized strategic road connections in Kyrgyzstan—Osh–Sarytash–Irkeshtam and Bishkek–Naryn–Torugart", Indeo 2018:139; see as well Mashrab Fozil, November 3, 2015; Megoran Nick 2017:105.

[30] Evidently, for many transports, train, even when obsolete, can be still useful and convenient, considering it can travel 24 h, in any condition.

[31] Reid Patryk 2017:21.

Fig. 11.9 Expo Astana (Nursultan) 2017, Turkmen pavilion representation of the TAPI project. (Photo: Igor Jelen)

two existing branches of the Soviet railroad system in CA with the circular railroad from Fergana to Tashkent out flanking the Trans-Alai mountain; furthermore, prolonging the track towards north-west in the Chuy Valley, reaching Bishkek.

In the south, the Soviet railway reached southern Tajikistan and the Afghan border close to Kulab which became the southern terminal of the SU railway system: today the trans-continental line Kulab–Ashkhabad–Moscow is still operating, but with incredibly slow time schedules; it crosses the five countries on its route (Turkmenistan, Tajikistan, Uzbekistan, Kazakhstan and Russia). The new Dushanbe–Qurghonteppa–Kulob railroad has recently been inaugurated.[32]

From this main track, a network of narrow-gauge railroads has developed, reaching industrial plants, mines and settlements, in fact extended networks of side branches which are mostly useless today. After independence this gigantic, and relatively efficient railway system was segmented by new borders, with secondary lines and only small residuals of rail lines remaining marginalized in lateral valleys or in abandoned industrial areas, which are often closed down.

[32] http://www.president.tj/en/node/12789#hisor, Inauguration of Dushanbe-Qurghonteppa-Kulob railroad, 24.08.2016, accessed 5.7.2018.

11.2.9 The Reconstruction of a Self-containing System

The post-independence priority is bypassing natural obstacles, as well as the settlement of political critical borders, in order to gain a better degree of accessibility to the international market. Usually this circuit (radial, linear or transversal, or whatever form it takes) evolves over the long term, consequently to development cycles of the country. But, and for the mentioned reasons, this has not happened in CA, so that at the moment of independence the different states had to deal with a set of incongruent infrastructures, connecting the different regions in a precarious way, evidencing a set of gaps (namely, the risk of marginalization).

The priority was to adjust these networks, connecting production regions, markets, urban areas, with the diverse centres and peripheries. In the case of Kazakhstan, the basic schema was always biased by connections with the Russian network, and today by the dualism between Almaty and Nursultan; the Tajikistan and Kyrgyzstan territories are heavily conditioned by the presence of geographical discontinuities, configuring disarticulated spaces, which are difficult to integrate into a uniform schema.

Similar situations can also be seen in Uzbekistan and Turkmenistan, where routes are often trespassing (crossing) borderlines, creating "cul de sac" situations, difficult to manage (prospecting regional and economic costs).

Considering this, the restructuring of the infrastructures seems too demanding—for the individual independent states—since too expensive, and also sometimes practically impossible, due to technical problems (mountain ridges, topographies, political unsolved questions, high distances).

The only possibility is to integrate the different national systems, elaborating mutual neighbouring planning; the alternative would be the risk of constructing self-referential, blind-street-like routes, impossible to manage economically.

Paradoxically, this apparent disadvantage could become the occasion for an advantage, representing for the NIS the opportunity to develop synergies, coordinate investments and improve the existing structures. In this way, it would be possible to elaborate a common schema for new connections, applying some general criteria (non-duplication, scale economics, integration, inter-modality, management efficiency) in order to avoid any waste of resources and political misunderstandings.

Otherwise, much of such infrastructures (like airport hubs, tunnels and highways, high speed trains, national flagship air companies) could simply become useless in a short time, like "side branches" without enough bulk (critical mass) to manage.

11.2.10 Further Modes and Further Infrastructures

The airports—usually the first image a foreign visitor has of a country—have been recently modernized, sometimes in a futuristic fashion. Any capital city tries to propose itself as a hub, building new infrastructures, like Nursultan and Dushanbe, with Almaty and Tashkent being the major hubs of the region. Air traffic is comprehensibly very important considering extension and distances (even when over the middle distances, it is challenged by high speed train new technologies, more convenient and possibly easier to manage particularly considering safety standards).[33]

Just after independence all the "stans", as usual for new independent states, tried to establish a national air company. To some extent this really was a necessity, considering provincial accessibility—for example, Murgab (Pamir) from Dushanbe, Khiva and Nukus from Tashkent, Osh from Bishkek over the Tien Shan, as well as most of the extended Kazakh territory.

Among them, Kazakhstan had success in developing the Air Astana, its flagship company with international standards; but more often than not such enterprises have proved to be very demanding, requiring not just investments, but "know how" and capability to access to markets as well.

Regarding maritime traffic, it is obviously limited by geographic and climatic aspects. Kazakhstan is investing in its maritime flotilla on the Caspian Sea (civil, military and mercantile), carrying forward a programme for the improvement of infrastructures for its internal water bodies (Aral and Balkhas especially) and rivers (considering also the potential of navigation of north-flowing rivers, like the Irtysh); to some extent Turkmenistan on its sea side as well.

Such questions have proved to be sensitive, with possible connections to the defence and security industries, and indeed the change of boundary status of the Caspian Sea started a kind of arms race that may have un-predictable consequences (representing one of the most critical aspects of the current geopolitical fragmentations, see later).[34] Kazakhstan recently ordered three units from Italian Trieste-based Fincantieri, to be equipped in a joint venture regime, partially with the intervention of local companies.[35]

A special case of infrastructure, inherited from the Soviet times, is that of the cosmodrome in Baikhonur, in central Kazakhstan, rented to Russia, still today the most important infrastructure in the world for spatial navigation.[36]

11.3 Infrastructures Driven Economics

11.3.1 Infrastructures as Semi-ideologies

Considering the geographic constraints (distances, natural barriers, landlockedness, continentalism), the question of infrastructures seems to be decisive to be able to permit orderly development; they are indeed so important that they assume, typically in CA, a cross-border scale, consequently a political character. For the same reason, they soon became an object of manipulation and were instrumental—from an internal point of view—to get consensus and externally to establish international relations.

Basically, their intention it that of pursuing development, finalizing public investments, feeding growth (indeed just for a while since often such investments have a limited Keynesian multiplier); but they also take on further meaning.[37]

In fact, they are instrumental for the autocrats in their aim to give purpose to their populations, with the intention of orienting and eventually diverting popular attention (something similar occurred with the Soviet economics, destined to fail); they are usually defined as "semi-ideologies"; they consist in large-scale projects (so not truly ideologies), capable of stimulating the imagination, like public works, spectacular urban realizations (skyscrapers, bridges), or some particular infrastructures (see the giant Roghun dam, Fig. 11.5).

[33]Gaynor Kelly Lee 2017.

[34]Silvestri Tommaso 2015–2016.

[35]Fincantieri, Singapore, 18 agosto 2016, presse communiqué reporting Kazakh National Maritime (Kazmortransflot, KMTF) procured 3 naval units Fincantieri that may change its impact and role, as well the strategic balance on this sea.

[36]Interview to Kazakh official, Nursultan, August 2017.

[37]Kudaibergenova 2015; Anceschi Luca 2017.

The recent publications report entire repertoires of such ideas, with different slogans and fantastic versions of projects, and prospected "routes", that the intention of pursuing some geopolitical intention is evident. Such projects often seem to be used in a kind of renewed "great game" based on communication manoeuvres, often just ephemeral proposals, playing with some trans-continental "variant" for pipelines, with highways crossing mountain ridges, extreme engineering works, with HSHC trains, or with SR or BRI segments and "land bridges", often with the exclusive motivation of re-orienting some decision.

11.3.2 Infrastructural Promise

All the "stans" soon started a programme of giant construction projects (following the wave of the transport revolution and the growth of international traffic), expecting it would exert a further push on local economics. Such programmes would further exert some response, above all consolidating pacification processes (considering the obvious trade-off between conflict and construction of big infrastructures on extended territories, crossing borders and delineating trans-national corridors).

They are usually presented in the periodical presidents' summits, as unrepeatable occasions for local development as well as for international economics, investors and banks (since it is commonly assumed, that in the landlocked CA "heartland", a well-planned infrastructural construction programme may realize enormous potential). In the current agenda of the NIS, such convincement assumes an official status: a long list of proposed infrastructures, like dams, canals, transport hubs, rail and roads, and especially pipelines for all kinds of commodities—that signify a strong "territorialization" effect of some politics.

11.3.3 List of Purposes

This is the case of irrigation channels and pipelines and of a list of small and big projects aimed at renewing, reconstructing and building; among them such mega projects like the Tajik dams,[38] western Kazakhstan "super-giant" Kashagan deposits and of the recently discovered Turkmenistan gas deposits.[39] It seems that the easiest and most direct way to pursue growth and to express value (currently unexpressed), comprehensibly is to open this space rich in all natural resources, renewable or not, to international investors and donors.

In particular, Kazakhstan has developed such infrastructure programmes prefiguring a new deal for long-term stabilization, with the project called Kazakhstan 2030, "Nury Zhol", widely propaganded by media and social events, for example, with mega screens in the country's squares on all occasions. It represents an extraordinary opportunity to invest, attracting foreign players to this central segment of the SR. But there are also risks: after a certain threshold, the multiplier effect of such policies can result as lower than expected (as indeed the experience of many countries shows). It is possibly the same mistake already made by gigantomaniac Soviet economics (the classic missed "lesson learned" from recent history).[40]

Besides these, CA authorities have elaborated a list of sustainable and socially inclusive projects, demonstrating an increased capability which corresponds to actual needs—overcoming a typical modernist quantitative threshold; this potentially being inspired by international agencies and institutions. It is the case of ecological recovery projects concerning the Aral Sea and other bodies of water, the re-forestation of extended areas, especially those surrounding the major cities, contrasting desertification (see dedicated chapter), so too for further projects elaborated on a local scale.

References

Anceschi Luca (2017) Turkmenistan and the virtual politics of Eurasian energy: the case of the TAPI pipeline project. Central Asian Survey 36(4):409–429. https://doi.org/10.1080/02634937.2017.1391747

Brill Olcott Martha (2012) Tajikistan's Difficult Development Path, Brookings Institution Press, 27 Nov 2012

Economist Intelligence Unit, "Uzbekistan: Economic Overview", Wikipedia, Uzbekistan. Accessed 29 Mar 2018. (citing eurasiacenter.org page not found)

Electronic Government of the Republic of Kazakhstan. http://egov.kz/cms/en/information/help/help-elektronnoe-pravitelstvo. Accessed 23 Apr 2018

Gaynor KL (2017) Transformations in Turkmen higher education: current opportunities and challenges at a new university. Central Asian Survey 36(4):473–492. https://doi.org/10.1080/02634937.2017.1367645

Global Witness, Germany must launch full inquiry into Turkmen funds in German banks, https://www.globalwitness.org/en/archive/germany-must-launch-full-inquiry-turkmen-funds-german-banks/, Press Release, July 3, 2007. Accessed 27 Mar 2018

http://gca.satrapia.com/+adb-suspends-its-participation-in-supporting-construction-of-the-tat-railway+. Accessed 29 Mar 18

http://gca.satrapia.com/+the-level-of-poverty-in-kyrgyzstan-will-continue-to-grow+. Accessed 27 Mar 2018

http://gca.satrapia.com/+unified-pension-fund-recommended-in-kazakhstan+, 23 January 2013. Accessed 11 Apr 2018

[38] https://web.archive.org/web/20080405065705/http://npdp.stanford.edu/damhigh.html, accessed 29.3.18; the second one is Nurek dam; both are on the water rich Vaksh river in Tajikistan.

[39] Anceschi Luca 2017:4.

[40] Klüter 1992.

http://news.tj/en/news/chormaghzak-tunnel-renamed-khatlon-tunnel-and-shar-shar-tunnel-renamed-ozodi-tunnel. Accessed 29 Mar 2018

http://whc.unesco.org/en/list/. Accessed 23 Mar 2018

http://whc.unesco.org/eng/list/603/documents/. Accessed 23 Apr 2018

http://www.imf.org/en/News/Articles/2015/09/14/01/49/pr03188. Accessed 27 Mar 2018

http://www.tajikistanmission.ch/business-and-investment/infrastructure.html. Accessed 29 Mar 2018

https://blogs.fco.gov.uk/leighturner/2013/05/07/trade-tunnels-transit-and-training-in-mountainous-tajikistan/. Accessed 29 Mar 2018

https://edition.cnn.com/travel/article/trans-siberian-road/index.html. Accessed 29 Mar 2018

https://web.archive.org/web/20080405065705/http://npdp.stanford.edu/damhigh.html. Accessed 29 Mar 2018

https://www.aljazeera.com/news/asia/2011/12/2011123015563314 6491.html, 30.12.2011, Accessed 6 July 2018, Las Vegas on the Caspian?, Luxury hotels in Awaza, built using gas revenues and aimed at foreign tourists, have yet to benefit ordinary Turkmens

https://www.cia.gov/library/publications/the-world-factbook/fields/2046.html. Accessed 27 Mar 2018

https://www.economist.com/news/asia/21582325-president-edifice-complex-screwing-motherland-folie-de-grandeur. Accessed 27 Mar 2018

https://www.economist.com/news/asia/21711943-authoritarian-president-pins-his-hopes-turkmen-las-vegas-shrinking-exports-spell. Accessed 22 Apr 2018, "Shrinking exports spell trouble for Turkmenistan", The Economist. 17 December 2016

https://www.economist.com/node/10181134. Accessed 29 Mar 2018

https://www.state.gov/e/eb/rls/othr/ics/2011/157382.htm. Accessed 27 Mar 2018

https://www.wto.org/. Accessed 29 Mar 2018

Indeo Fabio (2018) The impact of the belt and road initiative on Central Asia: building new relations in a reshaped geopolitical scenario. In: Center for Energy Governance and Security, Hanyang University, Zhang W et al (eds) China's Belt and Road initiative, Palgrave studies in internationalization in emerging markets. Springer International Publishing, Cham

Thorez Julien (2018) Central Asia: the state of the research, Asiac annual conference 2018, 5–7 December 2018, University of Trieste – Gorizia campus (manuscript)

Klüter H (1992) Russland und die Auflösung der Sovietunion, "Geographische Zeitschrift", Jahr 80, Heft 1

Kudaibergenova DT (2015) The ideology of development and legitimation: beyond 'Kazakhstan 2030'. Cent Asian Surv 34(4):440–455. https://doi.org/10.1080/02634937.2015.1115275

L. P. (2018) "Sorpreso dalla tempesa di sale, turista si ammala. Un 75enne in visita all'Uzbekistan con altri 11 friulani. Ha respirato la polvere pericolosa per la salute", "MessaggeroVeneto", p 29, venerdì 22.6.2018

Marchi M, Tonini C., a cura di (2009) Da Berlino a Samarcanda. Città in transizione, Quaderni Dip. Disc. Storiche, Univ. Bologna, Carocci, Roma

Mashrab Fozil (2015) Bishkek Puts Brakes on China–Kyrgyzstan–Uzbekistan Railway, Eurasia Daily Monitor 12-199, November 3, 2015, https://jamestown.org/program/bishkek-puts-brakes-on-china-kyrgyzstan-uzbekistan-railway/. Accessed 22 Apr 2018

Ministero degli Affari Esteri e della Cooperazione Internazionale, Dei – Kazakhstan: Italcementi firma MoU per stabilimento Shymkent, https://www.esteri.it/mae/it/sala_stampa/archivionotizie/approfondimenti/2015/04/dei-kazakhstan-italcementi-firma.html. Accessed 22 Apr 2018

Nick M (2017) Nationalism in Central Asia: A Biography of the Uzbekistan-Kyrgyzstan, Boundary, Central Eurasia in Context, 2017, University of Pittsburgh Press, https://doi.org/10.2307/j.ctt1vjqrk6

Paskaleva Elena (2015) Ideology in brick and tile: Timurid architecture of the 21st century. Cent Asian Surv 34(4):418–439. https://doi.org/10.1080/02634937.2015.1118207

PWC (2010) Doing business in Turkmenistan 2012–2013, https://www.pwc.com/uz/en/assets/pdf/doing-business-guide-in-turkmenistan-2012-2013.pdf. Accessed 20 May 2018

PWC (2011) Doing business guide Tajikistan 2012–2013. http://www.eurasianbusinessbriefing.com/wp-content/uploads/2015/04/pwc-doing-Business-in-tajikistan.pdf. Accessed 20 May 2018

PWC (2012–2013) http://www.eurasianbusinessbriefing.com/wp-content/uploads/2015/04/pwc-doing-business-in-Kyrgyzstan.pdf. Doing business guide in Kyrgyz Republic 2012–2013. Accessed 20 May 2018

Rasanayagam Johan (2014) The politics of culture and the space for Islam: soviet and post-soviet imaginaries in Uzbekistan. Cent Asian Surv 33(1):1–14. https://doi.org/10.1080/02634937.2014.882619

Reid Patryk (2017) 'Tajikistan's Turksib': infrastructure and improvisation in economic growth of the Vakhsh River valley. Cent Asian Surv 36(1):19–36. https://doi.org/10.1080/02634937.2016.1204533

Scharr K, Steinicke E (hg) 2012, Vom euphorischen Aufbruch in die Realitaet des Alltages, 1898–2000 – zwei Jahreszenten Transformationsforschuing, University of Innsbruck Press

Secchi Roberto, Spita Leone, a cura di (2018) Architettura tra due mari: Radici e trasformazioni architettoniche e urbane in Russia, Caucaso, e Asia Centrale, Macerata, Quodlibet

Silvestri Tommaso (2015–2016) Le strategie della federazione russa nel teatro del mar Caspio dopo la caduta dell'Unione Sovietica, Master thesis, University of Trieste, academic year 2015–2016

The Republic of Uzbekistan Accepts Article VIII Obligations, Press Release, November 11, 2003. https://www.imf.org/en/News/Articles/2015/09/14/01/49/pr03188. Accessed 6 July 2018

Tonini Carla (2017) Samarcanda, Bukhara, Khiva. Città dell'Asia Centrale, Città dell'Asia, Marzia Marchi a cura di. Bononia University Press, Bologna, pp 29–59

Wikipedia, "Transports in Turkmenistan". Accessed 29 Mar 2018

Wikipedia, Tajikistan. Accessed 23 Apr 2018

Wikipedians, Countries and territories in the world (no year) https://books.google.it/books?id=bR0hIC0Xhb0C&pg=PA120&lpg=PA120&dq=1216.3+kg+of+heroin+and+267.8+kg+of+raw+opium&source=bl&ots=XxL5vwuI0M&sig=g_CH8ED7D09ZwhA8qN4LH44Odxs&hl=it&sa=X&ved=0ahUKEwjr8JKwvP_YAhVDUBQKHfMlBegQ6AEIKDAA#v=onepage&q=1216.3%20kg%20of%20heroin%20and%20267.8%20kg%20of%20raw%20opium&f=false. Accessed 27 Mar 2018

World Bank, Tagikistan. https://data.worldbank.org/country/tajikistan. Accessed 27 Mar 2018

World Bank Group (2015) Kazakhstan: low oil prices; an opportunity to reform, Kazakhstan Economic Update Spring 2015, https://www.worldbank.org/en/country/kazakhstan/publication/kazakhstan-economic-update-spring-2015. Accessed 23 Apr 2018

12 Economics: From Micro to Macro

Abstract

The main characteristic of the economic situation is the coexistence of diverse and divergent tendencies: on the one side, the renovated top-down initiative, leveraging on the extraordinary wealth represented by domestic natural resources; on the other side, a kind of self-organized society, otherwise defined as a spontaneous "bazaar" or independent economy, using economic instruments mainly for survival, namely self-consumption purposes, rather than for accumulation; a situation that prospects further division and which the new powers have to recover as soon as possible.

Keywords

Current economic processes · Soviet legacy · Infrastructures · Mobility and accessibility · Urbanistic an territorial organization · Resources exploitation-based ("commandeering") economics · Bazaar and popular economics · Post-modern opportunities

12.1 Rapidly Changing Economic Setting

12.1.1 From the Post-Soviet Era to the End of Transition

Twenty-five years after the Soviet collapse, the transition cycle supposedly exhausted, the economic situation still has an ambiguous character, with volatile statistics and still unsettled questions.[1] This is especially evident in the material–economical aspects of the current reality, in which it is difficult to distinguish what is real development from what is "artificial", what is "easy" money (just quantitative induced growth), from what is sustainable, socially inclusive and self-alimenting economics.

In some areas signs of rapid growth, of globalizing economics and of the spreading of wealth are appearing; other areas appear to be caught in "traps", depending on the circumstances of a territorial, economic or cultural character.

It is difficult to make any predictions; the narratives diffused by the media, the reports from international organizations, as well as analyses published by institutions are often in contradiction with each other. Some reports inspire an image of growing economies, sometime comparing the CA systems to a kind of "new tiger"; such contingencies derive from favourable quotations for exported commodities (HC, cotton, gold and minerals) and to some extent for manufacturing production (Fig. 12.1).

Uzbekistan, even if still considered a poor country, is one of the fastest growing economies in the world,[2] exerting favourable repercussions on neighbouring Tajikistan and Kyrgyzstan, considered as even poorer countries, needing international assistance; Kazakhstan and Turkmenistan, defined as HC rentier states, after prolonged periods of growth, seem to be currently stabilizing (Tables 12.1, 12.2, and 12.3).

Other sources, contrarily, present the images of a fictitious growth or even of a regression; further sources underline the risk of excessive dependency on HC and commodity export, namely the risk of pursuing a fundamentally inflated growth, influenced by top-down spending.

The key question of the current situation is evident everywhere, namely the local institutions—still consolidating themselves and often pursuing neo-mercantilist policies—coming up against outside globalization pressure. In these circumstances the post-Soviet elite proved to be much more pragmatic than expected, using the "political capital" acquired (indeed in disputable terms, managing national

[1]Marzhan, 2015:459.

[2]PWC 2010:6, 2016, 2017; but in CA there is a general question of data accessibility and reliability, Anceschi Luca, 2017; Kazakh statistical agency obtained certificate ISO 9001 in 2008.

I. Jelen et al., *The Geography of Central Asia*, World Regional Geography Book Series,
https://doi.org/10.1007/978-3-030-61266-5_12

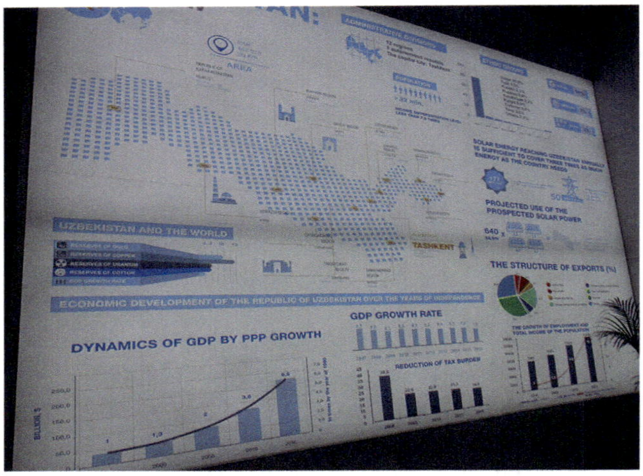

Fig. 12.1 Expo Astana (Nursultan) 2017, Uzbek pavilion celebrating the economic growth of the country. (Photo: Igor Jelen)

Table 12.1 Central Asia—current account balance (percentage of GDP)

	2013	2014	2015
Kazakhstan	0.4	2.8	−2.6
Turkmenistan	−7.2	−6.7	−12.7
Uzbekistan	2.9	0.7	0.0
Tajikistan	−2.9	−9.7	−10.2
Kyrgyz Republic	−15.0	−16.7	−14.7

Data source: Exchange Rate Developments and Policies in the Caucasus and Central Asia, International Monetary Fund, Middle East and Central Asia Department; https://www.imf.org/external/pubs/ft/dp/2016/mcd1602.pdf. Accessed 8/7/2017

Table 12.2 Central Asia—GDP growth (annual %)

	2012	2013	2014	2015	2016	2017	2018
Kazakhstan	4.8	6.0	4.2	1.2	1.1	4.1	4.1
Turkmenistan	11.1	10.2	10.3	6.5	6.2	6.5	6.2
Uzbekistan	7.3	7.5	7.1	7.4	6.0	4.4	5.1
Tajikistan	7.4	7.4	6.7	6.0	6.8	7.6	7.3
Kyrgyz Republic	−0.2	10.9	4.0	3.9	4.3	4.7	3.5

Data source: Data Bank, World Development Indicator, The World Bank; https://databank.worldbank.org/reports.aspx?source=2&series=NY.GDP.MKTP.KD.ZG&country=#. Accessed 20/05/2020

monopolies) in transition periods in order to consolidate themselves.

However, the moment is to be considered favourable; economic stabilization is occurring with the background of an overall, not obvious, period of peace, representing a unique opportunity to begin facing the other problems that affect the area—social justice, political asymmetries, ecology and many others—which just a few decades ago might have appeared practically without solution.

12.1.2 Diverse Options

As appeared almost immediately at the moment of independence, opportunities for the CA countries derived mainly from the availability of natural and energy resources (Fig. 12.3): a situation that seemed at that moment to render manifest the economic destiny of the "stans" as international suppliers of such goods. The local elites immediately acquired awareness of this and started on the reconstruction of top-down guided economies (to some extent giving continuity to the previous power), using the enormous opportunities offered by such resources: these were apparently easy to manage in a centralistic way, so ideal to be used both for investments and for maintaining power.

In reality this type of project proved to be more complicated; resources demonstrated to be relatively difficult to produce, transported and finally exported (this for HC, minerals, water for energy and irrigation, and also cotton and other crops, namely agricultural soil). It was necessary to fill this gap, to optimize organization capability, "know-how", capital, technology and above all social confidence (in terms of legitimization and eventually consensus) to be able to pursue such projects and to keep on with long-term programmes.

It must be considered that the sudden collapse of the Soviet Union (SU) created a situation of "de facto" annihilation of industrial capacity (besides some points that will be discussed later), which made it necessary for the whole system to reboot from a level close to "zero", and the whole development pattern needed re-thinking. The inherited industry often consisted in rusting bulks of machinery and installations which needed to be removed and cleaned up, representing a burden (rather than an asset) for the NIS (even when they can be considered as deposits of materials to be re-used, prospecting a new "circular economics").

The Soviet centralistic regime had the pretension of ruling (and planning) the entire reality (basic needs, supply, mobility, society and culture, politics and economics); so its sudden collapse configured an even worse vacuum on all levels, a void that populations, societies and the new elite had to fill—a task that was performed in diverse ways.

In that moment, the population found itself in a disorienting situation. Its initial reaction was not always rational, sometimes the people panicked (such as, in several circumstances, in Tajikistan), sometimes there was a search for an arrangement in the new situation, using the few instruments that were possible to access.

Often the reaction consisted in a kind of regression to an elementary status, with the populations often remaining without any resources, in empty kolkhoz and sovkhoz, just trying to achieve basic survival. This meant regression to a precarious way of operating, often returning to the steppe and mountain villages, adapting traditional—wherever pos-

Table 12.3 Central Asia—GDP per capita (current US$)

	2012	2013	2014	2015	2016	2017	2018
Kazakhstan	12,386.7	13,890.6	12,807.2	10,510.7	7714.8	9247.5	9812.6
Turkmenistan	6675.1	7304.2	7962.2	6432.6	6389.5	6587.0	6966.6
Uzbekistan	2137.5	2281.4	2492.3	2615.0	2567.7	1826.5	1532.3
Tajikistan	969,2	1048.2	1104.1	929.0	802.5	806.0	826.6
Kyrgyz Republic	1177.9	1282.4	1279.7	1121.0	1120.6	1242.7	1281.3

Data source: Data Bank, World Development Indicator, The World Bank; https://databank.worldbank.org/reports.aspx?source=2&series=ny.gdp.pcap.cd&country=#. Accessed date: 20/02/2020

sible—activities, like cattle breeding and caravanning operated with archaic, but self-organized (and inexpensive) methods.[3]

It signified the paradoxical regression to a self-consumption agriculture on a domestic scale, the reorganization of village economics and of traditional bazaars as well as the recovery of traditional mobility and of ancient caravan routes over the mountain passes (transversal to down-valley road networks).

In some cases such regression was made easier since the population was (to some extent, non-intentionally) prepared for such a possibility, practising the already described semi-legal or "bazaar economics" on a local level. They did so because the official system, in the last few decades of the Soviet epoch, was no longer capable of ensuring the basic needs for production, food and services.

Therefore, the question for the new governments was (as typical indeed for any transition country), what kind of strategic choices to undertake, namely the search for a development method—and for a new ideology—to give purpose to their populations: it had to be communicated as a kind of target, simplifying such questions as "where are we going" and "what are we working for".

It meant the difficult choice between reciprocally alternative options (a kind of epochal divide) such as openness and prosperity, centralization and safety, neo-nationalism or integration in IC, consumerism or "sacrifices" for the benefit of a future generation. Options which gave priority to immediate prosperity to the detriment of long-term development (or the reversal), inspired by a certain idea of the "rentier state" (HC export-based, supposedly): these countries would specialize in supplying raw materials to advanced economies (assuming a certain idea of the international labour division), or looking for other (more elaborated) policies, more or less based on public intervention, or on socially diffused initiatives.

Such policies sometimes coincided with the elaboration of what is possible to define as "semi-ideologies,[4] (see later), typical of semi-democratic regimes, easily communicated with simple statements to a large population, seeking a popular end. They usually concern some big work purposed as an epic challenge, comprehending some high-impact vision (the construction of a new town, of a new railway, of a high bridge etc.) with slogans like "new silk road", "energomost" ("energy bridge"), "golden age lake" and similar.

Usually such strategic-systemic decisions were not a matter of democratic discussion, but were taken directly, as happened in transition times, in situations of urgency, or of presumable necessity: the collapse of the Soviet paradigm provoked the breakdown of heavy industries (the main source of employment), planning practices and a centralistic supply of energy, heating systems, food and of most functions, leaving the population in a situation of desperate need.[5]

Often such circumstances signified truly catastrophic conditions, with special regard to social and ecological issues (especially critical in this continentally locked region), further aggravated by geopolitical stress, namely pressure from outside. Such emergencies were evident especially in the situation of political unrest, sometimes of war, that characterized some areas of CA for nearly the whole transition time.

12.1.3 Elements of Continuity

Soviet economics saw its heyday in the context of a mechanical paradigm—typical of the period of late modernity, before the affirmation of electronic and ICT—relatively easy to organize in a Fordist environment, with standardized instructions to be spread from central areas to the peripheries.[6] It was based on an extended heavy industrial sector, consisting of metallurgical, chemical, energy and construction and infrastructure industries to be consequently divided

[3] Jelen Igor 2002, 2003.

[4] Kudaibergenova Diana T., 2015.

[5] Brill Olcott Martha, 2012:283; it was the case of children and women without social assistance, elderly peoples without pensions, minorities deprived of citizenship, eventually persecuted, and further weak categories were particularly hit by such changes.

[6] The more demanding creative functions were devoted to eventually copying and reproducing innovations originating in advanced countries, with the s.c. reverse engineering; see as well Hughes Thomas P., 2004.

into production segments, to be managed and implemented with hierarchical commands.[7]

On the contrary, it mostly ignored light industries and large-consumer production: any need ("demand") was to be considered in a general planning schema, excluding the formation of market-oriented production from the start. Consequently, the "offer" was standardized in industrial monopolies, excluding the formation of a small-middle industry sector, capable of rapidly adapting to markets (indeed of critical importance for any consumerist-based economy).

In many cases this heavy industrial sector broke down with the same collapse of the SU. Indeed, at that time, it had already lost much of its original capability, appearing obsolete, polluting, inefficient and often simply counterproductive. However, in some cases it survived the transition passage, being renovated and re-launched in the new context.

This thanks to a set of reasons. Above all it is a question of "entry barriers", namely of investment costs in such heavy industrial sectors that are usually very high, discouraging construction of new plants from scratch, to the benefit of the older and pre-existing industries. In fact, the old plants continued to operate, notwithstanding such externalities.

This occurred in several cases, especially in scarcely inhabited regions, where opposition and the social cost of such activities are usually lower than in densely settled regions (because of environmental and soil-consuming costs, energy and raw material supply possibility). Here, such factories can benefit from a favourable cost-opportunity trade-off (from the fact, as said, that construction from scratch of new plants would be disproportionately expensive).[8]

Further, it can be said that such policies can be facilitated by the continuation of territorial organization, and a legacy of the previous regime. They rely on a culture of planning and territorial control in all senses: regional spaces are coordinated in a relatively efficient way, with inter-modal transport facilities organized on the base of effective zoning. This regarding both industrial and civil-residential areas, and in a wider sense, for regional and rural territories.

So too, for infrastructure networks characterizing the whole CA territory, built up in the Soviet period on a large scale (railways and highways, ID facilities, energy power plants and grids, pipelines, ports and airports). Today they are still often operating, although in poor maintenance conditions, representing a useful schema for further development.

In fact, the new power structures cannot simply give up on such "hardware", because they have to maintain their industrial potential and because it is necessary to give continuity to the entire (economic and social) pattern they represent. Thus, the new elite simply appropriate these industrial sectors, constructing new monopolies on them.

12.1.4 Energy, HC, Infrastructural Sector, Heavy Industry

At that time such industries represented an available asset and yet they often represented activities of mere exploitation of natural resources, heavily impacting natural and social environments.[9]

It is possible to list a sequence of such cases, starting from the HC extraction sector, and in general from the energy industry, the most important asset in the area. It began to be developed in late Soviet times with discoveries and research carried out by Soviet explorers and scientists, but which the regime did not subsequently have time to develop.

This is the case, in Kazakhstan, of the rich HC fields, the so-called supergiant Kashagan and Tengiz deposits; so too, for the Turkmen gas, with research finally leading to the discovery of the Galkynysh giant gas field, that is utilized by organizing the pipeline passing through Uzbekistan and Kazakhstan, to China and which has been operative since 2006 (see chapter 9).

It is also the case of other industrial areas, with coal mining, extraction, metallurgic and refinery industries.[10] This is especially pertinent for the coal-alimented plants in the Karaganda region (operated by Indian Mittal). A further case is that of Kyrgyz Kumtor gold mine,[11] discovered in late Soviet times, and of other minerals, coal and mineral mining, thermo- and hydro-energy production, located in remote areas, that represent a basic asset, even when needing to be continuously modernized, potentially with foreign capital. Among the hydroelectric power plants, the Nurek dam in Tajikistan must be mentioned (Fig. 12.2), and other dams on the Amu Darya and the Toktogul dams in Kyrgyzstan on the Syr Darya.

The Nurek hydroelectric power plant aliments the TALCO factory, processing bauxite that is imported from abroad

[7]This is the reason why in some case the industries started in Soviet times, even when obsolete, keep continue, representing a substantial asset for the NIC, that could eventually think about a further development, predisposing renovation and maintenance investments.

[8]It is a question of externalities, but also of the immaterial patrimony and of the "cultural capital" that they could represent in terms of "know how", of local work, that cannot be neglected, and that would represent an advantage for current necessities.

[9]That probably in more advanced societies, characterized by higher sensitiveness about such impact, would not be tolerated; but it is also to consider the current availability of updated technology mitigating such impacts.

[10]ArcelorMittal 2013.

[11]Lonely Planet, 2014.

Fig. 12.2 Tajikistan, Nurek dam reservoir, August 2017. (Photo: Igor Jelen)

nowadays;[12] planned and constructed in late Soviet times as a pure example of territorial Fordism; today it is still one of the biggest aluminium plants in the world. Set up in the mid-1970s, it has now been reconditioned and is today state-owned, currently representing the main income asset in Tajikistan.[13] It is the cause of many problems in terms of pollution in a politically sensitive border area with Uzbekistan.

A further case is that of the Tchaklov airplane industry in Tashkent, with the Uzbek government engaged in an attempt to rehabilitate the factory in the context of a neo-mercantilist policy. In its heyday, before "Perestroika", the factory had about 30,000 employees, and it was one of the largest airplane plants in the SU, serving a large market (both military and civil), relying on the knowledge and expertise of engineers and technicians arriving from all over the Soviet space.

Today, after a sequence of attempts and failures, it seems to have given up on its original production, stating other types of productions.[14] In general, the Uzbek management has tried to reconvert its precious industrial heritage for the actual needs of the country and for large-consumer standards, for example transport, automotive industries, domestic appliances and durable goods.[15]

The nuclear sector has been generally dismantled and abandoned; nuclear-fuel waste management—inherited from Soviet times—still represents a current issue, but it also represents the opportunity to elaborate new treatment techniques on the wide surfaces of these countries. The Uzbek government recently seems to be motivated in the recovery of a kind of nuclear industry in this country (Fig. 12.3).[16]

12.1.5 The Trend

The current situation evidences a split in two directions: a set of heavy industries concentrated in the hands of public authorities (or of some "oligarchs' clan"), and a "popular" economy, often in its initial stage defined as bazaar or kiosk economy. The first economy mostly manages commodities for export in hard currency productions; the second assumes the role of sustaining the wider population; it is scarcely institutionalized or totally submerged.[17]

International agencies and consultancy firms define this as "small business and private entrepreneurship", accounting usually for more than half of the economy.[18]

[12] Brill Olcott Martha, 2012; https://www.economist.com/news/asia/21582325-president-edifice-complex-screwing-motherland-folie-de-grandeur. Accessed 27.3.2018.

[13] Brill Olcott Martha, 2012:283:53.

[14] PWC 2016:4; but see "Uzbekistan to create its Aerospace industry", http://tashkenttimes.uz/national/1921-uzbekistan-to-create-its-aerospace-industry, 2018-01-11. Accessed 4.7.2017.

[15] Fazendeiro Bernardo Teles, 2015:490.

[16] Uzbekistan, 2018.

[17] This definition is rather difficult, among informal, shadow, submerged, grey or kiosk-economics; it could be considered as an intermediate economic step between pre-modern and advanced structured economics, or as a kind of local self-referential way of living; Abazov R., 1999; Martha Brill Olcott, 2012:283; this assuming a certain theoretical development path, from the quantitative-commandeer economics, to a more culture-intensive and transformative one.

[18] For the especially relevant Uzbek case see PWC 2016:6.

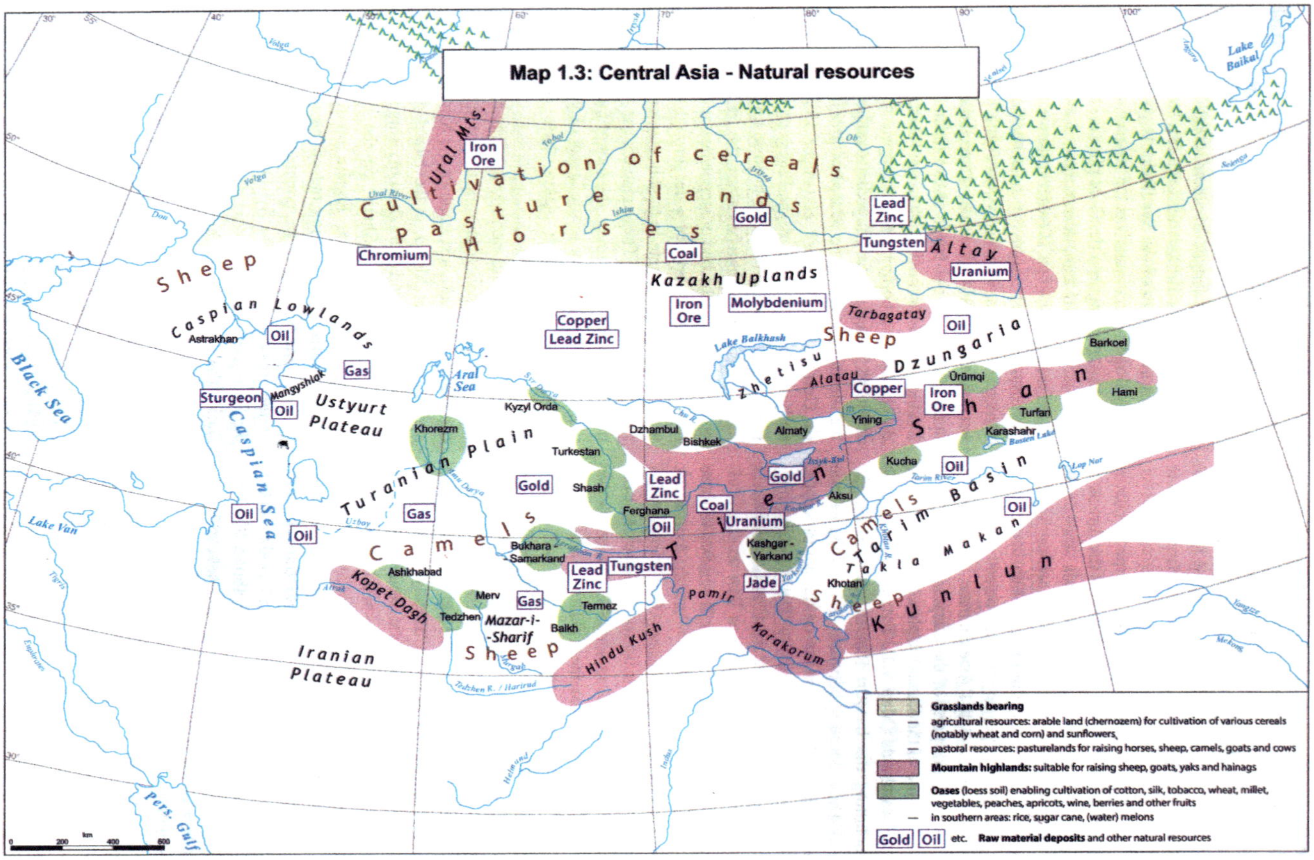

Fig. 12.3 Central Asian natural resources. (Courtesy Barisitz Stephan, 2017, p. 7)

These two levels, elite and popular economies, do not really dialogue, configuring a fundamental asymmetry that may influence any development path negatively. Such a gap has been covered in transition times by a consistent in-flow of money from HC and other commodities easily exportable, financing welfare and social spending, with the aim of maintaining consensus and social equilibria.

But today—after a bonanza of 20–30 years of high HC prices, and high revenues—the "stans" are aware that the change in energy model may be closer than expected, and that it is necessary to foresee new forms of development. Therefore, it is necessary to start up new policies, beyond the moment that HC will stop representing an easy source of revenue (indeed a situation of extra-profits in oil and gas economies that have financed the rentier states, especially Kazakhstan and Turkmenistan, in the last few decades).

Such an economic turn (coinciding with a kind of post-modern passage, overcoming the Fordist condition) would probably rely on innovations, technology- and culture-intensive activities, light- and market-oriented production, and high value-added industries (today in CA virtually non-existent). And furthermore, on advanced tertiary functions, such as social and personal care, education, tourism and amenity activities, assuming a qualitative use of the territory. It should probably organize the evolution in a new form of economy (green, circular and sustainable), pursuing continuing improvement in standards of living (and corresponding expectations), that in advanced economies represent a fundamental driving factor for the whole system.

Such economies would configure a diffused, rather than centralistically led economics, relying on accessible resources, rather than on political monopolies (Tables 12.4, 12.5, 12.6 and 12.7).

12.2 A Transition Phase

12.2.1 An Initial Stage: Trying for Survival

Soviet modernization provoked the destruction of traditional material culture—namely economies, infrastructures, housing—and of traditional means of livelihood, which were substituted by industrial technology, pursuing scale economics. However, this new form of economics proved to be inadequate in just a few decades, becoming obsolete, sometime self-referential—a kind of dramatic

Table 12.4 Central Asia—GNI per capita, PPP (current international $)

	1990	1991	1992	1993	1994	1995	1996	1997
Kazakhstan	5120	4690	4510	4240	3830	3640	3770	3950
Turkmenistan	2710	2600	2440	2200	1860	1680	1570	1430
Uzbekistan	–	–	1290	1260	1200	1190	1200	1260
Tajikistan	2210	2080	1480	1240	960	840	680	720
Kyrgyz Republic	1810	1700	1480	1280	1030	980	1050	1140

Data source: World Bank International Comparison Program database; Myant, Martin R., 2011, Appendices

Table 12.5 Central Asia—GNI per capita, PPP (current international $)

	1998	1999	2000	2001	2002	2003	2004	2005
Kazakhstan	3990	4110	4460	5260	5940	6530	7240	7830
Turkmenistan	1490	1710	1930	2430	2820	3380	4000	4350
Uzbekistan	1310	1360	1420	1490	1570	1650	1830	2000
Tajikistan	760	790	800	940	1040	1140	1310	1430
Kyrgyz Republic	1140	1170	1250	1370	1380	1500	1620	1670

Data source: World Bank International Comparison Program database; Myant, Martin R., 2011, Appendices

Table 12.6 Central Asia—GNI per capita, PPP (current international $)

	2006	2007	2008	2009	2010	2011
Kazakhstan	14,220.0	15,600.0	15,840.0	16,380.0	17,110.0	18,210.0
Turkmenistan	6140.0	7150.0	8040.0	8400.0	8920.0	10,130.0
Uzbekistan	3720.0	4320.0	4800.0	4980.0	5150.0	5520.0
Tajikistan	1600.0	1760.0	1920.0	1960.0	2590.0	2860.0
Kyrgyz Republic	2180.0	2420.0	2570.0	2640.0	2560.0	2610.0

Data source: Data Bank, World Development Indicator, The World Bank, https://databank.worldbank.org/reports.aspx?source=2&series=NY.GNP.PCAP.PP.CD&country=#. Accessed 20/02/2020

Table 12.7 Central Asia—GNI per capita, PPP (current international $)

	2012	2013	2014	2015	2016	2017	2018
Kazakhstan	19,380.0	21,290.0	22,380.0	23,620.0	22,930.0	23,610.0	24,450.0
Turkmenistan	11,290.0	12,530.0	13,780.0	15,070.0	16,450.0	17,280.0	18,490.0
Uzbekistan	5950.0	6420.0	7130.0	7460.0	7810.0	8250.0	4050.0
Tajikistan	3070.0	3400.0	3380.0	3410.0	3530.0	3710.0	3780.0
Kyrgyz Republic	2850.0	3050.0	3220.0	3320.0	3380.0	3550.0	8810.0

Data source: Data Bank, World Development Indicator, The World Bank, https://databank.worldbank.org/reports.aspx?source=2&series=NY.GNP.PCAP.PP.CD&country=#. Accessed 20/02/2020

illusion. It often produced more damage than not. Above all it separated the population from its history and its territory, and from any traditional skills, that have been completely forgotten in the meanwhile, finally configuring a situation of alienation.[19]

So, at the moment of the change in regime, people could not find any workable technology (the Soviet one no longer being usable, and the traditional one destroyed or forgotten): this situation proved even more difficult considering that, until that moment, most of the people had depended on the central power for all necessities (safety, food, mobility, health).

This new society realized that Sovietization had provoked the decay and loss of natural resources, soil consumption, over-building, an excess of construction and infrastructures (usually becoming obsolete and useless, especially considering the Soviet attitude to building in a disproportionate way,[20] without thinking about subsequent maintenance and depreciation).

Mostly the changes in the Soviet era proved to be irreversible; for example, the population forced to sedentariness lost their knowledge in preparing basic housing like yurts—a very efficient and inexpensive technology to build temporary homes in highland pastures. So too, for much practical knowledge, for cattle breeding, transhu-

[19] Jelen Igor, 2003.

[20] Klüter H., 1992:20–38; Hughes Thomas P., 2004:264.

Fig. 12.4 Kyrgyzstan, Lyailiak valley, July 1994, repairing a bridge. (Photo: Igor Jelen)

mance organization, food, milk and meat processing and conservation. So too, for artisan skills, use of local natural resources, for example finding water in arid environments, drying mud for making building bricks, or manure for making combustible in wood-less environments. So too, for the necessary expertise for survival in remote areas of the mountains or steppe.

The populations had no alternative but to try to adapt the precarious means they had at their disposal, for example arranging living quarters with residual industrial materials, used pneumatics or metal scraps. So too, for any activity, productive or cultural, and for any technology, self-producing food and energy, instruments for mobility and communication. So too, for the rehabilitation of local infrastructures (e.g. of building precariously roads and bridges and other rural infrastructures), ID channels and further essential facilities (Fig. 12.4 and 12.5).

12.2.2 A New Condition of Poverty

In some cases, the population maintained some basic knowledge (e.g. building the old-style yurts close to the modern settlements and manufacturing textile products using traditional techniques), or at least the ability to work the land and breed animals.

The population had to recover traditional ways of life based on hand work and animal labour for transporting, ploughing and other work, that in some circumstances resulted, especially considering the mountain topography, relatively efficient (also considering the poor standard of residual Soviet agro-machinery, and in general the inaccessibility of agro-industrial means, such as agro-chemicals and fuel).

It meant regression to survival economics, the only possible option, also considering the virtually non-existent connection with any wider market. The population of the kolkhoz regressed to transhumance and seasonal farming in highland pastures. This happened in a similar way on the plains and in the valleys, where the resident populations recovered traditions of self-consumption agriculture, adapting irrigation infrastructures, as well as other traditional activities.

The populations tried to recover their skills, their land use patterns and technologies as well as the solidarity habits of the past. The Kyrgyz occasionally went back to "jailoo", or highland transhumance, while Turkmen and Kazakh shepherds went back to the desert and steppe, considered as a kind of refugee. The Uzbek and Tajik went back to agriculture, applying the irrigation techniques they could assemble on a domestic scale. However, most of them remained entrapped in empty kolkhozes, without many opportunities (a situation indeed that was a non-secondary cause of conflicts in 1989–1991, and then in Tajik civil war).

This situation meant a kind of (not just territorial, but also cultural) regression as a consequence, to once abandoned peripheries, where it was possible to live in direct contact with nature in a situation of relative freedom, namely the freedom to access those resources. At the same time, this regression signified the re-discovery of other features which had once characterized the community-scale economy, like organizational local modes.

It meant the recovery of a condition of subsistence economics and of traditional technology, to some extent efficient and just sufficient enough for the survival of a small group from one season to the other. It meant the recovery of a basic solidarity, in communitarian terms, of a sense of protection and safety (the first need for any human being). And it meant in modernistic terms, the regression to a condition

Fig. 12.5 Turkmenistan, Kara Kum desert, 2000, village of yurtas and houses. (Photo: Igor Jelen)

of "poverty", but indeed a condition of rural poverty that was very different from the one typical of modernized areas (e.g. in the kolchoz), where it often meant marginalization and ghettoization, and precariousness on all levels.

Above all the urban "poor" had to survive far from any natural sources, almost "entrapped" in a concrete and asphalt neighbourhood, where they were deprived of everything, depending on public supply for basic needs (water, food), risking to be the object of some political blackmail.

In rural areas poverty is very different from that of an "urban" character: it means a lack of modern services (modern healthcare, transport, technology etc.), but it also signifies the recovery of social-communitarian solidarity that may substitute needs on other levels (in crisis time it is always a possibility).

Official data reports that poverty in rural areas apparently persists especially in marginal non-oil republics Tajikistan and Kyrgyzstan, partially Uzbekistan. It mentions, according to some sources in 2011, that one-third of the population live below the poverty line, with this percentage continuing to grow up to the current period.[21]

However, such data is difficult to interpret. The rural/urban diversity is hardly considered in terms of international accountability standards; on the contrary, community life is characterized by mainly non-monetary measurable benefits (e.g. of belonging to a solid community, the use of traditional medical remediation, the access to natural resources, basic freedom), even when marginalized from the wider economic circuits.

Finally—and ironically—at the end of transition, a neo-rural style of life would be reconsidered, rather than a true regression to a primitive stage, as the founding base for a new growth cycle. In fact, sometimes it would even be assumed as a starting point for a growth path, trying to enhance a basic territorial network, that often means the strengthening of rural family and communitarian economics, considered as a basic unit—a policy that has actually proved to be very wise in such circumstances. It has characterized particularly remote areas in Tajikistan and Kyrgyzstan, where such rural economics represents a basic foundation stone for any further development.[22]

12.2.3 Spontaneous Economics (Bazaar or Kiosk Economics)

In some cases the regression connects (and implements) the different forms of parallel economies that were already widespread in the previous regime: what in the Soviet times was the second, actually the most important job, carried out illegally at home or at the official workplace, could often become the primary occupation (e.g. small trade organized at the office desk, a "work" station becoming an artisan lab, a semi-legal collective taxi, a mahalla non-official kindergarten, a solidarity network organized thanks to the help of a neighbour in a "chruščëvka" quarter).

In the new circumstances these jobs slowly developed into new networks, connecting a certain economic area, for example a mountain valley to urban centres, a village network or a set of neighbourhoods in a certain city, creating a

[21] https://www.cia.gov/library/publications/the-world-factbook/fields/2046.html. Accessed 27.3.2018; http://gca.satrapia.com/+the-level-of-poverty-in-kyrgyzstan-will-continue-to-grow+. Accessed 27.3.2018.

[22] Megoran Nick, 2017; Dagiev Dagikhudo, 2014; Brill Olcott Martha, 2012; Esenova Saulesh, 1998:458.

spontaneous informal approach, what some economists have defined as "bazaar economics".[23]

In fact, all these melded into the vivacious environment of the bazaar, once again assuming the role it had had once upon a time, the place where everything "is possible" (negotiating, solving problems, meeting people, looking for help, buy and sale of home-made products), which was characteristic of traditional society (then complementary to the rigid family Islamic order).[24]

It should be considered as the early step for the formation of a structured market economy. However, its role in this progression (from a spontaneous to a structured economy) is questionable. Some authors think its role has even been negative since it used to become self-referential, impeding the formation of other more structured forms of economy (a kind of economy "trap"). Occasional dealers and small entrepreneurs do not invest in structures, developing more complex initiatives, capable of attracting investments, pursuing self-sustaining growth, but usually remain what and where they are (small activities).

Such economies tend to go "underground"—even when assuming a communitarian inclusive aspect—and invisible to institutions, eventually developing in opposition with the official economy (e.g. with operators not paying taxes, not adapting to regulations, not following policies). It usually remains on a local scale, with the bazaar recovering the role it had in traditional culture, but incapable of dialoguing with the official economic institutions.

It is a question, not just of trade at the bazaar, but also of other functions that may develop in opposition or independently from the official ones—evidently not efficient enough, and not capable of deserving the confidence of the wider population. This is the case of transport, with the spontaneous development of bus and "maršrutka" (collective taxi) stations, often developing close to the official stations, with the main railway or bus stations generally neglected or even deserted. This happens, for example, at Garm Dok close to the grandiose Asian Express Terminal in Dushanbe, and at the railway station in Nursultan and Almaty, and in most places.

This kind of mobility develops at the expense (detriment) of the train system in particular—the basic means of transport in Soviet times—which is becoming extremely obsolete, practically useless for local transport (also because the lines are fragmented by new borderlines, with customs, frontier controls etc.), sometimes simply surviving as a kind of relic of Soviet times.

And so on for many basic functions, for supply and information, personal and technical services, and even for health and education facilities. All this may signify the start-up of a new economy, meaning the early stages of a free market, but confusion and risk of speculation as well.

It is an economy without any accountability, that cannot grow, since it is impossible for businesses without proper bookkeeping to obtain finance resources from banks or credit institutions, or even undertake new activities, founding new units, or take on employees (paying salaries); then they remain incapable of planning investments, or innovations and further development.

Therefore, the challenge is to make the bazaar a real start-up business for more advanced economies, which may function as an instrument for redevelopment, consolidating the whole system.[25]

With this intention in particular, some governments have elaborated appropriate regulations, credit and fiscal apparatuses, and particular microfinance laws in order to reduce the spread of popular (or bazaar) economies, integrating them into a wider system.[26]

12.2.4 The Restart: Liberalism and Anti-prohibitionism as a Reconstruction Policy

The de-structuration of the Soviet standard, initially, coincided—intentionally or not—with the affirmation of a radically free market and of freedom in all senses, with the purpose of de-structuring what still remained of the previous centralistic system on the one hand, while on the other hand, setting the "underground" economy free, which would use the capital hidden "under the mattress". This would make the start of a new development wave possible.

Above all, the new policy tried to induce a certain confidence in the new economy, intended as independent from the power structure (the reversal of what communism had done three or four generations before).

It came to fruition in the anti-prohibitionist and liberal wave, signifying the explosion of capabilities as well as of tensions since it also meant decreased control activity by the authorities.

This is the case of the development of a kind of naive consumerism, and of many activities, also "borderline" ones, namely sectors previously strictly forbidden. It is the case of commercial liberation, including the import/export of any

[23] Abazov R., 1999; Brill Olcott Martha, 2012; Lamy Frederick, 2013:149.

[24] Jelen Igor, 2002, 2003.

[25] It is thinkable to apply microcredit strategies to develop such kiosk-economies; indeed, besides a percentage of small firms, often family-run, to be considered as "expected", the bazaar economies proved to be a barrier for development, often "looking" in on itself in local circuits; then, paradoxically, statistics register that the spontaneous economic initiative is still low; it possibly reflects cultural element inherited form soviet times; Brill Olcott Martha, 2012:83, 93fs.

[26] PWC 2012–2013:12.

Table 12.8 Central Asia—natural gas rent (% of the GDP)

	2011	2012	2013	2014	2015	2016	2017
Kazakhstan	1.4	1.4	1.2	1.01	0.9	0.8	1.1
Turkmenistan	28.4	26.1	22.9	15.2	13.0	8.2	12.3
Uzbekistan	14.1	13.3	13.1	8.5	5.2	3.1	6.6
Tajikistan	0.03	0.02	0.01	0.03	0.02	0.01	0.04
Kyrgyz Republic	0.02	0.06	0.04	0.004	0.003	0.001	0.002

Data source: Estimates based on sources and methods described in "The Changing Wealth of Nations: Measuring Sustainable Development in the New Millennium" (World Bank, 2011). https://databank.worldbank.org/reports.aspx?source=2&series=NY.GDP.NGAS.RT.ZS&country=#. Accessed 20/02/2020

Table 12.9 Central Asia—oil rent (% of GDP)

	2012	2013	2014	2015	2016	2017
Kazakhstan	17.2	14.2	13.6	6.7	7.1	10.1
Turkmenistan	12.7	10.5	8.9	5.0	3.7	5.0
Uzbekistan	2.8	2.4	1.8	0.6	0.4	0.8
Tajikistan	0.09	0.08	0.06	0.03	0.03	0.06
Kyrgyz Republic	0.3	0.3	0.2	0.1	0.08	0.1

Data source: Estimates based on sources and methods described in "The Changing Wealth of Nations: Measuring Sustainable Development in the New Millennium" (World Bank, 2011). https://databank.world-bank.org/reports.aspx?source=2&series=NY.GDP.NGAS.RT.ZS&country=#. Accessed 20/02/2020

kind of commodities, which took advantage of the abolished restrictions on transport communication.

So similarly, for new activities in culture, communication, trade, and to some extent, in land use, currency-exchange and finance and credit dealers, even if sometimes in improvised and precarious ways (as usual in these circumstances). This all took place with the use of devices made progressively available by new electronic technology.

Furthermore, it is the case of betting, gambling, casinos and of any similar kind of entertainment, particularly connected with transgressive behaviour bordering illegality (for light drug use, contraband and porno-material trafficking).[27]

Indeed, this period did not last long and after this wave, reflecting a kind of de-institutionalization, and of inebriation of freedom, there was a reaction from the state and other institutions (religious and traditional) recovering and restricting such (deemed as borderline) activities.

In fact, it soon became evident that such an approach would present the danger of exaggerations, for speculation and for deviation of any kind, even for the proliferation of lobbies and gangs, for crime and organized mafias, eventually taking root in some areas, monopolizing (or racketing) some activities. Society proved not resistant enough to deal with such "wild" liberation, with the emergence of the "new rich" and "new poor" classes, dismantling and privatizing the welfare public organization—or what it was left of it.

[27] See also Botoeva Gulzat, 2014.

12.2.5 A New Top-Down Economics

In such situations, a new phase in which the state tried to regain control (again) of the economic initiative started, but this occurred on the basis of principles different from those inspiring the previous system. In fact, the new republic had had the chance to start a consolidation phase, relying on the improved economics (mainly deriving from HC and natural capital export), starting a reversal policy, recovering its role in the economy and in society (Tables 12.8 and 12.9). Such intervention continues right up to the current day, although with somewhat different intensity.

From this moment forward, the extractive industry became the main source for the accumulation of the capital to re-invest in other aspects of development. The intention was to modernize the economy (assuming it was too weak and incapable of producing a "first push" by itself, namely an early capital accumulation); it resulted mainly in a kind of artificial reconstruction of the economic circuit, with massive inflow of capital supporting public spending. It meant the start of a kind of merely quantitative—as it can be defined—phase of growth, based, besides natural resource exploitation, on investments in hardware, namely in the reconstruction of a whole strata of structural landscapes.

There were also investments in sectors such as building and urbanistic projects, infra-structural and other industries which can easily be guided top-down, as an instrument to induce a push-pull effect on the wider economy. Sectors easily controlled and guided politically, which were perceived as essential for civic, public and economic needs (structural renovation obviously being necessary for any further activity).[28]

Such schemes have been followed by all the republics, to a greater degree by Kazakhstan and Turkmenistan, that have had to re-invest the high shares of HC revenues; less so for Uzbekistan, and even lesser so for Kyrgyzstan and Tajikistan (not to be considered as rentier states).

[28] Brill Olcott Martha, 2012; construction sector is rapidly expanding in Russia and Kazakhstan, attracting workforce form all the FSU space; it is expected, it will develop rapidly also in further CA countries.

All these applied policies, based on top-down investments, started the reconstruction of much of the public landscape, including new neighbourhoods and entire new towns, the creation of artificial lakes, of parks and of all kinds of facilities, especially in urban areas. These would soon be dominated by sparkling presidential palaces (the "face" of the new power), usually with magnificent "domes", and characterized by other infrastructures, by public and private housing, apartment and office buildings, malls, leisure, sport and entertainment facilities.

Such policies had some success, with the building sector, synergically with commodities exports, used as the driving force for a further economic wave, inducing rapid growth. The effects were soon evident. Just after the difficult transition decades, Kazakhstan reached a notable per capita PPP of $24,000, which is higher in comparison to others in CA and CIS countries (that started from the same initial level) and even to Russia.[29] Turkmenistan had similar possibilities, even if with far less growth, possibly because it had suffered from isolationism during the Niyazov government, combined with problems of accessibility for gas exports.

So too, to some extent for Uzbekistan, and for the poorest NISs, then Tajikistan and Kyrgyzstan, which applied similar policies—as start-up development devices—benefiting from the public investment push and registering encouraging results. This especially so for Tajikistan, considering the large-scale destruction provoked by the war, and the consequent push deriving from reconstruction, supported by international donors.

Such economies spread relatively rapidly, with some observers defining them as "post-Soviet new tigers", echoing the definition of south east Asian countries in the previous decades; but they spread in an uneven manner, sometimes just limited to the presidential apparatus (following a clan logic), in which the revenues (HC profits, royalties etc.) were usually redistributed.[30]

Such policies worked for a while, but then, after a certain point, proved to have less multiplication effect than expected, exerting low diversification effects. The same can be said of the energy and HC industries, which evidenced high revenues, even when demonstrating (beyond the effects of the booming consumerism) low social impact, if not negative ones, depressing the capability of autonomous initiatives.

[29] Özcan Gül Berna, 2015:409ss.

[30] Babajanian Babken, 2015:514; The Economist, https://www.economist.com/news/asia/21582325-president-edifice-complex-screwing-motherland-folie-de-grandeur. Accessed 27.3.2018.

12.2.6 A Social Question

Such policies, based on top-down induced changes and on a new consumerist culture, resulted as partially efficient but hardly socially inclusive. This was also because of subjective difficulties, namely because most of the population did not have the means to integrate themselves into the new circuits due to the described "cultural gap" (actually, the jump on the scale configured by the post-Soviet passage).

In fact, a wide majority of the population seem to be, rather than excluded, "self-excluded" from the circuit, not because of intentional reasons, but because they belong to an elderly generation, living in remote areas, often illiterate (due to "return" illiterates or to the new "digital" ones), characterized by poor expectations or by scarce confidence in such programmes.

This also because assets—for example a property, a farm with fields, an artisan lab, a space functional for some activity—are not available because owned by the state (as usually environment resources, land and real estates). Private housing investment, as well as land, real estate and property rights, is still an open question (due to incomplete privatization operation, see later).

On the contrary it is possible to observe the formation of a new emerging urban class, inebriated by the new money, used to luxury cars, costly shopping habits, "nouveau rich" restaurants and casinos, branded clothes and shoes, expensive travel configuring an initial wave of consumerism—usually to be considered as self-referential, often financed by bubble-like consumer credit.

For these reasons, in these rapidly enriched countries the social impact—namely the inclusive growth—is difficult to estimate. The poor are still numerous, even when it is difficult to distinguish whether they are truly poor or just economically "regressed" (as previously defined), or self-excluded peoples, namely members of a generation "left behind" because of the rapid changes (since incapable of culturally adapting); in this case the generational passage should greatly improve the situation.[31]

12.2.7 Looking for Further Economic Drivers

Whatever the strategy applied has been, the CA countries are enjoying a relatively long period of growth, which has pulled the local economics out of the stagnation that characterized the early transition time. Whatever the initial push has been

[31] Boboc Cristina, 2017a, b; it will take time before it forms a middle class practicing a balanced kind of life, intended as a spontaneous dynamic of private revenues, savings, consuming, investing in durable goods and fixed assets, then confident in their own country's future; in that moment the middle class will start a kind of stabilization effect with decisive character on the whole system.

(public or private investments, inside or outside demand), the growth lets us imagine that the local economies have a great deal of potential, starting the discussion about the possible strategies, in order to give to this growth a sustainable and durable character.

After the transition decline (after 2000), growth accelerated, with CA republics recovering and assuming the physiognomy of emerging or even booming countries. The issue concerns the element that induced such recovery. For Kazakhstan and Turkmenistan, the answer is usually the favourable situation for commodity prices (especially for HC).

So too, partly for Uzbekistan (essentially for cotton, gas, gold and minerals), even when this country has demonstrated potential in other sectors, developing manufacture, trade and services (tourism, folklore and archaeological heritage). Possibly also for this reason Uzbekistan has undertaken a neo-mercantilist route that should, in principle, facilitate internal diversification, but that presents some risks—namely involution in a protectionist "trap".

The case of Tajikistan and Kyrgyzstan is interesting, since these countries, poor in HC resources, have benefited from the general contingency, giving value to their environmental resources (water, essentially, for any purpose, energetic, irrigation, civil and industrial resources). They have demonstrated their intention to apply a "minimalist" strategy, at first incentivizing a widespread self-sustainment agricultural base (which seems to be a wise method in a country characterized by hardly integrated peripheral areas).

Tajikistan showed a growing rate of an average 9.6% yearly in the period 2000–2007,[32] after the war, evidencing the effects of post-war reconstruction as well. However, it has continued to grow, even accelerating its growth, meaning that pacification, once started, induced benefits, self-alimenting, then leaning towards long-term investments, with improvement in all economic indicators.

Kyrgyzstan has shown the same trend, benefiting from the decreased geopolitical risk affecting the region which has made the realization of infrastructural and industrial projects possible—that have a return in the long term.

Such a widespread (minimalist) human-resource development pattern tendentially induces a more inclusive form of economy, also inducing the strengthening of a nucleus of what could become a rural (or urban) middle class over time, relying on property, interested in such values as stability and participation.

[32] World Bank, Tajikistan.

12.3 True Development?

12.3.1 The Risk for Fictitious Loops

Today, after the modernity cycles, it is evident that economic progress is not linear—in CA as anywhere indeed, there are always risks of regression or of "vicious circles". Development should be represented as a "road map",[33] on the basis of a complex definition; certain mistakes in certain critical moments of the path can negatively influence the whole path—"road" to follow (as many cases of fictive growth in the world indeed demonstrate).

Recent tendencies show many examples of this, with investments often resulting as unprofitable and with gigantic infrastructures appearing in practice as useless; with "grey economies" flourishing, then aggravating the effect of the separation between society and politics, with a fundamental popular mistrust in the institutions arising.

So too, for many public functions, like mobility, housing, welfare and services (e.g. spreading perceptions of uncertainty in law and justice systems which seem hardly to function). The society demonstrates the tendency to organize itself in segmentary terms (on a local and community scale), bypassing any dialogue (and the further integration) with the public organisation. A dangerous trend, that can prefigure a kind of loop (e.g. of expectation of periodical juridical remission for tax evaders, or just speculating on the inefficiency of the public administration). Moreover, for many other "bubbles" in bureaucracy, credit and finance, real estate and the consumer market, an effect that as experience teaches can be especially dangerous, even if "invisible"[34] up to a certain point.

Such tendencies impact the top of the economic pyramid as well. The need to re-invest HC-rich profits is limited by the scarce efficiency of the public apparatus; this particularly as regards investing in high value-added initiatives (since limited by bureaucratically conservative attitudes). The easiest way is to invest in low-risk and state-controlled building industry, even when, after a certain point, it proves to be no longer remunerative. Finally, such investment risks financing the building of empty skyscrapers and useless infrastructures, fundamentally configuring real estate speculation and capitalist bubbles; in essence the wrong way to pursue development (paradoxically prefiguring the repetition of Soviet

[33] Marzhan Thomas, 2015.

[34] In fact, for any of such "inefficiencies", possibly causes of biased development in the long term, it is to expect the arising of a lobby, defending some particular interest, but biasing the development path; Brill Olcott Martha, 2012:91.

mistakes, as well as the imitation of the worse aspects of capitalist speculative economics).[35]

12.3.2 The Risk for Bias and Diversions

Probably, nowhere in the world so much as in CA, is the connection between politics and economics so evident—because of the possibility the power structure has, through HC and other natural resource exploitation, to condition the whole system. Indeed, the intervention of the public in these circumstances seems to be necessary, especially considering the mass of money suddenly put into circulation, as well as potential problems affecting transition, above all the disequilibria such rapid growth induces in society and on the territory.

Nevertheless, a major risk in such circumstances is for politics, with the formation of a model that has been defined as a patrimonial-authoritarian regime, to rely essentially on the administration of some monopoly, alimenting clientelistic relations.[36]

Evidently, it is impossible to have a functioning market in just a few decades and it usually takes generations (since it requires the formation of a set of internal controls, of adjusting mechanisms between demand and offer and much more). An operation which is even more difficult to pursue in the current situation—given that the contiguity between power and economics would impede all such evolution.

The news reports ongoing critical situations, evidencing the risk of such deviations; among them, there are continuous rumours of recycling and money laundering, mismanagement and corruption, which do not spare anyone from local governments. This is due to the cited confusion between the public and private, but also due to other situations, for example, the underground struggle between lobbies, bureaucratization affecting long chains of the public economy and similar (indeed already well-known in the Soviet era).

Free media, international agencies and independent observers report on cases involving the HC industry; this mainly for Kazakhstan and Turkmenistan, where the HC revenues are particularly evident, consisting in huge amounts of money managed by a relatively un-experienced public apparatus.

It is possible to cite several cases: in 2011 US authorities confiscated, due to an investigation in Switzerland, $48 million (and much more involved), accusing Kazakh officials of bribery;[37] in 2006 some NGOs investigating corruption and laundering and similar phenomena reported that Turkmenistan monetary reserves were being managed out of an account in a German bank (but possibly they are just reserves in currency to be managed).[38]

Rumours about money laundering regard Tajik industries, concerning export profits, especially for state-owned TALCO, in the shape of hundreds of millions of dollars in some foreign tax paradise.[39]

The building industry is especially exposed to such risks, which consist of a set of cost variables which are easy to manipulate in the different phases of executive projects, therefore predisposed to re-invest profits from illegal activities.

It is a major risk for such systems without effective internal controls, without effective segregation of duties, with justice apparatus depending on executive power. In the meanwhile, while waiting for such internal evolution, the only possibility relies on external controls that represent an independent view, producing analysis and criticism of local governments: UN or IMF agencies, NGO or IO observers, consulting and trust companies, which represent a kind of moral persuasion, sensitizing public opinion—but they have no political capacity.

This is possibly just the tip of the iceberg (also considering that findings have been made mainly by civil society actors and not by professional investigators): data about economic crimes and similar deviations are in principle uncertain, rarely based on material evidence. Furthermore, they do not consider contextual questions and the precariousness of the apparatus (namely the uncertainty of the procedures), which seem alone to give incentive to such behaviours, especially mismanagement, bribery, export of capital and further administrative and financial illegalities (Table 12.10).

Furthermore, a certain level of manipulation (obviously not to be encouraged) is considered unavoidable, namely connected with the same prerogatives of such young sovereign states, not yet stabilized in their procedures: they have to compete in a lawless international arena, providing security against secret and hostile forces, eventually using hidden reserves and "slush funds" registered in some "fiscal paradise"; thresholds and limits of such actions are not always

[35] Brill Olcott Martha, 2012:118fs.

[36] Brill Olcott Martha, 2012; Babajanian Babken, 2015:514; The Economist, https://www.economist.com/news/asia/21582325-president-edifice-complex-screwing-motherland-folie-de-grandeur. Accessed 27.3.2018; this involution is evidently linked with the failure—to some extent induced—of the privatization operation, occurring in early transition times, that in principle should have induced the formation of a market to some extent independent of the power.

[37] OECD, http://www.oecd.org/daf/anti-bribery/kazakhstan-should-build-on-efforts-to-fightcorruption-and-push-forwar-with-reforms.htm. Accessed 27.3.2018.

[38] See the London-based non-governmental organization Global Witness, https://www.globalwitness.org/en/archive/germany-must-launch-full-inquiry-turkmen-funds-german-banks/. Accessed 27.3.2018.

[39] The Economist, https://www.economist.com/news/asia/21582325-president-edifice-complex-screwing-motherland-folie-de-grandeur. Accessed 27.3.2018.

Table 12.10 Central Asia—percentage of households who paid a bribe when accessing basic services

Kazakhstan	29%
Turkmenistan	–
Uzbekistan	18%
Tajikistan	50%
Kyrgyz Republic	38%

Data source: "People and corruption: Europe and Central Asia" Global Corruption Barometer (Transparency International, 2016), https://www.transparency.org/whatwedo/publication/7493. Accessed 24/10/2017

obvious. Evidently such biases have to be faced on the geopolitical scale, improving wider international governance.

The same for other forms of submerged economies practised on all levels, but sometimes induced by the same lack of efficiency of public procedures, sometimes by scarce confidence in the power, specifically from situations that can be considered typical for "young" apparatuses.[40]

12.3.3 Risks Affecting Rapid Growth: The Cycle of Illegal Activities

A further issue is that of deliberate criminal activities, connected with international traffic: the position on the major routes connecting the east and the west, the north and the south of the Eurasian continent, makes the CA countries exposed to such risks—which in the transition period were greatly increased by the weakness of internal law-enforcement structures.

Illegal trafficking was alimented in those times by the opening of the borders, by instability (namely the insider risk conflict), as by-products of civil conflicts (alimenting themselves): drugs, weapons, poaching and human trafficking (for a kind of new slavery, namely for illegal migration and for any kind of exploitation) carried out by organized crime, organizing routes from Afghanistan towards Russia and Europe and anywhere.

In particular post-war Tajikistan seemed to be exposed to such risk; the country was recorded in 2006 as the third major country for the confiscation of raw opium and heroin, registering "1216.3 kg of heroin and 267.8 kg of raw opium in the first half of 2006".[41]

One must consider that, just after the end of the civil war, some "warlords" and local commanders were inclined to undertake other "activities", sometimes dedicating themselves to crime and drug trafficking.[42]

Some observers report that currently drug trafficking is decreasing, and that it does not represent a strategic risk anymore, capable of determining a systemic corruption of apparatuses, alimenting organized crime.

This is due to general improvement in law enforcement agencies and to the policies applied by local governments (in the case of Tajikistan, the Tajikistani Drug Control Agency), in collaboration with international agencies (UNODC). And thanks to the improved geopolitical situation, especially in the SCO (see later), that exerts a general control on the border areas (even when opium production in Afghanistan and neighbouring countries continues to represent a serious problem).

But the situation is still critical; trafficking proliferates on various scales, starting from local levels, then involving gangs controlling territories, using violence, rackets and extortion, that may degenerate, finally influencing even at governmental (both, local and nation) levels.

An especially complicated task is that of preventing "white collar" crime, proliferating also due to the spread of "easy money"; it is more difficult to recognize—because the criminals are usually inside the apparatus—and it can only be contrasted with a substantial improvement in internal control systems (not as usually occurs in post-Soviet countries with just a periodical "purge" of officials).

Among other measures the establishment of a special fund for accumulation of HC and other monopoly revenues can be put into action, to be administered with special regulations; this could be even more useful considering its role as a possible stabilization device in the long term (as happens for other commodity exporter countries, possibly under the auditing of professional agencies, IO or trust companies).

12.3.4 Foreign Investments (FI)

The "stans" have been characterized in the last decade by the booming of FI, either directly or in an indirect manner, with different degrees of involvement by local governments. It is evident that CA economies need foreign intervention, to recover the epochal economic delay, to get the technology for many demanding projects in the infrastructure, extraction and construction industries.

The list is rather long, starting from ID facilities like dams, channels, irrigation appliances, desalter plants with sophisticated technology, and then HC economics (the major

[40] Warf Barney, 2016; Brill Olcott Martha, 2012:53fs.

[41] Wikipedians, Countries and territories in the world (no year), https://books.google.it/books?id=bR0hIC0Xhb0C&pg=PA120&lpg=PA120&dq=1216.3+kg+of+heroin+and+267.8+kg+of+raw+opium&source=bl&ots=XxL5vwuI0M&sig=g_CH8ED7D09ZwhA8qN4LH44Odxs&hl=it&sa=X&ved=0ahUKEwjr8JKwvP_YAhVDUBQKHfMlBegQ6AEIKDAA#v=onepage&q=1216.3%20kg%20of%20heroin%20and%20267.8%20kg%20of%20raw%20opium&f=false. Accessed 27.3.2018; Brill Olcott Martha, 2012: 76ss; De Danieli Filippo, 2013:146 citing UNODOC estimations.

[42] Dagiev Dagikhudo, 2014; Brill Olcott Martha, 2012; Megoran Nick, 2017.

fields being located in difficult offshore areas and distant from the markets).[43] Further technology is needed for HC processing, distilleries, refineries and transport.

The inflow of capitals is especially important when considering the incumbent change of the model, from hard industry and HC extraction to post-industrial oriented economics and renewable energies; it means that, in the near future, the main value-added source will not rely on an HC heavy industry, but mainly on other activities such as culture-intensive and environment-quality functions.

Therefore, it is necessary to predispose investments in consumerist, "amenity" and "happiness" economies, in digital and high-tech equipment, sustainable mobility, renewable energy, waste management and recycling. This is in the context of the circular economy, the cleaning up of areas with high ecological risks (especially the abandoned Soviet industrial areas, see later), improving the environment and living standards.[44]

Evidently, the development of FIs has as a prerequisite the stability of the area, reducing the country-associated risk, as well as the improvement of law and organization systems, for example, demanding some reassurance from central government (preventing interference, successive rising of taxes or oppressive regulations), but the juridical internal frame is sometimes confused, representing an occasion for further mismanagement.

It could easily become matter of further disputes, eventually in an instrumental manner. This is the case of ecology or safety (often the object of complicated internal regulations).

This happens especially in HC-rich states, Turkmenistan and Kazakhstan, that periodically try to re-negotiate the international consortia interventions—with whom they signed long-term contracts for extraction and research. This happened recently in Kyrgyzstan, in the case of the Kumtor mine, with protests that some observers defined as an attempt at "racketing" the company by local worker associations.[45] This has also happened in other cases of public contracts and big works (dams, infrastructures, IR facilities), of large-scale interventions, especially in urban areas (in the restructuring of urban city quarters and in the setting up of new commercial facilities).

12.3.5 Local National and Popular Attitude

The attitude (political and popular) towards FI, and in general towards foreign interventions, is ambiguous; they are contested because they consume natural resources usually managed (disposed) under sovereign control. Such resources are often exploited by MNCs, which appear to local population as without scruples and connected to some foreign government, namely suspected of representing, even when non-intentionally, some political interest.[46]

This is also considering that the perception towards such situations can change depending on international contingencies or on government attitudes. This is possibly one of the reasons why some governments have undertaken neo-mercantilist policies.

Uzbekistan is particularly characterized by a very low FI index per capita, the lowest in CIS. The government deliberately did not make the currency convertible (until 2003 when it was obliged to by the IMF to ensure full convertibility) and used further expedients to limit outside influences (but with dubious results).[47] These practices present the risk of degenerating into protectionist policies, especially when pursued for long periods, paradoxically damaging the internal industries (as possibly happened for the Uzbek Chkalov plant, and for Tajik TALCO, see later).[48]

The exact opposite happens for Kazakhstan, which has an open policy, trying to give incentive to exchange at all levels, creating favourable conditions for FI, even when directly managed by foreign companies, therefore integrating itself further into international economics.

Often FI are facilitated by the mediation of local ethnic minorities, if for nothing else, for their linguistic capabilities, as in the case of Korean and German (but this, in the new

[43] Babaev Agajan G., 2012; PWC 2011, 2016.

[44] For example, it is easy to assign a contract for building hotels, university campuses, research centres or hospitals, not so to make them work with a desired quality standard; it is not sufficient to apply a Fordist sequence of instructions, but it is necessary to intervene on a set of qualitative variables.

[45] Lonely Planet, 2014; indeed governments need direct FI and multinational companies because they need management, commercial, and R&D capabilities difficult to obtain otherwise; just such corporations can devote significant budgets to such functions, at the same time using such capabilities with the necessary flexibility.

[46] It is to consider as a further Soviet cultural heritage in economics, a culture contrary to outside intervention, and to innovation in general; it concerns a certain attitude in social behaviour namely a diffuse sentiment against the individual initiative and in general against any innovation, that in the previous system was in principle considered as a transgression; even now the "new rich" are considered per default as "špekuljant" or tout court as "criminals" or "mafia" members; it means in general the necessity that the people start to accept the new reality as a "game" of opportunities (and eventually of risks), not as a sequence of planned commands; evidently a critical passage—the complete change of a way of thinking—that cannot occur but in the times of the generation change; significatively the post-Soviet lexicon about the deviant behaviours still comprehends foreign-origin words like "strafe", "contrabandist", "špekuljant", "brakonir" etc.

[47] http://www.imf.org/en/News/Articles/2015/09/14/01/49/pr03188. Accessed 27.3.2018; https://www.state.gov/e/eb/rls/othr/ics/2011/157382.htm. Accessed 27.3.2018; and similarly with other expedients in order to instrumentally obstruct the FI companies; se as well PWC 2016:3.

[48] Among the other requirements for direct FI in Uzbekistan, there is the substantial state participation share; Bernardo Teles Fazendeiro, 2015; PWC 2016.

Table 12.11 Central Asia—foreign direct investment (inward) as a percentage of GDP: Flows

	2011	2012	2013	2014	2015	2016	2017	2018
Kazakhstan	7.1	6.5	4.2	3.3	3.5	12.5	2.8	0.1
Turkmenistan	11.6	8.9	7.3	8.8	8.4	6.2	5.4	4.8
Uzbekistan	2.8	0.8	0.9	0.9	0.08	2.0	3.03	1.2
Tajikistan	2.2	3.1	3.3	3.5	5.7	3.4	2.5	2.9
Kyrgyz Republic	11.06	3.9	8.3	4.5	17.1	9.0	−1.3	0.5

Data source: Data Bank, World Development Indicator, The World Bank, https://databank.worldbank.org/reports.aspx?source=2&series=BX.KLT.DINV.WD.GD.ZS&country=#. Accessed 20/02/2020

globalization context, is often not much more than a folkloristic historic-emotional reminiscence, not economically very relevant).

12.3.6 Major Case of Investments

The direct economic intervention of foreign actors is considered as a good indicator of the capability of the country, not just to attract capital, but in general to represent an economically inviting environment (Table 12.11). Nonetheless, the attitude of the countries is different. Some of the CA countries apply regulations in order to maximize such capabilities, as in the case of favourable taxation, of prescriptions regarding re-investment (or of export) of profits, eventually redistributing benefits for the local population, while in other cases the governments participate directly, establishing partnerships and joint investments, sharing revenues. In other cases, opposite attitudes are rather apparent, with local institutions in fact limiting FDI, for example, prescribing complicated regulations or the compulsory establishment of a domestic company as partner for foreign corporations—with the real intention of controlling them.

The issue is even more sensitive in CA since often FDIs concern the most profitable sectors, those directly accessing natural resources (soil, water, HC and minerals) perceived as "national" common assets. From another point of view, it must be considered that, without the ready-to-use technology that such corporations can provide, much of the same resources would remain a kind of "lost opportunity".[49] This is the case of most of the valuable industrial productions, regarding heavy industry, extraction, mining and mineral processing, as well as thermo- and hydro-energy production. They are often located in remote areas, and need to be continuously reconditioned, precisely with foreign financial and technological capital.

Comprehensibly such interventions regard above all HC. It is the case of the rich Kazakh oil fields Kashagan and Tengiz, that from the beginning were participated in by foreign MNCs (multi-national companies), like Eni, Chevron and Exxon, incorporated in some Caspian Consortiums; and of the Turkmenistan giant gas field, that has been exploited with the collaboration of Chinese CNPC, organizing pipelines oriented through Uzbekistan and Kazakhstan to China, operative since 2006.

It is the case of the Kyrgyz Kumtor mines, where gold deposits were discovered in the 1970s, but which have been developed since independence with the application of updated technology. Currently they are operated by Canadian MNC Centerra Gold, which has invested substantial capital into upgrading extraction and production methods, as well as for mitigating environmental impact and enforcing worker safety (but object of periodical protestations by local workers, eventually inspired by expropriated authorities).[50]

Another case is that of the Kazakh Karaganda industrial area, with coal mines, extraction, metallurgic and refinery industries, with the nearby metallurgic industry that has been modernized with the investment of the Indian MNC Mittal;[51] further relevant investments concern the southern Kazakhstan Shimkent concrete plant operated by Ital-Cementi.[52]

There are also the hydroelectric power plants—among the most important in the world—the Amu Daria (in Tajikistan) dams (Nurek, Sangtuda) and the Syr Daria Toktogul dam in Kyrgyzstan must be mentioned: constructed in Soviet times, they need international help for an array of functions, like maintenance and safety monitoring; there is special concern regarding silting.

Among the contracts, we must mention the ones on the Rasht valley river of a dam constructed by an international consortium guided by MNC Salini-Impregilo (the cited Roghun dam, under construction, the highest in the world,

[49]As said, investments in technologies and capabilities for extracting, processing and transporting the commodities from remote areas; indeed, some authors consider oil industry already today something obsolete, while natural gas possibly will have still few decades of full development; in such circumstances it is even more difficult to make profits.

[50]Lonely Planet, 2014.

[51]ArcelorMittal, 2013.

[52]10.4.2015, https://www.esteri.it/mae/it/sala_stampa/archivionotizie/approfondimenti/2015/04/dei-kazakhstan-italcementi-firma.html. Accessed 22.4.2018, Ministero degli Affari Esteri e della Cooperazione Internazionale, Dei—Kazakhstan: Italcementi firma MoU per stabilimento Shymkent.

once finished, Fig. 11.5, chapter 11), and a long list of large projects in transport, communication and urban facilities.[53]

The Nurek hydroelectric power plant aliments the TALCO factory, processing bauxite that is imported from abroad nowadays (Tajikistan being deprived of this mineral). It has a complicated post-Soviet history of foreign participation and of a consequent legal struggle that has been settled with a final paradoxical re-nationalization.[54]

It is the main income asset in Tajikistan (as said, together with cotton exports and remittances from emigrants), started in the mid-1970s, modernized and today state-owned, producing “hundreds of millions of dollars in profits”.[55]

Further investments regard construction materials, like cement (concrete, beton), chemicals for all applications (fertilizers and pesticides in agriculture, HC refineries, special cements and steel for the building industry). So too, for other industrial sectors not strictly connected with strategic questions (light engineering, market-oriented industries and manufacturing), where the opportunities for FI are virtually unlimited (considering the growth of a local middle class characterized by increasing spending capability and tendentially imitating the western-globalized style of life).[56]

Uzbekistan, in particular, with its tradition of manufacture, can attract MNCs in a wide series of initiatives: representing the most diversified economics in the area, it can potentially exert a role of local hub for FIs in the whole area; but the country should definitively change its politics, pursuing a growth strategy, based on widespread initiatives, predisposing for the opening of the market and in general of its economies.

In fact, as it has been possible to observe recently—with the Mirzyioyev “new deal”—the country is changing its attitude, trying to integrate the internal production chain, avoiding the massive export of raw material (considering that the country is the seventh largest cotton producer in the world) and opening its internal market. The government recently started to incentivize the local processing of such resources, for example promoting the garment and textile industry and a market-oriented economics, in order to improve the value produced internally in the country, attracting foreign capitals.[57]

So for several further sectors, to be reorganized as consumer-oriented, not just as top-down managed economics. A paradigmatic case is that of the cited Tchaklov airplane industry, and of further cases of “heavy” industrial production, that have given up on their original production, starting other types of activities.[58] This is the case of the automotive (in joint ventures with US–Korean GM Daewoo),[59] and train industries, used by Spanish Talgo, operating the high speed Tashkent–Samarqand, and Bukhara, line, which functions more coherently with the country’s current needs (mass mobility, regional and urban transport).

A special, but significant case is that of the Cosmodrom in Baikhonur, central Kazakhstan (essential in order for Russia to pursue its space programmes), that has been rented to the Russian Republic until 2030 for an amount corresponding to US$150 million per year, used for space international programmes.[60] Recent purposes evidence that this infrastructure is suitable for development in wider terms, prefiguring some possibility of “space travel” for private purposes, not just reserved for strategic or military intentions.

12.3.7 Investments in Environmental Rehabilitation and in Circular Economics

Further sectors interested in FI are those suitable for the application of scale-economy production, for example agro-industrial resources like cotton or large-scale cultivation (rice and other cereals, exotic fruit, precious race breeding). In these sectors, FI contribute to the upgrading of technologies, representing valuable margins of improvement, for example considering the possible reduction of the waste of water, soil and energy, and the recycling of materials. These investments regard (as typical for post-Soviet times) the restructuring of the local environment, in order to reduce externalities, to make production convenient (namely safe and healthy).

Then the FI apply advanced techniques of ecological remediation; it is an interesting business. It is necessary to dismantle abandoned plants and factories (both civil and military), making it necessary to re-convert some obsolete infrastructures and to recycle used materials.

Such interventions need sophisticated technology, representing a kind of reversal of the same industrialization pro-

[53] Brill Olcott Martha, 2012.

[54] Brill Olcott Martha, 2012:53.

[55] Brill Olcott Martha, 2012; The Economist, https://www.economist.com/news/asia/21582325-president-edifice-complex-screwing-motherland-folie-de-grandeur. Accessed 27.3.2018.

[56] Particularly China in few years has become a main player in the region, with trade between China and the five CA republics, risen from US$1.8 billion in 2000 to US$34 billion in 2015, Indeo F., 2018.

[57] https://iteca.uz/oilgasconference/eng/PressCentre/index.php?ELEMENT_ID=44375, 17 October 2019, For the sake of workers, Uzbekistan is privatising its cotton industry. Accessed Dec. 5, 2019.

[58] PWC 2016:4; but see “Uzbekistan to create its Aerospace industry”, http://tashkenttimes.uz/national/1921-uzbekistan-to-create-its-aerospace-industry, 2018-01-11. Accessed 4.7.2017.

[59] Fazendeiro Bernardo Teles, 2015:490.

[60] Interview with Kazakh official, Nursultan, August 2017.

cess which occurred two or three generations ago. One must keep in mind the consolidation of new circular economies, possibly the most profitable activity in these circumstances—the whole post-Soviet space representing a huge deposit of materials which can easily be recycled.

The US- and IC-sponsored intervention in bonification of the Semipalatinsk polygon is especially relevant as ecological renewal investments.

And so too, for collaboration in further large-scale projects affecting what can be considered as the most relevant ecological disaster in the whole area, namely the drying up of Lake Aral; such collaboration includes further special interventions that can be considered decisive for the success of the whole rehabilitation project; it is the case of the clarification of the former chemical site of Aral lake island (renamed in Soviet time as Vozroždenie (Возрождения), namely "renaissance" island, that indeed in the last decades became mainland), representing "per se" an ecological disaster.

12.3.8 A New Urban Environment for the New Consumerist Economics

FI are even more relevant and impact activities such as construction, with the restructuring of entire new environments, both urban and regional. This especially occurs in major metropolitan areas, where interventions signify the reconstruction of a new landscape, suitable to the start-up of a new consumerist economy and of a new lifestyle.

It is the case of malls and shopping centres, and of commercial facilities and infrastructures, adapted for retail and large distribution, for logistic and social events, creating an environment of "globalized" appearances (eventually in opposition or close to the Soviet- or traditional city quarters).

The commercial reorganization of these city quarters (connected with general-purpose urbanistic planning) represents a valuable opportunity: with stabilization, CA populations and society have changed their habits, shopping in modern malls, with tourism in new resorts even abroad, driving Japanese or German cars, travelling on high speed trains, wearing Italian designer clothes and opening up a series of opportunities in all sectors.

In fact, several MNCs obtained contracts from the local authorities in order to reorganize market and distribution activities, planning large consumer and urbanistic restructuring, services and communications, intervening directly at the local market level and restructuring large surfaces of urban areas. It sometimes has meant the acquisition of a monopoly on commercial and cultural landscapes, inspiring the formation of a new form of society and a new style of life.

Major shopping centre and retail brands are continuously seeking new opportunities in order to be present in changing post-Soviet urban areas—Dushanbe, Almaty, Tashkent and others—to have the chance to occupy the best places (in perspective of long-term growth) and to set up new urban trade centres and entire new CBDs.

12.3.9 Scramble for Resources

Needless to say, companies have found several opportunities in the area. After independence, and taking advantage of a favourable international contingency, a scramble for such "newly discovered lands" began, countries which were literally unknown in western societies.

MNCs, foreign institutions, companies operating in territorial-environmental services, all took part in this scramble to acquire an advantageous position, trying to build up monopolies in HC and mineral resources, uranium, "rare earths" especially used today in the electronics industry; so too in other sectors in rapid growth—reflecting the expectations of a growing middle class (such as the building and constructing industry, large consumption, welfare, durable goods and light engineering, digital and culture-intensive activities).[61]

In fact, this "scramble" had the aim, supposedly, of acquiring some advantages, namely privileged positions for contracts in HC extraction and pipeline projects, in sectors exerting geo-strategic relevance, usually controlled by the public, but also in building industry and urban planning (e.g. for construction and setting up of malls and cultural-communication facilities).

12.4 Macroeconomic Policy Trends

12.4.1 On a Macro Scale

Even considering the neo-nationalist drift, it is not possible to say that the macro-economics of the individual CA countries are definitively diverging, or on the contrary converging; they seem rather to be subject to oscillatory politics that derive from different contingencies and by both internal and international situations (supply necessity, industrial traditions and commodities prices).

Indeed, the macro-economics of the NIS are to some extent "made" to run in a parallel way, rather than assuming

[61] https://blog.nationalgeographic.org/2013/08/20/race-for-rare-earths-in-central-asia/. Accessed 27.3.2018, citing Peyrouse Sebastien, 2013; about Kazkh HC, see the Tengiz and Kashagan Chevron, ENI, Exxon, Total and others participating in international consortium; Turkmenistan signed a similar convention with Chinese companies, see Silvestri Tommaso, 2015–2016.

a competitive attitude: because of basic material constraints, they are almost forced to find a way to collaborate in a bi- or multi-lateral manner. In fact, any function in such an environment immediately assumes a cross-border scale; this is the case of transport, trade, migrations, fiscal and custom policies, as well as civil rights, minorities and security questions. This is especially so for macro-economics, which in such circumstances cannot be organized in a truly independent manner. In fact, in CA, there is no measure that can be implemented without producing effects (intentional or not) on the other side of a border, involving neighbouring countries.

However, it took some time before the NIS would realize this. The early stages of their history were inspired by conflicting motivations, namely by the same necessity to consolidate the internal machine, potentially excluding neighbours. In this phase, economic geography was the instrument such "young" countries used for consolidating themselves in their own territory, even when they had an absolute need of foreign collaboration and of international trade at the same time. Thus, a situation that could not but result in contradictory policies (Tables 12.12 and 12.13).

Table 12.12 Central Asia—exports of goods and services (% of GDP)

	2012	2013	2014	2015	2016	2017	2018
Kazakhstan	44.1	38.6	39.3	28.5	31.8	33.5	37.5
Turkmenistan	58.8	49.8	46.7	35.6	22.1	22.4	22.6
Uzbekistan	21.3	20.8	17.7	15.3	14.8	21.8	29.1
Tajikistan	17.2	11.0	9.1	10.4	12.9	15.7	–
Kyrgyz Republic	44.4	42.2	37.4	35.1	35.8	34.2	32.7

Data source: Data Bank, World Development Indicator, The World Bank. https://databank.worldbank.org/reports.aspx?source=2&series=NE.EXP.GNFS.ZS&country=#. Accessed 20/02/2020

Table 12.13 Central Asia—imports of goods and services (% of GDP)

	2012	2013	2014	2015	2016	2017	2018
Kazakhstan	29.6	26.7	25.6	24.5	28.4	25.6	25.2
Turkmenistan	50.5	48.7	44.3	45.6	39.8	31.1	12.4
Uzbekistan	20.1	20.2	18.2	15.1	14.8	23.8	38.7
Tajikistan	65.8	60.6	45.4	42.2	42.0	40.9	–
Kyrgyz Republic	95.2	91.7	87.6	75.7	69.9	66.3	68.3

Data source: Data Bank, World Development Indicator, The World Bank, https://databank.worldbank.org/reports.aspx?source=2&series=NE.IMP.GNFS.ZS&country=#. Accessed 20/02/2020

Uzbekistan soon elaborated an evidently neo-mercantilist approach, in a neo-nationalist ideology, with the intention of protecting its domestic industry and agriculture (that relies on relevant assets tracing back to Soviet and colonial times, as well as to even earlier SR handicraft and manufacture traditions). As usual it led to a situation of mono-production, namely of excessive dependency on one exclusive resource and to a protectionist attitude to maintain that monopoly.

Kazakhstan on the contrary opted for integration into global economics, being aware of the benefits of participating in wider economic processes, looking for a place in the international schema of division of labour and furthermore developing this role in terms of policies, possibly to render its progress irreversible (rapidly involving a number of politically relevant international actors, avoiding the emergence of some protectionist lobbies).

Indeed, all the "stans" declared the transition to a market economy which was internationally integrated as a priority on official occasions (with exception of Turkmenistan in Niyazov era). It must be taken into consideration that Kyrgyzstan had already been admitted to the WTO in 1998, Kazakhstan and Tajikistan in recent years (2015 and 2013, respectively); Uzbekistan has the status of observer and Turkmenistan is not a member—but both the latter have a good chance of being admitted soon.[62]

But this intention was achieved with several contradictions. The governments often evidence a "lobbyist" attitude, maintaining and incentivizing sectors that are suitable to being centralistically controlled (showing a difficult attempt to sustain a double standard, liberal in international economics but rather authoritarian internally). This occurs in the cited case of heavy industries, energy and natural resources, agricultural production (considering that environmental resources are usually national property).

12.4.2 Some Cases

Uzbekistan's basic economics consists in commodity production, raw material, agro-industry (cotton), minerals (gold—4th largest deposits in the world, copper, uranium), gas (11th in the world) administered by national monopolies (it must be considered that, for some observers, the government deliberately pursues a policy opposed to spontaneous development);[63] the private sector is limited to bazaar economies and to small businesses.

The condition of HC rentier states may possibly develop further since Uzbekneftgas has recently started the exploitation of significant deposits of gas with the collaboration of foreign corporations such as the Chinese CNPC, Petronas, Luckoil and Korean oil.

Tajikistan presents a similar dependency since it relies on a small amount of natural capital which is exploitation based

[62] https://www.wto.org/. Accessed 29.3.2018; Megoran Nick, 2017:95.

[63] As said, according to the Economist Intelligence Unit, "the government is hostile to allowing the development of an independent private sector, over which it would have no control", cited by Wikipedia, Uzbekistan. Accessed 29.3.18.

and export oriented: TALCO aluminium production, cotton (60% of the agricultural output), as well as on environmental resources (above all water), difficult to evaluate in monetary terms (with the significant exception for hydropower production). The government started a policy to foster local-scale agricultural activities, considered as the base on which it is possible to stimulate development; this policy seems to be appropriate, considering the population scattered over wide territories and to consolidate community ties in such extreme conditions.

To some extent, there is a similar situation in Kyrgyzstan; the gold of the Kumtor mines represents 12% of exports, with hydroelectric power state-managed representing a further portion, but the government seems to give priority to agricultural projects, elaborating a minimalist way (already described), as a widespread schema for development.

The rural community base demonstrates a certain potential for diversification of economics, comprehending several forms of niche tourism (consider that a small rate of tourism growth is sufficient in order to have benefits for the whole population in such low-income remote areas). Furthermore, such economics proved to be seminal in order to promote other activities synergically like agro-food, landscape and local services, cultural events and similar.

Kazakhstan is the clearest example of such contradictions—even when its economics is to be considered, generally speaking, successful. The government rules in a tendentially autocratic manner, leveraging on HC revenues to develop further sectors of economics, but it has to limit its repressive attitude because of international relations, in order to maintain a respectable reputation.

The preoccupation of the leader to elaborate an acceptable perspective is evident, so after oil exploitation, and the re-investment of revenues in the building industry (with the epitomical case of Nursultan and of the Caspian coast cities), the government started to elaborate an "enlightened" vision, a second phase (the "bright path" strategy). It is based on a trans-continental infrastructure plan (the new SR-BRI possibly to be implemented in collaboration with powerful neighbours, China at first), and so could therefore continue to make investments, alimenting the economic circuits.[64] Thus, the government can maintain strict control over the economy at the same time.

In fact, the contradiction between internationalization and neo-centralization is apparent, since no policy is really possible without the collaboration of MNCs (e.g. for infrastructure construction,[65] HC exploitation and redistribution, technology development as well as urban restructuring), namely with the collaboration of relatively few but big corporations, with whom the governments can directly discuss agreements and policies (eventually avoiding any uncontrollable liberalization). This is therefore a policy made, not by the market, but by large-scale "contracts" (for public works and services, between government and few MNC directors, prospecting the risk for market distortive practices), that actually characterize the main aspect of current CA economics.

12.4.3 Limits of Neo-mercantilist Politics

Uzbekistan seems to be particularly refractory, even rejecting—as observed by many analysts—FI (when not under control of the local politics), elevating barriers of all kinds; but rather than a deliberate neo-mercantilist approach, it is possibly a device to cover up the "jealousy" of the internal elite, who have already transformed into a new nomenclature, trying to maintain exclusive control of national assets.

In such policies the government usually applies measures of a macro-political character, including monetary and devaluation manoeuvres on currencies. Eventually, it maintains certain international openness, but this in just "one direction", to be pursued with typical instruments such as import-substitution policies and specialization in key productions, usually those easily traded on international markets (deriving from massive exploitation of natural capital).

Such protectionist attitudes can last for a while, until local economies become sufficiently strong to function in an autonomous way and to open up to the free market. In fact, at the same time globalization processes can make such "neo-sovereigntist" politics, after a certain threshold, ineffective. It is the case of the metropolitan areas becoming "de facto" global cities; in Uzbekistan it is not just Tashkent, but to some extent Samarqand and Bukhara too, as well as the Fergana cities (the latter due to their lying on the trans-border area) increasingly open to international integration and in general of the increasing influence of the international civil society, consolidating—besides the eventual government obstruction—on all levels.

Besides governmental intentions, this practice results in a contradictory policy. The balance of international exchange steadily expands, and this in a reciprocal direction, considering export/import (of immaterial culture, technology, know-

[64] Lanzetti Eleonora, 2017; as said the current trains—a standard train indeed, using existent lanes—need about 15–18 days from Mortara, close to Milan, to Shangai, for 10.800 km, instead of 40 days of maritime travel.

[65] Swiss based, owned by Kosovo politician and business man Behgjet Pacolli, who founded Mabetex group build up nearly 40% of the whole new Nursultan city side, wiki, Mabetex Group, accessed 29.3.18; in practice all the CA major cities have been involved in their large-scale reconstruction, including building of shopping centres, public facilities and infrastructures, negotiating public authority construction permissions.

how, essential for local development), and also remittances from workers abroad (vital for Uzbek rural areas).

The best example of this situation is the industrial politics, with a mercantilist attitude evidencing its "double edge" aspects: the government maintains the high tech industry under its exclusive control (in the case of the cited Tchaklov plant relying on Russian collaboration in research, know-how, expertise and raw material supply), trying to keep it competitive; but finally it has to give up on such a project, transforming it with the help of foreign MNCs and adapting it to the country's actual needs (e.g. car and trains, instead of airplanes and armaments).[66]

12.4.4 State Administration and Budget, Currency and Monetary Politics

The construction from scratch of a new macroeconomics represents an extraordinary task. It signifies the construction of an entire system, including its cultural component, relying on the confidence the population has matured concerning issues such as tax collection, public spending capability, credit regulations, finance and currency institutions.

This task has been carried out in CA with the substantial help of the different international agencies, such as the IMF, and credit institutions, like the World Bank (WB) and the Asian Development bank (ADB), and today it is possible to say that it has been successful. However, at the beginning it had to face a set of problems, both economic and social. Among them can be considered the collapse of subsidized factories, the crisis of welfare, public and private services, often with conflict putting into doubt the legitimization of the ruling authorities.[67]

Just after independence, NIS governments established new public financial institutions, actually symbols and instruments of sovereign politics (central bank, ministers and regional offices, a state budget and social spending, a centralized administration and financial control function) and issued new national currencies, creating the premises for a new market economics.

Furthermore, they introduced a value-added tax, organizing a system for fiscal collection usually a severe problem in such situations (since "kiosk economics", spontaneously spread out, is hardly suitable for formal accountability), finally making it possible to draw up a state budget. But all this happened with several ambiguities.

Initial times were characterized by many inefficiencies—inevitably. In particular, the early transition times, 1991–1995, were characterized by hyper-inflation with the usual corollary of monetary mass restriction measures especially cuts in salaries and in employees in the public sector, as well as in services and facilities, supported with "wild" denationalization. It was also characterized by further situations typical of such circumstances, e.g. weak internal controls, inexperienced management and untested regulations, with, as expected, a sequence of cases of mismanagement.[68]

Situations that evidence the weakness and sometimes the complete failure of privatization operations that started soon after independence and which were considered an essential instrument to render the political change irreversible at that time; it produced contradictory results that actually did not prove successful in creating a functioning market.[69]

The most interesting case is that of monetary units that represent the same intrinsic confidence the new state must inspire, and which have to be restructured or invented from scratch. The fragmentation of the CA region into sovereign states signified the introduction of new and different monetary systems, that found some difficulties along the way, but which were finally able to consolidate themselves with the support of international financial institutions (like World bank and IMF).

In such situations of high spread among economic systems—among the different republics—the fact of using different currencies appears as an appropriate instrument in order to steer the differentials occasionally or structurally arising between them. Individual economics pursues different strategies, but has the same necessity of exchanging their products, namely, to access the international market.

In fact, the CA economies are in some cases complementary, but in others they find themselves in a situation of competition. On the one hand, they have to coordinate infrastructure projects, safety regulations and supply chains; on the other hand, they are competitors on the same international commodity market, especially when attracting foreign investments.

This is the case, specifically, of the two leading economies of the area that are characterized by their diverse approaches—Uzbekistan, assuming neo-mercantilist economics, and Kazakhstan which has pursued full integration into international economics since the beginning.

They developed different monetary policies such as in the case of Uzbekistan, applying devaluation competition, with convertibility of the currency used as a device to preserve internal capability, but possibly as an artificial manipulation

[66] Fazendeiro Bernardo Teles, 2015:485; in fact mercantilist policies can be considered necessary to protect a first phase (the start-up, the "take off") of internal developing economics; but it is important to understand when to say "stop" to such politics, before this degenerates in a distorted kind of protectionism.

[67] Brill Olcott Martha, 2012:66fs.

[68] Marzhan Thomas, 2015:471.

[69] Marzhan Thomas, 2015.

(as remarked on several occasions by the IMF). This country often seems to use such variables deliberately (customs, administrative procedures, taxation) in order to manipulate FI,[70] to be considered as a matter of continuing negotiation with the foreign corporations, assumed sometimes as "predators" or "invasions", rather than normal economic operators. An attitude that also spread on popular levels, degenerating into a sense of diffidence against foreign companies, used sometimes as a target for expressing social frustration.[71]

Indeed, the stability of the currency is a good indicator of the reliability of a certain system, and of its efficiency. It leads to further steps (that possibly only Kazakhstan has already undertaken), namely that of achieving self-alimenting growth and of involving wider population strata in the benefits of the growth (beyond the class of oligarchs, usually prone, among other things, to export profits and capital).

It means the formation of a middle class, capable of consuming and spending, in general of participating in development, with the promotion of individual and family consumption and accumulation, pursuing further privatization of some assets (relying on an effective land and propriety reform), promoting the formation of small private property.

Furthermore, it is the case of the diffusion of long-term social-financial services, like pension funds, health insurance and investments in advanced functions (e.g. planning the education for children or just getting a bank loan to buy a home or durable goods). In fact, such functions would enable the middle class to start a planning practice with regard to entire life cycles (relying e.g. on the expectation of a pension, of a health programme for the family), consequently developing a stabilization effect. This is possibly a key element for the passage towards a more stable advanced society, definitively eschewing under-development conditions.

Evidently, the capability of the system to "democratize" finance (bank stability and accessible stock exchange, financial state budget efficiency, savings and investment policy, eventually implementing auditing functions independent for the power) can reduce such a gap, incentivizing the population to participate in the wider economic circuits, overcoming the original "kiosk economics" condition (that otherwise could become permanent).[72]

Indeed local economics still evidences a strong dependency on natural resource exploitation and export, on international quotations, with the consequent risk of "importing" instabilities; the recent monetary crisis in most post-Soviet countries is the evidence of such situations. The Turkmenistan Manat devalued in 2015 by 18.9%, fundamentally demonstrating non-performing economics, consequent to the post-Soviet country (Russian mainly) monetary crisis which occurred in the same period.

Moreover, Kazakhstan had similar problems as a rentier state, incurring in a stagnation phase. The severity of the crisis is demonstrated by the astonishing news about strikes in Turkmenistan and Kazakhstan, evidencing the fragility of such models based on rentierism.[73]

12.4.5 Applied Macroeconomics

Uzbekistan seems to use macroeconomics and monetary instruments deliberately to protect both its own production system (also considering that the country has a manufacturing tradition, possibly the only one in CA space),[74] and its own "nomenclature".

On the contrary Kazakhstan seems to favour the internationalization of internal economics, possibly incentivizing the development of high value-added activities (establishment of foreign corporations, research for HC new fields), as well as implementing a systematic FI policy. Kyrgyzstan and Tajikistan (and to some extent Uzbekistan) with some difficulty may develop an autonomous macroeconomics. Their public and private budget depends on remittances (40% of the Kyrgyz GDP in 1998, about 800,000 migrated workers), and partially in the export of commodities. One must consider that the situation has improved over the last few decades, with economic indexes growing rapidly.

In fact, these countries have no alternative but the promotion of a bottom-up economics (not induced by the government), based on primary activities and local-scale work units (family, rural village, farms), with the possibility of materializing their work in some material assets (a property, a piece of land, a farm).

Indeed, some authors deem such an approach wise and economically efficient, avoiding the formation of asymmetries (as in any rentierism). This must be understood considering the multifaceted aspect of the situation, susceptible to unexpected evolutions; for example, the migrants returning to a definitively stabilized country can bring capital and knowledge with them, as usual in such situations, potentially starting up new initiatives.

[70] PWC 2016; as said Uzbeks opposed problems in convertibility; see FI Press Release: The Republic of Uzbekistan Accepts Article VIII Obligations; Imf.org.

[71] Indeed today, from the neo-mercantilist point of view; but possibly an obsolete question, since globalization tendencies can render protectionist policies inconsistent; Bernardo Teles Fazendeiro, 2015:488; PWC 2016.

[72] http://gca.satrapia.com/+unified-pension-fund-recommended-in-kazakhstan+, 23 January 2013. Accessed 11.4.2018; PWC 2016.

[73] Anceschi Luca, 2017; Institute for War & Peace Reporting, 2012, Daring to Strike in Turkmenistan, 29 Aug 2012, https://iwpr.net/global-voices/daring-strike-turkmenistan. Accessed 24.04.2018.

[74] Fazendeiro Bernardo Teles, 2015:488.

12.4.6 Risk for a "Monoculture" Organization of the Economics

In fact, the disadvantages deriving from "mono-production" are evident. At first, dependence on a single resource means the risk of being subject to the risk of market "bottlenecks" induced by international quotations, eventually manipulated by different speculators. Furthermore, such economics leads to concentration, inducing monopolies to suffocate the spontaneous emerging of a bottom-up economics: monopolies which are usually inefficiently managed, since the managers are appointed because of loyalty, not capability.[75]

The current major risk is that the governments would rely completely on this kind of economics, diverting investments from sectors with higher value-added potential. This is the case of cotton for Uzbekistan: for this country cotton means about the 60% of agricultural production, giving work to 75% of the rural population, organized in state-owned farms, and using almost the totality of irrigation water on 45% of arable land; obviously, the dependency on such a commodity can be extremely dangerous (namely involving the whole society).[76]

So similarly, for cotton production in Tajikistan; and so too, for the extraction activity in Turkmenistan and Kazakhstan, which besides the easy money earned through HC export, involves complex industrial economics with investments planned over the long term, that have to consider among other things environmental impacts, innovation and periodical upgrading costs, namely variables which cannot easily be estimated. So too, for Tajikistan, which relies on aluminium, indeed a sector exposed to competition on a worldwide scale market, and on water-hydropower production; so too, to some extent for Kyrgyzstan considering the revenues of Kumtor gold mine, the main and sometime unique source of hard currency of the country.

12.4.7 A Tendency to Structural Opening: Risk of De-territorialization

This "opening up" appears today as a structural force, potentially limiting centralization and neo-mercantilist tendencies. However, the excessive exposure, namely the indiscriminate "opening up" to globalization, must be considered as it may represent the risk of outflow of capital and of resources, as well as of out-migration (namely a de-territorialization risk). This is particularly due to the geopolitical location of CA countries, since capital is always seeking "safe ports" and better investment opportunities in a borderless environment, where it is sufficient to "click" on a mouse to move capital anywhere anytime (this especially for "young" economies, not yet fully established).

This considering the risk associated with these countries that have only recently become independent remains relevant: the concentration of power and wealth in a locked "club" (popularly defined as "oligarchy") is a potential risk from many points of view.

The monopolistic tendency consequently extends, from that of commodities (in the frame of a renovated "oriental despotism"), to further sectors, to communication and culture, seeking to control the media, both "social" and "mass" media (this even when in a digital post-modern environment it is more complicated for any power to extend totalitarian power directly over society as in the mechanic–Fordist paradigm, in previous times). The excessive pressure in a highly mobile world, without true barriers (as in a modern environment), can induce the loss of capital as a reaction as well as of persons and capabilities, towards more stability-perceived lands.

This risk is especially high for post-Soviet societies, inspired by a structural fear of power (as indeed any society that has experienced totalitarian degeneration), without much confidence in rule-of-the-law principles; so, for example for individual savings, which many try to accumulate and "export" abroad (in countries where they would eventually try to migrate to), and so in general for private and family saving and investing.

There is a "narrow" path between stability (and preservation of national interest) and integration into an open society, that in fact seems to be unavoidable—considering the geographical permeability induced by new technologies. Possibly, the only way is that of fostering the establishment of a domestic (national) middle class (namely territorialized, rooted, but to some extent independent from the power) that could give stability to the system.

12.5 Post-modern Tendencies and Resources

12.5.1 New Qualitative Activities Instead of Scale Economics

CA economics is experiencing a big change, which means new opportunities in the digital era, but also some risks for politics. Among these, as already considered, those concerning the flight of resources, above all human and cultural-financial resources: current technology accelerates mobility, rendering any asset a kind of "volatile" wealth (corporate profits but also private savings).

[75] In general, in such situations, the political-controlled management is less efficient—since the politics has different interests that the exclusive efficiency of the management, then much more exposed to risk of bureaucratization and arbitrariness.

[76] Wikipedia Tajikistan, accessed 23.4.2018; Black Sea Agro, 2013.

Fig. 12.6 Tajikistan, Rash valley, biker, August 2017. (Photo: Igor Jelen)

But also several opportunities: the affirmation of borderless economics means integration in an enormous, unlimited borderless world-market, delineating new activities, mainly culture-information based, such as digital manufacture, tourism, services, education and amenities (easily combined with the recovery of some traditional activities), with companies moving continuously from centres to peripheries; in principle anywhere and anyone can acquire assets usually excluded from the wider market, which, thanks to globalization and liberalization, are nowadays accessible.

This change configures a new paradigm: beyond appearances, resources usually considered precious (like highly "territorialized" HC, natural raw material, minerals, agrarian soil, with correspondent industrial capacity) seem to lose some value, making the weakness of rentier economics evident. On the contrary, an updated concept of development seems to be more oriented to culture-intensive factors today, and to environmental conditions (ensuring attraction in all senses, also politically).

A change which prospects a set of further adaptations: after the era of the heavy industry, and of the Fordist method, economics has to develop new activities like "high tech" and digital innovations, high value-added and advanced services; potentially, at the same time, it can recover (once neglected) traditional productions, like quality agriculture and handicraft productions, that may be traded on world markets, benefiting from the transport revolution, delivered with ICT organized cargos, relying on "low cost" organization.

So too, for outside demand as well as for income mobility: what once were considered just remote areas can easily become accessible ones, sometimes transforming themselves into natural, amenity and economic (tax) paradises.

This denotes a new prospective for the CA geography, that is characterized by natural and cultural peculiarities: remote untouched places, that once were the centres of tensions, today are entering Baedeker and guide-books; sometimes, they become tout-court "packages" offered by tour operators, registering a kind of boom in such activities—biking, trekking, canyoning etc. (Fig. 12.6). Often such areas are reconverted into hubs for new "free economic zones",[77] becoming the target for investments: things which had just been unknown assets until that moment (an abandoned mine, a nomad horsemen festival, a remote valley landscape, but also a Soviet-style university full of talents) may suddenly become the site of new activities, bringing capital and jobs from everywhere.

12.5.2 Cultural Heritages

In such contexts a small activity is sufficient to have a significant return in these scarcely inhabited countries.[78]

It is the case of traditional folklore, archaeological, cultural and natural landscapes, namely assets to be discovered, preserved in their authenticity, and economically recognized,

[77]PWC 2011:12.

[78]It is to say that the post-modern paradigm induces a kind of inversion of scale economics, privileging quality as quantity, improvement instead of quantitative growth, particularities instead of standardization.

Fig. 12.7 Transition times folk singer performing traditional epics, Bishkek, 1997. (Photo: Igor Jelen)

in synergy with local productions. It is the case of agricultural production, for food (vegetables, exotic fruits and further special production) and manufacturing: in times of anti-plastic (and further HC derivates) revolution, the production of natural fibres must be reconsidered (cotton, kenaf, hemp) possibly biologically cultivated, and socially and ecologically certified; such productions are increasingly appreciated by advanced markets, prospecting unexpected development for peripheral areas.

So too, for traditional art and artisan skills, to be organized in the frame of companies possibly collaborating with tourism agencies. This is the case of textiles, manufactured goods, Bukhara carpets and embroidery. It is also the case of music and dance, of traditional instruments, decorations, jewels, pottery, manufacture, clothing; local art marks a sophisticated expertise that also proves to be precious in identity terms. Such high value activities had almost disappeared in modernist Soviet times, when they were considered as a reminder of "feudal" times; but they currently represent a resource for any qualitative development (much more than niche economics).

It is the case of cultural productions, like literature and other artistic performances, like Manas and other traditional literature oral singing (visiting the amazing "Manasci" show, a unique experience) and similar traditional folkloristic uses: all these have relevant potential and are obviously precious because they are unique and irreproducible (not suitable of being copied or de-localized) resources (Fig. 12.7).

So too, for other nomadic life particularities, like horse and camel riding, connected with sports, tourism and similar activities and also, consequently, for the re-evaluation of the traditional zootecnic patrimony (Akhal-Tekke horse, Bactrian camels, Kara Kum dromedaries, Karakul sheep and many others) from all points of view (tourism, wool and food production etc.).[79]

And so, in a similar way, for apparently non-important activities, but which are essential for symbology and self-identification processes, like falcon hunting, horse and camel riding festivals and other traditional sports. So too, in similar way, for bear or bird watching, and for other precious wild animals, like Marco Polo sheep, antelope and mountain goats, Tien Shan bears, red wolves, snow leopards and other endangered species, which may be useful for specifically requested activities such as photo safaris. These and other opportunities for a nature experience may become, along with the new technologies (remote observation outposts, photo traps and specially equipped parks), extremely interesting, moving a significant number of visitors.[80]

12.5.3 The Landscape as an Asset

Besides popular and traditional activities, there is also the conventional "high culture" (archaeology, historical documents, artistic elements, theatres, monuments, museums) which represent an important resource: a long list of sights,

[79] PWC 2012–2013.

[80] All these resources prefigure diffused and accessible assets, also useful for fostering a sense of identification in the geography, of sensitivity of the territorial wildlife and nature. And much more, since such activities may attract and entertain visitors: the "preparation" of the "cultural product" to be offered to tourists exerts an cultural feedback, meaning the improvement of the domestic quality of life—e.g. improving accessibility and facilities, landscape quality, or simply a better knowledge of one's own environment. All these resources represent significant unexpressed value, that need management capabilities and projects tailored at a local scale in order to be translated into economic value (creating jobs, profits, and further self-alimenting investments).

Fig. 12.8 Tajikistan, Kulab, city park, August 2017. (Photo: Igor Jelen)

even when often of an auto-celebrative character that deserve a much more extended worth.

The CA cultural stratification evidences a unique mix of elements, western and Asiatic, Soviet and Islamic, nomadic and sedentary, that alone represent precious assets, visible in many dimensions—architecture, artistic products, performing arts.

Evidently, also the Soviet "layer" (artistic production, architecture and landscape, literature) begins, after a quarter of century or more, to assume a certain significance, opening up new possibilities: mosaics, underground stations and urban decorations, architecture and engineering structures from Soviet times are witness to an epoch, and besides any interpretations are of important aesthetic as well as functional evidence.[81]

The local masterpieces on "offer" are Samarqand, Bukhara, Khiva and Shakhrisabz (as said, the four Uzbek UNESCO World Heritage List sites), and the other ancient cities on the SR, representing a large spectrum of marvels which can potentially give value to any itinerary, where cultural tourism is increasingly reaching mass tourism scales.[82] Leaflets with images of Uzbek towns and itineraries have been in the shop windows of tourist agencies in the whole world for some time now, with Samarqand and Bukhara becoming top travel destinations.

So too, for other CA countries; in Turkmenistan, the archaeological complexes of Nisa and Merv (Mary) represent in world imagination the charm of ancient civilizations abandoned and ruined in the dust. Furthermore, the country offers wide natural and landscape resources. Among them the Darvaza site, an example of a non-intentional effect of Soviet engineering, since it was produced by an error in their own activity, provoking a terrain break down, while searching for gas, characterized by spectacular burning (a kind of "gateway to inferno"). So too, for other sights, of artificial or natural origin, that could represent a significant experience and knowledge source for visitors.

The Kazakh territory, because of its extension, represents a remarkable variety of environments. It is possible to list, among others, the extraordinary patrimony represented by the natural reserves at Charyn Canyon, at the Korgalzhyn and Nauryzumsky Reserves, UNESCO heritage sites, as well as the Bigach Crater. The western desert, with the surprising underground mosque complex and its endless horizons, is particularly fascinating. Other attractions are the mausoleum of Sufi Master Khoja Ahmed Yassaui at Turkistan, the petroglyphs at Tamgaly (Fig. 12.9), both recognized by UNESCO with "world heritage" status, and many others. Nursultan and Almaty represent diverse attraction targets for different reasons (ancient colonial districts, "chruščëvka" colourful neighbourhoods, utopian urban architecture).

The cultural offer in Tajikistan considers the Bronze Age site of the Andronovo culture of Sarazam, classified as a UNESCO heritage; other sights in the south are the magnificent Kulab mausoleums and parks (Fig. 12.8). Together with Kyrgyzstan, the country is famous for its mountain landscapes, with forests, glaciers and spectacular "north faces" (Trans Alai, Pamir, Tien Shan), which should be rendered

[81] Kluczewska Karolina, 2017.

[82] Paskaleva Elena, 2015:427, 436; http://whc.unesco.org/eng/list/603/documents/. Accessed 23.4.2018.

Fig. 12.9 East Kazakhstan, Tamgaly, July 1997, petroglyphs dated Skyft time. (Photo: Igor Jelen)

accessible by organizing an adequate tourist offer,[83] with the preparation of such routes as the Pamir highway (equipped with adequate public transport) characterized by fantastic views, ethnographic interest and natural reserves. Furthermore, Issyk Kol can be considered very popular. It was already used in Soviet times for organized tourism from the close urban areas in Kyrgyzstan and Kazakhstan, as a "kurort" and "sanatorium" with great potential that attracted about a million visitors in 2006 and 2007.[84] A special attraction is the "sacred" Arslanbob walnut forest, a unique case of such a forest, in the south of the country.

Such itineraries have to be considered as forms of sustainable development, relying on appropriate uses of natural resources,[85] qualitative agriculture and bio-farming synergically connected with green economic methods; the same obviously for environmental resources organized in "zapovednik" (natural reserves), representing reserves of space, clean air and water, offering endless horizons and landscapes (what in principle tourists coming from polluted and crowded urban areas are seeking).

12.5.4 Tourism, Amenity and Happiness Economics, and Social Feedback

All this means a turn in economics; in fact such evolution gives priority to bottom-up initiatives, namely on development resources widespread and accessible for the wide population (e.g. on niche agro-touristic and handicraft activities carried out in rural settlements), making new opportunities possible, and of new ways for doing tourism: desert and steppe trekking, biking and alpinism adventures, ethnographic and cultural experiences mean in general high added value but avoiding—here in CA—the annoyances characteristic of mass tourism.[86]

Such activities require a high professional skill in order to manage such resources in a market-oriented way, making them profitable (besides the obvious necessity of conservation). This is the case of rural areas (in order to organize rural-life experiences), of natural parks and bio-diversity areas (appropriately equipped for visitors), of the traditional and wild animals which can be observed in their habitat (without disturbing them), of ethnographic museums and similar facilities.

All these are important assets, to be consequently implemented (contributing to the spread effect of "authenticity", the most important characteristic in the post-modern era). One must also consider the amazing impact produced by having on a tiger or a snow leopard on a "leaflet" for visitors—animals on the "red list", risking extinction—[87] as well as the breeding of precious wool sheep or a Pamir yak, or some other exotic element. So too, as an attraction device for any kind of original landscape, for Lake Aral, the Balkhash swamps, and Tugay jungle, or the starry nights in the desert and on the steppe: the local environment offers an amazing catalogue of sights, potentially attracting tourists from all over the planet.

[83] Jelen Igor, 1994; Jelen Igor, 1995.

[84] Wikipedia, Kyrgyzstan. Accessed 23.4.2018.

[85] See IUNC—International Union for Conservation of Nature and Natural Resources, http://www.iucnredlist.org/details/22732/0. Accessed 23.4.2018.

[86] Paskaleva Elena, 2015: 427, 436; http://whc.unesco.org/en/list/. Accessed 23.4.2018.

[87] International Union for Conservation of Nature and Natural Resources, http://www.iucnredlist.org/details/22732/0. Accessed 23.4.2018.

It represents much more than a niche economy, namely a particular segment of tourist ("backpack-keepers", sport & alpinism, bicycle), nature friendly, sensitive to local populations, usually with high spending capacity.

It is therefore necessary to elaborate a new management scheme, combining a sense of hospitality and economic initiative while overcoming any negative attitudes that derive from the past culture. This might be the case of the "špion syndrome", namely a cultural aversion to foreigners, possibly induced in Soviet time by some isolationistic ideology: it is to assume the sense of hospitality is integral to any civilization (indeed such kind of ideologically induced "aversion" regarded in those times any kind of spontaneous initiatives and innovation).

There is the need for new expertise and for new education programmes, possibly for new university courses, but above all for a new sensitivity to such kinds of development (exactly contrary to that prospected by infrastructure-driven economics and natural capital exploitation prevalent today in the leading oil states).

Such economics is not necessarily in contradiction with the conventional forms of tourism that local governments have begun to develop in some areas particularly suitable for mass-scale amenity economics. This is the case of the new settlements on the Caspian, and possibly, in the future, also on Lake Aral (the only consistent water body in Uzbekistan) and in other amenity locations, often close to lakes, rivers and mountains. The new amenity resort centres Aktau in Kazakhstan and Avaza in Turkmenistan show relevant potential and might surely become favourite investment areas for the respective national economics.[88]

12.6 Development Options

12.6.1 Time Changes and the Arising of New Significances

All these possibilities can delineate a road map which should be developed on the basis of a plural and post-modern definition of progress, multi-dimensional and adaptability.[89]

The new economic paradigm is based on soft economies (instead of quantitative-commandeering extraction industries), on urban and ethnographic landscapes (the medersa of Samarqand as well as the colonial time wooden homes of Almaty, the shepherds yurts and the "palatke" of the Pamir), on natural resources and renewable energy sources (wind, soil, sun, water and glaciers, water steam etc.) and on assets such as culture, environment and landscapes.[90]

Further resources to be considered in these terms are the revival of traditional cultures and of local nature (destroyed or simply ignored in the modernist era) that can contribute to a new territorialization process, namely to the rooting of the population to a certain territory. As said, such activities have a high cultural-strategic relevance, since they induce feedback for domestic inhabitants, regarding the care of landscapes, and, in general, improving lifestyles. All this is important, above all, for preventing mass out-migration tendencies (always a risk for such peripheral regions).[91]

12.6.2 The Re-start as a Unique Chance: A Road Map for Development

This recovery, indeed, has as a premise the operation of cleaning up the areas contaminated by Soviet modernity, with wide surfaces sometimes literally covered by a bulk of ruins in whatever form: a long list of interventions, considering the clarification of any contamination in the ground, water and air, starting from the cleansing of nuclear testing sites (circumscribed to the north-eastern Kazakh region), already being undertaken meritoriously by the Kazakhstan government; so too, for abandoned industrial sites and post-Soviet "ghost cities".

This is especially relevant for Lake Aral and other internal waters which are currently the object of large-scale recovery projects, showing some encouraging success (see chapter 6). It is the case of the large-scale project in reforestation, fighting desert invasion or simply trying to till the steppe, improving land use patterns and contrasting climate change. It is the case of the area surrounding Ashgabat, where a reforestation programme is being carried out and the excavating of artificial lakes to improve the ID capability and to contrast cli-

[88] In 2007 it started the Aktau city project for tourism and investments; similarly for Avaza in Turkmenistan, https://www.economist.com/news/asia/21711943-authoritarian-president-pins-his-hopes-turkmen-las-vegas-shrinking-exports-spell. Accessed 22.4.2018; https://www.aljazeera.com/news/asia/2011/12/2011123015563314649l.html, 30.12.2011. Accessed 6.7.2018; the President Berdimuhamedov announced the creation of a "free tourist zone" in Avaza, PWC 2011:12; security questions are critical elements in this sense; evidently in such areas, recently included in international circuits, any event can have a worldwide echo; great impression was made by the case of western cyclists assaulted in Tajikistan in 2018, that cannot appear but as a terrorist attack.

[89] Marzhan Thomas, 2015.

[90] Times are changing and the changes can disclose new significances; what was until a point in time a residue could become suddenly a monument or a testimonial worthy of preservation; it is the case of an entire layer of social landscape, abandoned cities, of artistic and cultural products form Soviet times, possibly to be re-evaluated; Kluczewska Karolina, 2017.

[91] It is evident today, the most valuable economic resource is the "person" that has confidence in their own future, for him/she and hers/his family, living in a territory perceived as their own; if the authorities ignore this, the populations, especially in remote and marginalized areas, will take advantage of the globalization chances, and will sooner or later migrate, abandoning their homes.

mate change (but risking over time to appear unsustainable). Such projects have to be considered in the frame of a transnational programmes and agreements because they concern the whole water circulation on an even wider scale, from the glaciers to the "sea"; some waters arrive from outside—like the Irtysh River from China, flowing further into Russia, already and increasingly polluted, potentially impacting the endorheic basin, then posing an even more complicated question.[92]

It is necessary to think about a better use, for example, substituting open air channels with pipelines and pumps (reducing waste due to evaporation or filtering down underground), carefully planning maintenance of infrastructures, programming uses and accumulation; it is necessary, above all, to consider that the "free" environmental water is the factor that permits the self-depuration of the whole environment, improving the cleanliness of soils, the productivity of vegetation and of trophic chains.

These interventions can represent a relevant source of business in the context of new green (and circular) economics, representing an opportunity for the development of new industrial recycling techniques; for used materials, metal scraps, rubber, pneumatics, building materials, chemical residuals—actually a whole "geologic strata" of waste that Soviet economics had accumulated over the decades of its gigantic industrialization programme.

Recent reports show that ecology has not been definitively compromised and that it will maintain a homeostatic capability in the long term, until a certain threshold is reached; ecosystems, soils and water—fortunately—retain a basic capacity of self-depuration; this also considering the low population density and the huge surfaces, therefore the relatively low impact of such contamination on human life.

12.6.3 Possible Mistakes

Development issues show several options: at the moment of independence the destiny of the NIS appeared highly uncertain; among various scenarios it was possible to list those of new "colonies" exploited by rapacious MNCs, of a "new Third World" characterized by stagnating economies, of populist and autocratic leaderships, of battlefields for the new superpowers, and of possibly "rentier state" consuming natural stock unscrupulously (until possible); or finally that of a self-sustainable balanced development path.

Whatever the opinion, it is important to avoid mistakes today (supposedly at the beginning of a development cycle) that can provoke "snow ball" effects over time, potentially blocking the same growth (like several examples of advanced democracies can today evidence).

[92] Marzhan Thomas, 2015:467.

Such mistakes can be classified as reversible and irreversible and are usually to ascribe to a distorted view, and a biased use, of modernist innovations. This is the case of widespread pollution (of water, air, soil and natural resources) and of proliferation of unsustainable lifestyle practices; among these it is possible to consider several cases of consumerism "bubbles", with the abuse of individualist practices to the detriment of public functions (e.g. ruining networks of public transport, spread of "grey" economies, representing from a public accountability point of view a fundamental tax evasion problem).[93]

It is the case of a general difficulty in distinguishing between public and private spheres of competences: evidently the state tendency to intervene can depress the formation of a functioning market, and of a local entrepreneurial class, capable of start-up initiatives. But the opposite is also true, namely that private bodies cannot substitute the public in a set of functions; this particularly so for the functions representing basic needs, because not profitable or because they require the use of "common" and natural resources (popularly perceived as public).[94]

12.6.4 The Risk for "Resources Damnation" Effect

The current CA situation is characterized by an opposite tendency towards centralization and neo-nationalism. The patrimonial-authoritarian[95] policies privilege the exploitation of natural capital (the national treasure), in order to achieve products useful to be converted on the international market, into foreign currency; so too, for the building and infrastructure industry (water channels, pipelines etc.), that are suitable to be controlled by the power structure in a centralistic (Fordist) way.

[93] For example, the dismantling of the public network of welfare, facilities and transport, to the benefit of individual care (eventually supported by an individual insurance programme) and mobility, may represent irreversible damage in terms of public organization capability; in general, this situation configures a trade-off between the public and private; a further example is that of underground economies and of tolerance towards tax evasion (by private bodies), and corruption (typical for a public institutions), that would hinder a development path, further aggravating the fracture between institutions and society); see as well Brill Olcott Martha, 2012:66fs.

[94] Similar basic questions affect society, economics, politics and infrastructures. Indeed, applying appropriate techniques, and efficient governances, resources which are usually considered limited, can become renewable in practice (indeed so virtually for all such resources). Water use improvement is just an example: from the general point of view water is not a limited resource, but it is necessary to apply some efficiency criteria in order to pursue sustainable use, recycling or purifying it. So, similarly, for soil, raw materials, energetic resources, and for any kind of industrial or civil waste.

[95] Babajanian Babken, 2015.

The dependence on a unique resource is the premise for the well-known problem defined as "malédiction des ressources naturelles" (resource damnation), extremely dangerous for entire systems, even when such danger is not often popularly perceived.[96]

This risk is becoming concrete with the incumbent change in paradigms: the reason "of the wealth of the nations" can suddenly change, rendering such countries even more vulnerable. This is the case for certain standards of political and economic safety, for social stability and other basic needs, that should derive from efficient governance, both internally and internationally, rather than the simple extraction and consumption of fossil resources.

12.6.5 Change of Energetic Paradigm

The change of the economic paradigm is in full view of everyone; the old one, based on HC resources, and highly territorialized systems, is to be considered obsolete, and it is necessary to prepare for a new one.

This change will be difficult especially in rentier countries, where the HC economics represents the opportunity to consolidate a kind of "oriental despotism" with oil and gas (substituting water and water management, originally assumed by that classic geopolitical theory) delivered with a system of fixed infrastructures. This signifies the necessity for strict controls and even the militarization of the territories, predisposing strategic "corridors" and "safety belts" for pipelines, refineries and oil terminals, and the setting up of networks connecting the central points of the power right up to the "last mile".

But today, this economics based on fossil resources is challenged by more flexible technologies. It is the case of liquefied gas transported via sea routes that tendentially align to free market conditions (on the contrary the extractive industry is usually controlled in a monopolistic way). It is the case of the slowly but effective growth in renewable sources (defined as a utopia until just a few years ago by many observers, analysts and decision-makers), with energy saving and ecological-friendly regulations and with increasing technological improvements.

Considering that the cycle of the energy infrastructure (from exploration, to extraction, treatment, to finally last mile distribution) generally develops over the long term (in practical in the generational passage), the investments in HC infrastructures (and connected activities) will probably slow down, assuming the effect of a structural decrease (impacting negatively the profitability of such entire economic cycle). All this contributes to overcoming the paradigm on which the recent geopolitics of CA (together with Russia, Gulf countries and other oil countries) has been founded for the last few decades.

Such changes represent a major threat to local rentier-ruling classes that may not be able to prepare for the change—as usually rentier-state elites—who react to the risk of losing status and power in a disproportionate manner.

Among the most ambitious plans are those of the Kazakh government, which even when rich in fossil resources is continuously elaborating plans in order to overcome such changes. Kazakhstan aims at targeting the passage towards green economy based on renewable sources by 2050, adhering to the most advanced standard internationally, realizing the practical questions of a new economy, and of its impacts on society. It is doing this in a rather concrete manner, for example setting target prices for renewable energy production or investing in innovation.

Finally, one must take note that the government is implementing further programmes, in a complementary manner; among these is a programme in an e-governance project, in environment and education improvement, in general to create a society aware and ready for future challenges.[97]

Finally also Uzbekistan—the most conservative among the NIS economies—elaborated system of e-government and further reforms in order to reduce bureaucracy and to increase system efficiency.[98]

References

Abazov R (1999) Policy of economic transition in Kyrgyzstan. Cent Asian Surv: 197–223

Anceschi L (2017) Turkmenistan and the virtual politics of Eurasian energy: the case of the TAPI pipeline project. Cent Asian Surv 36(4):409–429. https://doi.org/10.1080/02634937.2017.1391747

ArcelorMittal (2013) Coal mines Karaganda Coal Basin, Kazakhstan, Pre-feasibility Study for Coal Mine Methane Drainage and Utilization

ArcelorMittal. Coal Mines Karaganda Coal Basin, Kazakhstan. http://aktau.arcelormittal.com/who-we-are/arcelormittal-kazakhstan.aspx?sc_lang=en. Accessed 23.4.2018

[96] Marzhan Thomas, 2015:467; Auty Richard, 1990.

[97] Electronic Government of the Republic of Kazakhstan, http://egov.kz/cms/en/information/help/help-elektronnoe-pravitelstvo. Accessed 23.4.2018; this also considering the wideness of the Kazakh territory; similar attitudes towards innovative and green economy standard, prescribed by 2015; Brill Olcott Martha, 2012:17; recently Kazakhstan's Supreme Court decisions oriented at improving transparency and of procedure at diverse levels, also using know-how of international organization, also of accounting in public spending, strengthening the state of the law.

[98] But—relying on the dominant neo-liberal economic paradigm—the demand induced top-down would not be sufficient since nothing can really substitute the impulse represented by the wider popular market, namely by the spontaneous participation of citizens in economics; and this also in political sense: the more sophisticated technology, even considering its perfect application, cannot result automatically, without a bottom-up popular control effect, in the quality of the governance.

Auty R (1990) Resource-based industrialization: sowing the oil in eight developing countries. Clarendon Press, Oxford
Babaev AG (2012) Desert problems and desertification in Central Asia: the Researchers of the Desert Institute, Springer
Babajanian B (2015) Promoting empowerment? The World Bank's Village Investment Project in Kyrgyzstan. Cent Asian Surv 34(4):499–515. https://doi.org/10.1080/02634937.2015.1095967
Barisitz Stephan (2017) Central Asia and the Silk Road, economic rise and decline over several millennia. Springer, Cham
Black Sea Agro (2013) Tajikistan: over 392.5 thousand tons of cotton picked in Tajikistan. http://bs-agro.com/index.php/news/other-countries/23909-tajikistan-over-392-5-thousand-tons-of-cotton-picked-in-tajikistan, 12.12.2013. Accessed 23.4.2018
Boboc C (2017a) 'Middle-class, limited-edition'? Middle class subjectivities in urban Azerbaijan. Caucasus Analytical Digest 95:11–14
Boboc C (2017b) Middle class subjectivities in urban Azerbaijan, ASIAC annual conference, November 6–7th, 2017. Dipartimento di Culture, Politica e Società dell'Università di Torino (oral presentation)
Brill Olcott Martha (2012) Tajikistan's difficult development path. Brookings Institution Press
Dagiev D (2014) Regime transition in Central Asia. Routledge, New York
Economist Intelligence Unit. Uzbekistan: economic overview. Wikipedia, Uzbekistan. Accessed 29 Mar 2018 (citing eurasiacenter.org page not found)
Electronic Government of the Republic of Kazakhstan. http://egov.kz/cms/en/information/help/help-elektronnoe-pravitelstvo. Accessed 23 Apr 2018
Esenova S (1998) "Tribalism' and identity in contemporary circumstances: the case of Kazakstan*. Central Asian Survey 17(3):443–462
Fazendeiro Bernardo Teles (2015) Uzbekistan's 'spirit' of self-reliance and the logic of appropriateness: TAPOich and interaction with Russia. Cent Asian Surv 34(4):484–498. https://doi.org/10.1080/02634937.2015.1114780
Filippo DD (2013) Counter-narcotics policies in Tajikistan and their impact on state building, in Heathershow, Heathershaw John, Herzig Edmund, 2013. Routledge, London and New York, pp 143–160
Frederick L (2013) Social networking practices: continuity or rupture with the soviet past? In: Akyildiz Sevket, Carlson Richard, edited by, 2013, social and cultural change in Central Asia: the soviet legacy. Routledge, pp 145–159
Global Witness. Germany must launch full inquiry into Turkmen funds in German banks. https://www.globalwitness.org/en/archive/germany-must-launch-full-inquiry-turkmen-funds-german-banks/. Press Release/July 3, 2007. Accessed 27 Mar 2018
Gulzat B (2014) Hashish as cash in a post-soviet Kyrgyz village. Int J Drug Policy 25-6, November 2014:1227–1234. https://doi.org/10.1016/j.drugpo.2014.01.016
http://gca.satrapia.com/+adb-suspends-its-participation-in-supporting-construction-of-the-tat-railway+. Accessed 29 Mar 2018
http://gca.satrapia.com/+the-level-of-poverty-in-kyrgyzstan-will-continue-to-grow+. Accessed 27 Mar 2018
http://gca.satrapia.com/+unified-pension-fund-recommended-in-kazakhstan+. 23 January 2013. Accessed 11 Apr 2018
http://news.tj/en/news/chormaghzak-tunnel-renamed-khatlon-tunnel-and-shar-shar-tunnel-renamed-ozodi-tunnel. Accessed 29 Mar 2018
http://whc.unesco.org/en/list/. Accessed 23 Apr 2018
http://whc.unesco.org/eng/list/603/documents/. Accessed 23 Apr 2018
http://www.imf.org/en/News/Articles/2015/09/14/01/49/pr03188. Accessed 27 Mar 2018
http://www.tajikistanmission.ch/business-and-investment/infrastructure.html. Accessed 29 Mar 2018
https://blogs.fco.gov.uk/leighturner/2013/05/07/trade-tunnels-transit-and-training-in-mountainous-tajikistan/. Accessed 29 Mar 2018
https://edition.cnn.com/travel/article/trans-siberian-road/index.html. Accessed 29 Mar 2018
https://iteca.uz/oilgasconference/eng/PressCentre/index.php?ELEMENT_ID=44375. 17 October 2019, For the sake of workers, Uzbekistan is privatising its cotton industry. Accessed 5 Dec 2019
https://web.archive.org/web/20080405065705/http://npdp.stanford.edu/damhigh.html. Accessed 29 Mar 2018
https://www.aljazeera.com/news/asia/2011/12/20111230155633146491.html. 30.12.2011. Accessed 6 July 2018. Las Vegas on the Caspian? Luxury hotels in Awaza, built using gas revenues and aimed at foreign tourists, have yet to benefit ordinary Turkmens
https://www.cia.gov/library/publications/the-world-factbook/fields/2046.html. Accessed 27 Mar 2018
https://www.economist.com/news/asia/21582325-president-edifice-complex-screwing-motherland-folie-de-grandeur. Accessed 27 Mar 2018
https://www.economist.com/news/asia/21711943-authoritarian-president-pins-his-hopes-turkmen-las-vegas-shrinking-exports-spell. Accessed 22 Apr 2018. Shrinking exports spell trouble for Turkmenistan. The Economist, 17 December 2016
https://www.economist.com/node/10181134. Accessed 29 Mar 2018
https://www.state.gov/e/eb/rls/othr/ics/2011/157382.htm. Accessed 27 Mar 2018
https://www.wto.org/. Accessed 29 Mar 2018
Hughes TP (2004) American genesis: a century of invention and technological enthusiasm 1870–1970, 2nd edn. The University of Chicago Press
Indeo F (2018) The role of Russia in the central Asian security architecture, policy brief #48, 2018., OCSE Academy Bishkek
Institute for War & Peace Reporting (2012) Daring to strike in Turkmenistan, 29 Aug 2012. https://iwpr.net/global-voices/daring--strike-turkmenistan. Accessed 24 Apr 2018
International Business Publication (2015) Uzbekistan Country Study, Washington DC – Uzbekistan, Guide Volume 1 Strategic information and developments
International Monetary Fund (2012) Kazakhstan Country Study Guide Volume 1 Strategic information and developments, IBP, Inc., Lulu.com
IUNC – International Union for Conservation of Nature and Natural Resources. http://www.iucnredlist.org/details/22732/0. Accessed 23 Apr 2018
Jelen I (1994) Spedizione Pamir-Alaj, Kirghizistan 1994 della SAF, in "In Alto", CXII, pp. 193–262
Jelen I (1995) Geografia e storia dell'Asia centrale in due storie di vita, pp.69–83, in "Alatau 95", Spedizione scientifico esplorativa Dzungarskij Alatau, Kazakistan, supplemento a "in Alto", CXIII, Società Alpina Friulana, 1995
Jelen I (2002) Tra i kirghisi del Pamir Alaj. Forum, Udine
Jelen I (2003) Ritorno a Ozgoruš. La condizione del sottosviluppo in una comunità kirghisa sul Pamir Alaj. Dai diari di viaggio della Società Alpina Friulana. Edizioni Università Trieste
Kluczewska K (2017) Project development in a Tajik NGO: everyday practices, instrumentalisation of donors and meaning of 'development work'. In: ASIAC annual conference, November 6–7th, 2017, Dipartimento di Culture, Politica e Società dell'Università di Torino (oral presentation)
Klüter H (1992) Russland und die Auflösung der Sovietunion, "Geographische Zeitschrift", Jahr 80, Heft 1
Kudaibergenova DT (2015) The ideology of development and legitimation: beyond 'Kazakhstan 2030'. Cent Asian Surv 34(4):440–455. https://doi.org/10.1080/02634937.2015.1115275

Lanzetti E (2017) La nuova «Via della Seta». Dall'Italia alla Cina in treno partendo da Mortara, "Corriere della Sera", 25.11.17

Lonely Planet (2014) Asia Centrale, EDT

Lonley Planet (1996) Central Asia

L.P. (2018) Sorpreso dalla tempesa di sale, turista si ammala. "MessaggeroVeneto", p 29, venerdì 22.6.2018

Marchi M, Tonini C (a cura di) (2009) Da Berlino a Samarcanda. Città in transizione, Quaderni Dip. Disc. Storiche, Univ. Bologna, Carocci, Roma

Ministero degli Affari Esteri e della Cooperazione Internazionale. Dei – Kazakhstan: Italcementi firma MoU per stabilimento Shymkent. https://www.esteri.it/mae/it/sala_stampa/archivionotizie/approfondimenti/2015/04/dei-kazakhstan-italcementi-firma.html. Accessed 22 Apr 2018

Myant MR (2011) Appendices

Nick M (2017) Nationalism in Central Asia: a biography of the Uzbekistan-Kyrgyzstan, boundary, Central Eurasia in context, 2017, University of Pittsburgh Press, https://doi.org/10.2307/j.ctt1vjqrk6

OECD. http://www.oecd.org/daf/anti-bribery/kazakhstan-should-build-on-efforts-to-fightcorruption-and-push-forwar-with-reforms.htm. Accessed 27 Mar 2018

Özcan GB (2015) Introduction: market adaptations, interventions and daily experience. Cent Asian Surv 34(4):409–417. https://doi.org/10.1080/02634937.2015.1103580

Paskaleva E (2015) Ideology in brick and tile: Timurid architecture of the 21st century. Cent Asian Surv 34(4):418–439. https://doi.org/10.1080/02634937.2015.1118207

PWC 2010. Doing business in Turkmenistan 2012–2013. https://www.pwc.com/uz/en/assets/pdf/doing-business-guide-in-turkmenistan-2012-2013.pdf. Accessed 20 May 2018

PWC 2011. Doing business guide Tajikistan 2012–2013. http://www.eurasianbusinessbriefing.com/wp-content/uploads/2015/04/pwc-doing-Business-in-tajikistan.pdf. Accessed 20 May 2018

PWC 2012–2013. http://www.eurasianbusinessbriefing.com/wp-content/uploads/2015/04/pwc-doing-business-in-Kyrgyzstan.pdf. Doing business guide in Kyrgyz Republic 2012–2013. Accessed 20 May 2018

PWC 2016. Guide to do business and investing in Uzbekistan 2016 edition. https://www.pwc.com/uz/en/assets/pdf/dbg_2016.pdf. Accessed 20 May 2018

PWC 2017. Doing business guide in Kazakhstan 2017. https://www.pwc.kz/en/publications/new-2017/dbg-2017-eng.pdf. Accessed 20 May 2018

Silvestri T (2015–2016) Le strategie della federazione russa nel teatro del mar Caspio dopo la caduta dell'Unione Sovietica, Master thesis, University of Trieste, Academic year 2015–2016

The Economist. https://www.economist.com/news/asia/21582325-president-edifice-complex-screwing-motherland-folie-de-grandeur. Accessed 27 Mar 2018

The Economist. 17 December 2016. Shrinking exports spell trouble for Turkmenistan. https://www.economist.com/news/asia/21711943-authoritarian-president-pins-his-hopes-turkmen-las-vegas-shrinking-exports-spell. Accessed 22 Apr 2018

The Republic of Uzbekistan Accepts Article VIII Obligations. Press Release, November 11, 2003. https://www.imf.org/en/News/Articles/2015/09/14/01/49/pr03188. Accessed 6 July 2018

Marzhan Thomas (2015) Social, environmental and economic sustainability of Kazakhstan: a long-term perspective. Cent Asian Surv 34(4):456–483. https://doi.org/10.1080/02634937.2015.1119552

Uzbekistan (2018) Russia agree on nuclear power station. The goal is to economize by using less gas and coal. https://eurasianet.org/uzbekistan-russia-agree-on-nuclear-power-station. July 12 2018. Accessed 27 Dec 2018

Warf B (2016) Global geographies of corruption. GeoJournal 81:657–669. https://doi.org/10.1007/s10708-015-9656-0

Wikipedia. Transports in Turkmenistan. Accessed 29 Mar 2018

Wikipedia, Kirgyzstan. Accessed 23 Apr 2018

Wikipedia, Tajikistan. Accessed 23 Apr 2018

Wikipedians. Countries and territories in the world (no year), https://books.google.it/books?id=bR0hIC0Xhb0C&pg=PA120&lpg=PA120&dq=1216.3+kg+of+heroin+and+267.8+kg+of+raw+opium&source=bl&ots=XxL5vwuI0M&sig=g_CH8ED7D09ZwhA8qN4LH44Odxs&hl=it&sa=X&ved=0ahUKEwjr8JKwvP_YAhVDUBQKHfMlBegQ6AEIKDAA#v=onepage&q=1216.3%20kg%20of%20heroin%20and%20267.8%20kg%20of%20raw%20opium&f=false. Accessed 27 Mar 2018

World Bank (2011) World Bank International Comparison Program database; Myant, Martin R., 2011, Appendices

World Bank Group (2015) Kazakhstan: low oil prices. An opportunity to reform, Kazakhstan Economic Update Spring 2015. http://documents.worldbank.org/curated/en/766781468182065283/pdf/96195-WP-PUBLIC-Box391443B-KAZ-Kazakhstan-economic-update-series-Bi-annual-Economic-Update-Spring-2015-Low-Oil-Prices-an-Opportunity-to-Reform-eng.pdf. Accessed 23 Apr 2018

World Bank, Tagikistan. https://data.worldbank.org/country/tajikistan. Accessed 27 Mar 2018

13 Institutions and Politics

Abstract

The agenda of the CA governments is impressive: politics has had to reconstruct the whole material background of local societies, as well as infrastructure, institutions, production centres and social and cultural networks; it has also had to develop the elements making independent states of what was once the province of an empire, then a self-functioning society, reconstructing international and internal relations, building a new sense of confidence among the social and economic actors.

Keywords

Current politics in Kazakhstan · Kyrgyzstan · Tajikistan · Uzbekistan and Turkmenistan · Formal and effective politics · Patrimonial-authoritarian politics · Problems and opportunities in political setting

13.1 The Transition and the Current Political Situation

13.1.1 Political Transition—Common Aspects

Legitimization processes, consensus acquirement and political consolidation are long-term processes, so it is hardly surprising that 25 years on from transition, the political CA elite are still engaged in a process of self-construction and self-consolidation.

In these times, the political narrative seems to be mainly influenced by a neo-nationalistic approach, considered in the context of a neo-organicist vision. This means that the political players—each sovereign state—is incorporated into, as classical geopolitics does, a kind of individual organism, motivated simply by its particular interest, engaged in the struggle for survival, competing for scarce resources in an open arena.[1]

However, this phase is possibly over. The CA countries are currently considered to be in a phase of stabilization from many points of view: internal apparatuses have been consolidated, territorial borders—a particularly sensitive argument—have been mutually recognized finally (if with some exceptions), infrastructures planned and sometime realized, international trade and supply chains (following a certain division of labour schema) have been reorganized.

Undeniably, the issue concerning political options is still open: the question was about the best political format to enact—as can be expected—just after the achievement of independence, essentially regarding what kind of regime should be established (among different options). However, soon such discussions were dismissed in the name of the generally recognized values of continuity and stability, necessary to maintain internal compactness.

This fact represents a motive for concern, with politics sometimes assuming a hybrid form in order to maintain its legitimization, adapting a government style based on a mix of new and old styles, Soviet habits—scarce transparency, actual discreteness—a clan-personalist way of doing things and potentially attitudes which have been recovered from local traditions.

This happens in the backstage of a general tendency to what has been defined as "patrimonial-authoritarian" politics,[2] based on the instrumental use of resources (intended as a kind of instinctive "commandeering" economics) in order to get consensus (or obedience) and leveraging on a limited circle of "stakeholders" (usually defined as the "oligarchy").

[1] Friedrich Ratzel, namely his nineteenth-century concept of sovereign state considered an absolute, evidently unsustainable today; Jelen Igor 2012; see also Megoran Nick 2017: 15, 135 fs; Lamy Frederick 2013: 146.

[2] Babajanian Babken 2015: 514.

I. Jelen et al., *The Geography of Central Asia*, World Regional Geography Book Series,
https://doi.org/10.1007/978-3-030-61266-5_13

These tendencies are the opposite of those leading towards a formal definition of the power administration, based on codified procedures, state of law and social participation, in a word—"democracy". This is certainly a word often used by CA leaders, but consequently rarely applied, sometimes appearing rather as a fig-tree leaf covering real authoritarian intentions.[3]

But it is not just something which has completely vanished. It is perceived, rather than as an ideal, as a long-term target, both at elitist and popular levels. However, the resulting political practice appears in rather ambiguous terms.

Indeed, none of the current leaders deliberately question the liberal-democratic theory (besides some attempts to accredit the grotesque theory of the "guided democracy").[4] This, if nothing else, for the purpose of complacency for the IC standard (inducing an opening tendency): besides any manipulation, power structures have to exhibit a certain degree of "democraticity", in order to appear as a reliable interlocutor on the international scene and to be able to enjoy the benefits the global market continuously offers.[5]

13.1.2 Continuity of Power—People and Elite

So far, the first characteristic of the transition is the search for the continuity of power, signifying alone, the limitation of the effects of the post-Soviet revolution. In fact, all the similar career elites (with partial exceptions of Tajikistan and Kyrgyzstan, for different reasons) were successful in this manoeuvre, maintaining control of the economies, even if in a different dimension—that of the independent state.

In these circumstances, leaders in Soviet times demonstrated the ability to adapt to change. A whole generation of "aparatčik", re-appointed as "presidents", who survived de-Sovietization, had the chance to establish a solid apparatus around themselves.

The five early independence "presidents" are examples of such situations, being characterized by a similar career: Niyazov (Turkmenistan), Karimov (Uzbekistan) and Nazarbayev (Kazakhstan) were already in power in late Soviet times; Kyrgyzstan's Akayev, who was not really a true politician but the representative of the science academy, was appointed in the early transition times in 1991 soon after independence. Tajikistan's Rahmon, originally a middle rank operator (manager in the cotton industry, the main state asset), was appointed as president of the Supreme Soviet in critical situations and then confirmed through elections after the explosion of civil war, in 1994.[6]

All the presidents share a similar basic Soviet style, demonstrating an undoubted ability. Even if they were taken by surprise by the changes, they were capable of grabbing an opportunity: in these circumstances they evidenced an attitude for a unscrupulous practice of government, even when pragmatic and to some extent oriented to mediation. It is possible to state today, at the completion of a transition cycle, that the tendency to power conservation (aborting de facto any revolutionary tension), combined with pragmatism in international relations, are the most remarkable aspects of this transition period.

It is to some extent understandable, considering the circumstances in which these countries realized their independence: once they took the power, reconverting from provincial party secretaries to presidents of new sovereign countries, the new leaders started a program to mitigate the tensions, both, inside and outside the country. Therefore, it was necessary to resist and extend control over possible oppositions; some of them had waited for decades or centuries for this moment, after generations of cruel Russian and Soviet repression. This is the case of religious leaders, of ethno-national minorities, of liberal oppositions.

After this first phase, the new presidents tried to consolidate their own power, improving political appearance, establishing new parties in order to configure a kind of internal dialectic. Furthermore, the new presidents soon learned to acquire control of the hydrocarbons (HC) and of further national assets, promoting what has been defined as a "patrimonial–authoritarian" regime,[7] inducing further centralization, using the financial resources acquired in this way politically.

In fact, they maintained strong personal control on any level of governance, eventually covering such a "party" with local lobbies and pursuing loyalty in a personalist or clanic way (in the absence of consolidated laws and procedures, which always need time to become efficient). This prefigures a kind of involution, with the political modalities regressing to pure arbitrariness whenever covered by a situation of "fictitious democracy", namely of the instrumental use of laws and procedures. In fact, while not deliberately denying democratic rules, the "presidents" were successful in establishing, or completely creating, a reality that just dissimulates the democratic game. This can be considered an extreme evolution of despotic systems.

[3]Namely based on a commandeer economics, simply extracting and exploiting natural capital, a very raw stage of economics, indeed, with scarce creation of value.

[4]Manifest of such idea is the letter sent to NYT by President Putin, 11 set 2013; Putin Vladimir 2013.

[5]a double standard, liberal (namely respectful of procedures) outside and authoritarian inside; in 1991 all the leaders declared the intention to consider democracy as a target of transition; then, it was widely assumed that the NIS would transform into democracies; Carlson Richard 2013: 127; Dagiev Dagikhudo 2014: 1.

[6]Dagiev Dagikhudo 2014: 216.

[7]Babajanian Babken 2015: 514.

13.1.3 From Personalization to Clan Politics

The increasing complexity of state governance, the ongoing diversification and the necessity to delegate some power—regionally, institutionally, economically—has induced further adaptations, from personalized power to a kind of oligarchy connected with the president and his government. Simply, the president cannot do anything alone, and some of the decisions in practice go outside of his power, configuring some more or less extended groups—who consequently have the possibility of "institutionalizing" themselves.

Such "inner circles" sometimes derive from the residual of the previous nomenklatura that had found continuity in the new regime. Sometimes they consist of the representatives of some regions and of the different "hard powers" (army, corporations, state monopolies, further institutions); often they are just relatives of the president who begin to manage the wealthy resources (HC, minerals, cotton, hydroelectric power) and appear as an unexpected "heritage" of the previously deceased system.

Such an adaptation means the emergence of a kind of clan politics, with different oligarchies engaging in a struggle, which sometimes escape [went out of] the control of the same president. It is possibly the most salient characteristic of post-Soviet times, dating back to the previous system: the "klanovost" (namely the tendency to organize in "clans") was already well known, especially by political police, although not widely recognized (being considered a kind of taboo).

It was, and still is, a further founding element of local political culture[8]; it can be realized in different forms of lobbyism, possibly echoing pre-modern tribalism (with the "power" intended as "possession", so legitimizing by itself). It can also manifest in a simply personalistic and paternalistic style of government, eventually manifesting the propensity for the formation of a new dynasty. In fact, the same power occasionally claims some particular ascendancy, and even some religious legitimization (even when the Islamic tradition does not really prescribe definitive criteria, with concern, among others, to such questions as legitimization, succession and government methods).

Furthermore it signifies a typical way of managing the power and any competition for it: it means a confusion of negotiations, menaces and reciprocal blackmails, even when not exhibited, with the final goal being to achieve a kind of absolute authority, namely the perfect arbitrariness. This occurs in order not to present an image of weakness, not giving pretexts to the outside (and foreign) enemies, considered per se as a threat (as is usual for previous Soviet habits).

[8]Ro'i Yaacov 1991.

It finally results in a sequence of rituals (evidencing obedience) and rhetoric, not excluding the use of repressive tools, but indeed rarely degenerating into violent struggles. In essence, it is a kind of police-state combined with welfare propaganda, pursuing formal consensus and persuasion, evidencing typical ambiguity—"good" or "bad", however, the power is without alternative; it penetrates deep into the economies and the society, spreading an ambiguous sentiment of loyalty and a mix of fear and consensus.

13.1.4 The Modalities of Power

Evidently, repression, threat and violence (the instinctive modalities of despotic power) are not enough to rule a complicated and articulated society, based on a modern economy, commodity production, investments and international commerce (the necessary way to produce profits for both the elite and the population).

In fact, this same necessity exerts itself as a self-limiting function for the powers. Repression usually maintains a character of "selective" persecution, relying on well-trained and efficient police (another heritage of the previous system), deeply infiltrated into the society. This is an important distinguishing feature, considering that the fine line between selective and mass repression (indiscriminate repression) usually means that such despotic powers become irreversible.[9]

This would actually be an authentic point of no return: a threshold well recognizable by international organizations (public or private) dealing with local governments: the new presidents need to maintain relations with them, if only to be able to continue to practise business and to export HC without disturbances (e.g. without incurring "democraticity" reports in NGOs, customer associations' accusations, media and journalist investigations).

Actually, the situation in late Soviet times in CA (and elsewhere in FSU in the 1970s) was already characterized by such an evolution of the power (just formal and fictive institutional mechanisms, characterized by arbitrariness substituting the principles of the "state of the law"), with the self-imposition of a clan-like "nomenklatura". This system was usually defined as an "oligarchy" by its criticizers, evidencing the formation of a self-referential power structure which had definitively become irresponsible towards the rest of the society. Such situations may be considered the con-

[9]Because obviously the repression indiscriminately hitting individuals or groups goes out of the control of the security apparatus easily, then triggering such effects as diffused hate and retaliation-vindictive intentions; it risks any time to degenerate in a purely totalitarian one, losing any capability of reversing in a less autocratic or even democratic power, and losing any possibility for dialogue (and development) capability.

tinuation of the kind of "tacit social contract" between institutions and the local society, characterizing the Soviet past (especially the late decades).[10]

This situation of compromise signified, and signifies even today, the renunciation of any pretensions of a more evolved society: a sign of ideological failure, with conformism simply maintaining the significance of a mask for the power of the new "president's party"—substituting in fact the "socialist" one.

13.1.5 Consequent Social Evolutions: The New Oligarch Class

In transition times, and following the consolidation of new leaderships, it is possible to register—as expected—an acceleration in social dynamics and the emergence of new classes.

In particular, it is possible to observe the formation of two new emerging groups, one deriving from the evolution and further consolidation of the old-style nomenclature around the new president, and the other from the new, sometimes chaotic, development of a new popular economics.

The upper class is made up of new oligarchs, military, bureaucrats and "aparatčik", members of the winning clan, or simply relatives of the president. It also includes diplomats, high-ranking managers, academicians and intellectuals, and even conformist religious leaders blandished by the new power. So, they can maintain their status, eventually acquiring new positions. Likewise, for technical staff, public career professionals and middle-rank officials, who usually took the place of the departed "vozvraščentsy", substituting them in assets and industrial management.

This new elite is often characterized by a multi-cultural attitude, since it derives from the Soviet melting pot (the s.c. "slijanje", namely the politics of ethnic confusion elaborated in late Soviet times especially in non-European provinces). Therefore it is possibly inclined towards internationalization originally, namely to the globalization of relations of any kind that characterizes independence times.[11]

This evolution is also urbanistically symbolized by the emergence in the major cities of new international central business districts (CBDs) and by the emergence of signs typical of new "global cities". What, until a few years ago, were boring provincial towns on the periphery of the empire had suddenly transformed into capital cities of new and rich independent "oil republics". These are the places where foreign people increasingly settle down, making them the headquarters of diplomacy, international organizations, multinational agencies, consulting firms (who sometimes collaborated in public treasury services during transition times),[12] oil corporations, global civil society (GCS) institutions and NGOs—obviously also for tourists, businessmen and visitors of any kind.

This process is facilitated by the fact that pluri-national and multi-linguistic CA societies were already prepared to some extent for international openness (the opposite happened in conservative circles, e.g. in Russia and in other post-Soviet or post-colonial countries in recent history).

13.1.6 The New "Parallel" Society

The other emerging new class derives from the popular economics (namely independently from public monopolies). It formed spontaneously in times of de-Sovietization and privatization, namely in the early stages of independence and in further liberalization circumstances—for example, internationalization of activities, trade and traffic development, migration remittances and the return of migrants working abroad.

Often this new class of "biznismeny" (small entrepreneurs, sometimes defined popularly "tout court" as "špekuljanti") had the chance to make easy money in the context of the informal economy, taking advantage of a vacuum in regulations for a set of activities, as is typical indeed of any transition. This is especially the case of the "wild" privatization of both small activities and "kombinat" assets, often in poor conditions which occurred in transition times.

It is the case of buildings and real estates, of machineries and instruments once properties of the collective factory; and it is the case of profitable assets such as mines and mineral productions easy to trade on international markets; so also for entire industrial sectors processing raw material, agro-food and fibres (cotton), as well energy sources, like water and HC. The same is true for any industry—usually characterized by low productivity since obsolete, deriving from a history of Soviet stagnation—where the managers offered the possibility of continuing production and of maintaining jobs eventually to the detriment of any residual corporate efficiency.

13.1.7 New Elites: Winners and Losers

Neither the new upper class nor the new "bizinismeny" are characteristics of the typical new "civil society" (which must be considered the real objective to obtain long term stability. The former group does not have these characteris-

[10] Black et al. 1991: 290; see as well Megoran Nick 2017: 78 fs.

[11] It is to be considered that about 40% of the Soviet population was already possibly ethnically mixed; Klüter Helmut 1993.

[12] Haghayeghi Mehrdad 1997: 337; including consulting and auditing companies like D&T, PWC, E&Y, that supervised the early attempts of privatization process in these countries; then auditing not just for HC, but also for further commodities; Heathershaw John 2013: 187.

tics because it depends directly on the power, the latter does not for opposite reasons, namely because "separated" and not participating in institutional life (as already considered in the previous chapter).

Indeed, they would possibly try to adhere to a "clan"—obviously in an informal manner—occasionally exerting pressure in order to pursue some interest, namely a kind of parallel society, "de facto" hiding themselves from the institutions (e.g. evading taxes and eluding regulations).

Nevertheless, they have to be considered the "winners" of the transition; consequently, they are trying to strengthen their position and legitimize their power (that they acquired fundamentally in an arbitrary way, incidentally, as an unexpected "heritage" of the previous regime), elaborating new ideological codes.

The new power is aware of this, because it needs the support of these two classes: neither the simple pragmatic approach (the fact it has demonstrated itself efficient, ensuring stability and also some prosperity) nor the repressive mode is enough to ensure the continuity of the government (beyond the supposed early emergence period).

In fact, the systemic transformation usually produces new societies continuously, new "winners" and new "losers", potential antagonists, that prove to be difficult to control definitively or which are eventually "paid" (politically) with HC revenues.

The new power needs both to elaborate a new ideology and to continue to reproduce consensus, remodelling the propaganda, eventually using updated "desinformacija" practices, applying well-used modalities deriving from the Soviet police traditions combined with current techniques.

13.1.8 The Personalization of the Power Struggle

Political evolution leads unavoidably to a kind of personalization of the power that seems to characterize simultaneously each NIS: a generation of local party secretaries, with more or less similar backgrounds and personal data, had the chance to accumulate unlimited power (becoming the head of the new sovereign state).

This fact produces a set of consequences, inducing a process of concentration of power, which—not being limited by procedures and rules—soon degenerates into the further personalization of governance, and eventually to a cult of personality (the extreme phase of this process; Fig. 13.1).[13]

Such "personalization" affects all aspects of political life; above all it induces, as a direct consequence, the personalization of the political struggle and of the opposition: often the only real opposition is represented by a president's personal enemies, often having become "enemies" just due to contingent questions, confusing personal or familiar issues with relevant political motivations.

The most typical case is that of Kazakhstan, with the former President Nazarbayev accumulating power but also made to articulate (and delegate) many of these functions in an increasingly complicated (and internationally integrated) system. In fact the same former president, who recently resigned, has retained some top national functions, appearing to create some confusion of powers (see later).

The consequences can be dramatic. For instance, this happened with the president's former collaborator and Prime Minister Akezhan Kazhegeldin, 1997–1999, who suddenly became his major opposition and who was finally accused of organizing a kind of insurgence.

Another example is the Rome kidnapping affairs (autumn 2013), with the Kazakh authorities allegedly instigating Italian police to arrest (actually kidnap) the wife and daughter of a high-ranking official and banker (supposedly managing public money bank accounts), who occasionally self-declared as a political opposer (indeed an affair apparently settled without major consequences).[14]

Another example is that of Altynbek Sarsenbayev, a former government official, who suddenly became a political opposer, where in 2006 officers of the country's security service were accused and convicted of his assassination: even if not directly instigated by the authorities, the assassination appeared as an "apparatus" manoeuvre.

The situation is even worse in other CA countries, where the political struggle usually provoked the police-individual persecution level, developing on a wider scale and often degenerating into mass repression or even massacres (a threshold that, as said, per definition means a point of no return for a despotic autocracy). These comprehend the disappearance and kidnapping of opposers, widespread practices of torture, arbitrary imprisonment and the whole repertoire of police violence repression.

Turkmenistan, from where indeed it is even now difficult to have reliable information, and Uzbekistan have recently experienced a change in regime, making it possible to imagine evolution towards a more advanced politics (that in the latter case has been defined also "Uzbek spring").

Tajikistan (after the pacification) and Kyrgyzstan show signs of relaxation, but the material questions (especially the common involvement in Fergana valley and in further border questions, that represent open issues) can at any time trigger escalation, then feed back into these countries and predispose them for an authoritarian regression.[15]

[13] Kudaibergenova Diana T. 2015: 447.

[14] https://www.theguardian.com/world/2013/jun/02/kazakh-oligarch-accuses-president-kidnapping, accessed at 29.06.2018; Scherer Steve 2013.

[15] Megoran Nick 2017.

Fig. 13.1 Tajikistan, Turzunzoda, 2017, ubiquitous poster President Rahmon. (Photo: Igor Jelen)

13.1.9 Further Consequences of Power Concentration

It must be considered that the accentuation of despotic characteristics, with the power structure becoming irreversible, always represents an incumbent risk. The new presidents demonstrate their capability, with great ease, of outflanking constitutional rules, either formally or informally (naïvely formulated in the early independency times), and also the eventual international pressure (elaborating a kind of double standard, liberal internationally, and illiberal internally).

Opposition and resistance occur in an occasional manner, rather than as in civic form (intended as rationale argumentation and debating). It is possible to register two kinds of popular reaction, respectively, practising indifference and eluding rules or possibly protesting with violence and confusion; the latter attitude leads to a reaction from the power structure in a kind of vicious circle, with the further concentration of power moving into the president's hands.

Then, some kind of extremization can be expected, with the risk of diffusion of subversive groups and the formation of secret associations inclined to conspiracy, possibly getting in contact with other extremist groups or with foreign powers interested in destabilizing the area, with a subsequent escalation of reciprocal accusations. This is the case of a long list of major and minor events characterizing all of the CA countries, and above all of the Tajik Civil War (while both post-Soviet rebellions in Kyrgyzstan, in 2005 and 2010, represent exceptions, which fortunately did not degenerate).

It is the case of the supposed assassination attempts on Uzbek former President Karimov (with a series of auto-bomb attacks) and of that of his Turkmen neighbour Niyazov in 1999, in specular manner, followed by reciprocal accusations.[16]

Such events immediately became the motivation for a Soviet-style "purge", followed by a sequence of reproaches, exile for opponents as well as former collaborators, firing and dismissals of public officials, and further large-scale repressive measures. It degenerated undeniably into a kind of pretextual police intervention, with a general limitation of freedom. There were also further restrictions on information media, travelling abroad, and "tout court" the blockade of the borders.[17]

A further effect of the personalist/clanic drift is the parallel involvement of family, with political reality often coinciding "tout court" (even tragically) with a "dynasty game". This usually means contrapositions and struggles among relatives, brothers-in-law betraying the older generation, daughters and sisters rebelling against "patriarchal" order, and similar issues, with family relations completely assimilated to state-political affairs.[18]

A further, often more relevant effect of such attitudes is the personalization of international politics. This means the emergence of inter-presidential rivalries becoming entirely the country's official politics, generating tensions against the neighbouring states (as happened on several occasions in the

[16] Hanks Reuel R. 2016.

[17] Megoran Nick 2017.

[18] World Report 2015: Uzbekistan, Events of 2014, https://www.hrw.org/world-report/2015/country-chapters/uzbekistan, accessed 23.4.2018.

Fergana three-border area region and, in 2002–2004, between Turkmenistan and Uzbekistan).[19]

The leaders reciprocally accuse and reproach each other of intrigues, or even support corresponding enemies, seeking retaliation "transversally", manoeuvring secret services, persecuting the respective national minorities or adopting economically obstructive measures (e.g. border traffic blockades). In this way international politics risks becoming another instrument the president uses to pursue personal targets or to solve internal problems (often indeed deriving just from personal or clan struggles).

13.1.10 Succession Rules

Succession issues represent other dangerous consequences in the personalization of power, and exert an obvious concern for future stability; this also considering that the "first presidents" do not (comprehensibly) elaborate rules concerning their future substitution, and that—these countries being structured as brand new institutions—usually there are no recognized procedures for regulating that discontinuity (to be considered physiological for any political system).

It is a complicated question (a kind of institutional time-bomb) that involves, besides functional questions, also legitimization criteria (both, internal and internationally) and popular sensitivity. All this obviously alongside the existence of reliable successors inside the family group, of sons or daughters, or of other relatives (considering the persistence of a dynastic schema as a possible legitimization device for the succession).

It particularly represents a critical aspect of CA countries: none of the existing political-cultural patterns may be taken into consideration to definitively regularize the succession right of a political leader. In fact, neither the Soviet standard—with changes orchestrated secretly in a "politburo"—nor the Islamic law, nor the traditional one (e.g. the archaic "warrior democracy" and tribal-family habit), predicts the succession criterion in a definitive manner.

Among such criteria, the simply "right" of the strongest must be considered (a kind of pragmatic attitude, typical indeed of the archaic Turchik as well of any ancient culture); then, several family succession criteria, including the right of primogeniture (eventually male) or the subdivision of power among sons. Then, we must consider the organization of elections, on a different level, that may easily degenerate into something fictitious in these circumstances, eventually reflecting some backstage competition. Another criterion is that of the direct nomination—even when dissimulated—of the president by a "hard power", like clans representing the secret police, the army or some economic monopoly.

This fact may be considered even more critical considering the similar age of the current leaders, namely the risk that the generational passage—or possibly the premature disappearance or retirement of some of the presidents—might occur in the same moment, possibly inducing a further shock-wave for the entire area. The personal accumulation of wealth and power, suitable to be simply inherited by anyone, represents by itself a temptation for any kind of pretenders, either inside or outside the apparatus.

13.1.11 The Cases

The description of the elite, the biographies of the current leaders, can be useful to understand what can be considered as the most effective sources of power (namely the instruments to produce power, for example, control of local elites, family ties, military or economic hard power). This in relation to institutional–juridical forms, to internal procedures and constitutional architectures: it is possible to say that succession will represent the true bench test for long-term stability of CA countries.

If the succession is to be considered a potential shock, in these case the risk is even higher because of the centralized leaderships (this fact, indeed, namely the possibility to codify succession procedures, can be considered a definitive advantage of state-of-law democracies); the question represents a source of preoccupation, inducing general prudence by the players (especially international investors), and in the economies in general.

The first cases of post-Soviet successions occurred in Kyrgyzstan (Akayev 2005 and Bakiyev 2010), but they may be considered as exceptions, since they represent cases of application of procedures in a tendentially democratic country, with a popular vote ratification. This is the case even when both have been characterized by uncertainties, street protestations and considerable use of violence.

The cases of presidential successions in Turkmenistan (2006) and Uzbekistan (2016), and possibly currently in Kazakhstan, might signify precedents in this matter; they occurred on the base of the apparent application of habits rather similar to that of Soviet origin.

The Turkmen case is particularly significant; the sudden death of Turkmen Niyazov in 2006 created a vacuum, with Deputy Prime Minister Berdimuhamedov (considered his "dauphin") becoming the interim president, and then elected with a landslide victory in the special presidential election in 2007. He was reaffirmed in 2012, with a nearly "soviet" style consensus of 97%.

[19] https://www.hrw.org/world-report/2014/country-chapters/turkmenistan, accessed 23.4.2018, World Report 2014: Turkmenistan, Events of 2013; see as well Megoran Nick 2017.

This case represented a precedent for the other CA countries, possibly approaching the same dilemma—the succession schema—which in that case was solved by applying the usual Soviet habit, with the departed president followed by his first follower in the party's ranking hierarchy.[20]

In fact, such customs limit uncertainty in transition methods, reducing the "right" of nomination to a small number of "insiders", usually members of the group; in the cited cases the deputy executive president was appointed, after a short period of secret politburo-level consultations, presumably consisting in a series of non-official negotiations among the different "clans".

Such negotiations were presumably influenced by outside powers, as is today the case of SCO (see later), the main guarantor of regional stability; it possibly signifies a mitigation of the scene, normalizing the passage—since in this moment destabilization is not in the interest of anyone, having the great powers China and Russia as neighbours, which have evidently reached a basic equilibrium in the wider geostrategic scenario.

In these cases, it seems that the succession is motivated at first by a need for continuity and the conservation of balance, both inside and outside the country. Consequently, the succession has to be supported by the most powerful lobbies, namely usually those ones relying on financial and economic assets (construction, HC or extraction industry, cotton or water monopoly).[21] The different cases that happened in transition phase evidence that the "dauphin" has usually been cooptated into the apparatus for a long time and assimilated to power rituals, therefore prepared in some sense by the "hard powers".

Such successions procedures were not really transparent or regulated, but they proved to be effective (as happened in Turkmenistan, Uzbekistan and eventually in Kazakhstan, in "controlled" Soviet-like circumstances).

13.1.12 Common Character of the Constitutional and Institutional Politics

All the CA republics promulgated the constitution soon after independence, with a characteristic urgency, as if considering the new constitution, by itself, an instrument for making change irreversible, with the preoccupation of covering the vacuum that the disappearance of the FSU had provoked unexpectedly. Such constitutions appear as technically well drawn up, almost perfect in their formalism, prescribing a juridical elaborated division of powers and responsibilities, elegant in their definitions of rights, and characterized by a peculiar completeness—to be envied by many advanced countries that have constitution papers written centuries ago and that have remained almost untouched since those times.

It is usually represented as a "book", possibly resembling the symbolism of a "sacred" document, again, mixing religious and secular elements (Fig. 13.2).

However, this "sacredness" is usually poorly respected, evidencing a gap between written prescriptions and practices, and possibly between the declarations and the true intentions of the governments.

In fact, the scarce respect of formal articulation, of roles and responsibilities (above all of constitutional laws), seems to represent an element characterizing each CA elite, actually a severe defect in the local politics. It derives from a general problem of personalization, thus centralization and "clanification", of power to the detriment of written law. It represents a severe obstacle for a general process of structuration ("depersonalization") of politics; the power practice should evolve independently from a specific person, namely from the "thaumaturgy" of the "founding father", or of the "first president", a figure evidently destined to disappear in the effectiveness of the state.[22]

13.2 Uzbekistan

13.2.1 Institutions and Political System

The current Uzbek state is the successor of the prestigious CA civilizations, reverberating today in a specific regional articulation; it comprehends the CA metropolis, Tashkent, the prestigious ancient cities Bukhara and Samarqand, the Fergana valley and the Khorazem (roughly corresponding to Karakalpakstan AO) region. All these represent also strong identities and are characterized by bi- or pluri-linguistic (or pluri-national) aspects, considering the presence of relevant minorities, especially the Tajik-speaking population.

The "material constitution"—namely the political practice—evidences a preponderant role of the presidential elite, claiming often exceptional powers, in order to face terroristic threat; much tensions have contaminated

[20] Article 61 of Turkmenistan Constitution: "The President may not transfer her or his executive powers to other organs or officials, except for the powers enumerated in parts 2, 9, and 11 of Article 57 of the Constitution, which may be transferred to the Chair of Parliament. If the President, for some reason, is not capable of meeting her or his obligations, until the election of a new President, her or his powers are transferred to the Chair of Parliament. In such a situation, a presidential election should be conducted no later than two months from the day of transfer of powers to the Chair of Parliament. A person meeting the obligations of the President may not be a candidate in the presidential election."

[21] Kudaibergenova Diana T. 2015: 447.

[22] Babajanian Babken 2015: 514; a political style defined as "patrimonial-authoritarian", with clan and lobbies, namely "informal power arrangements" forming in an un-transparent manner, connecting members of associations, hard powers, corporation, social or regional group finding occasionally a common interest, trying to condition the "visible" politics.

Fig. 13.2 Tajikistan, 2017, poster representing the celebration of the constitution as a "book". (Photo: Igor Jelen)

by contiguous Tajikistan, through porous Fergana borderline, that has been the centre of several tensions and outburst of conflicts; some observers indeed think the transition instabilities, and the consequent impossibility of establishing a stable order, are consequent to the same incapability of the government to periodically face these tensions, namely to the disproportionate reaction (often violent) of the power, inducing in-it-self the radicalization of the positions.

Uzbekistan is the cultural and political "hub" of the area, representing the example towards other countries are continuously looking; this not just for its geopolitical setting but also because it represents a "normal" situation, not biased by "easy money" (like rentier states Kazakhstan and Turkmenistan), facing the "normal" problems a country has to face daily; therefore its situation is even more significant for the destiny (evolution) of whole territory.

In general, it is to observe a long-term tendency of restricting the pluralism,[23] as well the freedom of individuals and of social players, especially NGOs, both, rooted or of foreign origins; international organizations define Uzbekistan as "an authoritarian state with limited civil rights".[24]

Such policy seems currently to evidence a turn, with some author speaking even about an "Uzbek spring", characterized by more relaxed policies; but it needs time to evaluate such change. Possibly it coincides with an attempt, made by the new president, of "normalizing" the inside situation—as indeed usual for new leaders, that have some chances to restart some politics at the beginning of their mandate. It is impossible to say, to which extent it is a true new politics or just a manoeuvre.[25]

13.2.2 Institutions and Policies

Uzbekistan is a presidential republic that declared its independence from the Soviet Union on 1 September 1991 (day currently celebrating as a national holiday). The latest constitution was adopted on 8 December 1992—the first since the previous Soviet Constitution of 1978—but it was often amended; it establishes the country as a republic dominated by a strong executive.

The constitution consists of six sections that include 26 chapters and 128 articles. On its geographical-institutional structure, the Republic of Uzbekistan is a unitary state; its administrative repartition relies in 12 provinces (Viloyatlar), one autonomous republic ("Avtonom Respublikasi" of Karakalpakstan) and one city (Shahar).[26] The relations between the Republic of Uzbekistan and the Republic of Karakalpakstan are regulated with treaties and agreements

[23] Fazendeiro Bernardo Teles 2015: 489; Warf Barney 2011: 14.

[24] Lillis Joanna, 3 Oct 2017; World Report 2015: Uzbekistan, Events of 2014, https://www.hrw.org/world-report/2015/country-chapters/uzbekistan, accessed 23.4.2018; sources report that occasionally repression use the forced internment in psychiatric hospitals, as happened to journalist in Uzbekistan, evoking Soviet habits.

[25] The progressive integration in international circuits, and especially current tourist boom, make more sensitive the local societies to stability and international reputation, especially in Uzbekistan; see "Meridiani" 2019.

[26] https://www.un.int/uzbekistan/uzbekistan/constitution-republic-uzbekistan/, accessed at 21.11.2017.

reached within the framework of the Constitution of the Republic of Uzbekistan.[27]

In May 2017, the constitution was amended for the last time, with a vote of the Senate about the terms of office and the requirements for candidates for judges. Now the Judges of the Constitutional Court are not allowed to be elected for two times, the age limits to be elected is increased from 30 to 35 years old and the age limit is 70 years old. Another amendment allows the Cabinet of Ministers and the Human Rights Commissioner of the Supreme Assembly (Oliy Majlis) of the Republic of Uzbekistan (ombudsman) to submit questions to the Constitutional Court.[28]

The procedure to amend the constitution is the following: the amendment is proposed by Supreme Assembly or by Referendum and required a two-thirds majority vote of both houses of parliament or a passage in a referendum.

The president is directly elected by an absolute majority in the election with a second round if needed, for a five years term; he can also be elected for a second term mandate. Shavkat Mirziyoyev won the last election on 4 December 2016. Previously, he was the interim president, being appointed after the death of the long-time President Islom Karimov, which occurred on 2 September 2016.

The cabinet minister is appointed by the president with an approval of the Senate chamber. The prime minister is nominated by the majority party who won the election; he is appointed along with deputy ministers by the president. Now Prime Minister is Abdulla Aripov. The results of the last election in December 2016 were the following: Miziyoyev (Liberal Democratic Party of Uzbekistan or O'zbekiston Liberal-Demokratik Partiyasi) got 88.6% of votes; Ketmonov (People's Democratic Party of Uzbekistan or Xalq Demokratik Partiyas) 3.7% and Otamuratov (National Revival Democratic Party of Uzbekistan or O'zbekiston Milliy Tiklanish Demokratik Partiyasi) 2.4%.

The Oliy Majlis is formed by Senate and the Legislative Chamber. The Senate has 100 members: 84 members elected indirectly by the regional governing council, 16 appointed by the president. They all have a five-year term mandate. The Legislative Chamber (Qonunchilik Palatasi) has 150 seats: 135 members directly elected in single-seat constituencies by absolute majority vote (with a second round if needed to reach the majority) and 15 indirectly elected by the Ecological Movement of Uzbekistan. The parliamentarians have a five-year term mandate.[29]

The main parties represented in Parliament are Ecological Movement of Uzbekistan[30] whose leader is Boriy Alikanov. The party was found in 2008 and registered to the Ministry of Justice as a political party on 20 September 2008. In 2008, a revised electoral law assigned 15 seats to this party. One legislator was elected from each territorial subdivision of Uzbekistan plus one member from the Executive Committee of the Central Council of the Ecological Movement. Delegates to the congress were elected in equal numbers at the conferences of each of the territorial branches of the Ecological Movement.[31]

About this, it is to consider a recent change; in November 2017 President Mirzoyev decided accordingly to his Action Strategy[32] to develop the political participation in his country and to abolish this rule in order to improve the direct election of all the parties.[33]

Another party that has 20 seats in Parliament is the Social Democratic Party of Uzbekistan. The Party was found in 1995, and it can be described to have a centre-left political position. The Uzbekistan Liberal Democratic Party won at the last elections 55 seats. Its leader is Abdulla Aripov, and it is considered a centre-right Party.

The Liber Party won the last election and is the party of the president in charge. The Uzbekistan National Revival Democratic Party has 35 representatives in the Legislation Chamber. It is mainly formed by intellectual and has a large proportion of female members. The party strongly opposes to the Russian influence in the region and desired a revival of Uzbek culture. There is also a representative of the past communist party; it is the People's Democratic Party of Uzbekistan, who is founded in 1991 after the independence. It has now just 27 seats; it is the main party, but—as observ-

[27]The Republic of Karakalpakstan is located in the northeast of Uzbekistan in the Southern shore of the Aral Sea; it is considered as an autonomous republic inside Uzbekistan with its own constitution, emblem and flag; the relationship between the two republics are based on treaty. http://www.uzbekembassy.org/e/investment_guide_by_karakalpakstan/ accessed at 29.06.2018.

[28]https://en.trend.az/casia/uzbekistan/2759942.html, accessed 9.5.2018.

[29]https://www.cia.gov/library/publications/the-world-factbook/geos/uz.html, accessed 9.5.2018.

[30]http://uza.uz/en/politics/ecological-movement-holds-conference-on-elections-to-parliament-27.12.2009-1160, accessed at 10.04.2018; the new amendments on the law of elections in 2008 provided such place for the party; the law reserved 15 seats for the Ecological Movement of Uzbekistan; such amendment was abolished by Mirziyoyev in order to put the countries parliamentary system in line with the OSCE 1990 Copenhagen's document which states that every member of a country parliament need to be directly elected, https://silkroadstudies.org/resources/pdf/SilkRoadPapers/1803-Bowyer-Uzbekistan.pdf accessed at 01.07.2018.

[31]https://en.wikipedia.org/wiki/Ecological_Movement_of_Uzbekistan, accessed 9.5.2018.

[32]On 7 February 2017, Mirziyoyev approved a major programme, the "Action Strategy on Five Priority Areas of the Country's Development for 2017–2021" (hereafter known as the "Action Strategy"); one of the five priority areas of the Action Strategy is improving the system of state and public construction; Bowyer Anthony C., 2018.

[33]Bowyer Anthony C., 2018.

ers' main opinion—it doesn't create a stronger opposition to the position of President Mirziyoyev.

There are two highest courts in Uzbekistan: the Supreme Court that consists of 67 judges divided for a specific branch of law: administrative, civil, criminal, economic and military. The second highest court is the Constitutional Court formed by seven judges. The president nominated the judges for these two courts and they are confirmed by Oliy Majlis. A judge initially is appointed for five years, subsequent to ten years and then can be subject to a lifetime term mandate. The subordinate courts are: regional, district, city, town court and economic courts.[34]

13.2.3 Current Politics

One of the most powerful institutions of the Uzbekistan is the National Security Services (the Uzbek KGB-like agency). The NSS is headed by a chairman who is answerable directly to the president; it deals with a broad range of national security questions, including corruption, organized crime and narcotics. The sources report that corruption among law enforcement personnel remained a problem. Another big problem for security forces is impunity, and officials responsible for abuses were rarely punished.[35]

The government continued to use an estimated 12,000 local neighbourhood committees as a source of information on potential extremists. Committees functioned as a link between local society, government and law enforcement.[36] Each neighbourhood committee assigns a "posbon"[37] (mahalla-neighbourhood guardian) whose job is to ensure public order and maintain a proper moral climate in the neighbourhood.

In practice, currently, this means preventing young persons in the neighbourhood from joining extremist Islamic groups. Neighbourhood committees also frequently identified by police those residents who appeared suspicious and, working with local NSS representatives, reportedly paid particular attention to recently amnestied prisoners and the families of individuals jailed for alleged extremism.[38] In these ways, the authorities can control in practice the whole Uzbek population.[39]

Despite the democratic constitution, Uzbekistan is strongly criticized by IC and especially by the United States and the EU for government abuses.

At the moment of its installation in 2016 as president, the former Prime Minister Shavkat Mirziyoyev failed to push in any meaningful improvements in Uzbekistan's minimal human rights record. Sources report that thousands of people are imprisoned on politically motivated charges and that torture is endemic in the criminal justice system.[40]

It seems that really the security forces are a problem, becoming increasingly self-referential organizations (as said, the reforms inaugurated by "Uzbek spring" are indeed too recent to be appropriately evaluated, and actually the power recent tradition does not permit excessive optimism).

Authorities continue to crackdown on civil society activists, opposition members and journalists. This with special concern for religious activists, either Muslims or Christians; in general, who practises religion outside strict state controls is persecuted, while basic freedom of cult and confessional expression is severely limited.[41] Authorities still deny justice for the 2005 Andijan massacre, in which government forces shot and killed hundreds of protesters (defined as extremist, suspected of organizing subversive actions).[42]

The situation in fact of civil, gender and sexual right—a taboo in as well in several post-Soviet countries—is particulary critical. LGBT people face deep-rooted homophobia and discrimination; in Uzbekistan, the sexual relationship among persons of the same gender is criminalized with a maximum prison sentence of three years.[43]

Article 18 of the Uzbekistan Constitution assured that all people shall have equal rights and freedom without discrimination, but it remains often only on the chart.[44] Authorities maintain a strict control of the population; thousands of people remain in jail due to political motivation charges.

Further critical aspects are related to economic monopoly, namely to country vital assets. The state every year forced

[34] https://www.cia.gov/library/publications/the-world-factbook/geos/uz.html, accessed 9.5.2018.

[35] The Mahalla is a traditional local government institution http://www.uzbekembassy.in/the-mahalla-a-unique-civil-society-institution-in-uzbekistan/ accessed at 10.04.2018

[36] Rasanayagam Johan 2014: 11ss.

[37] https://www.hrw.org/reports/2003/uzbekistan0903/5.htm, accessed 9.5.2018, Role and functions of mahalla committees.

[38] https://www.globalsecurity.org/military/world/centralasia/uzbek-inoyatov.htm, accessed 9.5.2018.

[39] As a form of local government they are also strictly connected with the Security System and other State services http://enews.fergananews.com/articles/2991 accessed at 10.04.2018

[40] Megoran Nick 2017.

[41] https://www.hrw.org/europe/central-asia/uzbekistan, accessed at 22.11.2017.

[42] Megoran Nick 2017: 34, citing HRW.

[43] Human Rights Watch, Report on Uzbekistan, January 2017, https://www.hrw.org/sites/default/files/uzbekistan_0.pdf, accessed at 22.11.2017.

[44] Article 18 Uzbekistan Constitution: "All citizens of the Republic of Uzbekistan shall have equal rights and freedoms, and shall be equal before the law, without discrimination by sex, race, nationality, language, religion, social origin, convictions, individual and social status. Any privileges may be granted solely by the law and shall conform to the principles of social justice."

one million people to harvest cotton under harsh condition. Only in 2015 and 2016, after a global campaign, authorities decided not to mobilized also children to harvest cotton. Instead, during these years, the number of adults forced labour is increased. Doctors, nurses, public servants are forced to harvest cotton under the threat of loss of work or welfare benefit and if they denounce such condition, they can be imprisoned.[45]

Sources report that police authorities and security forces use torture and other illegal treatment to people to do a confession. The EU Human Rights Court condemns Uzbekistan for torture and ill-treatment to prisoner that seems to be a typical procedure of security forces and prison staff. In July 2015, the UN Human Rights Committee urged the Uzbekistani authorities to ensure "that the prohibition of forced confessions and the inadmissibility of torture-tainted evidence are effectively enforced in practice by law enforcement officers and judges".[46]

International NGOs report that security forces usually persecute the family of the prisoners in order to force them to confess. They frequently beat male and female relatives of suspects, threaten them with rape or the murder of their children, sexually humiliate them, form committees to expel them from their homes and force employers to dismiss them from their jobs. They do not hesitate to resort to other physical and psychological abuse amounting to torture or other ill-treatment in order to trace and secure the conviction of a suspect.[47] During the last years, Uzbekistan's authorities don't do anything to improve the condition of prisoners and to reduce the use of torture.

Very important is the relationship with Russian's security forces. Instead of the proof of Human Right's abuse, Russia cooperates in a security operation in order to save the good relation with this country. Russian authorities often arrested Uzbekistan refugees in Russia, forcing them to return in the hand of Uzbekistan's Security Forces. When they arrive in their home country they are put in jail where they are tortured to force them to confess, they face unfair trials that result in long prison sentences served in cruel, inhuman and degrading conditions.[48]

[45] Human Rights Watch, Report on Uzbekistan, January 2017, https://www.hrw.org/sites/default/files/uzbekistan_0.pdf, accessed at 22.11.2017.

[46] Concluding observations of the Human Rights Committee, Uzbekistan, UN Doc. CCPR/C/UZB/CO/4, para 14, available at http://tbinternet.ohchr.org/_layouts/treatybodyexternal/Download.aspx?symbolno=CAT%2fC%2fUZB%2fCO%2f4&Lang=en accessed at 22.11.2017.

[47] Amnesty International 2016.

[48] Amnesty International 2016.

13.2.4 The New Leadership

The President is Shavkat Mirziyoyev who won the last election on 4 December 2016 and succeeded the deceased President Karimov, the first President of independent Uzbekistan.

Islom Abduganievich Karimov graduated in Mechanical Engineering and Economy from Central Asia Polytechnic. He became secretary of the Communist Party in Uzbekistan in 1989, and in 1990 he was elected president of the Uzbek Soviet Socialist Republic. When Uzbekistan became independent in 1991, he was elected president of this newly independent country.

Despite the ban in the constitution for a president of more than two mandates in 1995 and 2002, a national referendum extended his mandate. During his government, Uzbekistan became an authoritarian country, scarcely demonstrating to respect Human Rights, using sometime as a pretext the political tensions and the extremist activists. Despite all the lack of democracy, Uzbekistan became, after 11 September 2001, a key allied for NATO and the United States for its strategic position near the border of Afghanistan.[49]

During Karimov presidency, one of the most powerful positions was the Chairman of the National Security Service, Rustam Inoyatov. He was seen as the power behind the throne in Uzbekistan and in fact he represented a key figure in the succession of the same Karimov.

He is the head of influential Tashkent clan, which is allied to the Fergana one (possibly the most powerful in the current clan political map); supposedly their main opponent is the Samarqand clan. These kind of lobbies try continuously to influence politics and economics, reproducing patron-client networks built on regional identities or institutions, manipulating the political game. In fact, Karimov with the help of Inoyatov was also able to create an apparatus of deputies who constantly informed the president about what's going on in the government, in the parliament and in any institutional place.

Islam Karimov died in September 2016 and the cause of death for authorities was a stroke attack. The current President, Shavkat Mirziyoyev, being former prime minister, was able to negotiate with Inoyatov, obtaining finally the nomination as interim president. This happened without any transparent procedure, as a typical Soviet style game, probably after a struggle in the backstage between Mirziyoyev and the former Minister of Finance Rustam Asimov. Finally, Mirziyoyev, the first declared president at

[49] https://www.britannica.com/biography/Islam-Karimov, accessed at 22.11.2017.

interim after the death of the former leader, won the election in December 2016.[50]

13.2.5 Unrests, Attacks and the Recodification of War in Terrorism

The death of Uzbek President Karimov happened in an unexpected manner, creating as a reaction, as typical for Soviet style establishment, a period of silence; it meant possibly a backstage work, with negotiations among clans, and eventually with other powers (probably comprehending as well some representative of neighbouring super-powers).[51]

Indeed, the Uzbek politics—above all the international one—is difficult to be defined, often evidencing incoherencies, periodically disorienting allies and neighbours, and the whole IC, influenced by different non-transparent elements.

In some periods, Uzbekistan is an important partner for western countries, for the United States (as it happened especially in times of Afghan war) and the EU (notwithstanding it poor human rights record); above all it seeks equilibrium towards China and Russia. In these difficult years Karimov demonstrated a certain mediation capability, preventing Uzbekistan from falling into the sphere of influence of the one or the latter.

In the last times, he either demonstrated some preference towards China, integrating in SCO and in SR initiatives, pursuing a more intensive cooperation in infrastructural projects linking Far East with western markets, crossing Uzbek territory. This tendency appears as quite appropriate, since the landlocked country depends on its neighbours, and especially with fast developing China, for many aspects, especially for manufacturing and international trade. For all these reasons, the most probable scenario for the near future will be that of continuity of the power, eventually repressive internally, but pragmatic ("free rider" like) in the international scenario.[52]

The critical points of Uzbekistan are the border area with Tajikistan and Kyrgyzstan and above all the contiguity with Afghan scenario on the southern side that makes the country exposed to some contamination, especially in terms of trafficking: the gangs controlling such traffic demonstrated sometime to be so powerful that they may spread out corruption methods and further crime activities deep inside the Uzbekistan state.

In fact, the Uzbek recent history is characterized by unrests, internal conflicts and subversive movements spread out and periodically escalating, that but—as many argue—have been to some extent alimented by the same state apparatus disproportionate (and inefficient) reaction.

The tensions dates to early transition times (since cited Fergana 1989–1991 clashes),[53] when they originated spontaneously, then escalating in conflicts. Initially Karimov tried to blandish the conservative official representatives of Islam,[54] in order to strengthen its consensus; when he had the impression that Islam had gained much power, aspiring to a new active role, he suddenly changed the attitude, beginning with mass-scale persecutions.

In this way, it caused the radicalization of the positions and the spread of Uzbekistan Islamic Movements (MIU), basing on jihadism; this movement has been founded in 1998 by Tahir Yuldashev and Juma Namangani, acquiring soon a revolutionary significance, with the declared (openly enunciated) target to get the power in order to build up an Islamist state based on Sharia (religious-inspired civil regulations).

In fact, Uzbekistan leadership proved often to be incapable of offering a solution for the problems characterizing the rapidly changing society, and the consequent formation of new emarginated groups; so, the rebel attitude rapidly spread out, especially in rural-overpopulated areas (namely the described "empty kolkhozes), where the impoverishment has been the base for recruitment of new militants.[55]

Later, the oppositions developed a further stage, terrorism-like, changing modalities and eventually assuming initiative, attacking directly security forces; they also targeted directly the president, for example, in the case of auto-bombing in 1999, in Tashkent, with a sequence of explosions making many victims (even when facts suitable of different interpretations).[56]

The government took immediately drastic measures, but as usual hitting indiscriminately as a kind of mass retaliation; it is the case of new laws against the liberty of expression and political participation, of arbitrary imprisonments of supposedly dissidents, of use of torture on prisoners. This provoked the effect of ironically alimenting further the social hate.

So the religious movements remained the only political opposition, with Islamist extremists legitimizing by-theirselves just because they have been recognized as the only true resistance to the "dictator"; therefore, the situation lost any possibility of normalizing.

Such attitude accelerated from 2001, with religious movement pursuing affiliation with Al-Qaeda and then sending militias to resist in Afghanistan against western war coalition. In more recent periods, these groups support and affili-

[50] Article 96 of Uzbekistan constitution prescribes a procedure in case the President is not fit to exercise his duties: "In case the President of the Republic of Uzbekistan fail to exercise his duties, the Chairman of the Senate of the Oliy Majlis of the Republic of Uzbekistan shall be vested with acting duties and powers by holding election of President of the country within three months with strict observance of the Law "On election of the President of the Republic of Uzbekistan"".

[51] Considering the heavily influence induced by SCO trans-continental agreement; CESI, newsletter Sept. 2016.

[52] Parenti Fabio Massimo, Adda Iacopo 2017: 347; Limes n.8/2014.

[53] Elebayeva A. B. 1992.

[54] CESI, newsletter Sept. 2016.

[55] CESI, newsletter Sept. 2016.

[56] Jonson Lena 2006; Hanks Reuel R. 2016.

Fig. 13.3 Administrative map of Turkmenistan. (Source: http://ontheworldmap.com/turkmenistan/large-detailed-political-map-of-turkmenistan.jpg, accessed at 01.07.2018)

ate to DAESH ("Islamic state"), with—sources refer—about four training fields in Syria, with possibly 500 fighters.

After 2005 rebellion of Andijan, the repression became obsessive, making more capillary the social control: the riots resulted in an uncertain number (possibly 800 deaths) of people killed and a number of imprisonment of suspected persons.[57] Charges, detentions and indiscriminate police interventions in this occasion have been heavily criticized by international observers, accusing the government for systemic human rights abuse, requiring independent investigations. In reciprocal way, the Uzbek government accused western powers, in particular the United States and the United Kingdom, of having orchestrated the unrest.[58]

13.3 Turkmenistan

13.3.1 Institutional and Political System

Turkmenistan (Fig. 13.3) declared its independence from the FSU on 27 October 1991; on 18 May 1992, it adopted its first, as independent state, constitution; its legal system follows the civil law system, but it apparently has much of Islamic influence in the law applied. Gurbanguly Berdimuhamedov is the President of Turkmenistan since 14 February 2007; his position represents not only the chief of state but also the head of government; president is elected directly by absolute majority popular vote in two rounds if needed, and its mandate has a seven-year term, without limits for the re-election.[59]

[57] Simple believers, practising religion outside the state control, were considered suspected of disloyalty towards the regime, investigated for terrorism etc.; CESI newsletter sept. 2016.

[58] World Report 2015: Uzbekistan, Events of 2014, https://www.hrw.org/world-report/2015/country-chapters/uzbekistan, accessed 23.4.2018.

[59] Article 70 of Turkmenistan Constitution: "The President of Turkmenistan shall be elected directly by the people of Turkmenistan for a period of 7 years, and shall enter office on taking the oath." https://www.osce.org/odihr/262336?download=true, accessed at 19.05.2018; see Hanks Reuel R. 2016.

The last elections were held on 12 February 2017, while the new election will be held in 2024. Berdimuhamedov was re-elected president with 97.7% of the vote; the other candidate got only 2.3% of the vote.[60]

The legislative branch is formed by a unicameral National Assembly (Mejlis). The National Assembly is formed by 125 parliamentarians directly elected from single-seat constituencies by absolute majority of the vote, members serve for a five-year term mandate. The last parliamentary election was held on 25 March 2018, and the next will be held in 2023. The results of the last parliamentary election were as follows: the Democratic Party of Turkmenistan got 56 seats (the ruling party, it is the successor of the former communist party), the Party of Industrialists and Entrepreneurs 11 seats and lastly Independent candidates had taken 48 seats. [61]

Originally there was another Assembly that in the Soviet Era was called the Supreme Soviet and after the independence became the People's Council (Halk Maslahaty). This assembly shared the power with the National Assembly but a constitutional reform in 2008 revoked the power of the council and it was taken by the National Assembly. These changes were explained by analyst as a way for the president to reduce the opposition of parliament.[62] In 2016, a constitutional committee has drafted some amendments to the constitution to lengthening the presidential term and otherwise enhancing the already-powerful presidency.[63]

The Juridical Branch of Turkmenistan is formed by the Supreme Court of Turkmenistan. It represents the highest court in the country, is formed by 21 judges and a court president, and has jurisdiction on the civil, criminal and military law. The Supreme Judges are appointed by the president for a 5 years mandate.[64] In this way, the president can have the control on the Juridical Supreme Branch through the nomination of all the judges, a procedure that doesn't respect the principle of separation of the powers typical of a "rule of law" condition.

Other subordinate Courts are the High Commercial Court, the appellate courts, provincial, district and city courts. There are also military courts which concern military law. Political parties in Turkmenistan must be authorized by the government in order of doing political activity. In order to bypass the authorization, small opposition parties exist abroad.

If anyone wants to register and compete in the presidential elections, he/she needs to be nominated by a political party or collect 50,000 signatures. In addition, he/she has to be native born, live in Turkmenistan for the past 15 years and have no criminal record. Furthermore, the Law on Political Parties requires a party leader to have over 1000 members, to represent his party in all five regions, reporting all party members' names to the Ministry of Fairness, and allowing the officials of this ministry to attend party meetings. A very strict legislation that doesn't allow to some opposition party to participate in the elections.[65]

OSCE criticized in his report about the last presidential election the absence of separation of power in the Turkmenistan Republic; it highlighted also the absence of the respect of basic fundamental freedom who permits to hold a free election and the lack of a proper political pluralism.[66] Turkmenistan is now considered by Human Rights Watch one of the world's most repressive countries.[67]

The Ruhnama is the self-defined "sacred book" written by former President Niyazov, in which he expressed some statement and his idea about the state. The book talks also about the respect guarantee in Turkmenistan of the international treaty and the human rights.[68] Despite the principles enunciated in this "sacred book", independent observers report that Turkmenistan doesn't respect human rights. Paradoxically also the constitution in some articles states about the protection of Rights and Liberties, indeed as expected,[69] but it is something that remains only on chart.

Furthermore, the government doesn't permit any Human Rights organization to control the real respect of these principles and international groups are not in principle allowed to enter in the Country. There is no freedom of media, the state controls journal, and internet and also foreign journalist often are not allowed to enter in Turkmenistan. The government has destroyed all the private satellite to control the use of the internet. Currently, in this nation, the access to the internet is one of the most expensive and difficult to access

[60] https://www.cia.gov/library/publications/the-world-factbook/geos/tx.html, accessed at 06.01.2018.

[61] http://www.turkmenistan.gov.tm/_eng/?id=3069, accessed at 06.01.2018; all the results are available in the OSCE Election report https://www.osce.org/files/f/documents/2/8/382915_0.pdf.

[62] https://www.rferl.org/a/1104107.html, accessed at 06.01.2018.

[63] https://www.rferl.org/a/1144502.html, accessed at 06.01.2018.

[64] https://www.cia.gov/library/publications/the-world-factbook/geos/tx.html, accessed at 06.01.2018.

[65] https://freedomhouse.org/sites/default/files/NIT2017_Turkmenistan.pdf, accessed at 19.05.2018.

[66] http://www.osce.org/odihr/elections/turkmenistan/291726?download=true, accessed at 06.01.2018

[67] https://www.hrw.org/europe/central-asia/turkmenistan, accessed at 06.01.2018.

[68] Article 6 of Turkmenistan Constitution: "Turkmenistan recognizes the primacy of generally recognized norms of international law, is fully invested with the rights of a subject of the world community, and adheres in its domestic policies to the principles of peaceful coexistence, rejection of the use of force, and non-interference in the internal affairs of other governments."; see as well Decaux Emmanuel 2003.

[69] Article 3 of the Turkmenistan Constitution about Human Rights, Articles 8 and 16 of Turkmenistan Constitution the protection of foreign citizen and stateless, Article 20 about respect of Human Life, Article 21 about criminal repression and legality, Article 108 on the legal process.

due to the strict government control. Human Rights Watch reports also that many freelance journalists are arrested without charge or are victims of assault and violence.[70]

The "dissidents"—as oppositors are usually defined, echoing the previous regime—don't enjoy freedom of movement and their right to travel abroad is strictly limited. It is difficult to count the number of people imprisoned due to political motivation, because the organizations monitoring such situations (especially NGOs) haven't a complete access in the country. Trials are closed in case of a sensitive matter and severe punishment prevents an independent monitoring of justice. Many people imprisoned by the former dictator Niyazov are victims of enforced disappearance. After the arrest the suspect disappears, his family usually can't see him and it's very difficult to understand if he is alive or not. Source report that in jail they often suffer torture that is a practice especially used in high-security prison.[71]

Other important rights are violated in Turkmenistan such as the freedom of religion; the Turkmenistan Parliament approved in March 2017 a new law about: every religious movement must have some requirement to be registered by the authorities, and every religious book must also pass a state censorship. There isn't still a choice for who are objectors to military service and everyone must do their part in the armed forces. Similarly, for the freedom about individual and private rights, or about the right to express their own sexuality; homosexuality is severely punished by the authority: law enforcement officials and medical personnel are authorized to pursue humiliating examination to prove this "crime".[72]

13.3.2 Leaders Biography

The nomination of Berdimuhamedov as President of Turkmenistan was followed by expectations of some relaxing of political pressure. His political career began in 1995 when, being the dentist of the former president, he was nominated Minister of Health; then, some years later in April 2001, he became prime minister of Turkmenistan with the mandate of education, health and science. On 21 December 2006, he was appointed as president by the security council, after the death of his predecessor. As said, the succession has been managed in a rather Soviet style (scarce transparency but discreteness, carried on in the backstage of the power), in an apparently ordinate way.[73]

This nominee was a surprise for the IC and for the citizens because everyone thought that the security apparatus would prefer anyone considered more politically strong. Anyway, his nominee was confirmed on 21 February 2007, when in the election he received nearly 90% of the vote cast. His election was firstly saluted as a sign of renovation and his first law about pension and education went in this direction; but soon further signs evidenced doubts that he really wanted to democratize the country.[74]

The figure of the precedent dictator represents possibly the most interesting in the transition period from many points of view: Saparmurad Niyazov ruled the country for over 15 years since the declaration of independence, and for few years before, he was appointed as Turkmenistan SSR leader in 1985.

He had a difficult childhood: his father was killed in action during WWII, his mother and two brothers died in the earthquake that devastated the Ashgabat region in October 1948. He graduated as an engineer from the Leningrad Polytechnic Institute, but soon he had been starting his political career in the communist party. In 1980, Niyazov was the head of the Ashgabat City Party Committee and in 1985, as said, Mikhail Gorbačëv appointed him as the president of Turkmen Republic Communist Party. From this position, he soon started a campaign against corruption. In January 1990, he was appointed chairmen of the Supreme Soviet and in October 1990 he was elected executive president with 98.4% of the vote.

Turkmenistan voted his independence in August 1991 and Niyazov remained the president of the new republic, seeking and gaining legitimacy with a sequel of referendum and elections, arguably of disputable character, finally being appointed as life president.

He adopted in 1994 the sonorous name of Turkmenbashi (Head of Turkmen) to underline his role in the government as the leader and arbiter of all Turkmen. The People's Assembly in December 1994 gave him the right to maintain the power as long as he desired. Then soon he started to accumulate power and to fire all the forms of opposition. In January 2003, a presidential decree stated that everyone who has questioned his power should be considered a traitor. In March 2002, he purged the majority of the National Security Council and used an alleged coup to imprisoned every form of opposition.[75]

Two particular aspects describe his presidency: the desire to create a national ideology and the cult of personality (Figs. 13.4 and 13.5). The creation of a national ideology was important for Niyazov in order to unite all the Turkmen

[70] https://www.hrw.org/world-report/2017/country-chapters/turkmenistan#c3ba16, accessed at 06.01.2018.

[71] https://www.hrw.org/world-report/2017/country-chapters/turkmenistan#c3ba16, accessed at 06.01.2018.

[72] https://www.hrw.org/world-report/2017/country-chapters/turkmenistan#c3ba16, accessed 06.01.2018.

[73] See the cited Article 61 of Turkmenistan Constitution.

[74] https://www.britannica.com/biography/Gurbanguly-Berdymukhammedov, accessed 06.01.2018.

[75] https://www.britannica.com/biography/Saparmurad-Niyazov, accessed 6.1.2018.

Fig. 13.4 Turkmenistan, Ashgabat, 1999, ubiquitous poster President Niyazov (1). (Photo: Igor Jelen)

Fig. 13.5 Turkmenistan, Ashgabat, 1999, ubiquitous poster President Niyazov (2). (Photo: Igor Jelen)

tribes and his nationalist principle was written in the cited Ruhnama, published in early 2000, that must represent the basis of education for every citizen. It supported the biggest project to represent his personality and to remind to all the inhabitants his role as a Turkmenbashi.[76]

13.3.3 Niyazov Extravagances

Turkmenistan formally is a presidential republic, but like other former Soviet states, it is in fact an authoritarian state, giving continuity to the politics that characterized the previous regime; after the collapse of FSU in 1991, it needs some new arguments for maintaining legitimization.[77]

The government realized soon it had to face new realities from a weaker position; then it opted for an isolationist politics, possibly fearing manoeuvres from neighbouring states, but mainly in order to be able to have "free hand" in inside politics. It is also to be considered that, beyond the apparent political absolutism, the internal society was (and still is) fragmented in clans (and further cleavages, national, religions, regional), even when "invisible", but latent to any situation.

Comprehensibly, the moment of the sudden acquisition of independence represented a critical moment, inspiring not really a sense of liberation but also of fear: the power had at first to consolidate itself—inside and outside—facing all the difficulties created by the SU collapse.

It is to say that Turkmenistan remained outside of any Perestroika reform movement, with local elite—represented by the secretary Niyazov, whenever appointed by Gorbačëv—manifesting at the moment of Soviet collapse, the preference of conserving the "ancient regime", possibly in order to preserve its own position.

By SU collapse, in October 1991, Niyazov organized a referendum in order to approve the independence (the latest and last among CA republics, with exception of Kazakhstan), organizing presidential election in fact for himself—being the only candidate—in 1992; further, he became a "life president" winning presidential elections in 1994, and the following parliamentary elections in 1999.

All this combined with an absolute political locking: by parliamentary elections no opposition party could participate, and no international organization were allowed to send observers; it happened similarly for any kind of monitoring to presidential, as well to parliamentary elections in 2004 and 2005. However, the IC maintained a minimal influence, since Niyazov did not leave the OSCE, which means that a series of observes could continue to reside in the country.

Turkmenistan was not prepared for any independence; Niyazov was actually forced to declare independency due to the desegregation of SU: it meant "tout court" the disappearance of the state, then a big "empty" that was to be filled up by a new practice and a new ideology. In that moment, the only possibility was that of adopting a neo-nationalistic (indeed neo-statalistic) narrative that became soon the base for a rather grotesque cult of personality: Niyazov transformed soon himself, from a grey "aparatčik", in a national leader, inaugurating a centralization politics. He slowly slid in a kind of political paranoia, evoking much effort in establishing an apparatus for defending himself from imaginary or real enemies, internal or external, degenerating in a pure terror regime.

Further events menaced state security, inducing a further aggravation of the situation, with the apparatus becoming even more repressive inside, and more isolate towards outside; the 2002 terrorist attack attempted to assassinate the president (perhaps an event presenting elements of uncertainness), which was followed by a sequel of restrictions for individual and social rights, and by correspondent, as expected, accusations to some oppositions. Soon after, Niyazov carried out a wave of "purge" in perfect Stalin style, with officials' and government ministers' dismissals, with limitations and restrictions of any kind. As usual such opressive measures were combined with further extravagancies, apparently senseless, in fact signifying correspondent demonstrations of pure arbitrariness (namely humiliation of the opposition). This in particularly using instrumentally censorship on media and new papers, orienting any information source in order to glorify himself.

His rule was based on a mix of threat and persuasion, and paternalistic reassurance, alternating repression and welfare public concessions, for a country extremely rich in HC resources; in his way he generously financed any kind of social needs, but imposing in the same time a sequence of gestures, namely rituals of obedience and of exhibition of power, conditioning any moment of social and individual life.[78]

Besides this, he will be remembered by his population by such bizarre gestures as the sudden closing down of all the hospitals and rural libraries outside Ashgabat,[79] the mandatory teaching of his own book (the Ruhnama, to which it was given in a sacrilege manner the same status as Quran),[80]

[76] https://www.britannica.com/biography/Saparmurad-Niyazov, accessed 9.5.2018.

[77] Anderson J. 1995.

[78] Fox News 2006, Turkmenistan's Leader Promises Citizens Free Gas, Electricity and Water Through 2030, http://www.foxnews.com/story/2006/10/25/turkmenistan-leader-promises-citizens-free-gas-electricity-and-water-through.html, accessed 23.4.2018.

[79] Marat Gurt 2016.

[80] All the religion and especially Muslim faced a lot of pressure to accept the book and the example could be found in mosque with quotation of Ruhnama, http://www.refworld.org/pdfid/46891911 2.pdf, accessed at 10.04.2018; see as well Corley Felix 2005b.

required, among the others, for having drivers licences, and his biography.

Further strange decisions regarded the prohibition of golden teeth or beards,[81] the renaming of days of the week, and of months of years, giving his name to many places and facilities in all the country; a policy that has been stopped just with his sudden death. Then, the new president started a reverse campaign to reduce the impact of this personality cult, but maintaining some of its predecessor's changes.[82]

13.3.4 Opportunistic Isolationism

The arbitrariness in internal affairs had an exact correspondence in foreign politics that appeared as a scarcely coherent and sometime provocative one; he maintained relations with Afghan Taliban in the 1990s, until September 11th, 2001, and he promoted the build-up of the (small, indeed, inaugurated 1997) HC pipeline, outflanking the IC sanctions, with Iran; further, he does not adhere to SCO—annoying both, regional superpower and his neighbours—and other IO, combining such unpredictability with isolationist tendencies.

Indeed, his neutralism and "not alignment" permitted to the regime to maintain an autonomous position regards the regional powers, with Russia in particular; a not obvious task considering the critical position of the country, which borders with Iran, for long time emarginated from the intentional context, and Afghanistan. But above all it permitted to President Niyazov to consolidate its internal rule.

It is to say that the country mitigated isolationist tendencies recently, starting also a diversification policy for economics, with the aim especially to open new routes for export—possibly a way-out from isolationism[83]; the TAPI project seems to be the key question, even considering the difficulty of such realization.[84]

13.4 Tajikistan

13.4.1 Institutional and Political System

Tajikistan was created as an autonomous republic only during the Soviet era (1929) before it was as part of Uzbek SSR. It became independent in 1991, like other former Soviet republics. After the declaration of independence, Tajikistan fell in a civil war between regional factions, protracting from 1992 to 1997. The internal security problems continued for the country also during the last years: several domestic incidents were reported in the post pacification time, until in 2010 and later, between government armed forces and criminal groups. Another issue is the situation created by the war in Afghanistan with the flows of many Islamic fighters into the country.[85]

In 1994, it took effect the first democratic Constitution of Tajikistan that has designated the state as a presidential republic. Unique among the others CA countries, its constitution originally provides a strong legislative character, rather than a dominant executive one; in the Tajik constitution, the president is the head of the state and the prime minister is the head of government.[86] The amendments to the constitution should be presented by the president of the republic or by at least one-third of members of both houses of parliament.

The amendments need, after the approval of the Supreme Assembly, a referendum that, in order to be hold, should be approved by the president or by at least two-thirds of the Assembly of Representatives. The referendum approval needs the participation in the vote of the absolute majority of eligible voters. The Tajik constitution can't be amended in what concern the form of government, its territory, and the democratic nature.[87]

Despite such a difficult procedure, the constitution has been modified several times: the first amendment was in September 1999, the second in June 2003 and the last one in May 2016. During the last referendum regarding such matter, they were proposed 41 amendments, among them, the ones inserted in the constitution were the elimination of time limit for the presidential mandate, lower the age eligibility for President to 30 and a ban for the political parties based on religion (among them, also the IRP, with whom, as discussed, the Tajik government signed the peace treaty in 1997; see later).[88]

---

[81] https://www.telegraph.co.uk/news/worldnews/asia/turkmenistan/1460852/A-peep-into-the-strange-world-of-Turkmenbashi-whose-every-word-is-law.html, accessed at 01.07.2018.

[82] https://www.britannica.com/biography/Saparmurad-Niyazov, accessed 9.5.2018.

[83] He promises openness as well on internet, Warf Barney 2011: 14; http://www.eurasianet.org/departments/insight/articles/eav030807.shtlm; 2007, "In Turkmenistan internet access comes with soldiers".

[84] Anceschi Luca 2017.

[85] https://www.cia.gov/library/publications/the-world-factbook/geos/ti.html, accessed at 17.01.2018.

[86] https://www.britannica.com/place/Tajikistan#ref73604, accessed at 17.01.2018.

[87] https://www.cia.gov/library/publications/the-world-factbook/geos/ti.html, accessed at 17.01.2018.

[88] https://www.rferl.org/a/tajikistan-referenmdum-approved-rahmon-increasing-power/27751364.html, accessed at 17.04.2018; http://factsanddetails.com/central-asia/Tajikistan/sub8_6d/entry-4889.html, accessed 19.05.2018; http://www.c-r.org/downloads/Accord%2010_15Key%20elements%20of%20the%20Takikstan%20peace%20agreement_2001_ENG.pdf accessed at 19.05.2018; see as well Hierman Brent 2017.

The constitution has the highest legal power and it has direct application in all the Tajikistan territory (Article 10)[89]; it defines the government, the rights and liberty of the citizens as well as the legislative, executive and juridical power.[90]

The president of Tajikistan is Emomali Rahmnon who represents also the chief of state (Fig. 13.1). He adopts decree and orders, appoints the prime minister and other ministers, as well as the executive authorities at the central and regional level. The president nominates the chairpersons and deputies of the Constitutional Court, Supreme Court, Supreme Economic Court and the prosecutor general for parliamentary approval (Articles 69–70 of the constitution).[91]

Tajikistan politics indeed evolved, from a parliamentary, to a strong presidential system in which the executive branch exercises wide authority relative to the parliament. It is a presidential republic in which the president exercises a great role. He appoints not only the prime minister and other ministers but also the executive authorities at a central and local level.[92]

The president is elected by simple majority in a single election for a seven-year mandate. The last presidential election was held on 6 November 2013, and the next election will be in 2020. Sources refer that the election lacked a real pluralism and genuine political debate. Rahmon won the last election with a majority of 84.23% of the vote.[93]

The Prime Minister is Qohir Rasulzoda, appointed since November 2013 after the presidential election. The cabinet is also decided by the president and needs the approval of the Supreme Assembly.[94]

The Legislative Branch is formed by the Supreme Assembly (Majlisi Oli) who consists of two chambers: the National Assembly (Majlisi Milli) and the Assembly of Representatives (Majlisi Namoyandagon). The National Assembly is formed by 34 parliamentarians: 25 members indirectly elected by local assembly or majlis, 8 appointed by the president and 1 reserved for the ex-president. The Assembly of Representatives has 63 seats: 41 members directly elected in single-seat constituencies by 2-round absolute majority vote and 22 directly elected in a single nationwide constituency by proportional representation vote. All the parliamentarian of both chambers serves for a five-year mandate.[95]

The last election for the lower chamber was in March 2015. Important in this country is the respect for minorities and, despite the protection ensured by constitution, the candidates in parties that represented Uzbek, the Kyrgyz and Russian minorities were few. The winning party was People's Democratic Party of Tajikistan (PDPT) with 65.4% of the vote, Islamic Revival Party of Tajikistan (IRPT, 1.7%), the Communist Party of Tajikistan (CPT, 2.2%), The Agrarian Party of Tajikistan (APT, 11.4%) and the Party of Economic Reforms of Tajikistan (PERT, 7.5%). Further parties are the Socialist Party of Tajikistan (SPT, 5.5%), the Democratic Party of Tajikistan (DPT, 1.7%) and the Social-Democratic Party of Tajikistan (SDPT, 0.5%).[96]

The Supreme Court in Tajikistan consists of chairmen, deputy chairmen and 34 judges organized in a civil, criminal and military chamber. At the top of the Juridical Branch there is the Constitutional Court that consists in a court chairman, a vice-president, and five judges.

Separated by this court is the High Economic Court Judges who are organized in 16 judicial positions. These judges represent the highest juridical courts; they are chosen and nominated by the President with the approval by the National Assembly. The mandate has a 10-year term limit but it could be renewed with no limits except the age limits of 65 years. The juridical system is divided into subordinate courts: regional and district courts, Dushanbe City Court, provincial level courts (viloyat courts), Court of Gorno-Badakhshan Autonomous Region.[97]

The situation of peaceful dissent of presidential politics—that characterized the early pacification period—is shrinking drastically. The Government justifies its limitation of freedoms and liberties invoking the national security issue and the fight against Islamic terrorism.[98] Security forces have arrested over 150 human rights activists since 2015 with political charges. The IRP has been banned by an amend-

[89]Article 10 of Tajikistan Constitution: "The Constitution of Tajikistan possesses supreme legal power, and its norms have direct application. Laws and other legal acts that are contrary to the Constitution do not have legal force. The government and all its organs, officials, citizens, and citizens' associations are obligated to comply with and execute the Constitution and laws of the republic. International legal acts recognized by Tajikistan are a constituent part of the legal system of the republic. In the case of a discrepancy between the laws of the republic and recognized international legal acts, the norms of the international legal acts are applied. Laws and international legal acts recognized by Tajikistan enter into force after their official publication".

[90]http://www.president.tj/en/taxonomy/term/5/28, accessed at 17.01.2018.

[91]As per Article 69 of Tajikistan Constitution.

[92]Article 69 Tajikistan Constitution.

[93]OSCE, Republic of Tajikistan, Presidential election, 6 November 2013, Warsaw, 5 February 2014, http://www.osce.org/odihr/110986?download=true, accessed at 17.01.2018.

[94]https://www.cia.gov/library/publications/the-world-factbook/geos/ti.html, accessed at 17.01.2018.

[95]https://www.cia.gov/library/publications/the-world-factbook/geos/ti.html, accessed 19.5.2018.

[96]OSCE, Republic of Tajikistan Parliamentarian Election, 1° March 2015, Warsaw, 15 May 2015, http://www.osce.org/odihr/elections/tajikistan/158081?download=true, accessed at 17.01.2018.

[97]https://www.cia.gov/library/publications/the-world-factbook/geos/ti.html, accessed at 17.01.2018.

[98]https://www.amnesty.org/en/countries/europe-and-central-asia/tajikistan/report-tajikistan/, accessed at 17.01.2018.

ment of the Constitution adopted by a referendum on May 22, 2016 which banned all religious parties in Tajikistan.[99]

In June Tajikistan Supreme Court sentenced the leaders of IRPT to life prison after an unfair trial on charges for overthrow the government. The defendants were scared with torture, preventive arrest and during the detention, they were subjected to ill-treatment. Since 2014, authorities have arrested or imprisoned at least six human rights lawyers and also sentenced to 20 years prison two of the attorneys who have defended IRPT members.[100]

Torture is a practice forbidden by the constitution and there aren't enough information to inquiry the real situation inside the prisons and about the security procedure of the police forces.[101] Other freedoms are restricted to Tajikistan citizens: the right of free assembly and the right of association. The Ministry of Justice has written a draft to control the activity of NGOs and to verify their respect for the regulations. The verify is necessary due to obtaining the official registration.

In August the Government has allowed a new decree about the state controls of the media. The state control already all the media through the State Broadcasting Committee. Despite the state control, some independent journalists have tried to inquire and to write about the Tajikistan situation but they have been all faced intimidation and public harassment from security forces for their journalistic activity. A governmental decree permits also to block the use of the internet and require that all internet providers must do their service through a unique channel owned by the government, the Tajiktelecom.[102]

[99]As said, the IRP signed the Tajikistan Peace Agreement of 1997 who established to share the political power with the opposition party; the ban of an opposition party could rise some question about the future of the Peace Agreement; http://factsanddetails.com/central-asia/Tajikistan/sub8_6d/entry-4889.html, accessed 19.05.2018; http://www.c-r.org/downloads/Accord%2010_15Key%20elements%20of%20the%20Takikstan%20peace%20agreement_2001_ENG.pdf, accessed at 19.05.2018; https://eurasianet.org/s/tajikistan-bans-islamic-opposition-party accessed at 01.07.2018; consider that one of the leaders of the party Kurbon Mannonov died in prison http://catoday.org/centrasia/28551-v-tadzhikskoy-tyurme-skonchalsya-aktivist-pivt-kurbon-mannonov.html accessed at 08.07.2018; see also Hierman Brent 2017.

[100]https://www.hrw.org/world-report/2017/country-chapters/tajikistan, accessed 19.05.2018.

[101]Article 18 of Tajikistan Constitution: "Everyone has the right to life. No one may be deprived of life except by the court verdict for a particularly serious crime. The inviolability of a person (личности) is guaranteed by the State. No one may be subjected to torture [and] cruel and inhuman treatment. Forced medical and scientific experiments on human[s] are prohibited"; see as well https://www.amnesty.org/en/countries/europe-and-central-asia/tajikistan/report-tajikistan/, accessed at 17.01.2018; https://www.state.gov/documents/organization/265762.pdf accessed at 10.04.2018. The U.S. State Department says that despite the introduction of the torture as a crime for the Tajik criminal code, there are reported form of coercions to bring prison to confession. Tajikistan created an office to inquiry about prison and torture but there aren't enough information about their real impact and activity.

[102]https://www.amnesty.org/en/countries/europe-and-central-asia/tajikistan/report-tajikistan/, accessed at 17.01.2018; Dagiev Dagikhudo 2014.

13.4.2 Leader Biography

The Tajikistan President Emomali Rahmon (once Rahmonov, de-Sovietized name) is the figure dominating the Tajik politics and his People's Democratic Party of Tajikistan; he arrived relatively early to the presidency, in war time, starting from a position of middle-rank typical Soviet-CA "aparatčik".

He was born to a peasant family in Dangara District, in the south of the country; after military service in the navy and a graduation in Economics, he started to work on a collective farm; soon he was elected chairman of the Union Committee of Lenin collective farm in Dangara; in 1990 he was elected as a People's Deputy at the Supreme Council of the Republic of Tajikistan. From this moment he started the political ascension, and he was elected in other important position, in 1992, as Chairman of the Executive Committee of the Council of the People's Deputies of Kulab province and then also in the same year as Chairman of the Supreme Council of the Republic of Tajikistan.[103]

After the independence the President in charge was the former Soviet President of the Republic of Tajikistan, Rahman Nabiyev. He won the first election as a President with over 227 former communist members elected in the unicameral 230-seat Supreme Soviet of Tajikistan. He but became very unpopular, was overthrown during the civil war and must escape from the county. During the conflict that followed, Rahmon cleverly emerged as a political leader and in November 1994 he was elected President by a nation-wide vote.[104]

The election of Rahmon represented a sign of continuity with the Soviet nomenclature[105]; whenever belonging to southern Tajikistan Kulab lobby, he evidenced good mediation capabilities, in a country divided potentially in many

[103]http://www.president.tj/en/taxonomy/term/5/33, accessed 19.05.2018.

[104]In summer of 1994, during the civil war, the government proposed to parliament a new constitution and the holding of presidential election for September 1994; Rahmonov under the pressure of Russia and Uzbekistan decided to postpone the election in November and release 1000 political prisoners (25 August Amnesty). The opponents of Rahmon was the powerful member of the Khujand clan and former prime minister Abdumalik Abdullojanov; the election had been won by Rahmon with 59.6% of the vote. See the Compendium of Reports on Technical Election Assistance to Tajikistan, Pre-election Assessment, Election Day Observations, Post-Election Technical Assessment, March 1995, International Foundation for Election System, p.19, https://books.google.it/books?id=QqjoCF3AEYEC&pg=PA15&dq=elections+in+tajikistan+during+civil+war&hl=it&sa=X&ved=0ahUKEwiQx9b3z7LaAhWJCCwKHZcLBTQQ6AEILzAB#v=onepage&q&f=false, accessed at 11.04.2018.

[105]"Rahmon was then elected by the members of the Supreme Soviet as its chairman—a post equivalent to that of president—and the head of government", "Emomali Rahmon"", official website of the President of the Republic of Tajikistan, http://www.president.tj/en/taxonomy/term/5/33, accessed at 17.01.2018.

counter-posed parts (plain and mountain, urban and rural) and different regional areas (Kulab, Khudjand and Fergana gravitating area, Dushanbe—global society, Pamirs and further remote mountain provinces).

He demonstrated to be an ingenious leader, learning how to deal with fractions and lobbies. He prefers to be surrounded only people who don't have presidential ambitions and prevent anyone to emerge; for this reason, apparently, there aren't any strong political successors. For example, he continues to nominee as Prime Minister Oqil Oqilov, who has reached the age limits and for this reason, can't be considered as a competitor for the presidency.[106]

Rahmon controls also the judiciary and economic power of the country. The president appoints judges with legislative approval but he can unilaterally remove them from their position. His control of the judiciary power can be seen on the ban of all political parties connected with the government of Nabiyev. Differently to other CA country and despite the fact that Tajikistan is the poorest country, the economic officials and advisors of the government are mostly conservative (Soviet-like thinker), and therefore they hardly contribute to improve the economic situation.

This particular choice is desired by President Rahmon who doesn't want to be advised by people who have the ambition to become more powerful. The only possible rival of the president is the former Dushanbe mayor Mahmadsayid Ubaidulloyev who was dismissed from the assembly by the president in 2009's pre-election congress of the People's Democratic Party of Tajikistan. In this occasion, he also appoints on the party's central committee his son Rastam Emmomali.

The inner circle is very important and very influential for the President and this appears also by the rumours that Rastam could be in future his successor.[107] The succession at the presidential seat was also prepared by the amendments to the constitution that have to reduce the age limits for becoming president.

To ensure continuity, possibly in dynastic proto-monarchic terms, Rahmon started a kind of mild way of cult of personality. It is evident in the iconography (representing the president, in a kind of subliminal representation, with crown and kingdom attributes; Figs. 13.1 and 13.6), in the communication and in further evidences of governance.[108] The appointment of his son as mayor of Dushanbe, supposedly, is a signal showing the intention of giving continuity to its power, starting the accreditation of a procedure of succession.

13.4.3 Peace, Prosperity and Confidence Building

The civil war had—as usual for civil wars indeed—a tormented evolution and it is difficult to find a schema for a coherent description.

Soon after independency, tensions exploded, with oppositions (mainly religious and liberal, indeed considered as

Fig. 13.6 Tajikistan, 2017, presidential poster in Rasht valley. (Photo: Igor Jelen)

[106] https://www.globalsecurity.org/military/world/centralasia/rahmon.htm, accessed at 17.01.2018.

[107] https://www.globalsecurity.org/military/world/centralasia/rahmon.htm, accessed at 17.01.2018.

[108] Paskaleva Elena 2015: 425.

moderate in that phase) claiming reforms[109]; the conflict (aggravated also by disproportionate reaction of the power) spread out in any province of the country, with exception of the eastern Pamir side (as described in other chapter). After the conflict, the constitution has been in great deal changed, namely emendated; the new constitution's aim is the definitive pacification of the country, articulating the power in regionalist terms (to sedate separatist expectations, even when the state power seems to be still divided in regional clan-like organizations), acknowledging to Gorno Badakhshan (Pamir) a special status.

Tajikistan was always considered the poorest country, the last one in any statistic, in the whole Soviet space. It is the example of a dis-homogeneous national-state configuration (Fig. 13.10): the mountains outback, practically impossible to control with centralistic devices, is populated by autochthonous populations; the Khudjand province, gravitating on Fergana; and the Kulab area, in the southern side, representing the most dynamic economic regions. All these are represented by influent clan, exerting eventually anti-centralistic attitude.

Furthermore, the national-territory is heavily exposed to outside scenarios, and especially to Afghanistan, from that it can always infiltrate any kind of insurgents (especially in the unquiet Rasht valley), contaminating the whole scenario. The fact Dushanbe is gradually assuming the role of city capital, and of a true globalized city, appears as the key element for overcoming the fragmentation tendencies; it is mediating and exerting a fundamental attraction role on peripheries, contextually predisposing opportunities, jobs and perspectives for population from all over the country (Fig. 13.11).

After the peace deal 1997–1998, and beside the original causes, tensions persist in many forms; the war becomes something else for disbanded groups, for whom the rebellion became a style of life, and eventually a pretext for continuing to practise violence, undertaking criminal activities such as drug trafficking, racketing, extortions and contraband.[110]

The main problem is that of the hate provoked by decades of bloody conflicts; the pacification process needs time to be settled, possibly—as usually for civil wars—the time of the generation passage: individuals, families and groups, who have been hit personally, humiliated and deprived of everything, cannot easily forget. Besides the appearance, the "underground hate" is a continuous source of preoccupation for government.[111]

[109] Hanks Reuel R. 2016.

[110] Dagiev Dagikhudo 2014.

[111] https://jamestown.org/program/resurgence-of-islamic-radicalism-in-tajikistans-ferghana-valley/, accessed at 17.04.2018; it requires further and more refined pacification strategies, in order to exit such vicious circles (defined as "inside effects") self-alimenting tensions; but instabilities continue; Khayrullo Fayz 2012; Dagiev Dagikhudo 2014; Megoran Nick 2017.

13.5 Kyrgyzstan

13.5.1 The Kyrgyz Exception

Kyrgyzstan considers itself as a "unitary parliamentary republic", then "unitary" intended as non-federal, and non-divisible, possibly in order to mitigate the separation tendencies that derive, by-their-selves, from its fragmented geography (Fig. 13.7). The disarticulated form is the principal characteristic of the country, being even more aggravated by a dualism contraposing the Fergana valley side to the Bishkek capital city area, in the north-east (as Osh 2010 bloody clashes evidenced even recently).[112]

Further geopolitical characteristic of the country is the presence of remote areas, namely of the spectacular mountain outback, that but from governance point of view represents a challenge.

For a while, and during transition, Kyrgyzstan was indicated as a model country, the most liberal, and guided by an elite sincerely engaged in democratization, with the civil society in Bishkek considered rather robust, then refractory to any "oriental despotism" drift. This attitude is facilitated—in paradoxical sense—by the absence of relevant resources suitable of being exploited by monopolies also for political purposes, then potentially biasing the democratization process (as it happened possibly for oil-rich republics).[113]

Today the reality evidences some changes in centralistic terms, even when some elements leave imagine the persistence of a basic democratic attitude; it is the case of the "conservative" communitarian habits that continues to characterize the rural peripheries (traditionally self-organized on the base of traditions evoking archaic democracy); and it is the case, in the urbanized areas, of a basic ethnic pluralism, that can in principle predisposes to political "openness".

Anyway, the recent 2017 election was the first peaceful handover of powers from one democratically elected to another candidate in the whole CA. The winner of the election was Sooron Jeyenbekov who became President.[114] Despite such important achievement, the Republic shows some contradictions; some sources refer that some apparatus branches do not respect important rights and repress journalists and political opponents.

[112] Megoran Nick 2017.

[113] A kind of "malédiction des resources naturelles" (resources damnation), dangerous for entire systems, and even more for rentier states; Thomas Marzhan 2015: 467; Auty Richard 1990.

[114] https://www.economist.com/news/asia/21735067-life-getting-harder-opposition-politicians-and-journalists-repression-kyrgyzstan, accessed at 24.01.2018.

Fig. 13.7 Administrative map of Kyrgyzstan. (Source: http://ontheworldmap.com/kyrgyzstan/kyrgyzstan-road-map.html, accessed at 01.07.2018)

13.5.2 Constitution and Institutional Politics

Kyrgyzstan had been part of Russia since 1876, firstly as a part of the Russian Empire then under the Soviet power, and became independent in 1991 as other former Soviet countries. The Akayev presidency carried a rather open politics, even when the situation in south continued to be critical, with contrapositions either of ethnic- and social character. The political sentiment changed in 2005, and then in 2010, with mass demonstration against the government that had, as a result, the ouster the two President in charge during the demonstrations, respectively Askar Akayev and Kurmanbek Bakiev.[115]

The country is considered one of the poorest of the area, and supposedly also for this reason, Kyrgyzstan is one of the latest among the CA countries to have structured a national apparatus; this also considering law enforcement agencies and even a national army. Indeed, also after these have been founded, they faced several problems such as the low budgeting and an uncertain chain of command; this fact impacted also on politics because, being poorly subsidized and structured, the military appears scarcely inclined to influence the power, and to stay out the political questions.[116]

However, Kyrgyzstan, whenever carrying a "not aligned" politics, has been forced to make some active decisions in international questions.

It was one of the first countries to support the US-guided international coalition in the military operation in Afghanistan in 2001; for this reason on 25 September 2001 President Akayev gave the permission to US forces to use the Kyrgyz airspace and in November of the same years, US deploy also some military aircraft in his territory.

115 https://www.cia.gov/library/publications/the-world-factbook/geos/kg.html, accessed at 24.01.2018; Megoran Nick 2017.

116 https://www.globalsecurity.org/military/world/centralasia/kyrgyz-intro.htm, accessed at 24.01.2018.

The first Kyrgyz constitution was approved in 1993 but it has been amended several times during recent years. The last constitution, adopted in 2011, prescribes in a curious manner, the prohibition of any constitutional changes until 2020.[117]

Besides this, the process of amending the constitution appears to be rather heavy: the proposal has to be made by the majority of the Supreme Council or by 300,000 voters, in the next step the draft will be a vote by at least two-thirds majority vote of the council membership in each of the least three reading separated by 2 months apart, then the draft has to be submitted to a referendum that has to be approved by the majority of the Supreme Council and, if the draft is voted by the majority of population, it would need the approval of the President to be effective.[118]

Nevertheless, there are signs of a change in conservative sense; in 2016 it was held a controversial constitutional referendum about strengthening the power of the prime minister and, furthermore, to adopt a legislation that bans the same-sex marriage. The majority of voters rejected in December this new constitutional measures.[119]

Kyrgyzstan has been a parliamentary republic since 2011 with a change in the power of the President in favour of Parliament that now results to be more powerful. The previous constitution in 1993 imagined a powerful executive power that has strengthened since 1993 in several constitutional amendments. The new constitution is the result of the mass protest of June 2010 in which President Kurmanbek Bakiev was toppled. Now Kyrgyzstan is the only country in CA that has a single term limit presidency of 6 years. The President is directly elected in a first ride election, the second round can be necessary if in the first anyone gave the majority. The last election was held on the 15 October 2017 and the future presidential elections will be in 2023.[120]

Sooronbai Jeenbekov won the election with the 54.81% of the vote, while his main adversary Omurbek Babanov had 33.5% of the vote. At least this election has represented an important moment for the stability of the Region and also Russia was apparently interested in a peaceful passage of power in order to continue the relationships between the two countries.[121]

The president is the head of the State and the higher official in the Republic; he has the power of appointing the prime minister and his cabinet appoints the governor of provinces, cities and "oblast".[122] He is the head of the National Security Service and he can establish the head of the National Guards, of the Security Council and of other correlated security bodies. For what concern the administration of justice, the President has the power to appoint Supreme Courts Judge or the chairman of the Supreme Court, with the consult of the Parliament, and also establishes the judge of the local courts.[123]

The Legislative Branch of the Republic is formed by the unicameral Supreme Council (Jogorku Kenesh) who has 120 seats; the parliamentarians are directly elected in a single nationwide constituency by proportional representation vote to serve for a five-year term.[124] Since October 2007 the Supreme Council has consisted in an unicameral legislature; in 2010, with the new constitution, it established an increased number of seats from 90 to 120; then the new Law on Elections was adopted which was still in force during the October 4, 2020 Parliamentary Election.[125]

The latest result of the Parliamentarian elections was: Socialist Party of Kyrgyzstan (SDPK) 27.56%, Respublika-Ata-Jurt 20.26%, Kyrgyzstan Party 13.07%, Onuguu-Progress 9.38%, Bir Bol 8.5%, Ata-Meken 7.7%, other 14.1%. The elections result translate in a seat at the Assembly are SDPK 38 seats, Respublika-Ata-Jurt 28 seats, Kyrgyzstan Party 18 seats, Onuguu-Progress 13, Bir Bol 12, Ata-Meken 11 seats.[126]

The two Highest Juridical Court in Kyrgyzstan are the Supreme Court and the Constitution Court. The Supreme Court is formed by 25 judges and the Constitution Court consists only of one chairperson, a deputy chairperson and 9 judges. The judges of both courts are appointed by the Supreme Council with the recommendation of the President. The difference is about the time limits of their mandate: the

[117]The Law on the Enactment of the Constitution of 2010 said that Article 114 of Constitution regarding the amendment of the Constitution will enter into force only in 2020; https://www.osce.org/odihr/261676?download=true, accessed at 19.05.2018.

[118]https://www.cia.gov/library/publications/the-world-factbook/geos/kg.html, accessed at 24.01.2018.

[119]https://www.globalsecurity.org/military/world/centralasia/kyrgyz-gov.htm, accessed at 24.01.2018.

[120]https://www.cia.gov/library/publications/the-world-factbook/geos/kg.html, accessed at 24.01.2018.

[121]https://thediplomat.com/2017/10/a-vote-for-stability-steers-southerner-to-success-in-kyrgyzstans-presidential-election/, accessed at 24.01.2018.

[122]Article 42 of Kyrgyzstan Constitution, http://www.refworld.org/pdfid/3ae6b5ae0.pdf, accessed 19.05.2018.

[123]Article 46 of Kyrgyzstan Constitution, http://www.refworld.org/pdfid/3ae6b5ae0.pdf, accessed 19.05.2018.

[124]https://www.cia.gov/library/publications/the-world-factbook/geos/kg.html, accessed at 24.01.2018.

[125]https://www.osce.org/files/f/documents/b/6/465735.pdf, accessed at 7.11.2020.

[126]https://thediplomat.com/2015/10/final-results-confirmed-in-kyrgyz-parliamentary-race/, accessed at 24.01.2018.

first serves for only 10 years, while the constitution judges for 15 years, being their age-limits for both 70 years.[127]

Besides the fact that the constitution describes the juridical system as independent,[128] in reality, it is strictly under the influence of the prosecutor office. Most cases are originated in the local courts and after the appeal in the municipal or region courts, it may be decided by the Supreme Courts. Many cases are already judged by the traditional elder's courts (aksakal) that are under the jurisdiction of the prosecutor's office. The juridical system is heavily influenced by the previous Soviet system and many rights for people remains only on the chart such as the right a legal assistance of an attorney[129] and the presumption of innocence.[130]

13.5.3 Current Politics

Regarding the situation of human rights, Kyrgyzstan made some democratic progress over the years (especially in early enthusiasm independence times), but then it has been eroded in the last years. The "Foreign Agents" bill, a law that would affect the work of NGOs in the country was rejected by the Parliament but other restrictive laws are under discussion.

In 2011 a constitution amendment introduced the clause of the "supreme state values", weakening in this way the supremacy of the international treaty stipulated versus the domestic law.[131] It was also introduced a law that bans the homosexual marriage.[132]

The southern tensions opposing Uzbek minority is the principal un-resolved problem for Kyrgyz republic, representing much interpretations and feedback also inside the same state institutions. The Fergana ethnic minorities, soon after the independence, were the subject of the sprout out of violence; they started as a sequence of spontaneous street clashes, that but possibly the disproportionate (and inefficient) security apparatus reaction alimented further, inducing a "vicious circle" that will have heavy consequence until recently—considering the Osh violence in 2010.

These events will provoke hundreds of deaths, consequent to street fighting and to successive repression, with cases of ill-treatment, torture and arbitrary detention: much of arrested and charged persons had no fair trial and most of the crimes against them remain unpunished. In March 2017 the UN Commission of Human Rights Committee, about these interethnic violences in Osh, determined that four Uzbeks are still unfair detainee.[133]

One of the main opponents still prisoner is the right defender Azimjon Askarov, sources refer, has been subjected to torture and arbitrary detention. His case was discussed also by the UN Human Rights Committee who requested the government to free the right defender. In May 2017 the Supreme Court ruled against another right defender Tolekan Ismailova who had sued the President Almazbek Atambayev for defamation after he publicly smeared them.[134]

Despite the intentions, torture and other cruel and degrading treatment are usual in the security system of the country. Authorities are continuing to fail the effort to investigate and punish this crime. The acts of violence are used against both ethnic groups, Kyrgyz and Uzbek, with the latter suffering the most important damage and injuries.

Further frequent reason of denounced violence in Kyrgyzstan regards the domestic violence and the forced marriage practices. According to the data of the National Statistic Committee, there were registered over 4960 cases of domestic violence between January and October 2017, but only 158 are denounced and persecuted in courts. The economic situation made even more difficult for a woman to escape from this violence or from the forced marriage.[135]

In March and April 2017, Kyrgyzstan general prosecutor accused the journalist of the free radio Zanova, Azattyk and the Kyrgyz branch of Radio Free Europe. They were all accused of discrediting the President and spreading false information. Some of them are banned and forced to leave Kyrgyzstan and their bank account were all frozen. Other journalists such as Ulugbek Babakulov or Zulpukar Sapanov are reported to faced unfair trial only to reduce their activity.[136]

The freedom of assembly—a main right in order to pursue a proper development of a democratic country—has

[127] https://www.cia.gov/library/publications/the-world-factbook/geos/kg.html, accessed at 24.01.2018.

[128] Article 80 of Kyrgyzstan Constitution http://www.refworld.org/pdfid/3ae6b5ae0.pdf, accessed 19.05.2018.

[129] Article 16.3 of Kyrgyzstan Constitution "Everyone has the right to liberty and security of person. Detention, arrest and committal may be appealed in court. Anyone who is arrested or detained shall be informed, at the time of arrest, of the reasons for his arrest or detention, told his rights and allowed to defend himself in person or through legal assistance of an attorney." http://www.refworld.org/pdfid/3ae6b5ae0.pdf. accessed at 03.07.2018.

[130] https://www.globalsecurity.org/military/world/centralasia/kyrgyz-gov.htm, accessed at 24.01.2018.

[131] Individual-, sex- and gender-related issues seem to represent something particularly sensitive in transition circumstances (then eventually subject to political manipulation); they strongly impact the standard of privacy, rather diverse in the diverse circumstances, often mainly inspired to a prevalent conservative communitarian-traditional culture.

[132] https://www.amnesty.org/en/countries/europe-and-central-asia/kyrgyzstan/report-kyrgyzstan/, accessed 19.05.2018.

[133] https://www.hrw.org/world-report/2018/country-chapters/kyrgyzstan, accessed at 24.01.2018; Megoran Nick 2017.

[134] https://www.hrw.org/world-report/2018/country-chapters/kyrgyzstan, accessed at 24.01.2018.

[135] https://www.amnesty.org/en/countries/europe-and-central-asia/kyrgyzstan/report-kyrgyzstan/, accessed at 24.01.2018.

[136] https://www.hrw.org/world-report/2018/country-chapters/kyrgyzstan, accessed at 24.01.2018.

been reduced. For example, in February 2017, a Bishkek Court imposed a three-week ban on the public assembly and as a result, in mid-March five protesters were detained after their participation in a march to support freedom of speech. Before the election of October, the Bishkek Court banned all the assembly to reduce possible protests.[137]

Talking about human rights, Kyrgyzstan (like Russia and other post-Soviet countries) had introduced a homophobic "propaganda" law. The result of this law is a worrying rise of hostility against the LGBT community. The idea (eventually prompt by Russia and follow also by Kyrgyzstan) is that LGBT rights are western values that somehow constitutes a threat to the national security, perhaps predisposing to a climate of ignorance and violence that is growing in this country.[138] These people face discrimination, extortion and ill-treatment from both Security Authorities and non-state actors. All the responsible of these abuses remain free.[139]

It is to say, not in many cases it is a question of strictly government "bad will", nor of popular attitude; it happens often because of capability, namely of insufficient organization and financial resources. For example, Kyrgyzstan has signed the UN Convention on the Right of Person with a disability but has not ratified the convention allegedly due to economic and social motivation. The rights to health in this nation are not properly ensured to all the population. Actually, rural people face numerous difficulties to access to quality healthcare facilities especially people living in poverty and people with disability. The medical personnel are underpaid so in all the country it is very common to pay directly for the medical treatment.[140]

13.5.4 Leaders Biographies and Continuity of Power

The political system in Kyrgyzstan was overthrown two times in 2005 and 2010. There are theories that hypotheses the pervasive corruption and the influence of the organized crime, beside a lack of trust in the national political leaders, that finally brought the population to the revolt against the government. The most trusted are the local leaders ("aksakal", namely "white beard") such as senior community elders. So similarly for the leaders of woman council, very important traditional institutions, namely local leaders often called for resolving community problems (usually for local conflicts, not for severe penal-code or politically relevant cases).[141]

The first president, Akayev—an ethnic Kyrgyz, academic physicist—tried to establish a strong government in order to lead the country after the independence.[142] His curriculum is rather different from the other CA leaders, and his political career did not start in the Soviet era—although he was a member of the academia. Being very committed for democracy and liberties, he was called by western observers "the Thomas Jefferson of Central Asia". He supported in 1991 Gorbačëv during the attempted coup, and was indeed also very cautious about the independence of his country, considering the geopolitical delicateness of the situation.

After the independence he was appointed president of Kyrgyzstan in 1990, then he was directly elected and confirmed in new two elections of 1995 and 2000; then he was the first president elected without being in the high position of the Communist Party, ruling the country for 15 years; however, despite his juvenile passion about democracy, yet in Soviet era, during his presidency apparatuses branches continued to use police methods against opposition.[143]

Initially he had the support of the United States, and of the majority of IC, for his political reforms which allowed the country to obtain millions of dollars in foreign aid. In fact, the main problem Akayev faced in his presidency was the economic problems and the growth of crime rate especially connected with the drug trade beyond Afghanistan and Tajikistan routes.

His main opponents in parliaments were part of the former communist party elite and from the local governments, especially contrasting him because economic reforms (trying in principle to avoid the formation of rent positions). Akayev has begun since 1992 to appoint the local administrators in order to assure him a loyal support. This system didn't work and was the source of continue scandal for the administration. Since 1994 after a referendum he has assured the popular support to get rid of the parliament and its opponents. His presidency was finally overthrown in 2005 by the so-called Tulip Revolution, one of the "revolutions" picturesquely

[137] https://www.hrw.org/world-report/2018/country-chapters/kyrgyzstan, accessed at 24.01.2018.

[138] https://www.amnesty.org/en/latest/news/2017/12/former-soviet-states-entrenching-homophobia-and-demoralizing-lgbti-rights-activists/, accessed at 17.01.2018.

[139] https://www.hrw.org/world-report/2018/country-chapters/kyrgyzstan, accessed at 17.04.2018.

[140] https://www.amnesty.org/en/countries/europe-and-central-asia/kyrgyzstan/report-kyrgyzstan/, accessed at 17.04.2018.

[141] As typical for Kyrgyzstan, as well as in Kazakhstan, as a legacy of nomadic culture, women and men had equivalent roles, even when specialized per function and spaces; Jelen Igor 2002; Valikhanov Č. Č 1985; Valikhanov capt., Veniukof M. et al 1865; Kazakhstan, Kyrgyzstan, Tajikistan, Turkmenistan and Uzbekistan: Country Studies, Federal Research Division Library of Congress Edited by Glenn E. Curtis Research Completed March 1996; this means that in modern times woman could can play any role; locally they eventually rule traditional community villages; http://www.unwomen.org/en/news/stories/2017/5/feature-kyrgyzstan-from-housewives-to-local-councils, accessed at 19.05.2018.

[142] https://www.globalsecurity.org/military/world/centralasia/kyrgyz-politics.htm, accessed at 17.04.2018.

[143] http://factsanddetails.com/central-asia/Kyrgyzstan/sub8_5a/entry-4747.html, accessed at 24.04.2018.

denominated with names of flowers or colours, evoking possibly a "reform" (spring) seasonality (perhaps a structural instability).[144]

This street revolution was the result of an escalation of public dissent about the political decision of Akayev and his increasingly despotic government. The protest had begun in 2003 when a large-scale protest followed the arrest of dissident parliamentary deputy Azimbek Beknazarov. In 2003 the protest continued after a decision of the president to expand his powers. Despite the protest in 2005, a popular referendum confirmed all his new powers and the parliament approved a law that gives him and his family a complete immunity.

13.5.5 Continuity of Power, Namely a Tormented Transition

A further element disturbing public opinion was the negotiations with China about the border acknowledgement: a question inherited from Soviet and colonial times that was yet mature, to be settled, but that impacted tremendously the population, very sensitive about the question (with Kyrgyzstan perceived itself as surrounded by aggressive, powerful and non-democratic countries). The signature of the treaty prescribed a minor cession of territory (actually a rectification) to the powerful neighbour.[145]

The major protest exploded in 2005 in the capital when the protesters stormed the White House, the headquarters of the president. The protest occupation of the palace was unprepared and after a few hours, it was ended. But then the mass protest was followed, possibly due to the absence of police, with violence, continuing after that day with—it was reported—deaths, rapes and shop raided and burned.

After the assault the President and his family flew in Russia, allowing opposition leader Kurmanbek Bakiyev to be appointed as President. In absence of a formal resignation by the former president (as said), on 28 March 2005, Bakiyev was voted as Prime minister and President. In early April 2005 in Russia Akayev finally signed his resignation as President. In the successive presidential election in July 2005, Bakiyev won with over 90% of the vote. International observer accepted the election as fair and democratic: the first time that in Central Asia the succession between two presidents happened with a popular vote (whenever not in a pacific mode).[146]

Bakiyev is a graduate electrical engineer; his political career began in 1990 when he worked in government post in the southern part of the country. Only in 2002, he gains a national role in the national politics when he joined a central party who defend the regional interest. In 2004 before the "Tulip Revolution" he was the head of the new founded opposition movement People's Movement of Kyrgyzstan. His rule as a President was not simple, and his presidency, in fact, was marked by a frequent conflict with the Parliament.

The conflicts with opposition parties weren't only in Parliament but continued with a mass protest in the street. The president responded to the opposition in order to strengthen his rule with a constitutional referendum. In 2007 the constitutional changes were approved and allowed him to dissolve the parliament and to call a snap election. At the polls of December 2007, his party won 71 seats of 90. In the period between 2007 and the next presidential election of 2009, he faced several accuse of corruption.

Soon after his victory at the parliamentarian election another time in the country the protest spread out in the street against his autocracy policies and the corruption. In April 2010 at a mass protest the police were unable to disperse the crowd and used firearms, the results of the security forced action were 80 dead in the streets. On 7 April 2010 Parliament declared the state of emergency while people continued to protest in the city. At the early hours of 8 April Bakiyev escaped from the country in Belarus leaving the country to an interim opposition government that have to use the force to restore order.[147]

This until the nomination of a transition government, under the guide of Roza Otunbayeva, former foreign affairs minister, who resigned after his mandate in 2011, and the election of the new President Almazbek Atambayev.

After these two "revolutions", the political situation in the country remained quiet and, differently to other countries in the area, the new leaders have been chosen by popular election. The last election in October 2017 was won by Sooronbai Jeenbekov and his Social Democratic Party (SDPK). The citizen has chosen peacefully and democratically in an election the successor of Atambayev the previous president. The new challenge for Jeenbekov now is to rule a country without destroying all the effort made to became a democratic country.

The last year—2019—proved to be critical in this sense, possibly demonstrating the country is on the edge between regression and autocracy—possibly being the victim of geopolitical manipulation—or consolidation of a local civil society. The 8–9 August 2019 tensions following the charges and arrest of the former President Atambayev possibly can be interpreted in both terms.

[144] http://factsanddetails.com/central-asia/Kyrgyzstan/sub8_5a/entry-4747.html, accessed at 24.04.2018; Babajanian Babken 2015: 514; see also Megoran Nick 2017.

[145] Megoran Nick 2017.

[146] http://factsanddetails.com/central-asia/Kyrgyzstan/sub8_5a/entry-4749.html, accessed at 24.04.2018.

[147] https://www.britannica.com/biography/Kurmanbek-Bakiyev, accessed at 24.04.2018.

13.5.6 Revolution as a Ritual?

The transition times President Akayev introduced new government structures and appointed a new government composed mainly of younger, reform-oriented politicians. He tried to establish collaborative neighbourhood politics, especially with Kazakhstan. Such intention has been symbolized in the way the ancient hordes were used to settle alliances, namely with the matrimony of the sons of the two leaders, that really happened in 1999 with Akayev daughter and Nazarbayev son getting married.[148]

Beside his supposedly good intentions, he found itself in the turmoil of the events characterizing the early consolidation stage of the Kyrgyz state (to be considered to some extent unavoidable), finally, as said, being removed with "Tulip revolution" in 2005.[149]

Indeed, the event has been often discussed, and interpreted, either than a genuine civil uprising, as something manipulated, demonstrating then a distorted way of manifesting, targeted to conquest of power, either than the claiming of reforms; it represented a precedent, starting a period of troubles, with systemic accusations of corruption, possibly configuring a kind of political ritual (then legitimizing the "revolution" as anything to be periodically repeated).

In fact, Bakiyev, his successor (until a further "spring" revolution in 2010, denominated "April revolution"), found itself in an even more complicated situation, with troubles and unrest deliberately orchestrated, possibly, by some clan or lobby.

Such events demonstrated the weakness of his power, and the possibility for an organized opposition (even when minority) to mobilize the "crowd", and to gain the power out of the procedures, configuring a basic distortion in representative procedures. Nevertheless, a necessary test for a functioning democracy, that has to be prepared to such events.

It is to consider that the April revolution has been the premise for the most dramatic moment of the recent Kyrgyz history, namely of the Osh clashes and massacres occurred soon after (June 2010), possibly consequent to the perception of power vacuum induced by the same "revolution". Fergana valley confirmed to be the weakest point of the country, a place from where troubles reverberate to the rest of CA, exposing the country to a persistent source of problems.[150]

13.6 Kazakhstan

13.6.1 President Nursultan Resignment

On 19 March 2019, President Nursultan Nazarbayev announced his decision to resign from the presidential office during a televised address.[151] The day after the Speaker of Senate Kassym-Jomart Tokayev swear as the interim president of the country. Still, is too soon to appreciate how the power will effetely swift in the future until the new election planned for 9 June.

The resignation took completely by surprise populations and governments, upsetting geopolitical analysis and forecasts: the event is still to be appropriately pondered, since the Nursultan retained a set of functions but he is still in charge has the chairman of the important Security Council.[152]

This represents a historical moment not only for Kazakhstan but also for the all Central Asia: Kazakhstan is the most rich and extended country, strategically crucial for the entire CA territory, bordering with superpowers (Fig. 13.8); it is considered as a solid and peaceful country, enjoying a regular steady growth, with the president enjoying a high percentage of consensus—although being a considered a kind of moderate despot, concentrating much of the government functions. He was (actually is) the last Soviet political leader still in office after the independence, so with his resignation it is possible to say, the area definitively passed over the phase of the transition.[153]

13.6.2 Constitution and Institutional Politics

The constitution of Kazakhstan defines a parliamentary republic, with parliamentary elections prescribed any five years, and the role of the president crucial for any procedure—appointments of local authorities, economics and control of national natural resources.

[148] about the geopolitical significance of the wedding of the year Akayev son and Nazarbayev daughter, http://news.bbc.co.uk/2/hi/asia-pacific/132890.stm, accessed 23.4.2018, Wednesday, July 15, 1998, "Fairy tale wedding starts in Asia".

[149] Babajanian Babken 2015: 514; then a drift configuring the formation of a "patrimonial-authoritarian" power, with clan, lobbies and "informal power arrangements" forming in un-transparent manner; see as well Esenova Saulesh 1998; Among the others, parliamentary election 2005 registered heavy internet censorship, "event-based filtering", Warf Barney 2011: 14, possibly an early experiments of such manipulating of digital practices.

[150] Elebayeva A. B. 1992.

[151] https://thediplomat.com/2019/03/is-this-the-end-of-the-nazarbayev-era/, accessed on 06.04.2019.

[152] https://zik.ua/news/2019/03/19/nazarbaiev_pozhyttievo_ocholyt_radu_bezpeky_kazahstanu_1532473 accessed 29.042019.

[153] The new Kazakhstan President Tokarev met with large success (71%) in presidential elections June ninth 2019, as it was easy to foresee indeed; Nazarbayev remained president of the National Security Agency but with a not well defined role; the elections saw much protestation, with about 5000 arrests, and were criticized by most of the international organizations.

Fig. 13.8 Administrative map of Kazakhstan. (Source: http://ontheworldmap.com/kazakhstan/kazakhstan-political-map.jpg, accessed at 03.07.2018)

Independent since 16 December 1991 (today national holiday), Kazakhstan is the last CA country to have declared the independency (declared respectively by Kyrgyzstan 31 August, Uzbekistan 1 September, Tajikistan 9 September and Turkmenistan 27 October), then just before the effective dissolution of URSS (December 1991, with establishment of CIS), and the establishment of Russian federation, with which Kazakhstan has about 6800 km common border.

The constitution has been amended in several occasions; in this period the (former) President Nazarbayev has renewed the presidency, and after a sequence of referendum, has been appointed for life enjoying a special status (as prescribed by the same constitution). The constitution formally allowed and promoted opposition, the role of minorities, and the respect of civil and human rights, but the political practice is rather diverse; it must be said the Nazarbayev presidency, although undisputedly authoritarian, tried to avoid repressive degeneration (as it happened often in neighbouring countries), promoting the openness of his countries from international and economic point of view.

The republic has a parliamentary system that has been dominated for the whole transition time by the very popular president, with any opposition that could represent any serious problem to his rule. The president was (actually is) surrounded by part of the former Soviet elite, some members of his family and local leaders. His mainly opposition members were actually former supporters and collaborators.[154]

The popular opinion says supporters of the presidency are divided into two parts, "the hawks" and the "the doves". The hawks represent the old Soviet guard and they believe that the country will be accepted by the IC only because of its vast natural resources. On the other hand, "the doves", the youngest ones, believe in a new image of Kazakhstan, in a modern and west orientated nation. The decision of president

[154] https://www.globalsecurity.org/military/world/centralasia/kazak-politics.htm, accessed at 10.03.2018; Carlson Richard 2013: 127; possibly it is too early to talk about failure in democratization.

Nazarbayev would depend on which of the two groups is used to hear.[155]

Kazakhstan's constitution was approved in August 1995; it has been amended several times since 1995 and the last time was in 2017. Basically, the procedure to amend the constitution starts from a proposal made by the president, the government or the parliament. After the draft proposal, the president has two options: he can either submit the draft to the parliament or announce a constitutional referendum. Regarding the passage to the parliament, the amendments need a three-fourths majority vote of both houses and the signature of the president to become effective. On the other hand, the passage with the referendum needs the absolute majority of the votes and by one-half of the voter in at least two-thirds of the oblast, major city and capital, followed after the result by the signature of the president.[156]

The constitution guarantees equal rights for all the inhabitants without exceptions of ethnic or religious distinction.[157] The two main subjects of the constitutional system are the president and the parliament (the former Supreme Kenges), both elected by universal adult suffrage. The executive branch is formed also by the council of the ministers who are led by the prime minister; the legislative branch is formed by two chambers.[158]

The president is elected for a five-year term mandate by the popular vote. Nazarbayev was the chairman of the Supreme Soviet and after the independence he became the president of Kazakhstan. He has been in power since 1991 and even when he has been confirmed for another five years mandate after the election of April 2016,[159] he resigned in March 2019 as a President of the Republic. It is to consider that the president is the head of the state and determines not only the internal politics but also the foreign policy of the country.[160]

The constitution gives the president a lot of powers such as the powers of veto on the proposal of the parliament, the right to appoint the prime minister and the council of ministers, and the power to direct their actions. The council of ministers has the executive and administration powers.[161]

The Kazakh constitution, Article 48,[162] foresees the cases in which the President is released from its office or in case of resignation. Moreover, in this case the acting President of the Republic ad interim is the Speaker of the Senate, who will be in office until the next presidential election. The presidential election were scheduled for 2020, however after in March 2019 Nazarbayev announced his resignation and on 9 April his successor Tokayev announced that a snap election would be held on 9 June to avoid "political uncertainty".[163]

The acting president, and now also the main candidate in the next election, is Kassym-Jomart Tokayev. Tokayev is the man chosen by Nursultan Nazarbayev to take over his heredity as President, a man with a strong experience in the foreign service and also as a president of the UN office in Geneva. He speaks fluently English, Russian and Chinese, but had to take an exam of Kazakh language to run for the presidency. Currently are running seven other candidates in the next election, but he still can be considered the probable winner.[164]

The prime minister is Askar Mamin who was appointed in February 2019 by the former President Nursultan Nazarbayev after some protests in the country for the worsening of the living conditions.

A new amendment to the constitution in May 2007 introduced limits for the presidency to a five-year mandate and established a two-consecutive-term limit. Nazarbayev did not evidently respect them because he is appointed as a "First President of Kazakhstan", a special and unique institutional figure. The approval of the action of the President is nearly total among the population and this can be seen also on the

[155] https://www.globalsecurity.org/military/world/centralasia/kazak-politics.htm, accessed at 10.03.2018.

[156] https://www.cia.gov/library/publications/the-world-factbook/geos/kz.html, accessed at 11.03.2018.

[157] Article 14 of Kazakhstan Constitution: "1. Everyone shall be equal before the law and court; 2. No one shall be subject to any discrimination for reasons of origin, social, property status, occupation, sex, race, nationality, language, attitude towards religion, convictions, place of residence or any other circumstances."

[158] http://factsanddetails.com/central-asia/Kazakhstan/sub8_4d/entry-4659.html, accessed at 11.03.2018.

[159] https://www.cia.gov/library/publications/the-world-factbook/geos/kz.html, accessed at 11.03.2018.

[160] Article 40 of Kazakhstan Constitution.

[161] http://factsanddetails.com/central-asia/Kazakhstan/sub8_4d/entry-4659.html, accessed at 11.03.2018.

[162] Article 48 Kazakhstan Constitution paragraph one: "In case of premature release or discharge of the President of the Republic of Kazakhstan from office as well as in case of his death the powers of the President of the Republic shall be transmitted to the Chairperson of the Senate of the Parliament for the rest of the term; if the Chairperson of the Senate is unable to assume the powers of the President they shall be transmitted to the Chairperson of the Majilis of the Parliament; if the Chairperson of the Majilis is unable to assume the powers of the President they shall be transmitted to the Prime Minister of the Republic. A person who has taken the powers of the President of the Republic shall correspondingly withdraw his powers of the Chairperson of the Senate, the Majilis, the Prime Minister. In this case filling of those state positions shall be carried out in the order, stipulated by the Constitution." http://mfa.gov.kz/en/hague/content-view/the-constitution-of-the-republic-of-kazakhstan, accessed at 04.05.2019.

[163] https://www.aljazeera.com/news/2019/04/09/kazakhstan-to-hold-early-presidential-election-on-june-9/, accessed 8.11.2020.

[164] https://eurasianet.org/kazakhstan-election-candidate-dropped-for-poor-kazakh-skills, accessed 04.05.2019.

percentage of the vote given to him, usually about the 98% of the vote.[165]

The bicameral parliament consists of two chambers: the Senate and the Majilis. The Senate is formed by 47 members with 32 indirectly elected by majority two-round vote by the oblast-level assemblies and the remain 15 members appointed directly by the president that in this way can control the assembly. The senators remain in charge for 6 years but the assembly is renewed one-half every 3 years. The other assembly, the Majilis, is formed by 107 members, 98 of them are elected in a single national vote to serve for a five-year term mandate, the other 9 indirectly elected by the Assembly People of Kazakhstan. This assembly is formed by 350 members and its unique role is to be an advisor body of the president and designed to represent the ethnic minorities. The last Senate election was in 2017 and the last Majilis election in 2016.[166]

One of the most important constitution body is the Security Council, an organ with consultative and deliberative powers that elaborates decision and collaborates with the Head of the State in the field of: national security, protection of the national sovereignty, independence and territorial integrity of the state, maintenance of the social political stability in the country and in guarantying constitutional rights and freedoms of citizens.

The constitutive law foresees, as represented in the council, the following figures: President of Kazakhstan, Chairman of the Majilis of the Parliament, Chairman of the Senate of the Parliament, Prime Minister, Head of the Administration of President, Secretary of the Security Council, Chairman of the National Security Committee, Minister of Foreign Affairs, Minister of Defence.[167]

Despite the prediction of the law, the former President Nursultan Nazarbayev will remain its chairman, arising questions about how the power will be shared with the new President.

The main party is the party of the president and it is the currently ruling party; at the last Senate election it had the 82% of the vote, that means in the Majilis the election of 84 seats. The main (apparent) opposition parties are the Ak Zhol and the communist party, who reach only the 7% of the vote.

All the political parties in the country are organized around some political leaders who address their political activity; in order to do their activity, they must be registered and authorized and to do so they must submit all personal information about their leader and members included addresses and employment. Obviously, this restriction law doesn't permit a lot of people to join the political activity. On the other hand, the authorities push people to join the main political party the Nur Otan and to become active in it. The law about political party permits a party to be registered only if they hold a congress with over 1.000 delegates from two-thirds of the oblast and the major cities. This numerical request for the opposition parties is very difficult to reach and of course, reduced their activity and the number of political parties registered. The law also contains prescription about the funds that are necessary to come from individuals and other organization, and with these they have to pay all the political campaign—being strictly forbidden the foreign aid.[168]

13.6.3 Current Politics

This restrictive law, and its strict application permit to explain the vote result and also the difficulty of Kazakhstan to become a complete democracy following what it is written in the constitution. The two chambers operate together in continuous sessions. A law usually is drafted and submitted by the government or the president.

The president has also the power of veto on the law approved by the parliament that can be overcome only with a vote of two-thirds of the two houses. The same majority is also necessary in the case the two houses don't vote for a minister proposed by the president.[169] Needless to say, such a high majority, and the control of the Nur Otan on both parliament houses, makes it very difficult to have a real parliamentarian activity.

The highest court of Kazakhstan is the Supreme Court of Kazakhstan; it consists of 44 judges nominated by the president after the recommendation of the Supreme Judicial Council and confirmed by the Senate. The Supreme Court represents the appeal court for the decision made by the lower juridical court levels (district court and oblast court). The judges remain in charge until the age of 65 but their activity can be extended until the age of 70.

The second most important court is the constitutional court that judges about constitutional matters. Their members are appointed by the president, the Senate chairman and the majilis chairman for a three-year term; they appoint each one another judge for a six-year mandate and the chairman of the court is appointed directly by the president. Civilians

[165] https://www.cia.gov/library/publications/the-world-factbook/geos/kz.html, accessed at 11.03.2018.

[166] https://www.cia.gov/library/publications/the-world-factbook/geos/kz.html, accessed at 11.03.2018.

[167] www.akorda.kz/en/executive_office/presidential_councils/security-council, accessed 29.04.2019.

[168] https://www.globalsecurity.org/military/world/centralasia/kazak-political-parties.htm, accessed at 11.03.2018.

[169] http://factsanddetails.com/central-asia/Kazakhstan/sub8_4d/entry-4659.html, accessed at 11.03.2018.

do not have a right to appeal the constitutional court decision and also its decision is subject to the presidential veto.[170]

In the last years, Kazakhstan has increased its international role: in 2017, it held the Expo, one of the most important events, where every country presented its idea about an important theme (it was the first time for a CA country); in the same year in the capital Nursultan the Kazakh government hosted a very demanding diplomatic negotiation meeting between the Syrian government and the opposition groups. Despite all these acknowledgements, human rights situation—observers refer—in Kazakhstan remains bad.

Starting from political rights, the right to vote freely is limited. The opposition parties face a lot of problems to express criticism and opinions. The constitution gives the right of assembly but the law on national security restricts very much the free meetings. Other restrictions on politics are related to the freedom of speech, of vote and of press.[171]

Actually, independent and opposition journalists face several restrictions such as physical attacks and criminal persecutions. For example, in May 2007 unknown assailants stabbed the independent journalist Ramazan Yesergepov, a crime that remains unsolved by the police. Several courts have tried in the last years to convict newspaper and journalist: in September, the Almaty Court convicted the editor of an independent newspaper and banned him from journalism; in November 2017 a court sentenced the president of the National Press Club to 6 years in prison. Also, the parliament has scheduled to discuss new media law that could again restrict the right.[172]

NGOs and other organization face several limitations in pursuing their objects. Actually, the right of free assembly is very restricted in Kazakhstan. Any public organisation set up by citizens, including religious groups, must be registered with the Ministry of Justice, as well as with the local departments of justice in every region in which the organization conducts activities. During this registration procedure NGOs reported several problems mostly determined by the number of documents to be submitted.[173]

The freedom of assembly, despite constitution previsions,[174] actually is not respected. A public demonstration needs to be authorized by the security forces and its violation can be punished with 75 days in prison. On 1 August 2017, for example, two peaceful protesters Askhat Bersalimov and Khalilkhan Ybrahamuly were detained and sentenced for taking part in an unauthorized demonstration.[175]

Some changes in allowing peaceful demonstration can be seen during the 2019 presidential political campaign after the resignation of Nursultan Nazarbayev; the oppositions organized some peaceful protest against the ruling party in some of the cities, protest that gathered circa 300 people in the capital, in what it can be considered one of the biggest opposition gathering organized in this period.[176]

In the last years, Kazakhstan introduced new amendments in its criminal code to punish the crime of torture: according to the criminal code, it needs to be investigated by a different agency; the officer responsible for the abuse should be signalled.[177]

Despite the law foreseeing torture as a crime, it seems such kind of practices remain a standard procedure for police and security forces; so similarly for further ill-treatments to the prisoner: such cases are reported by the human rights ombudsman and the National Preventive Mechanism. Just in 2013, these two committees reported 43 cases of torture, and in 2014 the number rose to 295. The action of the two human rights authorities is very limited; they can only express to the security forces some recommendations that have no legal value.[178]

The police forces use not only torture but also arbitrary arrest and detention especially for crime related to politics. Only in 2012 security forces detained unlawfully 209 people. These procedures are often used for political opponents who are arrested or detained for minor offences for several months without any charges.[179]

Other important personal rights are restricted in the country. In January 2017, an independent organization of trade

[170] https://www.cia.gov/library/publications/the-world-factbook/geos/kz.html, accessed at 11.03.2018

[171] http://factsanddetails.com/central-asia/Kazakhstan/sub8_4d/entry-4666.html, accessed at 18.03.2018.

[172] https://www.hrw.org/world-report/2018/country-chapters/kazakhstan, accessed at 25.03.2018.

[173] https://www.amnesty.org/en/countries/europe-and-central-asia/kazakhstan/report-kazakhstan, accessed at 25.03.2018.

[174] Article 32 of Kazakhstan Constitution: "Citizens of the Republic of Kazakhstan shall have the right to peacefully and without arms assemble, hold meetings, rallies and demonstrations, street processions and pickets. The use of this right may be restricted by law in the interests of state security, public order, protection of health, rights and freedoms of other persons", https://kz.usembassy.gov/wp-content/uploads/sites/46/Kazakhstan-2018-Human-Rights-Report.pdf, accessed 13.11.2020.

[175] https://www.amnesty.org/en/countries/europe-and-central-asia/kazakhstan/report-kazakhstan/, accessed at 25.03.2018.

[176] https://eurasianet.org/kazakhstan-anti-government-rallies-signal-emboldened-protest-mood, accessed 06.05.2019.

[177] Article 146. Torment "1. Intentional infliction of physical and (or) mental suffering, committed by crime investigator, person, conducting an investigation, or other civil servant or other person with their incitement or with their consent or acquiescence, in order to obtain details or recognition from tortured or other person or punish him (her) for the actions, which he (she) or another person is committed or in commission of which they are suspected, as well as intimidate or coerce him (her) or third person or by any reason, based on discrimination of any nature, [...]"

[178] http://factsanddetails.com/central-asia/Kazakhstan/sub8_4d/entry-4666.html, accessed at 25.03.2018.

[179] http://factsanddetails.com/central-asia/Kazakhstan/sub8_4d/entry-4666.html, accessed at 25.03.2018.

union was closed by a court due to the absence of the proper authorization; also religious movements faced the same difficulties in doing their activity and in the last years; several Muslim, allegedly extremists, and Jehovah's witnesses were arrested.

A further sign of involution is the restrictive turn in private rights matter; former President Nazarbayev in July 2017 signed new laws that have decriminalized the domestic violence against women. The situation for homosexual is not good; often they hide their sexual orientation; actually, transgender people can face invasive and humiliating procedure such as psychiatric diagnosis and coerced sterilization.[180]

13.6.4 Kazakh Leadership

Nursultan Abishevich Nazarbayev was born into a family of Kazakh peasants. He studied in Ukraine and Kazakhstan, becoming an engineer. His political career began in fact in the 1960s as Party Secretary of Karaganda's metallurgical Kombinat, going through the ranks of the communist party. From 1977 to 1979 he served firstly as secretary, then as a second secretary of the party committee in Karaganda, then he became Secretary of the central committee of Communist Party Kazakhstan. Before the fall of the Soviet Union he was appointed prime minister of the Soviet Republic of Kazakhstan. When the country declared its independence, Nazarbayev presented himself at the first independent election of Kazakhstan.

He won the presidential election in December 1991 with 98.7% of votes, and thanks to the result he was appointed president of Kazakhstan for the first time. He won the presidential elections in 1999, 2005, 2011 and 2016, before announcing his resignation in 2019. These two last elections were strongly criticized by international observers due to lack of fairness.

In 2007, the parliament approved a constitutional law that restricts the presidential term limits to 5 years and limits the consecutive terms to two mandate as a president. Despite this limit he was appointed in June 2014 by a constitutional law as president of Kazakhstan and Leader of the Nation. The title permits Nazarbayev to have immunity from all the investigation and prosecutions; it attributed to him the power of veto to the parliament and the power to speak to the assembly even if he is not the president. Nazarbayev has two daughters and a son; his family in the last year doesn't appear so much in the political scene despite being parliamentarian and member of the main political organism in Kazakhstan.[181]

On 19 March 2019, Nazarbayev resigned in a television speech to the nation, in a date chosen just before the traditional Naurys (New Year) celebration. Nazarbayev, who is now, 78 has "reigned" for the last 29 years and he remains the only president known by most of the Kazakh. Despite the fact that he will still be considered Leader of the Nation and Chairman of the Security Council with the total immunity, this event represents an important legal transition to power in a region with a scarce tradition of democratic processes.[182]

13.6.5 General Situation

Nursultan demonstrated on several occasions a sincere interest in modernizing his country; nevertheless, he demonstrated a strong centralization and personalization attitude that is evident in daily governance style.[183]

The former president was the previous party secretary, and he had the chance to learn the functions of the administrative Soviet apparatus; then he took the advantage of the power vacuum, ruling the country with authoritarian style, giving deliberate priority to stability and economic development (eventually blandishing the oppositors with HC revenues), even at the expenses of individual rights.

In fact, his authoritarian rule is deemed by some observers unavoidable (or necessary), considering the condition of the country in those years of transition, and above all the risk for geopolitical manoeuvring from outside. Many signs evidence the intention of pursuing a socially inclusive development—combining market economy with some public monopolies—eventually to be considered as a premise for a further stage of political reforms.

For the entire Nursultan presidency, the HC monopoly represents the predominant aspect of economics; then today the major question is about such centralization of power, and about the correlated question—succession dilemma, bureaucratization of society, economic stagnation and inefficiency. It diffused the opinion that its strong presidency for so many years had a negative impact on the possibility of emerging new leaders and on the renovation of the political leaders.

Such questions are object of speculations of any kind; the presidential extraordinary accumulation of wealth and power represents a motive of general preoccupation. The constitution predisposes a criterion for organizing the transi-

[180] https://www.hrw.org/world-report/2018/country-chapters/kazakhstan, accessed at 25.03.2018.

[181] https://www.globalsecurity.org/military/world/centralasia/nazarbayev.htm, accessed at 25.03.2018.

[182] https://www.aljazeera.com/news/2019/03/kazakh-president-nursultan-nazarbaev-announces-resignation-190319132408987.html, accessed at 06.05.2019.

[183] The Kazakh chronicle has been characterized in the recent past by dramatic events as assassinations of opposition leaders as Altynbek Sarsenbayev in 2006, possibly not directly involving the regime; the attitude towards the media appears as scruples less; a journalist who accused the government to protect killers of the opposition leader was given two years in prison; Warf Barney 2011: 14.

tion to the future in case of premature release of the president[184]; but it is to consider there is not an obvious successor. Nazarbayev had success in building up a political construction around himself, taking benefit from a set of favourable circumstances, using more or less wisely the countries natural wealth (HC, minerals like uranium, extensive agricultural resources); he combined this practice with the diffusion of a mild neo-nationalist method without alarming neither the internal minorities nor the powerful neighbours.

One of the reasons of the country's success is wisely staying out from troubles that affected further CA countries: Nazarbayev evidenced a prudent attitude since early independence time, carrying a friendly and respectful politics to Russia and other neighbours. It is to underline the fact he took during that time some critical decisions, in absolute loneliness, as that, to be recognized today as a historical merit, of renouncing unilaterally to nuclear warheads, immediately stopping the atomic programmes, elaborating a policy looking at environmental clean-up and green technologies.[185]

The Kazakh territory is evidently exposed to geopolitical tensions, but—in some terms—it is "too big to fail", signifying that any tension here would have consequence on continental scale (therefore carefully avoided). Sometimes, indeed, Nursultan exhibited a kind of acrobatic capability, ruling a country exposed to different scenarios, extremely rich but demographically weak, characterized by environmental problems and geo-economic disequilibria. These, in particular, are arising mainly between the western Caspian provinces and Nursultan, rapidly growing, on the one side, the former capital city, with the north-eastern Russian-speaking provinces from the other; and finally a third region comprehending the southern Kazakhstan conservative centres, echoing the close Fergana influences. Such internal differentials indeed overlap further regional and cultural cleavages (presence of minorities, religious traditions), then resulting in something especially dangerous.

It must be said, such politics proved to be relatively easy in times of high oil quotations, but it will be presumably with difficult reproduced in the future (assuming the oil prices tendential reduction); probably Kazakhstan will face tensions deriving from the fact the new societies—mainly composed of urban class and oligarchies depending from HC extra profits—must re-dimension their standard of living.

The government is aware of this, and it seems to be committed in preventing such drift; it will be especially difficult to develop, in such public monopoly dominated regime, a truly open economics, connected intrinsically with the consolidation of a middle class. In such situation it is difficult to establish a well recognizable "line" between public and private limits with regard to a long list of situations (e.g. distinguishing between social benefits and individual profits, welfare and corporate activities, state-owned or open-market resources).[186]

The chronicle refers continuously about such kind of situations deriving from such kind of misunderstandings; it is the case of accusations about suspicious dealing, reproaching ministries and officials for not transparent administration of HC revenues[187]; some parties claim that such money should be administered, diversely as corporate profits, with a special status, being properly audited. In particular, it would be useful to predispose a kind of HC- or commodity-based stabilization fund, accumulating export revenues (indeed obtained exploiting the national natural capital stock possibly to detriment of future generations), to be administered with special regulations (a policy that has been already experimented in several case in the world, especially useful for rentier states).

Some recent signs evidence that the government made some concrete effort (at least considering the scarce records characterizing the other CA countries), making some progress in this path; furthermore, it is to consider the recent decisions undertaken by the Supreme Court, to modernize and render more effective the country legal system, eventually with the help of foreign experiences, in particular for reforming the public accountability system.[188]

[184] Articles 48 of the Kazakhstan Constitution: "In case of premature release or discharge of the President of the Republic of Kazakhstan from office as well as in case of his death the powers of the President of the Republic shall be transmitted to the Chairperson of the Senate of the Parliament for the rest of the term; if the Chairperson of the Senate is unable to assume the powers of the President they shall be transmitted to the Chairperson of the Majilis of the Parliament; if the Chairperson of the Majilis is unable to assume the powers of the President they shall be transmitted to the Prime Minister of the Republic. A person who has taken the powers of the President of the Republic shall correspondingly withdraw his powers of the Chairperson of the Senate, the Majilis, the Prime Minister. In this case filling of those state positions shall be carried out in the order, stipulated by the Constitution. 2. A person who has taken the powers of the President of the Republic of Kazakhstan, on the basis and in the order stipulated by Paragraph 1 of the present Article, has no right to initiate amendments and additions to the Constitution of the Republic of Kazakhstan." http://www.parlam.kz/en/constitution. accessed at 04.07.2018.

[185] Kazakh government promoted the entitlement of the holyday 29 August as the day against nuclear testing celebrating the closure of the Semipalatinsk Nuclear Test site 29 August 1991.

[186] U.S. Mission Kazakhstan, 23 February 2015.

[187] "In 2013, Aftenposten quoted the human-rights activist and lawyer Denis Jivaga as saying that there is an 'oil fund in Kazakhstan, but nobody knows how the income is spent'", http://www.almaty-kazakhstan.net/kazakhstan/economy/, 2017, accessed 6.7.2018; Claire Milhench 2018; see as well Heathershaw John 2013: 187 about EY auditing, not just for HC but also cotton, aluminium, and further commodities.

[188] Fenner Zinkernagel Gretta and Attisso Kodjo, no year.

13.6.6 Change of Capital City as Geopolitical Manoeuvre (1997)

One of the main problems Nursultan had to face—at the moment of independence, suddenly acquired, of such huge country—was that of reconciling different and possibly not compatible tendencies, between stability (necessary for the new leadership) and growth (usually predisposing to the opening of the society, but necessary in order to make the society better governable).

The solution Nazarbayev found, in order to propose a solution for such incongruences, was that of moving the capital status from Almaty to the new and recently renamed capital Nursultan (formerly Astana, Akmola and Tselinograd): the same city which in the 1960s was elected as centre for the "Virgin Lands" campaign, but then regressed to a status of provincial town, without big expectations.

With this manoeuvre Nazarbayev could solve the problem of conciliating two opposite tendencies; the building up a new capital town should symbolize the overcoming of such duality (then stability vs. opening).

Almaty, the CA metropolis, is characterized (still now) by a vivacious inter-ethnic society (resulting both, from generations of immigration from all over the SU and from a rapid internationalization occurred in transition times), while Nursultan would represent a brand-new capital city, the place where to build up a new country reference, showing a new perspective.

At the same time, the manoeuvre pursued different targets. Moving in the middle of the steppe, Nursultan removed the country barycentre from sensitive borders, getting closer to the Russian minority territory (it is to consider that the decision has been made in a period of relatively weakness of international role for Russia). Then in this way he could unify and gather the two components of the Kazakh nation (the Russian north-east and the Kazakh centre), centralizing the government practice.

Furthermore—as a not secondary motivation—with the same manoeuvre, he could renovate the old style Soviet bureaucracy, moving the state apparatus in new almost unknown location, then cutting-off the powerful administrative apparatus from its background. Then it also represents an easy way for investing huge HC revenues (indeed an operation without alternative, considering the insufficiency of both local demand and local entrepreneurial capability necessary in order to make the economic circuit functioning), then configuring an artificial (Keynesian) way for pursuing growth.

In this way, above all, Nazarbayev could build up a landscape suitable of representing materially a stabilization process, to keep distant the political risk, represented in that period by unpredictable street rebellions, taking the form of "coloured revolutions" (the nightmare of the local despots).

Nursultan represents also environmentally a disadvantaged place for such risks; it is cold in winter (and hot in summer), and its urbanistic—predisposing extended distances and empty urban spaces—represents a disincentive for any kind of protestations; at the same time it became the shop-window of the new power as well as new apartment building-city to attract Kazakh immigrants, modernizing their life with malls, auditoriums, facilities, presenting a set of perspectives attracting high qualification personnel, for example, top research institutions, medical and scientific faculties.

The city is dominated by an futuristic skyline, with original architectures combining with impressive urban symbols, with new city-satellite quarters continuously expanding—especially the futuristic Expo 2017 area—and with urban highways representing a kind of car-race lane for the increasing private traffic.

So, with the transfer Almaty-Nursultan, the latter would maintain the functions of a world city that may develop distant from the power, without disturbing too much the elite; while Nursultan became a kind of "fortress in the steppe", re-orienting the development axis of the country, in the core of its territorial "body", less exposed to outside influences.

Whatever it would be the main motivation, Nursultan will remain the symbolic place of its first president—comprehending as well "museum of the first president" and further monuments and symbolic constructions: it cannot but appear as a pharaonic representation of the Nursultan power.[189]

The story of the capital-city change is still yet to end and after the resignation of President Nursultan Nazarbayev, its name was changed to "Nur-Sultan" in order to celebrate the long-standing former president.[190]

13.7 The Big Divide—New Despotism or Democratization as a True Target?

13.7.1 A Double Standard

Neo-isolationist tendency has some limits; in fact, the same elite of the CA countries demonstrated to appreciate—and to need—the benefit deriving from participating in the world economics and politics. Then, since the indiscriminate opening cannot be tolerated, it is arising a kind of double standard, maintaining inside strong control on society, but without suffocating it, and without overcoming the threshold it would annoy the IC. A situation that can be maintained just in the frame of a kind of "patrimonial-authoritarianism",

[189] Carlson Richard 2013: 133.

[190] https://www.theguardian.com/world/2019/mar/23/kazakhstan-renames-capital-nur-sultan, accessed 29.04.2019.

financed with rich HC extra-profits, or with other public monopolies.[191]

For pursuing such contradictory targets (maintaining control inside, but developing integration in international economics), local governments have to elaborate new legitimization schema; beside the case of the former Turkmen President Niyazov, usually local governments do not deny democracy as an ideal.[192] This also considering that pluralism and openness (democracy's pillars) reflect the principles of the IC anyone adheres (above all UNO and OSCE), from which any country derives substantial benefits (considering international trade, tourism etc.) (even when this happens often in the frame of a manipulation).[193]

But this was realized in a rather fictitious way, leveraging on the manipulability of the population, that, having experienced the colonial and totalitarian power, is cauterized by a sense of deference for the power, that is either similar to that of fear and mistrust: such sentiment impedes to large population strata to adopt a more free attitude, namely autonomous towards the power.

Furthermore, it is to consider that the new "middle class" is mainly dependent on salaries paid by state-owned companies, exposing it to a kind of blackmail by the power. Evidently, it is a material and cultural question: the population need time to get accustomed to the democracy, as well as to market economy and to social participation (in order to control bottom up the functioning of welfare functions) and to economic initiative, then in general to a flexible idea of the existence as a dynamic of elements and responsible decisions (it takes time, possibly the generation passage).

13.7.2 Rhetoric of Guided Democracy

In order, possibly, to justify local failures in application of democratization standard, the local elite elaborated a hybrid category, the "guided democracy" (that, despite the recent origin, inspired already plenty of literature).

Such category has been often re-purposed by post-communist leaders who have to maintain an acceptable appearance, in order to justify the progressive closing in-itself of the system they are representing. It is often justified by the rhetoric of the "weakness" of the democracy, by the supposed fact that democracy cannot assure the necessary stability, namely that it cannot face the international chaotic environment.

Such interpretations have been eventually inspired by regional incumbent "great powers" (China and Russia), that have a specific interest in diffusing an ideology justifying a contraposition towards the "west" and its supremacy, that would realize through the globalization, considered as a pretext, in order to impose some subalternity (possibly a legacy of Soviet anti-western diffidence).[194]

Such interpretation would also derive from cultural-historical elements, namely from a certain delay, that characterizes the non-western countries, that would better be represented by collectivistic centralistic institutions, eventually sacrificing the individual and "bourgeoisie" rights (in order to permit the realization of the collective interest).[195]

Finally, such delay would be justified by a kind of rhetoric of the "first president": it means that, in such exceptional circumstances (in the early stages of the foundation of a nation-state), with the country engaged in the difficult transition, it would be necessary to have a strong guidance, eventually emendating the usual rights and procedures.[196]

13.7.3 Administrative Level Democratization as Intermediate Target

This model of "guided democracy" would represent an intermediate target, since the "tout court" "democracy", whenever representing a long-term target, would be not yet adequate for local societies that were not yet "ready" for such ideal.

Democracy seems to be, indeed, popularly considered as an ideal, or a long-term target, and it is recognized as the ideology characterizing the most advanced, prosperous and peaceful countries in the world. Therefore, whatever are the true intentions, or the geopolitical contingencies, the local leaderships need to purpose credible methods in order to demonstrate they are not against such principle.

One of them relies on the diffusion of such practice on the territory, at first in the peripheral functions, then, progressively, in environments closer to sensitive high-political

[191] Babajanian Babken 2015: 514; Megoran Nick 2017: 34 fs.

[192] Megoran Nick 2017: 44; Carlson Richard 2013: 127; Dagiev Dagikhudo 2014: 1.

[193] Carlson Richard 2013: 127.

[194] See the cited letter of Putin to NYT. Following the Wallerstein's schema, the CA would be collocated in the semi-periphery, then in the area where the regimes, since engaged in a competition with the "centre", defending itself from neo-colonialist exploitation attempts, cannot in principle "afford" the "privileges" of liberalism and democracy.

[195] Hanks Reuel R. 2016.

[196] Indeed the biggest present the current "first president" could do to their people—after a difficult transition, considering the un-doubtful merit of having maintained the stability—it that of having resigned, possibly exerting then, from a defiladed position, a role of "controller", using moral suasion instruments, for the period necessary in order to consolidate a state of the law, possibly originating with the time a true a democracy.

functions. It is based on a principle of gradualism and relies on a first experiment of de-centralization. This in particular in occasion of administrative elections, to be carried at the local level (then possibly meaning less political significance): this as an attempt in order to get the population accustomed with a political model based on initiative and critics (even when just at local level).

In theory after these experiments, the democratic method should be extended progressively (from the periphery to the centre), and with prudence, in further dimensions of the public life (then it is a model that assumes as "taken for granted" the "good will" of the ruling elite). Concretely such experiments concern administrative level of politics, appearing as an intermediate stage in the direction of the diffusion of a representative democracy; until now they have produced uncertain results. Usually administrative geography in CA is articulated on base NUTS 3 or 4 (state, oblast/region, rayon/province and micro-rayon, following roughly the Soviet schema) with big cities ascribing to a particular status, with governors being appointed centralistically.

Such decentralization has been applied in an experimental way in Kazakhstan and in Kyrgyzstan, with few examples also in Tajikistan; sometimes indeed it seems either an attempt for the actual power to get further legitimization, conciliating its role with the local traditions in order to get a further consensus effect.

An example of this tendency is the Kyrgyz regulation recognizing the role in the villages of the "aksakal" committee, prescribed even in constitution.[197]

The country is subdivided in seven oblasts (regions) with Bishkek and Osh being administrated with special status, similar to that of the region; the reforms promulgated in (early) transition times (in the 1990s) prescribed a kind of "local self-governments" in the frame of a decentralization process, establishing 475 villages councils, with the local community "head" ("aiyl okmotu") elected by the village council for four years.[198] The situation represents actually rather advanced results (in comparison with other CA countries), possibly predisposing to a further form of basic democracy, to be extended with the time from the local level to a properly more political one.

Kazakh Republic is subdivided in 14 oblasts (NUTS 2), governed by an "akim" (governor), plus 2 provincial status cities (Nursultan and Almaty), with the "akim" appointed by president himself[199]; so as well for the mayor of further metropolitan areas, appointed by the central government, eventually confirmed by bottom up elections. Lower level municipal "akims" (town mayor, apparently NUTS 4) are nominated by the regional governor.

Uzbekistan represents possibly the main exception in this order; it is characterized by a heavy structuration of a power, that flows from the centralistic level down the territorial hierarchy, till intra-urban level (it should be considered the situation of the country, characterized by relevant conurbations; Fig. 13.9).

It consists in forms of institutional control at mahalla (city neighbourhood) level, relying on a supervisor for local order (see the dedicated chapter).

It is to consider that such local committee, and similar neighbourhood-scale councils (for "chruščëvka" neighbourhoods, villages and kishlaks, kolkhoz and further community-scale settlements) exist anywhere and are rather of spontaneous character, either than top-down induced (reflecting possibly a basic community-democracy sentiment); sometimes they are just informal communities, to some extent recognized by wider authorities.[200]

13.7.4 Democratization Question

Democratization (and pluralism) issue is inevitable question, even when postponed periodically on the governments' agendas because of incumbent emergences; in fact, the increasing complexity and diversification, as well the inclusion in world economics, must theoretically induce the regimes to decentrate some capability, delegating some power on the territory, in the society and in the economics. Without some kind of pluralism—and political redistribution—the development path will encounter soon or later a limit, bringing to the conflict (since the "dictator" and the dictator's clan cannot just accumulate unlimitedly power).

Indeed, none of the CA republics until now has undertaken irreversibly such "road map" (with possible exception of Kyrgyzstan); after few experiments, in the early post-Soviet opening enthusiasm, the regimes demonstrated to prefer the "continuity" as the priority target, undertaking a reverse tendency.[201]

[197]Eventually considering as well a local religious leader guide; Brill Olcott Martha 2012: 24 fs.

[198]Babajanian Babken 2015: 514; Bhuiyan Shahjahan H. 2010.

[199]Kudaibergenova Diana T. 2015: 446.

[200]https://www.hrw.org/reports/2003/uzbekistan0903/5.htm, Role and functions of mahalla committees, accessed at 04.07.2018; in 1999 the Parliament voted the Law of Mahalla transforming this traditional social structure in an administrative unit inside the State. Its main powers are: control on education, employment, organize the use of land, organize volunteer service to provide the maintenance of public land. Since 2000 the committee has had also a function in the criminal prevention and with the Posbon Law in which designated the mahalla assembly to elect a posbon who has to verify the social and moral environment of the neighbourhood and all the information collected during its work are kept in a special book by the local police; see as well Hanks Reuel R. 2016; Brill Olcott Martha 2012: 27.

[201]Babajanian Babken 2015; Dagiev Dagikhudo 2014.

Fig. 13.9 Administrative map of Uzbekistan. (Source: https://commons.wikimedia.org/wiki/File:Uzbekistan_regions_map.png, accessed 5.7.2018)

For a while the pluralistic expectations may be delayed and covered by the HC (and further monopolies) revenues invested in terms of social spending (and opposition blandishing), but soon or later it will be evident that the politics will face some dilemma.

The diversification appears as a non-renounceable process, and also a convenient one; anyway, the current globalization tendencies (connected with ICT diffusion and transport revolution) makes it unimaginable a regression to a hermetic "locked in" regime like in Soviet times (since any political barrier is today technologically permeable).[202] It is connected with a societal development, bringing usually to the formation of a middle class (to some extent autonomous from the power), that has to be configured also institutionally (even when today the local society appears either fragmented in clan, clienteles and lobby).[203]

13.7.5 Democratization as Fictitious Game—As a Diversion

In the search for a political compromise, namely a "third way" between democracy and stability (a kind of political "shortcut"), the local governments elaborated (as said) an idea of "guided democracy".[204]

In this way the local "presidents" demonstrate an ultimate capability, namely the capability of conditioning the whole

[202] Indeed it is possible to observe some examples of societies, like Russia and Turkey, that historically represent important examples for local populations, that currently seem to regress to some autocratic pattern.

[203] Such evolution means the reorganization of some principles such as "rule of the law" (to be considered as an intermediate phase on the path to a "true" democracy); then it means the establishment of some independent powers, inside the apparatus, "balancing" the central government; it is the case of independent bureaucracies, of justice apparatus, of social and economic associations, universities and research centres, and others.

[204] That, besides the inconsistent theoretical bases, relies on the peculiar capability of the post-Soviet state of organizing a sophisticated manipulation of the democratic game (indeed a capability inherited by the previous totalitarian system). It consists in the deformation, and also "tout court" creation of the elements forming the political scenario, then rules, procedures, roles and actors, as governments and oppositions parties, rituals, as elections, organizing fictive parliamentary session and discussions, and so on; these autocracies are capable of manipulate all the variables, using media, ICT, economic monopolies, constitutional procedures—to be emptied of genuine sense—and institutional organs in order to produce such fictive political game.

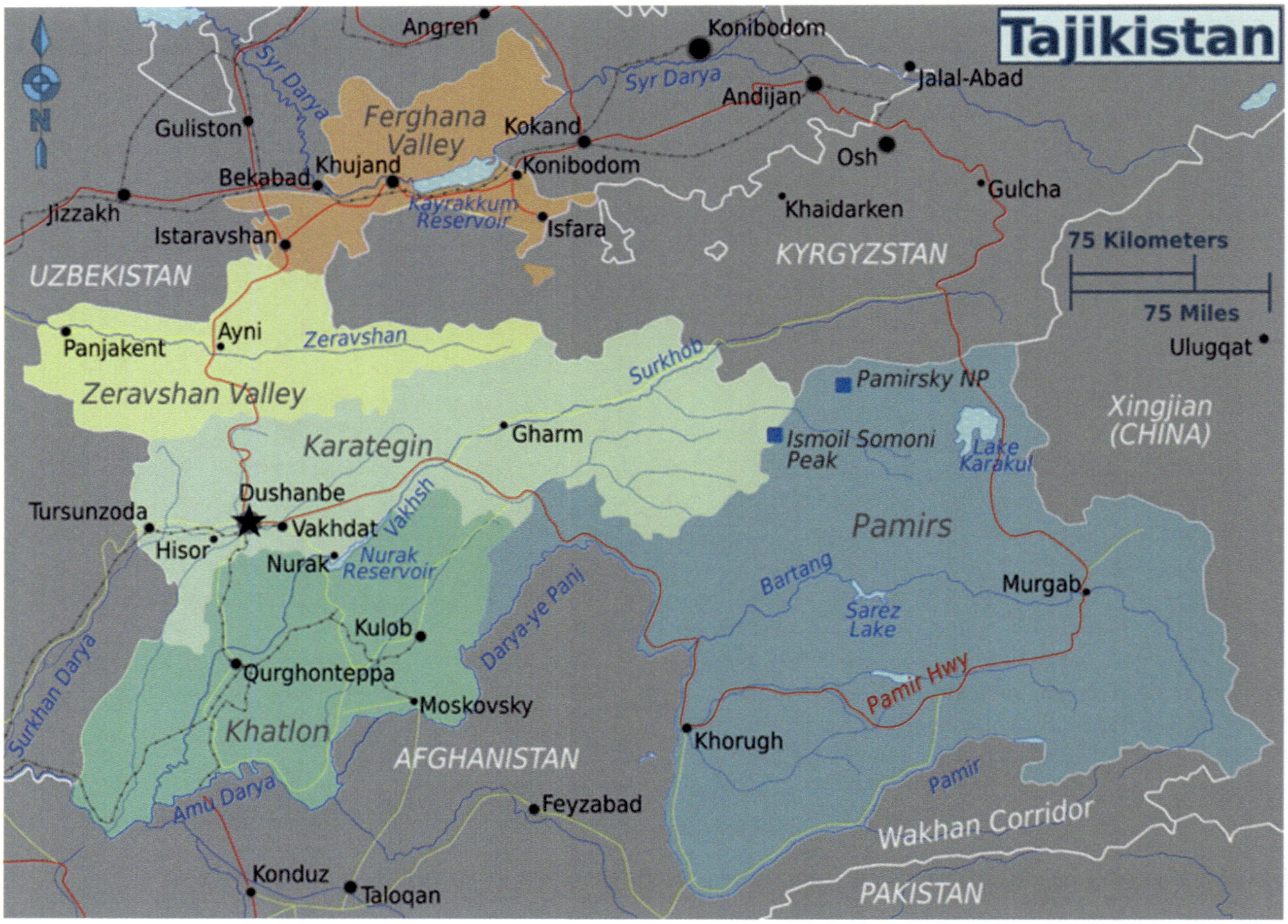

Fig. 13.10 Administrative map of Tajikistan. (Source: https://commons.wikimedia.org/wiki/File:Tajikistan_regions_map.png, accessed 5.7.2018)

political life of the country: they respect formally the political architecture, namely a constitution perfectly engineered by wise experts, in order to save the appearances, offering to IC always good arguments to appear as anything acceptable, and to have always some motivations to reject the reports of the independent organizations.

This can happen on the base of the elaboration of a sophisticated propaganda and manipulation techniques (relying on an expert "ideological" police, also this inherited by the previous regime); it is necessary to establish a kind of apparently real political game (in some sense, inducing the population to "believe" in it).

All the CA transition countries applied such techniques (again, with partial exception of Kyrgyzstan); the political-parties' screenplay is always similar, demonstrating the more or less intensive capability of the current presidents to manipulate the society.

In the case of Turkmenistan, Saparmurad Niyazov did it deliberately, establishing a heavy repressive regime, inspired by surveillance techniques, capable of controlling completely the Mejlis, the Turkmen parliament, and "de facto" of "kidding"—dramatically—the whole society (a policy in part continued by his successor Berdimuhamedov, even when some recent signs may evidence an evolution). He demonstrated the capability, indeed, not just to control, but to condition and directly "create" the political game and the entire society.

In other cases, it happened on more subtle manner. In Kazakhstan, much more exposed to international opinion, the presidential circle elaborated a more sophisticated game; it is the case of Majilis (Kazakh parliament, resembling the name of the ancient tribal assembly), with 2004 elections dominated by president's Otan party, with some residual seats conquered by nominal oppositions; in this game, in fact, the opposition is not completely banned or disappeared, but it results nevertheless incapable, being indirectly manipulated with tacit agreements, or other situations (it is notable that one of the opposition parties has been founded by former president daughter).

In this situation it appears the described phenomenon of personalization of the opposition, and of the political struggle in general: being the power personalized, the only true

Fig. 13.11 Tajikistan, Dushanbe, 2017, "bulevar" in the government district. (Photo: Igor Jelen)

opposition—namely any critical position—is made by individual president's enemies, that had supposedly betrayed the president loyalty, becoming then "dissident" (also this a typical figure inherited by the Soviet system).[205]

This "game" indeed has some limits; it cannot continue very long, and risk anytime toy degenerate "tout court" in a mass-repression dictatorship; in this perspective, the power can just to take time, hoping in some contingency, trying not to appear too "evil" (since in these times, it cannot imagine to rule just with physical repression).[206]

13.7.6 The Search for New Ideologies

Beside procedural and formal aspects, the NIS governments evidence the need for strengthening their legitimization, renovating the ideological "cement" at the base of the societies: repression and pragmatic approach (with a strong power appearing as the guarantor of order and peace) are not sufficient in order to maintain compact the society. It is necessary to elaborate as well a sense of persuasion, searching for consensus, actually the more convenient governance method, with population ideally adhering to government practices (evidently the mere repression has an intolerable, often incommensurable cost).

The Soviet collapse and the weakening of traditional and religious political-cultural roots evidence a vacuum that the NIS has to fill: after the failure of communism, of Soviet statism and of Russian colonialism, the local elite has to look for new ideologies.

The presidents need a new re-foundation mythology, elaborating a new inventory of ideologies; among them, observing the current evolution, it is possible to enlist consumerism, a kind of neo modernism based on ICT, neo-statism, based on a generous welfare conceded by public hand, financed by economic monopolies, and especially a neo-nationalist approach. This indeed relies on a diversified set of ideas, icons and values, evidencing a wide range of images, used as legitimization schemas, to be found in the ancient tradition, or sometimes being fully invented, manipulating symbols and myths, aesthetics and narratives.

Some of them assume respectively opposite significances, characterizing the current dialectic among the states, as happens respectively in the case of Tamerlane (Timur), assumed to be the founding father (the prototype of the power) for the Uzbek state; of the Manas epos, the primordial heroic warrior, for Kyrgyz fatherland; of the "batyr" tales for Kazakh traditions; and of the Somonid "arian" dynasty for the Tajik state.[207]

Further ideas trace to Jadid reformers of the nineteenth century, to Samarqand Medieval scientists, and to further

[205]In other words the opposition the current power occurs not for true political reasons, but for personal motivations; the oppositions are not—usually—structured groups, neither ideologically nor socially motivated, but they are often former president collaborators, possibly knowing his secrets (maybe bank accounts on some fiscal paradise).

[206]Brill Olcott Martha 2012:91; then a vicious circle, possibly predisposing to the definitive degeneration in an authoritarian state, that would encounter soon or later the risk of destabilization.

[207]Yilmaz Harun 2013:53; as said, this image has been eventually actualized with that of the "golden man", Kudaibergenova Diana T. 2013:164.

protagonist of the glorious cultural past; rulers sometimes claims to descend from Genghis Khan, while the Pamiri peoples legend refers to ascendency from Alexander the Great veterans settled down in the area; for post-Soviet Russians the figure of Lenin and even Stalin still represent a reference. All these are indeed ambiguous figures, alternatively considered as exterminators or wise rulers, exhibited in times of changes with intentions to "menace" or to reassure.[208]

Such signs of power are repurposed insistently by the public communication, adapting semblances and significances; the figures of Timurid and Samanid kings, that took the place on the city capital squares of Lenin, are represented either with European or Asiatic appearance, either with turban or crown, scimitar or sword, depending on circumstances.

So, for a long sequence of new figures, repurposed as the symbols legitimizing the actual power; and so also for religion, using symbols and images, hierarchies and "holy books", interchangeable with politics, even accrediting the ascendancy from Prophet Muhamad[209] as a legitimization right—as usual for who is aspiring to monarch status in Islamic countries.

In fact, the last trend seems that of the purpose of a monarchical feature, with the appearance here and there of kingdom signs on billboard, giant video screens and further communication devices, that seem to evidence a subliminal diffusion of such intention (being possibly the dynastic pretension a direct consequence of the entitlement as life president).[210]

13.7.7 Geopolitical Ideologies

Such images were mixed in a kind of popular geopolitics, in a game of symbols, territories and ideals, in the context of controlled mass media communication. Often they are manipulated in order to evidence the necessity of isolationist policies, of "fear of the foreigners", a sense of menace (separatism, terrorism, extremism, the "three evils" as defined by of the SCO statute, see later), eventually ethnic or religion-inspired, and finally justifying a kind of cult of personality: the new power needs some ideology (actually a soft- or a semi-ideology), easy to be assimilated, communicated and "consumed" by any social groups (with different level of education and sensitiveness).

This happens on an international scale, using geopolitical ideologies, with which the government can justify the belonging or the pursuing of an international order. It is the case of pan-Iranian culture for Tajikistan, of the revival of old religion peculiarities (eventually pre-Islamic, contraposing the Iranian origin to the Turchik one),[211] a mix of epical myths and ancient legends recovered as a legitimization tool by respective governments, tribes and clans.

It is the case of the "historical" empires, and obviously of "ancient" borders (evoking much larger sovereign territories), with figures as Genghis Khan or Tamerlane (or eventually Lenin and Stalin) used as proxy symbols, accrediting some "nobility" of origins, predisposing the transmission of the power from the "first president" to his descendants.

Such images would eventually be used in order to purpose representation of the mythical past, of the "historical" Uzbekistan or of the "great" Tajikistan, eventually of territorial claims, alimenting disputes over territories and resources. It is the case of the Uzbekistan claims over the whole Fergana, of the Tajiks on Samarqand and Bukhara, eventually considering the presence of rooted minorities: claims that were usually pretextually opposed and elaborated on order to maintain an "exchange" diplomatic-political money, to be spent on international disputes.[212]

Just a particular case is that of the Soviet "nostalgia" (the obsession of the "ranše"),[213] that but seems to lose appeal in the last decades (also considering the absence of celebration on occasion of the centenary of the Bolshevik revolution), as indeed many "nostalgia" feeling around the world; but it is still actual for Russian minorities in border areas, mainly in Kazakhstan and Kyrgyzstan (indeed becoming either a kind of private memory cult, than a public ideology).

13.7.8 Neo-consumerist and Globalization as Semi-Ideologies

Ideology is an especially relevant question for centralistic and tendentially despotic countries: the governments continuously try to elaborate what it is possible to define as semi-ideologies in order to regenerate motivations, to make periodically evident a target to their citizens and for the society in its whole.

So, after the season of twentieth-century modernist hard-ideologies, it is the time now of a diverse concept of social

[208] Forbes Andrew 2012; Totten Samuel and Bartrop Paul Robert 2008.

[209] Rasanayagam Johan 2014: 7, 12; Megoran Nick 2017; Yilmaz Harun 2013.

[210] Paskaleva Elena 2015: 425; Kyrgyz Republic National Academy of Sciences 1995; Glavnaja Redakcija Kirgizskoj Enciklopedii 1995; Abetekov 1995.

[211] Dagiev Dagikhudo 2014.

[212] Dagiev Dagikhudo 2014; Kudaibergenova Diana T. 2013:168.; namely, the key issue of international politics is that to find a common measure, a common convertible political "currency" in order to mutually exchange advantages and renounces in some scenario: this is possible if the international politics is organized in multi-lateral and institutional manner.

[213] Jelen Igor 2002, interviews.

Fig. 13.12 Tajikistan, 2017, Dushanbe, travel agency. (Photo: Igor Jelen)

coordination, searching for some popular idea easy to be reduced to slogan and easy to be communicated to wide populations.

It is the case of such programmes as the construction of new cities the setting up of giant infrastructures, emancipating the CA spaces, perceived popularly as landlocked also in economic and cultural terms.

It is the case of programmes such as "Nurly Zhol" (Bright Path), the TAPI "infrastructural promise" and the "energo-most" ("energy bridge" crossing this spaces), to cite just few of them; and it is the case of the same SR repurposed in terms of Belt and Road Initiative (BRI), as a device that would produce and redistribute peace and prosperity, mitigating popular anxieties.

Such devices finally result in a kind of giant-scale (even when usually ineffective) programmes resembling to some extent Soviet-like planning; indeed, they have further side motivations, beside the economic enunciated target, like that of keeping mobilized the apparatus (that otherwise risks a bureaucratization effect), and above all that of diverting the popular attention.[214]

But, possibly, all such ideological devices are losing relevance in the current days: the spread out of global practices, forming eventually an international middle-class standard (with expectation, style of life, values, comfort and capabilities standardized, due to imitation effect, similarly for the whole world) represents currently the most important reference (target) for millions of peoples (Fig. 13.12).

In fact, all these manipulations lead finally to a general confusion: such categories as nationalism, state-confessionalism or neo-communism (in the form of new-state monopoly) are here simply not applicable, if not with a renovated repressive politics.[215]

References

Abetekov A (1995) The culture of Kyrgyzstan in the age of "Manas", Kyrgyz Republic National Academy of Sciences: Manas 1000

Action Strategy on Five Priority Areas of the Country's Development for 2017–2021, Tashkent Times, http://tashkenttimes.uz/national/541-uzbekistan-s-development-strategy-for-2017-2021-has-been-adopted-following. Accessed 7.6.2018

Amnesty International (2016) Fast-track to torture, abduction and forcible return from Russia to. Uzbekistan, London

Anceschi L (2017) Turkmenistan and the virtual politics of Eurasian energy: the case of the TAPI pipeline project. Cent Asian Surv 36(4):409–429. https://doi.org/10.1080/02634937.2017.1391747

Anderson J (1995) Authoritarian political development in Central Asia: the case of Turkmenistan. Cent Asia Surv 14(4)

Babajanian Babken (2015) Promoting empowerment? The World Bank's village investment project in Kyrgyzstan. Cent Asian Surv 34(4):499–515. https://doi.org/10.1080/02634937.2015.1095967

Balletta Ruggero (2016) CESI, newsletter, 07.11.2016, Le incertezze dell'Uzbekistan dopo la morte del Presidente Karimov, https://www.cesi-italia.org/articoli/633/le-incertezze-delluzbekistan-dopo-la-morte-del-presidente-karimov. Accessed 5 July 2016

Barney W (2011) Geographies of global internet censorship. GeoJournal 76:1–23. https://doi.org/10.1007/s10708-010-9393-3

Bhuiyan Shahjahan H (2010) Decentralisation and local governance in developing and transitional countries. Int J Public Adm 33(12–13):658–672. https://doi.org/10.1080/01900692.2010.514445

[214]Kudaibergenova Diana T. 2015; Anceschi Luca 2017.

[215]Finally, the ideology is reduced to a technology of the power, intended as a mere political practice self-justifying, losing any ideological and social character, becoming a mere instrument used by the elite for its conservation.

Black CE et al (1991) The modernization of inner Asia. M.E. Sharpe, Armonk N.Y

Bowyer AC (2018) Political reform in Mirziyoyev's Uzbekistan: elections, political parties and civil society. Central Asia-Caucasus Institute & Silk Road Studies Program, Lithuania. Accessed at 19 May 018

Brent H (2017) Russia and Eurasia 2017–2018, 8th edn. Rowman & Littlefield, Baltimore

Brill Olcott Martha (2012) Tajikistan's difficult development path. Brookings Institution Press. 27 nov 2012

Compendium of Reports on Technical Election Assistance to Tajikistan, Pre-election Assessment, Election Day Observations, Post-Election Technical Assessment, March 1995, International Foundation for Election System, p 19. https://books.google.it/books?id=QqjoCF3AEYEC&pg=PA15&dq=elections+in+tajikistan+during+civil+war&hl=it&sa=X&ved=0ahUKEwiQx9b3z7LaAhWJCCwKHZcLBTQQ6AEILzAB#v=onepage&q&f=false, accessed at 11.04.2018

Corley Felix (2005a) Turkmenistan: President's personality cult imposed on religious communities. Forum 18. http://www.forum18.org/archive.php?article_id=522, 1 March 2005. Accessed 11 Apr 2018

Corley Felix (2005b) Turkmenistan: 2004, the year of demolished mosques. Forum 18. http://www.forum18.org/archive.php?article_id=481. Accessed 23.4.2018, 4 January 2005

Dagiev Dagikhudo (2014) Regime transition in Central Asia. Routledge, New York

Decaux Emmanuel (2003) OSCE Rapporteur's report on Turkmenistan, 12 March 2003, ODIHR.GAL/15/03 12

Elebayeva AB (1992) The Osh incident: problems for research. Post-Soviet Geogr:78–88

Lamy Frederick (2013) Social networking practices: Continuity or rupture with the Soviet past? In: Akyildiz Sevket, Carlson Richard, edited by, 2013, pp. 145–159

Rahmon Emomali (2018) official website of the President of the Republic of Tajikistan. http://www.president.tj/en/taxonomy/term/5/33. Accessed at 17 Jan 2018

Esenova S (1998) 'Tribalism' and identity in contemporary circumstances: the case of Kazakhstan. Cent Asia Surv 17(3):443–462

Fazendeiro Bernardo Teles (2015) Uzbekistan's 'spirit' of self-reliance and the logic of appropriateness: TAPOich and interaction with Russia. Cent Asian Surv 34(4):484–498. https://doi.org/10.1080/02634937.2015.1114780

Fenner Zinkernagel Gretta, Attisso Kodjo (no year) Returning stolen assets – learning from past practice: selected case studies. Basel Institute on Governance. https://www.baselgovernance.org/sites/collective.localhost/files/documents/131024_selected_case_studies.pdf. Accessed 23 Apr 2018

Forbes A (2012) Timur's legacy: the architecture of Samarkand. CPA Media. http://www.cpamedia.com/article.php?pg=archive&acid=120613150427&aiid=120613151433. Accessed 23 Apr 2018

Fox News (2006) Turkmenistan's leader promises citizens free gas, electricity and water through 2030. http://www.foxnews.com/story/2006/10/25/turkmenistan-leader-promises-citizens-free-gas-electricity-and-water-through.html. Accessed 23 Apr 2018

Haghayeghi M (1997) Politics of privatization in Kazakstan. Cent Asian Surv 16(3):321–338. https://doi.org/10.1080/02634939708400994

Hanks RR (2016) Narratives of Islam in Uzbekistan: authoritarian myths and the Janus-state syndrome. Cent Asian Surv 35(4):501–513

Harun Y (2013) The soviet construction of Kazakh batyrs. In: Sevket A, Richard C (eds) Social and cultural change in Central Asia. The soviet legacy. Routledge, London and New York, pp 45–62

http://enews.fergananews.com/articles/2991. Accessed at 10 Apr 2018

http://factsanddetails.com/central-asia

http://news.bbc.co.uk/2/hi/asia-pacific/132890.stm

http://uza.uz/en/politics/ecological-movement-holds-conference-on-elections-to-parliament-27.12.2009-1160

http://www.almaty-kazakhstan.net/kazakhstan/economy/ (2017)

http://www.c-r.org/downloads/Accord%2010_15Key%20elements%20of%20the%20Takikstan%20peace%20agreement_2001_ENG.pdf

http://www.eurasianet.org

http://www.osce.org/odihr/

http://www.president.tj/en

http://www.refworld.org

http://www.turkmenistan.gov.tm/_eng/?id=3069

http://www.unwomen.org/

http://www.uzbekembassy.in/

https://en.trend.az/casia/uzbekistan/2759942.html

https://en.wikipedia.org/wiki/Ecological_Movement_of_Uzbekistan. Accessed 9 May 2018

https://freedomhouse.org/sites/default/files/NIT2017_Turkmenistan.pdf. Accessed at 19 May 2018

https://jamestown.org/

https://thediplomat.com/

https://www.amnesty.org/

https://www.britannica.com/

https://www.cia.gov/

https://www.economist.com

https://www.globalsecurity.org/

https://www.hrw.org/

https://www.osce.org/odihr/

https://www.rferl.org/

https://www.state.gov/documents/organization/265762.pdf. Accessed at 10.04.2018

https://www.un.int/uzbekistan/uzbekistan/constitution-republic-uzbekistan/ Accessed at 21 Nov 2017

Human Rights Committee, Uzbekistan, UN Doc. CCPR/C/UZB/CO/4, para 14. Available at http://tbinternet.ohchr.org/_layouts/treatybodyexternal/Download.aspx?symbolno=CAT%2fC%2fUZB%2fCO%2f4&Lang=en. Accessed at 22 Nov 2017

Human Rights Watch (2017) Report on Uzbekistan, January 2017. https://www.hrw.org/sites/default/files/uzbekistan_0.pdf. Accessed at 22.11.2017

Jelen Igor (2002) Tra i kirghisi del Pamir Alaj. Forum, Udine

Jelen Igor (2012) Appunti di geografia politica ed economica. Aracne, Roma

Jonson Lena (2006) Tajikistan in the New Central Asia: Geopolitics, Great Power Rivalry and Radical Islam, I.B.Tauris

Kazakhstan, Kyrgyzstan, Tajikistan, Turkmenistan and Uzbekistan: country studies, federal research division library of congress edited by Glenn E. Curtis Research Completed March 1996

Khayrullo Fayz (2012 24 July) Tajikistan clashes: 'Many dead' in Gorno-Badakhshan. http://www.bbc.com/news/world-asia-18965366. Accessed 10.4.2018

Kudaibergenova DT (2013) National identity formation in post-Soviet Central Asia: the Soviet legacy, primordialism, and patterns of ideological development since 1991. In: Akyildiz Sevket, Carlson Richard, edited by, 2013, Social and cultural change in Central Asia. The soviet legacy, Routledge, London and New York, pp 188–200

Kudaibergenova DT (2015) The ideology of development and legitimation: beyond 'Kazakhstan 2030'. Cent Asian Surv 34(4):440–455. https://doi.org/10.1080/02634937.2015.1115275

Kyrgyz Republic National Academy of Sciences (1995) Glavnaja Redakcija Kirgizskoj Enciklopedii, 1995

Lillis J (2017, 3 Oct) Are decades of political repression making way for an 'Uzbek spring'? There is a mood of optimism in the air, but some fear that President Mirziyoyev's reforms are merely cosmetic tinkering. https://www.theguardian.com/world/2017/oct/03/are-decades-of-political-repression-making-way-for-an-uzbek-spring. Accessed 23 Apr 2018

Limes, n.8/2014, Cina, Russia, Germania unite da Obama

Marat Gurt (2016) Human rights watch, turkmenistan, events of 2016. https://www.hrw.org/world-report/2017/country-chapters/turkmenistan. Accessed 23.4.2018

Meridiani (2019) Special edition Uzbekistan, XXXII/248

Najibullah Farangis (2010). https://www.rferl.org/a/UN_Cites_Evidence_Kyrgyz_Clashes_Orchestrated_While_OSCE_Official_Fears_Ethnic_Cleansing/2072710.html. Accessed 23 Apr 2018, UN Urges Kyrgyz Government To Act Decisively Amid Turmoil, June 16, 2010, RFE/RL Kyrgyz Service. Accessed 5.7.2018

Nick M (2017) Nationalism in Central Asia: A Biography of the Uzbekistan-Kyrgyzstan, Boundary, Central Eurasia in Context, 2017, University of Pittsburgh Press, https://doi.org/10.2307/j.ctt1vjqrk6

OSCE (2015) Republic of Tajikistan Parliamentarian Election, 1° March 2015, Warsaw, 15 May 2015, http://www.osce.org/odihr/elections/tajikistan/158081?download=true. Accessed at 17.01.2018

OSCE (2013) Republic of Tajikistan, Presidential election, 6 November 2013, Warsaw, 5 February 2014, http://www.osce.org/odihr/110986?download=true. Accessed at 17 Jan 2018

Parenti FM, Adda I (2017) Are we going to live in a post-NATO world? A critical perspective on trends, obstacles and possibilities. GeoJournal 82:345–354. https://doi.org/10.1007/s10708-015-9693-8

Paskaleva E (2015) Ideology in brick and tile: Timurid architecture of the 21st century. Central Asia Survey 4–4

Putin V (2013) A plea for caution from Russia, 11 set 2013. https://www.theguardian.com/commentisfree/2013/sep/12/russia-putin-syria. Accessed 23 Apr 2018

Rasanayagam J (2014) The politics of culture and the space for Islam: soviet and post-soviet imaginaries in Uzbekistan. Cent Asian Surv 33(1):1–14. https://doi.org/10.1080/02634937.2014.882619

Richard C (2013) The failure of liberal democratization in Kazakhstan: the role of international invetmnet and civil society in impeding political refomrm. In: Sevket A, Richard C (eds) Social and cultural change in Central Asia: the soviet legacy. Rotledge, London and New York, pp 127–144

Ro'i Yaacov (1991) Central Asia riots and disturbances, 1989–1990: causes and context. Cent Asia Surv 3:21–54

Scherer Steve (2013) Kazakh oligarch's family accuses Italian officials of kidnap, World News, HTTPS://WWW.REUTERS.COM/ARTICLE/US-ITALY-KAZAKH-KIDNAP/KAZAKH-OLIGARCHS-FAMILY-ACCUSES-ITALIAN-OFFICIALS-OF-KIDNAP-IDUSBRE98O0UD20130925, September 25, 2013. Accessed 23 Apr 2018

Totten S, Bartrop PR (2008) Dictionary of genocide. Greenwood Publishing Group. http://31.210.87.4/ebook/pdf/Dictionary_Of_Genocide.pdf. Accessed 5 July 2018

U.S. Mission Kazakhstan (2015, 23 February) Signing of a mutual legal assistance treaty between the United States and Kazakhstan, https://kz.usembassy.gov/signing-of-a-mutual-legal-assistance-treaty-between-the-united-states-and-kazakhstan/. Accessed 23 Apr 2018

Valikhanov ČČ (1985) Sobranje sočinij v pjatu tomach. Glavnaja Redakcija Kazachskoj Sovetskoj enciklopedii, Alma-Ata

Valikhanov capt, Veniukof M., et alii (1865) The Russians in Central Asia (trans: John, Robert Mitchell, Edward Stanford). London

World Report (2014) Turkmenistan, Events of 2013. https://www.hrw.org/world-report/2014/country-chapters/turkmenistan. Accessed 23 Apr 2018

World Report (2015) Uzbekistan, events of 2014. https://www.hrw.org/world-report/2015/country-chapters/uzbekistan. Accessed 23 Apr 2018

14 Political Geography and Geopolitics

Abstract

The populations, in order to acquire a better degree of organization, aim at giving shape to the processes which characterize their life; they establish borders and further territorial elements, in order to give continuity to their actions, creating nations, states, infrastructures, regulations and whatever may be useful to stabilize a certain political configuration. Unfortunately, the geopolitical containers they use may have become obsolete—since reality continuously changes—making territorial states something to be considered as imperfect in principle. Therefore, societies are forced to pursue some new kind of dialogue continuously, on all levels, inside the sovereign state, transversally and internationally, in order to find a solution for problems which are ongoing and, which would otherwise be impossible to solve.

Keywords

Borders · Territorial disputes · Soviet legacy in geopolitics · International relations at regional scale · International relations at wider scale · Globalization influence

14.1 Boundaries, Territorial and Material Questions in Politics

14.1.1 New Territorial State and New Apparatuses

Similar to other social phenomena, politics needs to identify itself in a certain sovereign perimeter (it needs to "territorialize" itself); usually it tries to brand itself as the ideal state, assuming a certain neo-nationalist tendency. This is comprehensibly a sensitive matter, especially for such "young" countries as the NIS, but it is also a question without definitive solutions.

In fact, several geographical processes (such as migration, communication flows and supply networks) mostly develop indifferently to national or other kinds of territorial borders. This also happens for the NIS that try to reorganize institutions in order to adapt themselves to the current situation; they pursue manoeuvres like the transfer of capital towns, the reorganization of administration perimeters, the construction of infrastructures, and in general new territorial planning, that can only represent an insufficient answer, soon becoming something obsolete.

In such situations there is the continuous need for new negotiations, which take time and "good will" in order to be settled and which will presumably monopolize the agenda of the CA governments for a long period.[1]

14.1.2 Borders: General Characteristics

The border configuration inherited from Soviet times is problematic from many points of view—an understandable question considering that such borderlines were never drawn with the intention of becoming anything solidly political (Fig. 14.1).

They appeared at the moment of independence just as a casual design of lines and areas (even if they had been roughly designed originally as ethnic circumscriptions, as said), and they often seem to be deliberately designed in order to create problems and mutual dependency. Sometimes they cut natural circumscriptions transversally (oases, valleys and rivers, mountain ridges, climatic and agricultural regions), as well as functional axes (centre–periphery, supply chains, traffic and commuting routes), as well as caravan and

[1]Barisitz Stephan 2018; Carile Alessando 2018; Silvestri Tommaso et al. 2018; Comai Giorgio and Sofie Bedford 2018; Jelen Igor 2002; Megoran Nick 2017; Damiani Isabella 2013.

I. Jelen et al., *The Geography of Central Asia*, World Regional Geography Book Series,
https://doi.org/10.1007/978-3-030-61266-5_14

Fig. 14.1 Borders in Central Asia. (Source: https://www.un.org/Depts/Cartographic/map/profile/centrasia.pdf, accessed 17.11.2020)

pilgrimage itineraries, or further identity-marker representations.

In fact, on becoming political (changing functionality and significance with independence) the borders fragment a space that from many points of view is organically unified in an unsafe way. Above all, they leave, on the other side of the barrier, communities of different nationalities, configuring exclaves and enclaves, which represent a permanently open question.[2]

Furthermore, it must be considered that at the moment of the Soviet collapse, the borders were not often marked properly—as usual for internal circumscriptions—nor marked on the map, nor evidenced with fences or boundary stones, nor exactly described. This was the case given that they often concern remote and uninhabited areas.

At first, it was necessary to elaborate a method for jointly marking such borders that had never been mutually recognized historically in order to create a premise for any further collaboration. On the surface this would seem to be an apparently easy operation, but the objective situation, with continuous overlapping of jurisdictions, interferences and cross-border problems, made it more complicated.

Today the significance of the border seems to have changed again. After the time of modern geopolitics, when the border signified an exact line drawn on the territory for the nations, closing each-other off hermetically, the forces of de-territorialization seem to prevail, prefiguring a dynamic idea of state and political relations (challenging that order).

In fact, the postmodern border is somewhat different (a belt instead of a line, often assuming a dematerialized form); it has to "filter", rather than to block the flows, elaborating a method in order to leave economics and society to develop transversally, even when continuing to carry out the usual

[2]Julien Thorez 2008; Necati Polat 2002; Megoran Nick 2017; Dagiev Dagikhudo 2014; Brill Olcott Martha 2012; Thorez Julien 2018.

function of security (a kind of "double function" of the borders, typical indeed for advanced neoliberal systems). The current autocracies cannot just hermetically close themselves in, in complete arbitrariness, as they did in the twentieth century (as the FSU did).

The current theory seems to have embraced the reality that a perfect borderline does not exist and that the same state-sovereignty theory would result in inconsistencies (a modernist myth). Thus, it is possible to observe a diffusion of situations managed with "modus vivendi" formulas, or simply neglecting or understating problems, leaving them in an indefinite suspension status.

Such situations can be evidenced by a set of examples, actually by the more relevant border issue currently in CA. This is the case of the enclave status of the Fergana valley and of the Caspian Sea subdivision, negotiating case by case, risking misunderstandings and escalations.

14.1.3 The Current Border Context: Rationale

Such situation has consequences for the local life on all levels (economic political and administrative). Even when borders have been more or less acknowledged, the cross-border "dynamic" of neighbouring societies always creates new challenges. Periodically, the governments try to settle the open question with new regulations, purposing adjustment and rectifications, but without solving the problems definitively, possibly just eluding the problems. In particular, the question of territorial changes has proven to be an especially sensitive one, since it can be easily instrumentalized, affecting such arguments as identity and national pride (above all for such young independent states).[3]

Further, such questions are intrinsically critical, since they rely on procedures, articulated in different stages, some of them necessarily to be kept secret, namely not diffused initially to the wider public (as is the prototypical case of the "pour parler", in order to "guess" the true intentions of the counterpart). The negotiations usually consider a wide set of options, including territorial swaps, long-term territory lending without cession of sovereignty and the joint construction of infrastructure (in the case of a problem of accessibility, for instance).

The recent public narrative is characterized by recurrent rumours, about such kind of inter-governmental agreements. Often such negotiations are not successful, resulting in all things illusive—as usual regarding geopolitical matters.

Sometimes such uncertainty provokes catastrophic effects, as it happened in Kyrgyzstan, after the ratification of the border agreement with China in 2005, signifying the mutual reorganization of the borderline, provoking tensions, which finally resulted in the "Tulip Revolution" and Akayev resignations.

So too, for other CA countries; each one has some territorial weak point, the object of some claim to be settled, thus potentially exerting consequences on a wider scale. Negotiations, necessary in order to achieve a solution for pending border questions, periodically occur, involving procedures with constitutional relevance, parliamentary passages and referendum ratifications, sometimes also impacting third parties and foreign countries. Furthermore, there are rumours regarding secret agreement for payments, exchange of resources and of populations and territories as well.

In practice, such situations sometimes provoke paradoxical effects: the necessity of maintaining a continuous dialogue about such open questions may, however, transform into an advantage, namely the basis on which it is possible (or necessary) to build a multilateral framework—characterized by continuous reciprocal consultation mechanisms—overcoming a very immature status of "free riding" politics. The current border setting is so complicated that the CA individual states seem to be reciprocally co-penetrated, so that they are to some extent obliged to restore a situation of reciprocal confidence (a situation of de facto inter-dependency).

14.1.4 Fergana Borders

The most critical situation—as said—is in the Fergana valley, where the border is characterized by a sequence of discontinuities, both political and territorial, and by further problems. Sometimes, the situation appears so complicated that is impossible to be settled with the available instruments of politics and diplomacy, without the mutual recognition of some "exception" in borderland practices.

In transition time, CA is periodically interested by clashes and tensions, often localized in border areas, with neighbouring countries engaging alternatively in negotiations and hostilities. In some circumstances the situation is worsened to the point the authorities have to proceed to the building of "new walls", dividing territories and populations, and impeding relations of any kind; sources report as well of the destruction of the entire border communities, which could appear eventually as located in a "wrong" place—on the other side of a certain "line".[4]

The recent history demonstrates that critical tensions can easily escalate, as it happened in occasion of the terrible massacres of Andijan and Osh, in 2005 and 2010, connected directly or indirectly with border questions (minorities, borderline recognition, border area mobility, surveillance, supposed presence of hostile activities and correspondent manipulations).

[3] Julien Thorez 2008; Necati Polat 2002; Megoran Nick 2017:114fs.

[4] Megoran Nick 2017.

Fig. 14.2 Exclaves and enclaves amongst Uzbekistan, Tajikistan and Kyrgyzstan. (Source: https://de.wikipedia.org/wiki/Liste_von_Exklaven_und_Enklaven#/media/Datei:Exklaven_von_Usbekistan,_Tadschikistan_und_Kirgisistan.png; accessed 17.11.2020)

Recently sources report about a mitigation of the tensions. In particular, Tajikistan and Uzbekistan—the most critical segments in this "puzzle", consisting in a 1312 km segmented line—reach a border agreement. It signifies a result of the intention of the Uzbek leader Mirziyoyev "new deal"; it regards in particular the sector close to Bekobod and includes the cession of territorial assets (arguably a gold mine), the contextual construction of a bridge and of further infrastructures.

But such an initiative is also the consequence of a wide-scale geopolitical improvement, considering the urgency of arranging a general plan of transcontinental corridor, which is not possible to be postponed. Evidently the first step of such politics consists of the establishment of a new good neighbourhood policy, changing the post-independence attitude.[5]

[5]Tajikistan, Uzbekistan reportedly reach border settlement: Tajiks are angered by reports that the government is giving away a gold mine. 5 November 5, 2019, https://eurasianet.org/tajikistan-uzbekistan-reportedly-reach-border-settlement, accessed Dec. 5, 2019.

14.1.5 Enclaves and Exclaves

The critical situation in Fergana regards above all the question of territorial discontinuities, namely status and administration of enclaves and, respectively, exclaves. The map of such territories is per definition a problematic issue in political geography and is the best representation of the fragmented character of this territory; their definition usually dates back to the time of Soviet border re-drawing in the 1920s and 1930s of the twentieth century (Fig. 14.2).[6]

The reasons why these exclaves were established at that time are difficult to understand, if not as a wider manipulation manoeuvre (as said); they are territorially almost insignificant, but politically very sensitive. The Fergana area appears especially subdivided like a puzzle, a fact that creates ongoing problems, but also paradoxically obliging the current power to negotiate permanently, on a base of reciprocity (since the change of the border proved to be impossible in practice).

It is the case of border controls, transit and residency visas, trade and supply chains, flows and security (e.g. the military status of these "islands" in foreign territory), of trans-frontier minorities and migrations, energy and food procurements, commercial and industrial relationships, mer-

[6]Dagiev Dagikhudo 2014; Megoran Nick 2017.

chant itineraries, which in principle do not consider those delimitations, sometimes simply ignoring them.

The situation is complex and actually probably impossible to settle with the usual diplomatic bilateral tools. This is especially the case in some areas—as the cited Fergana valley—where the management of the borders necessarily involves more nations (not just the usual bilateral confrontation); this valley indeed has become the area in which the tensions that characterize the whole CA are reverberating, and this should induce a general principle of prudence by the respective governments.

There is one Kyrgyz enclave in Uzbekistan, the village of Barak (population 627), in the Fergana Valley, a few kilometres from Osh (Kyrgyzstan). Then, there are four Uzbek enclaves within Kyrgyzstan, the town Sokh (area 325 sq km), population of 42800 in 1993 (but probably much higher), with 99% Tajik population; Shakhimardan, area 90 sq km, population 5100 in 1993, with 91% Uzbeks, and remainder Kyrgyz; then the small territory of Chong-Kara and Jangy-ayyl, on the Sokh River, between the Uzbek border and the Sokh enclave, which is about 60 km east of Batken. Finally, the Tajik enclaves in Uzbek territory: Vorukh (area between 95 and 130 sq km, population about 23,000–29,000, mainly Tajiks, distributed among 17 villages), 45 kilometres south of Isfara on the right bank of the Karafshin River, and in Kyrgyz territory, a village close to Kairagach.[7]

14.1.6 Further Border Segments

Other critical situations can be found at the borders in the Glissar valley, which divides Tajikistan from the Tajik-speaking towns of Samarqand and Bukhara. Then there are some open questions between Kyrgyzstan and Tajikistan in the remote Pamir area (Pamir Alay, Great and Little Pamir, Piotra Velikogo mountains), and between them with Afghanistan and China, with borders that have only recently been recognized and marked. In these regions groups of shepherds practice transhumance (mainly Kyrgyz, who since ancient times have adapted to such conditions, exploiting the seasonal pasturages) crossing the poorly watched borderlines in remote areas.[8]

The northern Kazakh border with Russia derives from Soviet and colonial times and runs along the belt stretching just south of the continental tree line (in the transversal strip occupied by the trans-Siberian railway), separating the open space of the steppe from the taiga forests geographically and culturally.

The Kazakhstan eastern border with China coincides similarly with the previous Soviet external borderline; therefore, it is rather ancient and can be traced back to the nineteenth century. This was recognized by China–Kazakhstan in a recent treaty, signed just after Kazakh independence, attributing some minor territorial changes to China.

The China–Kyrgyzstan and China–Tajikistan mountain borders also correspond to the former external Soviet borderline and have recently been regulated, prescribing some territorial cessions, of limited dimensions, possibly to be defined as territorial rectification.[9] The Tien Shan and Pamir borders have not been recognized for a long time because they are difficult to mark, due to their remoteness and because of morphologic difficulties, namely to changes provoked by geo-glaciological action.

The Kazakhstan, Turkmenistan and Uzbekistan borders, in some parts of their territory, assume the geometric form (Kazakhstan–Uzbekistan and Uzbekistan–Turkmenistan segments) characteristic of colonial circumstances, usually crossing desert areas. They derive from internal Soviet regional subdivisions, they have been recognized, but not yet fully marked and some questions are still open due to border interpretation.[10]

Some border segments result to be critical, not just for strictly geographic motivations—to be treated with the usual delimitation methodology, but because they intertwine with further and pre-existent problems. It is the case of the "new" territory generated by the drying up of the Aral Lake, which is the object of a dispute between Kazakhstan and Uzbekistan, especially regarding the settlement and the bonification of the Soviet era bioweapon laboratory located in this area. So similarly for the appearance of new portions of territories occasionally deriving from the melting of glaciers, or from the changes characterizing river flows and further hydro-geological activities (that will possibly become more relevant in times of climate changes).

The Uzbekistan–Turkmenistan borderline presents some critical aspects, with the intersection of communication lines that do not exactly correspond to the border drawn. For a while some infrastructures were impossible to use without continuously trespassing the borderline, interrupted by exclaves, posing continuous problems for customs, safety and mobility in general. This borderline sometimes corresponds to the Amu Darya flow, then zigzagging over it; therefore, it is necessary to determine a number of adjustments.

The southern Turkmen, Uzbek and Tajik border with Afghanistan is uncontested, since delimitated by Amu Darya

[7] Necati Polat 2002; Thorez Julien 2008; Wikipedia, Central Asia.

[8] Shahrani 1979.

[9] "In 2011, Tajikistan ratified a 1999 deal to cede 1000 km^2 (390 sq mi) of land in the Pamir Mountains to the People's Republic of China, ending a 130-year dispute, and the relinquishing of China's claims to over 28,000 km^2 (11,000 sq mi) of Tajikistani territory; "Tajikistan cedes land to China". BBC News. 13 January 2011; "China's area increases by 1000 sq km". Times of India. 12 January 2011.

[10] Necati Polat 2002; Thorez Julien 2008.

and its tributaries. Nevertheless, it represents a problem from many points of view, especially in the Rasht valley, representing a kind of intrusion into the Tajik territory, this especially considering the risk for infiltration of Afghan "mujahidin" from Badakhshan.

The Turkmenistan–Iran border coincides with the Kopet Dag mountain edge, and since it was inherited from the Soviet external border, it is well defined and marked.[11]

14.1.7 Problems: Caspian Border Delimitations

The Caspian maritime borderline represents a particular problem, originally deriving from an apparently banal question, namely whether it is to be considered as a lake or a sea (besides the possibilities offered by the Russian controlled Caspian Volga–Don channel, linking the Caspian to the Black sea) (Fig. 14.3).

Depending on classification, it can appear as a unified (inseparable) body of water or as a "commodity" easy to be equivalently subdivided; in general, it must be decided what kind of border delimitation criteria have to be applied, as well as other similar questions affecting international law, but involving all aspects of the local life.

It represents a kind of geopolitical exercise, considering the set of possible interpretations, associated with visible or invisible interests and corresponding manipulations; depending on circumstances, the sea-bed median line, the distance from coast or, on the contrary, the sea must be considered as an "organic body" (essential from military, safety and ecological points of view).

The metric division of maritime territory is necessary to settle HC fields or other natural resource claims. This must occur to be able to plan infrastructures (notably pipelines crossing the seabed), ports and other facilities, accessibility and in general to regulate navigation. Other questions are about the interpretation of rules, application of international rights, material border delimitation and questions regarding the use of water and ecologic problems (especially for fishing and for the unique possibility of breeding the sturgeon).

It is a sensitive area from many points of view, political, military, economic (touristic) and ecological, especially considering the risk deriving from treatment of oil and corresponding pipeline construction: the Caspian is the place of the "super-giant" oil fields (often off-shore, located in disputed areas), discovered and since then exploited in diverse circumstances also by international consortia (especially the Kazakh and Azeri deposits).

This has occurred despite the Russian (and Iranian) aversion to foreign intrusion, trying to maintain this area and its HC treasure under its strategic control. This is possibly the key question, namely the fact that Iran and Russia have no interest in changing the original decision (when these two were the only sovereign states in the area, the latter under Soviet semblances), with possibly opposing policies regarding newcomers (namely the NIS) and their claims.

The recent development in negotiations concerns warfare and defence purposes, the Caspian being used by Russia for military offensive purposes, for launching strategic missiles in the Syrian war theatre in 2016. It represents an innovative use of the sea, even when such use was performed possibly either for confirming a right (namely for creating a precedent), or just an experiment, rather than for effective military purposes.

Having said this, the situation remains critical, creating much concern; the uncertainty of the Caspian Sea status has triggered a kind of political (and military) competition among the different coastal countries in the last few decades, preoccupied for the safeguard of the respective HC and other maritime prerogatives.

Kazakhstan and Turkmenistan, in particular, and Azerbaijan on the western side, have shown some ambitions of creating a significant maritime apparatus; Kazakhstan has recently ordered three transportation ships 113 × 22 m from the Italian Fincantieri to be positioned on the Caspian, to ensure "safe and convenient" maritime transport (therefore, Kazakhstan has established consulates in maritime foreign cities like Venice and Trieste).[12]

On 9 December 2017, press agencies referred that Foreign Affairs Russian Republic Minister Lavrov had announced that they had found a solution, which would be soon publicized.[13]

Despite this recent announcement, these issues are the object of continuing negotiations at the level of an established international commission searching for solutions, pending

[11] Megoran Nick 2017:107.

[12] Fincantieri, Singapore, 18th August 2016, presse communiqué reporting Kazakh National Maritime (Kazmortransflot, KMTF) procured to Fincantieri 3 naval units, which may change its impact and role, as well as the strategic balance on this sea.

[13] http://www.mid.ru/en/foreign_policy/news/-/asset_publisher/cKNonkJE02Bw/content/id/3103635, accessed 23.4.2018, 1 March 2018, "Communique of the 50th meeting of the Special Working Group on the development of the Convention on the Legal Status of the Caspian Sea at the level of deputy foreign ministers of the Caspian littoral states"; then "[P]recisely, to avoid further conflicts and regulate the water in August 2018 the Parties agree to sign the Convention on the Legal Status of the Caspian Sea [….] Still the situation taking into consideration a strictly legal point of view remains unclear and the last convention doesn't solve the problem of the legal status of what is called the <Caspian Sea>", Chiavon 2018.

Fig. 14.3 Caspian Sea with littoral states. (Source: https://nationsonline.org/oneworld/map/Caspian-Sea-map.htm, accessed 20.11.2020)

some arbitrage, presenting some risk for armament escalation.[14]

14.1.8 Regional Interaction

CA geography configures a situation of mutual dependency that seems to have become permanent: the "stans" depend on each other for vital resources like water, ecology, energy, migration and commerce, supply and trade lines and above all safety. This is the consequence, to some extent an intentional one, of colonial and Soviet territorial engineering, which today means it is necessary to establish a mutual collaboration method.

The aim was evidently that of establishing a "divide et impera" form of geopolitics, rendering the periphery structurally dependent on the centre and the peripheries dependent among themselves, making it impossible for the individual states to become self-sufficient (Fig. 14.4). The territorial repartition should be considered as fictitious, without any political significance, not becoming the reference for the consolidation of any identity.[15]

[14] Silvestri Tommaso 2015–2016; Necati Polat 2002; Caspian question seems to function indeed as a kind of "exchange currency" for the whole area, e.g. in negotiations for pipelines or infrastructure planning.

[15] As said, the soviet geopolitics, as indeed any totalitarian regime, had the intention of reducing the reality, then also territories and peoples to a set of standard elements, easy to manipulate and arbitrarily re-constructed, namely instrumentally used, as a kind of geopolitical LEGO.

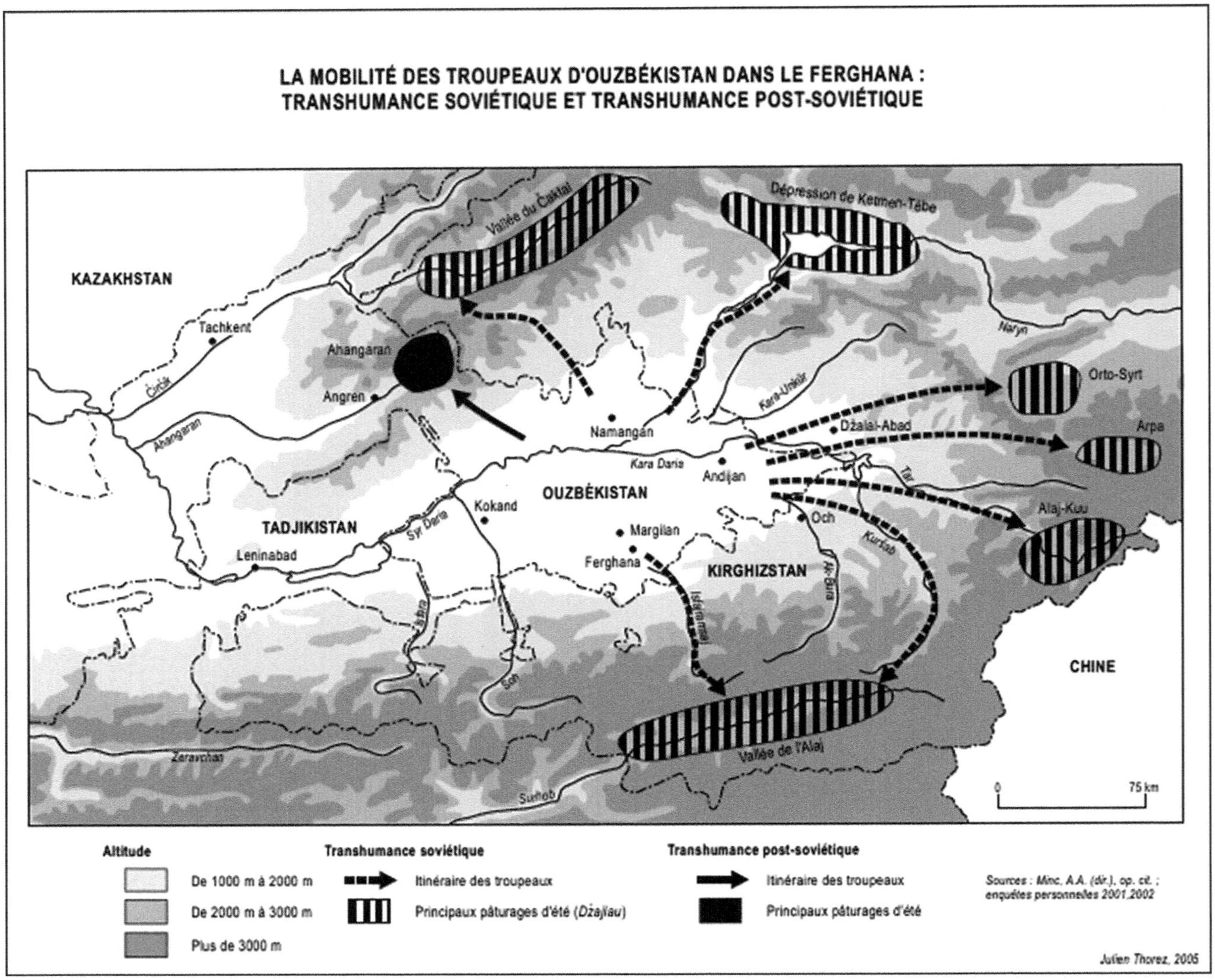

Fig. 14.4 Schema of mobilities in Fergana valley in transition post-Soviet period. (Source: Giraudo Gianfranco, a cura di, Integrazione, assimilazione, esclusione e reazione etnica, Editura Muzeului Ţării Crişurilor, 2012:373)

Among other consequences, one must consider the emergence of a progressive transnational condition, weakening original sovereignty pretensions. This means a situation of continuity between internal and international politics, namely the overcoming of the typical modernist "double standard" (arbitrary politics inside but respectful of regulations on the international scale, fostering multilateralism).[16]

In such situations it is simply impossible to distinguish the two spheres, considering that such a divide is neither evident nor easy to distinguish. It means the need for a new regionalist approach substituting the individual "free riding" politics with its multilateral approach. The sudden imposition of new borderlines on a pre-existing social and territorial body (comprehending assets, circuits of persons, freight and identities) represented a shock, provoking a sequence of contentions, often bringing the neighbouring countries close to war.

In fact, any decision one country takes, inevitably has consequences across some border, exposing the entire community to the risk of escalation and of manipulation (potentially also induced by lobbies, hidden powers, and also outside powers interested in local tensions, for bargaining, namely gaining some advantage from the induced conflicts).

The situation of mutual dependence is evident just on looking at the cartography and the basic data: Uzbekistan exports food and gas to Kyrgyzstan and Tajikistan pays with water and electrical energy. Kazakhstan and Turkmenistan supply oil and gas to all the republics, having in return a

[16] As described before, typical as well for CA regimes, that are authoritarian inside but that wanted to appear more pluralist towards the IC.

regulation of water sources, namely in electric production; Tajikistan and Kyrgyzstan, the poorest and marginalized countries, in fact control the upper flow of Syr Darya and Amu Darya, namely glaciers and snow covered peaks, the original richness of CA (obviously even more important considering the endorheic character of the basin).

Three countries share the agricultural production of Fergana and of the oases on the downstream flow of the major rivers, diverting their course in order to feed channels for thousands of kilometres and square kilometres of irrigated land (especially, the Karakum canal and the Great Fergana canal). And so on, with a long list of evidence of such interdependence.[17]

Finally, it must be considered that the increasing network of pipelines—for HC, and any other use—covers the whole territory and can only be used on reciprocity base, crossing the neighbouring countries, to reach international markets and customers. This network of infrastructures evidences poor efficiency; its basic schema can be traced back to Soviet times and urgently needs restructuring, which indeed for a significant part has been already started. They are now segmented by different borderlines, so that they often appear as useless. It is the case of railways, roads, canals, pipelines and energy lines, and any kind of natural and economic circumscriptions (traffic gravitation areas, metropolitan areas, supply basin), which are continuously overlapping (Fig. 14.5).

This must also take into consideration the demographic situation and migration policies, pilgrimage itineraries (especially, in the Osh area), transhumance and commercial routes, family and ethnic ties.

Therefore, it is necessary to consider the disequilibria that periodically occurred between the different regions, as in the case of the rapidly growing Kazakh cities, attracting the workforce from surrounding regions, but sometimes incurring a crisis of the different border areas, where people occasionally profit from the differentials in terms of taxes, job opportunities or commercial prices and organizing business.

It is the case, especially, of the overcrowded Fergana and of the surrounding Kyrgyz and Tajik mountain areas: this is the typically “contended space”, transversally divided by a new border that separated an organic unit, in which populations organized into “new” nations claim the right of the entire province.

The problem typically derives from fractures perpetrated in the past by Soviet geographers: in the case of Osh, the rural Kyrgyz considers the city area as their “own” city, belonging to their regional territory; vice versa, the Uzbek urban residents consider Osh as their national town and the Kyrgyz as recently immigrated outsiders, so this centre-periphery typically overlies further divisions such as mountain-plain, farmer-shepherd and oasis-desert.[18]

14.1.9 Contextual Situation

Depending on perspectives, the CA region can be considered as a set of individual countries or as a unique body. However, it is characterized by inseparable interests as in the case of geostrategic and defence capabilities, ecological questions and many other aspects, which exert cross-border impacts unintentionally.

The common condition of landlockedness that amplifies such effects is another element making the need for collaboration evident.[19] This does not only just occur between the CA states but also towards outside regional powers, like Russia, China, Iran and the Southern Asian context; particular concern is arising for security questions regarding mobility in border areas, concerning terrorism and trafficking and also for ecological questions, for fluid and aerial environments, for water and soil contamination and even for climatic changes.

It is the case of the rivers inflowing from China and flowing further to Russian Siberia (the Irtysh) or emerging in the Balkhash closed basin (like the Ili crossing the Zhetysu region), which are increasingly polluted; it is the case of the Aral, divided into two or more small lakes, the terminal of all the waters of the area and of the other dramatic ecological issues. Another example is that of long-range accessibility, in particular regarding the southern opening—towards the Indian ocean ports in the frame of the new transcontinental planning of SR infrastructures.

The only solution is the acceptance of this status quo, of sharing the same territory and the consequent common organization of practical functions, essential for the life of the populations. It is difficult for nation-states that have only just reached independence,[20] but there are no alternatives: rivalry in such locked scenarios risks triggering escalations, mutual retaliation and conflicts.

With time, the society may be able to elaborate better tools, namely supra- and inter-national regulation, organizing cross-border governance. In circumstances of stabilization and of low geopolitical pressure, the international relations game usually becomes a positive one (namely

[17]Megoran Nick 2017.

[18]Dagiev Dagikhudo 2014; Megoran Nick 2017; Brill Olcott Martha 2012.

[19]In order to access international markets, simply water resources for instance, each country needs to cross borders, and this fact configures a set of reciprocities that cannot but induce structural collaboration for pipelines, water and energy, but also for minority rights, visa regulations or communication networks, Thomas Marzhan 2015:467.

[20]Megoran Nick 2017; evidently new borders, superimposing a pre-existing integrated reality, are always problematic.

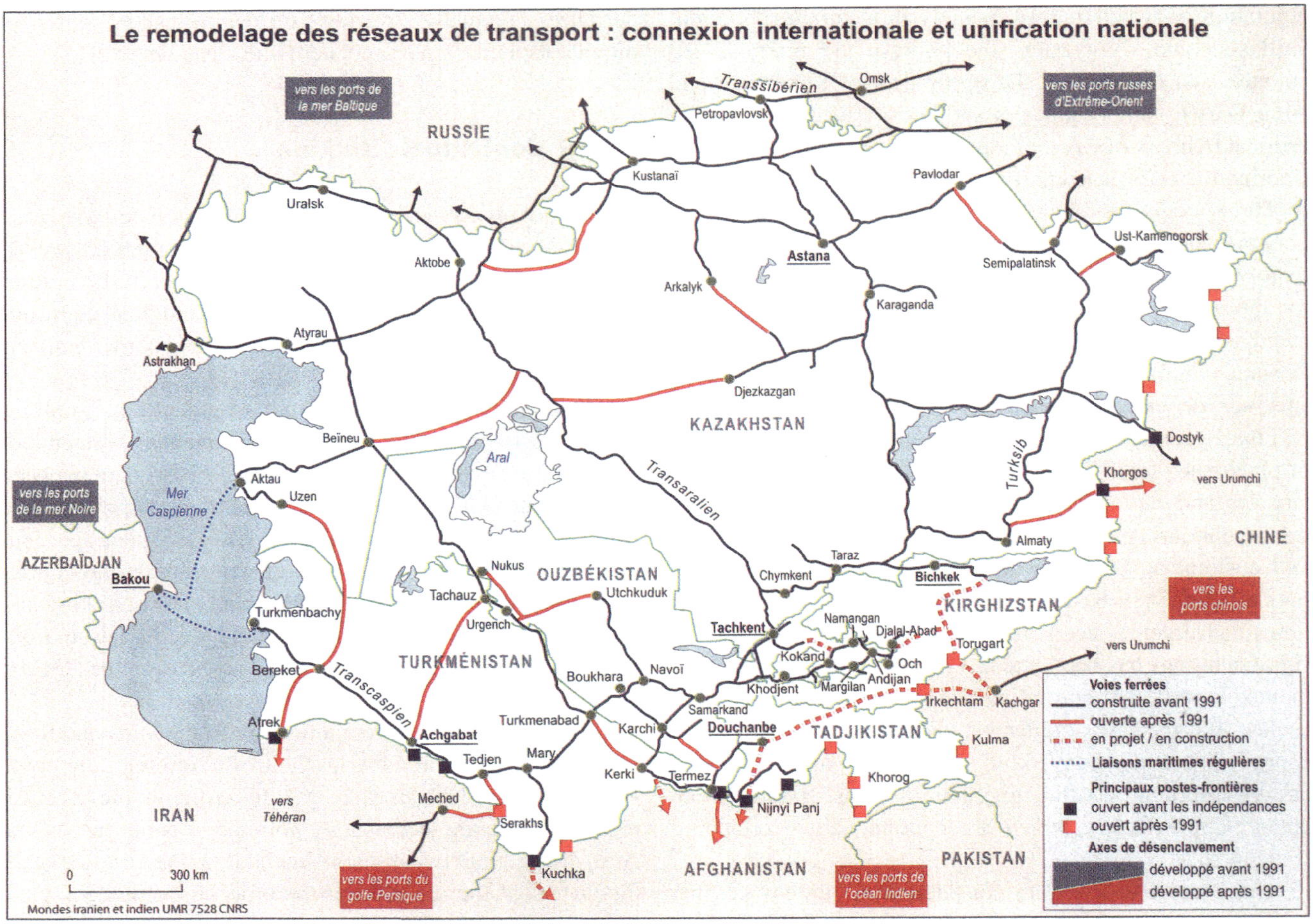

Fig. 14.5 Restructuration of railways, courtesy by Thorez Julien (sous la direction de), Asie centrale, 2015, Paris, Ellipses, p.V

based on collaboration), spreading the benefits over the whole population and over the whole territory.

14.1.10 Defence and Security Questions

Classic geopolitics assumes a "zero-sum game" scenario, a situation "made" by individual states (that is indeed very common for NIS at the beginning of their life cycle); it represents (as usual in such circumstances) a kind of chessboard (namely a "friend and foe" schema), with the neighbour assumed "per default" as an enemy (because in a new and unknown scenario, the first necessity is that of ensuring the preventive defence against a potential lethal enemy).

It is a sensitive question, since in such "locked" scenarios the question of security can soon initiate the questioning of mutual confidence, emulation and escalation policies. This is the case of a dangerous local "arms race", regarding not just weapons, but the whole defence apparatus, from industrial production to tactical military disposition (Table 14.1), reverberating in all aspects of civilian life.[21]

The security apparatuses of the NIS are generally poor, as far as military equipment and corresponding industrial production is concerned, the local states having just recently become independent, without having inherited any relevant weapons, either conventional or non-conventional (Kazakhstan also having renounced its atomic potential).

In this regard, the establishment of the SCO has proved to be essential. All these relations are included in the new frame designed by this international treaty that has proved capable of managing the delicate equilibrium between the different local states of central Eurasia. The SCO—and the wider IC—builds or actually imposes confidence between countries, impeding the escalation of uncontrollable tensions, limiting manipulation and taking out foreign powers (including themselves), which cannot easily intrude into these spaces.

[21] Above all, distracting resources that in this stage of their geopolitical life, could be very useful for the NIS, for starting up development and consolidation strategies, possibly alimenting aggressive attitudes.

Table 14.1 Central Asia: Military expenditure (% of GDP)

	2010	2011	2012	2013	2014	2015	2016	2017	2018
Kazakhstan	1.0	0.9	1.0	1.1	1.0	1.1	1.0	0.9	1.0
Turkmenistan	–	–	–	–	–	–	–		
Uzbekistan	–	–	–	–	–	–	–		3.6
Tajikistan	1.0	1.1	1.0	–	1.1	1.2	–		
Kyrgyz Republic	1.6	1.4	1.6	1.6	1.7	1.8	1.8	1.7	1.6

Data source: Stockholm International Peace Research Institute (SIPRI), Yearbook: Armaments, Disarmament and International Security. https://data.worldbank.org/indicator?tab=all, accessed 26/11/2018

However, there are further elements for preoccupation outside of the range interested by such a treaty; it is the case of Afghanistan, considering the porous border and the instabilities that still affect this country, but also other scenarios (Caucasus, Middle East, Xinxiang).

14.2 International Politics on a Regional Scale

14.2.1 Structural Element Influencing International Politics

> "Since their independence, Central Asian states have showed limited interest to develop regional cooperation or to bolster economic integration, primarily because they have privileged the achievement of national interests, which were based on strengthening political sovereignty and economic independence."[22]

But soon the situation of mutual interdependence induces all countries to pursue some kind of neighbouring politics, beyond the true intentions of the national leaders. In fact, local politics seems to be characterized by spontaneous integration, with tensions and contentions exerting mainly an instrumental significance (possibly for internal political purposes).[23]

All the NIS (included the post Niyazov Turkmenistan) are active members of IC, showing a diverse degree of internationalization, that is manifest in different schema. It is the case of CIS, as the FSU successor, currently made up of Russia, Belarus and Kazakhstan, and of the Eurasiatic Community, possibly a new edition of the one represented by the CIS. It is the case of the mentioned SCO, to which all the CA countries adhere (with the exception of Turkmenistan).

It is the case of OSCE that represents the basic intergovernmental organization in the area, besides UNO and UNO agencies, unifying the local post-Soviet countries with western European ones in a regional framework: it is a typical continental organization and it is still the major specialized regional security organization in the world (58 members). It is scarcely visible at the level of "high politics", but it is rather effective in day-to-day reality, exerting a monitoring role, sending observers and ensuring a kind of continuity of dialogue on a multilateral level.

It is considered obsolete by some observers (as well as NATO, a kind of relic of the "cold war"), because of the disappearance of the "iron curtain" that represented the major motivation for which this organization was once founded.[24] However, abandoning the OSCE would be too expensive in political terms, considering the potential need to re-establish a whole set of bilateral relations as well as dangers in terms of discontinuity of political relations.

It must be recalled that all of the NIS are members of major international organizations, like UNO, Organization of Islamic cooperation, International Organization of the Turkic culture (TÜRKSOY), as well as many regional associations (like ECO with other Asiatic neighbours).

So too, for the list of major IOs, comprehending IMF, World bank, Asian Development bank and other private or public institutions, sovereign wealth funds, international banks and MNCs. All these are comprehensibly attracted by the opportunities offered by local economies, which are potentially rich but present a relevant gap in terms of structures and infrastructures, which would be capable of developing much added value in principle. It must be considered that besides Kyrgyzstan, that had already accessed the World Trade Organization (WTO) in 1998, Kazakhstan and Tajikistan have also recently accessed this organization.

14.2.2 The First Step: The "Near Abroad" (Regional) Scenario

At the moment of independence, the NIS had to start the construction of the international framework on all scales from zero. This meant, at initial stages, the elaboration of a bilateral relations schema—negotiating the questions one by one—signifying as said a regression to individual state politics that is usually very expensive and inefficient.

Not surprisingly, the early steps in transition times were characterized by the emergence of rivalries, occasionally by reciprocal annoyances, sometimes degenerating into retalia-

[22] Indeo (manuscript, in press)

[23] Carlson Richard 2013:133.

[24] Parenti Fabio Massimo and Adda Iacopo 2017.

tions and disproportionate reactions. This is the case of the erecting of barbwire fences, the blocking of transit visas, commercial unilateral obstruction policies, systematic uses of pretexts (e.g. of controlling terrorist passage), and finally military training and exhibition of army capability ("muscle flexing").[25]

Actions, as usual for such novice diplomacy scenarios, are heavily conditioned by the personalist component, namely by neonationalist leaders in search of legitimization (since they have suddenly acquired huge power, almost casually). They sometimes exhibit regional "parades", asserting and representing respectively the brotherhood of Turkish or Iranic peoples, loyalty to a common Islamic faith, the awareness of the same Soviet past (meaning the interiorization of a certain culture, namely a secularized attitude) and others.

But beyond appearances, often nothing is concrete or stabile. Periodically the leaders build up alliances presented as eternal, based on consanguineous associations, or on pan-Turkish or Eurasian brotherhood, "Iranian" solidarity opposed to "Turanian", but which soon reveal themselves lacking in any real consistency. This is the case of solemn declarations of perennial friendships, celebrated in a family-like manner (as happened with the cited case of the Kazakh and Kyrgyz presidents' sons' wedding in the 1990s); it is the case of the signature of "definitive alliances", which changed the attitude in an unpredictable way.

Therefore, local "international" politics swings periodically between Russian and US friendships, or to some other geopolitical affiliation, potentially claiming Eurasian or total European integration, considering that some CA countries sometimes define themselves as a European appendix, which would be a title for adhering to the EU (e.g. as Kazakhstan states that 7% of the Kazakh territory lies west of the Ural River, or consequently some other interpretation of the geopolitical map).

A type of politics based on proclamations (with an evident demagogic or tactical character) that alternates public accusations, suspicions and reproaches, for example, occasionally not paying invoices for the import of gas or electricity, the threat of blocking water flow from dams and reservoirs or accusations of carrying out manoeuvres manipulating internal enemies or respectively persecuting national minorities. In such situations, it is difficult to elaborate a precise idea of one's national interest and of the collocation in the wider order.[26]

14.2.3 Shanghai Cooperation Agreement (SCO)

The SCO was founded in 1996 as the "Shanghai agreement" following negotiations that had started soon after 1990 and the FSU collapse in 1991, between Russia and China. They were accustomed to considering themselves as competitors in Eurasian spaces, but in the new circumstances they changed such traditions, realizing that they had complementary interests in this region.

It is made up of some permanent structures (a secretary based in Shanghai), in peripheral agencies and in a periodical sequence of summits and consultations, fostering neighbourhood relationships and confidence, always critical in such scenarios, characterized by long borderlines and a wide continental outback (representing sometime critical aspects). Another implicit motivation would be that of keeping the USA and western powers outside the area.

The SCO is possibly the main (or only) reference which has proved to be able to represent something solid in this situation, like an umbrella organization covering this side of the continent. It represents a rather conservative alliance (it has been compared to the "holy alliances" in Europe of the XIX Ct.[27]), with possibly repressive intentions, but it can also reassure the fragile CA countries in any confrontation with China and Russia, namely preventing any change in the geopolitical status quo. It derives from a basic geostrategic agreement between these two incumbent powers (above all from the necessity of the mutual recognition of borders), making the CA a kind a buffer and (what is more important for local societies) a pacified and resource-rich equilibrium area.[28]

The main stated targets of the SCO, in fact, have been summed up in the fight against the "three evils" affecting the "transition" phase, namely separatism, terrorism and religious extremism. To these a fourth one has been added, namely the prevention of trafficking, drug trafficking in particular: a problem in the meanwhile, which has increased until possibly reaching a level of strategic importance (impacting politics).

Therefore, they are a set of targets that should induce a confidence effect over borders and territories alone that have often been disputed in the past. This can be stated on the grounds of a continuative mutual exchange of information of

[25] Rotar Igor 2006; Megoran Nick 2017; it is especially the case of the Fergana, but there are in the area many other such critical areas, where the tension could easily escalate; the same politics could be interpreted as a part of a structured strategy, with the aim of maintaining the borderland and the neighbouring states, in a situation of artificial instability, in order not only to gain some advantages from their weakness, but also to justify an internal permanent mobilization, namely a kind of geopolitical blackmailing.

[26] Megoran Nick 2017:75.

[27] Hessbruegge Jan Arno 2004.

[28] De Bonis M., "Russia e Cina: terra in cambio di affari", Limes, 25/07/08.

programmes of joint military training and exercises, and of border-police cooperation.

The SCO-depending regional anti-terrorism structure (RATS), based in Tashkent, is especially relevant, with duties of antiterrorist prevention, with special concern for border areas, where such events may always provoke incidents and misunderstandings among frontier guards.[29]

In this way the CA countries try to play a difficult role of balancing elements between these two regional powers, which are indeed distracted by further and wider Eurasian-scale affairs and interests (as is the case for China in Xinxiang, Tibet and eventually Inner Mongolia, which represents its continental outback, saying nothing about the Russian appendices stretching along the whole Eurasia). The SCO has proved to be useful in order to build up neighbouring confidence, avoiding escalation and reciprocal instrumentalization of problems.

It has a significant role in this "zero-sum", namely a typically conservative scenario in which any unilateral initiative is to be considered as dangerous. Critical analysists report it may be anachronistic, considering current globalization, "de facto" preventing any political evolution (namely liberalization) of the scenario.

14.2.4 The Recent Russian Change

The world has been disappointed by the recent change in Russian politics: some authors have observed it means the risk of involution into an authoritarian state, then implicitly the incapability of Russian society to undertake pluralistic processes.[30] Indeed, it is marginally comprehensible, since democratization processes need time: people need to get accustomed to the values of a dynamic organization of life, to understand the political game as an effective (apparently unstable) "game" of democratic tensions and interests, not the sterile stability that characterizes a monolithic autocracy or communist planning.[31]

Besides this, with the recent invasions, Russia risks losing the capital of political confidence it gained in times of unilateral disengagement from CA (and other former FSU territories) in the early 1990s, which remains a historical merit: it allowed Russia to avoid a set of decolonization wars (as did many other European colonizers just a couple of decades earlier) and to maintain a material long-lasting presence in the area.[32]

In fact, it seems to have lost out in some geostrategic initiatives (but not in military capability, indeed), with its political action that derives mainly from exploitation of occasional chances (a kind of passive geopolitics), based on the exploitation of contingencies.

The basic data of Russia is similar to those of other middle-rank powers, so too for its demography, economics and commercial capability, but it inherited a geopolitical charge, a militaristic attitude and a nuclear arsenal. In particular, it inherited the capability of opposing nuclear blackmail and of resisting a nuclear attack thanks to its objective extension, making it tangibly a great power (in fact nobody could imagine invading Russia). In fact, Russia has practically been forced to pursue a world global politics, even when it no longer has the power to do so (therefore, it possibly tends to react in an instinctive-defensive manner, in an organicist way).

A further motivation for the persistent Russian influence in CA is the constant and relevant minority that in some cases, especially in northern Kazakhstan, is territorially concentrated in a border area, evidently indicating a critical situation. In the whole of CA there are not much more than 5 million Russians—depending on the sources and on the recording criteria—that account for less than the 10% of the population; eventually adding the several millions of eastern European of Slavonic origin to them (mostly Ukrainian), or nonetheless Russianized people, therefore speaking Russian or culturally Russian assimilated (finding it difficult to identify in any other nationality, Russian being the identity recognized per default as the successor identity of the Soviet one). Similar effects are also due to intermarriages, migrations and in general of the mixing cultural processes.

This population of "ethnic Russians",[33] or rather as "cultural-Russians", does not show a clear intention to migrate. It is a question not only concerning civil and national statuses but also of economic perspectives and political circumstances; in some cases (as in Turkmenistan) they may maintain a double citizenship, and everywhere they may enjoy a situation of "de facto" bilingualism.

But this is possibly a question which is losing importance since CA societies are stabilizing, with minority protection

[29] See Nicklas Norling 4/5/2006; RATS carried out exercises in order to protect from eventual attack to atomic reactor in Tashkent physics nuclear institute; the latest proposals about the revival of the nuclear industry in Uzbekistan, in such a fragile geopolitical context, are disquieting; Uzbekistan 2018.

[30] Indeo 2018.

[31] Ott Stephanie 2014.

[32] Also thanks to this "confidence", Russia has maintained the attitude of a world power, engaged directly on many world scenarios; this is because of its territorial extension (from Arctic to central Eurasian, from the Atlantic to Pacific ocean, from Black Sea to Middle-Eastern and Far Eastern countries, exposed to many scenarios) and of the imperial heritage (e.g. military prestige, but also the attitude to use force), as well as of its residual influence capability on peripheral areas (eventually leveraging and seconding anti-western Third World attitudes).

[33] Dunlop 1994:208.

becoming a less urgent question; furthermore, local governments seem to have the intention of developing bi- and multi-lingual policies (especially in Kazakhstan and Kyrgyzstan), which are acceptable and even desirable in the current situation (like indeed many countries in the world seem to demonstrate).

14.2.5 Powerful Neighbours: China

The influence of Russia in the CA is still to be considered dominant and still maintains structural characteristics, but it is easy to foresee that it will be challenged by China in a few years: "China has become the main economic power in the region. Trade between China and the five Central Asian republics has risen from US$1.8 billion in 2000 to US$34 billion in 2015, while the five states' trade with Russia amounted to only US$23 billion [...] [m]oreover, China is the most important creditor in the region".[34]

Indeed, China's position is perceived as an epochal chance. It appears perfectly complementary towards the countries of CA in all aspects, demographic, economic, commercial and strategic. It offers everything the CA countries need, namely human and economic capital, technology and turnkey-plants (ready-to-use), industrial heavy and light engineering, infrastructures, instrumental and consumer goods. And it does not have what it can get from CA neighbours, also energy (also in the contest of a transition toward the post-fossil sources that is beginning to exert some effects), raw materials, agrarian soil (both, irrigated or for extensive exploitation), therefore good potential for market development, economic expansion spaces and investment opportunities.

The list of major foreign investments in infrastructures, with the major strategic goal of opening the "treasure chest" of CA economies, demonstrates such interest and such complementarities. Above all, China has the appearance of non-colonial, disarmed power (to some extent), therefore it is perceived as less dangerous by its neighbours (but the question of the Uigurs, Moslem and Turchik population of Xinkiang must be considered, potentially disrupting the scenario).

In fact, it has some confidence capital to invest in this area, since it may play a role in counterbalancing the traditional influence of Russia (with whom China seems to have no immediate interest to enter into conflict, having settled their border questions in 2004–2008 and SCO treaty), playing a role in local equilibrium against any intrusion or internal degeneration.

But China's presence is also perceived as incumbent and indeed ambiguous, if not else for the imbalance in demography, economic capacity, commercial and financial power—not just because of an organicist perception: it is simply too big, and it can easily inundate local economies with quantities of products characterized by a wide range of quality. It must also be considered that China traditionally demonstrates a neo-mercantilist attitude, falling back on commercial "dumping" practices, devaluation policies and unscrupulous competition (although recent decisions may evidence a possible change in this attitude).[35]

The question is whether such economics can be converted in politics, eventually hegemony, and when this might occur; it has recently demonstrated, as said, some good will, settling some contentious border disputes that were left over from the Soviet and also from pre-revolutionary times.[36]

Furthermore, we must take into consideration that China, as well as Russia, do not have a particularly urgent interest in this quadrant (Russia much more in Eastern Europe-Black Sea "amenity places", while China seems interested in claiming spaces on the other side of the continent, on the sea which are areas considered as strategic gateways for its expansionism, contrasting the other Far East countries perceived as its true competitors). Therefore, CA countries may enjoy a period of relatively low tension.

In fact, there are no traces of the proactive (but not really aggressive) politics that China carries out on the other continental side (notably in the South China Sea). Here it seems much more conservative, respecting the status quo.[37]

China seems to have more motivation in the conservation of its wide continental outback, which is essential as the source of its major rivers (Tibet) and for other reasons (Xinxiang-Taklimakan for HC, military experiments, settlements possibilities). In general, it represents a strategic continental buffer, essential to China for playing its role as a "superpower". Nevertheless, it is subject to the risk of escalation effects and of independence claims, as has happened in the recent past with Soviet Asiatic republics (from this point of view, China is possibly to be considered as the "last surviving empire", even if it has never had, paradoxically, a colonial attitude).

Having said this, it is evident that China's politics is influenced by open economics—and manufacture-oriented economics, which (the reversal of Russia) should work well towards international integration and for tendentially peaceful politics, at least with its customer and trade partners. In fact, China has greatly benefitted from the international market, protected by IC institutions, on which the country

[34] Indeo 2018.

[35] It is probable that Chinese activities will assume consistency with time, developing more open forms; major Chinese actors in this area are China National petroleum Corporation CNPC, supported by the ubiquitous China Commercial Bank.

[36] Necati Polat 2002.

[37] If it were deprived of its continental outback, China would have to re-scale its status of world power.

depends for its development. Therefore, it is probable that Chinese policies will not degenerate into direct aggressive behaviour.

Possibly because of this, China also seems to prefer spaces that are not directly involved in political or military competition, located in relatively marginal situations for expansion areas (a kind of "understatement" geopolitics). For instance, China already promoted relations with Turkmenistan in Niyazov's time, promoting the Turkmenistan pipeline, possibly evidencing a kind of pragmatic attitude.[38]

14.2.6 Other Players: Neighbours

A considerable number of regional players are making calculations about the region for different motivations, some are attempting to gain a position in the new scenario (a "piece of the cake") in the new and supposedly rich but technology-hungry CA countries. In fact, the unexpected sudden opening up of a continent-wide market (after the Soviet empire collapse) started a race among corporations, trying to occupy what at first glance seemed to be a privileged position, possibly ensuring a long period of business in a market largely controlled by public authorities (relatively easy to penetrate), thus for both export and import, as well as for the numerous opportunities the post-Soviet reconversion prefigured.

Indeed, such economies were scarce at that moment in many sectors (technology, infrastructures, distribution networks), but nevertheless it includes a population with relatively high-standard education, suitable culturally to increase their level of consumerism, characterized by modern expectations (for commodities, communication, education, mobility, leisure, tourism) and lifestyle. Therefore, a high potential for growth, as well as a critical gap between economic expectations and cultural preparation, is possibly the main problem in circumstances of such rapid development.

14.2.7 Turkey

The 2016 political crisis of Turkey disoriented the populations and elite of CA. It had always been considered as a benchmark by other Turkic-speaking countries, since the beginning of time—from the times of the Osman empire, Enver Pascha pan-Turkish visions, until today, sided by the Western and NATO institutions. It was usually considered a competitor to Russia in these territories and a protector against anti-Turchik attitudes (e.g. today in Chinese Xinkiang). This is particularly so after the collapse of the Soviet empire, when Turkey purposed itself as the true partner of CA cultures, considering the Russian side as a residual of a colonial epoch.

Furthermore, it is especially important considering the geopolitical setting and the continental position, configuring a kind of gateway from CA towards the western economies, namely a terminal on the Mediterranean for traffic, pipelines and infrastructures of the Eurasian corridors.

Turkey entered this space and took advantage of the existing vacuum especially in some sectors like services and light industry, based on initiatives of private and middle/small companies, that were usually underdeveloped in the area (because the Soviet economics privileged heavy and planned industries). It purposed itself as a partner in sectors such as large consumer goods manufacture (textiles, food and leisure) and the building industry, international trade, retail organization, mobility and tourism on all levels (e.g. for air connections with western countries, considering the role the Turkish flagship company exerted in the period), and other market-oriented activities. For all of these, the Turkish economy represented an example and also a possible way to access to the wider international context.

Turkey based this intervention on further favourable circumstances such as the supposed attitude of local societies, characterized by language and ethnic affinities with local Turkish "brothers" in a revival of pan-Turkish suggestion. This was also achieved by promoting a massive public subsidization of any kind of relationship, like university student exchanges and common cultural initiatives, lobbying for the adoption of the Latin character alphabet (therefore making communication and literature reciprocally permeable).

In fact, the Turkish situation may represent an example to local realities, since it acquired a recognized position, relatively prosperous as advanced country (successor of a glorious Eurasiatic empire), relying exclusively on human resources, not on natural stock exploitation or on other rent positions (with the exceptions of the Mediterranean climate for the tourism and eventually migrant remittances sent from Western European countries); therefore, it represented the example of development which was completely different from that of the HC rentiers.

In general, the Turkish economy proved to be the symbol of the possibility of modernizing (and innovating) without losing peculiar traditional and religious characteristics. This was also considering the Turkish attitude to pragmatic adaptations (also regarding religion institutions, which usually do not have the capability to rapidly adapt to changes). It represented the example of a democratic way of pursuing development for a big Turkic Muslim country, attracting the attention of not just liberal advocates of the area, but of the whole population (a role possibly compromised by recent involutions).

[38] Anceschi 2017.

14.2.8 Iran

This country, neighbouring the scenario, represents an *incognita*: it has high potential, but is just coming out from a long period of sanctions and is only now starting a period of foreign politics and of communication in wider international scenarios. Notwithstanding the recent evolutions, it remains a clerical country characterized by a Shi'a organization—by a hierarchical internal religious structure, unusual for CA predominantly and Sufi influenced Sunni Muslims.[39]

It is characterized by a strategic alliance with Russia, probably based, among others, on the border divide agreement on the Caspian—namely by the common interest of maintaining the status quo, considering the NIS as outsiders on this scenario. Moreover, to some extent, also for the tactical reasons of dealing with regional rivalries on the Middle Eastern scene, trying to marginalize the Anatolian Turks and Saudi Arabs (the latter being competitors on the same energy world market) or others potentially. It is a question not only of ancient rivalries (that can be traced back to Osman, Arab and Persian empire times, far back in history) but also of current negotiations about material decisions like pipeline planning, Middle Eastern war intrusions and the support of some factions.

A situation that currently demonstrates the difficulty of escaping the trap of such a primitive way of diplomatic thinking (the already mentioned "chessboard" schema) and the urgent need to build an international and possibly supragovernmental framework (or at least to integrate the local regional rivalries into a wider UNO framework).

This country has yet to demonstrate its capability of integration with the global economy. The local natural reference of Iran is Tajikistan's close affinity in terms of language and ethnic background, which was considering the adoption of the Arab Persian alphabet after the war, recovering its historical pre-Soviet and pre-colonial traditions, which is prevalently Sunni. Such structural diversity is more than a simple curiosity. Iran's recent history is complicated and it is considered by the secular despotic CA leaderships rather as a threat, namely an example of theocracy, self-defining democratic and republican, originated from a (spontaneous chaotic) street revolution, diffusing the idea that religion inspired politics can become a political pattern.

14.2.9 Other Middle Eastern Countries: Saudi Arabia and the Gulf States

For the whole transition period Saudi Arabia (SA) tried to foster its position in CA in a peculiar way, namely consolidating the recovery of religious institutions and practices, contrasting what they defined as a prevalent secularist (or even not Islam-orthodox) attitude.

This also with material projects; SA is the base for many financial institutions, among others the IDB (Islamic development bank), based in Jeddah, engaged in a wide project of reconstruction for infrastructures, and especially for the elements of the religious landscape that for the whole period of the FSU were not properly maintained and often completely destroyed. This means mosques, minarets, Qoranic schools and mederse as well as Islamic charity, education and religious-cultural foundations, and banks and credit institutions managed in a way which is compatible with Islamic religious prescriptions.

This occurs even when local Islam does not always appear to conform to SA expectations, often being articulated in practices such as local pilgrimages, visits to saints' tombs, mausoleums and other cults and sacred manifestations (religious phenomenology), which demonstrate the Sufi and other influences in the region. Such practices mediate the special relation with (contradicting the principle of direct access to) the divinity, which as prescribed by fundamental Islam, must inspire individuals and inspired Wahhabism, originally a SA faith.

It only had relative success, considering that in transition times it was to register an increase in religious practices and in the establishment of believer organizations.[40]

However, the open proselytize practices would soon be considered with suspect, as an expression of a fundamentalist way of thinking and often directly fought against by the local authorities. It is significant that the term "wahhabism" became a common place term for defining subversion, especially in areas that had recently suffered conflicts, such as extremist proliferation.[41]

In the background of these movements the issue of the division and relations between power and religion that in Islamic cultures is mainly an open question has not yet been settled. In fact, from many points of view the investments of SA in CA have to be considered as an attempt to gain influence through the re-Islamization (and re-Arabization) of the much diversified, tendentially secularized local society. SA has often supported political religious parties but seems today to be incapable of purposing an economically "positive" pattern (like secular systems as Turkey, China or India).

Besides this, many conflicts in the recent past (possibly today overdue) can be interpreted as the effects of a proxy war between Russia and the Gulf oil-states, competing for influence in the energy market. The stake in this competition was supposedly the acquisition of a strategic supremacy in

[39] Golnaz Esfandiari 2018.

[40] Muminov Ashirbek 2014; Hanks 2016.

[41] Rasanayagam Johan 2014:9.

gas or/and oil, and all the consequent decisions in terms of the placement of pipelines and of connected long-term investments.

14.3 Further Elements in International Policies

14.3.1 International and Intergovernment Collaboration

The question of the cooperation among the CA countries represents a challenge. At the beginning of their independent life, local governments were comprehensibly motivated by the consolidation of their apparatuses, meaning, from the regional point of view, a kind of neo-nationalistic, often hostile external politics.

Then, in a further stage, until possibly current times, with the overcoming of a first generation of presidents, the situation evolved in a more relaxed situation. In fact that first generation of leader heavily conditioned the dialogue among the neighbouring states, with nationalistic politics often degenerating in personalistic contrapositions (among Niyazov, Karimov and Nazarbajev, especially).

Originally, the only multilateral institution of cooperation exclusively composed by Central Asian states was the Central Asian Economic Cooperation, renamed then Central Asian Cooperation Organization (CACO) in 2001.[42] At the time of its foundation its members were Kazakhstan, Kyrgyzstan, Tajikistan and Uzbekistan, while Turkmenistan refused to join it originally on the ground of its neutrality policy.

Such organization had the declared intention of pursuing a harmonization of trade and tourism and, in general, mobility of goods and people; it demonstrated just occasionally some intention for pursuing collaboration at wider scale and even integration intentions. It dissolved in 2006, possibly to be replaced by Russia-led Eurasian Economic Community (EurasEC), from which Uzbekistan withdrew in 2008; indeed, after Russia's and Belarus' adhesion, this multinational frame has lost to some extent its regional peculiarity.[43]

The "community" and the "communitarian" frame at regional scale have been eventually repurposed by the "union", then making it the Eurasian Economic Union (EAEU), which derives from a treaty that came into effect in 2015; it was purposed originally by former Kazakh President Nazarbayev, who first launched the idea in 1994, later suggesting, among the others, the creation of a single currency; in that time it was possibly an opportunistic manoeuvre in order to mitigate the growing influence, in those years, of Russia and China.[44] Currently the members of EAEU are Russia, Belarus, Kazakhstan, Kyrgyzstan and Tajikistan.

Today the most typical regional organization is the Central Asian Union, made up of Kazakhstan, Kyrgyzstan and Uzbekistan, and joined in March 1998 by Tajikistan; it may represent in the intentions an "umbrella" organization, like the EU or other similar regional organizations.[45] It remained at the state of idea in transition time, but it has been recently re-purposed; it is possibly the most interesting geopolitical experiment currently in the area.

14.3.2 The Search for a New Regional Multilateral Equilibrium

After transition, and once internal consolidation had started, the CA countries evidenced the intention of partaking in the benefits of international economics and of globalizing cultures.[46]

Therefore, the local governments undertook the construction of a sociology of international relations (on a regional scale), then went a step further with initial "free riding" (individualist) politics, in order to prevent that the different problems would become a matter of permanent contention.

The main target of such policies would be the creation of a common codex for the resolution of problems, which in the bilateral (reciprocity based) schema may not always find a satisfactory solution. It was necessary to avoid misunderstandings, accidents and instrumental intrusions, as had happened in this sensitive border area in the previous few decades. Indeed, this is the major risk when considering not only the tensions deriving from the contiguous scenario (besides Afghanistan, Xinxiang, Kashmir tensions), potentially manifesting a domino effect, but also in underground invisible interferences.

It is difficult to have an updated record of the affiliations since they change frequently, suddenly changing dramatically from the title of "eternal friend" to that of "definitive enemy". It is a sensitive question, and is usual for "young" susceptible states, usually governed with personalist-paternalistic methods. However, the CA states demonstrate a good attitude in participating in regional organization which,

[42] Indeo Fabio (manuscript, in press).

[43] Indeo Fabio (manuscript, in press); for the special Uzbek case see Hanks 2016:7.

[44] Parenti Fabio Massimo and Adda Iacopo 2017:350.

[45] Megoran Nick 2017:112; the EU process represents a reference, and even an example for such regional integration politics; in fact, the issues and the lexicon, such organizations recurrently adopt are those originally purposed by EU treaty, such as the necessity of "convergence" and "harmonization" of national (internal) legislations, the freedom of movement for persons, as well for goods, capital and services, the freedom of establishment and others similar expressions.

[46] Indeo 2018.

to some extent, is mandatory considering the described interdependency induced by the same interwoven geography.

Besides SCO (a decisive effort in the direction of a multilateral concept), it is possible to recognize other organizations, reflecting the precariousness of alignments in this phase of local policies.

The Collective Security Treaty Organization (CSTO) established in 1992, de facto Russian led, includes Russia, Belarus, Armenia, Kazakhstan, Tajikistan and Kyrgyzstan with part-time membership of Uzbekistan and has been defined as a kind of "answer to NATO", having a strategic character.[47]

14.3.3 Volatile Affiliations

The current situation is characterized by a "double standard",[48] with national and international politics divided by a structural threshold—conservative inside but ephemeral in international relations—with the latter sometimes being carried out simply for opportunistic reasons, for contingent purposes, potentially to influence public opinion. In fact, the game of international adhesions reflects the dynamic situation of the wider international arena, possibly the pressure of some local powerful neighbour.

The volatility of Uzbekistan is well known. In 1999 Karimov adhered to the Georgia, Ukraine, Azerbaijan and Moldova (GUAM) alliance (representing a group of former FSU peripheral countries), formed in 1997, then making it GUUAM. However, then it rejected it in 2005, possibly because the association was perceived as under an obsolete militaristic Russian influence.[49]

After the 2001 Twin Towers attack, Uzbekistan conceded the air base of Karshi-Khanabad in southern Uzbekistan to the USA, becoming a hub for the Afghan war. But soon after the Andijan massacre in 2005, and the USA reaction, Uzbekistan asked for a withdrawal of the American contingent.[50] Furthermore, it also quit the Eurasiatic Economic Community in 2008 and similarly in 2006 exited GUUAM, adhering to CSTO, that lasted until 2012, when Karimov suspended membership, without explaining this decision.[51]

This volatility is connected with the country's neomercantilist attitude, evidencing a tendency to use international politics as an economic instrument (and the reverse), but it is also connected to the fact that Uzbekistan has no border in common with the superpowers (therefore it is less exposed to the risk of military pressure). International affiliations are similar to a kind of "sliding door" geopolitics, depending on wider equilibria, possibly on the relative Russian influence or the need to avoid finding themselves in the middle of a China–Russia rivalry.

The case of Kazakhstan is different, as it seems to seek integration within IC coherently—considered possibly as a means to diminishing exposure to influential neighbours. In 2010 Kazakhstan was appointed as OSCE yearly president, an important acknowledgement, contradicting the discussed internal political attitude. So too, in the case of Kazakhstan again, which was elected in 2012 as a member of the UNO Human Rights Council. This was a decision criticized by some because of its poor domestic record in human rights. However, Kazakhstan, on this occasion, also demonstrated an important capability in accrediting the image of a reliable member of IC.

14.3.4 A Crowded Society: Super-national, Non-institutional Player and Others: Under the Threshold of Officiality

The effectiveness (namely the "power") of the IC can be discussed but it cannot be ignored. It exerts an impact as a whole, with a kind of characteristic concentric effect, from its innumerable institutions, down to the individual states, diffusing its standards and regulations. It promotes, perhaps unintentionally, a kind of transversal civil society, characterized by such capabilities as dialogue and transparency of administrative work. Elements in contradiction with the closure tendency of the autocratic oligarchy-based systems.

14.3.5 Corporate Multinational Companies (MNCs)

MNCs often prove to be more efficient than single states. This is the case of very structured corporations, especially of those dealing with HC and further extraction activities,

[47] Parenti Fabio Massimo and Adda Iacopo 2017, evidences a US-EU-NATO "expansionist process" XIX CT like imperialism; it may appear in these terms in a "zero sum game" scenario especially by non-democratic countries; CSTO does not include China.

[48] Intended as a different politics, carried out respectively internally and externally, international politics being assumed per default as chaotic and anarchic, where it prevails the use of the force.

[49] Parenti Fabio Massimo and Adda Iacopo 2017:347.

[50] The Uzbek government accused western powers of having orchestred the unrest in particular USA and UK; World Report 2015: Uzbekistan, Events of 2014, https://www.hrw.org/world-report/2015/country-chapters/uzbekistan, accessed 23.4.2018.

[51] Fazendeiro Bernardo Teles 2015:485; Fumagalli 2007; "Uzbekistan quits Russia-led CSTO military bloc", 28 June 2012, https://www.rt.com/politics/uzbekistan-quits-pro-russian-bloc-996/, accessed 23.4.2018; Parenti Fabio Massimo and Adda Iacopo 2017:347; Limes n.8/2014.

which sometimes make politics openly capable of negotiating directly with local governments (e.g. concerning obtaining of concession, exclusive exploitation and exploration of HC deposits, orientation of the pipelines and consequent realization of entire investment cycles).

The influence of powerful HC corporations in CA is particularly evident (Gazprom, Exxon, Eni, Petronas and many others), although in the last few years they appear depotentiated because of the expectations about energy paradigm change to some extent, which will possibly break down their geopolitical monopoly.

In fact, MNCs dealing with HC have balance sheets which are worsening due to long-term general depressing quotations that seem to have annulled extra-profit margins, which are the base on which such monopolies are able to maintain themselves. In general, the decrease of investments in the area must be registered, also considering that this sector, characterized by long-term return on investments (ROI), namely the time of construction of huge infrastructures, requires stability on extended surfaces (then a good standard of both national and international governance). [52]

Another case is that of advanced industries (high tech, innovation, digital electronic, mobility etc.), exerting strategic importance, and of related advanced services.

So too, in general for tertiary activities, comprehensibly obsolete in the post-Soviet phase, now recovering; consulting and trust agencies, in particular, had a special role as guarantor of many operations regarding development and setting up of such new sovereignty-states' apparatus and of their functioning. This is the case in early transition in particular, in operations such as privatization and denationalization, and in general in the reconstruction of macroeconomics of new states (currency, treasury services, market institutions, tax regulations, welfare administration). They have sometimes assumed a political significance in CA (Deloitte, PWC, KPMG and similar auditing companies, as well as international banks and consultancy specialized firms).[53]

14.3.6 Non-governmental Organizations and Further IOs

NGOs now represent effective players on all levels, filling the gap that arose in transition times, due to the fact the new apparatus was often too weak to execute many essential duties (e.g. wealth and sanitary, welfare and social assistance, international cooperation).[54]

Beyond the original statute and intention, their actions appear to be a network of multinational connected associations, financed by donors and subscribers, usually based in advanced countries. Their action is perceived, even if unintentionally, as politically oriented. In fact, they operate in sensitive situations where they are basically inspired by a liberal-democratic paradigm (individualist rights against collectivism, secular instead of religious attitudes, tendentially independence from the local elites), they can induce some ideological influences.

Therefore, they are targeted by periodical accusations of representing some particular interests, even under the semblance of assistance and charity, playing some decisive role in critical areas (conflict settlements, post-war pacification, development, basic needs).

In any case, they unavoidably represent a globalized (rather than westernized) style of life, by themselves interfering in local dynamics. For example, their "modus operandi" can appear as incompatible with local clan or patriarchal hierarchies, and with religious institutions; simply the fact that NGOs' actions try to address the war and the promiscuous violence in human relations can appear as interference in local questions.[55]

They often engage in a kind of "arm wrestling" with local, mainly governmental organizations or directly with the public administration; this is especially in CA, because here "foreign" NGOs started their activities when it was really needed, in early transition times, and were then joined by national organizations once the central public welfare recovered in the role of social assistance.

Their action is currently slowing down and becoming less relevant (at least in basic needs and assistance). The situation is no longer as critical and local governments are increasing their welfare capability, limiting foreign NGO action (e.g. in Tajikistan after the war).[56]

The dimensions of NGOs operating in CA are diversified. There are not only big multinational organizations but also a plethora of small-scale organizations, active on a small-local

---

[52] Similarly for further industries dealing with environmental and territorial resources, the necessity of permits to be obtained with the mediation of local governments; it is not only the case of minerals but also of building industry, such as cement, steel, energy production and distribution.

[53] Heathershaw John 2013:187; Jelen Igor 2007.

[54] Kluczewska Karolina 2017; just in Dushanbe there are 3000 NGOs, and of them about 1000 actives being subsidized by donors from all over the world, although some are forbidden, because accused of interfering in political question, since their action always appear politically biased.

[55] They do not necessarily appear as neutral, then mitigating causes that usually aliment the conflicts further, e.g. inside effects like "war entrepreneur" and speculators of any kind, addressing warlords and local gangster; Jelen Igor 2012:551; indeed the same difficulty of vigilance on such fragmented borderlines is by-it-self a cause of uncertainness; Megoran Nick 2017:126.

[56] Some situation very sensitive for politics and for popular sentiment, like children's and orphan protection, http://www.larca.org/, accessed 23.4.2018; Jelen Igor 2007.

scale, sometimes with little coordination, but which may nevertheless play an effective role. Local governments have proved to be sensitive to these elements, often contrasting them openly, often trying to establish a platform of domestic NGOs, namely GO-NGO promoted by the same governments. In recent times, and supposedly following the Russian (and Chinese) example, lists of banned NGOs have been drawn up (e.g. Endowment for democracy and some others). The question is about local civil society and its connection with the "global civil society", and about its capability of representing a base of autonomy from local power.[57]

The role of NGOs was exceptional in the Tajikistan civil war, playing an effective part, applying mediation and interposing actions. Their presence on the field, applying tactics in order to mitigate counter positions, was an extraordinary experiment. The view of blue helmets, white uniforms, red cross and/or crescent symbols, and humanitarian personnel "inundating" the battlefield with basic needs products (like packed food, ready-to-eat kits, water bottles, sanitary pharmaceuticals, dis-armed assistance personnel), alone exerted a pacification effect.

This in particular when displaced persons and civilians and other weaker categories (elderly people, children, disable, injured and ill people) were assisted in preparing the programme for refugees—considering that the return home of displaced people is to be considered the critical passage signifying in principle the end of the war.[58]

It has sometimes been defined with the slogan "the first war won by NGOs".[59] Indeed, this intervention has been made possible by exceptional circumstances, because of the low level of interferences exerted in that circumstance by other regional powers (possibly because Tajikistan lies in a marginal context, has no strategic resources like oil and also, as said, because of the relative weakness of neighbours' power in that period, like Russia and China).

14.4 Geopolitical Wide View

14.4.1 Regional–Continental Context

The collapse of FSU disclosed once again the "mackindaresque" relevance of the Eurasian heartland and not just in terms of HC resources, a true prize for world politics, starting a potentially predatory struggle. But soon the situation changed, prefiguring the change in the energy paradigm, becoming more flexible and even de-territorialized. The integration in the IC, in fact, realized its chance to use and exchange energy as convertible (no longer subject to the logics of zero-sum games, "fixed" in territorialized networks and geopolitically controlled transcontinental pipelines).

This fact should theoretically mitigate the predatory tendency, namely the tendency to engage in competition to gain the lion's share in the global market while potentially applying aggressive politics. The area could change its role from a precious heartland (for which it is worth organizing a war) to a new (not necessarily weak) periphery, or a battlefield for some proxy war: a change that has often happened in the history of CA countries.

In fact, the "heartland" of Eurasia periodically finds itself in the centre of geostrategic calculations, yet in some other occasions it has simply been exploited as a periphery. It depends on system-transformation elements such as the involution of Russia, the contiguity with Middle Eastern scenario, the pressure and the growth of China, the influence from the Western culture. It also depends on regional players like Iran, Saudi Arabia, Turkey and their role and interferences to spontaneous development.

From the geostrategic point of view, one of the major questions is the—originally planned for 2016 and continuously postponed (delayed)—retreat from Afghanistan of the USA-guided international coalition, which represented a potential threat for regional stability. In fact, the withdrawal would configure a new vacuum, and potentially escalation effects that could reverberate throughout the whole CA area and also wider. But probably—considering the evolution and the stabilization of the last few decades—this issue is a little obsolete. Several sources confirm a relative stabilization, considering that Kabul is becoming a big metropolitan area, and a hub for regional functions, where the activities of Taliban and other potentially subversive forces seem to be slowing down in this period. The main problem is gradually becoming socio-economic, possibly considering the difficulties of reconverting poppy production and trafficking that has become the cause of the continuation of the tensions.[60]

Whatever it may seem, it must be considered that a few thousand military, comprehended in a multinational force, are able to ensure a minimum compactness in a country char-

[57] Jelen Igor 2007.

[58] The high number of refugees and displaced persons must be considered, one million (on a population of about 4 million), besides the migrations of any kind affecting the area and the neighbour countries.

[59] Author's interview.

[60] Although, as said, the terrorists' activity in Kabul maintains a dangerous character, and the Taliban seem to be capable of occupying occasionally entire territories in north-eastern Afghan provinces, characterized by prevalent Tajik and Uzbek ethnic populations; in early May 2018 they occupied provinces of Badakhshan contiguous to Tajik border; indeed the Taliban seem today much more difficult to define than in the past, possibly a militia supported by not clear interests; it is to consider that Nov. 9th, 2018, a major sequestration of more than 300 kg of heroin, arriving from Afghanistan, through Iran, was made by Genoan Italian custom, signifying possibly the opening of a further itinerary in such kind of trafficking. Possibly the true question is today the definition of the "taliban" role becoming a kind of "proxy" army, manipulated occasionally by some hidden power.

acterized by dramatic history and internal differences, with a topographic territory as wide as half of Europe—however stabile enough to make the realization of important investments possible.[61]

14.4.2 Current Tendencies

The consolidation of a basic IC schema—on a Eurasian scale—is to be considered essential, in order to build confidence as well as communication channels between the governments. This is also because in recent times the governments have shown a certain nervousness, possibly a consequence of the s.c. Arab springs (or of similar reformist movements), namely the same idea of the street revolution, itself unpredictable and difficult to control (mainly characterized by civic motivations, by its high legitimization potential).

Indeed, assuming a farsighted policy, many of the tensions can be settled with the generational passage, with the populations—above all the victims of conflicts and persecution, their relatives or descendants—slowly overcoming physical and moral injuries. The CA republics are enjoying a period of peace and development, but they still present some alarming evidence, often consequences, direct or indirect, of the conflicts that characterized the transition.

Among these, perhaps the fact that these republics are the most important "suppliers" (country of origin) of "foreign fighters" and terrorists, possibly former fighters of the local conflicts (in Tajikistan, essentially), incapable of reintegrating into pacification programmes, engaging (and being manipulated) in further conflicts or terrorist attacks elsewhere in the world. Recent terrorist attacks have Uzbek or Tajiki exiled as protagonists (as in New York, 1 November 2017; Istanbul, 2015; and others), often trained in Syrian, Afghan or Iraq conflicts, possibly enrolled as mercenaries, with good military background, deriving from Soviet-like organization.

They are possibly individuals coming out from dismantled armies, frustrated or difficult to re-integrate (even when the identity records of such persons are usually not so clear). Often, they are "veterans" of the local wars (Afghanistan, and then in sequence Chechnya 1992–1994, Georgia, Eastern Ukraine and Syria), representing a kind of disbanded "army" of soldiers, recruited by "war agencies", exposed to manipulation.[62]

14.4.3 The Rhetoric of the Transition

CA populations and elites have to know that much of their future will depend on current decisions; apparently little mistakes today—at the initial stage of a theoretical development path—may have irreversible consequences in the long term. In this context the need for stability and continuity could easily become a pretext for not carrying out reforms, namely for accepting totally despotic politics.

Such attitudes would weaken the rule-of-law principle, relegating the written law as a "chiffon de papier" (XIX Ct.-like politics), and the constitution as completely able to be manipulated. At the moment it seems that any president may change it at will, eventually playing with variables such as social groupings and economic resources. Evidently the major aim of the rule-of-the-law order—namely to render power reversible—does not function: leaders tendentially become life dictators and behave in an irresponsible manner, as leaders that well know that they will never be called upon to answer for their actions.

In many cases they obtain such results with disputable referendums or ambiguous parliamentary acts that ratify virtually lifelong presidency proclamations, attributing to them a constitution-like legitimization, so for the deceased Karimov and Niyazov, and so similarly for the "life-presidents" Rahmon and Nazarbayev (the latter wisely resigned recently from many of his duties).

The current material politics, besides the announcements, cannot result as anything but contradictory, without much logic, tending to disorient social groups and wider populations, as well as IC and other interlocutors. Constitutional power and the same democracy are to be considered as middle-term targets, rejecting a naïve definition of democracy, since it has to be put into action in a difficult environment, in which there are lobbies, hidden powers and secret struggles. It requires a road map, to organize societies and economies, as well as a cultural attitude to accepting politics as a game of interests and of competencies: the game cannot be "planned" on all scales and societies have to be culturally accustomed to dynamic dialogues.

In this way, any kind of power can achieve legitimization: a situation which is sometimes preferred by the same IC paradoxically, which usually has no instruments to intervene in local politics, but just "moral persuasion", namely soft power and cultural-economic instruments. In the meanwhile, the same power has the chance to build up a game of reforms, a method for accrediting a fictitious democracy—with false elections, false opposition and so on—in order to appear as

[61]The Associated Press, The New York Times, Aug. 21, 2017, Full Transcript and Video: Trump's Speech on Afghanistan, https://www.nytimes.com/2017/08/21/world/asia/trump-speech-afghanistan.html, accessed 23.4.2018; possibly a question evoking disproportionate attention; see as well Specogna 2014.

[62]As said, the ban of an opposition religious-inspired party could rise some question about the future of the Peace Agreement, eventually alimenting in those period also the DAESH recrudescence, http://factsanddetails.com/central-asia/Tajikistan/sub8_6d/entry-4889.html, accessed 19.05.2018; http://www.c-r.org/downloads/Accord%2010_15Key%20elements%20of%20the%20Takikstan%20peace%20agreement_2001_ENG.pdf accessed at 19.05.2018; http://factsanddetails.com/central-asia/Tajikistan/sub8_6d/entry-4889.html, accessed at 19.05.2018; Hierman Brent 2017.

an acceptable compromise as a "transition country", and not being cut off (at the same time) from the benefits of the IC.

References

Anceschi L (2017) Turkmenistan and the virtual politics of Eurasian energy: the case of the TAPI pipeline project. Cent Asian Surv 36(4):409–429. https://doi.org/10.1080/02634937.2017.1391747

Barisitz Stephan (2018) Myth and actuality of the silk road: geo-economic opportunity vs. and geo-political convenience, Asiac annual conference 2018, 5–7 December 2018, University of Trieste – Gorizia campus (manuscript)

BBC news. http://news.bbc.co.uk/

Brill Olcott Martha (2012) Tajikistan's difficult development path. Brookings Institution Press. 27 nov 2012

Carile Alessando (2018) The issue of security, with particular regard to Uzbekistan-EU relationships, Asiac annual conference 2018, 5–7 December 2018, University of Trieste – Gorizia campus (manuscript)

Chiavon F (2018) The Legal Status of the Caspian Sea, Asiac annual conference 2018, 5–7 December 2018, University of Trieste – Gorizia campus (manuscript)

Comai Giorgio, Sofie Bedford (2018) Victims of double standards: double victimhood and changing narratives in Azerbaijan's public rhetoric, Asiac annual conference 2018, 5–7 December 2018, University of Trieste – Gorizia campus (manuscript)

Dagiev Dagikhudo (2014) Regime transition in Central Asia. Routledge, New York

Damiani Isabella (2013) Géopolitique de l'Asie centrale – Entre Europe et Chine: le coeur de l'Eurasie. Presses Universitaires de France, Paris

Dunlop JB (1994) Will the Russian return from the near abroad? Post-Soviet Geography 34:204–215

Eurasianet.org. https://eurasianet.org/tajikistan-uzbekistan-reportedly-reach-border-settlement. Accessed 5 Dec 2019, Tajikistan, Uzbekistan reportedly reach border settlement: Tajiks are angered by reports that the government is giving away a gold mine. Nov 5, 2019

Fazendeiro Bernardo Teles (2015) Uzbekistan's 'spirit' of self-reliance and the logic of appropriateness: TAPOich and interaction with Russia. Cent Asian Surv 34(4):484–498. https://doi.org/10.1080/02634937.2015.1114780

Fincantieri, Singapore, 18 agosto 2016, presse communiquè

Fumagalli M (2007) Alignments and realignments in Central Asia: the Rationale and implications of Uzbekistan's Rapprochement with Russia, "International Political Science Review", June 1, 2007. http://journals.sagepub.com/doi/abs/10.1177/0192512107077098. Accessed 5 July 2018

Golnaz Esfandiari (2018) Clashes highlight tensions between dervishes and Iran's Establishment, February 20, 2018. https://www.rferl.org/a/iran-dervhishes-sufi-explainer-tabandeh/29051140.html. Accessed 23 Apr 2018

Hanks RR (2016) Narratives of Islam in Uzbekistan: authoritarian myths and the Janus-state syndrome. Cent Asian Surv 35(4):501–513

Heathershaw John (2013) In: John H, Edmund H (eds) Tajikistan amidts globalization: state failure or state transformation? pp 177–198

Hessbruegge Jan Arno (2004) The Shanghai cooperation organization: a Holy Alliance for Central Asia?, "Al Nakhlah, The Fletcher School Online Journal for issues related to Southwest Asia and Islamic" Civilization, Spring 2004, Article 2

Hierman Brent (2017) Russia and Eurasia 2017-2018, 8th edn. Rowman & Littlefield, Baltimore

http://factsanddetails.com/central-asia/Tajikistan/sub8_6d/entry-4889.html

http://www.c-r.org/downloads/Accord%2010_15Key%20elements%20of%20the%20Takikstan%20peace%20agreement_2001_ENG.pdf. Key elements of the Tajikistan peace agreement. Accessed 6 July 2018

http://www.larca.org/. Accessed 23 Apr 2018

http://www.mid.ru/en/foreign_policy/news/-/asset_publisher/cKNonkJE02Bw/content/id/3103635. Accessed 23 Apr 2018, 1 March 2018, "Communique of the 50th meeting of the Special Working Group on the development of the Convention on the Legal Status of the Caspian Sea at the level of deputy foreign ministers of the Caspian littoral states"

Indeo F (2018) The role of Russia in the Central Asian Security Architecture, Policy Brief #48, 2018, OCSE Academy Bishkek

Indeo Fabio (manuscript, in press) New Trends in Central Asian Connectivity, in: Frappi Carlo and Indeo Fabio, edited by, Monitoring Central Asia and the Caspian Area. Development Policies, Regional Trends, and Italian Interests, Eurasiatica 13, Edizioni Ca'Foscari, Nato Defense College Foundation

Jelen Igor (2002) a: "Kirghizstan: Le contraddizioni della liberalizzazione", in: OS (Osservatorio Strategico), n.3, pp. 31–35; "Tagikistan: esperimenti di normalizzazione", in: OS/5, pp. 36–40; "Kazakistan: la difficile maturazione di una società moderna", in: OS/7, pp. 48–53; "A un anno dall'11 settembre. Asia Centrale: interpretazioni di contesto", in: OS/8-9, pp. 65–69; "La delusione uzbeka", in: OS/11, pp. 33–37; "Asia Centrale: quali prospettive?", in: OS – Prospettiva 2002–2003, n.12, pp. 63–66; CeMiSS, Roma

Jelen Igor (2007) La società civile nel Kazakistan della transizione post sovietica. Unica politica possibile o ultima ideologia rimasta?, in "Bollettino della Società Geografica Italiana", 2:373–392

Jelen Igor (2012) Appunti di geografia politica ed economica. Aracne, Roma

Thorez Julien (2008) Fermer les frontières, construire les nations. Le territoire comme fondement des nouvelles identités républicaines en Asie centrale post-soviétique. In: Giraudo G, Pavan F (eds) Integrazione, assimilazione, esclusione e reazione etnica. Naples, Scriptaweb, pp 247–273

Thorez Julien (2018) Central Asia: the state of the research, Asiac annual conference 2018, 5–7 December 2018, University of Trieste – Gorizia campus (manuscript)

Kluczewska Karolina (2017) Project development in a Tajik NGO: Everyday practices, instrumentalisation of donors and meaning of 'development work', ASIAC annual conference, November 6–7th, 2017, Dipartimento di Culture, Politica e Società dell'Università di Torino (oral presentation)

Limes, n.8/2014, Cina, Russia, Germania unite da Obama

Megoran Nick (2017) Nationalism in Central Asia: a biography of the Uzbekistan-Kyrgyzstan, Boundary, Central Eurasia in Context, 2017, University of Pittsburgh Press. https://doi.org/10.2307/j.ctt1vjqrk6

Muminov Ashirbek (2014) Traditional and modern religious-theological schools in Central Asia, CA&CC Press AB Sweden, CA&CC Press AB Publishing House (Sweden). http://www.ca-c.org/dataeng/09.muminov.shtml. Accessed 5 July 2018

Necati Polat (2002) Boundary issues in Central Asia. Transnational Publishers, Ardsley

Nicklas Norling 4/5/2006, RATS exercise in Tashkent: concern over nuclear terrorism? CACI Analyst, https://www.cacianalyst.org/publications/analytical-articles/item/10769-analytical-articles-caci-analyst-2006-4-5-art-10769.html. Accessed 23 Apr 2018

Ott Stephanie, Russia tightens control over Kyrgyzstan. https://www.theguardian.com/world/2014/sep/18/russia-tightens-control-over-kyrgyzstan#img-1. The Guardian, 18 September 2014. Accessed 5 July2018

Parenti Fabio Massimo, Adda Iacopo (2017) Are we going to live in a post-NATO world? A critical perspective on trends, obstacles and possibilities. GeoJournal 82:345–354. https://doi.org/10.1007/s10708-015-9693-8

Rasanayagam Johan (2014) The politics of culture and the space for Islam: Soviet and post-Soviet imaginaries in Uzbekistan. Central Asian Survey 33(1):1–14. https://doi.org/10.1080/02634937.2014.882619

Rotar Igor (2006) Resurgence of Islamic Radicalism in Tajikistan's Ferghana Valley, the Jamestown Foundation, Terrorism Focus, 3/15, April 20, 2006. https://jamestown.org/program/resurgence-of-islamic-radicalism-in-tajikistans-ferghana-valley/. Accessed 22 Apr 2018

Shahrani MNM (1979) The Kirghiz and Wakhi of Afghanistan: adaptation to closed frontiers. University of Washington Press, Seattle

Silvestri Tommaso (2015–2016) Le strategie della federazione russa nel teatro del mar Caspio dopo la caduta dell'Unione Sovietica, Master thesis, University of Trieste, academic year 2015–2016

Silvestri Tommaso, Specogna Erica, Chiavon Francesco (2018) Geopolitical status of the Caspian Sea, Asiac annual conference 2018, 5–7 December 2018, University of Trieste – Gorizia campus (manuscript)

Specogna E (2014) Le potenzialità di sviluppo dell'industria mineraria afghana: tra "maledizione delle risorse" e resilienza, pp 185–190, in AA.VV., Memorie Geografiche – Oltre la globalizzazione. Resilienza/Resilience, a cura di Capineri C., Celata F., De Vincenzo D., Dini F., Randelli F., Romei P., Firenze, 12/2014

The Associated Press, The New York Times, AUG. 21, 2017, Full Transcript and Video: Trump's Speech on Afghanistan, https://www.nytimes.com/2017/08/21/world/asia/trump-speech-afghanistan.html. Accessed 23 Apr 2018

Marzhan Thomas (2015) Social, environmental and economic sustainability of Kazakhstan: a long-term perspective. Cent Asian Surv 34(4):456–483. https://doi.org/10.1080/02634937.2015.1119552

Richard C (2013) The failure of liberal democratization in Kazakhstan: the role of international invetmnet and civil society in impeding political refomrm. In: Sevket A, Richard C (eds) Social and cultural change in Central Asia: the soviet legacy. Rotledge, London and New York, pp 127–144

Times of India, China's area increases by 1000 sq km, 12 January 2011

Uzbekistan (2018) Russia agree on nuclear power station, The goal is to economize by using less gas and coal, https://eurasianet.org/uzbekistan-russia-agree-on-nuclear-power-station, Jul 12, 2018. Accessed 27 Dec 2018

"Uzbekistan quits Russia-led CSTO military bloc", 28 June 2012., https://www.rt.com/politics/uzbekistan-quits-pro-russian-bloc-996/. Accessed 23 Apr 2018

Wikipedia, Central Asia

Final Comments

15

Abstract

In the twenty-first century, more scientific papers on the various research topics, prepared by a younger generation of researchers in CA countries as well as in Russia and in further FSU countries, have been published in international literature. At the same time, researchers from other countries have become more and more involved in different international scientific projects on the important ecological or resource management orientated problems of CA countries. Thus nowadays, the region and its diverse geographic objects, components and processes have become more open, understandable and clear, not only to the world's academic society but also for the ordinary people travelling and looking for unique places, new impressions and experiences. The geographic knowledge and its sharing among the countries unites people and assists them to find the sound and more balanced solutions for common timeline-evoked problems. With comprehensive focus on the CA region the attempts were done to present, understand and foresee its development.

Keywords

Natural constraints · Eco-system managements · Good governance practices · Human development · Economic and social progress · Political tendencies · Challenges and risks

15.1 Natural Constraints and Territorial Management

The knowledge of the CA region assumes increasing relevance, and this not just in academic and scientific circuits, but in the wider audience, recovering in the world representation the place and the role—to some extent mythical—it had in the past. And even more, considering the geography of this area is a kind of central tile of a global mosaic, necessary to describe and to understand the wide systemic equilibria: geological features and paleogeographic history, biomes and ecosystems, climate changes, and the correspondent impacts on soil formation processes, water and agricultural resources; and finally what is in principle even more important, the anthropogenic impact on the environment, including traditional and innovative activities and unexpected adaptations: all these aspects of a changing reality, that have been and will be increasingly renewed and systemized.

Water is an extremely valuable resource in arid CA; however, it is distributed unevenly across the five countries and that leads to the need for coordinated transboundary agreements, policies and actions on a more sustainable use of river runoffs in different years, wet and dry, normal and extreme. Water security and sustainable usage mean rational and coordinated development of irrigation and drainage, pastoralism and recreation, fishery and hydroenergetics, the rich biodiversity and the promotion of a sound environment. The rational use and development of water and hydroenergy resources in the Amu Darya and Syr Darya basins are still the most complex regional challenges for CA countries.

Rivers in CA mostly begin in the mountain areas; they are fed by both precipitations and glacier water. Thus, glaciers are the basis of the welfare of all of the main streams in the region; however, as ice monitoring shows, the glaciers in the CA mountains, as in other regions of the world, have been considerably reduced and degraded. With global climate warming the decrease in freezing periods and total reduction of winter ice cover has been observed in the north of the Caspian Sea over the last decade. Other research has shown that recent Caspian Sea level rises seem to be due to a combination of higher Volga discharges and lower evaporation from the Caspian Sea surface, in relation to changes in the persistence of southwestern circulation patterns.

I. Jelen et al., *The Geography of Central Asia*, World Regional Geography Book Series,
https://doi.org/10.1007/978-3-030-61266-5_15

Within the Aral Sea basin, the local impact on the microclimate can be observed due to the considerable drying up of the sea, and the impact of global warming. The trend towards increases in temperatures is explicit in the main countries, for example in Uzbekistan and Kazakhstan, directly surrounding the Aral Sea. Monitoring of global processes in different geospheres, such as the atmosphere, the surface of lithosphere—pedosphere, hydrosphere, cryosphere (in the mountainous regions of CA) in the crossover between the twentieth and twenty-first centuries—has shown that natural cycles have often been disturbed or worsened under the pressure of anthropogenic activities. First, the irrational use of water resources for irrigation in agriculture has to be mentioned as well as attitudes to drainage water management. The excess in irrigation water causes secondary salinization, which adds to desertification in the vast lowlands, like a new Aralkum desert.

An analogous to the Aral Sea region situation, though to a lesser extent is taking place in the Lake Balkhash basin of Kazakhstan, where after the construction of the Qapshaghay hydroelectric-power station on the Ili River, and with more water used for irrigation, these actions have caused the evident reduction in water levels and the volume of water in the lake. Thus, being the largest in area country, Kazakhstan has more numerous and complicated problems with water resource management than other CA countries. However, this also gives Kazakhistan's water specialists a challenge to seek for measures and solutions to stop the desiccation of the Aral Sea, at least its northern part, the Small Aral, or to find original solutions for the Lake Balkhash basin. One of such measures is to adopt more rational irrigation water-saving technologies. Another measure is a more rational use of groundwater for many purposes, including agriculture.

A few positive attempts which have already started to be implemented by CA countries have to be mentioned. Kazakhstan is planning a second phase to the Small Aral restoration project, which would bring the sea water back to the town of Aralsk, the former main port and transshipment point at the northern end of the Aral Sea. These attempts will reinstate traditional and recreational fishery in the region as well. However, large projects, which require much investment, have to be discussed and evaluated according to feasibility studies or environmental monitoring data. So far, there have been unconvinced opinions among the academic society on the necessity of such costly projects like establishing the Turkmen Lake in the Karakum desert in Turkmenistan. Some experts believe that runoff will be insufficient to fill the lake, as the drainage water will evaporate or seep into the desert through unlined feeder canals. This criticism has reasonable grounds: the areas along the Karakum canal, one of the longest irrigation canals in the world, are nowadays covered by numerous swamps and ponds, which have exacerbated salinization.

CA countries have rich deposits of the most known metals, minerals and other resources. Kazakhstan was the leading producer of oil, and Turkmenistan was the leader in gas production at the beginning of twenty-first century. At the same time Kazakhstan remains the largest, and Uzbekistan the second largest CO_2 emission producers. With economic development the CA countries will need to generate more energy; however, uneven geographic distribution of energy resources makes some of them more dependent on imported energy sources or electricity.

The exploration of renewable resources is weak at the moment because of high fossil fuel subsidies and low electricity prices, a lack of investments and a limited number of new technology developers, little public awareness etc.

According to Köppen, there are seven climatic zones in the CA region. The region is distinguished by strong continentality of the climate with wide amplitudes between the minimum and maximum values of air temperature, precipitation and evapo-transpiration index changes while going from north to south and according to the territory's altitude. Two main soil classifications are used in CA countries: national, based on the previous Soviet or Russian soil classification system, and the FAO-UNESCO classification. All five CA countries continue to explore resources such as agricultural land with different types of soil, of which the most fertile are Chernozems and Kashtanozems or Chestnut soils, suitable for the growing of different crops. Less fertile soils, naturally covered by grasses and shrub species adopted for arid environments, are favourable for pasture. The main threats to soils in the west lowland and plain zone of CA region are water and wind erosion and carbon mineralization. The grassland-covered soils, which are mostly used for grazing pastures, are a significant storage of carbon, and they have to be managed in order to maintain the organic soil matter level and structure.

Agriculture remains one of the most important sectors of CA countries' economics. However, since independence in 1991 more problems have occurred for farmers with land reforms, economic transition and maintenance of old degraded irrigation systems. Increased salinization was followed by a corresponding drop in crop yields. Soil salinity inhibits the growth of plants when the osmotic pressure of the soil-water solution in the root zone inhibits the ability of plants to absorb water. Among crops grown in CA, the most salt-tolerant crops are barley and sugar beets; moderately tolerant crops are alfalfa, rice, cotton, wheat, corn, potatoes, carrots, onion, cucumbers, pomegranates, figs, melons and grapes. The least salt-tolerant crops are stone fruits, almonds, peas and beans. To stop or reduce soil salinity on arable irrigated lands, conservation agriculture measures can be applied. Reduced tillage, establishing permanent raised sowing beds, enriching of land with plant residues under the cotton-wheat-maize rotation were studied and have shown positive results.

The recent ecological situation with vegetation complexes, biomes and ecosystems in the different parts of the CA region is closely related to the whole history of landscape changes, driven by many physical, biological and anthropogenic factors over the whole of the Eurasia continent and globally on Earth. The mountains of CA are one of the four centres of biological diversity within the territory of the former Soviet Union. The flora of the CA mountains comprises more than 5000 species, while the surrounding deserts accommodate about 2,000 species. Notwithstanding their relatively small coverage area, forests in CA play a key role in regional hydrology by helping maintain a steady river discharge from the high mountains to the irrigated lowlands. Land cover change in the boreal forest has a significant influence on the climate system mostly through changes to the surface radiative energy. In the grassland/steppe biome moderate surface albedo increases cause a reduction in net radiation. It was estimated that in winter and spring this cooled the surface, while in summer it warmed it. Since only 3.4% of the world's temperate grasslands/steppes are protected, it is important to pay more attention to this biome in CA countries.

In Tajikistan, the total area of natural protected areas makes up 22% of the territory of republic—this is the largest protected area in the CA region. The other countries protect much lesser areas: Kyrgyzstan 7.4%, Kazakhstan 8.8%, Uzbekistan 5.0% and Turkmenistan 4.0% of the country's total area. Though forest resources are not extended widely in CA countries, forests are recognized as very important and responsible for rich ecological functions and services, including the conservation of biological diversity and tackling desertification.

Over the last 15–20 years due to climate warming, more and more extreme situations are registered in CA countries. According to the national environmental reports, thousands of extreme situations and different accidents caused by natural and technogenic factors were registered in the region. Many of these disasters were followed by human losses. Natural disasters have made up about 1/5 of all accidents. Of the natural disasters the most damage was made by heavy rains, floods in the mountain valleys, landslides, mud and stone flows, fires and so on. Many ecological and social-health problems occur during some anthropogenic activities, for example, in previous urban mining or weapon test polygons, or in the Aral Sea region, in both the Kazakhstan and Uzbekistan territories, where dust storms carry the soil and the sea bottom particles, salts, other chemicals hundreds of kilometres from the sea. Radioactivity monitoring results in Semipalatinsk region of Kazakhstan in 2009 revealed that the northern part of the region has become safe, and the growing of agricultural crops and maintaining of cattle has returned. But several areas with elevated residual radioactivity were not recommended to be settled or used for agriculture.

In order to come to an overall conclusion on the discussed region and its natural processes and relationships between them and human activities, the region seems to be searching for its own path for further and more sustainable development. And at the same time, it is opening up for the exchange of different practices and experiences with other world regions and organization initiatives.

15.2 A Set of Structural and Cultural Changes

Any rapid change is in principle a risk: changes, neither in nature nor in human reality, are linear; often they prospect unforeseeable catastrophes. Therefore, a caution principle is to be assumed, and paradoxically this is also for policies disposed for environment improvement. Similarly, for human processes and for innovation-induced changes: globalization phenomena deeply impact everyday life at any scale, on politics, economy and society, and on the formation of cultural processes and attitudes, so, it is necessary to elaborate a correspondent technology, promoting in the same time social organization and cultural awareness in order to deal with such impacts.

Such changes are evident in the fragile endorheic CA environment, as well in the whole oecumene: a thin film of biosphere is consuming and saturating like a gas chamber, with rotten marshes, waste dumps full of residuals. Sometimes they are just subsequent to the "bubbles" created by a biased interpretation and application of the modernist economy, risking deflation at any time: in fact, all civilizations (and also the Soviet one) "let" their residuals—material and cultural—lie in a particular layer of the landscape.[1]

Such changes are even more violent in peripheral and remote areas, more sensitive to the changes, where the equilibrium between humanity and natural conditions is even more precarious. In particular transformation configures a tendency for re-thinking reality, involving the political level: the post-modern passage also means the crisis of the nation-states, based on exclusive belongings and on hermetic borderline. of Such kind of states—structured and hardly territorialized—had evolved from modern times, assuming a sequence of monopolies on basic functions (indeed the reason for their original legitimization, and of their current weakness).

The fracture of such monopolies is evident for a set of functions such as energetic supply, production, communica-

[1]Scharr K., Steinicke E., (hg.), 2012; Barisitz Stephan, 2018; Comai Giorgio, Sofie Bedford, 2018; Thorez Julien, 2018; Carile Alessando, 2018; Burri Ezio, Del Bon Andrea, Ferrari Angelo, Ragni Pietro, 2018.

tion, energy, transport and mobility, education and culture, communication, innovation and technology, and this means the weakening of the state sovereignty, as well as of its political significance.

Such evolutions are in contrast with the persistence of a monopolistic role of power, that, due to such transformations, can no longer exert a truly sovereign power.

Political configurations turn out to be permeable today, sometimes precarious; they cannot simply be hermetically cut off from the rest of the world: it is simply impossible to distinguish between internal and international levels, and the political units cannot simply function as "free riders", if not at the expense of the whole equilibrium and of their individual efficiency; they may no longer ignore each other.

All these effects are even more relevant for the new independent "stans" that are forced to search continuously for interstate collaboration for mobility, supply, minority questions, commercial activities and internal migration regulations.

This situation evidences a contradiction between maintaining the benefits of the free market (worldwide) economy and the conservation of self-referential elites; this especially in times of the ICT, with new social media representing something virtually impossible to annihilate; it is the case of the fear for street revolutions (Arab springs-like waves, often becoming, due to an escalation effect, uncontrollable), demonstrating the use of new technologies which may potentially disrupt local politics (if not at the cost of indiscriminate violent repression). It is the case of a general worldwide public opinion manifesting and protesting in unpredictable ways.

It must be considered that no hermetically locked regime may survive in times of globalization, as could have happened previously in Soviet and modern times; the NIS have to elaborate new schemas—among them that of "guided democracy", of a rather ephemeral character—trying to maintain an acceptable "business card" for the international society. However, the CA human rights record remains poor.

The most relevant characteristic of current politics is the search for a compromise between continuity and the conservation of international respectability (and the access to international markets). Any gesture, any action pursued by elites from the time of the post-Soviet passage is oriented in these terms.

This has happened in a context of manoeuvres, with the elite obsessively exhibiting an appearance of solidity; sometime leading to a recovery of despotism, potentially trying to promote family dynasties, or clan settings—even if in a dissimulating manner.

The highest risk in such situations is that of a possible fracture that would affect the power generational passage, the fear for the sudden disappearance of a charismatic historic leader, that has accumulated huge centralistic power, personalizing politics (evidently posing a problematic question for the succession).

15.3 Innovating in Continuity

It is necessary to identify the critical items that in a context of changes can contribute to strengthening local apparatuses, making society sensitive to a resource that the population sometimes has "in front of their eyes", but that they nevertheless were not able to notice.

This is the case of both social and natural resources, of mountain glaciers as well as of village communitarian life, of uncontaminated environments and of ethnographic resources. This is especially so for the marginalized outback: glaciers and snow-covered mountain slopes represent a reserve of water to be used with continuity in all seasons, making extended surfaces cultivable, as well as small village orchards (in fact, the presence of perennial snow and glaciers is not obvious at such latitudes, and can change consequent to wider climate changes). But such resources cannot be appropriately used, if not by an efficient basic organization and a functioning rural community, rooted in its territory.

Therefore, it is necessary to determine a renewed economic cycle, starting a new governance, at the same time consolidating a new local entrepreneurial class, independent from the power, which recognizes the freedom of initiative and all the connected kinds of freedom.

It is especially the case of HC monopolies, sometimes operating as political players: it is necessary to be aware of the inescapable surmounting of the HC paradigm and of the consequent overcoming of those monopolies (and of rentier positions for the elite) that such a passage will induce.

In fact, HC revenues should be used to promote further economic diversification (high-tech, high value-added, environment-quality functions, bio-organic brand-certified productions, good functioning social services etc.) to develop high value-added activities. This is the case of amenity functions like tourism on the Caspian Sea (Aktau, Atyrau in Kazakhstan, Avaza in Turkmenistan), on lakes and rivers, mountains, in the desert, steppe and anywhere else: a significant possibility which could potentially induce a turn in local economies.[2]

This strategy has to be implemented on the grounds of a wise use of HC revenues, investing such capital—the only available cash—to adjust local ecosystems, especially the hydrography (sparing water, rendering the ID systems more efficient, carefully managing the mountain resources) and the Aral in particular (as a water reserve, mitigating climate and supporting economics). A number of projects have been established and also implemented, hopefully inverting this tendency, bringing effective improvement to the situation of this and of other CA water basins.

Local governments show that they are highly aware of such situations, but they evidently have different intervention

[2] See cited IUNC programmes.

capabilities. The Kazakh government is particularly active in this matter; it aims at the passage towards a green economy, based on renewable sources by 2050, adhering to international advances standards, realizing the practical questions of further development of a new economy, as well as setting target-prices for renewable energy production.

Another opportunity is the project regarding new transcontinental connections (whatever it is named, SR, BRI, "bright path", "energomost" and others) opening a China-Europe pathway through Dzungarian gate, central Kazakhstan and southern Russia, possibly constructing a land bridge; it may signify the solution of the principle problem of such a continental region, namely its difficult accessibility to communicate with the outside world.

And it may also improve internal CA economics; in fact, internal connectivity typically occurs over middle-range distances, easily integrating into the wider corridor programmes (namely as a segment of the same infrastructures); this makes evident the necessity for the reconstruction of a regional networks for transports and communication, relying on sustainable technologies. This for goods and persons, energy flows and culture; it is the case of high speed trains and of further public transport technologies, that seem to be the most ecological and convenient over such distances. However, it must be taken into consideration that de-territorialization of economies, rapid technological development and political evolution will hardly justify the huge investments have been made in the past in such initiatives. Probably the better investments today regard immaterial activities, as it is the case of common programs in education at any level, involving visitors and students from any side of the border, as well of tourism, trade, social projects, human rights, in a cross-border cooperation context.

The huge availability of money—easy money from oil rent—risks becoming a paradoxical danger: it may set the concentration of power, the strengthening of a despotism, a kind of authoritarian-paternalistic economics, considering relations with the foreign countries as a risk and any innovation-development as a risk.

The question is, then, what to do with this money, also considering the risk that such improvised wealth may induce (over-complacency effects, financial "bubbles", corruption, attractiveness to foreign speculators): local governments started a kind of artificial economy that should be able to incentivize self-alimenting growth. This financing a more efficient welfare system, promoting advanced services and high value-added activities, and financing sectors such as environmental recovering, building (urban and infrastructural) industry, as well as further industries that are usually under the control of the established power (energy and hydrocarbons, military apparatus and armaments. This evolution presents evidently some risks, and especially the risk to becoming something self-referential, namely a dramatic consumeristic, real estate and financial "bubble", empty skyscrapers, "pharaonic" but useless infrastructures (with high depreciation rates), and finally a situation of dangerous abundance of arms.[3]

15.4 A Question of Governance

CA economies are growing, sometime booming; they are interdependent, and therefore they have to elaborate a mutually self-sustaining (and compensating) mechanism at a regional scale, overcoming aggressive competition (which in such a closed context could be destructive).

Kazakhstan is the most performing country, rapidly changing its social landscape, but the future of the whole region will depend on the performance of other CA countries, especially the demographically decisive Uzbek republic.

The same Kazakhstan will possibly encounter a short period of stagnation, consequent to cyclical settling after a long period of growth, potentially aggravated by a fall in oil prices (2013) and Crimean and Eastern Ukrainian crises (and consequent sanctions to main Kazakh partner, namely Russian republic); but its figures remain stable; this is also thanks to Kazakh conservative macro-economics, based on deficit control and fiscal accuracy, and especially thanks to accumulating reserves from hydrocarbons in an Oil Fund (named Samruk-Kazyna).[4]

In such situations it is possible to start advanced functions (like pension funds, social insurances programmes, private savings-services investments) that give stability in the long term, relying on a kind of middle-class interest (a further step to the evolution consolidation, strengthening the general situation). Kazakhstan is probably the only CA country currently capable of developing such functions (that require programmes in social participation, a solid financialization and diffuse investment capabilities); other republics are far behind.

[3]With the formation of a new "rich" middle class, but characterized by inadequate attitudes, therefore booming of consumerist attitudes, and disproportionate spending capacity; The Gazette of Central Asia, 23 January 2013; "Kazakh TV – Kazakhstan enters top 50 most competitive countries". Kazakh-tv.kz. 6 September 2013; Yessenov M.N., Beisakhmet A.A., Kalimbetov E.A., Artemiev A.A., 2017; Kazakhstan attractiveness survey 2013. EY.com, 8 March 2014; but such favourable statistics have been a little disappointed by the current crisis induced by low international HC prices, possibly anticipating the long-term transition to de-carbonized economics. In fact Kazakhstan was the best performing economic regime in the area, as said, growing in average by 8% yearly until 2013, then encountering a stagnation and slowdown; Kazakhstan was the first former Soviet Republic to repay all of its debt to the International Monetary Fund, Kazakhstan Country Study, 2012.

[4]"Kazakhstan's fiscal situation is stable. The government has continued to follow a conservative fiscal policy by controlling budget spending and accumulating oil revenue savings in its Oil Fund – Samruk-Kazyna", World Bank Group, 2015; PWC 2016, 2017.

The same for advanced functions, such as high tech and innovative industries, services like banking, finance and insurance, attracting foreign investments; but centralized management, even when representing an element of stability, can also become a systemic risk that the Kazakh government must be aware of.

Tajikistan and Kyrgyzstan are eschewing a difficult past of poverty; Turkmenistan–the other HC rentier state—is engaged in the post Niyazov transition, while Uzbekistan, possibly the CA country with biggest human and cultural potential, is still wondering about its economic model, more or less open to innovation or to conservation of internal equilibria.

The most important element of these times is the awareness of the need to overcome "productivist" economics, specializing in postmodern functions: attractiveness, quality of life and landscape for both tourists and residents, minorities and majorities, private parties and companies, to avoid the younger generation escaping and eventual mass migration tendencies of the more productive strata of the population (as well as outflow of capitals and further resources) that would exert a distortive effect on the whole society.

It is necessary to be aware that growth in an authoritarian context (without economic diversification, profit redistribution, cultural improvement and consciousness) means socially "zero" and could suddenly degenerate and collapse (as in several cases of apparently indestructible "superpowers" which have suddenly collapsed in the recent past); in fact society diversification means resilience, while economic diversification—in order not to depend on international quotation of oil—means a better possibility of durable growth.

All this happens in a kind of divide—between past and future—made evident by a set of paradoxes. Degraded soils, abused natural resources, endangered ecosystems, obsolete politics: it impossible to say how long it will take to recover; however, these problems also represent an opportunity for development at the same time, to elaborate new technologies, fostering circular and green economics: possibly, the big business today in CA is the recovery of the entire territory and urban stratifications contaminated in modernist times.

There is the need for a new theory of peripheral-qualitative development that can overcome the limits of productivism, and the forces of de-territorialization—indeed the true risk that local institutions will face in the near future. And there is a need for a new practice of land ownership—empowering the local communities—and a new ethic of the use of natural resources, in a frame of renewed sustainability—apparently a naïve policy, in fact preventing the damages of de-territorialization connected to the globalization drift.

The heritage of communism and colonialism can no longer represent a motivation for justifying any kind of transition. Many of the problems currently affecting CA republics can find a solution simply through good governance (e.g. improving the use of resources, setting up of appropriate infrastructures, inclusion of "drop out" social groups), with programmes carried out in the middle-long term. Further questions have proved to be simply overdimensioned for individual states (not just for CA new republics) and need to be settled in a context of international "good will".

References

Barisitz Stephan (2018) Myth and actuality of the Silk Road: geo-economic opportunity vs. and geo-political convenience, Asiac annual conference 2018, 5–7 December 2018, University of Trieste – Gorizia campus (manuscript)

Burri Ezio, Del Bon Andrea, Ferrari Angelo, Ragni Pietro (2018) Analisi ambientali e valorizzazione culturale dei qanat di Yazd, patrimonio dell'Umanità (UNESCO), Asiac annual conference 2018, 5–7 December 2018, University of Trieste – Gorizia campus (manuscript)

Carile Alessando (2018) The issue of security, with particular regard to Uzbekistan-EU relationships, Asiac annual conference 2018, 5–7 December 2018, University of Trieste – Gorizia campus (manuscript)

Comai Giorgio, Sofie Bedford (2018) Victims of double standards: double victimhood and changing narratives in Azerbaijan's public rhetoric, Asiac annual conference 2018, 5–7 December 2018, University of Trieste – Gorizia campus (manuscript)

International Monetary Fund, Kazakhstan Country Study, 2012.

Kazakh TV – Kazakhstan enters top 50 most competitive countries, 2013, The Gazette of Central Asia, Kazakh-tv.kz., 6 September 2013

Kazakhstan attractiveness survey 2013. EY.com, 8 March 2014

PWC (2016) Guide to do business and investing in Uzbekistan 2016 edition, https://www.pwc.com/uz/en/assets/pdf/dbg_2016.pdf, accessed 20.5.2018

PWC (2017) Doing business guide in Kazakhstan 2017, https://www.pwc.kz/en/publications/new-2017/dbg-2017-eng.pdf. Accessed 20 May 2018

Scharr K., Steinicke E., (hg.), (2012) Vom euphorischen Aufbruch in die Realitaet des Alltages, 1898-2000 – zwei Jahreszenten Transformationsforschuing, University of Innsbruck Press

The Gazette of Central Asia, 23 January 2013 Thorez Julien, 2018, Central Asia: the state of the research, Asiac annual conference 2018, 5–7 December 2018, University of Trieste – Gorizia campus (manuscript)

World Bank Group (2015) Kazakhstan: Low Oil Prices; an Opportunity to Reform, Kazakhstan Economic Update Spring 2015, http://documents.worldbank.org/curated/en/766781468182065283/pdf/96195-WP-PUBLIC-Box391443B-KAZ-Kazakhstan-economic-update-series-Bi-annual-Economic-Update-Spring-2015-Low-Oil-Prices-an-Opportunity-to-Reform-eng.pdf. Accessed 23 Apr 2018

Yessenov MN, Beisakhmet AA, Kalimbetov EA, Artemiev AA (2017) Tourism Opportunities for Kazakhstan on the Great Silk Road. J Geogr Environ Manag 4(47):180–189

Index

I. Jelen et al., *The Geography of Central Asia*, World Regional Geography Book Series,
https://doi.org/10.1007/978-3-030-61266-5

M

N

O

P

Q

R

S

T

GPSR Compliance

The European Union's (EU) General Product Safety Regulation (GPSR) is a set of rules that requires consumer products to be safe and our obligations to ensure this.

If you have any concerns about our products, you can contact us on ProductSafety@springernature.com

In case Publisher is established outside the EU, the EU authorized representative is:

Springer Nature Customer Service Center GmbH
Europaplatz 3
69115 Heidelberg, Germany

Batch number: 10755972

Printed by Printforce, the Netherlands